BIOLOGY
the essential principles

Tom M. Graham
University of Alabama

 SAUNDERS COLLEGE PUBLISHING
Philadelphia New York Chicago
San Francisco Montreal Toronto
London Sydney Tokyo Mexico City
Rio de Janeiro Madrid

Address orders to:
383 Madison Avenue
New York, NY 10017

Address editorial correspondence to:
West Washington Square
Philadelphia, PA 19105

This book was set in Baskerville by Clarinda Company.
The editors were Michael Brown, Lloyd Black, Patrice Smith, Bonnie Boehme.
The art & design director was Richard L. Moore.
The text design was done by Adrianne Onderdonk Dudden.
The cover design was done by Richard L. Moore.
The artwork was drawn by J & R Technical Services, Inc.
The production manager was Tom O'Connor.
This book was printed by R.R. Donnelley & Sons.

Cover credit: Flamingos in Bolivia. Photograph © by M. Phillip Kahl.

BIOLOGY: THE ESSENTIAL PRINCIPLES   ISBN 0-03-057838-8

© 1982 by CBS College Publishing. All rights reserved. Printed in the United States of America.
Library of Congress catalog card number 81-53071.

234 39 98765432

CBS COLLEGE PUBLISHING
Saunders College Publishing
Holt, Rinehart and Winston
The Dryden Press

This book is dedicated to my mother

GEORGIA INEZ GRAHAM

preface

For many students the exposure to college biology is a brief, one-time experience. For this to be a positive experience, both the professor and the textbook are responsible for presenting the scope, history, and flavor of scientific endeavor in a coherent and interesting manner. This is best accomplished using basic biological concepts supported by accurate examples and illustrations.

This text is an outgrowth of classroom experiences encountered in 8 years of teaching more than 10,000 nonscience majors from every region of the country and many foreign nations. In a book of this scope, a great deal of material has been, and should be, omitted, particularly that involving excessive scientific detail. A single college experience in biological science should not burden the student with the task of assimilating a massive number of facts. It is preferable that the discoveries and theories arising from scientific inquiry be presented in the form of essential principles to which the student can relate. To achieve this goal many of the essential principles of biology are presented from a human viewpoint. This method provides a familiar frame of reference from which many biological concepts can be developed.

In following the standard, or classical, approach, the book presents first the basic principles of chemistry, energy, and cell biology and progresses to the biology of organisms, ecology, and evolution. The second part emphasizes human structure and function and includes a chapter on behavior. It has been my experience that the vast majority of students are extremely curious about human function.

Each chapter is introduced with a brief statement, along with a list of learning objectives and a preview of the more difficult or pertinent terminology. These words, together with other important terms, also appear in boldface italics within the body of the text. Since many students are put off by terminology, much attention has been given to assure that terms are clearly defined and accessible. Illustrations in the form of drawings and photographs are important learning aids, and many of their captions provide additional explanatory information. Marginal notes have been provided throughout the text to reinforce important terms and concepts. At the end of each chapter are a concept review in summary form and a list of discussion questions. Complementing the many two-color drawings are 126 full color photographs contained in 32 pages; these are distributed in eight sections so as to appear as close as possible to corresponding text areas.

Considering the varied and immense scope of modern biology, a textbook on general principles must necessarily be selective. In using this text, the instructor may choose a sequence of chapters that better suits his topic preference or his desire to present a shorter course of study. For example, the chapters covering evolution and behavior could be omitted in a short course. Similarly, the chapters on viruses, animals, plants, and ecology could be rearranged to fit a different topic sequence without harm to the book's integrity. There is some evidence that an increasing number of institutions are offering a short one-term course in biology for biology majors who subsequently take a rigorous botany or zoology course. The text's emphasis on fundamentals and its flexibility as a teaching tool (which is further discussed in the accompanying Instructor's Manual) makes it an excellent candidate for this curriculum requirement.

Accompanying the textbook are two ancillaries. The Instructor's Manual highlights points in each chapter, explains major concepts in more detail, offers suggestions as to topic rearrangement for different course structures, and acts as a general aid to the instructor. The Student Guide offers study aids in the form of brief chapter summaries, terms to define, and sample test questions (true-false, multiple choice, sentence completion). Answers for these items will appear at the end of the guide.

I have reserved this paragraph to thank a group of extraordinary people at

Saunders College Publishing; my indebtedness to them is staggering. Not only are they highly competent professionals, but they have been enthusiastic supporters, dispensing invaluable suggestions and personal kindnesses whenever the need arose. The Publisher, Don Jackson, initiated this project by providing me the opportunity to write for Saunders. His confident support is acknowledged with deep appreciation. Mike Brown, Editor, actually is Robinson Crusoe's Friday in disguise. To recount his deeds of professional and personal service on my behalf would require the assistance of a small computer. For his counsel, friendship and innumerable courtesies, I shall always be grateful. Patrice Smith, Project Editor, provided me more encouragement and reinforcement than there is space here to relate. From beginning to end, she has been an indispensable guide and a fount of inspiration. Lloyd Black, Developmental Editor, directed the illustration program and provided enormous assistance in many areas. His diligence and concern have been a boon throughout, and I sincerely thank him.

Most of the full color photographs were supplied by Bruce Russell of Biomedia Associates and by Carolina Biological Supply Company. For their artistic contributions, an especial thanks.

Although many people have been involved in the development of this book, I want to pay special tribute to Doris Helms of Clemson University. Her incisive and creative suggestions proffered with great good humor were invaluable. In addition, I want to thank the following reviewers for their many helpful corrections and suggestions: William Bessler, Mankato State University; George Bleekman, American River College; Gwendolyn Burton, Community College of Denver, North Campus; Bruce Chorba, Mercer County Community College; Sid Crow, Georgia State University; Gerald Dotson, Community College of Denver, North Campus; Charles Granger, University of Missouri, St. Louis; Joseph Hindman, Washington State University; David Husband, University of South Carolina; Clarence Wolfe, Northern Virginia Community College.

Finally, to all of those behind the scenes at Saunders College Publishing and to my typist Sharon Creitz, thank you for everything you have done.

Tuscaloosa, Alabama
November, 1981

TOM M. GRAHAM

contents overview

CHAPTER 1	AN INTRODUCTION TO BIOLOGY		2
CHAPTER 2	SOME BASIC CHEMICAL PRINCIPLES		24
CHAPTER 3	LIFE: THE ESSENTIAL MATERIALS		46
CHAPTER 4	ENERGY AND LIFE		72
CHAPTER 5	CELLS AND CELLULAR ACTIVITY		82
CHAPTER 6	THE ESSENTIAL PROCESSES: PHOTOSYNTHESIS, GLYCOLYSIS, AND CELLULAR RESPIRATION		116
CHAPTER 7	DIVIDING CELLS: MITOSIS AND MEIOSIS		142
CHAPTER 8	THE BASIS OF HEREDITY		170
CHAPTER 9	VIRUSES: KINGDOMS MONERA, PROTISTA, AND FUNGI		200
CHAPTER 10	BIOLOGY OF PLANTS		248
CHAPTER 11	BIOLOGY OF ANIMALS		294
CHAPTER 12	GENERAL ECOLOGY		376
CHAPTER 13	ORIGINS AND EVOLUTION		420
CHAPTER 14	THE NERVOUS SYSTEM		460
CHAPTER 15	SKIN, SKELETON, AND MUSCLES		498
CHAPTER 16	RESPIRATION		524
CHAPTER 17	CIRCULATION AND IMMUNITY		544
CHAPTER 18	DIGESTION AND EXCRETION		584
CHAPTER 19	THE ENDOCRINE SYSTEM		620
CHAPTER 20	REPRODUCTION		648
CHAPTER 21	DEVELOPMENT		676
CHAPTER 22	ANIMAL BEHAVIOR		704

contents

1 AN INTRODUCTION TO BIOLOGY 2
  The Essential Objectives 2
  Preview of Essential Terms 3
  Biology: The Beginnings 4
  The Nature and Philosophy of Science 5
  The Scientific Method 6
  The Essential Principles 9
  The Cell Theory 12
  Unity and Diversity 20
  Review of Essential Concepts 22
  Applying the Concepts 23

2 SOME BASIC CHEMICAL PRINCIPLES 24
  The Essential Objectives 24
  Preview of Essential Terms 25
  The Structure and Chemistry of Matter 26
  The Role of Electrons 31
  The pH Scale 38
  Review of Essential Concepts 43
  Applying the Concepts 44

3 LIFE: THE ESSENTIAL MATERIALS 46
  The Essential Objectives 46
  Preview of Essential Terms 47
  The Elements of Life 48
  The Role of Carbon 49
  The Compounds of Life: Carbohydrates 50
  Lipids 54
  Proteins 57
  Nucleic Acids 61
  Water: The Liquid of Life 66
  Review of Essential Concepts 70
  Applying the Concepts 71

4 ENERGY AND LIFE 72
  The Essential Objectives 72
  Preview of Essential Terms 73
  Measuring Energy 75
  Adenosine Triphosphate (ATP) 76
  Energy Efficiency 78
  Energy Flow 78
  Review of Essential Concepts 80
  Applying the Concepts 81

5 CELLS AND CELLULAR ACTIVITY 82
  The Essential Objectives 82
  Preview of Essential Terms 83
  The Cell Theory 84
  Cellular Organization 85
  The Eukaryotic Cell 89
  Some Activities of Living Cells 101
  Review of Essential Concepts 112
  Applying the Concepts 114

6 THE ESSENTIAL PROCESSES: PHOTOSYNTHESIS, GLYCOLYSIS, AND CELLULAR RESPIRATION 116
  The Essential Objectives 116
  Preview of Essential Terms 117
  Photosynthesis 118
  Light & Life 121
  The Light Reactions of Photosynthesis 122
  The Dark Reactions of Photosynthesis 125
  Glycolysis and Cellular Respiration 128
  Glycolysis 129
  Cellular Respiration: The Krebs Cycle 133
  Cellular Respiration: The Electron Transport System (ETS) 135
  Review of Essential Concepts 138
  Applying the Concepts 141

7 DIVIDING CELLS: MITOSIS AND MEIOSIS 142
The Essential Objectives 142
Preview of Essential Terms 143
Introduction to Mitosis 144
The Scheme of Mitosis 147
Introduction of Meiosis 153
The Scheme of Meiosis 157
Meiosis I 158
Meiosis II 160
Meiosis in Human Beings 161
Abnormal Cell Division 164
Review of Essential Concepts 168
Applying the Concepts 169

8 THE BASIS OF HEREDITY 170
The Essential Objectives 170
Preview of Essential Terms 171
DNA: The Hereditary Material 172
Ribonucleic Acid (RNA) 173
The Genetic Code 175
Protein Synthesis 177
Mutations: The Genetic Errors 179
Some Essentials of Genetics 180
Monohybrid Cross 184
Test Cross 184
Dihybrid Cross 185
Sex Linkage 188
Incomplete Dominance 190
Multiple Alleles 192
Multiple Gene Inheritance 193
Gene Linkage 194
Review of Essential Concepts 197
Applying the Concepts 198

9 VIRUSES: KINGDOMS MONERA, PROTISTA, AND FUNGI 200
The Essential Objectives 200
Preview of Essential Terms 201
Taxonomy: Historical Background 202
Modern Taxonomy 202
Viruses 204
Kingdom Monera 211
Kingdom Protista 223
Kingdom Fungi 236
Review of Essential Concepts 243
Applying the Concepts 246

10 BIOLOGY OF PLANTS 248
The Essential Objectives 248
Preview of Essential Terms 249
Kingdom Plantae 250
The Algae 251
Introduction to the Land Plants 259
Nonvascular Plants 260
Vascular Plants 263
Division Tracheophyta 272
Plant Hormones 283
Nonhormonal Activity 288
Review of Essential Concepts 290
Applying the Concepts 293

11 BIOLOGY OF ANIMALS 294
The Essential Objectives 294
Preview of Essential Terms 295
Kingdom Animalia 296
The Invertebrates 303
Introduction to the Chordates 345
Review of Essential Concepts 368
Applying the Concepts 374

12 GENERAL ECOLOGY 376
The Essential Objectives 376
Preview of Essential Terms 377
The Biosphere 378
Introduction 378
Terrestrial Biomes 380
Aquatic Ecosystems 387
Water Pollution 389
Ecosystems: Basic Structure and Function 392
Food Chains and Webs 393
Trophic Levels 395
The Cycles of Life 399
Biological Magnification 408
Regulation of Population Size 409
Ecological Succession 414
Review of Essential Concepts 418
Applying the Concepts 419

13 ORIGINS AND EVOLUTION 420
The Essential Objectives 420
Preview of Essential Terms 421
Introduction 422
The Origin of The Earth 422
Chemical Evolution 424
The Origin of Life 426
Early Forms of Life 429
The Primitive Atmosphere 430
Early Eukaryotic Cells 431
Evolution: An Historical Background 433

Darwinian Evolution 436
The Theory of Natural Selection 437
The Modern Theory of Evolution 439
The Basis of The Synthetic Theory of Evolution 440
Evolution in Action 443
The Hardy-Weinberg Law 445
The Evolution of Man 449
The Genus HOMO 455
Review of Essential Concepts 457
Applying the Concepts 459

14 THE NERVOUS SYSTEM 460

The Essential Objectives 460
Preview of Essential Terms 461
Introduction to The Human Body 463
Introduction To The Nervous System 466
Neurons 466
The Reflex ARC 468
The Nerve Impulse 470
The Synapse 472
Divisions of The Nervous System 474
The Central Nervous System: The Spinal Cord 478
Enkephalins and Endorphins 481
Nervous Systems in Lower Animals 484
The Special Senses 486
Nervous System Disorders and Diseases 495
Review of Essential Concepts 496
Applying the Concepts 497

15 SKIN, SKELETON, AND MUSCLES 498

The Essential Objectives 498
Preview of Essential Terms 499
Introduction 500
The Skin 504
The Skeletal System 506
The Divisions of The Skeleton 507
The Muscular System 513
Diseases and Disorders of The Skin, Skeleton, and Muscles 521
Review of Essential Concepts 522
Applying the Concepts 523

16 RESPIRATION 524

The Essential Objectives 524
Preview of Essential Terms 525
Introduction 526
The Human Respiratory System 528
The Mechanics of Breathing 531
Respiratory Gas Exchange 533
Transport of the Respiratory Gases 535
The Control of Breathing 537
Respiratory Problems and Disease 540
Review of Essential Concepts 542
Applying the Concepts 543

17 CIRCULATION AND IMMUNITY 544

The Essential Objectives 544
Preview of Essential Terms 545
Introduction 546
The Human Circulatory System 547
Blood 547
The Cardiovascular System 557
Cardiovascular Disorders and Disease 570
The Lymphatic System 571
Immunity: Biological Defense 576
Organ and Tissue Transplants 580
Review of Essential Concepts 580
Applying the Concepts 583

18 DIGESTION AND EXCRETION 584

The Essential Objectives 584
Preview of Essential Terms 585
Introduction to Digestion 586
The Products of Digestion 588
The Physiology of Digestion 589
Digestive System Disorders 603
Nutrition and Diet 603
Introduction to Excretion 606
The Human Urinary System 607
Physiology of The Kidneys 608
The Kidneys and Regulation of Blood Pressure 613
Diseases and Disorders of The Urinary System 614
Review of Essential Concepts 615
Applying the Concepts 618

19 THE ENDOCRINE SYSTEM 620

The Essential Objectives 620
Preview of Essential Terms 621
How Hormones Work 624
Pituitary Gland 625
The Adrenal Gland 631

The Thyroid Gland   635
The Parathyroid Glands   637
The Pancreas   639
The Pineal Gland   643
Prostaglandins   643
Review of Essential Concepts   644
Applying the Concepts   647

20 REPRODUCTION   648
The Essential Objectives   648
Preview of Essential Terms   649
The Male Reproductive System   653
Male Sex Hormones   655
Male Sexual Disorders   657
The Female Reproductive System   658
The Female Sex Hormones and Menstrual Cycle   660
The Sexual Act   663
Fertilization and Preparation of the Uterus   664
Contraception   667
Sterilization   670
Abortion   671
Preview of Essential Concepts   672
Applying the Concepts   675

21 DEVELOPMENT   676
The Essential Objectives   676
Preview of Essential Terms   677
Cleavage   679
Implantation and The Fetal Membranes   680
Embryonic Development: First Month   683
Embryonic Development: Second Month   685
Fetal Development: Third Month   686
Fetal Development: Fourth, Fifth, and Sixth Months   686
Fetal Development: Seventh, Eighth, and Ninth Months   687
Birth   688
Control of Development   691
Aging and Death   699
Preview of Essential Concepts   701
Applying the Concepts   703

22 ANIMAL BEHAVIOR   704
The Essential Objectives   704
Preview of Essential Terms   705
Types of Animal Behavior Innate Behavior   707
Group Behavior   715
Review of Essential Concepts   724
Applying the Concepts   725

INDEX   727

BIOLOGY
the essential principles

an introduction to biology

THE ESSENTIAL OBJECTIVES

You have understood this chapter when you are able to:

1. Discuss the meaning of "science" in terms of its philosophy, potential, and limitations.
2. Discuss the steps of the scientific method and how these relate to solving problems.
3. Name the seven life principles and summarize their essential points.
4. Explain the relationship between unity and diversity in living things.

Biology is the science that attempts to understand and explain the workings of the living world. Its principles, established through objective observation and experimental evidence, attest to both the unity and the diversity of living things.

PREVIEW OF ESSENTIAL TERMS

science An organized body of knowledge acquired through observation, experimentation, and reason.

mechanism The doctrine that the universe is fully explainable in terms of physical and chemical laws.

vitalism The doctrine that, in addition to the laws of physics and chemistry, there is a supernatural "life force" that directs the universe.

scientific method A reasoned step-by-step procedure involving observation and experimentation in the solving of problems.

hypothesis An "educated guess" or temporary explanation that must be tested by experimentation to determine its validity.

theory A working hypothesis that, based on accumulated evidence, has a high degree of probability.

natural law A theory with such a high degree of probability that it is considered universally valid.

metabolism The sum of all the chemical reactions that occur in living things.

homeostasis Maintenance of a stable internal physiological balance or equilibrium within an organism or ecosystem.

cell theory All living things are composed of cells, and the cell is the structural and functional unit of life.

biogenetic law In the sphere of life, living things arise only from other living things.

BIOLOGY: THE BEGINNINGS

The roots of biology as an organized study extend far into the distant past. Its seed existed in the minds of Anaximander, Empedocles, Democritus, and other pre-Socratic Greek philosophers who first dared to make nature intelligible to mankind (Fig. 1.1). In the Chinese and Egyptian cultures prior to the Greek era, "science" was practically indistinguishable from theology; every operation in nature was believed to be overseen by some impelling supernatural force. With the early Greeks came the attempt to give natural explanations for the seemingly magical and mysterious workings of a fearsome universe. In time, however, the pursuit of scientific endeavors by the early Greek scholars began to wane as the demands of politics turned the thoughts of men elsewhere. For many years the seed of Greek science lay dormant; then, long after the death of Socrates (469?–399 B.C.), it was revitalized virtually by one man—the philosopher Aristotle (384–322 B.C.) (Fig. 1.2).

Presumably through the generosity of his pupil, Alexander the Great, Aristotle acquired a vast collection of zoological and botanical specimens from every region of Greece and Asia. Amidst these surroundings, Aristotle undertook the research that culminated in the monumental achievements that changed the course of science and history. As a biological scientist, he perceived the close relationship between structure and function in living things; contributed to the fields of experimental genetics, animal behavior, marine biology, and nutrition; and founded the science of embryology. Through his own work, Aristotle rescued the work of his predecessors, bringing it all together in a comprehensive body of knowledge. Moreover, he

Figure 1.1 Science was born in the minds of the early Greek philosophers. (From Holten, G., Rutherford, F. J., and Watson, F. G.: *Project Physics.* New York: Holt, Rinehart and Winston, 1981, p. 37.)

Figure 1.2 Aristotle, the founder of biology.

> Before Aristotle, biology as a science did not exist.

imparted structure to this knowledge by classifying each science by its principles and particular subject matter. Truly, it is little wonder that it was said of Aristotle that he "gave science to mankind."

Although practically nothing was known of the physical and chemical theories on which modern science is based, the illustrious accomplishments of Aristotle, superimposed on those of his predecessors, provided the groundwork for 2000 years of scientific progress. Nowhere is this more evident than in the science of biology, the study of living things. It was Aristotle—the founder of biology—who finally removed science from the realm of the mystical by replacing rumor and speculation with careful and continuous observation.

THE NATURE AND PHILOSOPHY OF SCIENCE

It has been over 23 centuries since Aristotle "gave science to mankind," but even today there is still uncertainty about exactly what it is he gave us. In other words, there is often misunderstanding or even ignorance concerning the meaning of "science." As an orientation to biology, it is essential that you, the student, understand what a science is, in terms of both its potential and its limitations.

The word "science" is derived from the Latin *scientia*, meaning "knowledge." In the strict sense, a science is an organized body of knowledge consisting of theories and natural laws subject to verification by observation, experimentation, and reason. It is important to point out that science refers to a body of knowledge related to *natural* phenomena and is concerned with discovering the principles or laws governing such phenomena. Accordingly, the philosophical attitude of science holds that the universe is fully explainable in terms of physical and chemical laws. This is referred to as the philosophy of **mechanism.**

The opposing view is that of **vitalism,** which proposes that there is "something" that controls natural events and directs them purposefully according to some divine plan. This includes the proposal that living organisms possess within them a vital principle or "life force" that cannot be explained within the limits of the laws of physics and chemistry. Vitalism, then, must appeal to the supernatural to account fully for the activities and processes associated with living things.

However, science, *by definition,* cannot concern itself with the supernatural. This is because scientific inquiry relies upon *experimentation* to verify its findings, and there is no known way to apply experimental analyses to the supernatural. It should be apparent, then,

An Introduction to Biology

> All scientific inquiry begins with objective observation.

that science *must* rely on the mechanistic approach, not in the sense that it necessarily refutes the supernatural, but in the sense that, scientifically, the influence of the supernatural can be neither refuted nor upheld. Quite simply, any account related to the supernatural is outside the domain of science. In biology, this means that we must deal with life in terms of what we can observe and in a manner subject to experimental validation.

Biologists, as with all scientists, must therefore substantiate their findings and ultimately can do so only in accordance with certain objective rules or standards. These standards have been outlined and organized into a reasoned step-by-step procedure generally known as the ***scientific method***.

THE SCIENTIFIC METHOD

While sounding rather overbearing, the scientific method is actually nothing more—or less—than an honest approach to solving problems. The steps in the approach are the following: (1) observation, (2) asking a question about or describing a problem related to the observation, (3) constructing a hypothesis, (4) experimentation, and (5) formulation of a theory (Fig. 1.3). As we comment on the scientific method, try to recall instances in your own life in which you have applied or could have applied these procedures.

1. It was pointed out that science is concerned with natural phenomena, which immediately restricts the scope of science by prohibiting inquiry into alleged supernatural or miraculous occurrences. This is understandable in that scientific investigation begins with objectively observable events. However, making an objective, unbiased observation is not always easy. The main obstacle is that we too often see only what we *want* to see or what we think we *ought* to see or what other people tell us we *should* see. In effect, there is a tendency to subject the observation to our opinions or emotions. Scientific observation tinged with bias, however, is a contradiction—a biased observation reflects an attempt at knowledge that may not be in accordance with reality. It is essential, then, to realize that the first step toward knowledge is to observe without prejudice.

2. In asking a question or proposing a problem related to the observation, the essential factor is that the question or problem be *testable*. For example, we might encounter an individual who claimed to be under the influence of evil spirits. Continued observation of his behavior might well suggest to some that he was indeed possessed. Scientifically, however, it is impossible to validate or invali-

date his claim because there is no way to test for the presence or absence of evil spirits. On the other hand, if we observe that grass appears green to us and ask why this is so, it is possible to approach the problem scientifically because grass and the human eye are subject to various testing procedures. Once it has been determined through repeated observations that grass does indeed appear green to human beings, the question "*Why* does grass appear green to human beings?" is a possible next step in the scientific method. In this instance, we can propose a testable problem directly related to an objective observation.

3. Next, one makes an educated guess, known as a **hypothesis**, concerning the possible answer (or answers) to the question or solu-

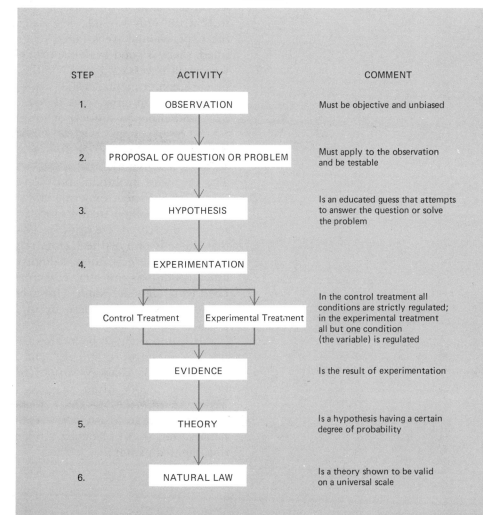

Figure 1.3 The steps of the scientific method.

> A scientific theory is always supported by established evidence.

tion (or solutions) to the problem. A hypothesis is a reasoned preliminary explanation of the possible outcome of the question or problem. However, such guessing, even though carefully reasoned, does not provide *evidence,* which is the result of the next step.

4. Through experimentation one attempts to determine the validity of the hypothesis. The experiment to be carried out is compared with a *control* treatment, in which all conditions affecting the experiment are strictly regulated. Consider, for example, the hypothesis that water is necessary for the normal growth of oak trees. The trees treated as controls would be supplied with adequate water, whereas the experimental trees would receive no water. All the other known factors that contribute to the normal growth of oak trees, such as adequate light and minerals, proper temperature, and so on, would be identical for both the control and the experimental trees, so that the only variable was water. If after repeated experiments the experimental trees failed to grow and the control trees flourished, there is good evidence to support the original hypothesis that water is a necessary factor for the normal growth of oak trees. In practice, experimental results are sometimes positive and sometimes negative, but in either case the results provide useful evidence that may tend either to support or to refute the hypothesis.

5. Accumulated evidence acquired through repeated experimentation makes possible the formulation of a ***theory***. A theory is simply a working hypothesis that has a certain degree of probability. The theory of evolution and the cell theory, for example, are supported by sufficient experimental evidence to give them a high degree of probability. In the case of the oak tree experiment just described, the supportive evidence allows the formulation of the theory that water is one of the factors required for the normal growth of oak trees. There is, then, a high degree of probability that, if all other conditions are met, oak trees will grow normally when supplied with adequate water. However, in the light of new evidence, it is always possible that these, or any, theories would have to be modified.

It is important to be aware of the distinction between the scientific use of the word "theory" and its use in everyday language. In science a theory is always supported by experimental evidence; used otherwise, theory often suggests only speculation or guesswork. Those scientific theories that repeatedly have been shown to be valid on a universal scale are called ***natural laws.*** The laws of thermodynamics, the laws of planetary motion, and the law of gravity are well-documented examples.

Figure 1.4 Although obviously quite different, a tiger and a tadpole are subject to the same basic life principles (see text). (*Tiger* from Ebert, J. D., Loewy, A. G., Miller, R. S., and Schneiderman, H. A.: *Biology.* New York: Holt, Rinehart and Winston, 1973, p. 356; *tadpole* from Silverstein, A.: *The Biological Sciences.* New York: Holt, Rinehart and Winston, 1974, p. 477.)

Even after the meticulous application of these steps, the pronouncements of science are not presented as absolute truths. Science must deal with the available evidence; it is simply not known if such evidence will be valid under every condition, everywhere in the universe, for all time.

The scientific method as a problem-solving instrument is basically an appeal to reason and honesty. In biology, as in other sciences, employment of the method is often exhausting and tedious, but as a guide to correct thinking, its application can be as appropriate in one's own life as it is in the life of the most eminent scientist.

THE ESSENTIAL PRINCIPLES

Biology is a scientific discipline founded upon certain essential principles from which a heightened understanding not only of our world but of ourselves is made possible. Biology, then, can be intelligible—and useful—to everyone if these principles are understood and applied. If the principles are used as basic guidelines, it is then possible to develop their associated concepts, or particulars, into an integrated and comprehensive view of the living world.

The basic biological principles that form the underlying framework of this book are applicable to every living organism on earth. It is obvious, however, that living things vary extensively in size, structure, nutritional requirements, geographic location, behavior, and so on. Nonetheless, any dissimilarity that distinguishes one form of life from another is essentially only a *particular* example of one or more of the major life principles. In other words, a tiger and a tadpole can be separated biologically in terms of particulars, such as size, structure, behavior, and so forth (Fig. 1.4). Fundamentally, however, the particulars do not alter the fact that these organisms are subject to the same life principles. It is the aim of this book to highlight and integrate the essential principles of biology while keeping their associated particulars to a minimum.

> All living organisms are subject to and united by the same life principles.

We shall list the principles, comment on each one briefly, and in subsequent chapters observe their applications in the living world. The principles are (1) **organization,** (2) **energy use,** (3) **growth and development,** (4) **sensation and response,** (5) **reproduction,** (6) **adaptation,** and (7) **homeostasis.** The coordinated interplay of these seven principles is responsible for the phenomenon we call "life," and all of them are characteristic of every living thing.

10
An Introduction to Biology

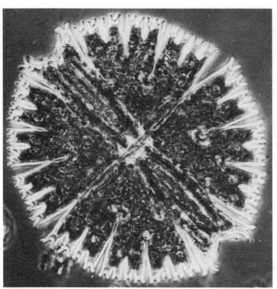

Figure 1.5 (a) This unicellular alga (shown here greatly magnified), as well as other single-celled organisms, never develops beyond the cellular stage. (b) Jellyfish are multicellular organisms that develop tissues, but they do not possess organs or systems. (c) Trees are examples of living organisms that develop beyond the tissue level to form definite organs and systems. (From Ebert, J. D., Loewy, A. G., Miller, R. S., and Schneiderman, H. A.: *Biology.* New York: Holt, Rinehart and Winston, 1973, pp. 578, 610, 642.)

> The cell is the structural and functional unit of life on earth.

Organization. Over 100 years ago it was determined that all living things are composed of one or more microscopic units called *cells* (Fig. 1.5A). The cell, as has since been discovered, is the fundamental structural and functional unit of life and is the smallest entity that truly can be called "alive." In the organizational make-up of living things, a group of similar cells forms a *tissue,* such as muscle tissue, nervous tissue, reproductive tissue, and so on (Fig. 1.5B). Tissues in turn are arranged to form *organs,* such as the heart, brain, ovaries, and so forth. Various organs can be arranged into organ *systems* (circulatory, nervous, reproductive, and so on), and various organ systems make up the complete individual *organism* (Fig. 1.5C). The simplest organisms—mainly the single-celled forms—do not attain even the tissue level of development; others—some of the multicellular organisms—may not develop beyond the tissue level, whereas still other multicellular forms may possess organs and systems.

Organization coincidental with life also exists *below* the level of the cell. Here, fundamental particles known as *protons, electrons,* and *neutrons* are organized into a number of different kinds of *atoms* (Fig. 1.6). Many of the atoms combine with each other to form *molecules* and various *compounds,* such as carbohydrates, lipids, proteins, water, and so forth, that constitute the physical building blocks

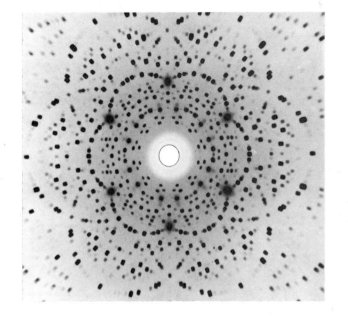
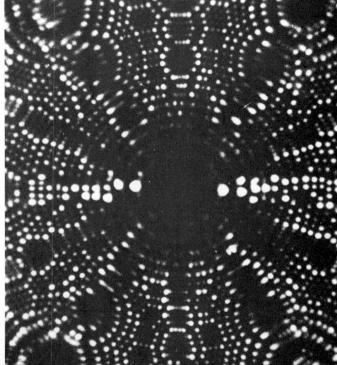

Figure 1.6 Electron micrograph of atoms arranged in a particular element or crystal. (From Holten, G., Rutherford, F. J., and Watson, F. G.: *Project Physics.* New York: Holt, Rinehart and Winston, 1981, pp. 558, 614.)

> The cell theory encompasses the concepts of cell division, metabolism, and heredity.

of living cells. At the other extreme, organization is evident *above* the level of the organism. Individual organisms of the same species, i.e., those that are genetically similar, together make up a **population.** Populations of different species living in proximity constitute a **community,** and any community plus its nonliving environment form an **ecosystem** (Fig. 1.7). The entire sequence of organization characteristic of living things is depicted schematically in Figure 1.8.

The Cell Theory

The early findings concerning the living cell have been incorporated into what is known as the **cell theory,** which states that all living things are composed of cells and that the cell is the structural and functional unit of life. Inherent in this statement are three concepts that are crucial to our understanding of the cell as the unit of life. The first of these is that *new cells arise only from pre-existing cells*. Except for the original cells on earth, all cells have arisen from living organisms by a process of cell division. All cells, therefore, have an ancestry that reaches back to the original cells. Second, *the chemical processes responsible for the life of the organism occur primarily within the cell*. These processes collectively are referred to as *metabolism*. This term simply refers to all of the chemical reactions that are carried out by cells and result in the building up or tearing down of bodily structures and materials. Third, *all cells contain genetic material that is passed to future cells*. This concept is the basis of the laws of heredity, which in turn account for the fact that like breeds like. Appleseeds always produce apple trees, and cats always give birth to kittens (Fig. 1.9).

Energy Use. Every living thing maintains its existence through its inherent capacity to use energy. By definition, energy is the capacity

Figure 1.7 Lake or marsh ecosystem. (From Turk, J.: *Introduction to Environmental Studies.* Philadelphia: Saunders College Publishing, 1980, p. 34.)

Figure 1.8 The sequence of organization characteristic of living things.

> Energy flow in living systems begins with the energy of sunlight, which is converted into other forms of energy by cellular processes.

Figure 1.9 In the sphere of life, like breeds like. (From Silverstein, A.: *The Biological Sciences*. New York: Holt, Rinehart and Winston, 1974, p. 390.)

Figure 1.10 Simplified model of a terrestrial ecosystem. *Arrows* indicate the direction of flow of energy.

to do work, i.e., the capacity to bring about some action or effect. An input of energy is required to build up the materials that sustain life, and, conversely, energy is released when these materials are broken down. In essence, this build-up and breakdown is what metabolism is all about. The ultimate source of energy for life on earth is the radiant energy of the sun. The conversion of this radiant energy into other forms of energy by living systems results in what is known as *energy flow* (Fig. 1.10). This refers to the passage of energy sources from one type of living organism to another type. For example, green plants, using the sun as a source of energy, produce carbohydrates that in turn are used as an energy source by animals that feed on the plants. These animals are then fed on by other animals, and so on.

Reduced to its essence, life is a continuous process of self-generating activity that depends upon an outside energy source. A basic knowledge of the sources, conversions, and flow of energy is therefore essential in understanding the phenomenon of life itself.

Growth and Development. Biologically, growth means that living cells increase in size, number, and complexity. For example, all human beings begin life as a tiny single cell. Through the process of cell division mentioned earlier, the fully formed human being be-

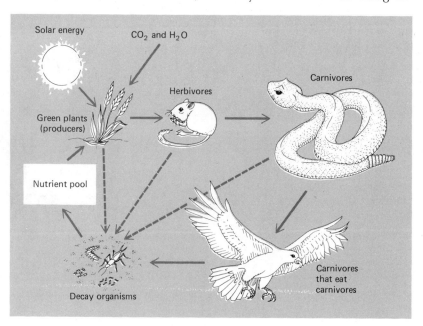

> Multicellular organisms grow through an increase in the number of body cells, not through an appreciable increase in the size of the cells.

comes a composite of trillions of cells. In the process of multiplying, the cells eventually become increasingly complex and specialized in order to carry out their specific functions. In more advanced animals, for instance, there are highly complex and specialized types of nerve cells. Some of these cells function in receiving stimuli; others, such as certain brain cells, store information; and still others are involved in controlling body movement. Various cells in both plants and animals increase in size as they divide and specialize, although not all do so. Any multicellular organism that increases in size does so primarily by an increase in the number of cells and not by a significant increase in the size of the cells. An elephant is larger than a mouse mainly because the elephant has more cells, not larger cells (Fig. 1.11).

Figure 1.11 An elephant is larger than a mouse because the elephant has more cells, not larger cells.

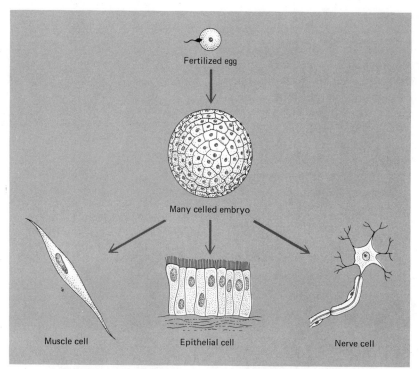

Figure 1.12 The process of development includes cellular differentiation. Many different types of cells may arise from a multicellular embryo.

Development in many living things involves cellular ***differentiation*** and ***morphogenesis***. Differentiation refers to the formation of different *types* of cells, such as skin cells, blood cells, muscle cells, and so on (Fig. 1.12). Morphogenesis is the process by which organs and other body structures acquire their particular shape or form. While certain aspects of growth and development do not apply to single-celled organisms, or to some multicellular types, both processes are nonetheless characteristic to some degree of all living things. Being unicellular, an amoeba, for instance, obviously does not grow by increasing its cell number; instead, it grows by increasing its size. Moreover, differentiation would not apply to the amoeba, since it does not form different types of cells. Some of the green algae, although multicellular, also show little tissue differentiation (that is, they do not possess distinguishable roots, stems, or leaves), but they grow by cell division (Fig. 1.13).

Sensation and Response. All living things are capable of sensing certain changes in their immediate environment. Any such environmental change is referred to as a ***stimulus***. Detection of the stimulus

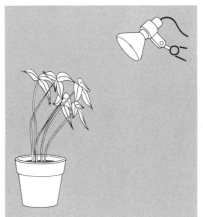

Figure 1.13 Although the body of this green alga is differentiated into parts, there is no tissue differentiation to form true roots, stems, or leaves. (From Villee, C. A.: *Biology.* 7th ed. Philadelphia: Saunders College Publishing, 1977, p. 168.)

Figure 1.14 Response in a living organism. Intense light (the stimulus) is a change in the plant's immediate environment. (From Applewhite, P., and Wilson, S.: *Understanding Biology.* New York: Holt, Rinehart and Winston, 1978, p. 9.)

by the organism is then followed by a certain *response* (Fig. 1.14). For example, animals follow the scent of food, and the roots of plants extend toward water. Among the animals, sensation and response are functions of specific nervous cells, tissues, or systems. In the plant kingdom, sensation and response are mediated by several different plant **hormones,** or growth-regulating substances. The ability of plants and animals to respond to various stimuli is intimately associated with their particular patterns of behavior. For the most part, the responses of living organisms tend to be useful in promoting their general welfare.

Reproduction. All living things have the ability to produce similar copies of themselves. The reproductive process may be as relatively simple as a protozoan splitting into two in a matter of minutes, or it may be as complex as the production and ultimate union of a sperm

> Fundamentally, evolution is the process whereby the genetic makeup of a population is affected.

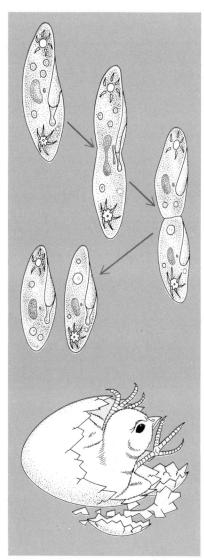

Figure 1.15 Reproduction results in the production of organisms that are similar copies of the parent or parents.

and an egg to produce a human being (Fig. 1.15). In the sphere of life, all living things arise only from other living things—a statement known as the ***biogenetic law.*** Through reproduction the perpetuation of life is made possible. Reproduction is also of tremendous consequence to the evolution of organisms. To a great extent, the number of individuals and the diverse genetic make-up of the offspring determine the course of the evolutionary process.

Adaptation. Biologically, adaptation refers to the ability of an organism to adjust to the changing requirements of the environment. As a consequence of adaptation, living things become suited to their surroundings. The forelimbs of birds, for example, have been modified for flight (Fig. 1.16); a pair of wings are of little use to a 1000-lb Kodiak bear. Adaptation may be short-term or long-term. Short-term adaptation involves the more immediate adjustments an organism may make when there are sudden changes in its environment. For example, if the air temperature drops sharply, the organism might respond by tightening its muscles and shivering in an attempt to keep warm. At the same time, certain internal adjustments are made that aid in maintaining the normal body temperature. Short-term adaptation, therefore, is a result of the sensation-response mechanism.

Populations of organisms may make adjustments over extended periods of time through long-term adaptation, more generally known as ***evolution.*** As environmental conditions and requirements change over thousands and millions of years, the continued existence of a population is dependent upon its capacity to adapt accordingly. The inability of individual organisms to make short-term adjustments often results in injury or death; for populations, the failure of long-term adaptation leads to extinction. Such failure is evident, for example, in the extinction of seed ferns and the fabled dinosaurs (Fig. 1.17).

Homeostasis. Homeostasis refers to the tendency of an organism to maintain a constant *internal* environment and to resist changes in that environment. The word "homeostasis" literally means "staying the same." Involved in homeostasis are such functions as maintaining proper water balance, temperature regulation, monitoring the composition of blood and other body fluids, and so on. What we generally describe as "being healthy" is essentially a reference to the fact that our bodies are in homeostatic balance. When this balance is upset, as when we have an elevated body temperature or become

Figure 1.16 The forelimbs of most birds are adapted for flight.

Figure 1.17 Extinct dinosaurs. These, like many other species of extinct organisms, failed to adapt over the long term. (a) *Tyrannosaurus;* (b) and (c) *Brontosaurus;* (d) *Stegosaurus;* (e) *Triceratops.* (From Villee, C. A., Walker, W. F., and Barnes, R. D.: *General Zoology.* 5th ed. Philadelphia: Saunders College Publishing, 1978, p. 808.)

> Homeostasis can apply to the functioning of an individual organism or to an entire ecosystem.

dehydrated, we are then ill. In effect, homeostasis is the result of proper functioning of the body's internal systems.

Homeostasis may also apply to groups of organisms, such as complex ecological communities that are considered to be relatively resistant to adverse environmental changes.

UNITY AND DIVERSITY

The discussion of the basic principles of life encompasses two interrelated concepts central to our understanding of the living world. As we have said, these principles apply to all organisms; it is apparent, therefore, that all living things are characterized by a fundamental *unity*, or sameness. At the same time, however, they exhibit considerable *diversity*. There is, for example, an enormous physical difference between a weed and a whale, but in terms of the seven essential life principles, they exhibit a remarkable unity.

The underlying unity of living things becomes more plausible in the light of our present knowledge concerning the origin of life on earth. Although incomplete, the available evidence strongly indicates that all living things are descendants of a common form of life—presumably some type of single cell. The unity, or sameness, that we find among living things can be explained, therefore, in terms of a common ancestry.

In the attempt to sort out and categorize the many types of living organisms, biologists have established a system in which each particular type of organism is classified according to certain distinguishing features. This systematic undertaking is known as the science of **taxonomy** (Gk. *taxis,* arrangement, + *nomos,* law).

As an illustration, all human beings are placed in a taxonomic group known as Primates, which also includes monkeys, gorillas, chimpanzees, orangutans, baboons, and a few others. As a group, the Primates are characterized by a large brain, three-dimensional vision with the eyes set in front of the skull, flexible limbs and digits, a shortened nose, upright posture, and so on. While every known primate possesses these characteristics, it is common knowledge that there is considerable diversity within the group. Baboons do not look exactly like human beings, nor are orangutans identical to chimpanzees.

With so many characteristics in common, why is there such diversity among the Primates? Basically, diversity among these organisms, as well as among other groups, has resulted from their innate capacity to adapt to the earth's many environments (Fig. 1.18). The

21
Unity and Diversity

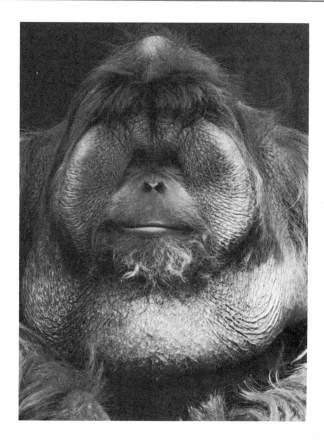

Figure 1.18 Primates. Although these animals belong to the same taxonomic group, their diversity is a result of their capacity to adapt to different environments. (From Villee, C. A.: *Biology.* 7th ed. Philadelphia: Saunders College Publishing, 1977, pp. 788, 789.)

adaptive changes in populations of organisms over long spans of time we have already described as evolution. Therefore, whereas unity among living things is an expression of their common ancestry, diversity results from their capacity to adjust or adapt to the demands of changing environments. It is the concept of evolution, then, that provides the unifying link between unity and diversity in living systems. In recognizing the unity and diversity of living things through their evolutionary relationships, our understanding of life on this planet is greatly enhanced and immensely simplified.

To close this introductory chapter and to state the guiding philosophy of this book, we will summon the words of Aristotle. He speaks to all of us when pointing out that much of our "knowledge" is nothing more than simple awareness of the existence of this or that thing. To have knowledge of them in the true sense, said Aristotle, we must know their causes, the principles behind them, and how they fit into the scheme of nature.

REVIEW OF ESSENTIAL CONCEPTS

1. The development of biology as an organized study is attributed primarily to Aristotle and the early Greek philosophers.
2. A "science" is a body of organized knowledge that depends ultimately upon the scientific method to verify its findings. Biology, the study of living things, is one of the disciplines of science.
3. Basically, there are seven essential principles that characterize all forms of life: *cellular organization, energy use, growth* and *development, sensation* and *response, reproduction, adaptation,* and *homeostasis*.
4. Through a common ancestry, all living things exhibit a fundamental unity; diversity among living things results from their capacity to adapt to different environments.
5. The adaptations made by populations of organisms over extended periods of time constitute *evolution*. Evolution is the unifying link between unity and diversity in living systems.

APPLYING THE CONCEPTS

1. Using specific concepts, explain why the term "science" would be inappropriate for use by cultures existing before the Greek era.
2. What do you think was the most important contribution made by Aristotle in establishing the science of biology? Why?
3. Is the philosophy of vitalism scientifically defensible? Explain.
4. Explain how the scientific method could be involved in a decision to buy a new car or to rescue a fading romance.
5. In what manner are metabolism, energy use, and the biogenetic law correlated with the cell theory?
6. Explain why an ant and a palm tree are characterized by a fundamental biological unity.
7. What is the relationship between evolution and adaptation?
8. Why is evolution considered the unifying link between unity and diversity in living systems?

some basic chemical principles

THE ESSENTIAL OBJECTIVES

You have understood this chapter when you are able to:
1. Describe the general structure of an atom.
2. Explain how atoms interact to form molecules and compounds.
3. Discuss the biological significance of the different electron energy levels.
4. Explain the formation of ions, ionic bonds, and covalent bonds.
5. Discuss the relationship between chemical bonds and the storage and release of energy.
6. Explain the concept of pH and discuss its biological significance.
7. Explain the role of buffers in living organisms and give an example of a buffer system.

All matter is constructed of one or more basic units called elements. Each element consists of atoms, which are composed of groups of electrical particles arranged in distinct configurations. Atoms interact to form the compounds and molecules that constitute the many kinds of living and nonliving matter. Even though immensely diverse, all forms of matter are fundamentally the same at the chemical level.

PREVIEW OF ESSENTIAL TERMS

matter Anything that has weight and takes up space.
atom The smallest part of an element that still retains the properties of that element.
compound A substance composed of two or more different atoms.
molecule The smallest part of a compound that still retains the properties of that compound.
oxidation The loss of hydrogen or electrons by an atom or molecule.
reduction The gain of hydrogen or electrons by an atom or molecule.
ion An atom or molecule that carries a positive or negative charge as a result of losing or gaining electrons.
chemical bond An attractive force between ions or atoms, resulting from the loss, gain, or sharing of electrons.
solution A homogeneous mixture of two or more substances, one of which is a liquid.
pH The symbol representing the relative hydrogen ion concentration of a solution. Acid solutions have pH values between 0 and 7, base solutions between 7 and 14. Lower pH values indicate a greater hydrogen ion concentration than do higher values.
buffer A chemical substance that aids in preventing drastic changes in the pH of solutions.

Some Basic Chemical Principles

Much has been written and stated and sung about "life," but like Plato's "Good," it has defied a comprehensive definition. The "Good," according to Plato, is a supreme principle of the highest value, an abstraction that cannot be put into words. We say that we have "life," but what exactly is it that we have? As yet, no one has been able to point a finger and claim unerringly: "There it is, *that* is life." Nonetheless, that life has a particular *quality* is obvious; we observe its manifestation in living things. Equally obvious is "nonlife," exemplified by a rock, mud, or a pane of glass. Life we envision as warm, responsive, dynamic; nonlife as cold, inert, static. It would appear, then, that life and nonlife are qualities at opposite extremes—separate, unrelated, and mutually exclusive. But as we shall see, appearances are sometimes deceiving.

THE STRUCTURE AND CHEMISTRY OF MATTER

Anything that has weight and takes up space—whether it is living or nonliving—is composed of one or more kinds of **matter**. Matter exists in one of three physical states: as a **solid**, a **liquid**, or a **gas**. In general, one state can be converted into another, primarily by raising or lowering the temperature (Fig. 2.1). For example, ice, a solid, can be converted to liquid water by raising the temperature of the ice. A further increase in temperature converts water into gaseous

Figure 2.1 The three states of matter (solids, liquids, and gases) may be converted in form by raising or lowering their temperatures. (From Whitten, K. W., and Gailey, K. D.: *General Chemistry*. Philadelphia: Saunders College Publishing, 1981, p. 4.)

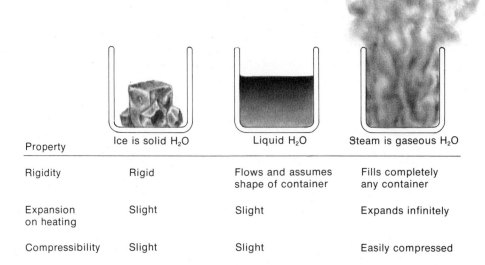

Property	Ice is solid H_2O	Liquid H_2O	Steam is gaseous H_2O
Rigidity	Rigid	Flows and assumes shape of container	Fills completely any container
Expansion on heating	Slight	Slight	Expands infinitely
Compressibility	Slight	Slight	Easily compressed

Table 2.1 COMMON ELEMENTS IN LIVING MATTER AND THEIR CHEMICAL SYMBOLS*

Carbon	:C	Sodium	:Na (for Latin *natrium*)
Oxygen	:O	Magnesium	:Mg
Hydrogen	:H	Calcium	:Ca
Nitrogen	:N	Chlorine	:Cl
Potassium	:K (for Latin *kalium*)	Sulfur	:S
Phosphorus	:P	Iron	:Fe (for Latin *ferrum*)

*From Clark, M.: *Contemporary Biology,* 2nd ed. Philadelphia, Saunders College Publishing, 1979, p. 27

steam, or water vapor. A sufficient lowering of the temperature will, of course, convert either water or steam into ice.

Each of the fundamental *kinds* of matter is called an **element.** There have been over 100 different elements discovered on earth, most of them occurring naturally, although a few have been produced in experimental laboratories. About 98 percent of all living matter is composed of but six elements: carbon, hydrogen, nitrogen, oxygen, phosphorus, and sulfur. The remaining 2 percent of living matter consists of the elements sodium, chlorine, magnesium, potassium, calcium, and others present in trace amounts (Table 2.1).

For convenience, each element has been given a different **chemical symbol** (Fig. 2.2). Often the symbol is simply the first letter of the element, such as C for carbon, H for hydrogen, N for nitrogen, O for oxygen, P for phosphorus, and S for sulfur. In combining the symbols of the six major elements of life, we come up with the acronym CHNOPS, which might help you recall them when necessary. Some of the other elements acquire their symbols from their Latin names. For example, the symbol for the element sodium is Na, taken from *natrium,* the Latin word for sodium, and the symbol for iron,

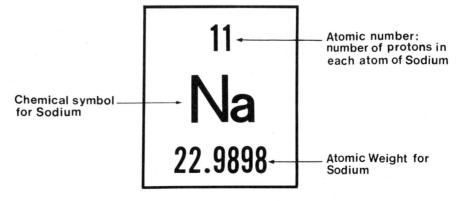

Figure 2.2 Sample box from the periodic table, showing the atomic number, symbol, and atomic weight of the element sodium. (From Peters, E. I.: *Introduction to Chemical Principles.* 3rd ed. Philadelphia, 1982.)

Some Basic Chemical Principles

> A *property* of a substance refers to any of its physical or chemical characteristics, such as density, odor, taste, boiling point, and so on.

Fe, comes from its Latin name, *ferrum*. Still other elements are designated by the first and one other letter of their English names, such as chlorine (Cl), magnesium (Mg), calcium (Ca), and so on.

Every element is composed of *atoms*, which are the smallest units of an element that have the same properties of the element. An atom in turn consists of particles called **protons, neutrons,** and **electrons.** The protons and neutrons together form the nucleus in the center of the atom. Each proton has a *positive* electrical charge of +1, and as the name suggests, neutrons have *no* electrical charge. Occupying orbitals or shells at varying distances outside the atomic nucleus are *electrons*, each of which has a *negative* electrical charge of −1. The number of electrons in an atom is the same as the number of protons. Structurally, then, an atom somewhat resembles a miniature solar system, with the atomic nucleus as the sun and the electrons representing the orbiting planets (Fig. 2.3).

Although a neutron has no electrical charge, it has roughly the same mass as a proton (mass refers to the quantity of matter in an object). The total number of protons and neutrons in the atomic nucleus is called the **mass number.** This number is usually written as a superscript at the upper left of the chemical symbol. For example, the element carbon usually has a mass number of 12 (actually

Figure 2.3 Diagrams of the atoms of the six major elements of life. Symbols: + = proton; ◯ = neutron; ⊖ = electron.

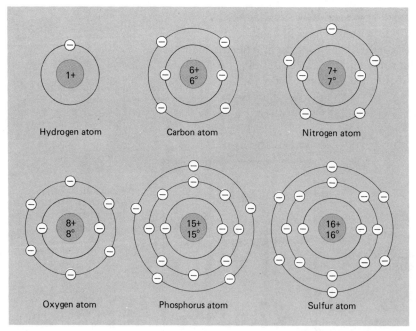

> The uniqueness of every kind of atom is attributable to its atomic number.

12.01115), i.e., the nucleus of a carbon atom is composed of 6 protons and 6 neutrons. The mass number of carbon, then, is designated ^{12}C. However, there are some carbon atoms that have *seven* neutrons and six protons in their atomic nuclei and still other carbon atoms that contain *eight* neutrons and six protons. Thus, the mass number of some carbon atoms is 13 (written ^{13}C), and for others the mass number is 14 (^{14}C), instead of the usual 12. Both ^{13}C and ^{14}C are called *isotopes* of carbon (Fig. 2.4). Isotopes of any element are atoms having the same number of protons and electrons but different numbers of neutrons. As another example, most nitrogen atoms have seven protons and seven neutrons—a mass number of 14. Some nitrogen atoms, however, have seven protons and *eight* neutrons, a mass number of 15. Thus, ^{15}N is an isotope of nitrogen because it has a different number of neutrons than protons or electrons. On the other hand, the number of protons, electrons, and neutrons of ^{14}N is the same.

Some isotopes are unstable, or *radioactive*, i.e., their nuclei emit radiation particles. With the use of special instruments, this radiation can be detected in biological systems. For example, by "tagging" red blood cells with a radioactive isotope, the course and velocity of the cells can be determined as they circulate through the body. In fact, our understanding of many biological reactions and processes, such as photosynthesis and the self-duplicating ability of DNA, has been greatly heightened by the use of radioactive isotopes.

Since the number of protons equals the number of electrons, an atom has a net electrical charge of zero. The number of protons and electrons in an atom of one element is different from the number in an atom of another element. For example, a carbon atom has six protons in its nucleus and therefore six electrons outside the nucleus; oxygen has eight protons and eight electrons; phosphorus has 15 protons and 15 electrons, and so on for each of the 100 or so elements. The number of protons in the nucleus of an atom is called the **atomic number.** Carbon, then, has an atomic number of 6, that of oxygen is 8, that of phosphorus is 15, and so forth. What makes each element unique, therefore, is that each one has an atomic number different from that of any other element.

The atomic number is written as a subscript at the lower left of the chemical symbol. For the element carbon, the atomic number is written $_6C$; for oxygen, $_8O$; for phosphorus, $_{15}P$, and so on. Each element, then, can be identified by a particular atomic number and, as discussed previously, by a particular mass number. Therefore, to designate both of these numbers, the element carbon is written $^{12}_{6}C$; oxygen, $^{16}_{8}O$; and phosphorus, $^{30}_{15}P$.

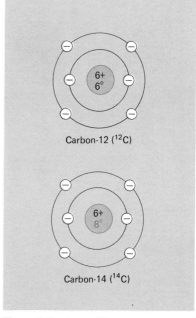

Figure 2.4 (Upper) Diagram of a normal carbon atom having a mass number of 12. (Lower) An isotope of carbon having a mass number of 14. Note that the only structural difference between the two atoms is the two additional neutrons in the atomic nucleus of ^{14}C. Symbols: + = proton; ○ = neutron; ⊖ = electron.

Some Basic Chemical Principles

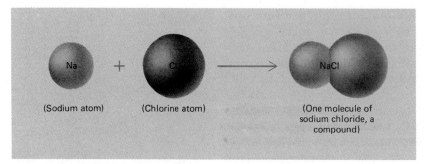

Figure 2.5 One molecule of a chemical compound is formed by combining two or more *different* kinds of atoms.

Knowledge of the basic structure of atoms is important in biology because many of the chemical processes that occur in living things are essentially interactions between atoms. For example, when two or more *different* kinds of atoms combine, they form a **compound** (Fig. 2.5). An atom of the element sodium (Na) combines with an atom of chlorine (Cl) to form the compound sodium chloride (NaCl), ordinary table salt. The smallest part of a compound that still has the properties of that compound is called a **molecule**. Thus, one atom of sodium and one atom of chlorine react to form one molecule of the compound sodium chloride. We could not break down this molecule of sodium chloride into any simpler form and still retain the properties of salt.

Molecules also can be formed by combining two or more atoms of the *same* element (Fig. 2.6). This is the case particularly in those elements that normally are found in the gaseous state. A molecule of oxygen gas, for instance, is written O_2, the subscript "2" indicating that there are two oxygen atoms in the molecule. Similarly, a mole-

Figure 2.6 (a) and (b) Molecules are also formed by combining two or more atoms of the *same* element.

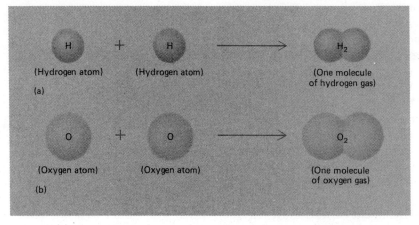

> Electrons at different energy levels have different amounts of potential energy.

cule of nitrogen gas, N_2, consists of two atoms of nitrogen, and hydrogen gas, H_2, is composed of two hydrogen atoms. One of the regions of the earth's atmosphere called the stratosphere consists in part of a layer of *ozone* molecules, O_3, which are formed by the complexing of *three* oxygen atoms. When it is necessary to indicate more than one molecule of a substance, we simply precede it by the appropriate number. Thus, three molecules of oxygen gas would be written as $3O_2$, six molecules as $6O_2$, and so on. In the same manner, two molecules of sodium chloride would be indicated as $2NaCl$, four molecules as $4NaCl$, and so on.

A compound or a molecule usually has properties that are different from the properties of the elements that compose them. For example, the common occurrence of rusting results from the combining of the element iron, a metal, with gaseous oxygen. The compound formed—iron rust—is a reddish, solid powder, almost totally unlike either of the elements that formed it. As another example, the liquid compound water, H_2O, certainly bears little resemblance, physically or chemically, to either the hydrogen or the oxygen elements that constitute it.

THE ROLE OF ELECTRONS

The tendency of atoms to combine, i.e., to undergo chemical reactions, results from the particular construction of each atom. It has been mentioned that the number of electrons (and protons) in the atoms of one element is different from the number in any other element. As a consequence, the overall arrangement, or configuration, of the electrons in their shells is different for each element. An electron shell, or orbital, is an imaginary path followed by an electron as it moves around the atomic nucleus. It would be more accurate to refer to these shells or orbitals as *energy levels*. This is because electrons have different amounts of energy, depending upon the shell they occupy. Electrons in the shell nearest the nucleus have the least energy, whereas electrons in each shell farther away from the nucleus have increasingly more energy. Thus, electrons in the last, or outermost, shell would have the greatest amount of energy.

Of what consequence is this to living things? Of major importance is the fact that electrons can be "boosted" from lower energy levels, i.e., from electron shells nearer the atomic nucleus, to higher ones farther away from the nucleus by an input of energy from some outside source. In the cells of green plants, for example, the energy for boosting electrons to higher energy levels comes from the

Some Basic Chemical Principles

sun. Electrons at high energy levels then become a source of **potential (stored) energy** that the plant cells can use. When the electrons return to lower energy levels, they release this stored energy for cellular work. To illustrate this principle, visualize a boulder lying at the foot of a steep hill (Fig. 2.7). An input of energy (such as that provided by your muscles) can "boost" the boulder up the hill. In moving up the hill to higher and higher levels, the boulder gains increasing amounts of potential energy. At the top of the hill, the boulder contains more potential energy than it has at any of the lower levels. If the boulder is then given a push (an input of energy), it releases its potential energy as it returns to lower levels of the hill. In rolling from the top of the hill to the bottom, the boulder would release more potential energy than if it began rolling only a few yards from the bottom. This could be verified to one's satisfaction by standing in its path.

In this example, the potential energy stored in the boulder becomes, or is converted into, **kinetic** (Gk. *kinetikos,* moving) **energy** as the boulder begins to roll down the hill. Kinetic energy is the energy for doing work. As pointed out, more potential energy would be released if the boulder began rolling from the top of the hill than if

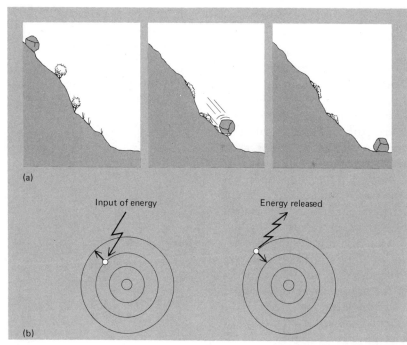

Figure 2.7 (a) When a boulder is up on the side of a mountain, it has high potential energy. As it rolls down the mountain, its potential energy is converted into kinetic energy. As it slows down and comes to rest, its kinetic energy is converted to heat and is given to the earth. The potential energy when the boulder is at the bottom of the mountain is lower than when it is at the top. (b) Similarly, an electron raised to a higher energy level gains potential energy. In returning to a lower energy level, the potential energy stored in the electron is converted into kinetic energy.

> Atoms tend to undergo reactions with each other because they have incomplete outer electron shells.

it started from some lower point. This means that the boulder could perform more work, since more potential energy would have been released and converted into kinetic energy.

Electrons in the atoms of living matter behave much like the boulder. Provided there is a sufficient input of energy, the electrons gain potential energy as they are raised to higher energy levels and release this energy as they return to lower levels.

Every atom is so constructed that its electron shells contain a limited number of electrons. The first shell, the one nearest the nucleus, can hold a maximum of two electrons; the second and third outer shells usually have a limit of eight electrons each. (Some of the largest atoms, with their high atomic numbers, may have as many as six or seven electron shells, but the more important elements found in living things have relatively low atomic numbers; consequently, their atoms rarely have more than three electron shells.) If the second or third shell of an atom is the *outermost* shell, and it contains *eight* electrons, it is said to be complete, or "closed." (This is true for all atoms except those few in which the *first* shell is the outer shell. In these cases, the first shell is complete with only two electrons; the helium atom with an atomic number of 2 is an example.) If the outermost shell has fewer than eight electrons, it is incomplete, or "open." Most atoms do *not* have complete outer shells, and this is the key to their ability to undergo chemical reactions.

An atom tends to fill its outermost shell with eight electrons by combining with other atoms. Consequently, reactions between atoms would not normally occur if their outermost shells were already complete. This is the case, for example, with atoms of such elements as neon, krypton, helium, and a few others. In contrast to all other atoms, these few types, with the exception of helium, have outer electron shells that always contain eight electrons. The outer (and only) shell of helium (atomic number of 2) is complete with two electrons. Accordingly, these atoms usually do not tend to undergo reactions with other atoms and are therefore referred to as **inert,** meaning nonreactive (Fig. 2.8). By comparison, an atom of oxygen—with an atomic number of eight—has two electrons in its first shell and only six electrons in its outermost shell. Phosphorus, with an atomic number of 15, has two electrons in its first shell, eight in the second shell, and five electrons in its outermost shell. Therefore, oxygen and phosphorus, having incomplete outer shells, tend to undergo reactions with other atoms to complete those shells.

Atoms acquire complete outer shells by either gaining or losing electrons, or by sharing electrons with other atoms. A gain and loss

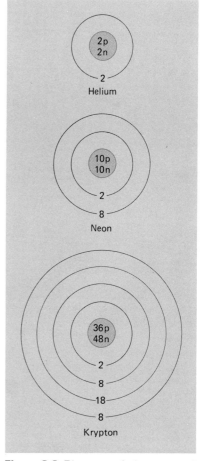

Figure 2.8 Diagrams of the atoms of three "inert" gases. Why do these atoms usually fail to undergo reactions with other atoms?

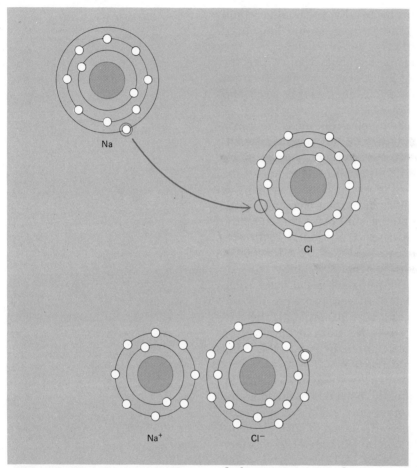

Figure 2.9 Reaction between sodium (Na) and chlorine (Cl) atoms to form the ionic compound sodium chloride (NaCl).

of electrons occurs, for example, in the reaction between sodium (Na) and chlorine (Cl) to form sodium chloride (NaCl) (Fig. 2.9). The sodium atom, with an atomic number of 11, has two electrons in its first shell, eight in the second shell, and one in its outermost shell. With an atomic number of 17, the chlorine atom has two electrons in its first shell, eight in the second shell, and seven in its outermost shell. Sodium, then, needs seven electrons to complete its outer shell, and chlorine needs only one electron. In this type of reaction, atoms with fewer than four outer shell electrons usually *lose,* or transfer, their electrons to another atom, whereas atoms with more than four usually *gain* electrons. Therefore, in this example, the sodium atom would lose its outer shell electron, which is taken up, or gained, by the outer shell of the chlorine atom. The sodium atom, then, is left

> Oxidation and reduction always must occur together.

with its completed *second* shell as the outermost shell; in gaining one electron, the outermost shell of the chlorine atom is also complete with eight electrons.

The loss of electrons by an atom is referred to as **oxidation,** and the gain of electrons is called **reduction.** Thus, in losing an electron, the sodium atom was oxidized, and the chlorine atom was reduced by gaining an electron. The terms oxidation and reduction also apply to the loss or gain of *hydrogen atoms*. A hydrogen atom consists of one proton and one electron. When a compound is oxidized or reduced, it loses or gains the hydrogen electron along with the hydrogen proton (H^+). When any compound is oxidized by losing electrons or hydrogen, another compound is reduced by gaining the electrons or hydrogen; this combination of oxidation and reduction constitutes one reaction known as an **oxidation-reduction reaction.**

In many of the chemical reactions vital to life, hydrogen atoms are removed from a compound and taken up by an oxygen atom to form water. In Chapter 8 we shall discuss the importance of the hydrogen atom in the oxidation-reduction reactions of photosynthesis and metabolism.

After the loss or gain of electrons, an atom is no longer electrically neutral. The sodium atom, for example, loses 1 electron and thus has a total of 11 protons, but only 10 electrons. Sodium, then, acquires a $+1$ (positive) charge, meaning that the protons outnumber the electrons by one. In chemical notation the charge on the sodium atom is expressed as Na^+. Chlorine, on the other hand, gains an electron, giving it a total of 18 electrons but only 17 protons. This gives the chlorine atom a -1 (negative) charge, written Cl^-. Any atom that carries an electrical charge is called an **ion.** Thus, Na^+ indicates a sodium ion, and Cl^- indicates a chloride ion. If more than one electron is gained or lost, the number is designated in the superscript. For example, in losing *two* electrons, the ionic form of the calcium atom (Ca) would be written Ca^{+2}, or Ca^{++}; in gaining *two* electrons, the oxygen ion would be written O^{-2}; for aluminum the ion is designated Al^{+3}, and so on.

In the reaction between sodium and chlorine, the ions are attracted to each other because they carry opposite electrical charges. The mutual attraction results in the formation of an **ionic bond** (Fig. 2.9). This, or any type of chemical bond, is simply an attractive force between atoms or ions.

Chemical reactions also occur in which the participating atoms *share* electrons in order to complete their outermost shells. Two atoms of chlorine, for example, interact to form a molecule of chlo-

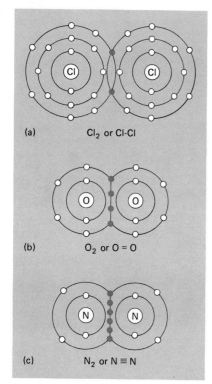

Figure 2.10 (a) By sharing one pair of electrons, the chlorine atoms are joined by a *single* covalent bond to form chlorine gas. (b) Oxygen gas is formed when the oxygen atoms share two pairs of electrons, forming a *double* covalent bond. (c) A *triple* covalent bond is formed when two nitrogen atoms share three pairs of electrons.

rine gas, Cl_2. With an atomic number of 17, the chlorine atom has two electrons in its first shell, eight in the second shell, and seven in its third, or outermost, shell. As we have seen, the chlorine atom requires one electron to complete its outer shell. In the reaction, the outer shells of the two chlorine atoms overlap so that each atom can share one of its seven electrons with the other. In this manner, the outer shells of both atoms are complete with eight electrons.

This sharing of electrons results in the formation of a **covalent bond** (Fig. 2.10). Since there is *one* pair of electrons shared between the chlorine atoms, the resulting bond is more correctly a *single* covalent bond. Covalent bonds are indicated by a dash (—) between the participating atoms; therefore, chlorine gas (Cl_2) can be written Cl—Cl. If atoms share *two* pairs of electrons, as in oxygen gas (O_2), for example, there is formed a *double* covalent bond. To indicate this, oxygen gas can be written O=O. If *three* pairs of electrons are shared, as in nitrogen gas (N_2), a *triple* covalent bond is formed. Nitrogen gas, then, would be written N≡N. Double or triple bonds are always stronger than single bonds joining the same two elements; in fact, the N≡N bond is the most stable bond we know about. Notice that the formation of covalent bonds does not involve the transfer of electrons and that ions are not formed. Being quite strong, covalent bonds are vital in the structural molecules of living things.

Hydrogen, one of the most important and abundant elements in living things, interacts with certain atoms, particularly oxygen and nitrogen, to form what are called **hydrogen bonds**. A hydrogen bond is formed between an already bonded hydrogen atom and a negatively charged atom of another element. In liquid water, for example, the oxygen atom of one water molecule is joined to the oxygen atom of another water molecule by a hydrogen bond (Fig. 2.11). Individually, hydrogen bonds are relatively weak; collectively, however, they are quite strong. Consequently, many of the molecules that make up living things have a great number of hydrogen atoms. This is important in that hydrogen bonds aid in maintaining the proper shapes of the molecules. In the following chapter, you will discover that many of the unique properties of water are attributable to hydrogen bonding.

How can the information concerning chemical bonds be applied toward a better understanding of living systems? We have said that chemical bonds are attractive forces between atoms. Accordingly, the molecules and compounds associated with both living and nonliving matter are "held together" by chemical bonds. When bonds form

37
The Role of Electrons

> Metabolism involves the storing of potential energy in chemical bonds, and the production of kinetic energy when these bonds are broken.

between atoms, chemical energy is stored in those bonds; when the bonds are broken, this energy is released. You, and every other living organism, require certain kinds and amounts of foods or nutrients. These are the compounds that supply the cells with chemical energy. This chemical energy is stored in the bonds and is thus a form of potential energy. Reactions within the cells break the bonds holding the food molecules together, releasing energy for cellular use. The released energy is kinetic energy—energy available for work (recall the boulder on the hill). Some of this kinetic energy is used in building up new molecules and compounds, and some of it provides heat energy for the maintenance of body temperature. Basically, this is what metabolism is all about—the making and breaking of chemical bonds. It should be apparent, then, that life is a saga

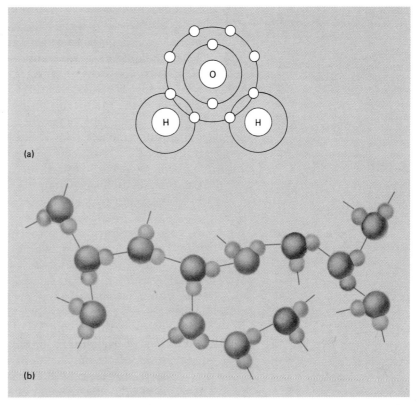

Figure 2.11 (a) Covalent bonding between one atom of oxygen and two atoms of hydrogen to form a molecule of water (H_2O). The larger oxygen nucleus attracts the hydrogen electrons, so that the oxygen atom acquires a weak negative charge. This leaves the hydrogen atoms with a weak positive charge. (b) Hydrogen bonds *(colored dashes)* between water molecules in liquid water. Each positively charged hydrogen atom is attracted to a negatively charged oxygen atom in the area.

> Acids release hydrogen ions (H^+) when they dissociate in water; bases release hydroxyl ions (OH^-).

of constantly changing energy forms. We shall pursue this essential biological concept throughout this book.

THE pH SCALE

The chemical reactions of life take place in **solutions** within and around the cells. A solution is a homogeneous, or uniform, mixture of two or more substances, one of which is a liquid. Sugar dissolved in coffee or tea is a common example. The dissolved substance is the **solute,** and the liquid is the **solvent.** In biological systems the liquid is always water. Many of the solutions in the human body, such as perspiration or tears, for example, are over 90 percent water; in fact, the entire body is nearly 70 percent water. Moreover, the living cells of every organism are continually bathed in a watery solution.

Almost all solutions are either acidic or basic. This is certainly true of the solutions in living systems. Everyone is familiar with the term **acid,** particularly as it refers to solutions that taste sour, such as vinegar, lemon juice, and other unsweetened fruit juices. While the term **base** is generally less well known, there are commercial bases familiar to practically everyone. For example, baking soda, bleaches, liquid ammonia, and scouring powders are bases used extensively in the home (Fig. 2.12).

Chemically, what is it that makes a particular solution either acidic or basic? Many substances found in living organisms *dissociate*, or "break apart," into ions when in solution. For example, when the salt sodium chloride (NaCl) is added to water, it dissociates into sodium ions (Na^+) and chloride ions (Cl^-).

Acids and bases also dissociate in water. An acid is a compound that releases hydrogen ions (H^+) when it dissociates. For instance, the dissociation of hydrochloric acid (HCl) in water releases hydrogen ions (H^+) and also chloride ions (Cl^-) (Fig. 2.13). Note that the hydrogen electron stays with the chlorine atom. A base is a compound that releases hydroxyl ions (OH^-) when it dissociates in water. A common base is the bleaching agent sodium hydroxide (NaOH), which dissociates into sodium ions (Na^+) and hydroxyl ions (OH^-) in water. Liquid water itself dissociates to some extent into hydrogen ions (H^+) and hydroxyl ions (OH^-) and therefore contributes these ions to the solution. However, since the amount, or concentration, of each ion is the same, the H^+ and OH^- ions of water itself do not affect the acidity or basicity of the solution.

The relative acidity or basicity of a water solution is measured in terms of **pH.** The symbol pH is derived from "*p*otential of *h*ydro-

The pH Scale

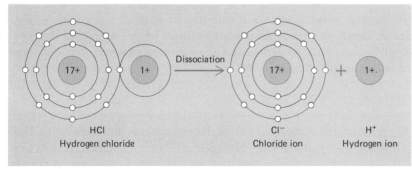

Figure 2.12 Some of these foods are acidic; others are basic. Can you tell which is which? (From Masterton, W. L., Slowinski, E. L., and Walford, E. T.: *Chemistry.* New York: Holt, Rinehart and Winston, 1980, p. 473.)

Figure 2.13 Dissociation of hydrochloric acid (HCl). All acids release hydrogen ions (H^+) when they dissociate in water.

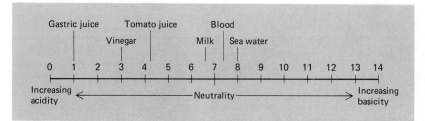

Figure 2.14 The pH scale. Acidic solutions (those with a pH below 7) contain more free hydrogen ions than hydroxyl ions. Basic solutions (those with a pH above 7) contain more hydroxyl ions than hydrogen ions.

gen" and refers to the concentration of free hydrogen ions in a solution. As stated, acids release hydrogen ions in water, and bases release hydroxyl ions. An acidic solution, therefore, contains more free hydrogen ions than hydroxyl ions, and a basic solution contains more hydroxyl ions than hydrogen ions.

The *pH scale* ranges from 0 to 14, with a solution of pH 7 being neutral (Fig. 2.14). A neutral solution, such as pure water, has an equal concentration of hydrogen and hydroxyl ions and therefore is neither acidic nor basic. Acid solutions have a pH between 0 and 7, and basic solutions have a pH between 7 and 14. Acidity increases in going from pH 7 down to pH 0, whereas basicity increases in going from pH 7 up to pH 14. For example, a solution of pH 2 is more acidic than one of pH 5, and a solution of pH 12 is more basic than one of pH 8. Actually, a change in pH from one whole number to another represents a 10-fold change in acidity or basicity. This means that a solution of pH 2 is 10 times as acidic as a solution of pH 3, and 100 times as acidic as a solution of pH 4. Similarly, a solution of pH 10 is 10 times as basic as a solution of pH 9, and 100 times as basic as a solution of pH 8 (Fig. 2.13).

A familiarity with the concept of pH is important because the normal functioning of living organisms depends upon the maintenance of a correct and constant pH of the body fluids. Human blood plasma, for example, is maintained at a pH value of about 7.4; if the plasma pH falls below about 6.9, the individual will usually pass into a coma. Moreover, a group of essential biological compounds called *enzymes* function only at certain pH values. In the following chapter we shall discuss the enzymes and their role as catalysts in the chemical reactions of life.

BUFFERS

The relative constancy of pH in living systems is essential to the maintenance of homeostasis. However, the metabolic reactions in liv-

> The bicarbonate buffer system is extremely important in regulating acid-base balance in the human body.

ing organisms continually add or remove hydrogen ions; if not countered, such fluctuations in hydrogen ion concentration tend to alter dramatically the pH of the body fluids. One of the fundamental mechanisms for maintaining a constant pH of the body involves chemical substances known as *buffers*. A buffer is capable of either bonding with hydrogen ions or bonding with hydroxyl ions in solution. If a fluid becomes too acidic, a buffer "takes up," or removes, the excess hydrogen ions responsible for the increased acidity. Conversely, a buffer removes excess hydroxyl ions if a fluid becomes too alkaline, or basic.

For example, one of the major buffer systems in human beings is the ***bicarbonate buffer system.*** All of the fluids of the body contain carbonic acid (H_2CO_3), a weak acid, and sodium bicarbonate ($NaHCO_3$), a weak base, in solution. If the acidity of the body fluids begins to increase, i.e., if the hydrogen ion concentration in the fluids increases, the sodium bicarbonate takes up the excess hydrogen ions. Assume, for example, that excess hydrochloric acid (HCl) enters the body fluids. You will recall that hydrochloric acid dissociates into H^+ ions and Cl^- ions in solution. The reaction between HCl and $NaHCO_3$ occurs as follows:

$$HCl + NaHCO_3 \longrightarrow H_2CO_3 + NaCl$$

The products formed are carbonic acid (H_2CO_3), a weak acid, and sodium chloride (NaCl), a salt. As you can see, the hydrogen ions released by hydrochloric acid are taken up by sodium bicarbonate. Consequently, the acidity of the body fluids does not increase significantly.

If a base such as sodium hydroxide (NaOH) is added to the body fluids, it reacts with the carbonic acid. Recall that sodium hydroxide dissociates into Na^+ ions and OH^- ions in solution.

$$NaOH + H_2CO_3 \longrightarrow NaHCO_3 + H_2O$$

In this instance, the hydroxyl ion (OH^-) of sodium hydroxide combines with a hydrogen ion from carbonic acid to form water. Removal of this hydrogen ion from carbonic acid forms the bicarbonate ion (HCO_3^-). When the sodium ion combines with the bicarbonate ion, the weak base sodium bicarbonate ($NaHCO_3$) is formed. By taking up excess OH^- ions, the bicarbonate ion prevents the body fluids from becoming too alkaline.

> All forms of matter are structurally the same at the chemical level. This is the key to the evolutionary relationship between life and nonlife.

If carbonic acid and sodium bicarbonate are always present in the body fluids, why do they not greatly affect the acidity or alkalinity of the fluids? The key to the answer is that H_2CO_3 is a *weak* acid and $NaHCO_3$ is a *weak* base. Acids are designated as either weak or strong, depending upon the degree to which they dissociate in water. Strong acids, such as HCl, dissociate more completely in water, whereas weak acids dissociate to a much lesser degree. Therefore, weak acids release fewer hydrogen ions in solution than do strong acids.

Similarly, strong bases form solutions that have a very high concentration of OH^- ions. These bases are corrosive and extremely irritating to human tissues. Weak basic solutions have low concentrations of OH^- ions. Solutions of sodium hydroxide (NaOH) and potassium hydroxide (KOH) are examples of strong bases; magnesium hydroxide ($Mg[OH]_2$), known as *milk of magnesia,* and aluminum oxide (Al_2O_3) are weak bases. Both of these weak bases are found in the antacids and buffered aspirin sold in drug stores. By taking up hydrogen ions, $Mg(OH)_2$ and Al_2O_3 help neutralize excess stomach acid. For milk of magnesia this neutralization reaction can be shown as follows:

$$Mg(OH)_2 + \underset{\text{(stomach acid)}}{2HCl} \longrightarrow \underset{\text{(magnesium chloride, a salt)}}{MgCl_2} + 2H_2O$$

The hydrogen ions released by the dissociation of stomach acid (HCl) are taken up by the OH^- ions to form water, and magnesium and chloride react to form the neutral salt magnesium chloride ($MgCl_2$).

In examining the basic structure of matter, we have uncovered another principle fundamental to a better understanding of living systems. All forms of matter—whether present in living or nonliving things—are structurally similar at the chemical level. Matter is composed of elements, which in turn consist of atoms, which are aggregates of subatomic particles—protons, neutrons, and electrons. The element carbon is still carbon whether it constitutes a lump of coal or a lock of your hair. This fundamental sameness of all matter is the key to the evolutionary relationship between the living and the nonliving. What this relationship suggests is that, in millenia past, an increasingly complex array of chemicals interacted to produce the building blocks of the first primitive cells, leading eventually to the progressive rise of a variety of plants and animals. In other words, chemical evolution led to organic evolution. We can appreciate this sequence only through an understanding of the basic structure of matter.

REVIEW OF ESSENTIAL CONCEPTS

1. Matter is composed of *elements*, which in turn consist of *atoms*; atoms of each element are composed of *protons* and *neutrons* that make up the *atomic nucleus* and of *electrons* located in shells or orbitals outside the nucleus.
2. Protons have an electrical charge of $+1$, and electrons have an electrical charge of -1; *neutrons* have no charge. The number of protons in an atom equals the number of electrons, giving the atom a net electrical charge of 0.
3. The number of protons in the atomic nucleus is called the *atomic number*; the number of protons plus the number of neutrons is the *mass number*. Atoms of any element having the same number of protons and electrons but a different number of neutrons are called *isotopes*.
4. Atoms may combine to form compounds or molecules; a *compound* is composed of two or more different elements. The smallest part of a compound that retains its properties is called a *molecule*; molecules may also be formed by the combination of two or more atoms of the same element.
5. Electron shells or orbitals of an atom are also referred to as *energy levels*; electrons in shells farther away from the atomic nucleus (high energy levels) possess more energy than electrons closer to the atomic nucleus (low energy levels). Through an input of energy, electrons can be boosted to higher energy levels and will release energy as they return to lower levels. Stored energy is called *potential energy*, which can be converted into the energy of motion, called *kinetic energy*.
6. The electron shells of atoms contain a limited number of electrons: The first shell contains a maximum of *two* electrons, and the second and third outer shells have a maximum of *eight* electrons each. Shells having the maximum number of electrons are said to be *complete* or *closed*.
7. The fact that most atoms do not have the maximum number of electrons in their outer shells accounts for their ability to combine, or react, with each other.

8. Reactive atoms acquire complete outer shells by either gaining or losing electrons, or by sharing electrons. Atoms with fewer than four outer shell electrons tend to lose them; atoms with more than four tend to gain electrons. The loss of electrons is called *oxidation;* the gain of electrons is called *reduction.* Oxidation and reduction also involve the loss or gain of *hydrogen* atoms.

9. An atom that has gained or lost electrons is called an *ion;* an ion, therefore, carries either a positive or a negative electrical charge. The attractive force between oppositely charged ions is called an *ionic bond.*

10. The sharing of electrons by atoms results in the formation of a *covalent bond;* there are single, double, and triple covalent bonds. *Hydrogen bonds* are formed by the sharing of hydrogen electrons between a hydrogen atom and another atom.

11. Individual *hydrogen bonds* are weak covalent bonds; collectively, however, hydrogen bonds are quite strong. Hydrogen bonds are important in maintaining the proper shape of molecules.

12. Solutions in living systems are either acidic or basic. An *acid* is a substance that releases hydrogen ions when it dissociates; a *base* releases hydroxyl ions. Acidity and basicity are measured in terms of *pH,* which refers to the relative concentration of free hydrogen ions in solution. Acid solutions contain more hydrogen ions than hydroxyl ions; basic solutions contain more hydroxyl ions than hydrogen ions. The *pH scale* ranges from 0 to 14, with pH 7 being neutral; acids have a pH between 0 and 7, and bases have a pH between 7 and 14.

13. A *buffer* is a chemical substance that aids in preventing drastic changes in the pH of body fluids. Buffers act by taking up excess hydrogen or hydroxyl ions and thereby control the acidity or alkalinity of the fluids.

14. Since all matter is composed of 1 or more of the more than 100 elements, the living and the nonliving are structurally similar at the chemical level. This fundamental sameness of matter is indicative of an evolutionary relationship between life and nonlife.

APPLYING THE CONCEPTS

1. Diagram the nuclei and electron shells of the following atoms: aluminum ($^{26}_{13}Al$), helium ($^{4}_{2}H$), magnesium ($^{24}_{12}Mg$), and sulfur ($^{32}_{16}S$).
2. What is the relationship between potential energy, kinetic energy, and chemical bonds?

3. In the chemical reaction between magnesium and oxygen to form magnesium oxide (MgO), which atom is reduced and which atom is oxidized? What are the electrical charges on each of the formed ions?
4. Explain the formation of the double covalent bond in oxygen gas.
5. Why is a knowledge of the concept of pH important in biology?
6. Why is pure water neither acidic nor basic?
7. If solution A has a pH of 5 and solution B has a pH of 3, how could you express the relative differences in hydrogen ion concentration between the two solutions?
8. Why are buffer systems in living organisms composed of weak acids and weak bases? How do weak acids and weak bases function in regulating the pH of a solution?

3 life: the essential materials

THE ESSENTIAL OBJECTIVES

You have understood this chapter when you are able to:

1. List the major chemical elements essential to living things.
2. Explain why the carbon atom is the basic structural element of life.
3. Describe the general structure and function of each of the compounds of life.
4. Explain with examples the reactions of hydrolysis and dehydration synthesis.
5. Discuss the biological importance and functional activity of enzymes.
6. Describe the general function of ATP, NAD, and FAD.
7. Describe the basic chemical structure of water and discuss its biologically important properties.

Living matter is constructed from molecules and compounds containing the element carbon. Carbohydrates, lipids, proteins, and nucleic acids compose the bulk of living matter, but a variety of elements and other compounds, including water, are equally essential.

PREVIEW OF ESSENTIAL TERMS

carbohydrate An organic compound consisting primarily of carbon, hydrogen, and oxygen atoms organized to form monosaccharides, disaccharides, and polysaccharides.

lipid An organic compound consisting primarily of carbon, hydrogen, and oxygen atoms organized to form fats, oils, waxes, and so on.

protein An organic compound composed of amino acids, which in turn consist of carbon, hydrogen, oxygen, and nitrogen atoms.

nucleic acid An organic compound composed of nucleotides; each nucleotide consists of a nitrogen base, a 5-carbon sugar, and a phosphate group.

hydrolysis A metabolic reaction in which a large molecule is split into smaller molecules by the addition of a water molecule.

dehydration synthesis A metabolic reaction in which a large molecule is formed from smaller molecules by the removal of a water molecule.

enzyme An organic catalyst; a protein that accelerates the rate of a chemical reaction.

coenzyme An organic compound such as nicotinamide adenine dinucleotide (NAD) or flavin adenine dinucleotide (FAD) that functions as an electron and hydrogen acceptor in an enzymatically controlled reaction.

adenosine triphosphate (ATP) An organic compound composed of adenine, ribose, and three phosphate groups; the major storage form of chemical energy in the cell.

Life: The Essential Materials

In Chapter 1 it was pointed out that living things are characterized by a fundamental unity, or sameness, while at the same time exhibiting extreme diversity. The unity of life through a common ancestry suggests that living things also exhibit unity through a common structural organization. This unity is evident at the chemical level in that all living matter is structured around the element carbon. With carbon atoms as the structural base, the bonding of hydrogen, nitrogen, oxygen, and other atoms to carbon forms a variety of *organic* compounds essential to the structure and function of living systems. The term "organic" is used in referring to living things in general, as well as to their carbon compounds. Nonliving matter is termed *inorganic*, although some inorganic compounds, such as carbon dioxide (CO_2), also contain carbon. There is, however, a variety of inorganic molecules found in living things; in fact, life would not be possible without them.

The chemical elements constituting the bulk of living matter also compose many forms of nonliving matter. The elements carbon, hydrogen, and oxygen, for example, make up about 92 percent of the body weight of a living animal, but these elements are also the constituents of many nonliving substances, such as alcohols, ether, various acids, nail polish remover, and so on. This is not meant to suggest, however, that living matter and nonliving matter always or even usually contain the same elements. What it does suggest is that there is a difference between the two forms of matter that goes beyond their basic chemical make-up. What accounts for this difference? Essentially, the answer lies in *how the elements are organized.*

THE ELEMENTS OF LIFE

There are four elements that constitute about 95 percent of all living matter: *carbon, hydrogen, nitrogen,* and *oxygen*. Adding two others—*phosphorus* and *sulfur*—the total approaches nearly 98 percent. Most of the remaining 2 percent consists of the elements calcium, chlorine, potassium, and magnesium. (Table 3.1). These elements are found in living things usually in the form of ions or in combination with other elements to form various insoluble compounds. For example, calcium ions (Ca^{++}) are necessary for normal muscle contraction; in addition, calcium is a constituent element of the inorganic compounds calcium carbonate ($CaCO_3$) and calcium phosphate ($CaHPO_4$), both of which are components of bones and teeth.

The *trace elements,* such as iron, cobalt, zinc, copper, iodine, fluorine, and so on, account for less than 0.01 percent of the elements

Table 3.1 ELEMENTS FOUND IN LIVING THINGS*

MAJOR ELEMENTS	TRACE ELEMENTS
Carbon (C)	Flourine (F)
Hydrogen (H)	Silicon (Si)
Nitrogen (N)	Iron (Fe)
Oxygen (O)	Cobalt (Co)
Phosphorus (P)	Copper (Cu)
Sulfur (S)	Zinc (Zn)
Calcium (Ca)	Tin (Sn)
Chlorine (Cl)	Iodine (I)
Magnesium (Mg)	Boron (B)
Potassium (K)	Vanadium (V)
Sodium (Na)	Chromium (Cr)
	Manganese (Mn)
	Selenium (Se)
	Molybdenum (Mo)

*A given group of living organisms may require many of these elements, but not necessarily all of them.

in living things. This is not to suggest, however, that these elements are anything less than essential. For example, copper and zinc are necessary for the proper functioning of various plant and animal enzymes. Iron is an element of central importance in hemoglobin, the pigment in red blood cells, and is also required for the synthesis of the plant pigment chlorophyll. Iodine is a constituent element of the hormones produced by the thyroid gland. The trace elements required by plants are taken up directly from the soil, whereas animals must obtain their requirements through a proper diet.

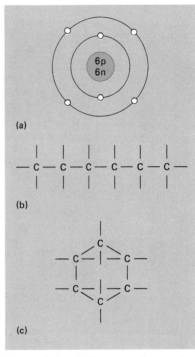

Figure 3.1 (a) Diagram of a carbon atom. With its four outer shell electrons, the carbon atom forms four covalent bonds simultaneously. (b) Carbon atoms are capable of bonding with each other to form chains or (c) rings—the structural "backbones" upon which biological compounds are built.

THE ROLE OF CARBON

The six major elements (CHNOPS) are organized and bonded together to form the fundamental structural and functional compounds that are components of all living things. Since their basic organization is structured around the element carbon, it is necessary that we first examine the role of the carbon atom in living systems.

Carbon is truly a unique element. With an atomic number of six, the carbon atom has four electrons in its outer shell and thus forms four covalent bonds simultaneously (Fig. 3.1). Moreover, these bonds are quite strong, providing a structural base upon which a great number of varied and complex compounds can be built. In addition, carbon atoms can bond with each other to form rings or short or long chains. No other earthly element is capable of this feat.

Life: The Essential Materials

> Carbon forms more than 3 million different compounds, representing more than 90 percent of all known substances.

The carbon chains or rings form the structural "backbone" of the biological compounds mentioned previously, as well as an enormous variety of other compounds. Carbon, then, forms strong bonds with hydrogen (C—H), oxygen (C—O), and nitrogen (C—N) atoms. One need only observe the great diversity of living things to appreciate the extraordinary versatility of the carbon atom.

Interestingly, there are atoms other than carbon with four outer shell electrons—for example, silicon (Si) and germanium (Ge). These atoms, however, present two major biological problems. First, having weak bonds, neither element is able to bond with itself to form long chains or rings; second, both atoms are quite heavy relative to carbon. Silicon is more than twice as heavy as carbon, and germanium is over six times heavier. If living matter were composed of these atoms instead of carbon, we would have a weight problem indeed. Except for oxygen, silicon is the most abundant element by percentage (about 26 percent) in the crust and atmosphere of the earth. It is interesting that carbon, in contrast, constitutes less than 1 percent.

Nonetheless, it is the uniqueness of carbon—with its four bonds, self-pairing ability, and lightness—that accounts for the basic structure of the four major types of biological macromolecules: *carbohydrates, lipids, proteins,* and *nucleic acids.*

THE COMPOUNDS OF LIFE
Carbohydrates

All carbohydrates contain the elements carbon, hydrogen, and oxygen, and a few also contain nitrogen. We generally think of carbohydrates as either sugars or starches. Certain carbohydrates are the principal outside energy sources for living cells, but various lipids and even proteins also are available as energy sources.

The simplest carbohydrates are the *monosaccharides* ("one sugar"), of which there are several of biological importance (Fig. 3.2). Some simple sugars may contain only three carbon atoms, while others may have five, six, or more. It is primarily the 6-carbon sugars that bond together to form the larger, more complex carbohydrates. A common 6-carbon monosaccharide is *glucose,* an energy source for living things unmatched by any other carbohydrate. In the cells of plants and animals, it is primarily glucose that provides the initial source of energy for cellular metabolism. Other important 6-carbon sugars include *fructose,* a very sweet sugar occurring naturally in a variety of fruits, and the sugar *galactose,* a constituent of milk.

Two 5-carbon monosaccharides of major biological significance are *ribose* and *deoxyribose,* both of which are found in the nucleic acids (Fig. 3.3). In addition, ribose is a constituent of the high-energy compound **adenosine triphosphate** (ATP).

The bonding together of two monosaccharides forms a disaccharide ("double sugar"). **Sucrose,** ordinary table sugar, is formed

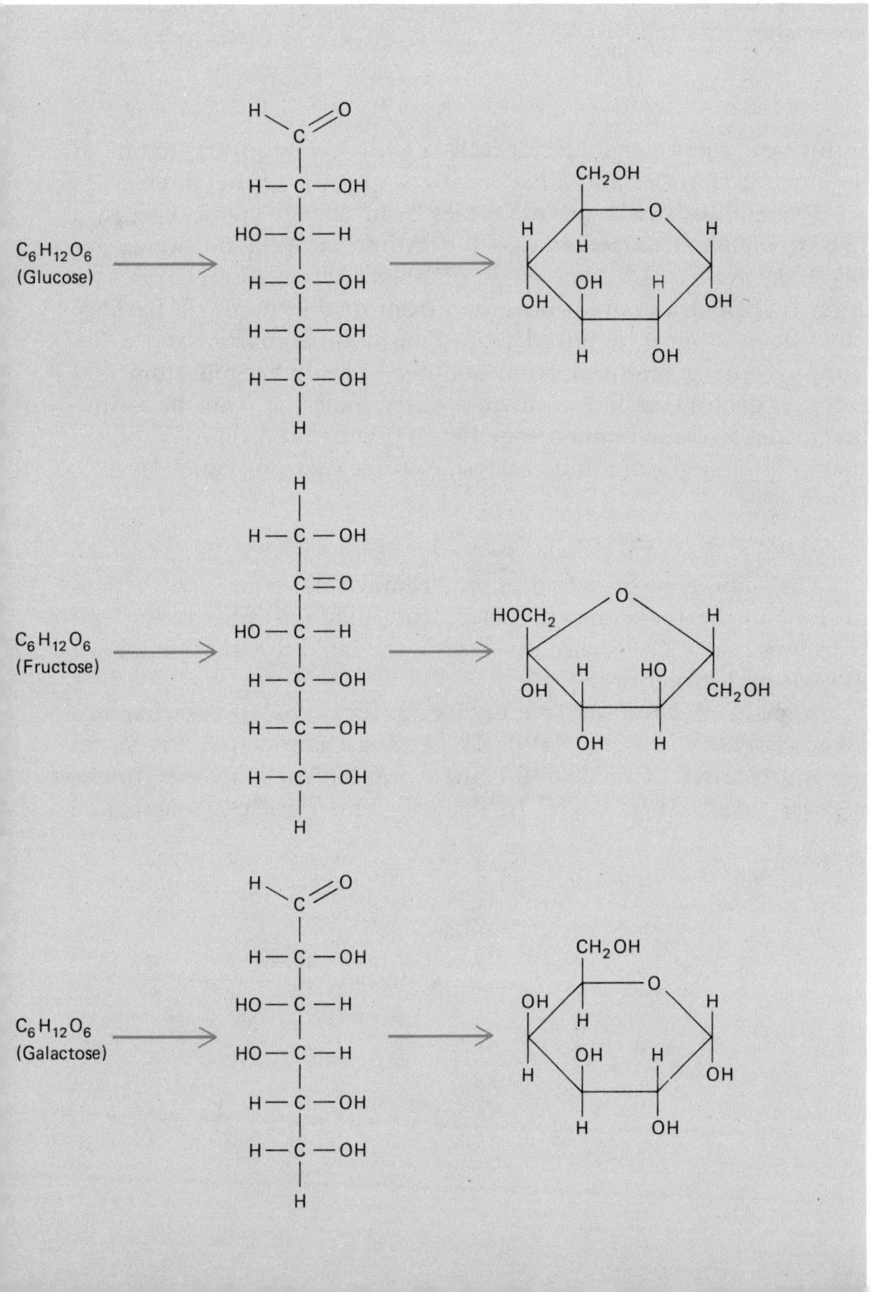

Figure 3.2 Three of the more common 6-carbon simple sugars (monosaccharides). Although each sugar is composed of the same number of carbon, hydrogen, and oxygen atoms ($C_6H_{12}O_6$), their differences result from the manner in which these elements are organized. These sugars may exist in the form of straight chains or as ring compounds. Carbon atoms are located in each corner of the ring structures.

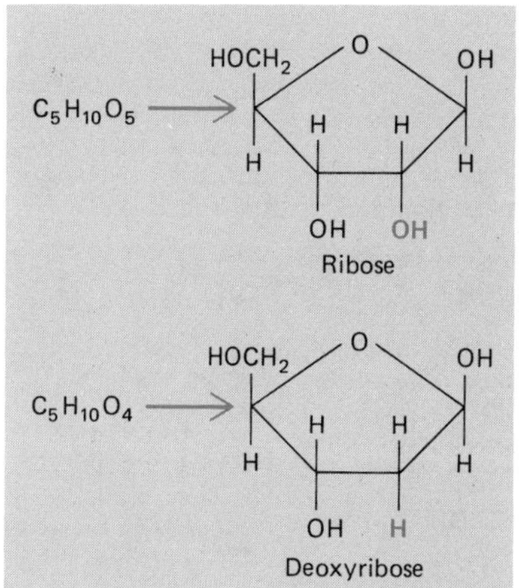

Figure 3.3 The two 5-carbon monosaccharides found in nucleic acids. Note that the two sugars are structurally identical except for one oxygen atom not present in deoxyribose.

> The six-carbon sugar glucose is the most common and most biologically important monosaccharide.

from the bonding together of glucose and fructose. **Lactose,** or milk sugar, is a disaccharide consisting of glucose and galactose, and **maltose,** or malt sugar, is composed of two glucose molecules. With the aid of specific enzymes released in the digestive system, each of these disaccharides can be broken down into its component monosaccharides. This breakdown is referred to as **hydrolysis** because water is *added* during the reaction. For example, the hydrolysis of sucrose can be shown as follows:

$$C_{12}H_{22}O_{11} + H_2O \longrightarrow C_6H_{12}O_6 + C_6H_{12}O_6$$
$$\text{(sucrose)} \quad \text{(water)} \qquad \text{(glucose)} \quad \text{(fructose)}$$

In this case, the water molecule adds a hydrogen atom (H) to one of the monosaccharides and a hydroxyl group (OH) to the other.

The build-up of larger molecules from smaller ones occurs by a type of metabolic reaction called **dehydration synthesis,** which involves the *removal* of a molecule of water. When disaccharides or larger carbohydrates are synthesized from smaller monosaccharides, a hydrogen atom is removed from one monosaccharide and a hydroxyl group is removed from another. The hydrogen atom and hydroxyl group combine to form a water molecule, and the monosaccharides become bonded together. To illustrate, consider the reaction between glucose and fructose to form sucrose (Fig. 3.4):

$$C_6H_{12}O_6 + C_6H_{12}O_6 \longrightarrow C_{12}H_{22}O_{11} + H_2O$$
$$\text{(glucose)} \quad \text{(fructose)} \qquad \text{(sucrose)} \quad \text{(water)}$$

The same type of dehydration ("removal of water") synthesis is involved in the formation of lactose and maltose. Dehydration and hydrolysis reactions are also involved in the build-up and breakdown of lipids and proteins.

A chain of 10 or more monosaccharides bonded together is a **polysaccharide** ("many sugars"). The polysaccharide **starch** is found in a wide variety of foods and is the storage form of excess glucose in green plants (Fig. 3.5A). In the liver and muscles of man and

Figure 3.4 Dehydration synthesis using structural formulas. Through the removal of a molecule of water, glucose and fructose bond together to form the disaccharide sucrose. In the reverse of this reaction (hydrolysis), a water molecule is added to sucrose, which then splits into glucose and fructose.

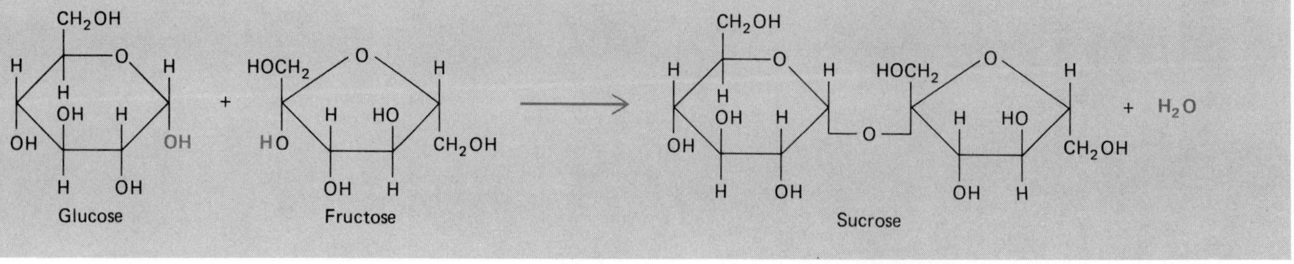

many other animals, glucose is stored in the form of **glycogen,** or "animal starch" (Fig. 3.5B). When the plant or animal requires an energy source for cellular metabolism, appropriate enzymes break down the stored starch or glycogen into individual glucose molecules.

Other than serving as storage forms, some polysaccharides are structural components of cells. **Cellulose** (Fig. 3.5C) forms part of the

Figure 3.5 General structures of three major polysaccharides. All three are composed of glucose molecules, but the organization of these molecules in each polysaccharide is different. Cellulose is a straight-chain polysaccharide, but starch has branching side chains, and glycogen branches more frequently than starch.

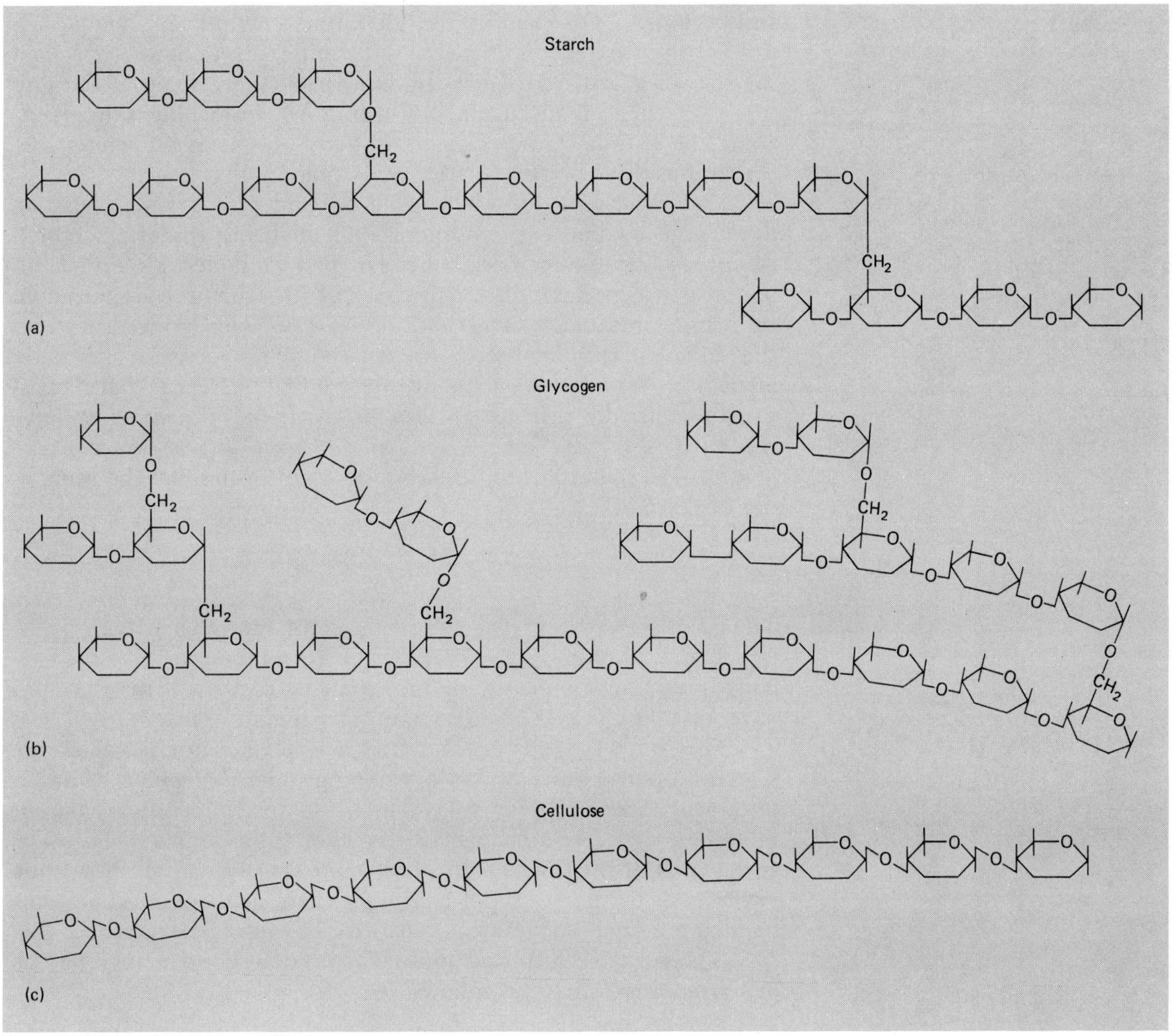

Starch

(a)

Glycogen

(b)

Cellulose

(c)

54
Life: The Essential Materials

> Knowing the relationship between structure and function is fundamental to an understanding of living systems.

Figure 3.6 Cellulose fibers in a plant cell wall. (From Camp, P. S., and Arms, K.: *Exploring Biology*. Philadelphia: Saunders College Publishing, 1981, p. 189.)

supporting structure of plant cell walls (Fig. 3.6), and **chitin,** which incidentally is one of those carbohydrates containing nitrogen, is a structural component of the tough outer skeleton of insects and crustaceans (crayfish, shrimp, lobsters, and so on) (Fig. 3.7). All of these polysaccharides—starch, glycogen, cellulose, and chitin—are *polymers,* or chains, of glucose molecules or glucose derivatives bonded together. Their functional differences result from the precise manner in which their glucose molecules are linked together and organized. For example, cellulose is a straight-chain polysaccharide, with the glucose molecules forming an unbranched chain suggestive of a string of boxcars linked together. Some of the glucose molecules of glycogen, however, form branching side chains along the length of the compound. Starch also has branching side chains, but these occur less frequently than in glycogen.

The structures of starch, cellulose, and glycogen demonstrate the crucial significance of organization in living things. Although composed of the same basic units, i.e., glucose molecules, each compound is functionally distinct because of its structural organization. In living plants and animals, each of these polysaccharides has a specific function related directly to its particular structure. This statement is representative of one of the essential biological concepts associated with the principle of organization: *In the make-up of living systems, there is almost always a direct correlation between structure and function.* We shall refer to this concept often throughout the remainder of this book.

Lipids

Figure 3.7 The tough outer skeleton of the crayfish is composed of the polysaccharide chitin. (From Barnes, R. D.: *Invertebrate Zoology.* 4th ed. Philadelphia. Saunders College Publishing, 1980, p. 728.)

The lipids include a variety of organic compounds, such as fats, oils, waxes, phospholipids, and steroids. These compounds, like the carbohydrates, consist mainly of the elements carbon, hydrogen, and oxygen, although some also contain nitrogen and phosphorus.

The simplest lipids are the **fatty acids,** which form part of the structural "backbone" of most of the other lipids. The most common fatty acids are composed of chains of 14 to 22 carbon atoms, although some carbon chains may be shorter or longer. The other component of the "backbone" of most lipids is a sugar alcohol called **glycerol.** Glycerol consists of a straight chain of three carbon atoms bonded together, each having a hydroxyl group (OH) attached.

Glycerol and fatty acid molecules become bonded together by dehydration synthesis, as follows:

> The carbon atoms of saturated fatty acids are fully occupied by bonded hydrogen atoms; in unsaturated fatty acids, double bonds prevent some of the carbon atoms from being completely occupied by hydrogen atoms.

$$\begin{array}{c}
\text{H} \\
| \\
\text{H}-\text{C}-\text{OH} \\
| \\
\text{H}-\text{C}-\text{OH} \\
| \\
\text{H}-\text{C}-\text{OH} \\
| \\
\text{H} \\
\text{(glycerol)}
\end{array}
\quad
\begin{array}{c}
\text{O} \\
\| \\
\text{HO}-\text{C}-\text{C}_{15}\text{H}_{31} \\
\text{O} \\
\| \\
+\ \text{HO}-\text{C}-\text{C}_{15}\text{H}_{31} \\
\text{O} \\
\| \\
\text{HO}-\text{C}-\text{C}_{15}\text{H}_{31} \\
\text{(terminals of 3 long-chain} \\
\text{fatty acid molecules)}
\end{array}
\longrightarrow
\begin{array}{c}
\text{H} \quad\ \text{O} \\
| \quad\ \ \| \\
\text{H}-\text{C}-\text{O}-\text{C}-\text{C}_{15}\text{H}_{31} \\
| \quad\ \ \text{O} \\
| \quad\ \ \| \\
\text{H}-\text{C}-\text{O}-\text{C}-\text{C}_{15}\text{H}_{31} \\
| \quad\ \ \text{O} \\
| \quad\ \ \| \\
\text{H}-\text{C}-\text{O}-\text{C}-\text{C}_{15}\text{H}_{31} \\
| \\
\text{H} \\
\text{(fat molecule)}
\end{array}
+\ \begin{array}{c}
\text{H}_2\text{O} \\
\text{H}_2\text{O} \\
\text{H}_2\text{O} \\
\text{(3 water} \\
\text{molecules)}
\end{array}$$

In this reaction, a hydrogen atom is removed from each of the three hydroxyl groups of glycerol, and an —OH group is removed from each fatty acid molecule. The hydrogens and —OH groups combine to form three water molecules, and the glycerol and three fatty acid portions become bonded together.

In the metabolic breakdown, or hydrolysis, of a fat molecule, three water molecules are *added,* and the fat molecule is split into glycerol and three fatty acids.

Fats and Oils. Fats and oils are lipids consisting of three fatty acid molecules attached to one molecule of glycerol. One fatty acid is bonded to each hydroxyl site of the glycerol molecule. Fats and oils differ in that fats contain *saturated* fatty acids, while oils contain *unsaturated* fatty acids.

Saturated fatty acids have only single bonds between their carbon atoms (C—C). The available remaining carbon bonds are "saturated," or fully occupied, by hydrogen atoms. In unsaturated fatty acids, some of the carbon atoms are joined by double bonds (C=C) instead of single bonds. Accordingly, some of the carbon atoms of unsaturated fatty acids cannot be fully occupied by hydrogen atoms. In addition, fats are solid at room temperature, whereas oils are usually liquid.

Common *examples of fats* include lard, butter, and animal fat; the *oils* include olive oil, peanut oil, corn oil, and other vegetable oils. Many plants store their lipids in the form of oils. The sebaceous glands of human beings produce an oily secretion called *sebum,* which aids in keeping the hair and skin soft and waterproof. Both fats and oils are also referred to as *triglycerides.*

Waxes. Like the fats and oils, waxes also consist of three fatty acids, but they contain alcohols other than glycerol. Waxes are solid lipids that are important structural components in many plants and animals. The waxy coating on the surface of leaves and other plant structures aids in preventing excessive water loss. Waxes also form a protective film over the body coverings of many animals. In human beings, the skin around the opening into the ear canal is supplied with numerous glands that secrete a waxy substance called *cerumen*. Cerumen aids in preventing foreign objects from entering the ear canal.

Phospholipids. These lipids are constructed like the fats and oils, except that phospholipids usually have only two fatty acids bonded to glycerol instead of three. The third fatty acid is replaced by a compound containing phosphorus or nitrogen or both. Phospholipids are important as structural components of cells, particularly the cell membrane.

Steroids. Unlike any of the other types of lipids, the steroids do not consist of fatty acids and glycerol but have a basic structure consisting of four rings of carbon atoms joined together (Fig. 3.8). The biologically important steroids include *cholesterol*, an important component of certain cell membranes and nervous tissue; the human *sex hormones*; certain adrenal gland hormones, such as *cortisone*; various *constituents of bile*; and *vitamin D*, which is necessary for proper bone formation.

Some of the lipids we consume in our diet are important as stored energy reserves for the body. If the body depletes its carbohydrate stores, it must then call upon its reserve lipids for a source of energy. These reserves are mostly fats, which are well-suited to their task, yielding about twice the energy of an equal amount of carbohydrate or protein. If we consume more carbohydrates than can be stored as glycogen, much of the excess is converted into fats.

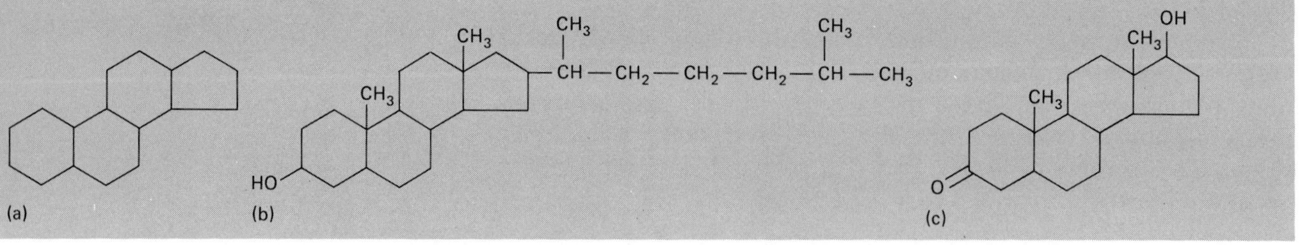

Figure 3.8 (a) The four interlocking rings of carbon atoms that form the basic structure of steroids. Steroids differ according to the particular side groups attached to the four carbon rings. (Carbon atoms are located in the corners of each hexagon and the pentagon.) (b) The steroid cholesterol, a component of cell membranes and nervous tissue. (c) The steroid testosterone, the sex hormone of human males.

A peptide bond is formed by removing an OH from a carboxyl group (—COOH) of one amino acid and an H from the amino group (—NH$_2$) of an adjacent amino acid. The OH and H then combine to form water.

The Compounds of Life

An excess of body fat results in overweight or obesity, both of which appear to be related to a variety of human health problems, such as heart disease and diabetes.

Proteins

The real keys to the physical make-up and much of the functioning of living cells are the proteins. All proteins include the elements carbon, hydrogen, oxygen, and nitrogen. Some proteins also contain sulfur and phosphorus, and a few also contain iron, magnesium, and copper. The carbon, hydrogen, oxygen, and nitrogen atoms are organized to form individual compounds, or units, called **amino acids,** the building blocks of all proteins. These units are bonded together in chainlike fashion to form the enormous variety of structural and functional proteins found in living things (Fig. 3.9). The linkage between one amino acid and another is a type of covalent bond called a *peptide bond.* Essentially, a peptide bond is a carbon-to-nitrogen bond between adjacent amino acids.

Peptide bonds are formed as a result of dehydration synthesis. The bonding of two amino acids together forms a **dipeptide;** when many amino acids are bonded together, the resulting structure is a **polypeptide.** A protein may consist of one or several polypeptides.

Figure 3.9 (a) The basic structure of an amino acid. Amino acids differ according to their side group, designated R. The C$\diagup\!\!\!\diagup$O$\diagdown$OH group represents an acid and is called a *carboxyl group.* The NH$_2$ is called an *amino group.* (b) Examples of different amino acids, showing some of the variations in side groups.

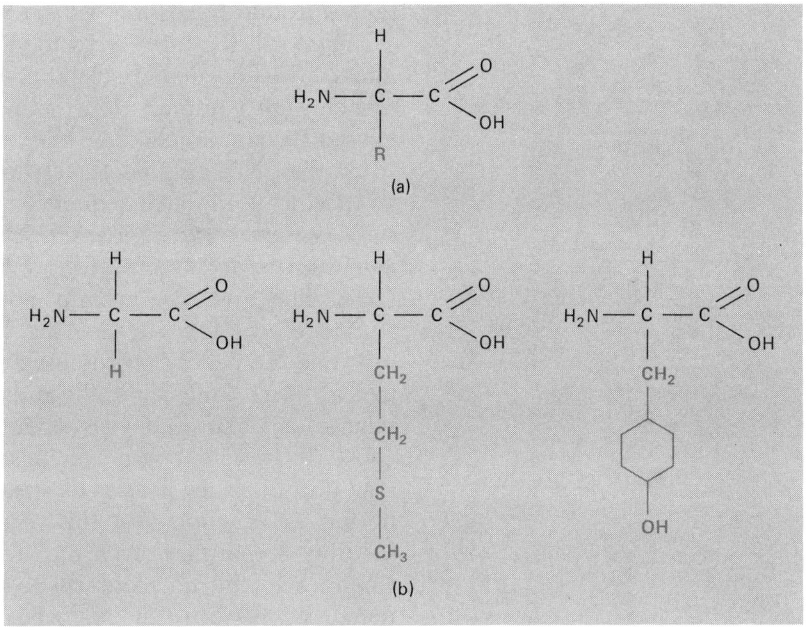

Life: The Essential Materials

> Twenty different amino acids can be sequenced to form an enormous number of different proteins, just as the 26 letters of the English alphabet can be sequenced to form a multitude of different words.

$$\underset{\text{(amino acid)}}{\text{H}-\text{N}(\text{H})-\text{C}(\text{H})(\text{H})-\text{C}(=\text{O})-\text{OH}} + \underset{\text{(amino acid)}}{\text{H}-\text{N}(\text{H})-\text{C}(\text{H})(\text{H})-\text{C}(=\text{O})-\text{OH}} \rightarrow \underset{\text{(dipeptide)}}{\text{H}-\text{N}(\text{H})-\text{C}(\text{H})(\text{H})-\text{C}(=\text{O})-\text{N}(\text{H})-\text{C}(\text{H})(\text{H})-\text{C}(=\text{O})-\text{OH}} + \underset{\text{(water)}}{\text{H}_2\text{O}}$$

In this reaction, as in all dehydration synthesis reactions, water is removed and smaller molecules bond together to form larger molecules. The two amino acids composing the dipeptide are joined by the carbon-to-nitrogen peptide bond. In the same manner that we have seen with carbohydrates and lipids, the breakdown, or hydrolysis, of a dipeptide, polypeptide, or protein into its smallest units, i.e., amino acids, involves the *addition* of water.

There are 20 different amino acids commonly found in proteins, making possible the building of enormous numbers of different protein molecules. This can be accomplished mainly because the 20 amino acids can be sequenced in an almost infinite variety of ways. The result is the vast physical diversity we observe in the living world. The specific sequence of amino acids in any one polypeptide is its *primary structure* (Fig. 3.10A). Sometimes several polypeptide chains can associate together to form a functional protein molecule. For example, hemoglobin, the pigment in red blood cells, is a functional protein consisting of four polypeptide chains.

The specific task performed by each different protein is determined by its particular structure. For example, your hair is a type of protein, but your muscles, too, consist of proteins. Obviously, these have different functions in the body and accordingly have different structures. Not only do the structures of individual proteins consist of specific amino acid sequences, but also each entire protein molecule is twisted into a helix, or spiral (called the *secondary* structure), and then folded into various shapes (*tertiary* structure) (Fig. 3.10B, C, D). This gives the protein molecule an overall three-dimensional structure.

The shape of a protein molecule is maintained by chemical bonds, which unite the folds and preserve the turns and twists in the molecule. If subjected to sufficiently high temperatures—usually above 50 to 55°C—some of these bonds separate and the molecule may unfold. Once it loses its structure permanently, the protein can no longer function. For this reason, we find very few living things existing at temperatures much above 55°C. There are occasions when a protein molecule will regain its normal structure if its functional temperature is restored. Extreme heat, however, usually

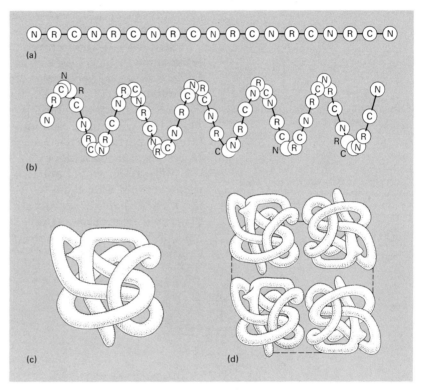

Figure 3.10 (a) The *primary* structure of a protein is determined by the sequence of amino acids in the chain. (C—N = peptide bond between amino acids.) (b) A common *secondary* structure of proteins is the formation of a helix. The turns in the helix are maintained by hydrogen bonds between amino acids. (c) A folding of the helix results in the *tertiary* structure of a protein. The folds are united by chemical bonds between two sulfur atoms. (d) Some proteins, such as hemoglobin, mentioned in the text, are composed of several folded polypeptide chains; this association of chains forms the *quaternary* structure of a protein.

causes the protein to *coagulate,* after which its original structure is lost forever. You have observed coagulated proteins many times in the form of scrambled eggs and shrunken bacon. Unfortunately, the chemistry of proteins prevents us from "unscrambling" the eggs or "unshrinking" the bacon, regardless of our expertise in the culinary arts.

One of the most serious effects of a bad burn on the body is the coagulation of body proteins. In addition, environmental factors other than heat, such as a significant change in pH, various chemicals, electricity, and others, may also disrupt the normal structure of a protein.

Enzymes: Nature's Catalysts. Some of the most important functional proteins are the **enzymes**. Found in all living things, the enzymes are defined as *organic catalysts.* A catalyst is a substance that speeds up the rate of a chemical reaction while at the same time remaining unchanged by the reaction. Enzymes are said to be *specific,* meaning that one enzyme usually catalyzes only one type of

Life: The Essential Materials

> In the absence of enzymes, it would require unreasonably large amounts of energy, principally in the form of heat, for chemical reactions to occur in living things.

chemical reaction. Therefore, each reaction of cellular metabolism requires its own specific enzyme.

Specificity is apparent in the precise manner in which an enzyme binds to the substance, or *substrate*, with which it reacts. The binding area on the enzyme molecule is a very small region called the ***active site*** (Fig. 3.11). During a chemical reaction, only a certain area of the substrate molecule binds with the active site of the enzyme. In this manner, each individual enzyme, with its unique structure and particular substrate, catalyzes only one type of chemical reaction specific to it alone.

Since an enzyme is not affected by a chemical reaction, it can be used over and over again to catalyze the same reaction. Enzymes do not *cause* a chemical reaction to take place; they *speed up* a reaction that could take place anyway, but if the catalyst is not present, the reaction would take place at a rate that would be useless to the organism.

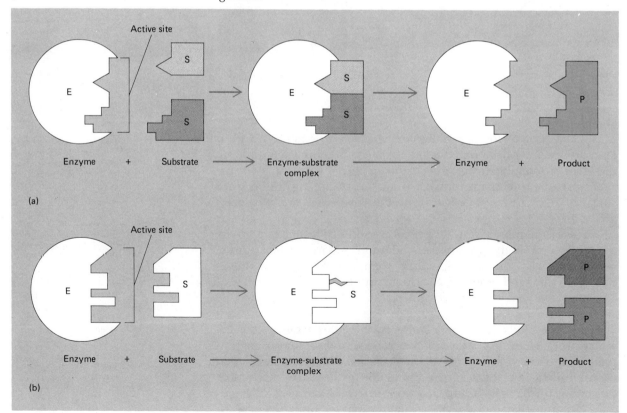

Figure 3.11 (a) Participation of an enzyme in a reaction in which a large molecule (product) is synthesized from two smaller molecules (the substrate). In binding to the active site, the substrate unites with the enzyme, forming a temporary *enzyme-substrate complex*. After the product is formed, the enzyme can be used again and again. (b) Participation of an enzyme in a reaction in which a large molecule (substrate) is broken down into two smaller molecules (product).

> The three most crucial factors affecting enzyme function are concentration, temperature, and pH.

Many enzyme-controlled reactions occur in your own digestive tract. For example, starch (the substrate) is acted on by the enzyme *ptyalin*, or *salivary amylase*, in your mouth to form molecules of the disaccharide *maltose* (the product). This, of course, is an example of hydrolysis, a chemical reaction in which bonds are broken, reducing a larger molecule to smaller ones. Enzymes also catalyze the dehydration synthesis reactions that bond together smaller molecules, such as glucose and fructose, to form the larger molecule, sucrose.

Factors Affecting Enzymes. Three major factors influence the functioning of enzymes. First, there must be a sufficient amount, or *concentration*, of substrate molecules. If there are more enzyme molecules than substrate molecules, the reaction slows down. One way to speed up the reaction is to add more substrate. By so doing, more of the enzyme molecules become involved in the reaction. Once a sufficient substrate concentration is reached that involves all the enzyme molecules, the reaction will then proceed at its maximum rate. Beyond this point, however, adding more substrate will not further increase the rate of the reaction.

Second, the *temperature* of the cellular environment is crucial. At very low temperatures the rate of a chemical reaction is greatly slowed. If the temperature rises too high, the bonds holding the enzyme molecule intact are disrupted. As with other proteins, disruption of its bonds causes the enzyme molecule to unfold and lose its activity permanently. Enzymes affected in this manner are said to be *denatured*.

Third, every enzyme works most effectively at a certain *pH*. In your stomach, for example, some enzymes function at a pH well below 2; in your mouth, however, the enzyme in saliva is active at a pH of about 7. If either of these environments became too acidic or basic, the respective enzymes would cease to function.

Nucleic Acids

The fourth major group of the compounds of life comprises the **nucleic acids**. These large compounds are composed of individual units called **nucleotides**, which consist of the elements carbon, hydrogen, nitrogen, oxygen, and phosphorus. There are two basic types of nucleic acids: **deoxyribonucleic acid**, abbreviated **DNA**, and **ribonucleic acid**, or **RNA**. These two nucleic acids are involved in the synthesis of enzymes and other cellular proteins. In the cells of all living organisms, the nucleic acids contain the hereditary information that is passed from parent to offspring, generation after gen-

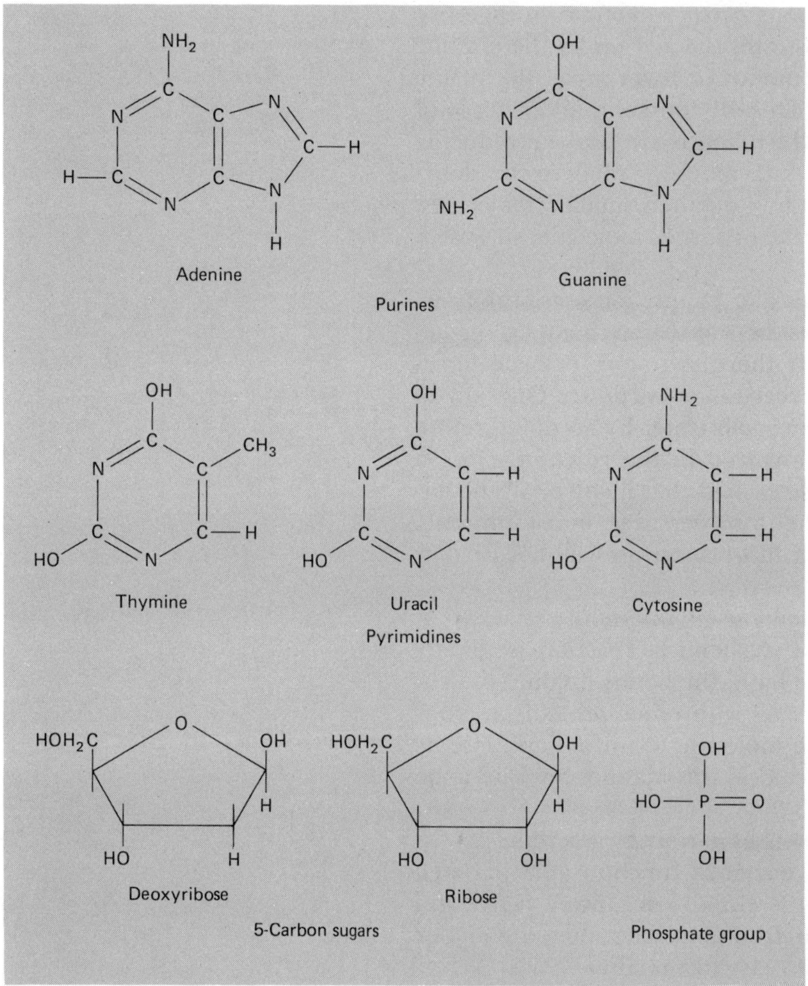

Figure 3.12 Constituents of nucleic acids.

cration. Much of what is known about the phenomenon of life itself is tied in with the structures of DNA and RNA.

Each nucleotide of DNA or RNA is composed of a carbon-nitrogen ring-shaped compound called a **nitrogen base,** plus a **5-carbon sugar** and a **phosphate group** (PO_4) (Fig. 3.12). A nitrogen base having only one carbon-nitrogen ring is called a **pyrimidine.** The pyrimidines in nucleic acids include *thymine, cytosine,* and *uracil.* The other type of nitrogen base, called a **purine,** has two interlocking carbon-nitrogen rings. The two purines in nucleic acids are *adenine* and *guanine.*

> Nucleic acids carry and convey the hereditary information within the cells of every living thing.

There are two slightly different monosaccharides constituting a part of each nucleotide. The 5-carbon sugar *deoxyribose* is found in DNA, and RNA contains the 5-carbon sugar *ribose*. Deoxyribose has one less oxygen atom than does ribose, and that is the only structural difference between the two. The phosphate group is the same in both DNA and RNA.

Each nucleotide contains one of the five nitrogen bases. The base thymine, however, is found only in the nucleotides of DNA, and the base uracil is found only in the RNA nucleotides. The other three nitrogen bases—adenine, guanine, and cytosine—are components of the nucleotides of both DNA and RNA.

In the make-up of a complete nucleotide, the 5-carbon sugar is bonded to a nitrogen base, and the phosphate group is bonded to the sugar. In the structures of DNA and RNA, the nucleotides are arranged in a linear sequence, held together by covalent bonds between the sugar of one nucleotide and the phosphate group of the adjacent nucleotide. There is formed, then, a long chain of alternating sugar-phosphate groups, which make up the "backbone" of the nucleic acid molecule. The nitrogen bases extend out perpendicularly along the length of the sugar-phosphate strand. Although composed of nucleotides that are quite similar, the complete structures of DNA and RNA are distinctly different.

The Structure of DNA. DNA is composed of *two* chains, or strands, of nucleotides, resembling a ladder (Fig. 3.13). The nitrogen bases along the two strands are oriented across from each other, forming **base pairs.** The sugar-phosphate backbone of each strand forms the sides of the "ladder." In DNA the pairing of the nitrogen bases is quite precise: Adenine (A) always bonds with thymine (T), and guanine (G) always bonds with cytosine (C). Thus, adenine and thymine are *complementary* bases, as are guanine and cytosine. These base pairs may occur in an enormous variety of sequences throughout the length of the DNA molecule. For example, if we use the initial letters of the nitrogen bases, a section of one strand of DNA might "read" TGACGGACT. The corresponding section in the other strand would then "read" ACTGCCTGA. Thus, the sequence of the base pairs would be T—A, G—C, A—T, C—G, G—C, and so on. Depending upon the species in which it is found, a molecule of DNA may contain thousands or millions of base pairs. In practically all species these can be sequenced in an almost infinite number of ways, providing each individual organism with its own unique DNA.

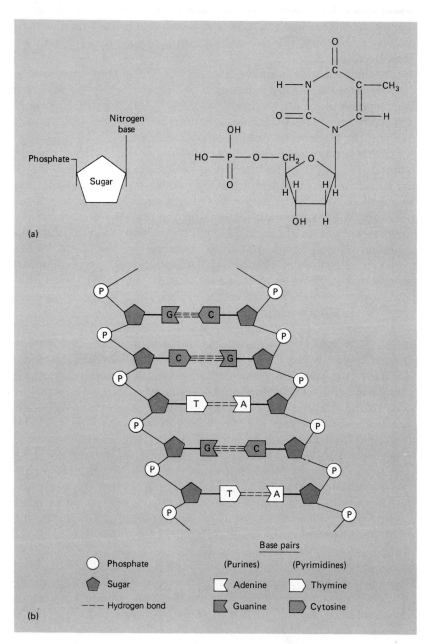

Figure 3.13 (a) General structure of a nucleotide and detailed structure of the nucleotide thymine. (b) Molecular structure of DNA, showing base pairing and orientation of sugar-phosphate "backbone."

Not only is DNA ladder-shaped, but also the entire molecule is twisted into a spiral, or helix (Fig. 3.14). DNA is the principal constituent of *chromosomes* and is the carrier of the hereditary units called *genes*. We shall look more closely at the genetic role of DNA in Chapter 8.

The Structure of RNA. In contrast to DNA, the RNA molecule is single-stranded; however, it can fold back on itself to form double-stranded sections. The nucleotides in a strand of RNA, as in DNA, can be sequenced in a multitude of ways. As mentioned, the nitrogen base uracil is substituted for thymine in the nucleotides of RNA, and the 5-carbon sugar of RNA is always ribose. Whereas DNA is found predominantly in the nuclei of cells, RNA is primarily located in the cytoplasm. RNA is actually made in the nucleus but is then transported through the nuclear membrane into the cytoplasm. There are three principal types of RNA, which, along with DNA, are involved in the synthesis of cellular proteins. We shall examine the types of RNA and their roles in protein synthesis in Chapter 8.

Other Important Nucleotides. In addition to those constituting DNA and RNA, there are several other nucleotides of vital importance in living things. These include **adenosine triphosphate,** or ATP (Fig. 3.15), the "energy currency" of all living cells; **nicotinamide adenine dinucleotide,** or NAD; and **flavin adenine dinucleotide,** or FAD. The latter two are termed coenzymes and are important in certain reactions of cellular metabolism.

In the earlier discussion of carbohydrates, it was pointed out that glucose is the primary outside energy source for cellular metabolism. In the cell, other carbohydrates, as well as various lipids and

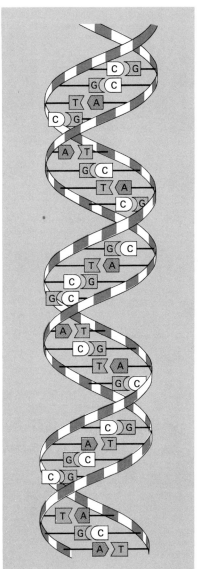

Figure 3.14 Model of double helix structure of DNA. A, T, G, C = nitrogen bases. Dark area on strands is deoxyribose; light area is the phosphate group.

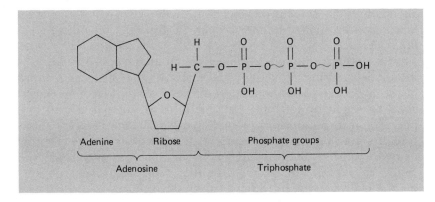

Figure 3.15 Generalized structure of the nucleotide adenosine triphosphate (ATP). The symbol ~ indicates a so-called high-energy bond.

> Adenosine triphosphate (ATP) is used by all forms of life as both a storage form of energy and a provider of immediate energy for cellular work.

proteins, can be converted into glucose and then broken down, releasing their chemical bond energy. Through a series of chemical reactions, some of this energy is used to synthesize the high-energy compound ATP. As the "energy currency" of the cell, ATP can be drawn on like money in the bank. When the body requires an immediate source of energy, it "spends" some ATP. Certain reactions of photosynthesis also produce ATP by utilizing light energy from the sun. ATP is then used by the plant cells as an energy source to produce glucose and other organic compounds.

The coenzymes NAD and FAD are essential to the normal functioning of living cells. Their primary function is to accept, or "pick up," electrons and hydrogen atoms that have been split from glucose and other molecules as these are being broken down within the cells. In accepting the electrons and hydrogen atoms, NAD and FAD are reduced to their high-energy forms, $NADH_2$ and $FADH_2$, respectively. In the earlier stages of cellular metabolism, most of the usable energy released from the breakdown of glucose resides in $NADH_2$ and $FADH_2$. In subsequent reactions, both compounds release this energy by transferring their electrons and hydrogen atoms to other molecules in the cell. By virtue of this transfer, $NADH_2$ is converted back into NAD, and $FADH_2$ becomes FAD. Both coenzymes can then be used again in the same manner when a new molecule of glucose is broken down.

WATER: THE LIQUID OF LIFE

On the earth, water and life are inseparable. Of all the liquids known, only water possesses those properties that have made the origin, evolution, and perpetuation of living things possible. Water covers over 70 percent of the surface of the earth and constitutes about the same percent of your own body. What is it about water that makes it essential to and inseparable from the phenomenon of life on earth?

Like other compounds, the key to the uniqueness of water is in its atomic structure. You will recall that the water molecule consists of two hydrogen atoms bonded to one atom of oxygen. The oxygen atom has eight protons in its atomic nucleus, compared with only one proton in the nucleus of each hydrogen atom. Oxygen, therefore, has a stronger positive charge that tends to attract the electrons of hydrogen over more to its corner of the water molecule. As a result, the oxygen atom acquires a rather weak negative charge, leaving the hydrogen atoms with a weak positive charge.

> The unique properties of water are attributable to hydrogen bonding between its molecules.

Since it forms these two opposite electrical poles, water is referred to as a *polar* molecule (Fig. 3.16). Its polar nature is one of the main reasons water is the ideal solvent for living systems. Many of the various organic and inorganic substances in living cells dissolve, or go into solution, because the charged water molecules tend to separate them into their constituent ions. Ordinary table salt, NaCl, readily dissolves because water separates the compound into Na^+ ions and Cl^- ions, which are then attracted to the respective poles of the water molecule. Many other cellular constituents, also in ionic form, have a similar attraction for the charged water molecules. In fact, more different substances will dissolve in water than in any other liquid known.

In the make-up of water, hydrogen bonds between oxygen and hydrogen atoms serve as "bridges," linking one water molecule with another water molecule. The tendency of water molecules to cling together through this type of hydrogen bonding accounts for its *cohesive* properties. Cohesion refers to the clinging together of molecules of the *same* substance. One of the results of this molecular interaction is known as **surface tension**, a property of water that makes it possible for a needle or small insects to float on its surface (Fig. 3.17). Cohesion is also important in the upward transport of water through plant tissues, a process we shall discuss later.

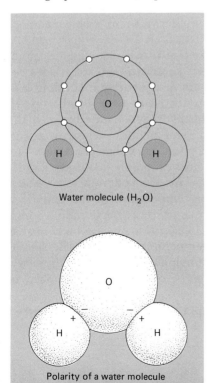

Figure 3.16 Water is a polar molecule because it forms oppositely charged electrical poles. The hydrogen electrons are pulled closer to the oxygen atom, which acquires a weak negative charge. This leaves the hydrogen atoms with a weak positive charge. Ionic compounds readily dissolve in water because their constituents are attracted to the electrically charged poles of the water molecule.

Figure 3.17 Surface tension. Unbalanced downward attractive forces at the surface of a liquid pull molecules into a difficult-to-penetrate skin capable of supporting small bugs or thin pieces of steel, such as a needle or razor blade. The bug literally runs on the water; it does not float in it. Molecules within the water are attracted in all directions, as shown.

> Having a high heat of vaporization permits a small amount of water to absorb a large amount of heat.

As a charged, or polar, molecule, water is strongly attracted to charged molecules of other substances. The clinging together of *different* substances is known as **adhesion.** Adhesion also plays a role in water transport within the plant, as well as in the movement of water through the soil.

Another property of water attributable to hydrogen bonding is its high **specific heat.** The specific heat of a substance refers to the amount of heat necessary to raise the temperature of that substance so many degrees. Water has a high specific heat, second only to that of liquid ammonia. This means that it takes quite a large amount of heat energy to raise the temperature of water so many degrees. The essential reason for this is that most of the heat energy applied to raise the temperature of water is used to break the hydrogen bonds between the water molecules. Because water has a high specific heat, its temperature does not tend to rise or fall rapidly. This is of crucial importance to living things because chemical reactions in the body are adversely affected by extreme temperature changes.

We have seen, for example, that enzymes function effectively only within a limited temperature range. Since body cells have a high water content, living things are able to maintain a more constant internal temperature than if some other substance were present in their cells. Moreover, the high specific heat of water prevents extreme temperature fluctuations in the oceans, seas, and other bodies of water. Consequently, aquatic organisms live in an environment where the temperature is relatively constant.

An additional property of water that makes it such an ideal liquid for living systems is its high **heat of vaporization,** i.e., it does not tend to evaporate readily. Here again, it is hydrogen bonding that accounts for this property. For water to evaporate—that is, to change from a liquid to a gas—the hydrogen bonds between the water molecules have to be broken. The amount of heat necessary to cause evaporation is much greater for water than for almost any other liquid. However, when water does evaporate, it absorbs a great deal of heat from the environment. This absorption factor is essential, for example, in regulating the temperature of the body if it attempts to overheat. During vigorous exercise, perspiration, which is mostly water, has a cooling effect by absorbing excess body heat as it vaporizes. Therefore, the body temperature does not rise significantly. By comparison, rubbing alcohol, although cooling the body, evaporates much too readily and must constantly be reapplied to *keep* you cool.

Finally, there is one other property of water that is truly striking. When cooled, water contracts like any other liquid; but upon

Water: The Liquid of Life

> Water is one of the few substances that is denser in the liquid state that in the solid state.

reaching a temperature of 4°C, water expands, becoming lighter and less dense. We know, for example, that a glass container of water or a bottle of soda will break if left in the freezer and that ice floats on the surface of water. The expansion of water at 4°C and its accompanying decreasing density are attributable to hydrogen bonding and the unique structure of the water molecules (Fig. 3.18).

In the make-up of ice, the oxygen atoms are linked together by hydrogen bonds to form a ring or, more correctly, a hexagon of water molecules (Fig. 3.19). The large open space in the center of each hexagon widely separates one water molecule from another. Consequently, the water molecules of ice are not packed closely together, which is another way of saying they are less dense. When ice melts, the hydrogen bonds are broken, and the water molecules lose their hexagonal arrangement. As a result, liquid water occupies less space than frozen water, even though the number of water molecules in both is the same.

What are the biological implications of this unique property of water? Since over 70 percent of the surface of the earth is covered by water, the implications are vastly significant. During the cold winter months, ice always forms on the *surface* of ponds and lakes when the temperature falls below freezing. If ice were more dense, it would sink and the ponds and lakes would freeze from the bottom

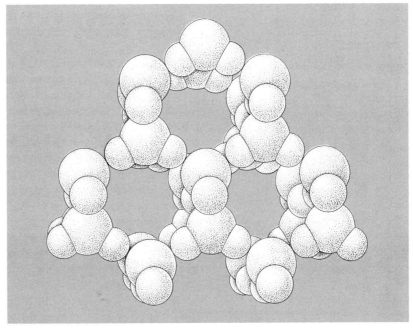

Figure 3.19 The structure of ice. Water molecules are held in a hexagonal shape by hydrogen bonds. This arrangement prevents the water molecules from becoming packed together, as they are in liquid water. Consequently, ice is less dense and lighter than water in the liquid state.

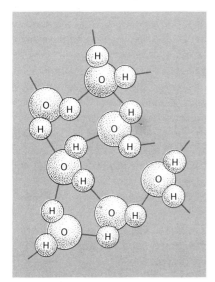

Figure 3.18 Water molecules united by hydrogen bonds *(colored bands)*. A single water molecule can form hydrogen bonds with four other water molecules.

up. Instead, ice on the surface acts like a layer of insulation, preventing the water below from freezing at all. If such were not the case, aquatic life in the cold waters of the earth would be impossible.

It is intriguing that of all the liquids on earth, only water has the properties necessary to sustain life. If there is life elsewhere in the universe, it seems likely that it exists in company with this unique and versatile liquid.

REVIEW OF ESSENTIAL CONCEPTS

1. Living matter is structured around the element *carbon*. The other major elements of life are hydrogen, nitrogen, oxygen, phosphorus, and sulfur. A variety of trace elements accounts for less than 0.01 percent of the elements in living things.
2. The element carbon can form four covalent bonds simultaneously; it also is capable of bonding with itself to form chains or rings. These chains or rings form the structural "backbone" of organic compounds.
3. The four major types of organic macromolecules are *carbohydrates*, *lipids*, *proteins*, and *nucleic acids*. The carbohydrates include *monosaccharides*, *disaccharides*, and *polysaccharides*. The lipids include *fats*, *oils*, *waxes*, *steroids*, and *phospholipids*. Most of the lipids consist of *glycerol* and *fatty acids*.
4. The splitting of a large molecule into smaller molecules involving the addition of a water molecule is known as *hydrolysis*. The formation of a larger molecule from smaller molecules involving the removal of a molecule of water is called *dehydration synthesis*.
5. Proteins are constructed of *amino acids*, 20 of which are commonly found in proteins. Protein molecules have specific shapes that are related to the function of the protein. Various environmental factors may denature a protein, resulting in total loss of function.
6. Enzymes are *organic catalysts*, specific types of proteins that speed up the rate of a chemical reaction. The substance on which an enzyme acts is called its *substrate*. During a reaction, the substrate binds with the *active site* on the enzyme.
7. Enzyme function is dependent upon enzyme-substrate concentration, temperature, and pH.
8. Nucleic acids are composed of *nucleotides*, which are units consisting of a *nitrogen base*, a *5-carbon sugar*, and a *phosphate group*. Nucleic acids are the carriers of hereditary information.
9. The two basic types of nucleic acids are DNA and RNA. The nucleotides of DNA consist of the bases *adenine*, *guanine*, *cytosine*, and *thymine*, along with the 5-carbon sugar *deoxyribose* plus a phosphate group. The nucleotides of RNA consist of adenine, guanine,

cytosine, and *uracil*, along with the 5-carbon sugar *ribose* plus a phosphate group.
10. DNA is usually double-stranded, the molecule assuming the shape of a double helix. The two strands of the DNA molecule are complementary, i.e., composed of complementary base pairs. DNA is the major component of chromosomes.
11. RNA is usually single-stranded and is located primarily in the cytoplasm.
12. The nucleotide *ATP* is the "energy currency" of living cells, providing an immediate source of stored cellular energy.
13. The *coenzymes NAD* and *FAD* are hydrogen and electron acceptors involved in the process of cellular metabolism.
14. Water is the most important liquid on earth. Water is a *polar molecule*, one of the main reasons it is an ideal solvent.
15. Hydrogen bonding accounts for many of the unique properties of water. These include its *adhesive* and *cohesive* properties, its high *specific heat*, its high *heat of vaporization*, and its expansion on reaching a temperature of 4°C.

APPLYING THE CONCEPTS

1. Differentiate between "organic" and "inorganic" compounds.
2. What factors make carbon the appropriate constituent of the "backbone" of living matter?
3. What are the most common monosaccharides associated with living things?
4. How are disaccharides and polysaccharides formed?
5. How do the comparative structures of starch and glycogen illustrate the significance of organization in living things?
6. Differentiate between a fat and an oil. What are some of the functions of fats and oils in the body?
7. Describe the general structure of a protein. What makes each different type of protein unique?
8. What is the significance of the primary structure of a protein?
9. Explain the general mechanism of enzyme function. How could this mechanism be disrupted?
10. What are the essential structural differences between DNA and RNA? In what ways are these two compounds similar?
11. What are the important functions of ATP, NAD, and FAD?
12. Explain why water is referred to as a polar molecule.
13. In your own words, explain what is meant by the following:
    a. Adhesion   c. Specific heat
    b. Cohesion   d. Heat of vaporization

4 energy and life

THE ESSENTIAL OBJECTIVES

You have understood this chapter when you are able to:

1. Explain, by examples, the first and second laws of thermodynamics.
2. Define calorie and kilocalorie and comment on their use in biology.
3. Discuss the biological significance of energy efficiency.
4. Discuss the concept of energy flow in living systems.
5. Describe the structure and function of ATP; using an example, explain how ATP stores and releases energy.

The existence and continuance of life depend upon a constant source of energy. The ultimate source is the radiant energy of the sun, which, through the metabolic activities of living things, becomes packaged in the form of cellular ATP.

PREVIEW OF ESSENTIAL TERMS

energy The capacity to do work, i.e., to bring about some action or effect.

thermodynamics The discipline concerned with the study of the relationships between heat, energy, and work.

calorie The amount of heat necessary to raise the temperature of 1 gram of water 1°C; a kilocalorie is equal to 1000 calories.

adenosine triphosphate (ATP) An organic compound that serves as the major source of cellular energy in living things. ATP contains adenine, ribose, and three phosphate groups (PO_4).

efficiency The ratio of useful energy to the total energy of a system.

energy flow The passing of chemical (food) energy from one organism to another along a food chain.

74
Energy and Life

Life, we have said, is activity—growing, developing, sensing, responding, reproducing, and so on. As a consequence of these activities, a price is exacted from every living thing. This price is *energy*. Energy is the capacity to do work, i.e., to bring about some action or effect. This capacity applies, of course, in the nonliving world as well as in living systems. The rush of waters, blowing winds, and rolling boulders are familiar examples of energy producers in the nonliving sphere. Moreover, the very source of energy for life on this planet comes from the radiant energy of the nonliving sun (Fig. 4.1).

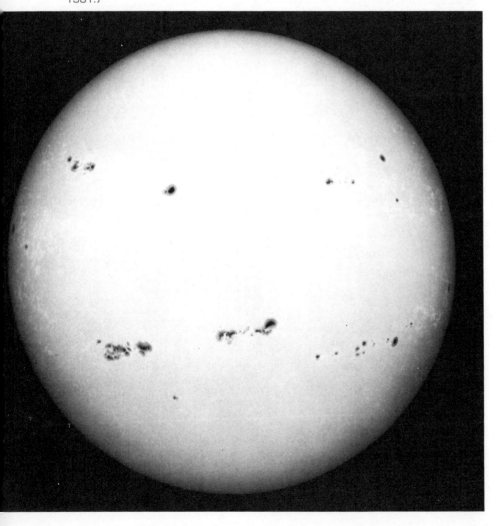

Figure 4.1 The sun is the ultimate source of energy for life on the planet Earth. (From Pasachoff, J. M.: *Contemporary Astronomy.* 2nd ed. Philadelphia: Saunders College Publishing, 1981.)

Inherent in the laws of thermodynamics is the fact that the *total* amount of energy in the universe is constant (first law) but the amount of *usable* energy is continually decreasing (second law).

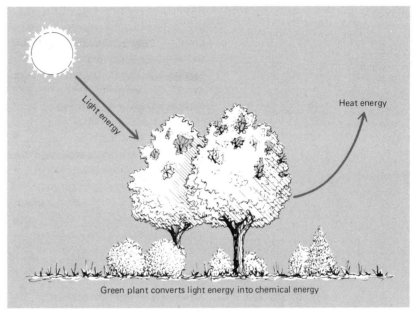

Figure 4.2 Green plants convert light energy from the sun into chemical energy. In the conversion process, some of the energy is dissipated to the environment as heat energy, which is unavailable to do work (second law of thermodynamics). However, the *total* amount of energy entering and leaving the system (green plant) is the same (first law of thermodynamics).

Energy is subject to a law of conservation that states that ***energy can be converted from one form to another, but it cannot be created or destroyed.*** This statement is also known as the ***first law of thermodynamics.*** Thermodynamics (literally, "heat movement") refers to the relationship between heat, energy, and work. In the conversion of light energy into chemical energy—as in green plants, for example—no energy is created in the conversion, nor is any of the energy destroyed. However, in any conversion of energy from one form to another, there is a corresponding decrease in *usable* energy. The *total* amount of energy, however, remains the same, but *some of it is dissipated into the surroundings as heat energy.* ***In all energy conversions, some heat energy is "lost" to the environment and cannot do work.*** This last sentence is a general statement of the ***second law of thermodynamics.*** Applied to living things, both the first and second laws of thermodynamics have far-reaching consequences, as we shall discuss (Fig. 4.2).

MEASURING ENERGY

In biology it is often important to measure the amount of energy expended by living things. For example, the bonding together of smaller molecules such as nucleotides to form larger nucleic acid

> The energy (measured in calories) required to run the chemical reactions of life is stored in and released from the phosphate bonds of ATP.

molecules requires an input of energy. The reverse process of breaking down large molecules into smaller ones releases chemical bond energy. The unit used to measure the amount of energy involved in these and other metabolic processes is the *calorie.* One calorie is the amount of heat energy required to raise the temperature of one gram, or one milliliter, of water one degree Celsius (centigrade). A thousand calories is a ***kilocalorie,*** written either Kcal or Calorie.

The energy contained in the foods we eat is usually expressed in Calories. There are about 113 Calories (113,000 calories) per ounce of carbohydrate or protein and about 255 Calories (255,000 calories) per ounce of fat. In the metric system this is about 4 Calories per gram of carbohydrate or protein and about 9 Calories per gram of fat (1 oz = 28.35 grams). These figures represent the heat energy released when each foodstuff is broken down during a chemical reaction.

ADENOSINE TRIPHOSPHATE (ATP)

Most of the energy requirements of living things are supplied by the high-energy compound ***adenosine triphosphate,*** or ***ATP***. As the "energy currency" of the cell, ATP can be drawn on like money in a bank account. Found in all living organisms, ATP must be synthesized continuously by the body to run the chemical reactions of life. Structurally, ATP is a nucleotide consisting of the nitrogen base adenine (the same purine found in nucleic acids), the 5-carbon sugar *ribose,* and three *phosphate* (PO_4) molecules (Fig. 4.3). Adenine and ribose bonded together form *adenosine,* and the three phosphates are bonded together in chain-like fashion, with the first phosphate bonded to ribose.

Most of the stored energy (chemical bond energy) in ATP utilized by living things resides in the high-energy bond between the middle and terminal phosphate groups. Therefore, the release of energy from ATP more often involves the breaking of this terminal phosphate bond. The bond between the middle and first phosphate groups also is a high-energy bond, but it is less often involved in the routine energy transformations of the cell. Both of the high-energy bonds in the ATP molecule are indicated by the symbol ~.

With the breaking of the terminal phosphate bond, one phosphate group is removed and ATP becomes adenosine *diphosphate,* or ADP (Fig. 4.4A). In breaking the second bond, ADP becomes adenosine *monophosphate,* or AMP. In cellular reactions the free phosphate group removed from ATP is often transferred to another

Adenosine
Triphosphate (ATP)

> In the living world, ATP is generated through the reactions of photosynthesis and cellular respiration.

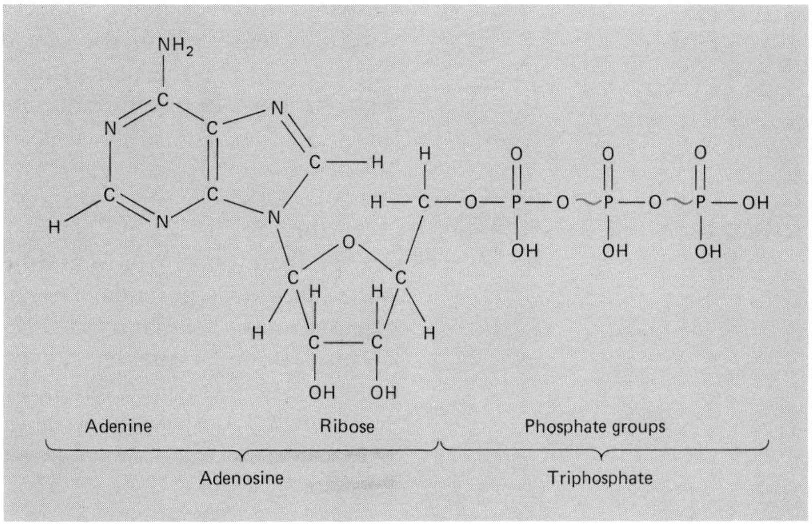

organic molecule. Thus, in acquiring the phosphate group, the molecule obtains some of the energy previously stored in ATP. Subsequent reactions remove the phosphate group from the molecule, releasing the energy stored in the phosphate bond. Most of this released energy is used to bond the molecule to another molecule. In this manner the cell uses the energy stored in the phosphate bonds of ATP to synthesize larger molecules from smaller ones. An example of this process is the synthesis of larger molecules such as disaccharides and polysaccharides from smaller monosaccharide molecules. The phosphate group released from the molecule can then be bonded to ADP to reform ATP (Fig. 4.4B).

What is the source of energy for bonding the free phosphate group to ADP? As referred to earlier, there are two sources. One is light energy used by green plants in photosynthesis. The other source is the chemical energy stored in the carbohydrates, lipids, and proteins that living things utilize every day. In the cells, enzymes break down these foodstuffs, releasing the chemical energy stored in their bonds. Essentially, then, the energy from sunlight or from the breakdown of food is required to bond the free phosphate group to ADP. As with other chemical reactions in the cell, both the breakdown and the synthesis of ATP are dependent upon the activity of appropriate enzymes.

It is characteristic of living things that they release energy in a slow, step-by-step fashion, instead of all at once. A sudden burst of energy lasts too short a time and produces temperatures that would

Figure 4.3 The complete structure of ATP, the "energy currency" of living cells. The high-energy bonds are designated by the symbol ~. Removal of the terminal phosphate group forms adenosine diphosphate (ADP); removal of the middle and terminal phosphate groups from ATP forms adenosine monophosphate (AMP).

Figure 4.4 (a) Breaking the terminal phosphate bond of ATP results in the formation of ADP and a free phosphate group (P). The energy released in this reaction is used by cells to synthesize organic molecules. (b) The synthesis of ATP from ADP and a phosphate group requires an input of energy from an outside source. In living systems, the source is either light energy from the sun or chemical energy stored in the bonds of organic molecules.

ATP $\longrightarrow$ ADP + P + Energy
(a)

ADP + P + Energy $\longrightarrow$ ATP
(b)

> The second law of thermodynamics prohibits any system from being 100 percent efficient.

destroy the cell. Equally important is that energy release has to be carefully regulated in the course of a complex chemical reaction. Each step in the reaction would be impossible to regulate if energy were supplied in one sudden burst.

ENERGY EFFICIENCY

According to the second law of thermodynamics, any conversion of energy from one form to another—as is the case in cellular metabolism, in which potential energy stored in chemical bonds can be released to do cellular work—results in a decrease in usable energy because some of the energy escapes as heat. Although the total amount of energy entering and leaving a system is exactly the same, only some of the energy actually does work. **The ratio of useful energy to the total energy is known as *efficiency*.** This can be shown as follows:

$$\% \text{ Efficiency} = \frac{\text{Useful energy}}{\text{Total energy}} \times 100$$

As an example, assume that your favorite chocolate candy contains 100 Calories. This is the total energy that would enter the body when you eat the chocolate. Certain tests might determine that only 40 Calories were actually used by the body during cellular metabolism. Thus, the percent efficiency of the body in this instance would be as follows: 40 Calories (useful energy) ÷ 100 Calories (total energy) = 0.40 Calories × 100 = 40 percent; the remaining 60 percent is given off as heat energy. Every energy-converting system releases some heat energy that is unavailable to do work; no system, therefore, is ever 100 percent efficient.

ENERGY FLOW

In all living organisms an outside source of energy is required to sustain life, and this energy must be converted in form to be used. The original source of energy for life on earth is light energy from the sun. Through photosynthesis, green plants convert this light energy into chemical energy that is stored in the bonds of ATP and other organic compounds. One of these compounds is glucose, which the plants use as an energy source for their own metabolism. Any excess glucose not used by the plant cells is stored in the form of starch, which has even more bonds because it is formed by linking many glucose molecules together. During cellular metabolism, some of the energy released from the breakdown of glucose is used im-

Energy, unlike matter, is not recycled; consequently, the flow of energy through the living world ultimately depends upon the sun as an outside energy source.

mediately by the plant cells, some of it is stored in the form of ATP, and some escapes as useless heat energy. Various animals feed on the plants, and these animals in turn are fed on by other animals. The compounds synthesized by and incorporated into the plant cells thus serve as initial sources of energy for cellular metabolism in animals. Again, some of the energy is used immediately by the cells of the animals, some is stored in the form of chemical bonds in ATP and other compounds, and some escapes as heat energy. Various soil and water microorganisms obtain their energy by decomposing the bodies of plants and animals that have died. Energy is used in essentially the same manner within the microorganisms as it is within plants or animals. This movement of energy sources among living things can continue as long as there is an outside energy source available. Since energy is not recycled, this outside energy source ultimately is the sun. **The passage of energy from one organism to another throughout the living world is known as *energy flow*** (Fig. 4.5). We shall examine this important concept more closely in Chapter 12.

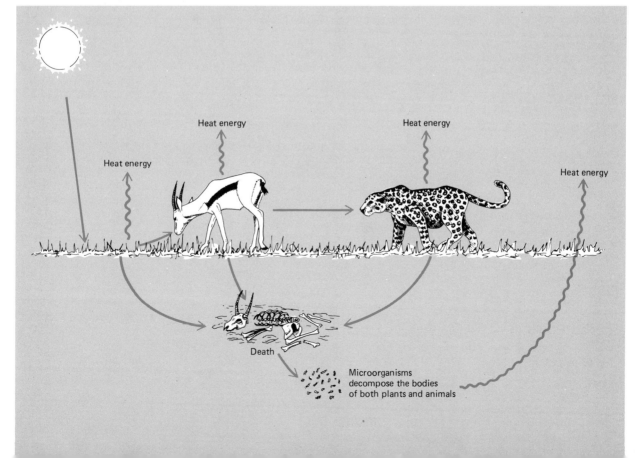

Figure 4.5 Energy flow on the earth. Utilizing light energy from the sun, green plants manufacture organic compounds taken into the bodies of animals that feed on the plants. These animals in turn are fed on by other animals. Microorganisms obtain energy by feeding on the bodies of plants and animals that have died. In the energy conversions that occur within the plants, animals, and microorganisms, some of the energy is used for cellular metabolism, and some escapes to the environment as heat.

Energy and Life

REVIEW OF ESSENTIAL CONCEPTS

1. Life is dependent upon a constant source of energy. *Energy* is the capacity to do work.
2. The *first law of thermodynamics* states that energy can be converted in form, but it cannot be created or destroyed.
3. Any conversion of energy from one form to another results in the escape of some nonusable heat energy, which cannot do work. This is a general statement of the *second law of thermodynamics*.
4. Energy is measured in *calories,* defined as the amount of heat energy required to raise the temperature of one gram of water one degree Celsius. A thousand calories is a *kilocalorie* (Kcal).
5. *Efficiency* is the ratio of useful energy to the total energy of a system. Efficiency is always less than 100 percent because of the escape of nonusable heat energy from the system.
6. *Energy flow* is the passage of energy from one organism to another. The ultimate source of energy on earth is the sun.
7. ATP is the storage form of cellular energy for living things. ATP is generated by certain reactions of photosynthesis and by cellular metabolism.
8. Energy release in living systems is a step-by-step procedure that allows for the regulation and control of chemical reactions.

APPLYING THE CONCEPTS

1. In what manner does a living organism demonstrate the first law of thermodynamics? How does it demonstrate the second law of thermodynamics?
2. What is the significance of energy efficiency as applied to living organisms?
3. What is the general application of the first and second laws of thermodynamics to energy flow in living systems?
4. Explain how energy is stored in and released from ATP.

5 cells and cellular activity

THE ESSENTIAL OBJECTIVES

You have understood this chapter when you are able to:

1. Explain what is meant by the cell theory.
2. Describe the general composition of a eukaryotic cell.
3. Discuss the function or functions of each of the cellular organelles.
4. Compare generally the structures of plant and animal cells.
5. Explain the significance of the surface area—to—volume ratio as applied to living cells.
6. Define diffusion and discuss by example its importance in living things.
7. Explain, using examples, the terms hypotonic, hypertonic, and isotonic.
8. Define osmosis and discuss by example its importance in living things.
9. Explain by example the cause and effect of plasmolysis.
10. Describe the processes of facilitated diffusion, active transport, phagocytosis, and pinocytosis and discuss their importance in living cells.

The cell is the structural and functional unit of life, a miniature factory that synthesizes vital materials and converts energy into forms useful to the organism. Through their diverse regulatory activities, cells contribute to the establishment and maintenance of homeostatic control.

PREVIEW OF ESSENTIAL TERMS

eukaryotic Pertaining to cells that have a membrane-bound nucleus and a variety of organelles.

prokaryotic Pertaining to cells that lack a distinct nucleus and membrane-bound organelles.

organelle "Little organ," such as a mitochondrion, lysosome, plastid, and so on, found in the cytoplasm of a cell.

diffusion The spontaneous movement of a substance from where it is in higher concentration to where it is in lesser concentration, until equilibrium is reached.

osmosis The movement of a solvent from an area of high solvent concentration to an area of lesser solvent concentration across a selectively permeable membrane, until equilibrium is reached.

hypotonic Pertaining to a solution that has a *lower* solute concentration than that of another solution with which it is compared.

hypertonic Pertaining to a solution that has a *higher* solute concentration than that of another solution with which it is compared.

isotonic Pertaining to a solution that has the *same* solute concentration as that of another solution with which it is compared.

plasmolysis The pulling away of the plasma membrane from the plant cell wall due to the loss of water.

facilitated diffusion The movement of a substance across a cell membrane—assisted by carrier molecules—from regions of high concentration to regions of lower concentration.

active transport The energy-requiring transport of a substance across a cell membrane by carrier molecules against a concentration gradient.

phagocytosis The engulfment of large particles by a cell; the particles are then taken into the interior of the cell and digested or destroyed.

pinocytosis The engulfment of very small particles or liquids by a cell; the process is similar to that of phagocytosis.

84
Cells and Cellular Activity

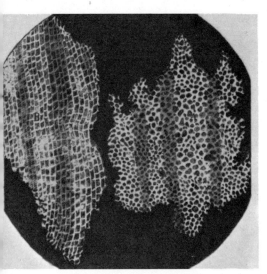

Figure 5.1 Drawing by Robert Hooke of the microscopic structure of a thin slice of cork. Hooke referred to the empty compartments within the cork as "cells." (From Villee, C. A.: *Biology*. 7th ed. Philadelphia: Saunders College Publishing, 1977, p. 11.)

Almost 2500 years ago the Greek philosopher Aristotle wrote: "Nature makes so gradual a transition from the inanimate to the animate kingdom that the boundary lines which separate them are indistinct and doubtful." What Aristotle meant, of course, is that there is an exceedingly fine line that separates "life" from "nonlife." At what level of organization, then, do we first encounter an entity that manifests "life"? Certainly the individual chemical elements are not alive. What about the major biological macromolecules—carbohydrates, lipids, proteins, and nucleic acids? Again, there is no "life" in any one of these individual compounds. Where, then, do we draw the biological line between life and nonlife? The answer is that *life is first encountered at the level of the individual cell.* No single element, compound, or groups of compounds below the organizational level of the cell can be considered a living thing. It follows, then, that any living organism must consist of at least a single cell. As a living entity, a cell is capable of metabolism, growth, response, reproduction, and adaptation—all of those essential characteristics attributable only to living things. In every living organism, the cell is the structural and functional unit of its existence.

THE CELL THEORY

Although he did not discover cells as such, the English scientist and architect Robert Hooke (1635–1703) is credited with the first use of the word "cell." By using a crude microscope, Hooke described the tiny empty compartments in a thin slice of cork as "cells," although dead cork actually has no living contents (Fig. 5.1). However, through Hooke's observations back in the mid 1600s, the word "cell" has now endured for 3 centuries.

The first person to observe living cells was Anton van Leeuwenhoek (1632–1723), a Dutch lens grinder and contemporary of Hooke's (Fig. 5.2). Using an improved microscope of his own design, Leeuwenhoek was able to observe many species of algae and protozoa, the organisms later to be described as unicellular forms of life.

The discovery of the cell eventually brought about a revolution in biology, opening up a new field of inquiry into the structure and function of living systems. From this early foundation, aided by the studies of many scientists over the years, the basis of the ***cell theory*** was formulated in 1839. The cell theory states that all living things are composed of cells and that the cell is the structural and func-

> The cells of most eukaryotic organisms have essentially the same basic architectural plan.

Figure 5.2 Anton van Leeuwenhoek fascinated seventeenth century scientists with his description of microorganisms seen through the lenses he ground as a hobby. (From Silverstein, A.: *The Biological Sciences.* New York: Holt, Rinehart and Winston, 1974, p. 65.)

tional unit of life. The theory was expanded when, in 1858, it was affirmed that all cells arise from pre-existing cells by a process of division. In the years since, all further pronouncements concerning the functioning of living things have been based on this original theory.

CELLULAR ORGANIZATION

Knowing that life has a cellular basis, we may next ask "What is a cell?" In a specific sense the question cannot be readily answered because cells come in a variety of shapes and sizes and have widely varying functions. In a general sense, however, the cells of most organisms known as *eukaryotic* organisms have the same basic architectural plan (Fig. 5.3). All eukaryotic cells have a nucleus and membrane-bound organelles (Fig. 5.4). For example, muscle cells, blood cells, bone cells, various plant cells, and so on are all eukaryotic cells with a fundamentally similar structure. However, the functions of each are quite different. Two groups of living organisms—the bacteria and blue-green algae—are prokaryotic organisms composed of **prokaryotic** cells, which have a less complex structure than do eukar-

86
Cells and Cellular Activity

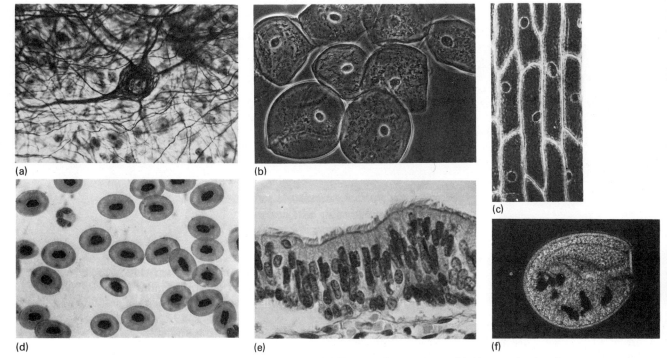

Figure 5.3 Diversity of eukaryotic cells. (a) Nerve cell, or neuron. (b) Human cheek cells. (c) Cells from an onion. (d) Frog red blood cells. (e) Cells lining the human windpipe. (f) A single-celled protozoan. (From Silverstein, A.: *The Biological Sciences*. New York: Holt, Rinehart and Winston, 1974, p. 77.)

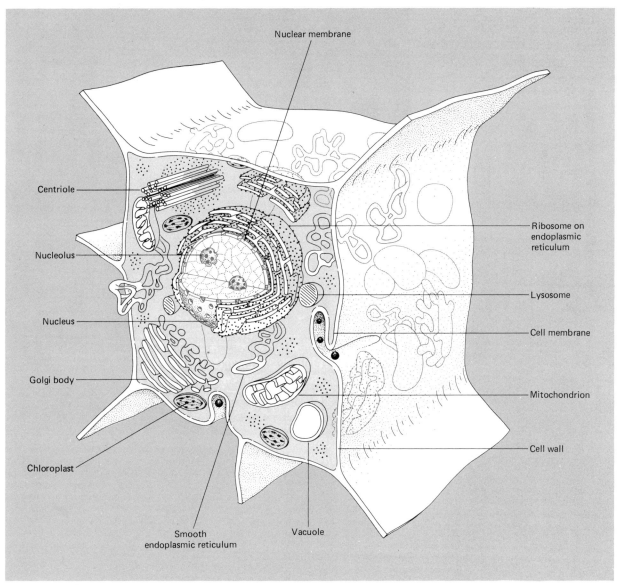

Figure 5.4 Diagram of a eukaryotic cell, showing a composite of plant and animal cell organelles.

Cells and Cellular Activity

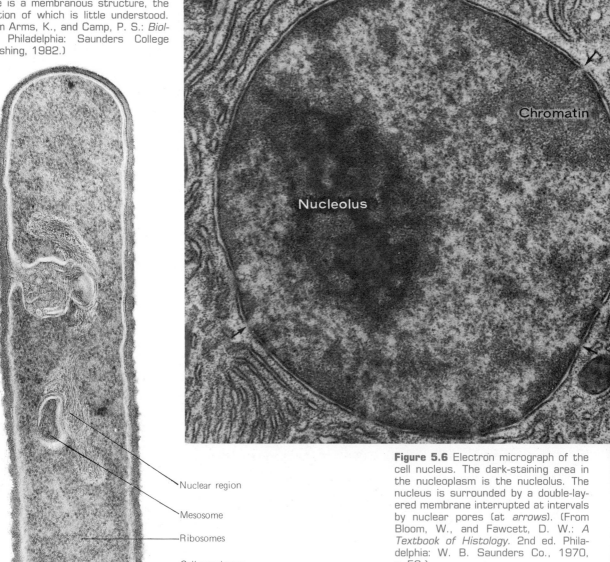

Figure 5.5 A prokaryotic cell of the bacterium *Bacillus subtilis*. The mesosome is a membranous structure, the function of which is little understood. (From Arms, K., and Camp, P. S.: *Biology*. Philadelphia: Saunders College Publishing, 1982.)

Figure 5.6 Electron micrograph of the cell nucleus. The dark-staining area in the nucleoplasm is the nucleolus. The nucleus is surrounded by a double-layered membrane interrupted at intervals by nuclear pores (at *arrows*). (From Bloom, W., and Fawcett, D. W.: *A Textbook of Histology*. 2nd ed. Philadelphia: W. B. Saunders Co., 1970, p. 56.)

> "Protoplasm" is a general term for all the functional constituents of the cell nucleus and surrounding cytoplasm.

yotic cells and are considered to be the more primitive types (Fig. 5.5). Prokaryotic cells lack a distinct nucleus and membrane-bound organelles.

The following discussion outlines the general structure and function of eukaryotic cells; in Chapter 9, we shall comment further on some of the basic characteristics of prokaryotic cells.

THE EUKARYOTIC CELL

Essentially, a cell is structurally and functionally much like a miniature factory. Just as a factory consists of machines, assembly lines, and control centers, all housed within concrete and steel, a cell, too, contains metabolic "machinery," molecular "assembly lines," and a control center, all housed within a surrounding cell membrane. Moreover, both factories and cells manufacture certain "products," utilizing energy in the process. Also like factories, cells are usually individual units of a larger structure. Many factories may compose an industry, and many cells make up the tissues of an organ. The heart, for example, is an organ consisting of tissue that is primarily a collection of individual cardiac muscle cells. Whether we speak of animals or plants, each constituent organ or structure is composed of specific types of cells.

Basically, every cell is an aggregate of *protoplasm* surrounded by a thin *plasma membrane.* "Protoplasm" (Gk. "first form") is a rather general term used for convenience to designate living matter, i.e., the constituents of a cell. In effect, protoplasm is the physical and chemical basis of life. Within the boundaries of the plasma membrane, the protoplasm consists of a semi-fluid *cytoplasm* and the *nucleoplasm* located in the *nucleus* (Fig. 5.6). The nucleus, the "control center" of eukaryotic cells, contains the genetic information in the form of deoxyribonucleic acid (DNA). With directions from the nucleus, the cellular machinery produces the enzymes and other proteins necessary for the structure and function of the cell. We shall pursue the role of DNA in Chapter 8. It has long been known that the nucleus is necessary for a cell to function properly over an extended period of time. Similarly, a factory will not function properly without a manager to give directions. In fact, a cell without a nucleus has but a very short life span (Fig. 5.7). For example, when red blood cells are in the formative stages within the bone marrow, they are all nucleated. Just before their release into the circulating blood, all normal red cells lose their nuclei. As a result, these cells exist in the blood for only about 120 days before they die. They are continually replaced, however, by new red cells formed by special tissues within the marrow.

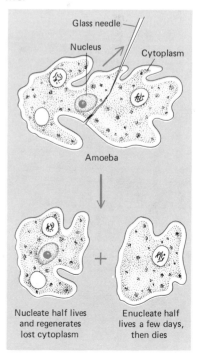

Figure 5.7 Experiment demonstrating the essential role of the nucleus in cell survival. An amoeba is cut in two with a glass needle. The half with the nucleus survives; the other half eventually dies. Transplantation of a nucleus into the enucleated half would allow it to survive.

Cells and Cellular Activity

> The cell nucleus houses the nucleolus, chromosome material, and nucleoplasm within a double-layered membrane.

In nondividing cells, the nucleus is usually the most prominent structure in the cytoplasm. It is a large, usually spherical body bounded by a double-layered *nuclear membrane.* The membrane is penetrated by small pores, each of which is covered by a single thin membrane. The pore membrane separates the cytoplasmic and nuclear contents and apparently regulates the passage of substances into and out of the nucleus.

Lying within the nucleus is a small, spherical body called the *nucleolus* ("small nucleus"). The nucleolus is composed of DNA and protein and is involved in the synthesis of a specific type of ribonucleic acid (RNA).

The DNA of the nucleus comprises most of the structure of tiny threadlike bodies called *chromosomes.* The inheritance of traits from one's parents is due to the passing of chromosomes from the parents' cells to the offspring through the production of eggs and sperm. Each species of plant or animal has a set number of chromosomes in its cells. You have 46 chromosomes, the normal number for all human beings; your dog has 78 chromosomes; certain roundworms have only 2; and there are species of protozoa that have hundreds of chromosomes.

Most of the remaining cellular machinery consists of several different organelles found in the cytoplasm. These are small, mostly

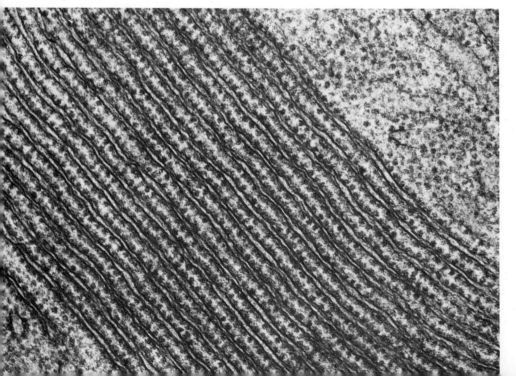

Figure 5.8 Electron micrograph of endoplasmic reticulum (ER). These ER membranes are studded with ribosomes (dark-staining dots). Endoplasmic reticulum having ribosomes attached is referred to as "rough" ER; if ribosomes are not attached, the endoplasmic reticulum is referred to as "smooth." (From Bloom, W., and Fawcett, D. W.: *A Textbook of Histology.* 2nd ed. Philadelphia; W. B. Saunders Co., 1970, p. 34.)

microscopic bodies, each having a particular role to play in carrying out the metabolism of the cell.

Throughout the interior of the cell there is a network of membranes that separates the cytoplasm into compartments. This network, referred to as the *endoplasmic reticulum,* or ER, divides the cytoplasm into separate areas, much like a factory is divided into sections (Fig. 5.8). Some ER membranes are studded with tiny spherical bodies composed of RNA and protein, called *ribosomes.* These organelles, also found lying free in the cytoplasm, are the sites in the cell where proteins are made. The endoplasmic reticulum extends from near the plasma membrane inward all the way to the nuclear membrane, forming channels or passageways through which cellular materials can be distributed.

In 1898 the Italian scientist Camillo Golgi first described a cellular organelle that has since been found in almost all cells. The organelle is called, appropriately enough, the *Golgi body* (Fig. 5.9). It is composed of several flattened, elongated sacs lying parallel to each other. It has been suggested that some of the proteins made on

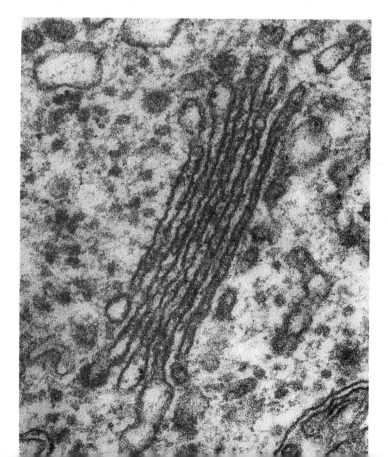

Figure 5.9 Electron micrograph of a Golgi body. This organelle is involved in cellular packaging and secretion. (From Villee, C. A.: *Biology.* 7th ed. Philadelphia: Saunders College Publishing, 1977, p. 68.)

> Various genetic disorders may result in the absence or improper functioning of lysosomal enzymes. The substances on which these enzymes act thus accumulate in the cells, resulting in *storage diseases*, such as those that depress skeletal muscle function.

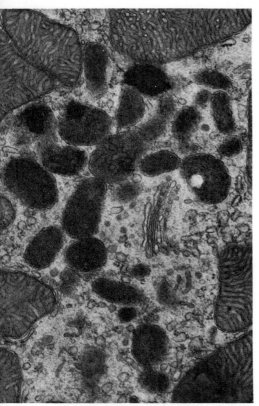

Figure 5.10 Electron micrograph of lysosomes. These membranous sacs contain digestive enzymes. (From Bloom, W., and Fawcett, D. W.: *A Textbook of Histology*, 2nd ed. Philadelphia: W. B. Saunders Co., 1970, p. 42.)

Figure 5.11 This plant cell contains several vacuoles, which are open spaces in the cytoplasm bounded by a cell membrane. (From Silverstein, A.: *The Biological Sciences*. New York: Holt, Rinehart and Winston, 1974, p. 103.)

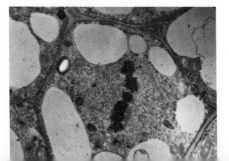

the ribosomes of the endoplasmic reticulum move into the Golgi body, where they are stored temporarily. The proteins are then packaged in tiny sacs produced by the Golgi body, and the sacs move to the plasma membrane, where they burst and deposit the proteins outside the cell. It has been further suggested that after releasing the proteins, the membrane of the ruptured sac becomes incorporated into the plasma membrane. Other cellular substances, including hormones, enzymes, pigments, and so on, also accumulate in the Golgi body. These, too, are packaged and released or secreted by the cell.

Found predominantly in animal cells are organelles called *lysosomes* (Fig. 5.10). A lysosome is simply a membranous sac containing any one or a few of a variety of digestive powerful enzymes. Although the function of lysosomes is not totally clear, it is believed that they are involved in digesting food particles and other substances that enter the cytoplasm. In human beings and other animals, disease-causing organisms that enter the body are digested by certain white blood cells, presumably through the action of their lysosome enzymes. The single-celled protozoa apparently use lysosomes to break down food particles contained within food vacuoles.

Vacuoles are open spaces in the cytoplasm bounded by a cell membrane. They may contain a variety of substances, including food, water, ions, pigments, wastes, and so forth. Vacuoles are typically more developed in plant cells than in animal cells. In fact, the vacuoles of many plant cells become so large that the cytoplasm is forced to the periphery of the cell. Among some protozoans, small specialized vacuoles called **contractile vacuoles** aid in maintaining the internal environment of the cells by pumping excess water to the exterior. **Food vacuoles** formed in many types of cells store and distribute nutrients. The lysosomes and food vacuoles may fuse, so that the contents of each mix and the food particles are digested. Lysosomes are also believed to be involved in the breakdown of worn-out cells or cells that are no longer of use to an organism. Such is the case, for example, when a tadpole absorbs its tail on the way to becoming a frog. Finally, lysosomes may be involved in the process of aging and in degenerative diseases of the heart, brain, and skeletal muscles. Through the release of the digestive enzymes, lysosomes may contribute to the breakdown of body cells and the subsequent destruction of vital tissues.

If we were to designate the major energy-producing "departments" of the cell, they would be the organelles known as **mitochondria** (sing. mitochondrion) (Fig. 5.12). These small bodies contain

> Mitochondria are the "powerhouses," i.e., the major sites of ATP production, in eukaryotic cells.

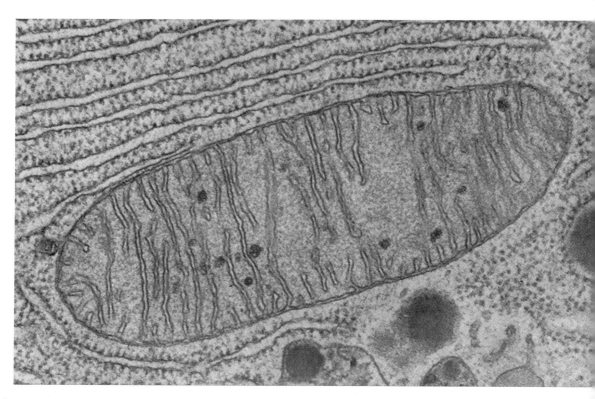

enzymes and electron carrier molecules that are involved in the major reactions of cellular metabolism. Cells that are actively metabolizing, and thereby requiring a good deal of energy, are abundantly supplied with mitochondria. The cells of the heart and skeletal muscles are good examples. Because of their role as the sites of energy production (mainly in the form of ATP), mitochondria often are referred to as the "powerhouses" of the cell.

Two types of organelles believed to play a role in intracellular support and movement are *microfilaments* (Fig. 5.13) and *microtubules* (Fig. 5.14). Microfilaments are long, threadlike structures found in several different cell types, including skeletal muscle cells, a variety of protozoans, and some plant cells. In skeletal muscle cells, the sliding action of thick and thin protein microfilaments is responsible for contraction of the muscle. Microfilaments also appear to be involved in the flowing motion of cytoplasm called *cytoplasmic streaming,* which occurs in various plant cells and in the pseudopodia of amoeboid cells. In some cells, microfilaments apparently act like an internal skeleton, providing strength and support.

Figure 5.12 Electron micrograph of a mitochondrion. The mitochondrion is surrounded by two membranes. The channel-like areas within the mitochondrion are formed by inward foldings of the inner membrane. These foldings are called *cristae.* Located on the cristae are some of the enzymes and electron carrier molecules involved in cellular respiration. Other enzymes also involved in cellular respiration are in solution within the mitochondrion. (From Gerking, S. D.: *Biological Systems.* 2nd ed. Philadelphia: Saunders College Publishing, 1974, p. 15.)

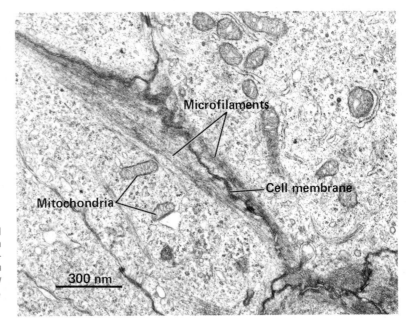

Figure 5.13 Microfilaments near the cell membrane. Microfilaments are thin protein fibers thought to be involved in cellular contraction and amoeboid movement. (From Arms, K., and Camp, P. S.: *Biology.* New York: Holt, Rinehart and Winston, 1979, p. 55.)

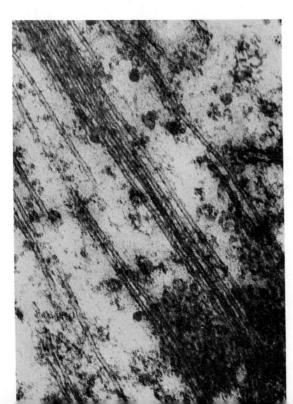

Figure 5.14 Microtubules in the cellular cytoplasm. Microtubules are involved in intracellular support and movement. (From Bloom, W., and Fawcett, D. W.: *A Textbook of Histology.* 2nd ed. Philadelphia: W. B. Saunders Co., 1970, p. 51.)

> Microfilaments and microtubules are organelles composed of specialized proteins that are capable of lengthening and shortening.

Also involved in intracellular support and movement are microtubules, which are long, hollow cylinders. Microtubules are involved in the beating of cilia and flagella, in giving shape and internal strengthening to the cell body, and in the movement of chromosomes during cell division. When cells are dividing, the chromosomes are attached to an arrangement of microtubules called the *spindle*. By shortening, the microtubules of the spindle move the chromosomes to different locations in the cell.

Another organelle, located just outside the nucleus of most animal cells, is the **centriole** (Fig. 5.15). Centrioles are usually found in pairs, lying at right angles to each other. A centriole contains nine groups of microtubules arranged in the shape of a cylinder; each group consists of three microtubules. Although they are associated with the spindle and appear to be involved in its formation, the centrioles apparently do not function in this capacity. In fact, their function is unknown. Centrioles are not present in most of the cells of higher plants, and yet these cells form spindles during cell division in essentially the same manner as animal cells.

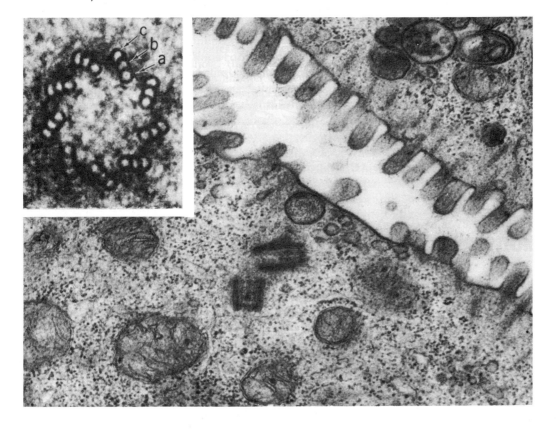

Figure 5.15 Electron micrograph of a pair of centrioles oriented at right angles to each other. The inset shows a cross section of one centriole, showing the nine groups of three microtubules. (From Bloom, W., and Fawcett, D. W.: *A Textbook of Histology.* 10th ed. Philadelphia: W. B. Saunders Co., 1975, p. 58.)

> Most plant cells differ from animal cells in having large vacuoles, plastids, few or no lysosomes, and a cell wall.

Some of the cells of the body, as well as those of some plants and many single-celled organisms, have tiny hairlike organelles called *cilia* and *flagella* (Fig. 5.16). Some of the single-celled organisms use cilia or flagella for locomotion. In man and other higher animals, cilia line the respiratory and reproductive tracts. In your windpipe, for example, cilia aid in preventing foreign particles from entering the lungs. In the fallopian tubes of human females, cilia that line the walls help propel the egg toward the uterus. The sperm cells of many organisms, including man, have a single long flagellum used for locomotion. Plants such as mosses have biflagellate sperm cells, whereas ferns and a few other plant groups have multiflagellate sperm cells. Actually, cilia and flagella in most eukaryotic organisms are structurally identical; the only essential difference is that flagella are longer. Cilia and flagella are somewhat similar in structure to centrioles; however, cilia and flagella have nine *pairs* of microtubules arranged in a cylinder, with two single microtubules in the center. This basic structure is the same in all cilia and flagella of eukaryotes, whether on the cells of protozoa, plants, or human beings.

As you would imagine, animal cells and plant cells are not structurally the same. As mentioned, many plant cells have large vacuoles; in addition, plant cells contain organelles called *plastids* and possess a *cell wall* outside the cell membrane. One of the vital energy-producing "departments" of green plant cells is a type of plastid called the *chloroplast,* an organelle not found in animal cells (Fig. 5.17). Chloroplasts contain *chlorophyll,* the pigment that gives

Figure 5.16 Electron micrograph of a cross section of a flagellum. All cilia and flagella in eukaryotes are constructed of nine pairs of microtubules arranged in a cylinder, with two single microtubules in the center. (From Loewy, A. G., and Siekevitz, P.: *Cell Structure and Function.* 3rd ed. New York: Holt, Rinehart and Winston, 1969, p. 62.)

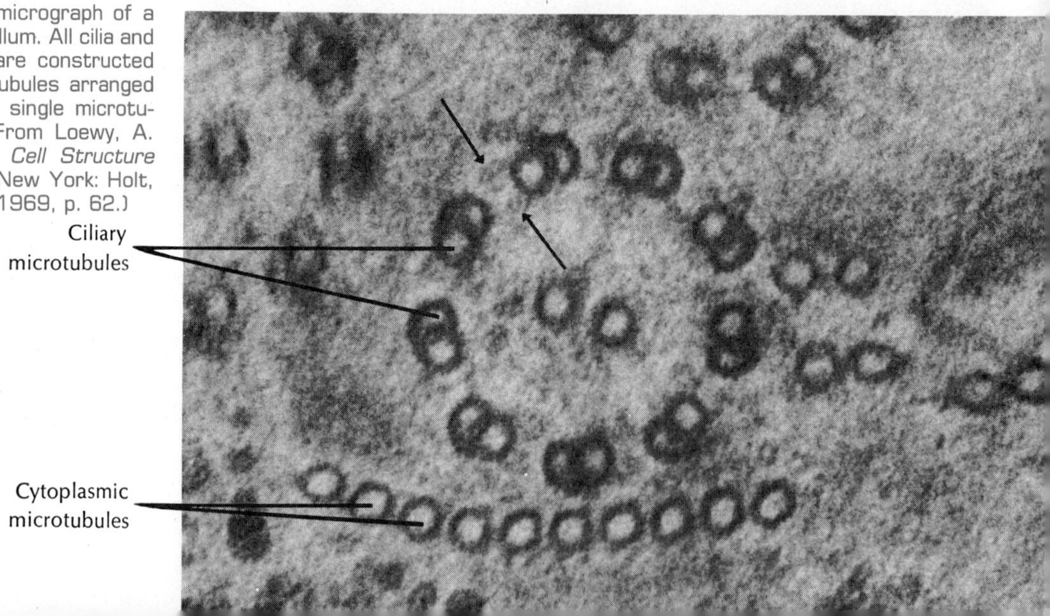

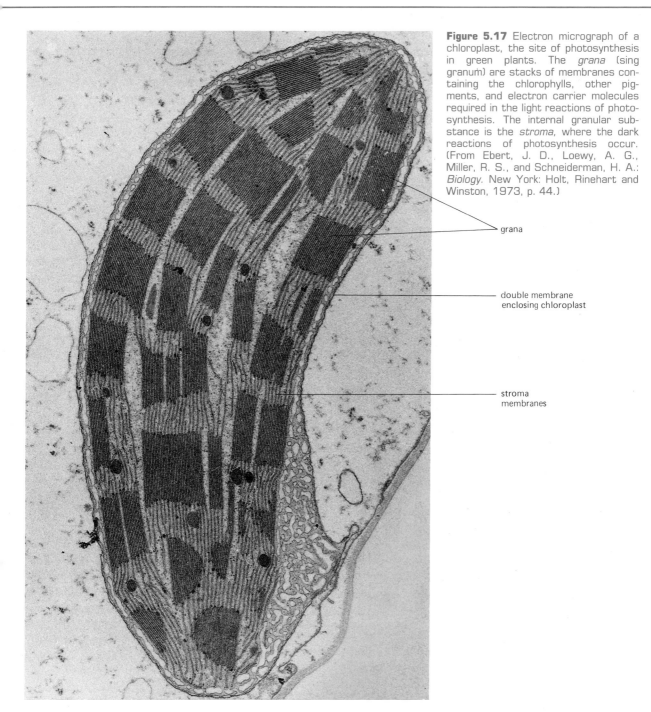

Figure 5.17 Electron micrograph of a chloroplast, the site of photosynthesis in green plants. The *grana* (sing *granum*) are stacks of membranes containing the chlorophylls, other pigments, and electron carrier molecules required in the light reactions of photosynthesis. The internal granular substance is the *stroma*, where the dark reactions of photosynthesis occur. (From Ebert, J. D., Loewy, A. G., Miller, R. S., and Schneiderman, H. A.: *Biology.* New York: Holt, Rinehart and Winston, 1973, p. 44.)

grana

double membrane enclosing chloroplast

stroma membranes

> The unique internal environments of the eukaryotic cell and its organelles are maintained by selectively permeable membranes.

green plants their characteristic color. More importantly, chlorophyll is necessary in the sugar-producing reactions of *photosynthesis*. Other plastids called **chromoplasts** store the red, yellow, and orange pigments displayed in the hues of autumn leaves and in the vivid shades of flowers and fruits. Proteins, lipids, and starch may be stored in colorless plastids called *leucoplasts*.

In addition to harboring the organelles mentioned previously, the cytoplasm is a reservoir for a variety of other materials. In general, the cytoplasm of most cells is an aqueous solution of proteins, inorganic ions, simple sugars, enzymes, and other organic materials, including ATP. Some of these substances are raw materials used by the cellular machinery during metabolism, and some are the products manufactured from these raw materials.

In many types of cells, the cytoplasm contains specific substances that are associated with the unique function of the cell. Your red blood cells, for example, contain the pigment *hemoglobin,* which is vital in the transport of oxygen by the blood.

Like any building, a factory, along with each of its individual departments, is partially isolated from the outside environment. So it is with a living cell. There is usually quite a difference between the physical and chemical events occurring within the cell and those occurring in the outer environment. Similarly, the activities within the organelles are usually quite different from those occurring in the cytoplasm outside. Like a walled room, the interior of each organelle, along with the cytoplasm itself, is a separate and distinct environment isolated, not by solid walls, but by surrounding membranes. In fact, because of their particular structures, the membranes surrounding each organelle serve to regulate and maintain these unique environments.

Basically, cell membranes are considered to be constructed of a double layer of phospholipids in which are embedded protein molecules of various sizes. Some of the proteins appear to "float" in the inner and outer surfaces of the double lipid layer, whereas other proteins extend all the way through the layer (Fig. 5.18).

It might appear that membranes form an impenetrable barrier between the inside of a structure and its surroundings. Such is not the case, however, in that the structure of membranes permits the passage of some substances while restricting the passage of others. Thus, membranes are said to be *selectively permeable,* i.e., permeable only to certain "select" substances. By regulating what gets into and out of the cell, the various membranes thereby maintain the cellular environment (Fig. 5.19). You may recall earlier references to the relationship between structure and function in living systems. As you will discover shortly, cellular membranes exquisitely demonstrate the significance of this crucial relationship.

In addition to a bounding plasma membrane, plant cells are encased in a *cell wall* lying outside the membrane. The cell wall is composed mainly of a tough, rigid carbohydrate called *cellulose*. Cellulose is a polysaccharide composed of a long chain of glucose molecules. The cell wall is quite porous, allowing molecules of almost any size to pass through with relative ease.

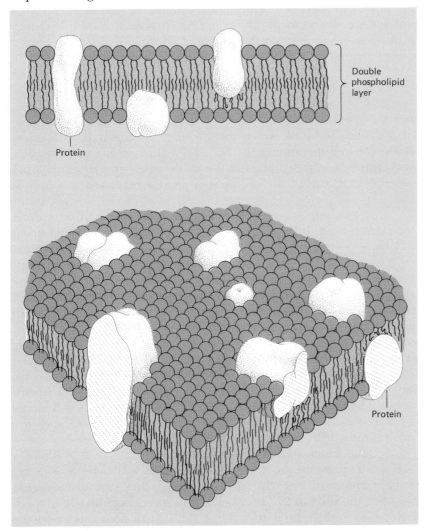

Figure 5.18 Basic structure of a cell membrane in cross-sectional and three-dimensional views. This is generally referred to as the *fluid mosaic model* of cell membrane structure. "Mosaic" refers to the inlaid protein molecules that form a pattern, and "fluid" is a reference to the mobility of the lipids (and to some extent the proteins), some of which are thought to move laterally to different positions in the membrane.

Figure 5.19 Electron micrograph of two plasma membranes separated by a narrow intercellular space.

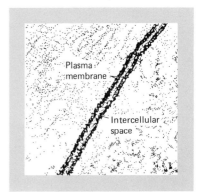

> The significance of an increase in the size of cells on the surface area-to-volume ratio can be appreciated when we consider that, as a cell grows, its surface area increases as the *square* of its diameter, while its volume increases as the *cube* of its diameter.

Any discussion concerning living cells would be incomplete without referring to the most vital substance of cellular composition—water. About 70 to 85 percent of protoplasm is water, and all living cells must be bathed in a watery medium. The cells of many lower organisms are surrounded by the fresh or salt water in which they live. In higher organisms, body cells are bathed in a watery **extracellular fluid** (fluid in the spaces outside the cells). There is a vital relationship, as we shall see, between this fluid and the **intracellular fluid** (L. *intra*, inside or within) of the cytoplasm. Moreover, the unique properties of water make possible the vast array of chemical and physical activities of living cells. A good deal of this activity is attributable to the fact that water, unlike many other substances, passes through biological membranes easily and rapidly.

SOME ACTIVITIES OF LIVING CELLS

In its role as the basic unit of life, a cell is involved in a broad range of activities, some of which include the movement of substances across the cell membrane. By the very nature of life as a dynamic yet carefully controlled process, the fluids, gases, and other materials entering and leaving the cell must be closely regulated as a part of that process.

One attribute common to almost all cells—plant or animal—is their small size. Why is this so? Given your present height and weight, why couldn't you just as easily be composed of a few dozen or a few hundred larger cells, instead of trillions of smaller ones? Basically, the answer involves the relationship between the *surface area* of a cell and its *volume*. The relationship is such that a smaller cell has more surface area exposed *in proportion to its volume* than does a larger cell. As a hypothetical example, consider two different cells—one in the shape of a 2-inch cube and the other a 3-inch cube (Fig. 5.20). The surface area of *one* side of the 2-inch cube would be 2 inches × 2 inches = 4 square inches; since the cube has six sides, its *total* surface area would be 6 × 4 square inches = 24 square inches. The *volume* of the 2-inch cube is 2 inches × 2 inches × 2 inches = 8 cubic inches. Therefore, there is a surface area–to–volume ratio of 24 to 8, or 3 to 1.

For the 3-inch cube, the total surface area is 3 inches × 3 inches × 6 = 54 square inches; the volume is 3 inches × 3 inches × 3 inches = 27 cubic inches. In this instance, the surface area–to–volume ratio is 54 to 27, or 2 to 1. Thus, increasing the size of a cell reduces its surface area–to–volume ratio. How does this affect the

Diffusion is a passive process, requiring no expenditure of cellular energy.

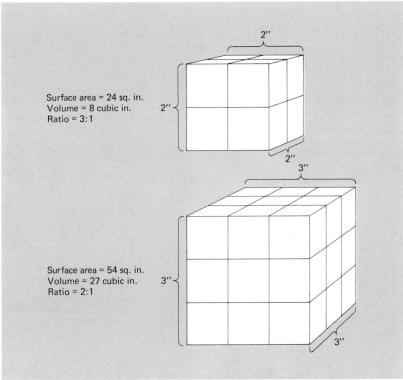

Figure 5.20 Relationship between surface area and volume in a 2-inch cube and a 3-inch cube. Like the cubes, living cells are three-dimensional structures. A smaller cell (represented by the 2-inch cube) has more surface area exposed in proportion to its volume than does a larger cell (the 3-inch cube).

functioning of living cells? To answer that question, we first need to become familiar with two of the basic processes involved in the movement of materials into and out of the cell.

The first of these is the process of *diffusion* (Fig. 5.21). By definition, diffusion is *the spontaneous movement of a substance from an area where it (the substance) is in high concentration to an area where it is in lower concentration, until equilibrium is reached*. To put it simply, a diffusible substance moves from where there is more of it to where there is less of it. This net movement continues until the substance is equally distributed throughout the area. We can illustrate diffusion with a familiar everyday example. When a substance such as perfume or shaving lotion is applied, it is in high concentration on the body. The substance begins to diffuse spontaneously out into the room, where it is less concentrated. Once there is an equal concentration of the molecules of the substances throughout the room, diffusion stops. The molecules may continue to move about, but there is no *net* movement in any direction. If the room happened to be large, it would take a relatively longer time for the molecules to

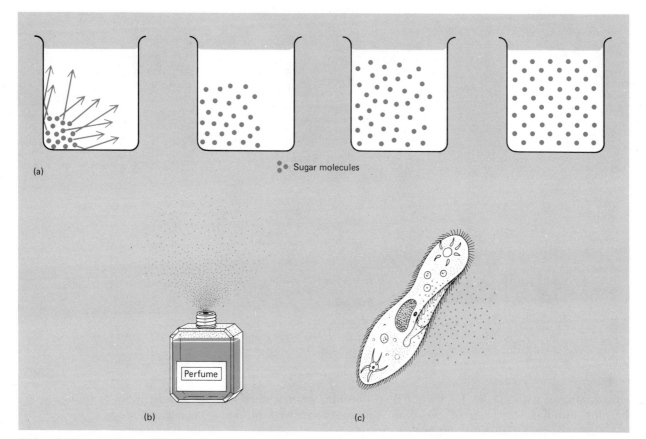

Figure 5.21 Examples of diffusion. (a) Diffusion of a dissolved substance in water (b) Diffusion of perfume molecules through the air. (c) Diffusion across a cell membrane. In each example, the molecules move from an area where they are in greater concentration to an area where they are in lesser concentration. How long will this process continue?

reach equilibrium than if the room were quite small. This is because diffusion occurs quite rapidly over very short distances.

With living cells, substances in high concentration outside the cell membranes would tend to diffuse to the inside and vice versa (providing of course that the substances are not too large, and so on). Two of the most vital substances exchanged across cell membranes are the respiratory gases—oxygen and carbon dioxide. Oxygen is required as a raw material by almost all living cells for the reactions of cellular metabolism. As the cells metabolize, they use up this oxygen and produce carbon dioxide as a by-product. Fresh oxygen, constantly resupplied by the blood, diffuses into the cells as the "old" oxygen is being used. As diffusion dictates, oxygen moves into the cell because it is in higher concentration in the blood than it is inside the cells. On the other hand, the concentration of carbon

> In the solutions found in living things, the solvent is always water.

dioxide increases inside the cells and thus diffuses outward into the blood, where it is in lower concentration.

The mechanism of diffusion, as described here for oxygen and carbon dioxide, also applies to other diffusible substances, such as water, ions, and dissolved nutrients that pass into and out of the cell. The essential points to remember regarding the diffusion of a substance across cell membranes is that there must be a difference in concentration of that substance inside and outside the membrane and that substances always diffuse from an area where they are in higher concentration to an area of lower concentration.

Returning to our original question, how does the surface area–to–volume ratio affect the functioning of living cells? One of those functions we have just described involves diffusion. Since diffusion occurs more rapidly over a short distance, a substance can move into and out of a small cell quickly. In the case of oxygen this is vital, since oxygen is used by the cell almost as rapidly as it enters. Moreover, once a substance is inside a small cell, it can move throughout the interior quickly. In a cell having a larger volume, a substance may diffuse around the interior for a relatively long period of time before it is distributed to where it is needed.

Apparently, cell size also affects the functioning of the nucleus. Being the control center of the cell, the nucleus can regulate only a certain amount of cytoplasmic activity. Part of this activity involves the movement of materials to and from the cell nucleus, as well as from one area of the cytoplasm to another. In small cells, nuclear control of such movement would tend to be more efficient than in large cells.

Finally, a greater surface area–to–volume ratio means that there is proportionally more membrane surface exposed for the effective exchange of materials. As a result, the small cell is a much more efficient unit for the passage and production of essential commodities.

It has been pointed out that the cytoplasm of living cells is an aqueous solution containing a variety of organic and inorganic materials. Any solution is composed of the dissolved solids, or *solute*, which is dispersed throughout the liquid, or *solvent*. In the cells of all living things, the solvent is water. Solutions may have varying amounts of solute, resulting in each solution having a different concentration. For example, if there were two glasses of the same size filled with iced tea (the solvent), we might add 1 tsp of sugar (the solute) to the first glass and 3 to the second glass. Thus, the solution in the first glass is *less* concentrated than the solution in the second

104
Cells and Cellular Activity

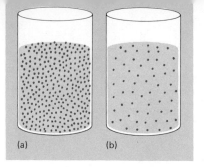

Figure 5.22 Two glasses containing solutions of different concentrations. The more highly concentrated solution in glass (a) is said to be *hypertonic* to the less concentrated solution in glass (b). It can also be said that (b) is *hypotonic* to (a).

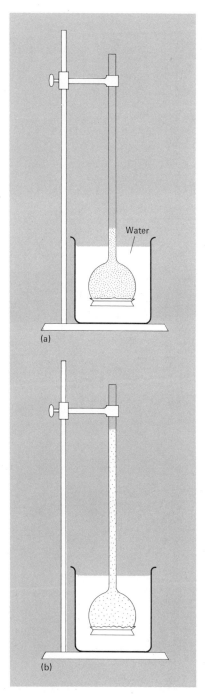

glass. Another way of expressing the comparative difference in concentration between the two tea solutions is to say that the first solution is **hypotonic** to the second solution (Fig. 5.22). Hypotonic simply refers to the fact that a solution has *less* solute than another solution of equal volume with which it is compared. Conversely, the solution with *more* solute than another is said to be **hypertonic.** If two solutions of equal volume have the *same* concentration of solute, they are *isotonic* to each other.

The fact that living cells may be exposed to solutions of varying concentrations presents some problems that affect the functioning and indeed the very existence of the cells. It was mentioned earlier that water moves freely through cell membranes, whereas many other materials cannot pass through. The movement of water across a membrane is the result of the process of **osmosis** (Fig. 5.23). Osmosis can be defined as *the movement of a solvent* (water) *from an area of high solvent concentration to an area of lesser solvent concentration across a selectively permeable membrane.* As with diffusion, osmosis continues until a state of equilibrium is reached, i.e., until the *concentrations* of the solutions on both sides of the membrane are equal.

Osmosis, a word derived from the Greek *osmos,* meaning "a push or thrust," can be better understood if we consider a few basic concepts involved in the process. As we have pointed out, the solutions inside and outside living cells are separated by a selectively permeable membrane, which readily allows water and other small molecules to move into or out of the cells but restricts the passage of larger molecules. Although some of the small molecules may pass through the membrane, the important consideration in osmosis is the movement of *water* molecules across the membrane. Often described as the "diffusion" of water across a membrane, osmosis is indeed that, but it is also something more.

The movement of water across cell membranes depends upon the number of particles (molecules or ions) of solute. Water has potential energy that is released when the water moves. The potential energy of a flowing river, for example, can be used to generate electricity in a hydroelectric plant. The potential energy of water, re-

Figure 5.23 Osmosis. The thistle tube initially contained a concentrated sugar solution (a). Water diffused through the semipermeable membrane into the tube (b), lowering the concentration of the solution and increasing the volume of liquid in the tube (b). The membrane does not permit passage of sugar molecules out into the beaker.

> Osmosis occurs when two solutions of different concentrations are separated by a selectively permeable membrane.

ferred to as *water potential,* is *decreased* when a solute is added; therefore, pure water (water having no solute) has a greater water potential than does water containing solute particles; similarly, water containing less solute has a greater water potential than does water containing proportionately more solute. Since osmosis involves the movement of water between regions of higher and lower solute concentrations, water will move from a region of high water potential (less solute) to a region of lower water potential (more solute) across a membrane until equilibrium is reached.

But what does it mean to "reach equilibrium"? It means that the potential energy of water, i.e., the water potential, on both sides of the membrane is equal. Stated more simply, equilibrium is reached when there is no net movement of water either into or out of the cell. How is this state of equilibrium established? As water moves across the cell membrane, the concentrations of the solutions inside and outside the cell begin to change. The solution with the *higher* solute concentration gains water, which makes the solution more dilute. In losing water that passes across the membrane, the solution with the *lower* solute concentration becomes less dilute, or more concentrated. At some point, the concentrations of the two solutions become equal, and the net movement of water stops. The two solutions are then said to be in *osmotic equilibrium.*

As an illustration of osmosis, assume that a hypothetical cell has a cytoplasmic concentration of 1 percent sugar. (A 1 percent solution contains 1 gram of solute dissolved in enough water to make 100 ml of solution; a 5 percent solution would have 5 grams of solute dissolved in enough water to make 100 ml of solution, and so forth.) If this cell were immersed in a 5 percent sugar solution, what would happen to the cell? Recalling the definition of osmosis, water moves across a membrane from a region of high water concentration to a region of lesser water concentration. In the example here, the 1 percent sugar solution contains more water than the 5 percent sugar solution. Therefore, water would move from the interior of the cell (where there is proportionately more water), across the plasma membrane to the outside (where there is proportionately less water).

Because of their size, the sugar molecules do not passively penetrate the membrane and so do not enter or leave the cell. The loss of water from the interior of the cell causes it to shrink. Such is the result of exposing most cells to a *hypertonic* solution (Fig. 5.24). If the concentrations were reversed so that the cytoplasm contained a 5 percent sugar solution and the cell was immersed in a 1 percent sugar solution, water would move across the plasma membrane *into*

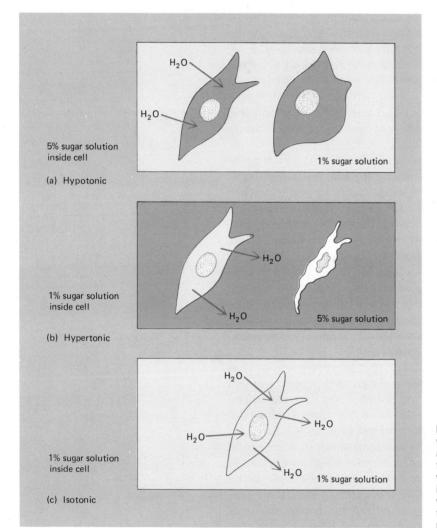

Figure 5.24 In a hypotonic medium (a), water diffuses into the cell; it swells and may ultimately burst. In a hypertonic medium (b), a cell tends to lose water by osmosis and shrivels. In an isotonic medium (c), the net inflow of water is equal to the net outflow, and the cell is at equilibrium with its surroundings.

the cell, causing it to swell. In this instance, the cell was exposed to a *hypotonic* solution. If the sugar solutions inside and outside the cell happened to be of the same concentration, i.e., if the solutions were *isotonic* to each other, there would be no net movement of water in either direction.

It is important to realize that the terms "hypertonic," "hypotonic," and "isotonic" are *comparative* terms and cannot be used to describe any one solution unless that solution is being compared with another. To say that a certain solution is hypertonic, for exam-

> Living cells react to changes in their osmotic environment by swelling or shrinking.

ple, makes no sense unless its concentration is being compared with that of a less concentrated solution. In our example, then, it would be correct to state that the 5 percent sugar solution is *hypertonic* to the 1 percent sugar solution or that the 1 percent solution is *hypotonic* to the 5 percent solution. It would be incorrect, however, to state that a 5 percent sugar solution sitting alone on a table is hypertonic. Compared with another solution, it *could* be hypertonic, hypotonic, or isotonic.

As a condition of their existence, all living cells are continuously bathed in a watery solution. As we have seen, the concentration of this solution determines the movement of water into or out of the cell. The water solution bathing the cells of the body is called *extracellular fluid*, which normally is isotonic to the *intracellular fluid* of the cytoplasm. The fact that these two solutions are isotonic minimizes the problem of appreciable water uptake or loss by the cells. It could also be stated that there is no net movement of water into or out of the cells because the intracellular and extracellular fluids are in almost constant osmotic equilibrium. There are, however, many organisms that do not live in constant osmotic equilibrium with their environment. Unicellular organisms such as many of the protozoa live in a freshwater environment that is hypotonic to their intracellular fluids. The osmotic problem confronting these organisms is that water continually tends to move into their cells. Unless counteracted, the inward movement of water would cause the cells to swell and eventually burst. This problem is taken care of by a tiny cytoplasmic organelle called a *contractile vacuole*, which pumps water to the outside.

The result of living in a hypotonic environment is somewhat different among plant cells. In most plant cells the plasma membrane is bounded by a rigid cell wall. As water accumulates in the plant cell, it exerts pressure against the inside of the plasma membrane, which expands and exerts pressure against the cell wall. The force exerted by the water is also known as **turgor pressure** (Fig. 5.25). This pressure maintains the leaves and other plant structures in a firm, upright position. In contrast to most animal cells, the turgid plant cell does not burst because of the rigid cell wall. If plant cells with thin cell walls are exposed to an isotonic solution, water ceases to enter the cells and the plants wilt. In hypertonic solutions, water is drawn out of the cells, and all turgor pressure is lost. In plant cells with thick walls, the plasma membrane pulls away from the cell wall and collapses. In this case, the loss of turgor pressure causes the cytoplasm and vacuole to shrink. The pulling away of the

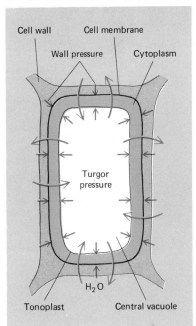

Figure 5.25 Contributions of the cell wall and central vacuole to plant cell shape. The central vacuole, surrounded by its membrane, takes up most of the volume of the cell. As the cell swells, its internal turgor pressure is opposed by the pressure of the cell wall, which resists stretching. When turgor pressure equals wall pressure, water is squeezed out by wall pressure as quickly as it enters the cell *(colored arrows)*.

plasma membrane from the cell wall due to the loss of water is termed *plasmolysis* (Fig. 5.26); a cell so affected is said to be plasmolyzed.

Although water, some gases, and many small ions readily pass through cell membranes, molecules such as sugars and proteins are too large to penetrate the membrane by simple diffusion or osmosis. However, these and other large molecules are vital constituents of the cytoplasm of cells. How do they get there? One of the processes involved is called *facilitated diffusion,* which is the principal means by which some of the sugars in particular cross a membrane (Fig. 5.27). By "facilitated" it is meant that the substances crossing the membrane are "helped" in some manner. Involved in the mechanism of facilitated diffusion are *carrier molecules,* possibly small proteins, that combine with the sugar or other substance and transport them across the membrane. The carrier molecules are located within the membrane, where they make contact with either the outer or the inner surface of the membrane. The carriers can thereby

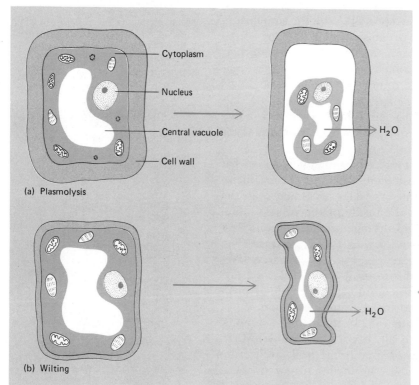

Figure 5.26 Plasmolysis and wilting in plant cells. (a) In plants with a thick cell wall (trees and shrubs), osmotic loss of cell water causes a shrinkage of cytoplasm and the central vacuole. The pulling away of cytoplasm from the cell wall is called plasmolysis. (b) In plants with a thin cell wall (lettuce and celery), loss of cell water causes a similar shrinkage of the cell and vacuole. In this case, however, the fall in turgor pressure also causes collapse of the cell wall, or wilting.

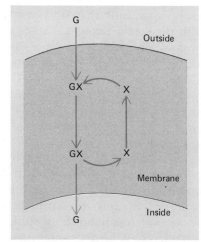

Figure 5.27 The facilitated diffusion of glucose through membrane. The theory suggests that the membrane contains specific carriers (X) to which the glucose is bound and transported through the membranes. A complex of carrier and glucose is formed in the outer membrane and cleaved at the inner membrane surface.

> Active transport—the process of moving a substance against a concentration gradient—requires the expenditure of cellular energy supplied by ATP.

transport substances into or out of the cell. After the substance is escorted across the membrane, it separates from the carrier, which can then pick up and transport another large molecule. So far as is known, there is a specific carrier molecule for each type of substance transported across the membrane. Facilitated diffusion, like ordinary diffusion, involves moving from an area where the substance to be transported is in higher concentration to an area where it is in lower concentration. The essential difference, however, is that facilitated diffusion requires carrier molecules, whereas ordinary diffusion does not.

There are conditions in living organisms that require the movement of substances from a region where they are in low concentration to a region of higher concentration across a cell membrane. By comparison, this is movement in a direction opposite to that occurring in diffusion. In effect, the movement of a substance from lower to higher regions of concentration is analogous to water running uphill. How is this possible? Just as it would require an input of energy to get water to run uphill (as with a water pump), the movement of large cellular substances "uphill" also requires an input of energy. Such movement occurs through the process of *active transport* (Fig. 5.28). In a manner similar to that of facilitated diffusion,

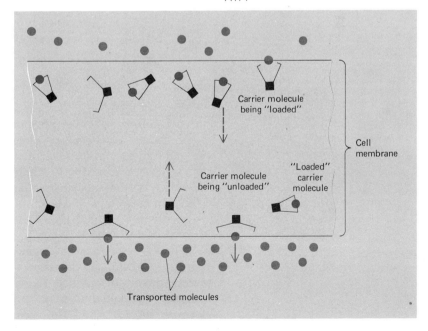

Figure 5.28 Active transport across a cell membrane. The molecules to be transported combine with carrier molecules at the surface of the cell membrane. The "loaded" carrier molecules then diffuse across the membrane from a region of lower concentration to a region of higher concentration, where they discharge their load into the cell. The carrier molecules then move back to the surface of the membrane. The entire process of active transport requires an input of energy derived from ATP.

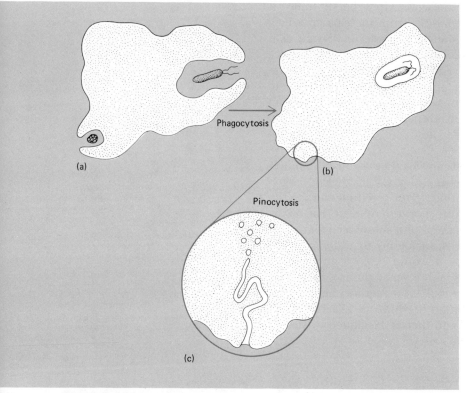

Figure 5.29 Endocytosis. (a) Phagocytosis involves the engulfment of a large molecule by cytoplasmic extensions of the cell, after which the molecule is contained within a cytoplasmic vacuole (b). (c) Pinocytosis involves the engulfment of small particles and liquids by inversion of the cell membrane.

the mechanism of active transport involves the use of carrier molecules. These molecules are located in the cell membranes and transport substances into and out of the cell. The energy for active transport is derived from the breakdown of ATP.

Some of the substances that are actively transported in living cells include some of the sugars, most of the amino acids, hydrogen ions, sodium ions, and chloride ions. Each of these apparently has its own specific carrier system in the cell membrane.

Finally, some large molecules do not enter a cell by any of the processes previously discussed. Instead, they may approach the plasma membrane, which then sends out cytoplasmic extensions, forming a pocket that engulfs the molecule. The membrane then closes to form a vesicle or vacuole with the molecule inside. The

> The exchange of materials across cell membranes illustrates one aspect of the maintenance of homeostasis at the cellular level.

vacuole finally pinches off and moves farther into the cytoplasm. The general term for this process is **endocytosis,** of which there are two types (Fig. 5.29). When large molecules or substances are engulfed in this manner, the process is referred to as **phagocytosis** ("cell eating"). The engulfment of bacteria and other foreign substances by your white blood cells is a common example of phagocytosis. Many protozoans, such as the amoebae, obtain their food by phagocytosis. In white blood cells, amoebae, and other cells, lysosomes often fuse with the vacuoles formed by phagocytosis, and the lysosome enzymes are released into the vacuole. The enzymes then destroy the foreign substance in the cell or act as digesting enzymes to break down the food particles.

The engulfment of very small particles or liquids by a cell is termed **pinocytosis** ("cell drinking"), a process essentially the same as phagocytosis, except that the membrane inverts and then closes around the engulfed molecule (Fig. 5.30). However, pinocytosis often involves the engulfment of particles by tiny vesicles ("little vacu-

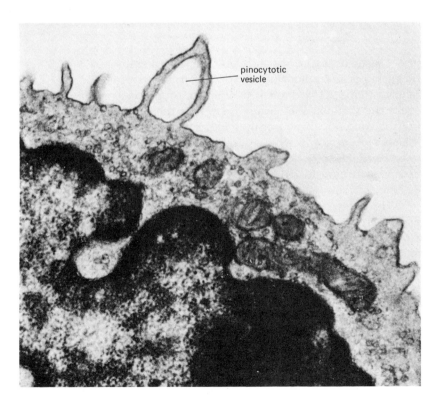

Figure 5.30 Electron micrograph of a pinocytic vesicle on the surface of an amphibian white blood cell. The vesicle contains fluid, which eventually will be transported to the interior of the cell. (From Ebert, J. D., Loewy, A. G., Miller, R. S., and Schneiderman, H. A.: *Biology.* New York: Holt, Rinehart and Winston, 1973.)

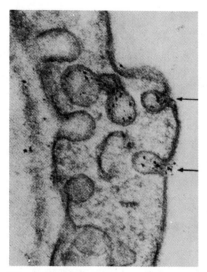

Figure 5.31 Two stages in the uptake of tiny particles by pinocytosis. The particles are visible at the surface of the cell membrane *(arrows)*. (From Bloom, W., and Fawcett, D. W.: *A Textbook of Histology*. 10th ed. Philadelphia: W. B. Saunders Co., 1975, p. 396.)

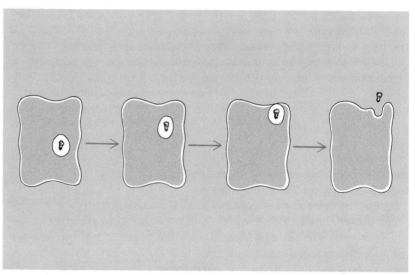

Figure 5.32 Exocytosis, the reverse of endocytosis. When the vacuole moves from the interior of the cell to the surface, the vacuolar membrane fuses with the cell membrane, and the contents of the vacuole are expelled to the outside.

oles") formed on the surface of the cell membrane (Fig. 5.31). Phagocytosis and pinocytosis may occur in reverse, i.e., vacuoles with their contents move from the interior of the cell to the surface, where the vacuolar membrane fuses with the cell membrane, expelling the contents to the outside. This process is referred to as ***exocytosis*** and is employed by many cells in releasing their secretions and waste products to the exterior (Fig. 5.32). Some of the vacuoles that transport substances from the interior of the cell to the cell surface are formed from membranes of the Golgi bodies.

With reference to the seven life principles, the exchange of materials across cell membranes illustrates one aspect of the maintenance of homeostasis at the cellular level. By controlling the movement of these materials into and out of itself, the cell maintains the internal environment of living things at a constant and stable level.

REVIEW OF ESSENTIAL CONCEPTS

1. The *cell theory* states that the cell is the structural and functional unit of life and that all cells arise from pre-existing cells.
2. Eukaryotic cells are found in all living organisms except bacteria and blue-green algae. Eukaryotic cells are bounded by a *plasma membrane*

and contain a variety of *organelles* dispersed in the cytoplasm. Plant cells have a *cell wall* surrounding the plasma membrane.

3. Protoplasm is the physical basis of living matter; within the plasma membrane, protoplasm consists of the *cytoplasm* and the *nucleus*.
4. The nucleus is bounded by a nuclear membrane; within the nucleus are the *nucleolus* and the *chromosomes*. Each species of organisms has a set number of chromosomes in its cells. Chromosomes consist primarily of DNA, the hereditary material.
5. The cytoplasm of the cell is separated into compartments by the *endoplasmic reticulum*. The sites of protein synthesis, the *ribosomes*, are found on some ER membranes and also are found free in the cytoplasm.
6. The *Golgi body* is an organelle involved in packaging and secreting cellular substances.
7. *Lysosomes* are membranous sacs containing digestive enzymes; lysosome enzymes digest food and foreign particles, break down cells no longer needed by the organism, and may be involved in the process of aging and in certain diseases.
8. *Vacuoles* are open spaces in the cytoplasm bounded by a cell membrane. Vacuoles are more developed in plant cells than in animal cells and contain food, water, ions, pigments, wastes, and so on.
9. *Mitochondria* are the major sites of energy (ATP) production in the cell. They contain the enzymes and electron carrier molecules involved in many of the reactions of cellular metabolism.
10. *Microfilaments* and *microtubules*, found in a variety of cells, function in cell support and movement. Microtubules are constituent organelles of the *spindle*, the structure involved in the movement of chromosomes during cell division. Microtubules also form the *centrioles*, paired bodies located just outside the nucleus of most animal cells.
11. *Cilia* and *flagella* are hairlike organelles involved in cell movement and filtering and are composed of microtubules.
12. *Chloroplasts* are plastids containing the pigment *chlorophyll* and are the sites of photosynthesis in eukaryotic cells. *Chromoplasts* store other plant pigments, and *leucoplasts* store proteins, lipids, and starch.
13. Cell membranes are composed of phospholipids and proteins; these membranes are selectively permeable and thereby regulate the internal environment of the cell.
14. The cells of all higher organisms are bathed in a watery *extracellular fluid*; the cytoplasm of cells contains an *intracellular fluid*, which is also predominantly water.

Cells and Cellular Activity

15. Small cells have more surface area in relation to their volume than do larger cells. Small cells, therefore, are more efficient in terms of the passage, distribution, and exchange of vital substances.
16. *Diffusion* is the spontaneous movement of a substance from an area where it is in higher concentration to an area where it is in lower concentration, until equilibrium is reached.
17. A *hypertonic solution* contains *more* solute than a solution of equal volume with which it is compared; a *hypotonic solution* contains *less* solute; and an *isotonic solution* contains the *same* amount of solute as another solution.
18. *Osmosis* is the movement of a solvent from an area of high solvent concentration to an area of lesser solvent concentration across a selectively permeable membrane, until equilibrium is reached.
19. Cells exposed to *hypertonic* solutions tend to lose water. In plant cells, the pulling away of the plasma membrane from the cell wall due to a loss of water is termed *plasmolysis*. Cells exposed to *hypotonic* solutions tend to gain water.
20. *Facilitated diffusion* is a process in which *carrier molecules* transport substances across cell membranes, from regions of high concentration to regions of lower concentration.
21. *Active transport* is an energy-requiring process in which carrier molecules transport substances across cell membranes *against* a concentration gradient, i.e., from an area of lesser concentration to an area of greater concentration.
22. Some large molecules enter cells through *phagocytosis*, a process involving engulfment of particles by the cell membrane. Small particles and liquids may enter the cell by *pinocytosis*.

APPLYING THE CONCEPTS

1. Why is the cell considered the fundamental unit of life?
2. What is the biological significance of the cell theory?
3. Name the major organelles of eukaryotic cells and give their general functions.
4. In your own words, explain why small cells are more efficient biological units than large cells.
5. Explain how the following processes affect the functioning of living cells: (1) diffusion, (2) osmosis, (3) facilitated diffusion, and (4) active transport.

Color Plate 1 Moss cells packed with chloroplasts. (Photograph by Bruce Russell, Biomedia Assoc.)

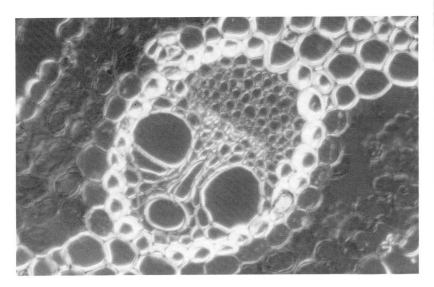

Color Plate 2 Vascular tissue cells from the stem of a monocot. (Photograph by Bruce Russell, Biomedia Assoc.)

Color Plate 3 The rectangular cells of the alga *Spirogyra*. (Photograph by Bruce Russell, Biomedia Assoc.)

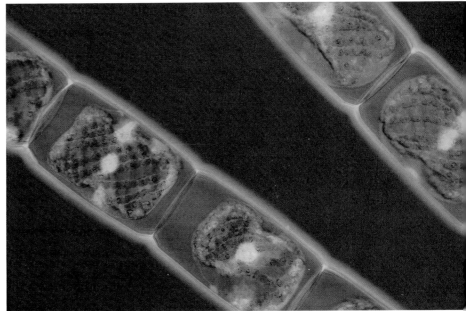

Color Plate 4 A cluster of single-celled paramecia. (Photograph by Bruce Russell, Biomedia Assoc.)

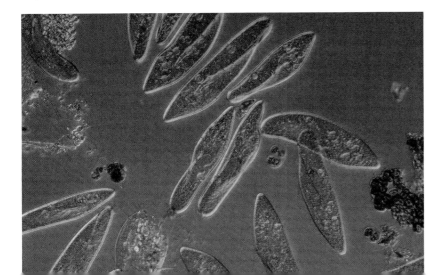

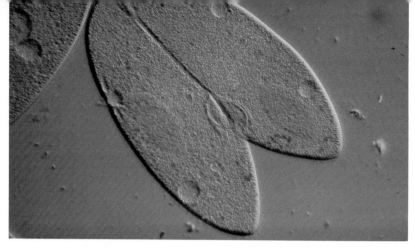

Color Plate 5 Conjugating paramecia. (Photograph by Bruce Russell, Biomedia Assoc.)

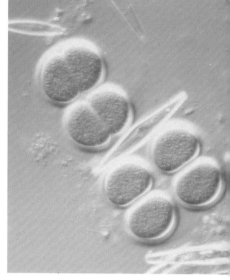

Color Plate 6 Blue-green algae dividing. (Photograph by Bruce Russell, Biomedia Assoc.)

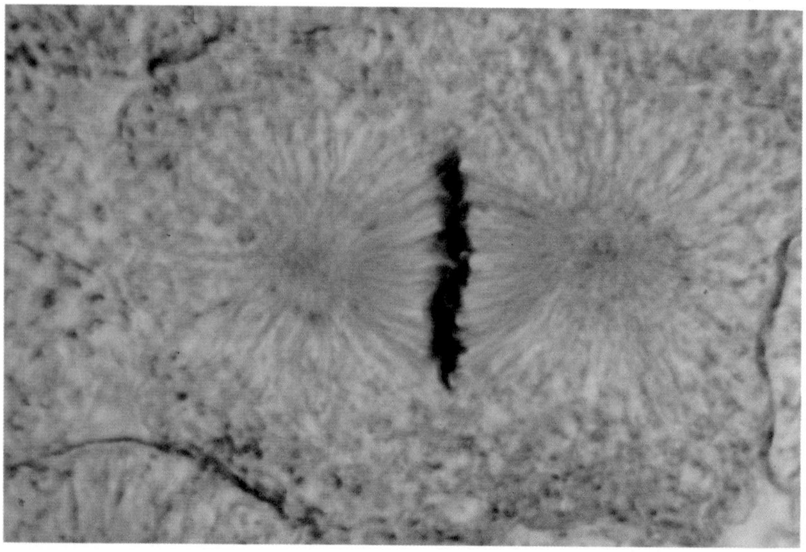

Color Plate 7 Cell division: Mitotic metaphase. (Photograph by Carolina Biological Supply Co.)

Color Plate 8 Cell division: Mitotic anaphase. (Photograph by Carolina Biological Supply Co.)

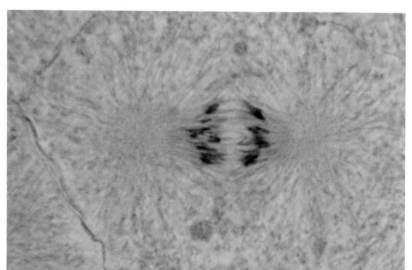

Color Plate 9 *Paramecium* stained to show food vacuoles. (Photograph by Bruce Russell, Biomedia Assoc.)

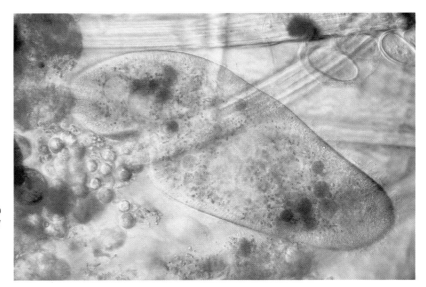

Color Plate 10 The contractile vacuole of *Paramecium*. (Photograph by Bruce Russell, Biomedia Assoc.)

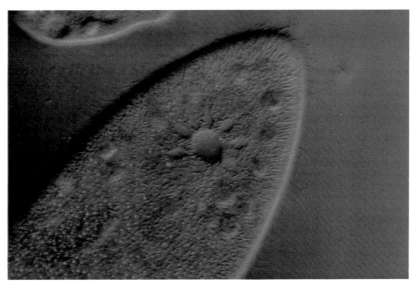

Color Plate 11 The ciliate *Euplotes* feeding on algal cells. (Photograph by Bruce Russell, Biomedia Assoc.)

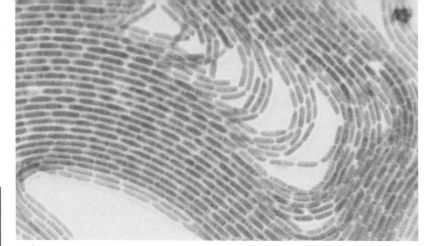

Color Plate 12 Chains of rod-shaped bacteria. (Photograph by Bruce Russell, Biomedia Assoc.)

Color Plate 13 Live spirilla bacteria (Photograph by Bruce Russell, BIOMEDIA Assoc.)

Color Plate 14 Filaments of blue-green algae. (Photograph by Bruce Russell, Biomedia Assoc.)

Color Plate 15 Filaments of blue-green algae. (Photograph by Bruce Russell, Biomedia Assoc.)

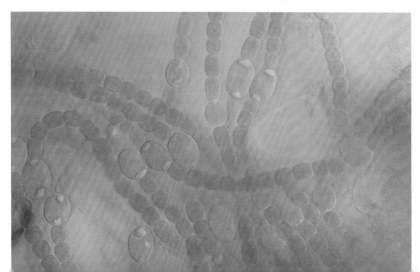

6. Using a specific example, illustrate the effect of exposing a living cell to a hypertonic solution.
7. What is *turgor pressure,* and what is its significance in plant cells?
8. Explain the general mechanisms of phagocytosis and pinocytosis.
9. Basically, how is the cell involved in maintaining homeostasis of an organism?

6 the essential processes: photosynthesis, glycolysis, and cellular respiration

THE ESSENTIAL OBJECTIVES

You have understood this chapter when you are able to:

1. Write and explain generally the summary reaction for photosynthesis.
2. Describe the basic structure of the green leaf.
3. Discuss the concept of light energy and its relationship to photosynthesis.
4. Discuss the basic steps in the light and dark reactions of photosynthesis.
5. Explain the significance of photosynthesis to the development of life on earth.
6. Write and explain generally the summary reaction for cellular respiration.
7. Outline the basic steps of glycolysis, the Krebs cycle, and the electron transport system.
8. Describe the process of fermentation.
9. Explain the cause of oxygen debt.
10. Compare aerobic and anaerobic metabolism in terms of energy efficiency.

By utilizing the radiant energy of the sun, photosynthesizing plants produce carbohydrate, the original source of nutrition for the living world. In all living cells, carbohydrates and other nutrients are broken down stepwise by a process called cellular respiration, releasing energy that drives the chemical reactions of life.

PREVIEW OF ESSENTIAL TERMS

photosynthesis The process by which chlorophyll-bearing cells, utilizing light energy, produce carbohydrates and oxygen from carbon dioxide and water.

chloroplast In green plant cells, the chlorophyll-containing organelle in which photosynthesis takes place.

light The movement of physical particles (photons) in streams of varying wavelengths.

photolysis The splitting of water molecules into hydrogen ions, oxygen atoms, and electrons by light energy.

nicotinamide adenine dinucleotide phosphate (NADP) A coenzyme that functions as an electron and hydrogen acceptor in the light reactions of photosynthesis.

cellular metabolism All the chemical reactions within a cell that synthesize or break down products; includes glycolysis, Krebs cycle, and electron transport system.

fermentation The anaerobic process whereby glucose is broken down into alcohol or lactic acid.

oxygen debt A deficiency of oxygen in the cells resulting from the lack of oxygen delivery to the body during exercise.

nicotinamide adenine dinucleotide (NAD) A coenzyme that functions as an electron and hydrogen acceptor in cellular metabolism.

flavin adenine dinucleotide (FAD) Another coenzyme with essentially the same function as NAD.

The Essential Processes: Photosynthesis, Glycolysis, and Cellular Respiration

PHOTOSYNTHESIS

The renowned biochemist and Nobel Prize winner Albert Szent-Györgyi once stated: "What drives life is . . . a little electric current, kept up by the sunshine." That "little electric current" is part of a never-ending drama between the energy of the sun and every green plant on earth. Through their capacity to trap and utilize the light energy of the sun, the green plants of the earth annually produce billions of tons of sugar—the original source of food for the living world.

The process by which green plant cells manufacture sugar (glucose) is called *photosynthesis,* literally meaning "putting together with light." The process may be summarized by the following reaction:

$$6CO_2 + 12H_2O + \text{light energy} \xrightarrow{\text{chlorophyll}} C_6H_{12}O_6 + 6H_2O + 6O_2$$

Actually, the reaction as written is a little misleading. It appears that carbon dioxide and water, in the presence of light, interact or combine *with each other* to produce the sugar and other products. As we take a look at the photosynthetic process, you will find that this is *not* the case. Basically, the reaction indicates that 6 molecules of carbon dioxide and 12 molecules of water are in some fashion involved in the formation of glucose ($C_6H_{12}O_6$) molecules and that 6 molecules of water and oxygen gas are also produced. Furthermore, light energy is necessary to start the reaction. Finally, there is the involvement of *chlorophyll,* the pigment that gives green plants their color. Chlorophyll is found in small cell organelles called *chloroplasts* (Fig. 6.1), which are the structures in the cell where the reactions of photosynthesis actually occur.

The principal organ of photosynthesis in green plants is the leaf (Fig. 6.2). The size and structure of leaves vary considerably according to the species; there is, however, a basic arrangement of

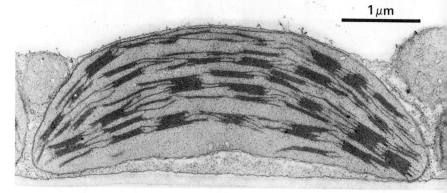

Figure 6.1 Electron micrograph of a tobacco chloroplast. The reactions of photosynthesis take place within the chloroplasts of eukaryotic cells. (From Arms, K., and Camp. P. S.: *Biology.* 2nd edition. Philadelphia: Saunders College Publishing, 1982.)

> Sunlight, the original source of energy on earth, is made available to living cells through the reactions of photosynthesis.

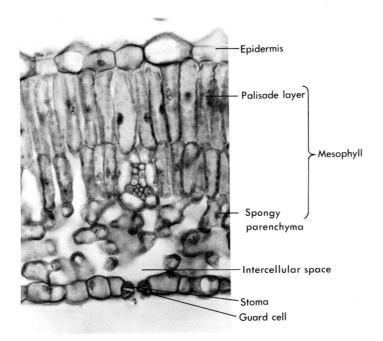

Figure 6.2 Cross section of a leaf. The upper surface of the leaf is at the top of the photo. (From Gerking, S. D.: *Biological Systems.* 2nd edition. Philadelphia: Saunders College Publishing, 1974, p. 243.)

leaf structures common to a great number of different plants. The upper and lower surfaces of the leaf consist of a layer of *epidermal cells,* covered by a waxy, waterproof *cuticle.* Between the epidermal layers is the *mesophyll,* composed of various chloroplast-bearing cells. Extending through the mesophyll are *veins,* bundles of vascular (vessel) tissue composed of *xylem* and *phloem.* Xylem functions in conducting water and minerals, and phloem mainly conducts dissolved food. The epidermal layers are perforated with tiny openings called *stomates* (Gk. *stoma,* mouth), which are bordered by a pair of *guard cells.* The guard cells, unlike the surrounding epidermal cells, are photosynthetic and regulate the size of the stomatal opening. When water enters the guard cells by osmosis, they become turgid and spread apart, thereby increasing the size of the stomate. As water is lost, the guard cells become more flaccid and compress in on each other, reducing the size of the stomate. Most of the photosynthesis occurs in the chloroplasts of the mesophyll cells (Fig. 6.3).

Photosynthesis is really two sets of reactions. The first set is referred to as the **light reactions,** so called because light must be available as an energy source. The second set is a series of **dark reactions.** Light is not directly involved in these reactions, although they may occur in either light or darkness, and energy is supplied by products formed during the light reactions.

Figure 6.3 How stomates work. (a) Epidermal cells of a leaf, showing the position of stomates among the epidermal cells. (b) Cross section through a stomate. Changes in turgor pressure alter the shape of the guard cells, opening and closing the stomate. (c) Scanning electron micrograph of an open stomate on the lower surface of a pear leaf (Magnification X 3700). (From Ebert, J. D., Loewy, A. G., Miller, R. S., and Schneiderman, H. A.: *Biology*. New York: Holt, Rinehart and Winston, 1973, p. 300.)

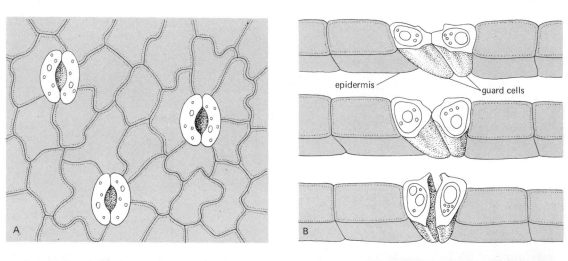

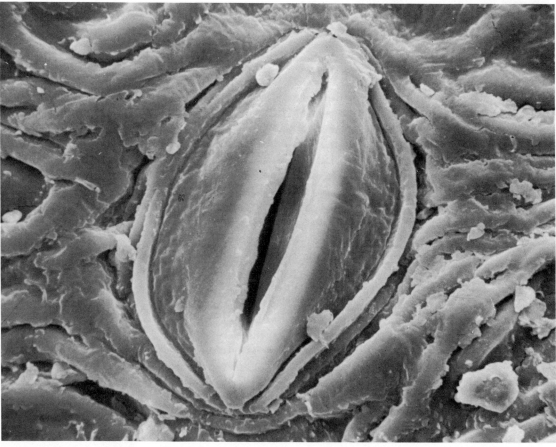

Life on earth is sustained by the energy in the wavelengths of visible light.

How then does light, a gas, some water, and a green pigment go about producing a sugar? Let's put it all together and see.

Light and Life

The existence and continuance of life on earth is possible because of the energy of light. The link between light and life is the green plant, which converts light energy into chemical energy through the process of photosynthesis.

Light is the movement of a wave or stream of physical particles called **photons**. There are different kinds of these particles, each having different amounts of energy and moving with a different frequency to produce wavelengths of light. Going from shorter to longer wavelengths, the energy content of photons decreases (Fig. 6.4). Thus, the photons in shorter wavelengths of light have more energy than those in longer wavelengths. Light, then, is more accurately defined as the movement of physical particles, or little packets of energy called photons, in streams of varying wavelengths.

Visible light is light that we can see. White light, as from daylight or a light bulb, is a mixture of all visible wavelengths, or colors. When the wavelengths of white light are separated out by passing

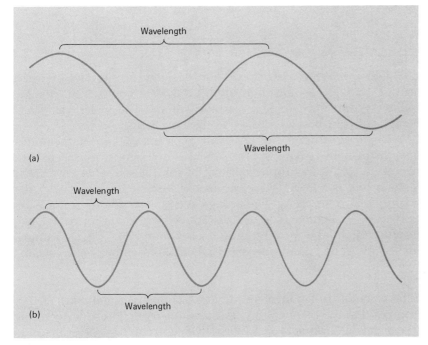

Figure 6.4 Light is the movement of photons in streams of varying wavelengths. Longer wavelengths of light (a) have less energy than do shorter wavelengths (b).

The Essential Processes: Photosynthesis, Glycolysis, and Cellular Respiration

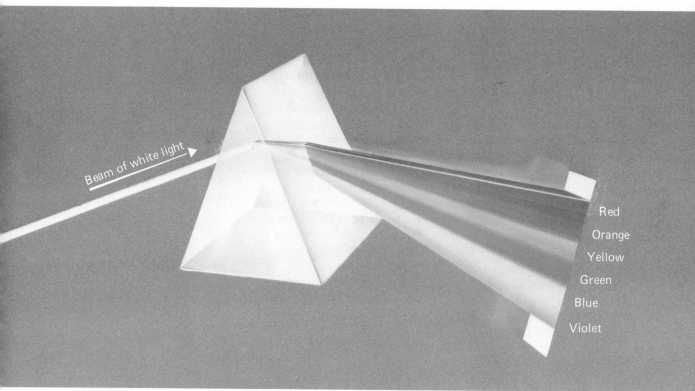

Figure 6.5 White light is a mixture of all visible colors or wavelengths. When passed through a glass prism, white light is separated into the visible spectrum. Having the longest wavelength, red light has the least energy in the spectrum; violet light has the shortest wavelength and therefore the most energy. (Abell, G. O.: *Drama of the Universe.* New York: Holt, Rinehart and Winston, 1978, p. 97.)

the light through a glass prism, our eyes interpret the wavelengths as one or some combination of the following colors: red, orange, yellow, green, blue, or violet. Going from red to violet, the wavelengths of light become progressively shorter (Fig. 6.5).

Depending upon the object or substance, some of the wavelengths of light striking it are absorbed and some are reflected. In the green plant, chlorophyll primarily absorbs red, blue, and violet light, while the green wavelengths are reflected. This is why the plant looks green to us. The orange and yellow wavelengths are absorbed only slightly. It is the absorption of the appropriate wavelength of light that introduces an energy source into the chlorophyll molecule. At this point, the reactions of photosynthesis have begun.

The Light Reactions of Photosynthesis

The essential function of the light reactions is to generate energy sources that the plant can use to make sugar. Light energy entering the plant is "trapped" by chlorophyll molecules and converted to

> In the light reactions of photosynthesis, ATP and NADPH$_2$ are generated by the kinetic energy of electrons.

chemical energy in the form of high-energy organic compounds. The conversion of energy is dependent upon chlorophyll because it is this pigment that "reacts" when stimulated by the appropriate wavelengths of light. The essence of the reaction is that electrons in the chlorophyll molecules are boosted to high energy levels and release their energy as they return to lower levels. The energy released goes into the formation of adenosine triphosphate (ATP) and a reduced hydrogen and electron acceptor called reduced **nicotinamide adenine dinucleotide phosphate.** In the spirit of charity it has been abbreviated *NADPH$_2$*.

You will recall that when an atom or a compound is reduced, it gains hydrogen or electrons or both. The reduction of NADP to form NADPH$_2$ occurs in the following manner:

$$\underset{\text{(oxidized form)}}{\text{NADP}} + \underset{\text{(two hydrogen ions, or protons)}}{2\text{H}^+} + \text{electrons} \rightarrow \underset{\text{(reduced form)}}{\text{NADPH} + \text{H}^+ \text{ or NADPH}_2}$$

As you will see, the H$^+$ ions are split from water, and the electrons are contributed by a chlorophyll molecule. Therefore, it is ATP and NADPH$_2$ that are the aforementioned sources of chemical energy generated in the light reactions. Basically, the light reactions of photosynthesis proceed according to the following sequence.

Light energy striking the green plant is absorbed by chlorophyll molecules, located within the grana of the chloroplasts. Actually, most plants contain two types of chlorophyll, or pigment systems, that can be designated as **chlorophyll I** and **chlorophyll II.** The essential difference between the two is that they absorb different wavelengths of light.

In some manner, light energy causes the electrons in the chlorophyll II molecule to become "excited" and to gain sufficient energy that they are boosted to an energy level high enough to make them actually leave the chlorophyll molecule. The released electrons are then "caught" by what is termed a **primary electron acceptor,** which simply "holds" the electrons at a high energy level. At the same time that the chlorophyll II molecule is being activated, light energy is performing its other photosynthetic function. One of the reactants—water—is split by light energy into H$^+$ ions, oxygen atoms, and electrons. This process is known as **photolysis.** With electrons missing, being held by the primary electron acceptor, "holes" are left in the chlorophyll II molecule. The electrons released from the splitting of water rush in to the molecule to fill these "holes." The oxygen atoms from water form oxygen gas, O$_2$, which diffuses

The Essential Processes: Photosynthesis, Glycolysis, and Cellular Respiration

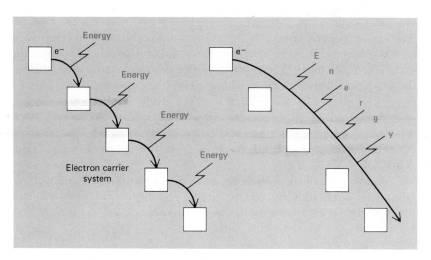

Figure 6.6 By descending stepwise down an electron carrier system, an electron (e⁻) releases its potential energy gradually. This permits the cell to utilize the energy efficiently in the performance of useful work. An electron descending abruptly down the carrier system releases its stored energy all at once. This energy cannot be efficiently controlled and produces excessively high temperatures.

through the stomates and out into the atmosphere. This is the oxygen that you and other animals breathe every day.

Meanwhile, the electrons held by the primary electron acceptor are released and passed "downhill" along a system of ***electron carriers.*** You might envision this as a bucket brigade system, with the person at the top of the hill passing a bucket down to the next person, who passes it down to the next person, and so on to the bottom of the hill. The people of the brigade represent the electron carriers, and the buckets represent the electrons. As the electrons are passed along the carrier system to lower energy levels, they release their stored energy. Some of this energy is used to bond free phosphate (PO_4) to ADP, thus generating the high-energy compound ATP.

You will notice in this scheme that the electrons are passed "downhill" in a step-by-step fashion (Fig. 6.6). Thus, the potential energy of the electrons is not released all at once, as would be the case if they were "dropped" suddenly from the top of the electron carrier system to the bottom. It is true that the total amount of energy released from the electrons is the same whether they descend from one energy level to another stepwise or descend abruptly. However, chemical reactions within the cell can be controlled and regulated if energy is released gradually. This permits the cell to perform useful work. Energy released in one sudden burst cannot be effectively controlled and produces excessively high temperatures that could damage the cell. You would experience little comfort if the burning log in your fireplace released its stored energy in one instantaneous flash and then died out. A more efficient use of the

> Utilizing the high energy compounds ATP and NADPH$_2$, the dark reactions of photosynthesis produce glucose, the original source of nutrition for the living world.

energy in the wood is to allow the log to burn slowly so that some of its energy (heat) can do useful work, such as keeping you warm. The gradual, step-by-step release of energy is the general rule not only in photosynthesis but also in almost all the major chemical reactions in living cells.

While chlorophyll II was being activated, light energy also was exciting electrons in the chlorophyll I molecules. These electrons, too, are boosted out of the molecules to a higher energy level, where they are accepted and held by another primary electron acceptor. This loss of electrons leaves "holes" in the chlorophyll I molecule. Up to now, these events essentially parallel those in the chlorophyll II molecule. However, the "holes" in the chlorophyll II molecule were filled by the electrons split from water; the "holes" in the chlorophyll I molecule are filled by the electrons that were passed down the bucket brigade of electron carriers.

From the primary electron acceptor, the electrons from chlorophyll I also are passed "downhill" along another system of electron carriers. At the "bottom" of the chain, the electrons are transferred to NADP; in addition, some of the hydrogen ions (H$^+$) split from water also are accepted by NADP. By accepting the electrons and hydrogen ions, NADP is reduced to its high-energy form, NADPH$_2$. With the formation of NADPH$_2$ and ATP, the light reactions are completed. The essential function of the light reactions is to produce these two high-energy compounds. From this point on, the remaining reactions of photosynthesis depend upon the chemical energy stored within the bonds of ATP and NADPH$_2$ (Fig. 6.7, and Table 6.1).

The Dark Reactions of Photosynthesis

The second set of photosynthetic reactions do not require light, although they can proceed equally well in light or darkness. The essence of the dark reactions involves the use of carbon dioxide, ATP, and the hydrogen atoms from NADPH$_2$ to form sugar (Fig. 6.8). The production of this one carbohydrate signals the introduction of nutrition into the living world. Directly or indirectly, life ultimately depends upon this carbohydrate for every morsel of food consumed.

Carbon dioxide from the atmosphere diffuses into the leaf through the stomates and enters the stroma of the chloroplasts. The carbon dioxide then bonds with a 5-carbon sugar called **ribulose diphosphate,** or RuDP, present in the leaf from previous photosyn-

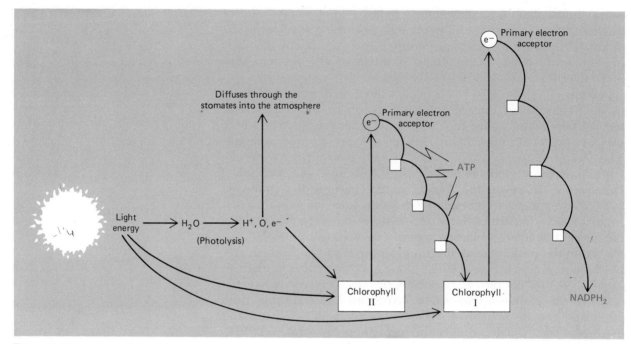

Figure 6.7 Summary of the light reactions of photosynthesis (see text).

thetic reactions. The bonding of CO_2 to RuDP forms an unstable 6-carbon compound, which almost immediately splits into two molecules of a 3-carbon compound called **phosphoglyceric acid,** or **PGA.** Using the hydrogen atoms and electrons from $NADPH_2$, and the energy supplied by ATP, each PGA molecule is reduced to **phosphoglyceraldehyde,** or **PGAL.**

This compound also has three carbon atoms and, except for the addition of hydrogen, is structurally the same as PGA. Three-carbon compounds such as PGAL are the simplest carbohydrates and con-

Figure 6.8 Generalized summary of the dark reactions of photosynthesis (see text).

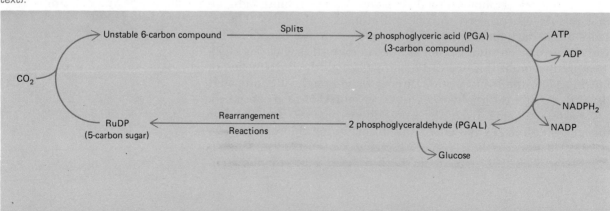

> With the rise of photosynthetic organisms some 3 billion years ago, the face and history of the earth were changed forever.

Table 6.1 THE BASIC CHEMICAL REACTIONS OF PHOTOSYNTHESIS, SHOWING ONLY RAW MATERIALS AND END PRODUCTS (The starting materials for one are often the end products for the other.)

	RAW MATERIALS	END PRODUCTS
Light reaction	Light energy Water ADP + P_i (from dark reaction) $NADP^+$ (from dark reaction)	Oxygen ATP NADPH + H^+
Dark reaction	Ribulose diphosphate Carbon dioxide ATP (from light reaction) NADPH + H^+ (from light reaction)	Sugars ADP + P_i NADP

sequently can be bonded together to form larger, more complex monosaccharide carbohydrates. This is basically what happens in the dark reactions of photosynthesis. Some of the 3-carbon PGAL molecules bond together to form 6-carbon sugars, such as glucose. Glucose molecules in turn join together to form the storage product starch. However, some of the PGAL, through a series of complex rearrangement reactions, is used to form new molecules of RuDP. This is necessary, of course, in that new carbon dioxide entering the plant must continually be taken up by new RuDP.

In addition to carbohydrates, the other product of the dark reactions is water. This "new" water is formed from some of the hydrogen released by the photolysis of water and from the oxygen removed from carbon dioxide when it bonds with RuDP.

Finally, after being used in the dark reactions, ATP has released its bond energy and becomes adenosine diphosphate (ADP); similarly, $NADPH_2$ has released its hydrogen and electrons to become NADP. Both ADP and NADP can then be recharged to high-energy forms by a new sequence of light reactions. Figure 6.9 shows schematically both the light and dark reactions of photosynthesis and the connection between them.

With the rise of photosynthetic organisms some 3 billion years ago, the face and history of the earth were changed forever. In the millions of years before their appearance, there was no oxygen-rich atmosphere, no oxygen-breathing organisms, no plant or animal capable of living on the land. But with the rise of **photosynthetic oxygen-producers**, probably the first of which were the prokaryotic blue-green algae, the atmosphere and the land were eventually al-

The Essential Processes: Photosynthesis, Glycolysis, and Cellular Respiration

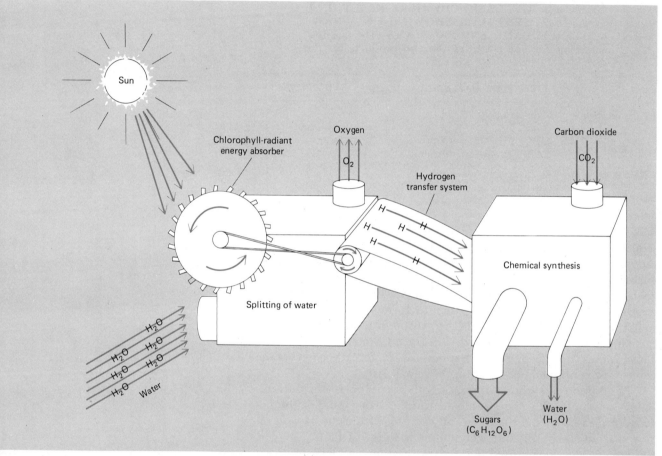

Figure 6.9 Photosynthesis: energy source, raw materials, and products. On the left side is the photo part of photosynthesis; on the right is the synthetic part.

tered dramatically. With the build-up of atmospheric oxygen came the formation of the protective ozone layer. No longer bombarded by the intense ultraviolet radiation of the sun, the land became habitable, providing the opportunity for terrestrial evolution. With free oxygen available, many organisms left the water to evolve into more active, more complex forms. And to this day, almost every form of life depends upon oxygen—and the production of oxygen is a consequence of photosynthesis.

GLYCOLYSIS AND CELLULAR RESPIRATION

In the final analysis, you or any other organism exists because you and they are capable of utilizing energy. More exactly, the cells of every living thing have the capacity to convert energy from one form to another. Your body requires a reasonably steady supply of nu-

> Metabolism is separated into two major sets of reactions: (1) *anabolism* and (2) *catabolism*. Anabolism refers to reactions that synthesize cellular materials, and catabolism includes reactions that break down cellular materials.

trients, primarily in the form of carbohydrates, lipids, and proteins. When taken into the body, these compounds are broken down into smaller molecules by the digestive process, but digestion alone does not provide the body with usable energy. These molecules must leave the digestive tract, enter the circulating blood, and then infiltrate every living cell. The real action, then, takes place within the cells, where these various small foodstuff molecules release the energy that drives the chemical reactions of glycolysis and cellular respiration, collectively known as cellular **metabolism.**

Almost every living organism utilizes glucose as its primary energy source. Our central concern, therefore, is to observe what happens to this simple sugar as it is metabolized by a living cell. The metabolic process involves a series of steps (reactions) in which glucose, in the presence of oxygen, is ultimately broken down, or oxidized, to produce carbon dioxide, water, and energy. However, as we shall discuss, there are a few organisms in which this complete scheme is not followed.

For most living things, though, the complete breakdown of glucose can be summarized by the following reaction:

$$C_6H_{12}O_6 + 6O_2 \longrightarrow 6CO_2 + 6H_2O + \text{energy}$$
(glucose)   (oxygen)   (carbon dioxide)   (water)

This overall reaction actually requires three distinct but continuous sets of reactions. The first set is called **glycolysis;** the second set includes the reactions of the **Krebs cycle;** and the third set constitutes the **electron transport system,** or ETS. The Krebs cycle and ETS together are termed cellular respiration. Thus, the complete breakdown, or metabolism, of glucose involves glycolysis and cellular respiration.

Glycolysis

The first series of metabolic reactions, called *glycolysis,* occurs in the cytoplasm of the cell (Fig. 6.10). Glycolysis literally means "loosening of glucose," which refers to the breaking of its chemical bonds as it is metabolized. The basic scheme of glycolysis involves the splitting of one molecule of glucose (a 6-carbon compound) into two molecules of a 3-carbon compound called **pyruvic acid.** Some of the chemical bond energy released as glucose is broken down goes into the formation of ATP. In the course of the breaking down process, some of the hydrogen atoms and electrons that make up part of the structure of glucose are removed. The hydrogen atoms and electrons are then taken up by the coenzyme NAD, nicotinamide adenine dinucleotide, which is thereby reduced to $NADH_2$. NAD and

The Essential Processes: Photosynthesis, Glycolysis, and Cellular Respiration

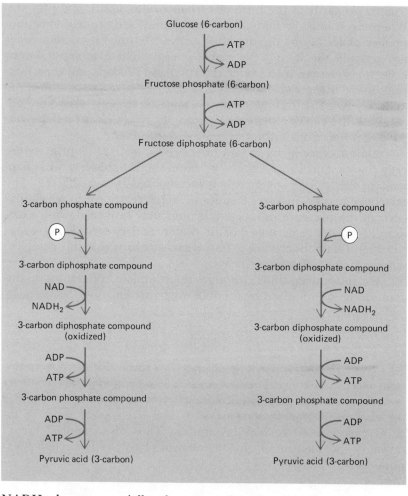

Figure 6.10

$NADH_2$ have essentially the same function as the NADP and $NADPH_2$ molecules we encountered in photosynthesis. The only structural difference is the phosphate group.

Glycolysis does not require oxygen and therefore can be an *anaerobic* (*an*, without; *aero*, air) process. The Krebs cycle and ETS, however, take place only in the presence of oxygen and thus are *aerobic* processes.

Glycolysis is actually a sequence of nine chemical reactions, resulting in the splitting of glucose into two molecules of **pyruvic acid** ($C_3H_4O_3$). The reactions are initiated by "energizing" the glucose molecule through the addition of a phosphate group from ATP. Af-

> In the step-by-step reactions of glycolysis, the energy stored in a molecule of glucose is released and utilized to form ATP and $NADH_2$.

ter the phosphate group is attached, the glucose phosphate molecule is rearranged by an enzyme and converted to the 6-carbon sugar *fructose phosphate*. Another molecule of ATP adds a phosphate group so that the fructose now has two phosphate groups bonded to it, forming fructose *diphosphate*.

Fructose diphosphate is then split into two 3-carbon compounds, each with a phosphate group attached. Keep in mind that two ATPs have been used thus far.

Next, the two 3-carbon compounds each receive another phosphate group, although these are inorganic phosphate ions present in the cell and are not donated by ATP. In addition, both compounds are oxidized, losing hydrogen ions (H^+) and electrons, which are accepted by NAD to form two molecules of $NADH_2$. At this point, then, glucose has been split into two 3-carbon compounds, each having two phosphate groups attached. In addition, both compounds have lost hydrogen ions and electrons that have gone into the formation of two molecules of $NADH_2$.

In the next reaction, a phosphate group is removed from *each* 3-carbon compound and transferred to each of the two ADP molecules to form two molecules of ATP. There have now been two ATPs used during two steps at the beginning of glycolysis and two ATPs now produced in this one step, so the *net* gain of ATP is zero at this point.

The remaining reactions involve a slight alteration in the structures of the two 3-carbon compounds plus the removal of the other phosphate groups. Each phosphate group is removed from its 3-carbon compound and transferred to ADP to form two additional molecules of ATP, to give a net gain of two ATPs. Each 3-carbon compound is now in the form of pyruvic acid, or *pyruvate*. At this point the reactions of glycolysis are completed.

In summary, one molecule of glucose is split into two molecules of pyruvate. In the process, there is a net energy gain of two molecules of ATP and two of $NADH_2$.

Many carbohydrates, after being converted into glucose, may enter the glycolysis pathway. Sugars such as the disaccharides sucrose and lactose, for instance, are converted into glucose and then broken down into pyruvate. Polysaccharides, too, enter the pathway after being broken down into simple sugars.

If oxygen is in sufficient supply, the pyruvate formed in glycolysis enters the next series of reactions, called the Krebs cycle. In those cells that normally metabolize in the absence of oxygen, pyruvate follows a different route. Yeast cells, for example, do not re-

The Essential Processes: Photosynthesis, Glycolysis, and Cellular Respiration

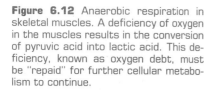

Figure 6.11 Anaerobic respiration (fermentation). In the absence of oxygen, yeast cells convert the pyruvic acid formed in glycolysis into ethyl alcohol and carbon dioxide.

quire oxygen; thus, instead of continuing into the Krebs cycle, pyruvate is converted into *ethyl alcohol* and *carbon dioxide*. This anaerobic process by which alcohol is formed from the breakdown of sugars is called **fermentation**—a process of great importance to the wine and liquor industry. In addition, in baker's yeast used in breadmaking, it is the carbon dioxide produced during fermentation that causes the dough to rise (Fig. 6.11).

In higher animals, anaerobic metabolism is common in skeletal muscles (Fig. 6.12). After strenuous and prolonged exercise, the

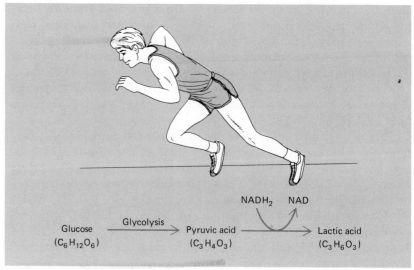

Figure 6.12 Anaerobic respiration in skeletal muscles. A deficiency of oxygen in the muscles results in the conversion of pyruvic acid into lactic acid. This deficiency, known as oxygen debt, must be "repaid" for further cellular metabolism to continue.

> Pyruvic acid can be converted to acetyl-CoA and enter the Krebs cycle only if sufficient oxygen is available.

muscles are often sore and fatigued. If the supply of oxygen to the muscle cells is inadequate, pyruvate cannot enter the Krebs cycle but is converted into **lactic acid** (human beings and other animals do not produce alcohol from pyruvate). Lactic acid is considered to be a contributing factor to muscle soreness and fatigue. During heavy exercise, lactic acid accumulates and the muscle cells build up what is known as an **oxygen debt.** This is simply an oxygen deficiency in the cells. After you cease exercising, you continue to breathe deeply and rapidly in an effort to "pay back" the oxygen debt. This, of course, results in an increased supply of oxygen to the cells and eventually a return to the normal rate of breathing. Physiologically, sufficient oxygen in the cells allows lactic acid to be converted into pyruvate, which can then enter the Krebs cycle to be completely metabolized.

After glycolysis, the next major sequence of reactions involved in the aerobic metabolism of foodstuffs is the *Krebs cycle* (Fig. 6.13). This pathway was named after the English biochemist and Nobel laureate

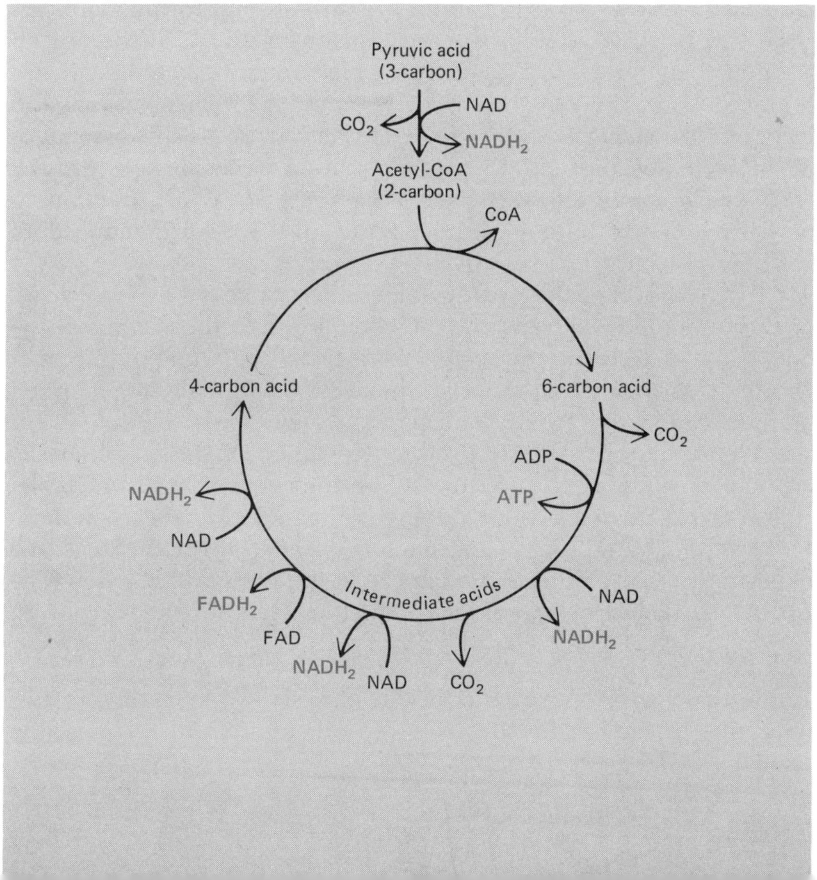

Cellular Respiration: The Krebs Cycle

Figure 6.13 Generalized summary of the transistion reaction and Krebs cycle. The energy yield for one molecule of pyruvic acid is shown in color.

> The transition reaction converts a broken down glucose molecule (in the form of two pyruvate molecules) into two acetyl groups, each of which bonds with coenzyme A to form two molecules of acetyl-coenzyme A.

Hans Krebs, who first described most of its reactions. Whereas glycolysis occurs in the cytoplasm of the cell, the Krebs cycle reactions in eukaryotes take place within the cellular organelles called *mitochondria*. As pointed out previously, mitochondria are the major energy-producing "departments" of living cells.

At the end of glycolysis the partially broken down molecule of glucose is in the form of two molecules of pyruvate. Before entering the Krebs cycle, each pyruvate molecule is converted into a 2-carbon compound called **acetylcoenzyme A,** or simply **acetyl-CoA.** This conversion is sometimes referred to as the **transition reaction.** In the mitochondrion, pyruvate loses carbon dioxide, thereby becoming a 2-carbon compound (acetyl). It then bonds with coenzyme A, which is present within the cell, to form acetyl-CoA. As a part of the conversion of pyruvate into acetyl-CoA, hydrogen atoms are removed from both pyruvate and CoA and taken up by NAD to form $NADH_2$. The carbon dioxide removed is exhaled from the lungs. It is acetyl-CoA that now enters the Krebs cycle.

At the beginning of the Krebs cycle reactions, coenzyme A releases the acetyl group and returns to pick up another acetyl group. The acetyl group is transferred to a 4-carbon acid, forming a 6-carbon acid. In the course of the cycle, involving the formation of an additional eight different acids, two molecules of carbon dioxide are removed from the 6-carbon acid. In addition, hydrogen ions and electrons are removed from the various acids at four different steps in the cycle and passed to NAD and another electron acceptor molecule called **flavin adenine dinucleotide,** or FAD. With the gaining of hydrogen ions and electrons, NAD and FAD are reduced to $NADH_2$ and $FADH_2$, respectively.

The removal of the hydrogen ions, electrons, and two molecules of carbon dioxide converts the 6-carbon acid into the original 4-carbon acid that accepted the acetyl group at the beginning of the cycle. In other words, this initial 4-carbon acid is continually regenerated or reformed so that the cycle may begin anew.

We can summarize the product formation of the transition reaction and Krebs cycle as follows: With the conversion of one molecule of pyruvate to acetyl-CoA, there is one molecule of carbon dioxide given off and one molecule of $NADH_2$ formed. Since *two* molecules of acetyl-CoA are produced in the transition reaction (one for each molecule of pyruvic acid), the total yield for the reaction is:

$$2\text{pyruvic acid} + 2\text{CoA} + 2\text{NAD} \rightarrow 2\text{acetyl-CoA} + 2CO_2 + 2NADH_2$$

In the Krebs cycle, for each molecule of acetyl-CoA that enters, two molecules of carbon dioxide are given off, whereas three molecules

> The most significant accomplishment of the Krebs cycle is the production of the high energy compounds $NADH_2$ and $FADH_2$.

of $NADH_2$ and one molecule of $FADH_2$ are formed. Finally, some of the energy released during the Krebs cycle is used to generate two molecules of ATP, one for each acetyl-CoA that enters the cycle. Thus the Krebs cycle reactions can be summarized as follows:

$$2\text{acetyl-CoA} + 3NAD + FAD + 2ADP \rightarrow 2CoA + 4CO_2 + 6NADH_2 + 2FADH_2 + 2ATP$$

Only some of the energy released in glycolysis and the Krebs cycle is packaged directly into ATP. Most of the energy resides in the coenzymes $NADH_2$ and $FADH_2$. The formation of these two compounds is the most significant accomplishment of the Krebs cycle. The carbon dioxide given off in the Krebs cycle, as well as that released in the transition reaction, is a waste product that we exhale from the lungs.

At this point, then, the original glucose molecule has been completely broken down, with most of its released energy stored in $NADH_2$ and $FADH_2$. We shall now see what happens to this stored energy in the final stage of cellular respiration.

Cellular Respiration: The Electron Transport System (ETS)

The electron transport system, like the Krebs cycle, requires oxygen and is found within the mitochondria. The theme of the ETS is a familiar one, involving the passage of electrons from higher to lower energy levels (Fig. 6.14). The electrons referred to are those held by $NADH_2$ and $FADH_2$. These two compounds pass their electrons to an electron carrier molecule that passes them to another carrier molecule and so on "downhill" along a system of these carriers. As the electrons are transported along the system, i.e., from high energy levels to lower energy levels, they release their stored energy, which is used to generate large amounts of ATP from ADP.

At the "bottom" of the transport system, the electrons are finally taken up by oxygen atoms. $NADH_2$ and $FADH_2$ give up their hydrogen ions, which combine with the oxygen and electrons to form cellular water. With this final step of the ETS, the available energy in the original glucose molecule has been completely utilized. NAD and FAD move out to pick up yet another group of hydrogen ions and electrons as a new molecule of glucose is being broken down.

When passing down the electron carrier chain, one pair of electrons from each $NADH_2$ molecule produced in the transition reaction and Krebs cycle yields three molecules of ATP. Since the transition reaction and Krebs cycle together form eight $NADH_2$ molecules, a total of 24 new molecules of ATP are produced. However, a pair of electrons donated by each molecule of $NADH_2$ from gly-

136

The Essential Processes: Photosynthesis, Glycolysis, and Cellular Respiration

In the electron transport system, large amounts of ATP are generated from the energy released as electrons pass from higher to lower energy levels.

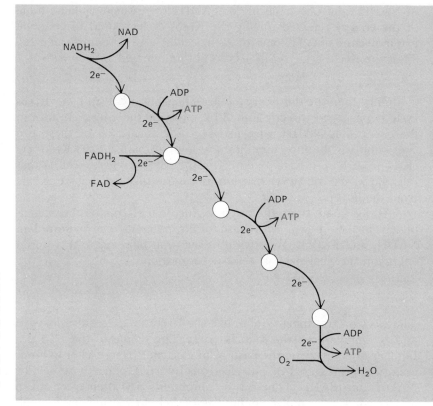

Figure 6.14 Summary of the electron transport system (ETS). $NADH_2$ and $FADH_2$ donate their electrons, which are passed "downhill" along a system of electron carrier molecules. In moving from higher to lower energy levels, each pair of electrons releases potential energy, which is used to produce ATP. At the "bottom" of the system, hydrogen ions from $NADH_2$ and $FADH_2$ combine with oxygen to form water (H_2O).

colysis yields only two molecules of ATP; the same is true for a pair of electrons donated by each molecule of $FADH_2$ produced in the Krebs cycle. Thus there is a total of four ATP molecules produced from the two $NADH_2$ molecules from glycolysis and four ATP molecules from the two $FADH_2$ molecules formed in the Krebs cycle.

Overall, then, the complete breakdown of one molecule of glucose produces 12 reduced coenzyme molecules (10 $NADH_2$ and two $FADH_2$), each of which donates a pair of electrons to the ETS. In passing down the system of electron carriers, these 12 pairs of electrons release energy that generates a total of 32 molecules of ATP. Therefore, with the two ATP molecules formed during glycolysis and the two formed in the Krebs cycle, the complete metabolism of one molecule of glucose yields a grand total of 36 molecules of ATP (Fig. 6.15). This represents the useful energy derived from cellular

Glycolysis and Cellular Respiration

> For most living things, the generation of sufficient amounts of ATP is dependent on an almost constant supply of oxygen.

metabolism. Only two of the ATP molecules come directly from glycolysis, however; the remaining 34 ATPs are generated by aerobic respiration in the mitochondria. It should now be obvious why you and most other living things depend upon an almost constant supply of oxygen.

In Chapter 4 it was pointed out that the efficiency of a given system is the ratio of useful energy to total energy. Moreover, it was also stated that no system can be 100 percent efficient because of the escape of useless heat energy (second law of thermodynamics). This principle applies, of course, to the metabolic activities of living cells. How, then, does cellular metabolism stack up in terms of efficiency? All we need to do to find out is to compare the useful energy derived from ATP with the total energy originally put into the cell. The chemical bond energy in one molecule of glucose is 686 Kcal. This is the total amount of energy originally supplied to the cell. The chemical bond energy (i.e., the energy stored in the terminal phosphate bond) in one molecule of ATP is approximately 7.3 Kcal. Since 36 molecules of ATP are produced during cellular respiration, there are 36 ATPs × 7.3 Kcal = 262.8 Kcal of useful energy obtained from the breakdown of one molecule of glucose. If the total energy put into the system and the useful energy produced are known, the percent efficiency can be easily determined:

$$\% \text{ efficiency} = \frac{262.8 \text{ Kcal (useful energy)}}{686 \text{ Kcal (total energy)}} \times 100$$

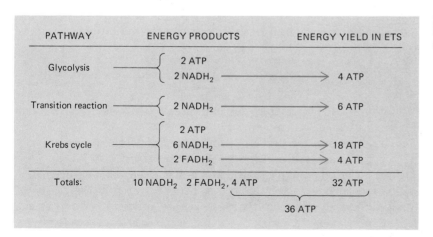

Figure 6.15 Energy yield from the aerobic metabolism of one molecule of glucose.

> The energy efficiency of any organism is limited by its available supply of ATP.

Therefore, the efficiency is 0.383 × 100 = 38.3 percent. By comparison, a simple steam engine may have an efficiency as low as 10 percent.

In those organisms that metabolize anaerobically, such as yeasts, efficiency is very poor when compared with that of aerobic metabolism. Fermentation or glycolysis yields only two ATPs; with each ATP molecule having an energy value of 7.3 Kcal, the total useful energy derived from anaerobic metabolism is only 14.6 Kcal. As stated, one molecule of glucose contains 686 Kcal of available energy. Therefore, the efficiency rating in this instance is only 14.6/686, or about 2 percent. The same result is seen in exercising muscles, which also work in the absence of oxygen. The limited supply of ATP explains why strenuous bursts of energy, such as the 100-meter dash, cannot be sustained for long. As we have seen, exercise causes the accumulation of lactic acid in the cells, preventing further cellular metabolism. By stopping and breathing deeply, you supply the cells with oxygen, the lactic acid is removed, and metabolism continues into the Krebs cycle and ETS. There is then a replenishment of the supply of ATP, which the body utilizes as its primary energy source.

Although our discussion has centered on carbohydrates, the other major foodstuffs—lipids, proteins, and nucleic acids—must also undergo metabolism in the cells. These macromolecules, after being broken down into their simpler forms by digestion, also enter the Krebs cycle and ETS, where they are metabolized to carbon dioxide and water. As with carbohydrates, some of their energy is released and stored in the bonds of ATP. The end result of the complete metabolism of any foodstuff, then, is the production and packaging of energy in forms that can be used to do the work of the cell.

REVIEW OF ESSENTIAL CONCEPTS

1. Green plants manufacture carbohydrates and release oxygen through the process of *photosynthesis*.
2. Photosynthesis occurs within *chloroplasts*, which contain the pigment *chlorophyll*.
3. The major organ of photosynthesis is the green leaf. Leaves consist of an upper and lower epidermis, between which is the *mesophyll* layer. Extending through the mesophyll are *xylem* and *phloem* tissues. Xylem functions in conducting water, and phloem conducts food.

4. The leaf epidermis is perforated with *stomates*, which are bordered by *guard cells*. Guard cells regulate the size of the stomate.
5. Photosynthesis consists of a series of *light reactions*, which require light as an energy source; there is also a series of *dark reactions*, which depend upon energy supplied by ATP and $NADPH_2$ formed in the light reactions.
6. Light is the movement of physical particles called *photons* in streams of varying wavelengths. Photons in shorter wavelengths of light have more energy than do photons in longer wavelengths. In the visible spectrum, the wavelengths of light become progressively shorter in going from red to violet.
7. Chlorophyll primarily absorbs red, blue, and violet light, while reflecting green light. This is why plants appear green to us.
8. During the light reaction of photosynthesis, water is split into hydrogen ions, oxygen atoms, and electrons. In addition, light energy boosts chlorophyll electrons to high energy levels, where they are held by electron acceptor molecules. These electrons are passed down a system of electron carriers, releasing energy in the process. The released energy goes into the formation of ATP. The electrons, along with the hydrogen ions split from water, are taken up by NADP to form $NADPH_2$. The oxygen split from water passes out of the plant into the atmosphere.
9. In the dark reactions of photosynthesis, CO_2 bonds with the 5-carbon sugar *RuDP* to form an unstable 6-carbon compound. This compound splits into two molecules of the 3-carbon compound PGA. Using ATP and $NADPH_2$, each PGA molecule is subsequently reduced to PGAL.
10. Some of the PGAL molecules bond together to form glucose and other intermediate sugars, and some PGAL molecules form new RuDP. Glucose molecules may be stored in the form of *starch*.
11. Some of the hydrogen ions released during photolysis, along with oxygen atoms from carbon dioxide, bond to form cellular water.
12. Photosynthesis is the source of atmospheric oxygen, including the oxygen forming the ozone layer.
13. Most living organisms utilize glucose for cellular energy. In the cells, glucose is metabolized to carbon dioxide and water; the energy released in this process is stored in the form of ATP. The complete process of cellular metabolism includes *glycolysis*, the *Krebs cycle*, and the *ETS*.
14. Glycolysis occurs in the cytoplasm and involves the splitting of glucose into two molecules of *pyruvate*. In the process, two molecules

of ATP are formed, and two molecules of NAD are reduced to form two molecules of $NADH_2$.

15. Yeast cells respire anaerobically, converting pyruvate into *ethyl alcohol* and *carbon dioxide*. However, in animal cells, anaerobic respiration converts pyruvate into lactic acid.
16. In the presence of oxygen, pyruvate is converted into *acetylcoenzyme A*; in the conversion, carbon dioxide is given off and NAD is reduced to $NADH_2$. Acetylcoenzyme A then enters the Krebs cycle.
17. The Krebs cycle occurs in the mitochondria and involves the interconversion of several acids, accompanied by the formation of two molecules of carbon dioxide, three molecules of $NADH_2$, one molecule of $FADH_2$, and one molecule of ATP for each pyruvate that enters the cycle. The original acid in the cycle is regenerated so that the cycle may continue. The carbon dioxide produced in higher animals is exhaled from the lungs.
18. The ETS occurs in the mitochondria and involves the passing of electrons and hydrogen ions to electron carrier molecules. As the electrons are passed from one carrier molecule to another, they release energy that generates ATP. Most of the ATP produced by the cell is formed in the ETS. The electrons are finally taken up by oxygen; the hydrogen ions combine with this oxygen to form water.
19. Glycolysis and cellular respiration result in a total net gain of 36 molecules of ATP from the breakdown of one molecule of glucose. This represents an efficiency rating of 38.3 percent. Anaerobic respiration, however, results in an efficiency of only 2 percent.

APPLYING THE CONCEPTS

1. Describe the general structure of the green leaf. How is the structure of the leaf related to photosynthesis?
2. What is the relationship between light wavelength and chlorophyll?
3. Outline the essential steps of the light reactions of photosynthesis.
4. What are the basic functions of the light reactions? How do these relate to the dark reactions of photosynthesis?
5. Why is photosynthesis such a basically important process?
6. What is glycolysis? What does it accomplish?
7. Compare anaerobic glycolysis in yeast cells and human muscle cells.
8. Why do some cells acquire an "oxygen debt"?
9. What is the relationship of the transition reaction to glycolysis and the Krebs cycle?

Applying the Concepts

10. What is the most significant accomplishment of the Krebs cycle? Why?
11. Discuss the basic scheme of the ETS. Why is oxygen essential to the ETS?
12. Why is anaerobic respiration considered an inefficient process?
13. Write the overall reaction of photosynthesis and, in your own words, explain the relationships between the reactants and the products.
14. Do the same as in Question 13 for the metabolism of a molecule of glucose.

dividing cells: mitosis and meiosis

THE ESSENTIAL OBJECTIVES

You have understood this chapter when you are able to:

1. Discuss the function of mitosis in plant and animal cells.
2. Describe the events that occur in each of the four phases of mitosis.
3. Explain the significance of meiosis and its relationship to gametogenesis.
4. Describe the events that occur in meiosis I and meiosis II.
5. Describe the phenomenon of crossing over and explain its genetic significance.
6. Explain what is meant by nondisjunction and cite an example.
7. Discuss the causes and symptoms associated with Down's syndrome, Turner's syndrome, trisomy-X, Klinefelter's syndrome, and the XYY-syndrome.
8. Describe the technique of amniocentesis and discuss its use in genetics.

Growth and repair in the living world are accomplished through the process of mitotic cell division. In plants, mitosis also gives rise to gametes. Meiotic cell division produces spores in many plants and produces gametes in the higher animals.

PREVIEW OF ESSENTIAL TERMS

diploid Having two sets of chromosomes; twice the haploid number of chromosomes.

haploid Having one set of chromosomes; half the diploid number of chromosomes.

sex chromosome A specialized chromosome associated with determination of the sex of an organism; in human beings the sex chromosomes are designated X and Y.

autosome Any chromosome except a sex chromosome.

chromatid One of the two strands of a duplicated chromosome; a chromatid is equivalent to a helix of DNA.

centriole A cellular organelle associated with formation of the spindle fibers in animal cells and a few other cell types.

centromere A small region that holds together the two chromatids of a chromosome.

homologue A chromosome that pairs with another similar chromosome in meiosis.

crossing over The exchange of genetic material between the chromatids of homologous chromosomes.

gametogenesis The production of gametes, or sex cells, by meiosis; divided into oogenesis and spermatogenesis.

nondisjunction The failure of homologous chromosomes to separate from each other during meiosis.

Dividing Cells: Mitosis and Meiosis

INTRODUCTION TO MITOSIS

Omnis cellula e cellula: "All cells come from cells." This now famous aphorism, coined by the German physician Rudolph Virchow, is one of the cornerstones of the cell theory. The inherent capacity of living cells to undergo division to produce others of the same kind is the basis for the processes of cell growth and repair in the living world. The process by which one cell gives rise to others like it is called *mitosis* (Gk. "thread condition").

The body cells of every eukaryotic organism contain a certain number of **chromosomes** located in the nuclei. The chromosomes are composed primarily of the hereditary material DNA. Each chromosome is a helix of *deoxyribonucleic acid* (DNA) (recall that a helix of DNA is composed of two chains of nucleotides bonded together and twisted into a helix). The units of heredity, called **genes,** are located along the length of the DNA chains. In designating the normal number of chromosomes in each body, or *somatic,* cell, such as 6 in the mosquito, 78 in your dog, 46 in you, and so on, we are referring to what is known as the **diploid** (Gk. "double") **number** of chromosomes.

The term "diploid" is used because chromosomes are usually found in pairs. Thus, the diploid number of chromosomes for human beings is 46, or 23 pairs, and for the mosquito the number is 6, or 3 pairs. In human beings, 1 pair of the 23 is the **sex chromosomes;** the other 22 pairs are called **autosomes.** The sex chromosomes of humans are designated X and Y. Normal human females have two X chromosomes (XX), one contributed by the mother and one by the father. Normal males have one X and one Y chromosome (XY); the X chromosome is contributed by the mother, and the Y chromosome by the father. In the case of the 22 pairs of autosomes, 22 individual chromosomes are contributed by the mother and 22 by the father (Fig. 7.1).

A term we will use later in this chapter is **haploid** (Gk. "single"), which refers to *one-half* the diploid number of chromosomes. Cells containing the haploid number of chromosomes contain only one of each of the paired body cell chromosomes. In man, then, the haploid number is 23, and in the mosquito it is 3. Among higher organisms, only the reproductive cells—eggs and sperm—are haploid. Many species of plants produce haploid reproductive cells called **spores.** As a convenience, the haploid number is also referred to as the n number of chromosomes. The diploid number is then referred to as $2n$, or twice the haploid number. Table 7.1 gives the diploid numbers for common plants and animals. In human beings, then, $2n = 46$ and $n = 23$.

Introduction to Mitosis

> Mitosis serves to apportion the hereditary material equally among dividing cells.

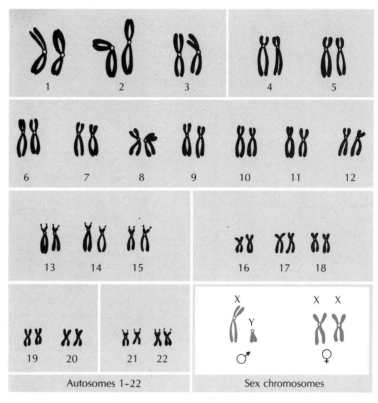

Figure 7.1 Human mitotic metaphase chromosomes arranged in pairs. There are 22 pairs of body, or somatic, chromosomes, and 1 pair of sex chromosomes, either XX (female) or XY (male). The arrangement of chromosomes as shown here is called the *karyotype*. (From Johnson, W. H., Delanney, L. E., Cole, T. A., and Brooks, A. E.: *Biology*. 4th ed. New York: Holt, Rinehart and Winston, 1972, p. 110.)

The essential accomplishment of mitosis (or nuclear division) is the formation of two identical **daughter cells,** i.e., the division of one cell into two, each having the same number of chromosomes and thus the same hereditary information, or DNA, as the original parent cell. This is absolutely necessary for the replacement of lost or injured body cells. If you scrape your knee, for example, the skin cells are replaced by mitosis—and each new skin cell is genetically the same as all the other skin cells. The same is true in regard to growth. Every organ or body structure increases in size by the addition of more cells, and all the cells making up the particular tissue of an organ or body structure are identical. Mitosis, therefore, serves to pass the genetic material (DNA) faithfully from cell to cell.

Mitosis actually refers to division of the *nucleus* to produce two new identical nuclei, i.e., nuclei having the same number of chromosomes as the original parent nucleus. Mitosis may or may not be accompanied by division of the cytoplasm. A variety of cells, such as those of some fungi and green algae, undergo mitosis without sub-

Table 7.1 CHROMOSOME NUMBERS IN VARIOUS SPECIES OF PLANTS AND ANIMALS

PLANTS	DIPLOID NUMBER
Corn	20
Tomato	24
Tobacco	48
Oats	42
Cabbage	18
Garden pea	14
Sunflower	34
Carrot	18
Animals	
Mouse	40
Amoeba	50
Frog	26
Monkey	42
Cat	38
Rabbit	44
Earthworm	36
Pigeon	80

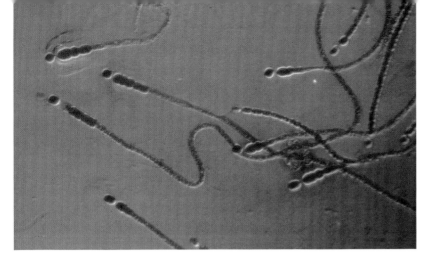

Color Plate 16 Filaments of the blue-green alga *Gloeotrichia*. (Photograph by Bruce Russell, Biomedia Assoc.)

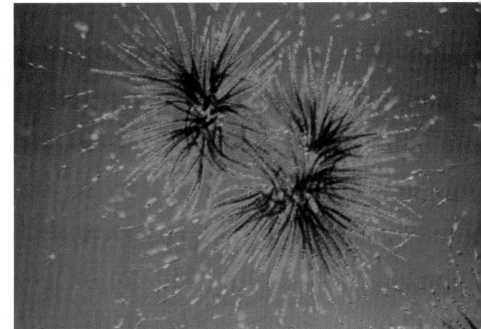

Color Plate 17 The filaments of this blue-green alga come together at one end, forming a rosette. (Photograph by Bruce Russell, Biomedia Assoc.)

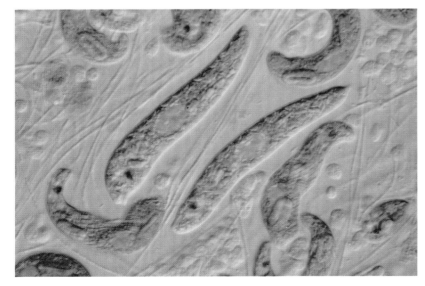

Color Plate 18 A group of green *Euglena*. (Photograph by Bruce Russell, Biomedia Assoc.)

Color Plate 19 Diatoms. Unicellular algae of the division Chrysophyta. (Photograph by Bruce Russell, Biomedia Assoc.)

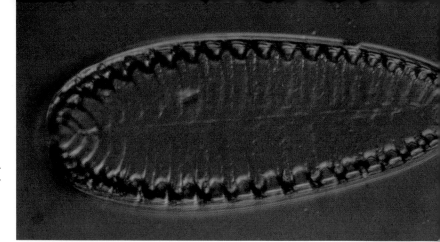

Color Plate 20 A test, or shell, of a diatom. (Photograph by Bruce Russell, Biomedia Assoc.)

Color Plate 21 A dinoflagellate. Division Pyrrophyta. (Photograph by Carolina Biological Supply Co.)

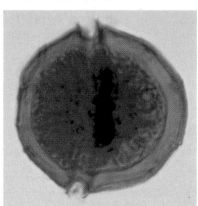

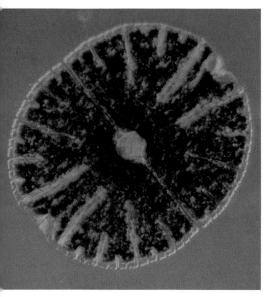

Color Plate 22 *Micrasterius,* a unicellular green alga. (Photograph by Bruce Russell, Biomedia Assoc.)

Color Plate 23 *Scenedesmus,* a green alga. (Photograph by Bruce Russell, Biomedia Assoc.)

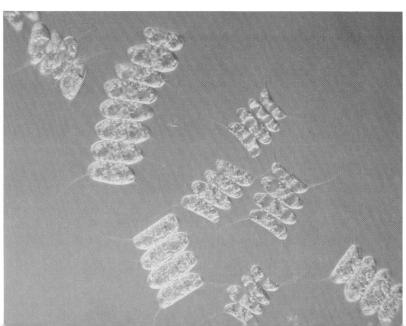

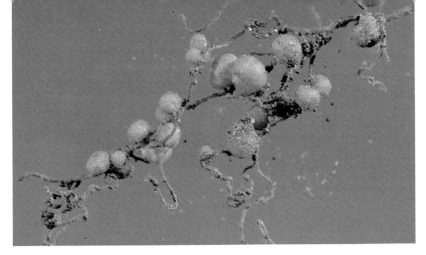

Color Plate 24 The root nodules of many legumes harbor nitrogen-fixing bacteria. (Photograph by Bruce Russell, Biomedia Assoc.)

Color Plate 25 A lichen is a mutualistic relationship between an alga and a fungus. (Photograph by Bruce Russell, Biomedia Assoc.)

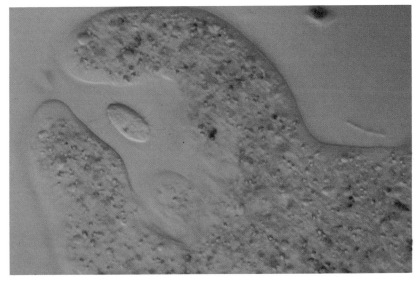

Color Plate 26 Phagocytosis in *Amoeba*. (Photograph by Bruce Russell, Biomedia Assoc.)

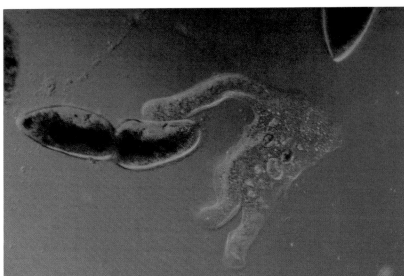

Color Plate 27 An amoeba about to feed on a dividing paramecium. (Photograph by Bruce Russell, Biomedia Assoc.)

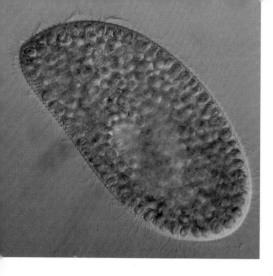

Color Plate 28 Symbiotic green algae within a paramecium. (Photograph by Bruce Russell, Biomedia Assoc.)

Color Plate 29 The filamentous green alga *Spirogyra,* highlighting its unique chloroplasts. (Photograph by Bruce Russell, Biomedia Assoc.)

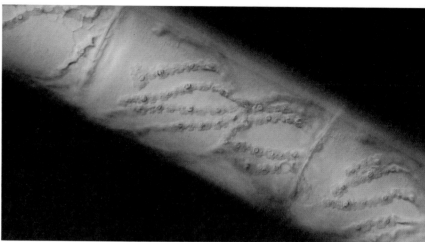

Color Plate 30 A brown alga, division Phaeophyta. (Photograph by Carolina Biological Supply Co.)

Color Plate 31 A red alga, division Rhodophyta. (Photograph by Carolina Biological Supply Co.)

> A duplicated chromosome consists of two double helices of DNA joined together at the centromere region.

sequent cell division. Consequently, each of these cells has several nuclei in its cytoplasm and is thus *multinucleate*. However, in the majority of living species, nuclear division (mitosis) is accompanied by cytoplasmic division, also known as *cytokinesis* (Gk. "cell motion").

It was not until the late 1800s that new staining techniques made possible the observation of chromosomes in living cells. Soon thereafter, a German biologist named Walther Flemming described in detail the events of mitosis using the body cells of a salamander. Basically, the scheme of mitosis is relatively simple. There is much to be learned about mitosis, of course, such as what exactly starts the process and what stops it. Nonetheless, as a physically observable process, the scheme of mitosis is elegant in its simplicity. When a body cell is ready to divide, the hereditary material (DNA of the chromosomes) is first duplicated so that each chromosome is composed of two helices of DNA joined together at some point along the chromosome. The result, then, is that the chromosomes essentially double their number. This, of course, is the same thing as saying that the diploid number is doubled. During mitosis, these duplicates of each chromosome separate from each other so that each of the two daughter cells receives the original diploid number of chromosomes (Fig. 7.2).

Although mitosis is a continuous, nonstop process, for convenience it has been separated into four distinct phases, making it easier to follow the progressive sequence of events (Figs. 7.3 and 7.4). The phases referred to are *prophase, metaphase, anaphase,* and *telophase.* In addition, there is a nondividing stage called *interphase,* which occurs prior to each mitotic division. During interphase the cell produces new organelles, manufactures materials for its own structure and function, and duplicates its DNA in preparation for division.

When viewing a particular cell under the light microscope, we usually see a cell in interphase. This is because cells spend only a short time actually dividing. The nucleus is quite often visible, but the individual chromosomes are not. During interphase the chromosomes are not fully formed but exist as a tangled mass of long, thin threads called *chromatin.* Just outside the nucleus of animal cells lie two pairs of tiny cylindrical organelles called *centrioles.* For some unknown reason the centrioles appear to be absent from the cells of higher plants.

THE SCHEME OF MITOSIS

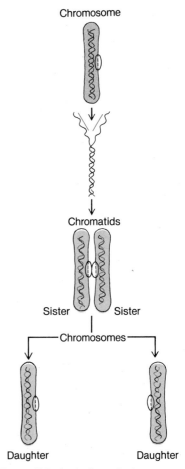

Figure 7.2 A duplicated chromosome consists of two chromatids joined at the centromere region. Each chromatid consists of identical helices of DNA.

148
Dividing Cells:
Mitosis and Meiosis

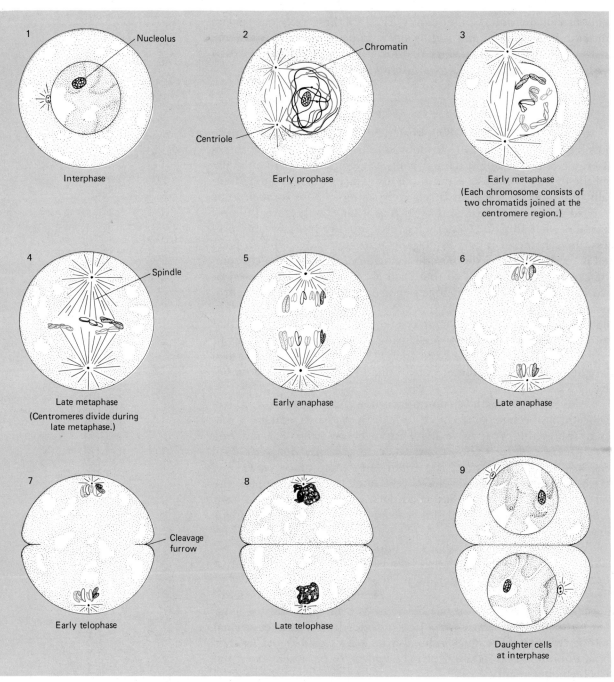

Figure 7.3 Mitosis in animal cells ($2N = 6$).

149
The Scheme of Mitosis

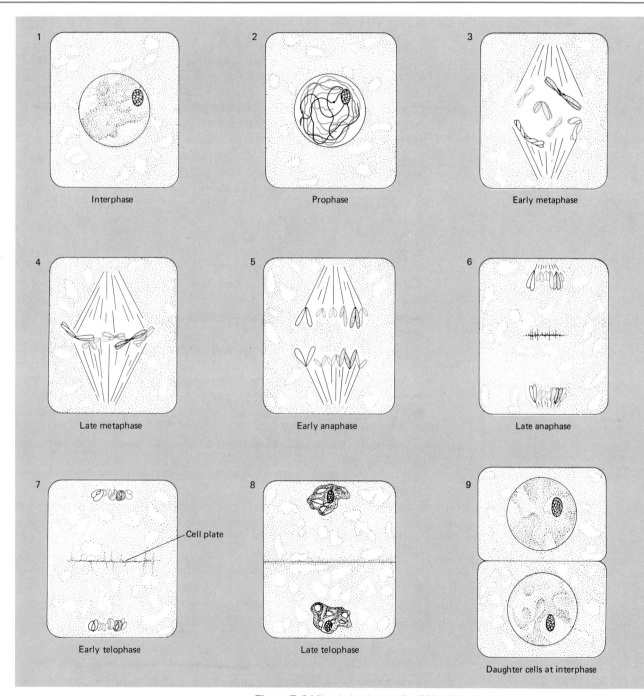

Figure 7.4 Mitosis in plant cells (2N = 6).

> During interphase the cell synthesizes enzymes and structural proteins; in addition, DNA replication occurs, forming double-stranded chromosomes.

For a number of years the term "resting stage" was applied to interphase, but as is now known, the cell is doing anything but "resting" at this time. As has already been mentioned, one of the essential events occurring in interphase is *duplication of the chromosome material,*" or *DNA.* Each soon-to-be chromosome makes a replica, or copy, of itself, forming a double-stranded chromosome. Each strand is referred to as a **chromatid,** and the two chromatids are held together at a region called the **centromere** (Fig. 7.5). The two chromatids making up one chromosome are genetically identical, i.e., they are composed of the same kind and amount of DNA.

As stated, the essential function of mitosis is to produce daughter cells in which the genetic material is equally distributed. Let us look, then, at how this division and distribution come about.

Prophase. By early prophase of mitosis, which immediately follows interphase, the double-stranded chromosomes are fully formed and distinctly visible under the microscope. The centrioles outside the nucleus migrate toward the opposite ends, or poles, of the cell. Later in prophase a system of long, thin **spindle fibers** extends from each centriole across the cell. Finally, the nuclear membrane and nucleolus disperse, leaving the chromosomes suspended in the cytoplasm.

Metaphase. At metaphase the chromosomes have moved to the center of the cell, where they line up on what is called the **equatorial plate.** The centromere region of each chromosome is attached to spindle fibers extending from the centrioles at either pole of the cell.

Late in metaphase *the centromere of each chromosome divides,* resulting in separate chromatids, each having its own centromere region. After the centromeres divide, each chromatid is then called a chromosome.

Anaphase. During anaphase the now separated "sister" chromosomes move toward the opposite poles of the cell, apparently pulled by the spindle fibers. Remember that the DNA of the chromosomes moving toward one end of the cell is the same as the DNA moving to the other end. By late anaphase each set of chromosomes arrives at its respective pole of the cell.

Telophase. In telophase (or late anaphase) the cytoplasm is separated. If present, one centriole from each pair duplicates itself to form two new centriole pairs. In animal cells a **cleavage furrow** forms on opposite sides of the cell, and the cell eventually pinches in two.

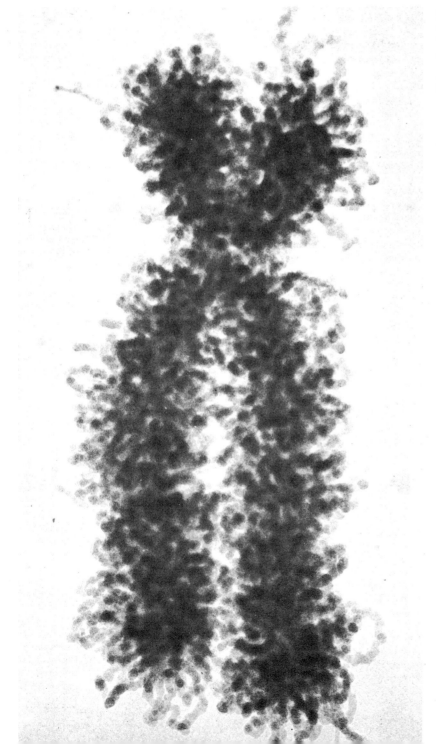

Figure 7.5 A human mitotic chromosome. The two "sister" chromatids (each composed of a DNA helix) are joined at the centromere region. (From Goodenough, U., and Levine, R.: *Genetics.* New York: Holt, Rinehart and Winston, 1974, p. 53.)

Dividing Cells: Mitosis and Meiosis

> Mitosis is one example of the control and regulation of living processes at the cellular level.

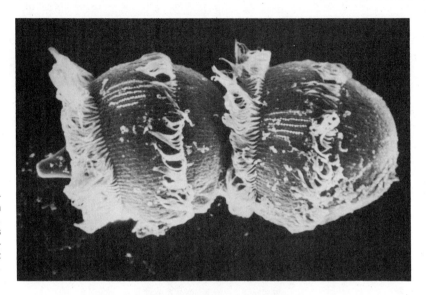

Figure 7.6 An electron scanning micrograph of an early stage of binary fission in the protozoan *Didinium nasutum*. This type of asexual reproduction is common among protozoans. (From Villee, C. A.: *Biology*. 7th ed. Philadelphia: Saunders College Publishing, 1977, p. 587.)

By division of the cytoplasm, each set of chromosomes is thereby contained within a separate *daughter cell*. In plant cells a **cell plate** forms that eventually separates the cytoplasm, resulting in two distinct daughter cells. As pointed out earlier, some cells do not undergo cytokinesis after mitosis and thereby become multinucleate cells.

After the two genetically identical daughter cells have been formed, the nuclear membranes and nucleoli are reformed in each cell. The chromosomes soon become less and less distinct as the cells prepare for the interphase stage.

One of the most distinctive characteristics of living things is their capacity to reproduce others of their own kind. Among some of the unicellular organisms—algae, yeasts, protozoa, and so on—mitosis is essential to the process of reproduction. By various asexual methods to be considered later, these organisms undergo mitotic cell division to produce offspring in basically the same manner as your skin cells produce other skin cells (Fig. 7.6).

Mitosis provides further evidence that life is regulated and controlled at the cellular level. Moreover, in observing mitosis, one can appreciate more fully the fact that the cell is the fundamental unit of all living things. Without the capacity of cells to make copies of themselves, life could not exist for long; in fact, it would never have progressed much beyond the original cells in the ancient seas (Figs. 7.7 and 7.8).

> Meiosis reduces the diploid number of chromosomes to the haploid number.

INTRODUCTION TO MEIOSIS

Not all of the cells produced by man and other sexually reproducing organisms are diploid. In the reproductive tissues of human beings—the *ovaries* in the female and *testes* in the male—cells arise that contain only half the diploid number of chromosomes. These cells are called *gametes* and include the eggs, or *ova,* of the female, and the *sperm* of the male. In every normal human gamete there are 23 chromosomes, which represent the *haploid,* or *n,* number of chromosomes. The haploid number, as mentioned, is always one-half the diploid number. Why would gametes be haploid while all other body cells are diploid? The answer is in the nature of the sexual reproductive process.

By definition, sexual reproduction involves the union of gametes from individuals of opposite mating types. The union of gametes (fertilization) produces a single-celled *zygote,* the beginning of a new living organism. All normal offspring (excluding some hybrids—a hybrid is an offspring of a cross between two organisms of

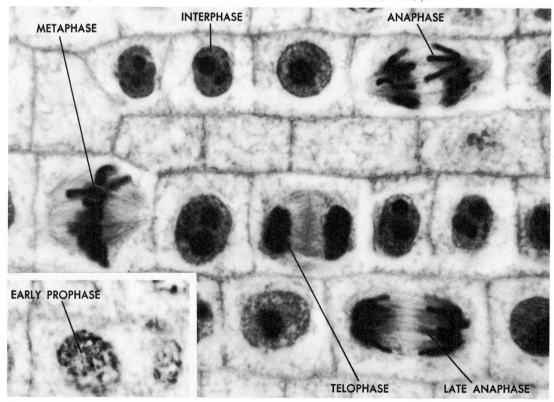

Figure 7.7 Photomicrographs of meristematic cells from the tip of an onion root showing stages in mitotic division. Note the characteristic thin walls, large nuclei, and absence of vacuoles. (From Johnson, W. H., Delanney, L. A., Cole, T. A., and Brooks, A. E.: *Biology.* 4th ed. New York: Holt, Rinehart and Winston, 1972, p. 117.)

154
Dividing Cells: Mitosis and Meiosis

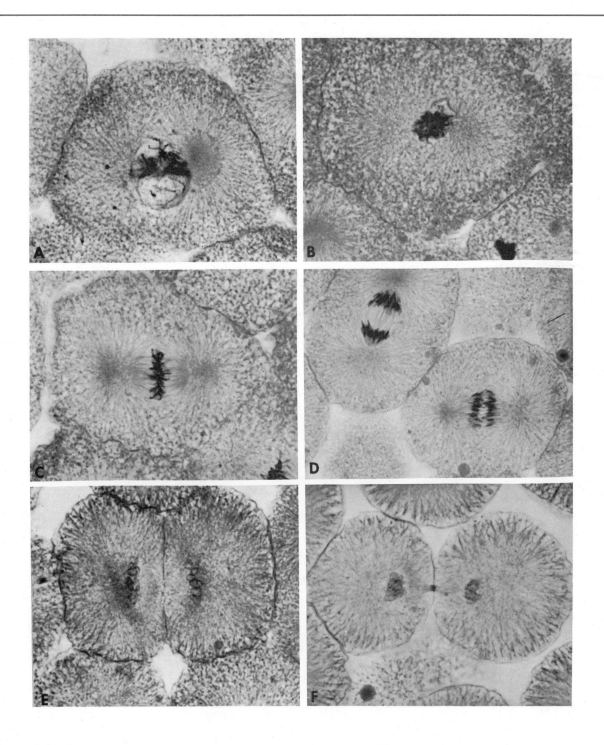

Figure 7.8 Photographs of stages in mitosis in the white fish blastula. (a) Early prophase. (b) Later prophase. (c) Metaphase. (d) Two cells in early and late anaphase, respectively. (e) Early telophase. (f) Late telophase. The dark spot connecting the two cells in (e) is the remainder of the spindle. (From Villee, C. A., Walker, W. F., and Barnes, R. D.: *General Zoology*. 4th ed. Philadelphia: Saunders College Publishing, 1978, p. 43.)

In higher plants and animals, sex cells, or gametes, are produced only in specialized reproductive tissue.

different, but closely related, species) possess the same number of body chromosomes as their parents. With each parent contributing a haploid gamete, the full chromosome complement, i.e., the diploid number, is restored in the zygote. If gametes were diploid instead of haploid, each succeeding generation would have twice as many chromosomes as its parents. In human beings, assuming that we started with *diploid* gametes containing 46 chromosomes, this would mean that the second generation would then have 92 chromosomes, the third generation would have 184, and the tenth generation would have 23,552 chromosomes!

The type of cell division that reduces the diploid number of chromosomes to the haploid number is called **meiosis** (Gk. "to diminish"). In higher animals, meiosis occurs *only* in the reproductive tissues. In human beings, these tissues include the ovaries of the female and the testes of the male. In the reproductive tissues of both sexes, there are specialized diploid cells that undergo meiosis to produce the haploid gametes. In the female these cells are called **oogonia** (Gk. "egg offspring"), and in the male they are called **spermatogonia** (Gk. "seed offspring"). The production of gametes is termed **gametogenesis,** which can be separated into **oogenesis** (egg production) and **spermatogenesis** (sperm production) (Fig. 7.9).

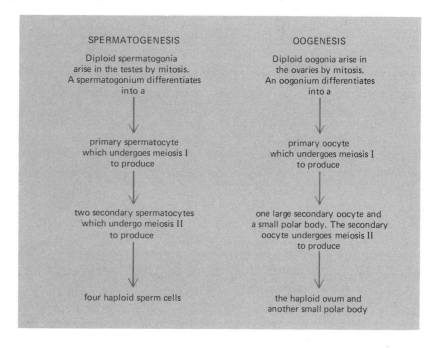

Figure 7.9 General scheme of gametogenesis in animal cells.

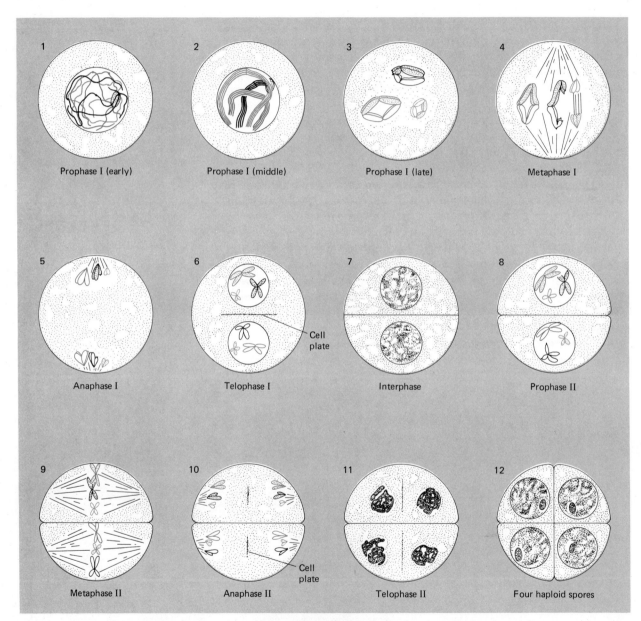

Figure 7.10 Meiosis in plant cells ($2N = 6$). The haploid spores divide by mitosis to produce the gametophyte of the plant.

Although meiosis results in the production of gametes in animals, this is usually not the case in most plants. Meiosis in plants usually gives rise to haploid reproductive cells called *spores* (Fig. 7.10). Unlike gametes, which unite to form a diploid zygote, a spore undergoes mitosis to produce, in some species, an entire haploid multicellular plant; in other species, spores undergo mitosis to produce small reproductive structures found on the plant. In the mosses, for example, the moss spore divides by mitosis to produce the leafy green organisms we recognize as moss plants. These are haploid multicellular plants known as *gametophytes,* since they eventually give rise to gametes. On the other hand, in the flowering plants, spores undergo mitosis to form small reproductive structures also called gametophytes, which produce the male and female gametes. The male gametophyte, or pollen grain, produces haploid sperm, and the female gametophyte within the ovule gives rise to the haploid egg.

You will notice here that the male and female gametophytes are haploid because they arise by mitotic division of a haploid spore. Furthermore, the gametes produced by the gametophytes are also haploid. Thus, in contrast to animal gametes, which are produced by *meiosis*, the gametes of most plants arise by *mitosis*. In the case of the mosses, the principle is the same—specialized cells on the haploid plant body undergo mitosis to produce the haploid male and female gametes.

At fertilization the haploid gametes unite to produce the diploid zygote, which develops into a multicellular *sporophyte.* In angiosperms, the plant or tree itself is the sporophyte; in mosses, the sporophyte is a stalklike structure growing out of the female moss plant. The sporophytes produce spores by meiosis, and the reproductive cycle of the plant starts over again.

It was pointed out earlier that chromosomes usually come in pairs. At fertilization, one chromosome of a pair is contributed by the female, and one by the male. The two chromosomes constituting each pair are usually the same size and shape and therefore are referred to as *homologous* (Gk. "same proportion") chromosomes. One chromosome of a pair, then, is the *homologue* of the other. Two of the distinguishing events of meiosis are the pairing and separation of homologous chromosomes (Fig. 7.11).

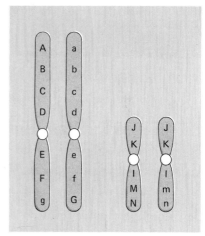

Figure 7.11 Two pairs of homologous chromosomes. Each chromosome of a pair separates from its partner during meiosis I.

THE SCHEME OF MEIOSIS

Meiotic cell division actually involves two stages, generally referred to as *meiosis I* and *meiosis II.* Whereas mitosis produces two genetically identical diploid cells, meiosis results in *four haploid cells* (gametes or spores) that are genetically different, i.e., the chromosomes of each cell contain different genes. The stages in the first meiotic division (meiosis I) are designated *prophase I, metaphase I, ana-*

> Crossing over occurs between pairs of homologous chromosomes, resulting in a "mixture" of male and female genes.

phase I, and *telophase I;* in the second division (meiosis II), the stages are *prophase II, metaphase II, anaphase II,* and *telophase II.* Meiosis I and sometimes meiosis II are preceded by an interphase stage.

Prior to meiosis I, the DNA in the parent diploid cell is duplicated, just as it is in mitosis. Homologous chromosomes then come together in pairs (this is a major difference between meiosis and mitosis), and eventually each homologue is separated from its partner. Two cells are then formed, each containing one set of the homologues. (In some organisms, the cells do not separate from each other at this point.) In meiosis II the events resemble those of mitosis. Each of the two cells formed in meiosis I undergoes division, which separates the *chromatids* that still compose each chromosome. This results in the formation of four cells, each containing one-half the number of chromosomes as the parent diploid cell.

Let's look first at a more detailed explanation of meiosis and then comment on its application to gametogenesis in human beings.

Interphase. At the first interphase the diploid cell nucleus contains chromatin in the form of a tangled mass of threads. Later in interphase there is DNA duplication, such that each of the individual chromosomes consists of two chromatids joined at the centromere.

Meiosis I

Prophase I. In late prophase I the duplicated chromosomes are clearly visible under a microscope. Homologous pairs of chromosomes come together and orient themselves next to each other. In humans this would mean that the 46 duplicated chromosomes are arranged in 23 homologous pairs. Next, a phenomenon occurs that has far-reaching genetic consequences. The homologous pairs exchange some of their genetic material, a process known as ***crossing over*** (Fig. 7.12). What actually happens is that a segment of a chromatid from one chromosome breaks off and is exchanged with a segment that has broken off a chromatid of the other chromosome. Crossing over, therefore, occurs between the homologous pairs and not between the two chromatids making up the same chromosome.

The result of crossing over is ***genetic recombination,*** i.e., a rearrangement of the genes carried by each chromosome. Since one chromosome of each homologous pair was originally donated by the male parent and one by the female parent, each chromosome, after crossing over occurs, is a "mixture" of male and female genes.

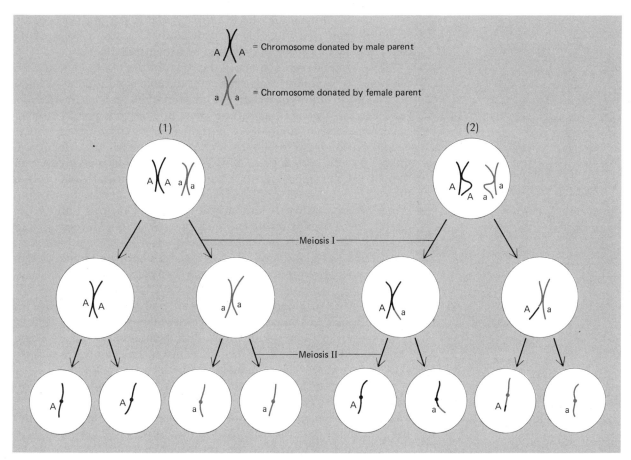

Figure 7.12 Comparative diagrams of meiosis showing (1) the four types of haploid cells formed when crossing over does *not* occur and (2) the "mixture" of male and female genes (*A* and *a*) in two of the four haploid cells as a result of crossing over.

Later in prophase I the spindle fibers are formed, and the centrioles, if present, migrate to opposite poles of the cell. Finally, the nuclear membrane and nucleolus disperse, marking the end of prophase I.

Metaphase I. In metaphase I the *pairs* of homologous chromosomes line up along the equatorial plate. In mitosis you will recall that *each* chromosome lined up, one under the other, along the equatorial plate. In meiosis, *pairs* of homologous chromosomes line up in this manner. The spindle fibers are attached to the centromere of each homologue. Notice here that the centromeres do *not* divide as they do in mitosis; each chromosome, then, still consists of two chromatids joined together at a single centromere region.

> Each of the two cells formed at the end of Meiosis I contains one chromosome of each homologous pair.

Anaphase I. During anaphase I the spindle fibers pull each of the chromosomes of a homologous pair toward opposite poles of the cell. Each chromosome *still* consists of two chromatids.

Like genetic recombination, the separation of homologous chromosomes at anaphase I also has considerable significance. As mentioned, a chromosome from the male parent and a chromosome from the female parent constitute each homologous pair. The manner in which each homologous chromosome separates from its partner in anaphase I is essentially random. This means that, depending upon how the homologous pairs were lined up at metaphase I, either of the chromosomes of a pair *could* move toward one pole of the cell or toward the opposite pole. In this manner all the chromosomes become assorted randomly at the end of meiosis I and subsequently in the gametes at the end of meiosis II. Moreover, since crossing over results in each of these chromosomes containing a combination of genes from the male parent and the female parent, this mixture of chromosomes is even more complex.

Telophase I. At telophase I each set of homologues has arrived at opposite poles of the cell, and a nuclear membrane begins to form around each set. The spindle fibers disappear, and the chromosomes begin to fade from view. In some organisms the cell divides into two cells; in others the cell remains undivided.

In summary, then, the chromosomes of a diploid cell have duplicated themselves; homologous pairs have exchanged genetic material; and each homologue has been separated from its partner. Keep in mind that each of these chromosomes is still composed of two chromatids joined at the centromere region. At this point, meiosis I is completed, and either two cells or a single cell containing two "nuclei" has been formed. If present, the centrioles duplicate in preparation for the second meiotic division.

Interphase II. At the end of meiosis I, there may be a period known as interphase II, in which the nuclear material disperses, but there is no duplication of chromosomal DNA as in the first interphase. In some cells telophase I is followed immediately by the second meiotic division.

Meiosis II

Prophase II. During prophase II the chromosomes of each of the two nuclei formed during meiosis I condense, new spindle fibers form, and the nuclear membranes break down.

> At the end of Meiosis II in higher organisms, every gamete is genetically different from every other gamete.

Metaphase II. At metaphase II the chromosomes in each nucleus line up along the equatorial plate. Spindle fibers from both poles of the cell emanate from the centrioles, if present, and are attached to each chromosome at the centromere region. Late in metaphase II the centromere regions of the chromosomes divide as they do in mitosis, thus allowing the two chromatids to exist independently.

Anaphase II. During anaphase II each *chromatid* (now called a chromosome) is pulled toward its respective pole of the cell.

Telophase II. During telophase II the separated chromatids, now called chromosomes, arrive at the poles of the cell, the spindle fibers disappear, and a nuclear membrane forms around each of the two sets of chromosomes from each nucleus, giving a total of four. The cytoplasm then divides, forming four haploid cells (gametes or spores).

MEIOSIS IN HUMAN BEINGS

In human beings (and other higher organisms), meiosis is the process that produces the male and female sex cells, or gametes. As mentioned, meiosis in humans is called gametogenesis and includes spermatogenesis in the male and oogenesis in the female. These processes are controlled and regulated by various hormones secreted by the reproductive organs and other endocrine tissues.

Spermatogenesis. The parent diploid cells in the testes of the male are called *spermatogonia*, which differentiate to form the **primary spermatocytes** (Fig. 7.13). These cells contain 23 pairs of chromosomes consisting of two chromatids each, since DNA duplication occurs prior to division. The primary spermatocytes undergo meiosis I to produce two **secondary spermatocytes.** Each of these cells contains 23 chromosomes; remember, however, that each chromosome still consists of two chromatids.

The secondary spermatocytes undergo meiosis II to produce four haploid **spermatids,** which are simply immature sperm cells. These cells contain the haploid number of single chromosomes. Eventually, each spermatid develops a flagellum and differentiates into a mature haploid gamete (sperm cell). In the male, each parent spermatogonium undergoing meiosis gives rise to four functional haploid sperm of equal size. Spermatogenesis normally results in the production of several hundred million sperm every day.

Figure 7.13 Diagram of spermatogenesis as it occurs in male human beings and higher animals (2N = 6).

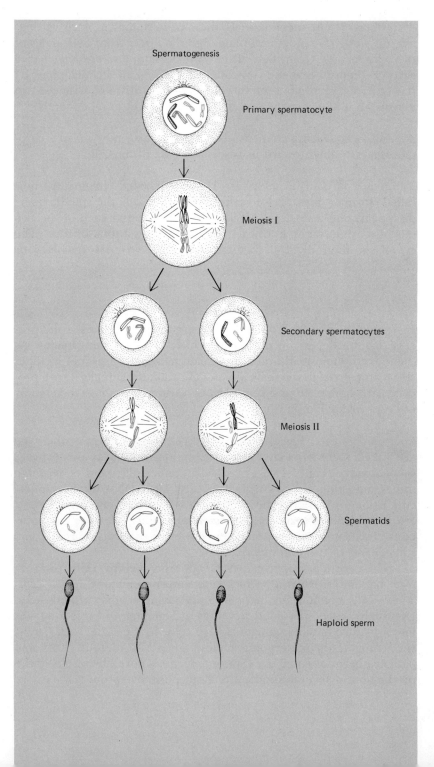

> In humans, spermatogenesis produces hundreds of millions of sperm every day; oogenesis usually produces only one ovum every 28 days.

Oogenesis. In the human female, the parent diploid cells in the ovaries are called *oogonia*. These cells differentiate to form the **primary oocytes,** which contain 23 pairs of chromosomes consisting of two chromatids each (Fig. 7.14). A primary oocyte undergoes meiosis I

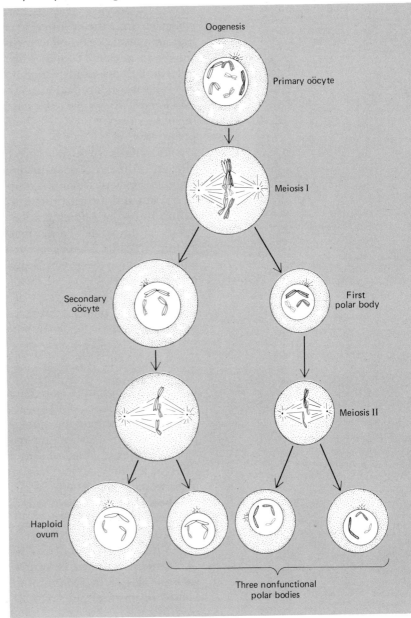

Figure 7.14 Diagram of oogenesis as it occurs in female human beings and higher animals ($2N = 6$).

> In human and other vertebrate females, an oocyte may be arrested in meiosis I for as long as 35–40 years; only after fertilization will the oocyte complete meiosis II.

to produce one large **secondary oocyte** and a very small cell called the *first polar body*. The difference in size results from unequal division of the cytoplasm. Each of these cells contains 23 chromosomes, each of which is still composed of two chromatids. It is the secondary oocyte that is discharged from the ovary (a process called *ovulation*) into the oviduct, or fallopian tube. The first polar body may or may not complete meiosis II. If it does, it divides into two similar-sized polar bodies. This is of little consequence, since all the polar bodies are nonfunctional and eventually disintegrate.

The secondary *oocyte*, however, completes meiosis II, *but in human beings this does not occur until after fertilization.* The division of the secondary oocyte produces the large haploid *ovum* and another small polar body.

Oogenesis, then, may produce four haploid cells, but only one of them—the ovum—is functional. Meiosis proceeds as usual during oogenesis except for unequal division of the cytoplasm. Most of the cytoplasm and its constituent organelles go into the formation of the ovum; the polar bodies receive very little cytoplasmic material. As a result of unequal cytoplasmic division, a human ovum is some 75,000 times as massive as an individual sperm cell. Through unequal division, the ovum acquires a large supply of nutrients and organelles, which will be needed during the early development of the zygote. Cellular division of the zygote is so rapid that the DNA does not have time to direct the production of new organelles and nutrients needed for development.

ABNORMAL CELL DIVISION

For reasons not fully understood, there are times when meiosis in human beings does not proceed in its normal manner. In these instances, the homologues may not separate from each other at anaphase I, or the centromeres may not divide at metaphase II. At the end of meiosis, then, one of the gametes is minus an autosome, whereas another gamete has one too many. In other words, one gamete would contain only 22 chromosomes, and another would have 24. The failure of chromosomes to separate properly is known as **nondisjunction** (Fig. 7.15). If the gamete with 24 chromosomes unites with a normal gamete, the resulting zygote would have 47 chromosomes instead of the normal 46.

The classic example of extra chromosome inheritance is *Down's syndrome* (Fig. 7.16). Most commonly, Down's syndrome involves the failure of chromosome pair 21 to separate during meiosis. Conse-

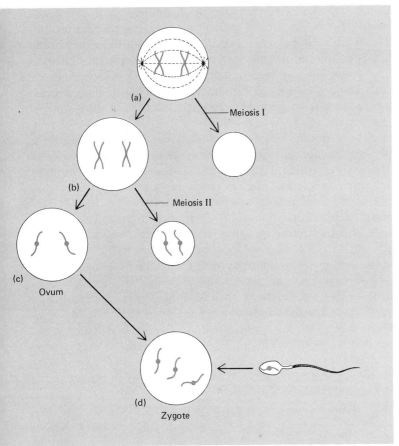

Figure 7.15 Diagram of nondisjunction involving one chromosome pair (Number 21), as occurs in Down's syndrome. (a) Primary oocyte contains two Number 21 chromosomes (the other 22 pairs of chromosomes have been omitted for clarity). (b) After meiosis I, failure of the chromosomes to separate from each other (nondisjunction) results in the secondary oocyte receiving *both* Number 21 chromosomes. The smaller first polar body normally receives one of these chromosomes. (c) After meiosis II, the ovum and small second polar body each contain two Number 21 chromosomes. (d) At fertilization, a normal sperm containing one Number 21 chromosome unites with the ovum, producing a zygote with three Number 21 chromosomes. Nondisjunction of chromosome pair Number 21 also may occur in spermatogenesis.

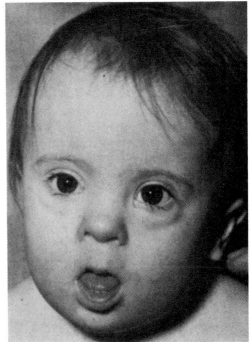

Figure 7.16 As a result of nondisjunction of the twenty-first chromosome during meiosis, this child received two of these chromosomes from its mother as well as one from its father. The extra chromosome produces mental retardation and a characteristic facial expression. (From Clark, M. E.: *Contemporary Biology.* 2nd ed. Philadelphia: Saunders College Publishing, 1979, p. 379.)

> Although human males produce new sperm throughout their lives, all the oocytes a female will ever have are present at her birth.

quently, one of the gametes contains two chromosomes 21. At fertilization, then, the zygote would contain an extra, or third, chromosome 21, instead of the normal two. Down's syndrome is one of the more common human genetic abnormalities, occurring in approximately 1 of every 700 live births. Symptoms of the condition include mental retardation, eyelid folds characteristic of Mongolian people, short stature, stubby fingers, and abnormalities of various internal organs.

The risk of giving birth to a child with Down's syndrome is about 10 times greater in women over 40 than in younger women. This apparently is traceable to the age of the oocytes. In humans and other mammals, all of the oocytes a female will ever have are present at birth. Evidence indicates that older oocytes have an increased tendency to undergo nondisjunction or to produce structurally altered chromosomes, resulting in the production of abnormal gametes.

Nondisjunction also may involve the sex chromosomes. An individual with *Turner's syndrome,* for instance, has only one X chromosome; the second X or the Y chromosome is missing (Fig. 7.17). In this case, then, the affected person has only 45 chromosomes instead of the normal 46. Such an individual is genetically XO and has the external features of a female. All XO women, however, are sterile, and their reproductive systems resemble those of juvenile females. Moreover, these women are abnormally short in stature, have poorly developed breasts, and do not menstruate. The incidence of Turner's syndrome is slightly less than 1 in every 1000 live births.

Females also are born who have three X chromosomes (XXX). In this case, an XX ovum is fertilized by an X-bearing sperm, and the individual thus has 47 chromosomes. This extra chromosome condition, known as *trisomy-X,* often results in sterility, although some of the women are fertile. Many XXX females are mentally defective, and some have poorly developed sexual characteristics. The frequency of trisomy-X per live births in the general population is a little higher than that for Turner's syndrome.

An abnormal chromosome condition associated with males is *Klinefelter's syndrome.* This condition is genetically designated XXY and is estimated to occur in 1 in 500 live births. Presumably, the XXY state can result when an XX ovum is fertilized by a Y sperm or when an X ovum is fertilized by an XY sperm. Individuals with Klinefelter's syndrome have the appearance of males, but they are sterile, have enlarged breasts and other female characteristics, and are usually mentally retarded.

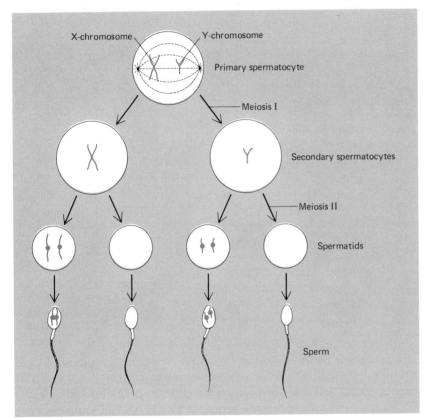

Figure 7.17 Nondisjunction of sex chromosomes in spermatogenesis in the human male (the 22 pairs of autosomes have been omitted). The abnormal sperm cells contain two X chromosomes, two Y chromosomes, or no sex chromosomes. If a Y-containing sperm fertilizes an XX ovum, the result is Klinefelter's syndrome; fertilization of a normal ovum by a sperm containing no sex chromosome results in Turner's syndrome (XO). If a normal ovum is fertilized by a YY sperm, the XYY-syndrome results. The diagram here illustrates nondisjunction in the second meiotic division. What would be the result of nondisjunction in the first meiotic division?

One other type of trisomy is the *XYY-syndrome*. It results when a normal X ovum is fertilized by a YY sperm. First observed among inmates in a prison in Scotland, the XYY condition produces males who are tall (averaging over 6 ft), below normal in intelligence, and often antisocial. Although several investigators have suggested that the antisocial behavior of XYY men is attributable to the extra Y chromosome, the evidence is inconclusive. Some XYY men appear to exhibit perfectly normal behavior.

In some cases, information concerning suspected chromosome abnormalities can be obtained before a child is born. This is possible using a technique called **amniocentesis** ("puncture of the amnion") (Fig. 7.18). The amnion is a fluid-filled sac that forms around the fetus developing within the uterus of a pregnant woman. The developing human fetus is suspended in and surrounded by fluid (amniotic fluid) within the amnion until birth. By inserting a syringe through the abdominal wall of the pregnant woman and into the

About one third of all human pregnancies spontaneously abort in the first 2 or 3 months, and most of the aborted embryos have the wrong chromosome number.

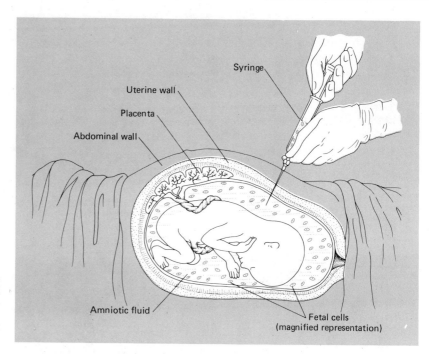

Figure 7.18 Amniocentesis ("puncture of the amnion") involves the withdrawal of amniotic fluid surrounding the developing fetus. (From Meeks, L., and Heit, P.: *Human Sexuality*. Philadelphia: Saunders College Publishing, 1982.)

amnion, some of the fluid can be withdrawn from the amnion. Cells sloughed off from the fetus are usually present in the fluid, and they can be examined microscopically for chromosome abnormalities. If the examination indicates any serious disorder, the parents may elect to terminate the pregnancy by abortion. Cells obtained from amniotic fluid may also be useful in revealing the sex of the fetus. The sex of a male fetus can be determined with virtually 100 percent confidence, but in the case of a female fetus, there is the chance that the cells withdrawn from the amnion are those of the mother.

Amniocentesis is not always advisable but may be recommended, for example, if a pregnant woman has previously given birth to a child with genetic defects or if she is having a child later in life. This latter point is important because, as mentioned, it has been shown that changes in the structure and in the number of chromosomes often result from the effects of aging.

REVIEW OF ESSENTIAL CONCEPTS

1. In the living world, cell growth and repair occur by the process of cell division called *mitosis*.
2. Mitosis involves the production of genetically identical *daughter cells*, all of which contain the diploid, or 2n, number of chromosomes.

3. Mitosis is separated into *prophase, metaphase, anaphase,* and *telophase.* A nondividing stage called *interphase* precedes each mitosis. During interphase, DNA is replicated, and the cell produces materials and organelles for its own structure and function.
4. In higher animals, gametes are produced by the process of cell division called *meiosis.* However, in plants gametes arise by mitosis.
5. Meiosis reduces the diploid number of chromosomes to the haploid, or *n*, number.
6. In higher animals, meiosis occurs only in the ovaries of the female and the testes of the male. In females the process of egg formation is called *oogenesis;* in males, sperm formation is called *spermatogenesis.*
7. In plants, meiosis usually gives rise to haploid *spores.*
8. Meiosis consists of two stages: meiosis I and meiosis II. Each stage is separated into prophase, metaphase, anaphase, and telophase. Meiosis I is always preceded by interphase.
9. *Crossing over,* which usually occurs in prophase I, involves the exchange of genetic material between homologous chromosomes. Crossing over results in *genetic recombination.*
10. Spermatogenesis results in the formation of four functional haploid sperm from each parent diploid cell (spermatogonium). Oogenesis produces one functional haploid ovum from each parent diploid cell (oogonium).
11. Abnormal cell division is a consequence of *nondisjunction,* in which chromosomes fail to separate properly. This may occur with the autosomes or the sex chromosomes.
12. Information concerning suspected chromosome abnormalities in the unborn can often be obtained by *amniocentesis.*

APPLYING THE CONCEPTS

1. Why does cell growth and repair occur by mitosis and not by meiosis?
2. Would mitosis be possible without an interphase? Why?
3. What is the functional relationship between centromeres and spindle fibers in cell division?
4. Differentiate between mitosis and cytokinesis.
5. Why is meiosis important?
6. What is gametogenesis, and what is its relationship to meiosis?
7. What differences are observed among plant and animal cells undergoing mitosis? Meiosis?
8. Diagram the chromosomal events of a cell ($2n = 4$) undergoing mitosis. Do the same for a cell ($2n = 4$) undergoing meiosis.
9. What is nondisjunction? What is its general effect?
10. What specific kind of information do you think could be obtained from amniocentesis?

the basis of heredity

THE ESSENTIAL OBJECTIVES

You have understood this chapter when you are able to:
1. Describe the self-reproduction of DNA.
2. Describe the formation, structure, and function of messenger RNA, transfer RNA, and ribosomal RNA.
3. Explain what is meant by the "genetic code."
4. Describe the processes of transcription and translation as applied to protein synthesis.
5. Discuss the possible causes and effects of gene mutations and chromosome mutations.
6. Discuss the contributions of Gregor Mendel to the science of genetics.
7. Explain the mendelian principle of segregation.
8. Determine the phenotypic and genotypic ratios from a given monohybrid cross.
9. Determine an unknown genotype by means of a test cross.
10. Determine the phenotypic and genotypic ratios from a given dihybrid cross.
11. Explain the mendelian principle of independent assortment.
12. Explain the inheritance of sex-linked traits in males and females.
13. Describe the phenomenon of incomplete dominance and discuss its application in sickle-cell anemia.
14. Illustrate inheritance of multiple alleles as revealed in the human ABO blood group system.
15. Discuss multiple gene inheritance, using an example.
16. Discuss the concept of linked genes and their effects on offspring phenotypic and genotypic ratios.
17. Discuss the significance of crossing over as it applies to the evolutionary process.

The genetic information specific for every living organism is contained within its nucleic acids. In accordance with its structure, each nucleic acid contributes to the synthesis of the cellular proteins unique to each organism. The passing of genetic information from parent to offspring conforms to specific principles that normally ensure the perpetuation of each kind of living thing.

PREVIEW OF ESSENTIAL TERMS

deoxyribonucleic acid (DNA) The hereditary material; two polynucleotide chains bonded together and wound into a double helix.

ribonucleic acid (RNA) A chain of nucleotides, usually single-stranded, that functions in protein synthesis; there are three major types of RNA.

transcription The copying of the nucleotide sequence of DNA into a complementary sequence of nucleotides in messenger RNA.

translation The sequencing of amino acids in a protein by transfer RNAs as directed by the nucleotide sequence of messenger RNA.

gene A specific sequence of nucleotides in DNA that determines the amino acid sequence of a polypeptide.

mutation An inheritable change in a gene; involves changes in individual nucleotides or nucleotide sequences in DNA.

allele One of the alternative forms of a gene; alleles occupy the same location on homologous chromosomes.

dominant Referring to an allele that masks or covers up the phenotypic effect of the allele with which it is paired.

recessive Referring to an allele that is masked or covered up by the phenotypic effect of the dominant allele.

phenotype The outward physical appearance of an organism.

genotype The kinds of genes present in an organism.

DNA: THE HEREDITARY MATERIAL

What is now known as deoxyribonucleic acid, or DNA, was first extracted from cell nuclei in 1869, but its complete structure was not determined until almost 85 years later. With the use of information accumulated by many scientists, this was accomplished by an American biologist, James Watson, and an English physicist, Francis Crick, in 1953. The structure of DNA, therefore, is referred to as the *Watson-Crick model.*

As described in Chapter 2, the structure of DNA consists of two strands of nucleotides, twisted into the shape of a double helix. The nitrogen bases along one strand are paired with complementary bases on the other strand (Fig. 8.1). This means that adenine (A) is always paired with thymine (T) and that cytosine (C) is always paired with guanine (G). The nucleotides of DNA may be sequenced in an almost infinite number of ways along any one of the strands, and the other strand is always complementary. As we shall discuss, it is the sequence of the nucleotides that explains how DNA carries its genetic information.

It was brought out in the discussion of mitosis and meiosis that prior to cell division there is chromosomal, i.e., DNA, duplication. This ensures that each cell formed receives its required amount of DNA. How does DNA accomplish its task of self-reproduction? Research has shown that the essence of the answer lies in the structure of the DNA molecule itself.

First of all, the two nucleotide strands separate or "unzip." Each of the separate strands now becomes a pattern for the formation of a new DNA strand (Fig. 8.2). The nucleotides needed for synthesis of the new strands are available in the cell. With the assistance of enzymes, these nucleotides are fitted into position according to the sequence of the complementary nucleotides on the "old" strands. This means that every cytosine (C) inserted into the new strand is paired with guanine (G) on the old strand, and vice versa. Similarly, every adenine (A) inserted on the new strand is paired with thymine (T) on the old strand. This procedure is somewhat like matching up pieces of a jigsaw puzzle—each piece has a definite place into which it fits—and the finished "puzzle" is two identical replicas of the original DNA molecule. The old strand serves as the "director" of the process, providing each new nucleotide with a specific location on the new strand and a specific nucleotide to bond to. Each molecule, then, is composed of one "old" strand of nucleotides bonded to one "new" strand of complementary nucleotides.

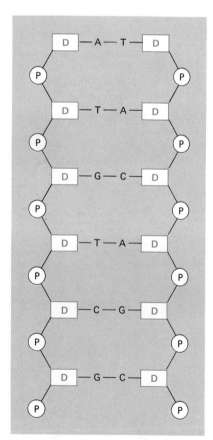

Figure 8.1 A section of the complementary strands of DNA. A = adenine; T = thymine; C = cytosine; G = guanine; D = deoxyribose; P = phosphate group.

Ribonucleic Acid (RNA)

> In all living organisms, the hereditary information is contained within a specific sequence of nucleotides in chromosomal DNA.

RIBONUCLEIC ACID (RNA)

After it was discovered that DNA was the carrier of hereditary information, there logically arose the problem concerning the mechanism by which this information could be used by the cell. The solution was found to involve the other major type of nucleic acid—RNA. Actually, there are three different types of RNA: *messenger RNA*, abbreviated *mRNA*; *transfer RNA*, or *tRNA*; and *ribosomal RNA*, or *rRNA*.

Let's look briefly at each of these and then observe the roles of all three types of RNA in conveying to the cell the information carried by DNA.

Messenger RNA. During interphase, after the chromosomes have duplicated, the cell must then synthesize the various enzymes and other proteins required for its proper maintenance and function. This, too, occurs in interphase, just prior to actual nuclear division (mitosis or meiosis). The first major step in the process involves the production of mRNA.

Essentially, a strand of mRNA is formed much as a new strand of DNA is formed during duplication. One strand of DNA serves as the pattern on which the mRNA strand is copied. In other words, DNA directs the synthesis of a complementary strand of mRNA. For example, if the sequence of nucleotide bases in a section of DNA were AGCA, the complementary section on the mRNA strand would

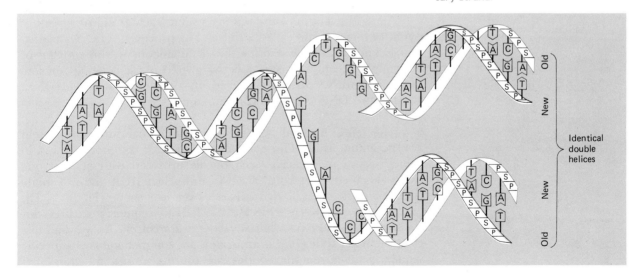

Figure 8.2 Double helix of DNA. The normal structure, on the left, shows the complementarity of base pairs: A always pairs with T; C always pairs with G. The two strands are thus "mirror images" of each other. To reproduce, the two strands unwind, as shown on the right. Each serves as a model or template for synthesis of its complementary strand.

> The essential function of RNA is to assemble the enzymes and structural proteins as directed by the information coded in DNA.

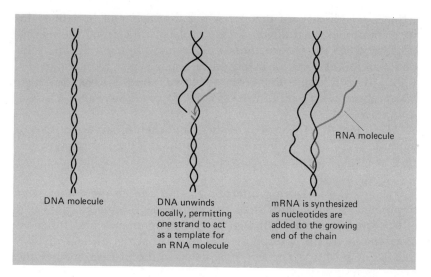

Figure 8.3 Synthesis of mRNA. One strand of DNA acts as a template for mRNA. After mRNA synthesis is complete, it unravels from the DNA and escapes from the nucleus into the cytoplasm. (Modified from Stahl, F. W.: *The Mechanics of Inheritance.* 2nd ed. Englewood Cliffs, N.J.: Prentice-Hall, 1969.)

be UCGU (recall that uracil is substituted for thymine when making RNA, and uracil pairs with adenine just as thymine pairs with adenine in DNA). The nucleotides for the synthesis of mRNA are found in the cell and, assisted by enzymes, are inserted along the mRNA strand according to the complementary nucleotide sequence of DNA. In effect, the genetic information contained in DNA has been copied by mRNA.

Along the length of the mRNA strand, each group of three nucleotides forms a unit called a ***codon***. The process of mRNA production is known as ***transcription*** and is analogous to the copying of information from a master sheet (Fig. 8.3). As it is being formed, the single mRNA strand is peeled off the double-stranded DNA, and the two DNA chains rewind into a double helix.

Transfer RNA. You will recall that the individual units of a protein are the amino acids. There are 20 amino acids commonly found in proteins, and in the cell, each one is provided with at least one tRNA. Each tRNA is synthesized from nuclear DNA and is a single chain of nucleotides folded into a clover leaf shape (Fig. 8.4). On one loop of the tRNA molecule is a specific sequence of three nucleotides that make up what is called an ***anticodon***. As we shall discuss, the nucleotides in this anticodon are complementary to specific sequences of three nucleotides, or codons, found in strands of mRNA.

Ribosomal RNA. The cellular organelles known as ribosomes are the sites where proteins are made in the cell. Much of the make-up of ribosomes consists of ribosomal RNA, which is also a single-stranded polynucleotide chain produced by transcription from DNA. The nucleolus synthesizes rRNA and assembles it into ribosomes by combining it with protein.

The real keys to the structural and functional uniqueness of any living organism are its proteins. These include the proteins that make up the actual physical structure of its cells, as well as the en-

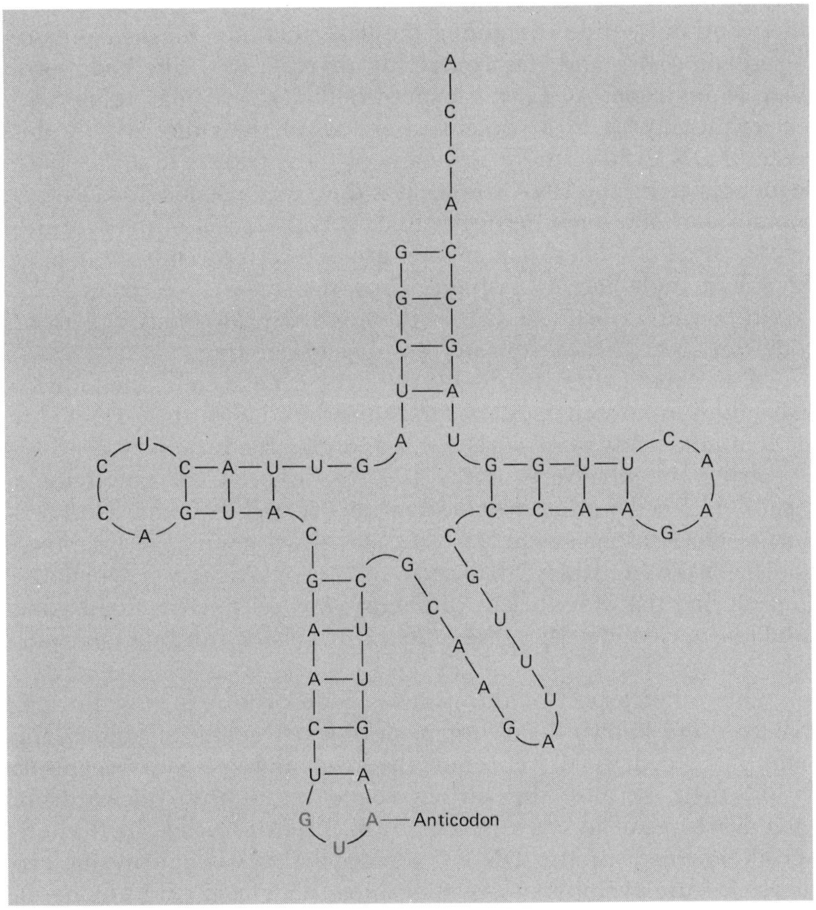

THE GENETIC CODE

Figure 8.4 Generalized structure of a transfer RNA molecule. The anticodon is complementary to a specific codon of messenger RNA. A = adenine; U = uracil; C = cytosine; G = guanine.

> The genetic codewords in DNA are transcribed into messenger RNA in the form of codons. Each codon is composed of three adjacent nucleotides that specify 1 of 20 amino acids.

zymes—the specialized proteins that catalyze the reactions of metabolism. Each different protein has its own unique sequence of amino acids, and every organism has proteins that are specific to it alone. But how would a cell "know" which amino acids it needs and in what sequences to arrange them to form its own specific proteins? This question was answered in 1961 with the discovery of what is now known as the **genetic code.**

Through the work of several investigators (including Francis Crick at Cambridge University, Severo Ochoa of New York University, and Marshall Nirenberg at the National Institutes of Health), it was determined that each amino acid had a "codeword" that consisted of a sequence of three nucleotides in the DNA strand. For example, such a sequence might include the following three nucleotides: one nucleotide containing the base adenine (A), one containing guanine (G), and one containing thymine (T). The codeword, then, is the triplet AGT, which specifies that a particular amino acid will eventually be in a protein whose synthesis is directed by this piece of DNA. Since there are many of these triplets in a variety of sequences along the DNA strand, it is therefore possible for DNA to contain the codewords necessary to construct a great number of different proteins. These codewords are transcribed into messenger RNA in complementary fashion. Thus, the triplet AGT from DNA would be transcribed as UCA on the mRNA strand. The triplets, or codewords, on mRNA are referred to as *codons.*

A few years after the discovery of the genetic code, the individual codons in messenger RNA (and thus the codewords in DNA) for all 20 amino acids were worked out and identified.

From the discovery of the DNA codewords, we now have a meaningful and workable definition for the unit of heredity carried on the chromosomes—the *gene.* Essentially, a gene is a certain sequence of DNA triplets that code for one particular polypeptide. Considering that the nucleus of one of your cells may contain some 5 billion nucleotide pairs, the amount of genetic information available is staggering.

One other aspect of the genetic "code of life" is most intriguing. In every known living organism on earth, plant or animal, the same DNA codewords represent the same amino acids. Organisms differ, then, because the various sequences of the codewords in their DNA result in the synthesis of different proteins. In the universal sameness of the DNA codewords, there is convincing evidence for the evolutionary relationships thought to exist among all living things.

> In all organisms except certain viruses, protein synthesis involves the replication of DNA, the transcription of RNA from DNA, and the translation of the RNA message into a protein.

PROTEIN SYNTHESIS

From the foregoing discussion it should be apparent that for any cell to make a protein calls for a rather vast array of cellular materials. The absolute essentials include DNA, mRNA, tRNA, ribosomes, several different enzymes, ATP, and the amino acids, which are found in the cytoplasm of the cell.

As stated, the synthesis of cellular proteins takes place during interphase, just prior to nuclear division. Since DNA is double-stranded, the initial event is the unwinding and separation of the two strands. One of the strands then becomes the pattern, or **template**, from which the complementary strand of mRNA is copied. When transcription is completed, the mRNA strand moves away from the DNA template and passes through the nuclear membrane into the cytoplasm. Back at the nucleus, the two strands of DNA rewind to form the double helix.

After entering the cytoplasm, ribosomes become attached to the mRNA strand and eventually form a row of ribosomes called a **polyribosome**, or polysome. The polyribosome-mRNA complex is where protein synthesis occurs. Dispersed throughout the cytoplasm are the various types of tRNAs, along with the amino acids. With the aid of certain enzymes, and energy supplied by ATP, each tRNA molecule becomes attached to its specific amino acid. Transfer RNAs with attached amino acids are said to be *charged* tRNAs. Recall that one of the loops on the tRNA strand contains a three-nucleotide codeword called an *anticodon*. *The tRNA anticodon is complementary to the mRNA codon.* For example, if the tRNA anticodon is GAA, the complementary codon on the mRNA strand is CUU. This is simply another case of matching complementary bases. Since each mRNA codon calls for a certain amino acid, then the appropriate tRNA, with its complementary anticodon, carries an amino acid specified by that codon. Through complementary base pairing, each tRNA can bring its attached amino acid to the correct location on the mRNA strand. The result is a sequence of amino acids that form the protein or polypeptide unique to that organism.

The tRNA carrying its attached amino acid moves to the ribosome, where the mRNA is anchored. The anticodon of the tRNA then attaches temporarily to the appropriate codon on the mRNA strand. The next tRNA carrying its specific amino acid moves to the mRNA strand and attaches to the next codon. A specific enzyme forms a peptide bond between the two amino acids so that the first amino acid loses its attachment with its tRNA and is now held as a dipeptide by the amino acid still attached to the second tRNA. The first tRNA molecule then separates from the mRNA strand. When the third amino acid is brought to the mRNA and bonded to the second amino acid, the tripeptide formed is held by the third tRNA and the second tRNA molecule separates from the mRNA strand. This continues for the subsequent amino acids that are carried to the mRNA by their respective tRNAs. Once a certain point on the mRNA strand is reached, protein synthesis stops. This is accomplished by what is called a *termination codon*—UGA, for example—which is a genetic signal for terminating the synthesis of a particular polypeptide. The polypeptide (formed of linked amino acids) is then released into the cytoplasm of the cell. The tRNAs are reusable and move out into the cytoplasm to await another round of protein synthesis. The mRNA strand may persist for weeks or months, after which it is broken down into its component nucleotides by a specific enzyme (Fig. 8.5).

The synthesizing of proteins along the strand of mRNA is called *translation,* i.e., the transcribed information coded in mRNA is

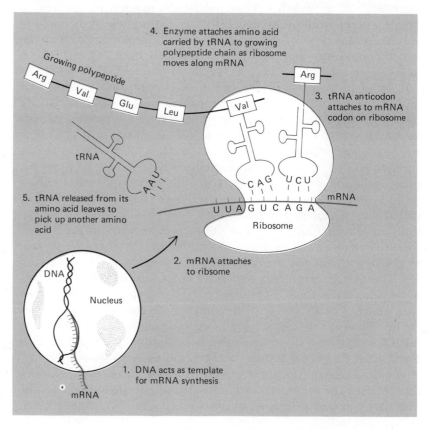

Figure 8.5 Basic scheme of protein synthesis. Within the nucleus, one of the strands of DNA synthesizes a complementary strand of messenger RNA. The mRNA passes across the nuclear membrane into the cytoplasm, where it becomes associated with ribosomes. Each type of transfer RNA picks up a specific amino acid and carries it to a ribosome that is moving along the mRNA strand. The anticodon of each tRNA bonds with its complementary codon on the strand of mRNA. One by one, the amino acids are brought to the mRNA strand and linked by peptide bonds. After its amino acid is linked to the growing polypeptide chain, each tRNA separates from the mRNA strand and moves out into the cytoplasm to await another round of protein synthesis. When synthesis is completed, the polypeptide is released into the cytoplasm of the cell.

> A mutation is a change in the structure of a gene, and the vast majority of mutations are harmful. Only rarely is a mutation beneficial to an organism.

"translated" into a specific kind of protein. It is the particular sequence of the codons of mRNA, then, that determines the sequence of the amino acids in a protein. But the codons of mRNA are copied from DNA; thus *DNA is the real determinant of protein structure.* Therefore, when a cell divides, it is really the unique nucleotide sequences of DNA (the hereditary information) that are passed by the chromosomes to future cells and to future generations.

MUTATIONS: THE GENETIC ERRORS

Ordinarily, the genetic machinery of the cell functions faithfully and accurately. There are instances, however, when the machinery goes awry as a result of genetic errors, or **mutations**. Various environmental influences are known to cause mutations—ultraviolet radiation, x-rays, and an assortment of chemicals are well-documented examples. In addition, for reasons unknown, some mutations occur spontaneously.

All types of mutations involve alterations in the nucleotides of DNA. Such alterations include the *addition* of a single nucleotide somewhere along the DNA strand, the *loss* of a nucleotide, and the *substitution* of one nucleotide for another. From the discussion of protein synthesis, you can see how many of these mutations could result in a change in the amino acid sequence of a protein. In many cases, such a change may have minor significance, but there are other cases in which the alteration of a protein has deadly consequences.

Such consequences may result, for instance, from a change in one of the body's most vital proteins—hemoglobin. Hemoglobin is the red, oxygen-carrying pigment found in red blood cells. Various inherited blood diseases involve mutations that change the amino acids or their sequence in the hemoglobin molecule. In some cases, a change in only one amino acid severely interferes with the ability of hemoglobin to combine with oxygen. One disease in particular in which this occurs is *sickle-cell anemia* (Fig. 8.6). This malady is characterized by severe anemia and sickle-shaped red cells that not only have poor oxygen-carrying capacity but also tend to clog up small blood vessels.

Sickle-cell hemoglobin is structurally identical to normal hemoglobin except for the substitution of one different amino acid. As a result of a mutation that changes one nucleotide in a segment of DNA, there is a single incorrect amino acid inserted among the 574 that make up the hemoglobin molecule. As a consequence of respi-

The Basis of Heredity

> Modern genetics received its greatest impetus from the work of Gregor Mendel. His now classic paper on inheritance was published in 1866.

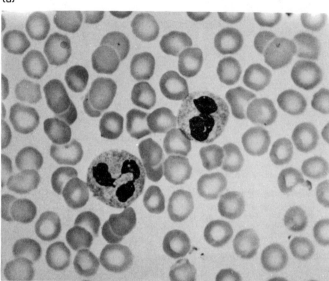
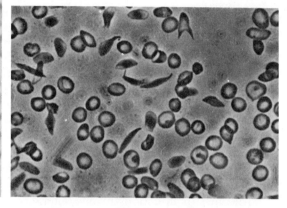

Figure 8.6 Photographs of (a) normal human blood, showing two white blood cells surrounded by red blood cells, and (b) blood of individual with sickle-cell anemia. Note the distorted shapes of many of the red blood cells. (From Silverstein, A.: *The Biological Sciences*. New York: Holt, Rinehart and Winston, 1974, pp. 276–277.)

ratory and circulatory malfunction, sickle-cell anemia is usually fatal—and all because of one amino acid.

Other types of genetic errors include **chromosome mutations,** in which extended segments of DNA are involved (Fig. 8.7). These mutations usually result from breaks in a chromosome; the broken segments may be lost entirely, they may rejoin the chromosome in an inverted position, or some segments may become attached to other chromosomes. Any of these chromosome alterations could change the nucleotide sequence in DNA. This, of course, changes the genetic information in the cell and may adversely affect the synthesis of cellular proteins.

SOME ESSENTIALS OF GENETICS

One of the more obvious attributes of living things is their capacity to reproduce others of their own kind. Such a capacity involves the passing of hereditary traits from the parents to their offspring. We can therefore predict within certain limits which traits to expect in the offspring if we know something about the genetic make-up of the parents. The science of *genetics* attempts to discover and explain the mechanisms that account for hereditary relationships and to determine how the genetic make-up affects the entire organism.

The individual whose work marks the beginning of modern genetics was an Austrian monk named Gregor Mendel (1822–1884) (Fig. 8.8). During the latter part of the nineteenth century, Mendel

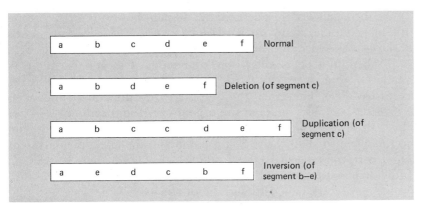

Figure 8.7 Diagram illustrating the types of mutations that involve changes in the structure of the chromosome.

Figure 8.8 Gregor Mendel, the "father of modern genetics." (From Silverstein, A.: *The Biological Sciences*. New York, Holt, Rinehart and Winston, 1974, p. 391.)

Figure 8.9 Alleles are found at the same location on homologous chromosomes.

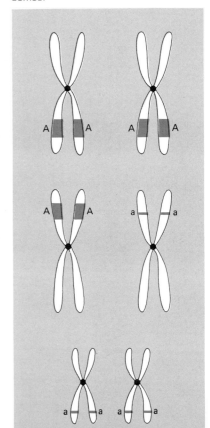

experimented with a variety of pea plants, crossbreeding them and meticulously recording his findings. He was the first to determine that hereditary characteristics are carried and transmitted by what he called "factors"—known today as *genes*. Moreover, Mendel suggested that these factors occur in pairs in the offspring—one factor contributed by one parent and one factor by the other parent. In other words, a new pea plant, for example, receives one factor for each of its characteristics (color, height, and so on) from the male structure and one factor from the female structure.

There are alternative forms of a gene, each of which is called an **allele** (Gk. "another form"). Alleles are located at the same position on homologous chromosomes, one allele on one chromosome and one allele on the other chromosome (Fig. 8.9). In meiosis, then, the alleles are separated from each other at anaphase I. If the two alleles for a given trait are the same, the organism is said to be **homozygous** for that trait; if the two alleles are different, the organism is then **heterozygous** for the trait. What is meant by "same" and "different" alleles? First of all, an allele may be designated as **dominant** or **recessive**. In genetics, the dominant allele is indicated by a *capital* letter, such as *A;* the recessive allele is indicated by a *small* letter, in this example, *a*. Therefore, there are three possible combinations of two alleles in a pair—*AA*, *Aa*, and *aa*. The *AA* alleles are the same,

The Basis of Heredity

> Although one organism homozygous dominant for a genetic trait and another heterozygous have different genotypes, they may have the same phenotype.

as are the *aa* alleles; the alleles in the *Aa* pair, however, are different. In this example, then, *AA* is designated *homozygous dominant,* *aa* is *homozygous recessive,* and *Aa* is *heterozygous.*

When an allele is said to be "dominant," it means that the trait carried by that allele "covers up," or "masks," the trait carried by the recessive allele. Therefore, the trait carried by the dominant allele is expressed, or observable, in the organism. As an example, consider the trait of hair color, in which the allele for dark hair is dominant to the allele for light hair. Being dominant, the allele for dark hair color is *D;* the recessive allele for light hair color is then *d*. Thus, an individual with the alleles *Dd,* i.e., heterozygous for hair color, would have dark hair, although he or she carries the allele for light hair color. An individual, of course, may be homozygous for dark hair, in which case the alleles would be *DD*. For the recessive trait—light hair color—to be expressed or observable, *both* alleles must be recessive. A light-haired individual, then, would possess the alleles *dd*. The outward appearance of a flower, an animal, or another human being is the result of its genetic make-up. The outward appearance of an organism, i.e., what it looks like, is referred to as its **phenotype.** On the other hand, the genetic make-up of an organism, i.e., the kinds of genes it has, is called its **genotype.** The gene (allele) pairs *DD, Dd,* and *dd,* therefore, are examples of genotypes. The outward expressions resulting from these genotypes, i.e., either dark hair or light hair, are the phenotypes.

Although we may know the phenotype of an organism simply by looking at it, we do not always know its genotype. In observing an individual with dark hair, we know that he or she has to possess the dominant allele *D*. What is *not* known simply by looking is whether the individual is homozygous dominant (*DD*) for dark hair or heterozygous (*Dd*). It is apparent, then, that two individuals could have the same phenotype but different genotypes.

With light hair, or other recessive traits, the situation is quite different. If an organism displays the recessive trait, we automatically know its genotype. With light hair, the genotype has to be *dd*. Obviously, if a dominant allele were present, one's hair would be dark, not light.

With Mendel's discovery that hereditary traits are carried by pairs of factors (genes), it later became apparent that these factors separate from each other during meiosis. This meant that one allele of a pair must be located on one chromosome and that the other allele must be on the homologous chromosome. Therefore, if a genotype were *AA,* for instance, one allele (*A*) would be on one chro-

Mendel's first law, or principle of segregation, states that genes usually occur in pairs and that members of each pair eventually pass into different gametes.

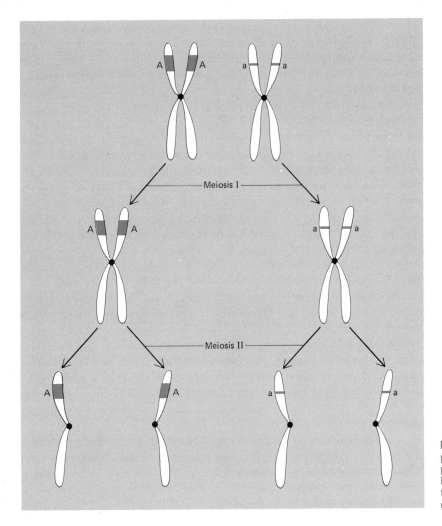

Figure 8.10 Illustration of Mendel's principle of segregation using the allele pair *Aa*. At the end of meiosis, each allele passes into a different gamete, so that each gamete receives only one of each kind of allele.

mosome, and the other allele (*A*) would be in the same location on the homologous chromosome. The same is true, of course, for the *Aa* and *aa* combinations. At the end of meiosis, then, each gamete would contain one allele of a pair. The separation of a pair of alleles into different gametes is known as Mendel's *first law*, or the **principle of segregation** (Fig. 8.10). Using the allele pair *Aa* as an example, the *A* allele and the *a* allele separate at meiosis and eventually pass into different gametes. Therefore, each gamete receives only one of each kind of allele.

> To ascertain the kinds of offspring from any type of genetic cross, the initial task is to determine the types of gametes the parent organisms can produce.

MONOHYBRID CROSS

By employing the genetic principles discovered by Mendel, it is possible to set up various crosses between parent organisms and then determine the types of offspring expected from such matings. The simplest of these involves only one genetic trait and is called a *monohybrid cross*.

To illustrate, let's consider the trait of color in a certain species of flowers. The allele R = red color, and r = white, with red dominant to white. One type of monohybrid cross might involve two homozygotes, such as a mating of parents with the genotypes RR and rr (the parent generation is designated P). In any type of genetic cross, the first requirement is to determine the kinds of gametes each parent can produce during meiosis. In this example, the RR parent can produce gametes containing only an R allele; the rr parent can produce only r gametes. At fertilization, all of the offspring (designated as the *first filial*, or F_1, *generation*) will have the genotype Rr, i.e., they will all be heterozygotes. Although they carry the allele for white color (r), phenotypically all of the offspring are red.

Suppose now that we used two members of the F_1 generation as parents, i.e., mated an Rr flower with another Rr flower. What would the members of the *second filial*, or F_2, *generation* be like phenotypically and genotypically? The first task, as always, is to determine the types of gametes the parent organisms can produce. In this case, each parent can produce an R gamete and an r gamete. In determining all the possible ways the gametes could combine to form the offspring, it is often convenient to use a **Punnett square,** as shown in Figure 8.11.

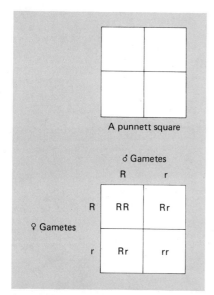

Figure 8.11 A Punnett square is used in "crossing" parental gametes to determine the possible types of offspring. This is an example of a monohybrid cross, since only one genetic trait (color) is involved.

The possible offspring genotypes resulting from a cross between the two heterozygotes are thus the following: 1 RR, 2 Rr, and 1 rr. This **genotypic ratio** is designated 1:2:1, i.e., one homozygous dominant (RR), two heterozygotes (Rr), and one homozygous recessive (rr). This also could be expressed as $\frac{1}{4}$ RR, $\frac{1}{2}$ Rr, and $\frac{1}{4}$ rr. Phenotypically, three-fourths of the F_2 offspring are red flowers and one-fourth are white flowers—or there are three dominants (red) to one recessive (white). This is referred to as the **phenotypic ratio** and is written 3:1.

TEST CROSS

If an organism displays a dominant trait, such as dark hair or red color, the organism may be homozygous for that trait, or it may be heterozygous. Is it possible to determine which is the actual genotype? Mendel asked the same question and, in an attempt to answer

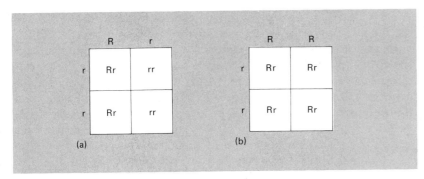

Figure 8.12 A test cross involves crossing an organism displaying a dominant trait (red) with a homozygous recessive individual (white). The test cross in (a) produces white offspring, indicating that the organism displaying the dominant trait is heterozygous. In (b) the organism displaying the dominant trait must be homozygous, since no white offspring are produced.

it, performed an experiment known as a *test cross* (Fig. 8.12). This involves crossing an organism displaying a dominant trait with a homozygous recessive individual. For example, a red flower may have the genotype *RR* or *Rr*. By crossing each of these with a homozygous recessive (*rr*), or white, flower, we would obtain the results shown in Figure 8.12. Crossing the *Rr* flower with the homozygous recessive (*rr*) gives a phenotypic ratio of two red to two white, or a 1:1 ratio. The fact that the test cross produces *white* flowers as well as red ones indicates that the parent *had* to be heterozygous (*Rr*). On the other hand, test crossing the homozygous dominant (*RR*) flower produces offspring that are all red. In other words, if no white flowers are produced from this test cross, the genotype of the red parent *has* to be *RR*.

A mating between organisms in which *two* different inheritable traits are involved is called a *dihybrid cross* (Fig. 8.13). Although there are a few new wrinkles to consider, the same essential principles apply to dihybrid crosses as to monohybrid crosses. Again, it was the work of Mendel that explained the basic mechanisms governing the inheritance of these traits.

Since we have been working with the trait of color in flowers, let's use these same flowers and add a second trait—height. In the following examples, then, tall (*T*) will be dominant to short (*t*). Thus, there are red (*RR*, *Rr*) and white (*rr*) flowers that are either tall (*TT*, *Tt*) or short (*tt*). What are the possible combinations of alleles for any one flower? Since each trait—color and height—is determined by a pair of alleles, the possible combinations would be as follows: *RRTT, RRTt, RRtt, RrTT, RrTt, Rrtt, rrTT, rrTt,* and *rrtt*. These allele combinations, of course, represent the genotypes, each of which determines a particular phenotype.

DIHYBRID CROSS

	♂ Gametes			
	RT	Rt	rT	rt
RT	RRTT	RRTt	RrTT	RrTt
Rt	RRTt	RRtt	RrTt	Rrtt
♀ Gametes rT	RrTT	RrTt	rrTT	rrTt
rt	RrTt	Rrtt	rrTt	rrtt

Figure 8.13 A dihybrid cross involving two organisms heterozygous for both traits. The phenotypic ratio of such a cross is 9:3:3:1.

> Mendel's second law, or principle of independent assortment, states that in a cross involving two or more pairs of genes the members of one pair segregate or separate independently of any other pair.

To illustrate one type of dihybrid cross, let's start as we did with the monohybrid cross, using the homozygous dominant (*RRTT*) and homozygous recessive (*rrtt*) plants as parents. At the end of meiosis, the *RRTT* parent will produce gametes containing the *RT* alleles; the *rrtt* parent produces *rt* gametes. At fertilization, the only possible genotype of the F_1 offspring is *RrTt*, i.e., heterozygous for both traits. Phenotypically, the offspring would be red and tall, as the dominant alleles would dictate. Note that in a dihybrid cross one allele for each trait is present in each gamete.

Using the *RrTt* offspring of the F_1 as parents, what could we expect in the F_2 generation? First, each F_1 parent could produce four different kinds of gametes: *RT, Rt, rT,* and *rt*. These can be set up on a Punnett square, and the possible F_2 offspring can be determined as shown in Figure 8.13.

In considering flower color only, when we count the number of red flowers in the F_2 and then the number of white flowers, there are 12 red and 4 white—a phenotypic ratio of 3:1. This is the same ratio obtained using heterozygous parents in a monohybrid cross. Notice, too, that there are 12 tall F_2 offspring and 4 that are short. Again, the phenotypic ratio is 3:1. Moreover, the principles of dominance and recessiveness still prevail. Therefore, there is nothing in this dihybrid cross that conflicts with the principles or results obtained from a monohybrid cross.

Although the same basic principles apply in both types of crosses, there is an interesting contrast in the dihybrid cross. In the F_1 generation (in this case, the parents), the red and tall alleles and the white and short alleles are all combined in one flower (*RrTt*). In the F_2 generation, however, these alleles are inherited in different combinations, as if each allele were totally independent of any other allele. This event was noted by Mendel, who then formulated his second law, known as the ***principle of independent assortment*** (Fig. 8.14). Essentially, this means that when one pair of alleles separates in meiosis, it is not affected by how another, different pair of alleles separates. For example, when the alleles *R* and *r* separate and move toward opposite poles of the cell, their direction of movement is not affected by the direction of movement of the *T* and *t* alleles.

By referring to Figure 8.13, it can be seen that the phenotypic ratio of the F_2 offspring is 9:3:3:1. There are nine offspring expressing the two dominant traits (red and tall), three offspring that express one combination of dominant and recessive traits (red and short), three offspring that express the other combination (white and tall), and only one that expresses both recessive traits (white and

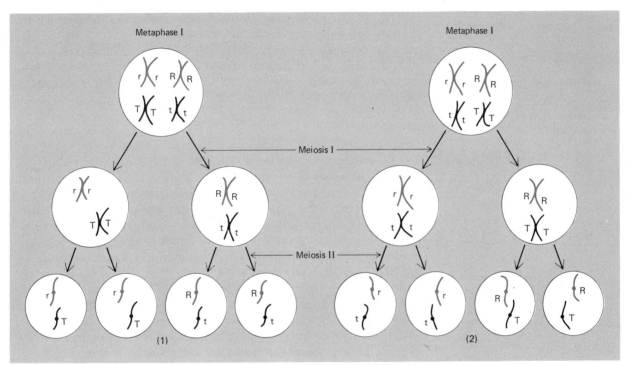

Figure 8.14 Diagrams illustrating Mendel's principle of independent assortment of chromosomes. The two sets of chromosomes may line up at metaphase I, as shown in (1) or as shown in (2). These chance chromosomal arrangements result in the formation of two different sets of reproductive cells. When the chromosomes with the R and r alleles separate from each other at anaphase I, their direction of movement is not affected by the movement of the chromosomes with the T and t alleles, i.e., each pair of homologous chromosomes separates independently of the other pair of homologous chromosomes. The number of genetically different reproductive cells possible at the end of meiosis can be determined by the expression 2^n, where n = the haploid number of chromosomes. In the diagrams here, there are four chromosomes, so the haploid number is 2. Thus, the number of genetically different reproductive cells possible is $2^2 = 4$. In an organism with eight chromosomes, the haploid number is 4. Therefore, there are $2^4 = 2 \times 2 \times 2 \times 2 = 16$ different reproductive cells that can be produced. In human beings with 46 chromosomes, $n = 23$; thus the number of genetically different gametes possible is 2^{23}, or 8,388,608. Actually, if crossing over is considered, the number of genetically different reproductive cells possible is much higher than the figures given here.

> In human beings, the X chromosome carries thousands of genes; in contrast, the Y chromosome carries but a few genes, most of which are concerned with male sexual development.

short). Such a ratio is the result of any dihybrid cross in which the parents are heterozygous for both traits.

To make sure you can work with a simple dihybrid cross, set up a Punnett square, if necessary, and determine the phenotypic ratio of the F_1 generation for the following crosses: (1) $RRTt \times Rrtt$; (2) $RrTt \times rrTt$; and (3) $rrTT \times RrTT$.

SEX LINKAGE

We have stated that the normal diploid number of chromosomes for all human beings is 46, or 23 pairs. More accurately, there are 22 pairs of body chromosomes, or *autosomes,* and one pair of *sex chromosomes.* Most of the genes responsible for one's physical make-up—height, shape, hair color, and so forth—are carried on the autosomes. But as stated earlier, whether one is male or female is determined by the one pair of sex chromosomes, designated as either XY or XX. Human males have an X and a Y chromosome (XY), and human females have two X chromosomes (XX). In female offspring, one X chromosome is contributed by the mother and one by the father. In males, only the mother contributes the X chromosome, and the father contributes the Y chromosome. It is the male parent, then, who determines the sex of the offspring in human beings.

Associated with the X chromosome are certain genetic traits—many of them undesirable or harmful—that more frequently affect males than they do females. The reason for this will become apparent shortly. A genetic trait carried by an allele on the X chromosome is said to be **sex-linked,** or **X-linked** (the Y chromosome carries very little genetic information). One such sex-linked trait is *red-green color blindness,* a form of color blindness in which there is difficulty in distinguishing red, yellow, and green. Color blindness is caused by a recessive allele, designated c; the dominant allele for normal color vision is then C. (In genetics, the alleles for sex-linked traits usually appear as superscripts next to the X chromosome, as shown in the following sentences.) Since a female has two X chromosomes, she may have the dominant allele on one of her X chromosomes (X^C) and the recessive allele on the other (X^c) and still have normal color vision. The reason, of course, is that the dominant allele (C) "masks" the effect of the recessive allele (c). Although having normal color vision, women who have the dominant allele on one X chromosome and the recessive allele on the other X chromosome are *carriers* of the color blindness trait. To be color blind, a woman would have to carry recessive alleles on *both* X chromosomes (X^cX^c).

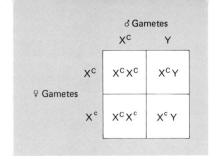

Figure 8.15 A mating between a female who is a carrier of the color blindness trait ($X^C X^c$) and a male with normal vision ($X^C Y$). What percentage of the offspring have normal color vision? What percentage of the male offspring are color blind? What percentage of the female offspring have normal color vision?

In human males, the story of the X chromosome takes on more unfortunate twists. In the case of color vision, the X chromosome of the male may carry the allele for normal color vision (*C*) or the allele for color blindness (*c*). There is no corresponding allele on the Y chromosome. If he has the allele for color blindness ($X^c Y$), he is then color blind. The lack of a second X chromosome in males makes them more likely than females to be afflicted with color blindness. In the following discussion, we will consider the mechanism of inheritance of sex-linked traits and illustrate why these traits are more frequently found in men.

The three possible genotypes associated with color vision in females are $X^C X^C$ (normal vision, noncarrier), $X^C X^c$ (normal vision, carrier of color blindness allele), and $X^c X^c$ (color blind). In males there are only two possible genotypes: $X^C Y$ (normal vision) and $X^c Y$ (color blind). In determining the possible offspring of matings between these males and females, the same principles apply as for a monohybrid cross. The first step is to determine the possible kinds of gametes that each sex can produce. Meiosis (oogenesis) in the female produces haploid egg cells, each having a single X chromosome. In the male, meiosis (spermatogenesis) produces haploid sperm cells, half of which contain an X chromosome and half of which contain a Y chromosome. Using this information, let's consider a cross between a female who is a carrier of the color blindness allele and a male who has normal color vision.

The genotype of the female in this cross would be $X^C X^c$ and that of the male would be $X^C Y$. At meiosis, the female could produce either an X^C egg or an X^c egg; in the male, one-half the total number of gametes would be X^C sperm, and one-half would be Y sperm. As we did previously, the gametes are aligned horizontally and vertically along a Punnett square and then crossed to determine the possible types of offspring. This has been done for you in Figure 8.15. All of the females from this cross have normal vision, although one-half of them are carriers of the color blindness allele. Only half of the males, however, have normal vision. As you can see, the color blind males inherited the recessive allele from their mother. Suppose that the mother had been color blind instead of a carrier. In working this out, you will find that there are still no color blind daughters produced, although *all* of them are carriers. Having inherited the dominant normal color vision allele from their father, the daughters cannot be color blind. On the other hand, *all* of the sons are color blind because both the mother's X chromosomes carry the color blind allele.

	♂ Gametes	
♀ Gametes	X^H	Y
X^H	$X^H X^H$	$X^H Y$
X^h	$X^H X^h$	$X^h Y$

Figure 8.16 A mating between a female who is a carrier of the allele for hemophilia and a normal male. What percentage of the offspring are hemophiliacs?

Color blindness affects about 1 male in 12, whereas only 1 female in about 250 is affected. For a woman to be color blind, not only must her father be color blind ($X^c Y$), but also her mother must be either color blind ($X^c X^c$) or a carrier ($X^C X^c$). Even if her mother is a carrier, a woman whose father is color blind still has a 50:50 chance of having normal color vision.

There are, of course, several other examples of sex-linked inheritance, one of the most serious of which is bleeder's disease, or *hemophilia*. This malady is characterized by the inability of the blood to clot. In normal individuals, blood escaping from a cut or wound will clot within minutes, thereby preventing excessive blood loss. Hemophiliacs, however, lack a substance in the blood necessary for normal clotting and may bleed for hours. Like color blindness, hemophilia results from the presence of a recessive allele on the X chromosome. The allele for normal blood clotting (*H*) is dominant to the recessive allele for hemophilia (*h*). As you can see in Figure 8.16, a cross between a normal male and a female carrier of the disease obeys the same principles of inheritance as those involving color vision. Here again, it is rare for a woman to be afflicted with hemophilia because her father would have to have the disease, and her mother would also have to have it or be a carrier. Hemophilia occurs in only 1 female in about 100 million but is estimated to occur in 1 male in 10,000.

INCOMPLETE DOMINANCE

Many organisms—both plant and animal—display shades of color that appear to be a blending of the colors of their parents. Crossing a particular species of red flower with a white one, for example, sometimes produces F_1 offspring that are pink. In this instance the *R* (red) allele is not totally dominant, nor is the *r* (white) allele totally recessive. Instead, there is a partial expression of *both* alleles in the offspring. This phenomenon is referred to as ***incomplete dominance***, or intermediate inheritance (Fig. 8.17). In a monohybrid cross between heterozygous parents—such as *Rr* × *Rr*—*complete* dominance results in a phenotypic ratio in the F_1 of three red to one white. The genotypic ratio is 1*RR*:2*Rr*:1*rr*. In the case of *incomplete* dominance, the two *Rr* flowers would not be red, but pink. Obviously, this changes the phenotypic ratio from three red to one white, to one red, two pink, and one white.

Earlier, we referred to the human genetic disease sickle-cell anemia. Incomplete dominance is also involved in the inheritance of this

Although lethal when homozygous, the sickle-cell gene provides a resistance to malaria among some African blacks who are heterozygous for the gene.

condition. The allele for normal hemoglobin is designated *A*, and the allele for sickle-cell hemoglobin is *S* (Fig. 8.18). Individuals with the homozygous *AA* genotype have normal hemoglobin. In those with the *SS* genotype, all the red cells are sickled, and such individuals therefore suffer from the disease. Anemia is so severe in these individuals that they usually die in their early years. Those who are heterozygous *AS* have *both* normal hemoglobin and sickle-cell hemoglobin. Although having the sickle-cell trait, the heterozygous individual does not have the disease and often leads a normal life. Therefore, neither the *A* allele nor the *S* allele is completely dominant; the heterozygote possesses equivalent amounts of both normal and sickle-cell hemoglobin, which apparently is sufficient for adequate oxygen transport.

It has been estimated that about 8.5 percent of the black population of North America carry the sickle-cell trait, whereas the incidence may be as high as 20 percent among the blacks of central and western Africa. Why should the incidence of the sickle-cell trait be

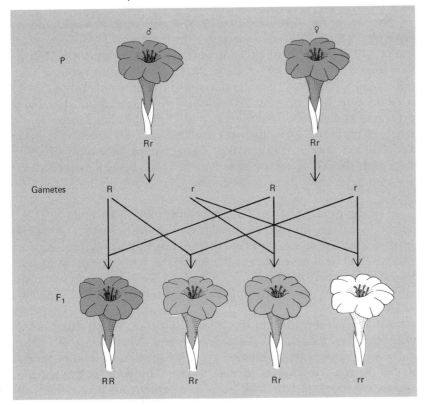

Figure 8.17 Incomplete dominance (*R* = red, *r* = white). Although the heterozygous parent flowers are red, the two heterozygous offspring are pink. The *R* allele in this instance is not totally dominant, nor is the *r* allele totally recessive.

Figure 8.18 Mating between individuals who carry the alleles for both normal hemoglobin (*A*) and sickle-cell hemoglobin (*S*). The probability that any of their offspring will have sickle-cell anemia (*SS* genotype) is one in four. Since neither the *A* nor the *S* allele is dominant, offspring with the *AS* genotype may lead normal lives.

so much greater in Africa than in North America? It has been found that individuals heterozygous for the sickle cell gene have a greater than normal resistance to malaria, a disease common in many parts of Africa. The heterozygous condition thus confers an adaptive advantage on individuals who are continually exposed to the disease. In North America, malaria is not a common disease, and the heterozygous condition affords no particular adaptive advantage.

MULTIPLE ALLELES

In our discussion thus far, we have considered only two possible forms of the alleles responsible for a certain trait. The R and r alleles in flowers are examples. In some populations, however, more than two alleles can be associated with the inheritance of a given trait. Still, no organism receives more than two of these alleles. As we have seen, the alleles may be the same, in which case the organism is homozygous for a trait, or the alleles may be different, and the organism is then heterozygous.

One of the best-known examples of **multiple alleles** is the human ABO blood group system. There are three important alleles in this system—A, B, and O—any two of which determine a person's blood type (Fig. 8.19). Only *two* of these alleles can be found in any one individual. Such allele pairs (genotypes) in man are as follows: AA or AO, BB or BO, AB, and OO. Both alleles A and B are dominant to O, but when found together, A and B exhibit incomplete dominance. Therefore, persons with the AA or AO genotype have type A blood; those with the BB or BO genotype have type B blood; those with the AB genotype have type AB blood (since neither allele is dominant); and persons with the OO genotype have type O blood.

The designation "A" or "B" represents the presence of a specific *antigen* found on the surface of the red blood cells; type A blood contains the A antigen, and type B contains the B antigen. Persons with type O blood have neither the A nor the B antigen on their red cells, whereas persons with type AB blood have both antigens. An antigen is a substance that stimulates the production of specific antibodies.

An *antibody* is a protein that reacts with the antigen that caused its production, usually attacking the antigen if it is harmful to the body. Associated with each blood type are specific antibodies found in the plasma. Because of certain antigen-antibody reactions, the transfusing of blood from one person to another must be done with caution. These reactions will be discussed in Chapter 17.

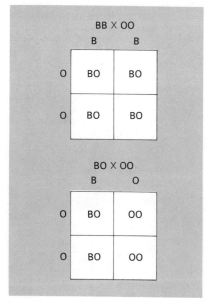

Figure 8.19 Matings between parents with blood types B and O. Parents with the genotypes BB and OO produce offspring with blood type B only. Parents with the BO and OO genotypes produce offspring with blood types B and O.

> The value of the human ABO blood group system in paternity suits lies in its use in determining who is *not* the father rather than who is.

The ABO blood group system sometimes proves valuable in paternity suits. However, its value lies not in determining who *is* the father, but who is *not*. By applying the same mendelian principles we have discussed, the inheritance of blood types can be determined. For example, assume that a man with type AB blood was involved in a paternity suit, in which it was claimed that he fathered a child with type O blood. Is the claim genetically possible? Since the A or B allele contributed by the father is dominant to the O allele, it would be impossible for the child to be type O, no matter what the genotype of the mother might be. Although this proves that a man with type AB blood could *not* have been the father, it could not prove who *is* the father, since any number of men would have the appropriate blood type to father a child with type O blood.

As another example, what blood types could be expected and what types would not be possible in a mating between parents who are types B and O? The possible genotypes of the parents are BB or BO and OO. By referring to the Punnett squares in Figure 8.19, you can see that the only possible blood types of the offspring are B and O. Therefore, these children could not have blood types A or AB.

Multiple allelism is also responsible for the various kinds of abnormal hemoglobins evident in certain blood diseases. As in sickle-cell hemoglobin, other variant hemoglobins usually differ from normal hemoglobin by the substitution of one amino acid.

MULTIPLE GENE INHERITANCE

Although many genetic traits depend upon a single pair of alleles on homologous chromosomes, there are traits that result from the contribution of several or many genes. The alleles responsible for the inheritance of these traits are found at several different locations on the chromosomes. Some of the human traits determined by multiple genes include height, weight, skin color, intelligence, and eye color.

In the case of skin color, for example, all persons except albinos possess varying amounts of a brown skin pigment called *melanin*. A light-skinned individual has mostly non–melanin-producing alleles, whereas the appropriate alleles of a dark-skinned person all produce melanin (Fig. 8.20). The varying shades of human skin are thought to result from the relative proportions of melanin in each person. Very high concentrations of melanin, for instance, cause the skin to appear black. These persons, then, have more melanin-producing alleles present than do lighter-skinned people.

In the final analysis, multiple gene inheritance is surely much more complex than we have indicated here. Various environmental

194
The Basis of Heredity

> Linked genes, i.e., genes located on the same chromosome, do not separate from each other during meiosis and thus are exceptions to Mendel's principle of independent assortment.

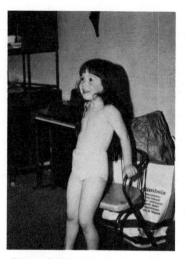  

Figure 8.20 Multiple alleles determine the inheritance of human skin color. The resultant differences in the amount and distribution of the dark pigment melanin produce a continuous gradation of skin tones. (From Silverstein, A.: *The Biological Sciences.* New York: Holt, Rinehart and Winston, 1974, p. 413.)

and other genetic factors are known to affect many human traits strongly. For example, although height is definitely influenced by multiple genes, such factors as diet and hormones also exert their regulatory effects.

GENE LINKAGE

It was pointed out previously that alleles are assorted independently if they are located on separate chromosomes. It is known, however, that genes may be found on the *same* chromosome and thus are inherited together, not separately. These genes are said to be **linked** and constitute a **linkage group.** Therefore, during meiosis, linked genes move together toward one pole of the cell and eventually into a gamete. Obviously, this would change the expected genotypic and phenotypic ratios of the offspring from those expected if the genes had assorted independently. This, then, is an exception to Mendel's principle of independent assortment.

As an example, let us again use the $RrTt \times RrTt$ dihybrid cross in flowers. You will recall that the principle of independent assortment would dictate the formation of four different types of gametes—RT, Rt, rT, and rt—from each parent if the genes were located on separate chromosomes. The expected phenotypic ratio from this cross is 9:3:3:1. Suppose, however, that the R and T alleles were linked, i.e., located on the same chromosome, and that the r and t alleles were linked on the homologous chromosome. What would then be the expected phenotypic and genotypic ratios?

In this example, each parent could produce only RT gametes and rt gametes. By using a Punnett square as shown in Figure 8.21, the types of F_1 offspring can be determined. Thus, the phenotypic ratio is three dominant to one recessive, and the genotypic ratio is one homozygous dominant to two heterozygotes to one homozygous recessive.

Quite often, linked genes are rearranged by *crossing over*. Crossing over occurs in prophase I of meiosis and involves the exchange of genetic material between chromatids of homologous chromosomes. In this manner, genes from maternal and paternal chromosomes can be interchanged. As the example in Figure 8.22 shows,

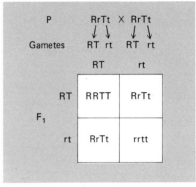

Figure 8.21 Dihybrid cross involving linked genes. The R and T alleles are located on the same chromosome, and the r and t alleles are located together on the homologous chromosome. Consequently, the alleles on each chromosome do not assort independently but move as a single unit.

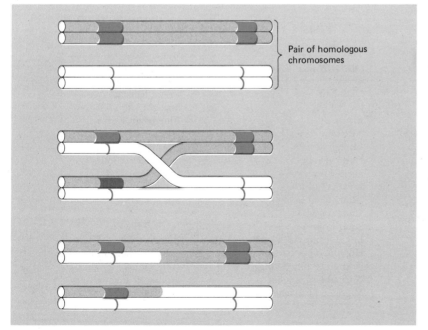

Figure 8.22 An example of crossing over in which there is an exchange of dominant (wide bar) alleles and recessive (narrow bar) alleles. As a result of crossing over, reproductive cells carry a new and different assortment of genes.

> Crossing over can affect the evolutionary process by providing favorable or unfavorable gene combinations that alter the gene pool of a population.

crossing over can result in an exchange of dominant and recessive alleles between homologous chromosomes. As a consequence of crossing over, gametes carry a different assortment of genes than if crossing over had not occurred.

Crossing over is tremendously significant in the evolutionary process. Through the exchange of genetic material, a chromosome might acquire an advantageous allele to go along with another advantageous allele already on the chromosome. In meiosis, these alleles would enter a gamete together; and at fertilization, the offspring receiving this allele pair would have a genetic advantage. In evolutionary terms, this means that these offspring would be favored to reach reproductive age and to pass their advantageous alleles to their offspring. On the other hand, offspring receiving the less advantageous alleles would be less likely to survive and reach reproductive age. Consequently, a high percentage of their alleles would not be passed to future offspring, thereby diminishing the frequency of the alleles in the population. Thus, crossing over could change the make-up of a population by producing individuals with different survival potentials. Those individuals with favorable gene combinations have an increased adaptive advantage and will survive in greater numbers. Those individuals with less favorable gene combi-

Table 8.1 SOME HUMAN INHERITED TRAITS

DOMINANT TRAIT	RECESSIVE TRAIT
Dark hair	Blond hair
Curly hair	Straight hair
Brown eyes	Blue or gray eyes
Nearsightedness	Normal vision
Farsightedness	Normal vision
Free ear lobes	Attached ear lobes
Large eyes	Small eyes
Long eyelashes	Short eyelashes
"Roman" nose	Straight nose
Short stature	Tall stature
Short fingers	Normal fingers
Blood groups A, B, AB	Blood group O
High blood pressure	Normal blood pressure
Normal blood clotting	Hemophilia
Normal hemoglobin	Sickle-cell anemia
Migraine headache	Normal

nations are more likely to be eliminated. The result is evolution, i.e., a change in the gene pool (total genetic material or kinds of genes) of a population over a period of time (Table 8.1).

REVIEW OF ESSENTIAL CONCEPTS

1. The replication of DNA involves the separation of the two nucleotide strands, each of which becomes a pattern for a new strand of DNA.
2. Messenger RNA is formed from a nucleotide strand of DNA. The nucleotides in a strand of mRNA are complementary to those of DNA. The copying of mRNA from DNA is called *transcription.*
3. Transfer RNA is synthesized from DNA and is shaped like a clover leaf. The three-nucleotide sequence on a loop of tRNA is called an *anticodon.*
4. Ribosomal RNA is synthesized by the nucleolus and is a component of ribosomes.
5. The *genetic code* of DNA consists of "codewords," or nucleotide triplets that specify a particular amino acid. The triplets are transcribed to mRNA, where they are called *codons.*
6. A *gene* is a specific sequence of DNA triplets that code for one particular polypeptide.
7. Protein synthesis involves the transcription of mRNA from a DNA template strand. Actual synthesis occurs on the ribosomes, to which the mRNA is attached. Amino acids are carried to the mRNA strand by specific transfer RNAs, each of which contains an anticodon complementary to a mRNA codon. Through complementary base pairing, each tRNA inserts its amino acid in the correct sequence along the strand of mRNA. The formed polypeptide is released into the cell, and the tRNAs are reusable. The synthesizing of proteins on the mRNA strand is called *translation.*
8. Genetic errors, or *mutations,* may occur through environmental influences or may arise spontaneously. Mutations involve alterations in the nucleotides of DNA (gene mutations) or extended segments of DNA (chromosome mutations).
9. Modern genetics began with the work of Gregor Mendel, who developed the basic laws of genetics. These include the concept of inheritance by genes, the principles of dominance and recessiveness, and the laws of segregation and independent assortment.
10. The outward appearance of an organism is called its *phenotype;* the kinds of genes it has is its *genotype.*
11. A monohybrid cross involves only one genetic trait; a dihybrid cross involves two genetic traits.

12. The proportions of different phenotypes in the offspring is called the *phenotypic ratio;* the proportions of genotypes is called the *genotypic ratio.*
13. A *test cross* involves crossing an organism displaying a dominant trait with a homozygous recessive individual. A test cross serves to determine if an organism having a dominant trait is homozygous or heterozygous for that trait.
14. A genetic trait carried on the X chromosome is said to be *sex-linked* or *X-linked.* Sex-linked traits include color blindness and hemophilia.
15. The partial expression of both the dominant and recessive alleles in the phenotype results from *incomplete dominance.* Incomplete dominance is involved in the inheritance of sickle-cell disease.
16. The term "*multiple alleles*" refers to the fact that more than two alleles can be responsible for a given trait. The human ABO blood group system is an example of multiple allele involvement.
17. *Linked* genes are found on the same chromosome and are inherited together. Linked genes may be rearranged by crossing over.
18. Through crossing over, a chromosome may acquire an advantageous allele that would provide an evolutionary advantage for an organism.

APPLYING THE CONCEPTS

1. How is transcription accomplished in the cell? What is its significance?
2. What is the difference between transcription and translation?
3. How do tRNAs function?
4. Assume that in a hypothetical segment of DNA the sequence of bases is AGGCTTGATCCG. What would be the sequence of bases in the transcribed mRNA strand? What would be the base sequences of the tRNA anticodons that would bind to this strand of mRNA? If the base (adenine) of the first nucleotide in the DNA segment were deleted through mutation, how would this deletion affect the synthesis of a polypeptide? What possible biological effect could this deletion have on a living organism?
5. In what manner do mutations affect protein synthesis?
6. a. Determine the phenotypic and genotypic ratios of the F_1 for the following (R = red color; r = white color):

    $Rr \times rr$ $RR \times Rr$ $Rr \times Rr$

    b. What would be the phenotypic ratios of these crosses if dominance were incomplete?

7. Determine the phenotypic and genotypic ratios of the F_1 for the following (R = red color; r = white; T = tall; t = short):
   a. $RRTT \times rrtt$ b. $RRtt \times rrTT$
   c. $RrTt \times RRTT$ d. $RrTt \times RrTt$
8. Assume that in corn kernels smooth texture (S) is dominant to wrinkled texture (s). You observe an ear of corn with only smooth-textured kernels. How could you determine if the smooth texture represents a homozygous or heterozygous condition?
9. One of the forms of nonlethal muscular dystrophy results from a sex-linked recessive gene. A man who has the disease marries a woman who is a carrier of the dystrophy gene.
   a. What percentage of their sons could be expected to have muscular dystrophy?
   b. What percentage of their daughters could be expected to be carriers of the dystrophy gene?
   c. What percentage of all of their children could be expected to be normal?
10. A man with blood type A is involved in a paternity suit in which it is claimed he is the father of a child with type B blood. Is the claim genetically defensible? Explain.
11. a. Explain what is meant by a linkage group.
    b. What is the significance of linked genes in determining the phenotypes of offspring?
    c. Explain how crossing over could affect linkage groups.

9 viruses; kingdoms monera, protista, and fungi

THE ESSENTIAL OBJECTIVES

You have understood this chapter when you are able to:

1. Describe the structure of DNA and RNA viruses and explain the method of infection and replication in each type.
2. Describe the characteristics of bacteria in terms of structure, methods of nutrition, and biological roles.
3. Describe the recombinant DNA technique and explain its practical application.
4. Describe the structure and biological roles of the blue-green algae.
5. Name the three major divisions of unicellular algae and describe the types of organisms in each division.
6. Compare the plasmodial and cellular slime molds in terms of structure and reproductive processes.
7. Name the four major classes of protozoans and describe the representative organisms in each class.
8. Name the four divisions of fungi and describe the methods of reproduction characteristic of each division.

The acellular viruses are a group of infectious agents requiring living host cells for their maintenance and replication.

The diversity of life at its simplest level is revealed in the prokaryotic monerans and the eukaryotic protists and fungi.

PREVIEW OF ESSENTIAL TERMS

taxonomy The biological science concerned with the systematic arrangement and classification of living things.
RNA transcriptases Enzymes that enable viral RNA to act as a template for the synthesis of new viral RNA.
reverse transcriptase An enzyme used by retroviruses to transcribe viral RNA into DNA.
binary fission A type of asexual reproduction in which an organism separates into two new individuals.
heterotroph An organism that must acquire food energy from an outside source.
autotroph An organism, such as a blue-green alga, capable of synthesizing its own food.
symbiosis An intimate biological relationship between two or more organisms of different species.
plasmid A small circular piece of DNA found in the cytoplasm of some bacterial cells.
restriction endonuclease An enzyme that cleaves short sequences of nucleotides from a DNA molecule.
phytoplankton microscopic photosynthetic organisms that float on the surface of bodies of water.
mutualism A type of symbiosis in which each organism benefits from the relationship.
life cycle A series of processes undergone by an organism in its development from one form to another and back to the same form again.
sporangium A structure in which haploid spores are produced by meiosis.
conjugation A method of sexual reproduction among lower organisms in which genetic material is passed between mating organisms.
multiple fission Mitotic cell division in which many individual cells are produced from a single cell.
zygospore A type of spore formed by the fusion of nuclei from two conjugating organisms.

Figure 9.1 Carolus Linnaeus, the eighteenth century Swedish physician and naturalist who developed the taxonomic system still in use today. (From Gerking, S. D. *Biological Systems.* 2nd ed. Philadelphia: W. B. Saunders, 1974, p.88.)

TAXONOMY: HISTORICAL BACKGROUND

In the following three chapters we will consider the biology of the major groups of organisms that inhabit the earth. Our primary concern will be the distinguishing characteristics of each group; at the same time we will emphasize their underlying biological unity.

To some extent we will need to rely upon the science of **taxonomy** in referring to the scientific names of some of the organisms and in placing them within a group of organisms of the same or related type. Taxonomy, a term meaning "law of arrangement," is the biological science that deals with the naming and classification of living things.

Biologists have recognized the need for a system of classification since the time of Aristotle. Through his exacting observations, Aristotle became convinced that living organisms could be arranged in a continuous series from lowest to highest, on the basis of their degree of complexity. This "ladder of nature" concept, although not entirely correct, was perhaps the first attempt at a comprehensive system of classification.

Although Aristotle used the terms "genus" and "species," which are derived from the Greek words *genos* (race, kin) and *eidos* (form), the language of taxonomy since the Middle Ages has been Latin. The taxonomic system in use during the Middle Ages included a generic name for a particular group of organisms, plus a rather long descriptive phrase identifying each specific kind of organism within the group. This somewhat cumbersome **polynomial** ("many names") **system** persisted well into the eighteenth century. There was, however, an innovative concept emerging in the work of a Swedish physician and naturalist named Carolus Linnaeus (1707–1778) (Fig. 9.1). Using ideas conceived by several European scientists, Linnaeus developed a taxonomic system that recognized two **kingdoms**, Plantae and Animalia. He further subdivided the two kingdoms into **orders, families, genera,** and **species.** Each kind of organism in this system was assigned only two Latin names—a generic name and a species name. This "two name" classification, or **binomial system,** a contribution of the Swiss naturalist Kaspar Bauhin (1560–1624), is still used today.

MODERN TAXONOMY

The largest and most inclusive taxonomic category is the *kingdom.* Currently, the most widely used scheme of classification recognizes five major kingdoms: **Monera, Protista, Fungi, Plantae,** and **Animalia.** Organisms in the kingdom Monera are all prokaryotes, i.e.,

> Taxonomy is the science that describes and classifies living things. A closely related science known as *systematics* is concerned with the evolutionary relationships of living things.

they are composed of cells that lack nuclear membranes and the membrane-bound organelles found in the cells of eukaryotic organisms. All organisms in the other four kingdoms are eukaryotes.

Below the level of the kingdom, taxonomy includes several less inclusive categories, which further distinguish one group of organisms from another. The taxonomic category below a kingdom is either the **phylum** (pl. phyla), which is applied to the animals and the animal-like protists, or the **division**, usually applied to the plants, the plantlike protists, the fungi, and the monerans.

Phyla or divisions are divided into **classes**, which in turn are composed of **orders**. Orders consist of **families**, which are groups of different **genera** (sing. genus). A genus includes a particular kind of organism, or **species**. Each of these seven major taxonomic categories may be divided into various others, such as subkingdom, superclass, superfamily, subgenus, and so on.

To illustrate the taxonomic scheme using the seven major categories, the biological classification of roses, for example, would be:

Kingdom: Plantae
Division: Tracheophyta
Class: Angiospermae
Order: Rosales
Family: Rosaceae
Genus: *Rosa*
Species: *virginiana, multiflora,* and others

In going from the most inclusive category—the kingdom—down to the least inclusive—the species—each category more narrowly and more exactly describes the organism being classified. For example, the kingdom Plantae includes every known type of plant, whereas the division Tracheophyta, although an extremely broad category, includes only those plants with vascular tissue, i.e., plants having a system of internal vessels that conduct water and nutrients throughout the plant. A narrower or more restrictive category is the class Angiospermae, which includes only vascular plants with flowers. The order Rosales is further delimiting by including flowering plants with a particularly unique arrangement of the floral parts (sepals, petals, and reproductive structures). These types of flowering plants with additional structural modifications are placed in the Rosaceae, the rose family. However, there are other members of the rose family, such as apples, peaches, cherries, and strawberries, that are quite different from roses. Just as apples, peaches, cherries, and strawberries have their own genera, roses, too, are separated from their family members and placed in the genus *Rosa*. Finally, all roses

Viruses; Kingdoms Monera, Protista, and Fungi

Table 19.1 THE CLASSIFICATION OF THE HUMAN SPECIES

Kingdom	Animalia	(animals)
Phylum	Chordata	(with notochord)
Class	Mammalia	(with hair and mammary glands)
Order	Primates	(with opposable thumb and binocular vision)
Family	Hominidae	(with erect posture)
Genus	*Homo*	(man)
Species	*sapiens*	(wise)

are not exactly the same but are represented by several different "kinds," or species. Thus, a particular species of roses may be referred to as *Rosa virginiana, Rosa multiflora,* and so on.

In the binomial system, the generic and specific names are always written in italics or underlined. In addition, the generic name is capitalized, whereas the specific name is not. Thus human beings are designated *Homo sapiens* ("wise man") (Table 9.1); the common house cat is *Felis domestica* ("domestic cat"); *Tulipa montana* is the mountain tulip; and the world's tallest species of trees is the redwood *Sequoia sempervirens,* named for a Cherokee Indian, Sequoyah, to which was added the species name, meaning "evergreen."

Part of the value of a taxonomic system is that it affords accurate communication between scientists anywhere in the world. Since Latin is an unchanging language, there is uniformity in referring to a given species, whether it is found in the United States or the Canary Islands. Essentially, then, taxonomy speaks a universal language and, in an applied sense, reflects the unity and diversity of the living world.

VIRUSES

The viruses constitute a rather controversial group of "organisms" that does not fit into any of the usual categories of biological classification. Essentially, the controversy centers on the question of whether viruses are really living organisms. First of all, viruses are not even cells; they have no organelles, nor do they have the cellular machinery to generate their own energy (adenosine triphosphate, or ATP). They cannot exist outside the body of another living thing and are therefore entirely parasitic. Structurally, they are the essence of simplicity, consisting only of a nucleic acid—either deoxyribonucleic acid (DNA) or ribonucleic acid (RNA)—surrounded by

> Salvador E. Luria, Nobel Prize laureate 1969, referred to viruses as "bits of heredity looking for a chromosome."

a protein coat. Finally, they do not meet one of the essential criteria of life in that they cannot reproduce on their own but must depend upon materials acquired from the host.

It would be most unusual to find anyone who has not had an encounter with a virus. The word "virus" is from the Latin, meaning "poison," and in many cases viruses are exactly that. Viruses are responsible for a variety of plant and animal diseases, including such human afflictions as influenza, measles, mumps, chicken pox, and the common cold (Fig. 9.2).

The methods by which viruses enter and infect the host cell vary with the type of virus in question. Some viruses—including many that infect both plants and animals—first become attached to the cell membrane and then move into the interior of the cell. In almost all DNA-containing viruses, the nucleic acid enters the nucleus of the host cell, where it takes control of the cell's metabolic machinery. The protein coats of the viruses are broken down by cellular enzymes. Within the nucleus of the host cell, the viral DNA becomes

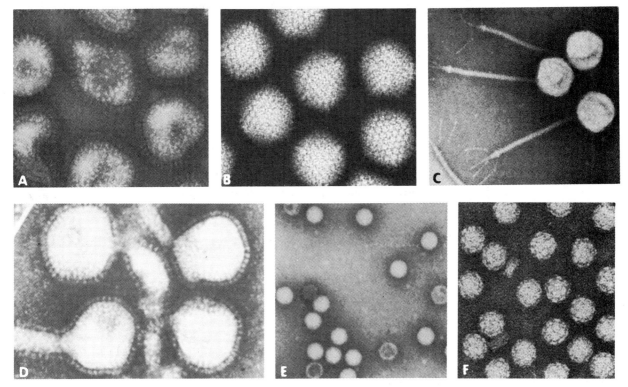

Figure 9.2 Some important viruses. In the top row are DNA viruses; below are RNA viruses. (a) Influenza virus ($\times$ 170,000). (b) Adenovirus, one of the cold viruses ($\times$ 180,000). (c) T5 bacteriophage ($\times$ 150,000). (d) Retrovirus, causing mammary tumors in mice, isolated from mouse milk ($\times$ 150,000). (e) Poliomyelitis virus ($\times$ 160,000). (f) Turnip yellow mosaic virus ($\times$ 250,000). (From Clark, M. E.: *Contemporary Biology*. 2nd ed. Philadelphia: Saunders College Publishing, 1979, p. 99.)

Viruses; Kingdoms Monera, Protista, and Fungi

> RNA viruses are unique in that they require highly specialized enzymes—either RNA transcriptases or reverse transcriptase—to carry out their replicative processes.

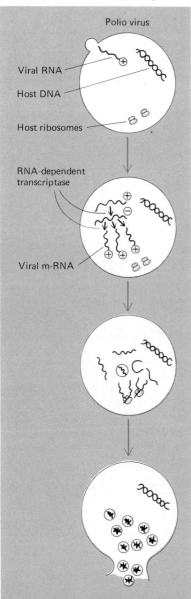

Figure 9.3 Conventional RNA viruses. A single strand of RNA, labeled ⊕, enters the host cell, along with the viral enzyme, RNA-dependent transcriptase. This enzyme first catalyzes the synthesis of a complementary ⊖ strand of RNA and then uses it to make many ⊕ RNA strands, the viral m-RNA. These parasitize the host cell ribosomes, making viral protein. The many new viruses then leave the cell, usually killing it.

the pattern, or template, from which new viral DNA, structural proteins, and enzymes are synthesized. As pointed out, viruses cannot reproduce on their own but must utilize materials derived from the host. Accordingly, under instructions coded in the viral DNA, the host cell manufactures new viruses, each with a DNA core and protein coat. These new viruses are then released from the host cell, either by disruption of the cell membrane or by extrusion, i.e., an outpocketing of the cell membrane that pinches off with the virus enclosed inside.

Among viruses in which the nucleic acid is RNA, reproduction within the host cell takes on some unusual twists. The RNA viruses do not enter the nucleus of the host cell but remain in the cytoplasm, where, in some cases, viral RNA functions as a template for the synthesis of new RNA. Obviously, this is in variance to the usual method of RNA synthesis. From the discussion in Chapter 8, it would appear that DNA is always required as a template for the synthesis of RNA. Since the RNA viruses do not introduce a DNA template into the host cell, how do they carry out their replicative process? The answer involves some unusual enzymes known as **RNA transcriptases.** These enzymes enable the viral RNA to act as a template for the synthesis of new viral RNA and hence of new viral particles (Fig. 9.3). The RNA transcriptases either are brought into the host cell by the virus or are synthesized in the cytoplasm of the cell. Synthesis of these enzymes is possible because viral RNA entering the host cell assumes the role of messenger RNA, which codes for the synthesis of the transcriptases by the cell. In some RNA viruses, then, reproduction of new viral particles is dependent upon unique enzymes—RNA transcriptases—which permit the synthesis of new RNA from the original viral RNA entering the host cell.

Another group of RNA viruses—known as *retroviruses*—employ an even more unusual method in reproducing (Fig. 9.4). These viruses bring with them into the host's cells an enzyme called *reverse transcriptase*. This enzyme, as its name implies, enables transcription to occur in reverse, i.e., it enables viral RNA to be transcribed into DNA. Newly formed viral DNA then attaches to the chromosomes of the host cell, where it replicates along with the host DNA. Although this replicating process produces no new RNA viruses, the viral DNA may be passed along to succeeding generations of the host's descendants. At times, however, the viral DNA does more than just replicate; it becomes active, transcribing new viral RNA and mRNA. Eventually, protein coats are assembled around the viral RNA molecules, forming new retroviruses, which are extruded from the host cell.

The *infectious cycle* of a bacteriophage begins with its attachment to the host cell and ends when new phages are released from the cell.

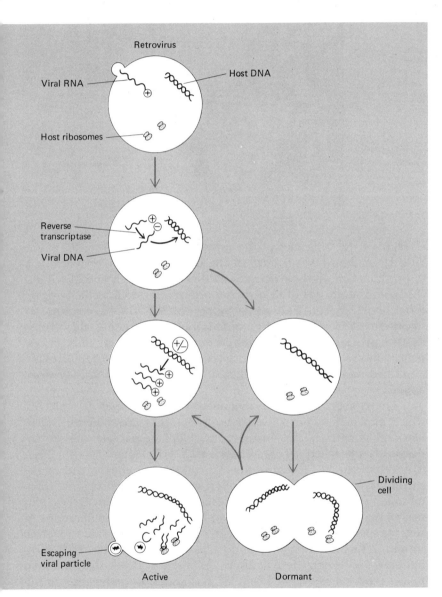

Figure 9.4 Retroviruses. The viral RNA that enters is first transcribed into complementary DNA by the viral enzyme, reverse transcriptase. This then becomes a +/− double helix that fuses permanently with the host DNA dividing each time the host cell divides. The viral DNA may remain dormant for long periods, only making m-RNA and new virus particles from time to time.

Bacteriophages

Certain types of viruses infect only bacteria and are referred to as *bacteriophages* ("bacteria eaters"), or simply as *phages*. Like some of the viruses previously discussed, phages consist of a nucleic acid (DNA) surrounded by a protein coat. Phages also possess an elongate tail region by which they attach to the wall of the bacterial cell. After attaching to the cell, a bacteriophage injects its DNA into the

Once introduced into the chromosome of the host cell, the DNA sequences of temperate phages eventually can be transmitted to billions of *E. coli* cells as the bacterium divides.

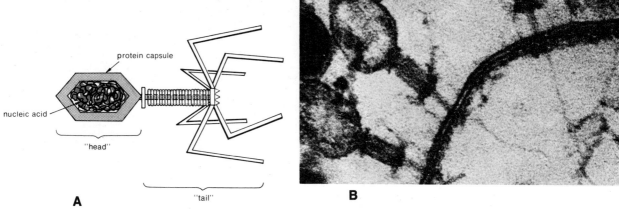

Figure 9.5 A DNA virus, bacteriophage T2, which attacks the human colon bacillus, *Escherichia coli.* (a) Diagram of phage virus. The "head," consisting of a protein capsule surrounding the nucleic acid core, is typical of all viruses. The protein "tail," used to insert DNA into the host cell, is characteristic of many phages. (b) Phage particles attacking an *E. coli* cell. The tail attaches to the cell wall, and the DNA is injected through a tube into the cell interior. Threads already visible within the cell are probably viral DNA. (From Clark, M. E.: *Contemporary Biology.* 2nd ed. Philadelphia: Saunders College Publishing, 1979, p. 99.)

interior of the cell, leaving the protein coat outside (Fig 9.5). The viral DNA then takes control of the bacterium's cellular machinery to produce new viral particles. Soon thereafter, the bacterial cell disrupts, and hundreds of new viruses are released to infect a new bacterial host (Fig 9.6). This entire process of bacteriophage reproduction is referred to as the **lytic cycle** (Gk. *lysis,* loosening, dissolving), i.e., the process of disintegration or destruction of cells. This process almost always kills the host cell; accordingly, lytic bacteriophages are described as **virulent,** a term meaning extremely injurious or damaging.

There are other bacteriophages, known as *temperate* phages, which may or may not kill their hosts, i.e., they are sometimes virulent and sometimes not. In the nonvirulent state, these phages undergo what is called a **lysogenic cycle** (Fig. 9.7). After the viral DNA is injected, the cycle begins when the DNA becomes joined to the chromosomal DNA of the bacterium. As a part of the bacterial chromosome, the phage DNA replicates along with the host DNA. This means that the phage DNA can be passed to many generations of cells as the bacterium reproduces. Occasionally, various external stimuli induce the phage DNA to separate from the bacterial chromosome and enter the lytic cycle. As just discussed, the separated phage DNA takes over the cellular machinery of the host cell to initiate production of new virulent bacteriophages.

An example of a virus that illustrates the lysogenic and lytic cycles is *Herpes simplex,* the cause of cold sores. Anyone who has been infected with the herpes virus knows that the cold sores persist for some time, clear up, and then reappear at a later time. Apparently, the virus is always present in the body but does not cause cold sores

Viruses

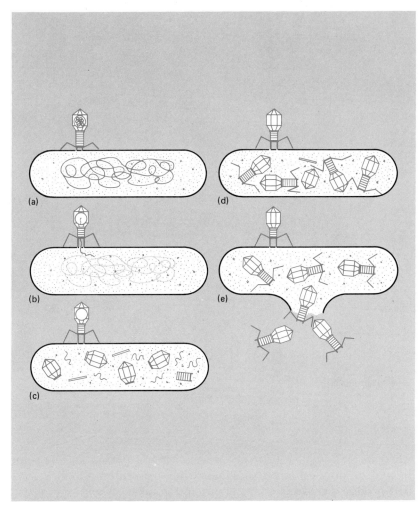

Figure 9.6 Phage infection. (a) Phage particle attached by its tail. (b) The cell wall is penetrated, the DNA is injected, and the protein coat is left behind. (c) Virus DNA takes over cell function and protein coats are synthesized. (d) DNA appears in some of the phage coats. (e) all phage particles are complete and the bacterium lyses. The whole process takes about 30 minutes. In some virus-bacterial systems, as many as 300 virus particles are reproduced from a single cell infection. (From Gerking, S. D.: *Biological Systems.* 2nd ed. Philadelphia: W. B. Saunders, 1974, p. 215.)

Figure 9.7 Reproductive cycle of a lysogenic, or temperate, phage. The host may survive for many generations with the phage genetic material incorporated into the host genome, until some condition triggers the phage to become lytic.

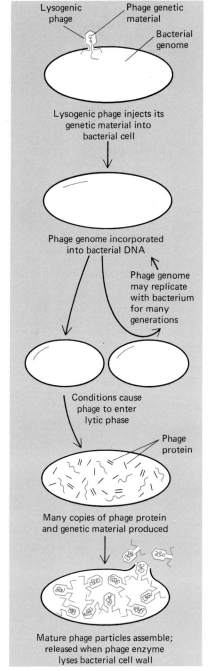

Viruses; Kingdoms Monera, Protista, and Fungi

> Cancerous cells can spread from one organ or tissue into many other areas of the body—a process known as *metastasis*.

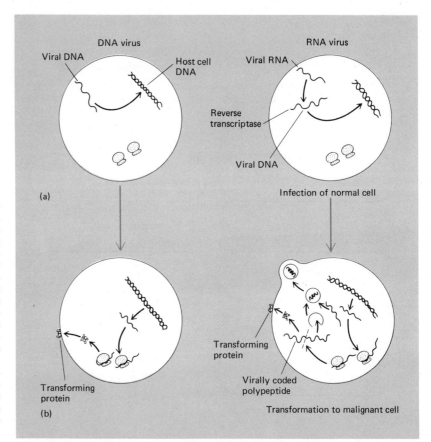

Figure 9.8 How viruses may cause cancer. (a) DNA viruses. The viral DNA fuses with that of the host and remains latent except for transcription of a small piece of m-RNA coded for transforming protein. This protein somehow converts the cell from a normal cell to a cancer cell. It is shown here acting at the cell surface, but it could be acting elsewhere in the cell, perhaps by altering the regulation of the host cell's genes. (b) RNA viruses. Only retroviruses that inject virally coded DNA into host chromosomes produce cancers. When viral DNA is active, both viral proteins and transforming protein are made, presumably by cleavage of one long polypeptide chain. The viruses are eliminated from the cell without killing it.

when in its lysogenic form. However, when a person suffers some mental or physical stress or illness, the virus enters the lytic cycle and produces cold sores on the lips and gums.

Interferon

Viral infections cause some eukaryotic cells to produce a protein called **interferon**. When released from the infected host cells, interferon interacts with neighboring cells and in some manner protects them from the viral infection. When a person is immunized against virus-caused diseases, the vaccine usually contains "killed" viruses or viral nucleic acid. When introduced into the body, the virus or nucleic acid induces the production of interferon by the cells. Should infection then occur by a live virus at some later time, the increased amount of interferon makes the body more resistant to the infection.

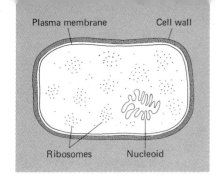

Figure 9.9 Generalized prokaryotic cell.

Viruses and Cancer

Over the past several years there has been considerable interest in the relationship between viruses and certain forms of cancer. Basically, a cancer is an abnormal growth, or *tumor* (L. "a swelling"), that invades normal, healthy tissue. A *benign* tumor is confined to one place and does not destroy healthy tissue; on the other hand, a *malignant* tumor, through uncontrollable cell division, invades and destroys surrounding tissues. The loss of control of normal cell division can result when an invading cancer virus takes control of the genetic machinery of the host cell (Fig. 9.8). The viral genetic material also transforms the host cells by altering their shape and metabolism. Cancerous cells are generally spherical in shape and, as a consequence of their rapid rate of division, have a higher metabolic rate than do normal cells. Thus, cancerous cells utilize the bulk of food energy in the organism, leaving the healthy cells without adequate nourishment. In animals, the result is often progressive weight loss and emaciation.

Although it is known that a variety of plant and animal cancers are caused by viruses, there is as yet no definite evidence that viruses are responsible for any known human cancers. Some human cancers, however, such as leukemia and breast tumors, are suspected of being virally induced.

KINGDOM MONERA

Only two groups of living organisms are placed in the kingdom Monera—the **bacteria** and the **blue-green algae**—both of which are microscopic in size. Most of these organisms are unicellular, and all are **prokaryotes.** Prokaryotic cells lack a nuclear membrane, although they do contain a region of nuclear material called the **nucleoid.** They have none of the membrane-bound organelles found in eukaryotic cells, but they do possess ribosomes (Fig. 9.9). The blue-green algae and photosynthetic bacteria do not contain chloroplasts or other plastids, although both groups contain some type of chlorophyll. The "chromosome" of a prokaryotic cell is a single loop of DNA, attached to the cell membrane. Prokaryotic organisms reproduce primarily by splitting in two—an asexual process known as **binary fission**—but neither mitosis nor meiosis occurs (Fig. 9.10). During cell division the DNA duplicates, and one DNA molecule goes to each organism (Fig. 9.11).

Bacteria (Division Schizomycetes)

In just about any environment—in the soil, in water, within other organisms, even in the human digestive tract—one would find some type of monerans called **bacteria** (Gk. *bakterion*, rod, staff). These

Viruses; Kingdoms Monera, Protista, and Fungi

> Placed end to end, half a million "average" bacterial cells would extend a distance of only about 1 inch.

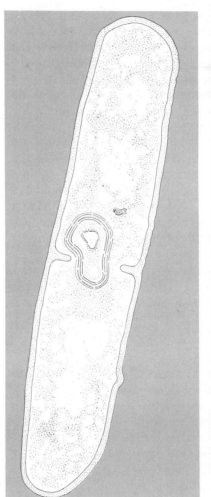

Figure 9.10 Binary fission in a bacterium.

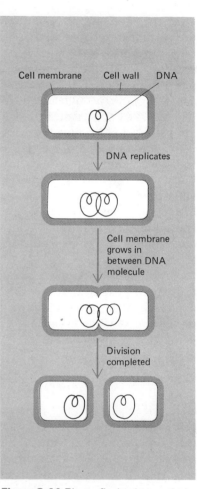

Figure 9.11 Binary fission in a prokaryotic cell.

primitive, one-celled microorganisms have been around for some 3 billion years and are involved in a variety of enterprises that belie their lowly status among living things.

Most of the bacteria can be divided into three major groups according to their particular shape. There are spherical bacteria, referred to as *cocci* (Gk. "berry"); rod-shaped bacteria, or **bacilli** (L. "stick"); and spiral-shaped bacteria, called *spirilla* (L. "coil") (Fig. 9.12). Sometimes the cocci bacteria are found in clusters, resembling a bunch of grapes. When grouped together in this fashion, they are referred to as **staphylococci.** On other occasions, cocci bacteria may

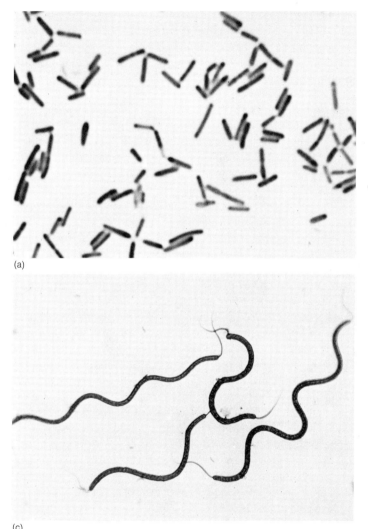
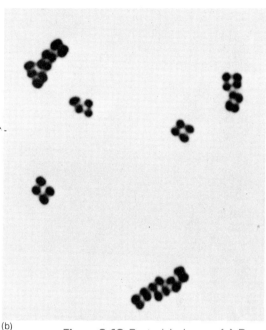

Figure 9.12 Bacterial shapes. (a) Bacilli, or rod-shaped. (b) Cocci, or spheres. (c) Spirilla, or spiral-shaped. (From Arms, K., and Camp, P. S.: *Biology*. 2nd ed. Philadelphia: Saunders College Publishing, 1982.)

form beadlike chains, in which case they are called *streptococci*. Accordingly, comments concerning a "staph" infection or a "strep" throat are references to bacteria arranged in either clusters or chains. In addition, some coccus cells, such as those of the bacterium that causes pneumonia, are found in pairs called *diplococci* (Gk. *diploos*, double).

Another important group of bacteria comprises the corkscrew-shaped *spirochetes* (Gk. "coil bristle"). One member of the spiro-

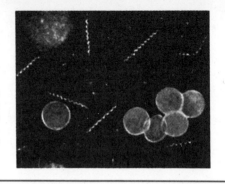

Figure 9.13 The spirochete bacterium *Treponema pallidum* is the causative agent of syphilis in human beings. (From Clark, M. E.: *Contemporary Biology*. 2nd ed. Philadelphia: Saunders College Publishing, 1979, p. 396.)

chetes, *Treponema pallidum*, is the causative agent of syphilis in human beings (Fig. 9.13). The bacterium is transmitted by sexual intercourse.

Many bacteria are capable of movement, either by means of an as yet unexplained gliding motion, or by long, rotating flagella.

Characteristics of Bacteria. Most bacteria are **heterotrophic,** i.e., they must acquire their food from an outside source. Of these, the majority are **saprophytes** (Gk. *sapros*, rotten), feeding on dead or decaying organic matter. The remaining heterotrophic bacteria are **parasites,** living within the body of the host organism. Nutrients already in solution enter the bacterial cell by diffusion; however, some foods must first be broken down by bacterial enzymes before they go into solution and enter the cell.

There are some bacteria capable of synthesizing their own food in a manner similar to that of green plants. Such **autotrophic,** i.e., "self-feeding," bacteria may be **photosynthetic,** although they do not possess the same type of chlorophyll found in higher plants. Instead, one group of photosynthetic bacteria—the *green sulfur bacteria*—contains **chlorobium chlorophyll,** a pigment very similar to a chlorophyll of higher plants. The other major group of photosynthetic bacteria, known as *purple bacteria,* contains **bacteriochlorophyll,** a pale green pigment somewhat different from the chlorophyll of higher plants. In further contrast to plants, the green and purple bacteria do not utilize wavelengths of light in the visible spectrum as the source of energy for photosynthesis. You will recall from Chapter 6 that green plants primarily absorb red, blue, and violet light; photosynthetic bacteria, on the other hand, absorb light in the infrared region of the spectrum. Wavelengths of light in the infrared region are longer than those in the visible spectrum and cannot be seen by humans. Thus, as far as human vision is concerned, bacteria photosynthesize in darkness. Moreover, bacteria do not utilize water in photosynthesis and thus do not release free oxygen. In fact, bacterial photosynthesis is a totally anaerobic process. The substitutes for water in bacterial photosynthesis include hydrogen sulfide (H_2S), fatty acids, and other organic compounds. These compounds serve to donate hydrogen and electrons just as water does in photosynthesis in higher plants.

The fact that bacterial photosynthesis occurs anaerobically and in darkness explains how some bacteria can survive in the mud at the bottom of a pond or lake. There is little or no oxygen in the mud, but it usually contains hydrogen sulfide, produced by the de-

> To a major extent, the process of decomposition or decay, i.e., the breaking down of large complex molecules of dead organisms into similar ones, is accomplished by enzymes secreted by bacteria.

composition of plant material. Moreover, only the longer wavelengths of light reach the bottom of the pond or lake, since visible light is filtered out by the surface waters.

A few autotrophic bacteria are **chemosynthetic**, using ammonia (NH_3), nitrate (NO_3), nitrite (NO_2), and other compounds as energy sources for metabolism. As we will discuss in Chapter 12, some of the chemosynthetic bacteria are essential in the cycling and recycling of nitrogen compounds through the living world.

Activities of Bacteria. Many species of bacteria are beneficial to human beings as well as to many other living things. We have already mentioned their role in the nitrogen cycle. In addition, they are also important in the process of decomposition of organic matter. Were it not for bacteria, dead plants and animals would not decay but would be left to accumulate. Moreover, decomposition by bacteria is an integral part of the cycling and recycling of many forms of matter on earth (see Chapter 12).

Other beneficial roles of bacteria include their use in the dairy industry, where they are involved in the production of buttermilk, various kinds of cheese, and other products. Decay-causing bacteria are used by sewage treatment plants in the purification of water. Quite importantly, many of the antibiotics, such as streptomycin and oxytetracycline, are produced by certain species of bacteria.

Many animals harbor **symbiotic** bacteria and other microorganisms in their digestive tracts. Symbiosis is an intimate relationship between two or more organisms of different species. Animals known as *ruminants* (L. *ruminare,* to chew over), such as cows, deer, sheep, goats, and so forth, depend upon symbiotic bacteria in the stomach for digestion of foodstuffs. In ruminants, the stomach is divided into four chambers—the *rumen, reticulum, omasum,* and *abomasum* (Fig. 9.14). Most of the bacteria (and other microorganisms) live in the rumen and reticulum. When food is swallowed, it first enters these two chambers, where carbohydrates, fats, and proteins are broken down by the microorganisms. Much of this digested food is regurgitated into the mouth, and the animal "chews its cud." On its second trip down, the food enters the omasum and then passes into the abomasum. Many of the bacteria and other microorganisms also move into the abomasum, where they are digested along with the food.

In human beings, symbiotic bacteria inhabit the large intestine, where they are involved in the synthesis of vitamin K, vitamin B_{12}, riboflavin, and thiamine. Each of these vitamins is required for nor-

Viruses; Kingdoms Monera, Protista, and Fungi

> The bacterium *Escherischia coli* is found in the gut of not only humans but also many other mammals.

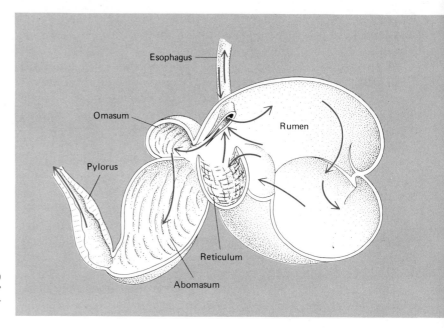

Figure 9.14 The stomach of a sheep with the wall cut away to show its four compartments. Arrows show the direction in which the food travels.

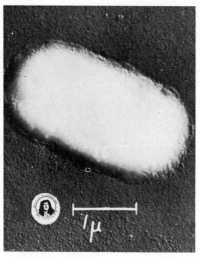

Figure 9.15 The human intestinal bacterium, *Escherichia coli*. (From Gerking, S. D.: *Biological Systems*. 2nd ed. Philadelphia: W. B. Saunders, 1974, p. 219.)

mal functioning of the body. Adequate vitamin K, for example, is necessary for normal clotting of blood, and vitamin B_{12} is necessary for normal body growth and maturation of red blood cells. In addition, intestinal bacteria release enzymes that break down roughage, such as cellulose, that human enzymes cannot digest. Using the products of roughage digestion, some bacteria synthesize amino acids, which are absorbed into the bloodstream through the intestinal wall.

The most notable of human intestinal bacteria is *Escherichia coli* (usually referred to as *E. coli*), a species widely used in biological and medical research (Fig. 9.15). There are, however, a number of other intestinal bacteria, which, like *E. coli*, are bacillus forms.

As pointed out, some bacteria are disease-producing, or **pathogenic**, causing such human diseases as tuberculosis, syphilis, gonorrhea, scarlet fever, diphtheria, whooping cough, and many others. A bacterium called *Clostridium* is responsible for **botulism**, an extremely severe form of food poisoning. The organism produces a toxin that interferes with nerve activity, often leading to muscle paralysis. Commercial canned goods are heat-sterilized to prevent growth of *Clostridium*, but home-canned foods that have been inad-

> Much of the research concerning the molecular biology of genes has employed species of bacteria as experimental organisms.

equately heated may provide a suitable environment for growth of the bacterium.

Most bacteria require oxygen to exist, but *Clostridium*, like the photosynthetic bacteria, cannot live in the presence of oxygen. Another anaerobic species of *Clostridium* is responsible for the disease **tetanus** (lockjaw), characterized by spasmodic contractions of various voluntary muscles, especially those of the neck and jaw. These and other bacteria incapable of existing in the presence of oxygen are known as ***obligate anaerobes***. Still other types of bacteria can live with or without oxygen and are called *facultative anaerobes.*

Bacteria and Research

For decades bacteria have been used widely in scientific research, primarily because they are relatively easy to grow in the laboratory and they reproduce quite rapidly. Some species may divide once every 20 minutes or so. Organisms that reproduce rapidly have proved to be invaluable in scientific research for two basic reasons: First, researchers can grow large numbers of organisms with which to work; second, many generations of offspring can be observed in a short span of time. Let us say, for example, that you have decided to investigate the immediate and long-term effects of a certain chemical that you suspect is mutagenic, i.e., capable of producing mutations in living organisms. As a part of your experiment, it may be desirable, or even crucial, to test the chemical on massive numbers of organisms, under widely varying environmental conditions. Moreover, you may also wish to know the effects of the chemical on future generations not directly exposed to it. Since bacteria can be grown in massive numbers and reproduce rapidly, your research could be greatly facilitated in that you could observe the effects of the chemical on numerous organisms over many generations in the span of a few hours. In fact, a single bacterial cell dividing once every 20 minutes could produce around 1 million descendants in about $6\frac{1}{2}$ hours. Think how long you would have to wait for Asian elephants, producing one offspring every 20 months, to give you experimental results—not to mention the problem with space. It is true, of course, that bacteria are not appropriate for many kinds of biological research; for one thing, they are prokaryotes and therefore different structurally, metabolically, and so on from eukaryotic cells. Nonetheless, bacteria have been and continue to be one of the most significant groups contributing to scientific knowledge and advancement. As we shall discuss in the next section, perhaps their greatest contribution is yet to come.

Viruses; Kingdoms
Monera, Protista, and
Fungi

> Through the use of bacterial plasmids, it is now possible to insert a variety of genes into an *E. coli* host.

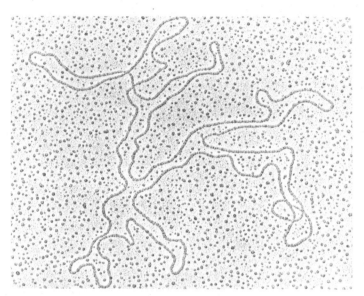

Figure 9.16 A circle of DNA from a bacterium. (From Camp, P. S., and Arms, K: *Exploring Biology.* Philadelphia: Saunders College Publishing, 1981, p. 257.)

Recombinant DNA

We have stated that a bacterial cell contains a single chromosome in the shape of a loop (Fig. 9.16). Genes are arranged in a certain sequence along the chromosome, which is composed of double-stranded DNA. In addition to the main chromosome, most bacterial cells contain small circular pieces of DNA called ***plasmids,*** which are separate from the main chromosome. By experimentally disrupting the bacterial cells, the plasmids can be removed and used in what is called the ***recombinant DNA*** technique. This procedure, also referred to as "gene splicing" and "genetic engineering," has most commonly employed the bacterium *Escherichia coli* (Fig. 9.17).

After being removed and isolated from the bacterial cells, the plasmids are precisely cleaved by specific enzymes called ***restriction endonucleases.*** These types of enzymes (of which there are many) have the ability to "recognize" short sequences of nucleotides on a DNA molecule and to cut the molecule apart at that point. Thus, the plasmids, which were originally circular, are split into small DNA segments. In addition, restriction endonucleases are used to cleave DNA extracted from the cells of some other species of organisms. This DNA, then, is also cut into small pieces. Next, fragments of plasmid DNA and foreign DNA are mixed together and sponta-

> The genetically identical copies of recombinant DNA produced by asexually dividing bacteria constitute a *clone*. In fact, any group of genetically identical forms produced by asexual division is called a clone.

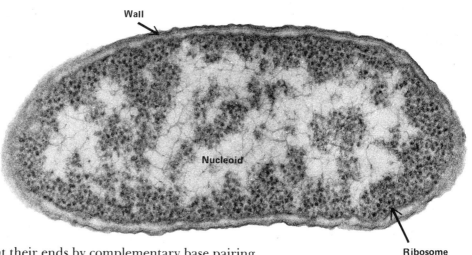

Figure 9.17 Structure of *Escherichia coli*. (From Goodenough, U.: *Genetics.* 2nd ed. New York: Holt, Rinehart and Winston, 1978, p.140.)

neously join each other at their ends by complementary base pairing. This process of joining, also known as *annealing,* is assisted by an enzyme called **DNA ligase** (L. *ligare,* to bind). The result is a group of circular "hybrid" plasmids composed partly of plasmid DNA and partly of foreign DNA—in other words, *recombinant DNA* (Fig. 9.18).

The newly formed plasmids are then added to a flask containing live *E. coli* cells. Some of the cells will take up the plasmids, which, because of the foreign DNA, transform the *E. coli* cells, giving them new genetic properties. When the *E. coli* cells undergo division, the new plasmids will also reproduce along with the cells. It is therefore possible to grow enormous quantities of *E. coli* cells containing some type of foreign DNA in combination with plasmid DNA.

What are the actual and potential applications of the recombinant DNA technique? First, this technique makes possible the production of numerous copies of the same gene (segments of DNA). This affords researchers the opportunity to study more thoroughly the structural and functional properties of a particular gene.

Second, cells transformed by the insertion of foreign genes are capable, in many cases, of synthesizing proteins coded for by the inserted gene. This is the case, for example, with the gene that codes for the protein hormone insulin. Dividing bacterial cells, in which the insulin gene has been inserted, produce numerous copies of the gene, making possible the synthesis of great quantities of insulin for treatment of diabetics. Currently, insulin for human use is extracted from pigs and cattle; however, once the procedure for bacterial production of human insulin is perfected, the hormone can be obtained

Viruses; Kingdoms Monera, Protista, and Fungi

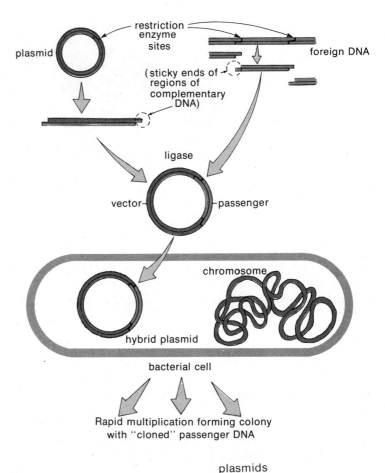

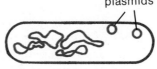

Figure 9.18 Summary of steps in the recombinant DNA technique. The passenger, or foreign, DNA combines with the vector, or plasmid, DNA to form recombinant DNA. As the bacterium multiplies, it produces a "clone" of foreign DNA. (From Clark, M. E.: *Contemporary Biology*. 2nd ed. Philadelphia: Saunders College Publishing, 1979, p. 650.)

more cheaply than animal insulin and in virtually unlimited quantities. Moreover, other human hormones, including growth hormone and thymosin (secreted by the thymus gland), as well as the protein interferon discussed earlier, have also been produced in bacteria.

Third, the future may find recombinant DNA technology applied in the area of human health through what has been called "gene replacement therapy." This would involve replacing defective genes in a chromosome with healthy genes. As an example, a variety

Kingdom Monera

> The blue-green algae are possibly the oldest form of life on earth. The oldest fossil known is a blue-green 3.5 billion years old.

of human genetic diseases result from the absence of certain enzymes or from defective enzymes. If the genes coding for the correct enzymes could be inserted into the chromosomes, the diseases could be cured.

Finally, it should be pointed out that recombinant DNA technology is not limited to the use of bacterial cells as vehicles for carrying foreign DNA. Several experiments have shown that DNA can be inserted into some eukaryotic cells, such as those of mice and monkeys. In these cells, as with bacterial cells, the foreign DNA replicates along with the host DNA. As a result, scientists are better able to study and understand the mechanisms of gene regulation in higher organisms.

Blue-Green Algae (Division Cyanophyta)

The other group of monerans are the blue-green algae, or simply "blue-greens." Some of these organisms are not actually blue-green at all but may be red, green, yellow, and other colors. Most blue-greens are photosynthetic and are found in moist soil and in fresh and salt waters. They are either unicellular or filamentous, i.e., constructed of individual interconnected cells that form a chain or filament (Fig. 9.19). All blue-greens, like the bacteria, are prokaryotes.

The various colors of the blue-greens result from the presence of several **carotenoid** (L. *carota*, carrot) pigments (red, orange, yellow), along with a chlorophyll found in higher plants; in addition, *phycocyanin*, a blue pigment, and *phycoerythrin*, a red pigment, are also found in blue-greens.

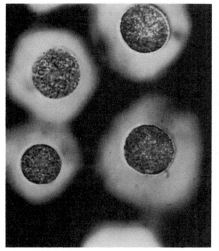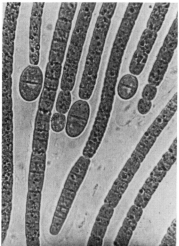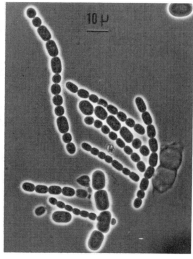

Figure 9.19 Some representative species of unicellular and filamentous blue-green algae. (From Gerking, S. D.: *Biological Systems*. 2nd ed. Philadelphia: W. B. Saunders, 1974, p. 224.)

Viruses; Kingdoms Monera, Protista, and Fungi

> Blue-green algae occupy some of the most inhospitable habitats on earth, including deserts, leached soils of the tropics, and hot springs.

Like the bacteria, blue-greens reproduce by binary fission; on occasion, however, filamentous species reproduce by *fragmentation,* in which the filaments break apart into smaller fragments, each of which grows into a new filament. Some blue-greens are capable of a gliding motion, but flagella are never present.

Blue-greens are important ecologically in that some species are capable of fixing atmospheric nitrogen (N_2), i.e., converting the nitrogen into forms that can be used by living organisms. Nitrogen fixation, which is also carried out by certain species of bacteria, is one of the steps in the cycling of nitrogen through the living world. We shall discuss the nitrogen cycle in more detail in Chapter 12.

In addition, the aquatic blue-greens constitute a large part of the *phytoplankton* (Gk. "drifting plants") of the oceans, seas, and fresh waters (Fig. 9.20). Being photosynthetic, these tiny, one-celled organisms—predominantly algae—are a major source of oxygen for many aquatic environments. Moreover, they may supply as much as half the oxygen in the earth's atmosphere.

Some species of blue-greens exist in association with a fungus to form a type of plant life called a *lichen.* This is a type of symbiosis known as *mutualism,* in which both species benefit. The blue-green provides food through photosynthesis, and the fungus retains water and provides protection.

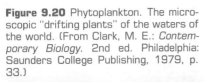

Figure 9.20 Phytoplankton. The microscopic "drifting plants" of the waters of the world. (From Clark, M. E.: *Contemporary Biology.* 2nd ed. Philadelphia: Saunders College Publishing, 1979, p. 33.)

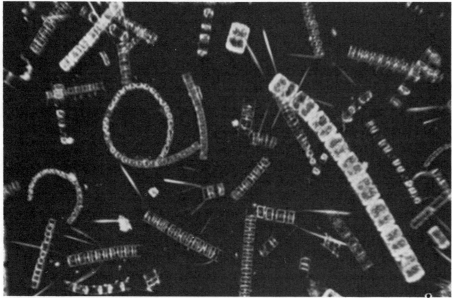

It is possible to destroy the chloroplasts of photosynthetic euglenas by keeping the organisms in the dark. The colorless strain produced does not die but shifts to a heterotrophic method of nutrition.

A most interesting aspect of the blue-greens concerns their apparent role on the primitive earth. It is thought that, through their photosynthetic activity, these organisms were responsible for much of the development of the oxygen-rich atmosphere formed around the earth perhaps 2 billion years ago. In all the ensuing centuries, the structural features of the blue-greens apparently have changed very little.

KINGDOM PROTISTA

In viewing the make-up of living organisms, we can note there are those that are too complex structurally to be included with the prokaryotic monerans and that are also difficult to classify as either plants or animals. These "intermediate" organisms have been assembled into a single group called the *protista,* a term derived from the Greek *protos,* meaning "first." The protists include the *unicellular algae* (except the blue-green), the fungus-like *slime molds,* and a group of unicellular or colonial organisms known as *protozoa.*

Unicellular Algae (Plantlike Protists)

There are three major divisions of unicellular algae—Euglenophyta, Chrysophyta, and Pyrrophyta. Although the organisms placed in these groups are referred to as "plantlike," many of them possess characteristics of animals as well.

Division Euglenophyta. The Euglenophyta (Gk. *eu,* good, + *glene,* eyeball, + *phyton,* plant), or "euglenoids," are microscopic unicellular organisms found mostly in fresh water; some species are photosynthetic, and some are not.

The most common genus in this division is the green *Euglena,* an elongate, somewhat tapered organism, that possesses one short flagellum and one long flagellum at its anterior end (Fig. 9.21). Also located near the anterior end is a tiny reddish-orange granule called the *stigma* (eyespot), from which the organism, and the division, got its name. The stigma is light-sensitive, permitting the organism to respond to light in accordance with its photosynthetic activity. Scattered throughout the interior of the cell are small chloroplasts, each containing an organelle called a *pyrenoid.* Pyrenoids produce a special kind of starch called *paramylum,* which the organism stores as a reserve food supply.

The nonphotosynthetic euglenoids closely resemble *Euglena,* but since they lack chloroplasts, they are essentially colorless. All euglenoids possess a single **contractile vacuole** that pumps water from the cell to the exterior. Although they are not known to reproduce sexually, euglenoids undergo asexual reproduction by mitosis, which divides the cell longitudinally.

Division Chrysophyta. The organisms found in the Chrysophyta (Gk. *chrysos,* gold) are generally yellow or brown in color, and all are photosynthetic. Movement is by flagella, although some species have no flagella. As with most unicellular algae, reproduction in the Chrysophyta is predominantly asexual.

Some of the most interesting—and beautiful—members of the Chrysophyta are the ***diatoms*** (Fig. 9.22). These algae are found in both fresh and marine waters, where they constitute one of the major groups of phytoplankton. Diatoms are constructed of two silicon shells that fit together like the top and bottom of a Petri dish. Many of these shells have minute perforations that form exquisite patterns (Fig. 9.23). Over millions of years, the shells of diatoms have accumulated on the floor of the ocean, forming deposits of "diatomaceous earth." This material is mined and then used in fertilizers,

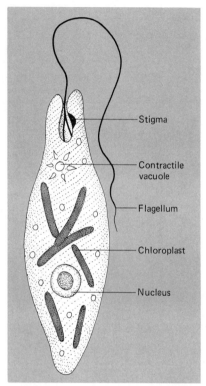

Figure 9.21 *Euglena*, a photosynthetic unicellular alga.

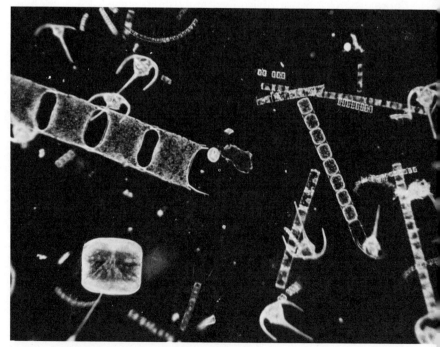

Figure 9.22 Various species of diatoms. (Division Chrysophyta) (From Ebert, J. D., Loewy, A. G., Miller, R. S., and Schneiderman, H. A.: *Biology.* New York: Holt, Rinehart and Winston, 1973, p. 582.)

225
Kingdom Protista

> So fine are the pores and surface structures of some diatoms, that slides of the organisms are used to check the lens systems of light microscopes.

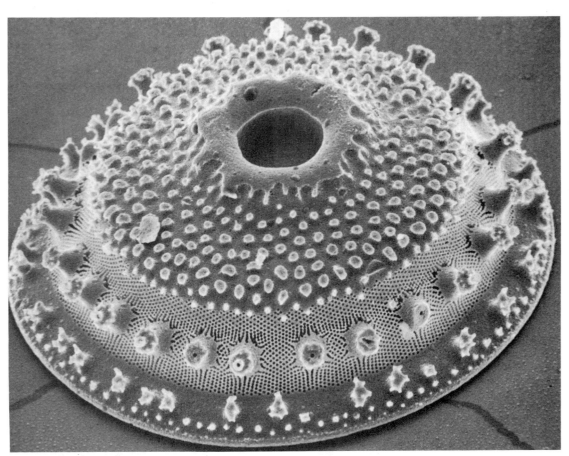

Figure 9.23 Electron micrograph of a diatom, a unicellular alga. (From Ebert, J. D., Loewy, A. G., Miller, R. S., and Schneiderman, H. A.: *Biology*. New York: Holt, Rinehart and Winston, 1973, p. 582.)

> Humans can become ill by consuming clams or oysters that have taken up dinoflagellate toxins released in the surrounding water.

detergents, polishes, toothpaste, scouring powders, and a variety of other products.

Division Pyrrophyta. Members of the Pyrrophyta (Gk. *pyrros*, reddish, flame-colored) are known as ***dinoflagellates*** ("spinning whips") (Fig. 9.24). Some species are photosynthetic, whereas others are either free-living or parasitic heterotrophs. Almost all of the dinoflagellates are marine and, like the diatoms, constitute one of the main groups of phytoplankton. Only some of the dinoflagellates are red in color; others may be greenish-yellow, brown, or colorless.

Some of the dinoflagellates are covered by a tough cellulose plating, but other members are naked. Almost all species have two unequal flagella, one encircling the middle of the organism and the other extending outward posteriorly. The beating of the flagella causes the organisms to spin like tops.

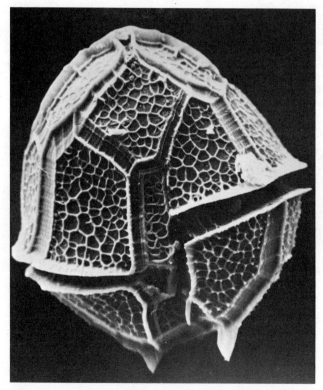

Figure 9.24 A scanning electron micrograph of a dinoflagellate taken at a magnification of more than 2,000 times. (From Camp, P. S., and Arms, K.: *Exploring Biology*. Philadelphia: Saunders College Publishing, 1981, p.13.)

> The slime molds basically are multinucleated naked masses of protoplasm. They are both plantlike and animal-like, but they have no apparent relationships to other living things.

Several species of dinoflagellates are able to give off flashes of light that illuminate the sea at night. On occasion, some species of red-pigmented dinoflagellates multiply into billions, coloring the water red and thereby producing what is known as a "red tide." Red tides, usually appearing on the Pacific Coast and in the Gulf of Mexico, are responsible for the death of countless fish, a consequence of an extremely potent nerve toxin produced by one species of dinoflagellates.

The Slime Molds

Division Gymnomycota. There are two major groups of slime molds, the *true,* or *plasmodial, slime molds* and the **cellular slime molds.** They are fungus-like in that both groups are nonphotosynthetic and are usually found growing in damp areas or on decaying organic matter. The division name, Gymnomycota (Gk. *gymnos,* naked, + *mykes,* fungus), alludes to the fungus-like characteristics of the group and to the fact that for the most part they are naked masses of protoplasm. Although the plasmodial and cellular slime molds are physically quite similar, each has a somewhat different life history, which we can use to illustrate the concept of a *life cycle.* A life cycle is a series of processes undergone by an organism in its development from one form to another and back to the same form again. Such forms may be quite diverse in a given life cycle, or they may closely resemble each other except for a few minor variations.

Plasmodial Slime Molds. A slime mold appears as a mass of naked protoplasm that creeps along like an amoeba. The mass, called a **plasmodium,** feeds on small particulate matter, which it engulfs by phagocytosis. The plasmodium is diploid and contains numerous nuclei (multinucleate). When unfavorable conditions develop, such as a lack of food or moisture, the plasmodium stops moving and separates into individual masses or stalks called *fruiting bodies.* At the tips of the fruiting bodies, structures called **sporangia** (Gk. *spora,* seed, + *angeion,* vessel) develop, in which meiosis occurs, producing haploid spores. When released from the sporangia, the spores germinate to produce flagellated gametes. Pairs of gametes fuse to form diploid zygotes, each of which develops into a mass of protoplasm. The diploid nucleus in each zygote divides repeatedly by mitosis to produce a multinucleate plasmodium, completing the cycle.

Cellular Slime Molds. The life cycle of the cellular slime molds is rather intriguing in that there is no sexual phase, i.e., no fusion of gametes, and the structures formed are all haploid.

Viruses; Kingdoms Monera, Protista, and Fungi

> Although most protozoans are unicellular, many of them are much more complex than any single cell of higher multicellular organisms.

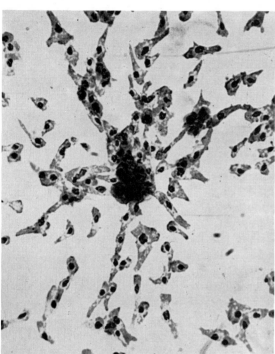
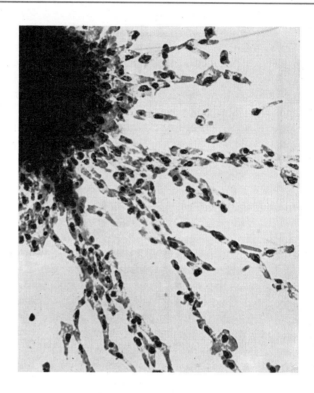

Figure 9.25 Reproduction in cellular slime molds. (Left) Individual haploid cells aggregating to form a pseudoplasmodium. (Right) Later stage of aggregation. (From Villee, C. A.: *Biology.* 7th ed. Philadelphia: Saunders College Publishing, 1977, p. 178.)

The protoplasmic mass in the cellular slime molds is called a ***pseudoplasmodium,*** which is formed from separate individual haploid cells (Fig. 9.25). When conditions are suitable, a pseudoplasmodium develops into a single stalked fruiting body with a sporangium at its tip. Within the sporangium, haploid spores are produced by mitosis. After their release, the germinating spores do not form gametes but instead develop into tiny haploid amoeba-like cells. These cells feed by phagocytosis, during which they undergo repeated mitotic divisions to produce numerous haploid offspring. As the food supply dwindles, groups of these amoeboid cells begin to migrate slowly toward different areas, where each group aggregates to form its own pseudoplasmodium. As mentioned, each pseudoplasmodium consists of individual haploid cells; the pseudoplasmodia persist for a time, after which another life cycle begins anew (Fig. 9.26).

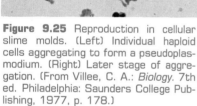

Phylum Protozoa

The phylum Protozoa (Gk. *protos,* first, + *zoion,* animal) includes the animal-like protists, all of which are either unicellular or colonial heterotrophs. In our survey here we will consider the four major classes of protozoans: Mastigophora, Sarcodina, Ciliata, and Sporozoa. As we shall discuss, protozoans are placed in each of the first three groups according to their method of locomotion. Adult organisms in the class Sporozoa are nonmotile. Although most protozoans are quite small, they often exhibit complex functional characteristics and an interesting assortment of life styles.

Class Mastigophora. These protozoans, also known as zooflagellates, constitute more than half of all protozoan species (Fig. 9.27). Most of them move by means of long, whiplike flagella, of which there may be one, two, or dozens, They are considered to be the most primitive of the protozoans, and it is possible that ancestral flagellates gave rise not only to the other groups of protozoans but also to the multicellular animals as well.

Zooflagellates reproduce by binary fission, in which the nucleus divides by mitosis. Although some species inhabit fresh and marine waters, most of these protozoans have taken up a symbiotic relationship with various species of plants and animals. For example, the parasitic genus *Trypanosoma,* which is responsible for African sleeping sickness, lives in the bloodstream of human beings and other higher vertebrates. The parasite is transmitted by tsetse flies. There are other zooflagellates that live symbiotically in the intestine of the common termite. The zooflagellates release enzymes that aid the termite in digesting wood. This is an example of mutualism, since both organisms in the relationship benefit. These mutualistic protozoans characteristically have the greatest number of flagella of any of the organisms in the Mastigophora.

Class Sarcodina. The Sarcodina, or sarcodines, a term meaning "fleshlike," generally move by means of ***pseudopodia***, or "false feet," which actually are cytoplasmic extensions of the body. The slow streaming motion of these organisms is called ***amoeboid movement.*** We have encountered this type of movement previously in our discussion of the slime molds. Pseudopodia also are used to surround and engulf prey, which are then digested by cellular enzymes (Fig. 9.28). This is another incidence of phagocytosis, a process also used by your white blood cells in surrounding and engulfing harmful substances in the blood.

The sarcodines are found in both fresh and salt waters. The freshwater group faces the problem of living in a hypotonic environment and, as a consequence, possesses *contractile vacuoles* that continually pump water out of the body. Almost all sarcodines are free-living, although some are parasitic, including one pathogenic form that causes the human inflammatory intestinal disease called amebic

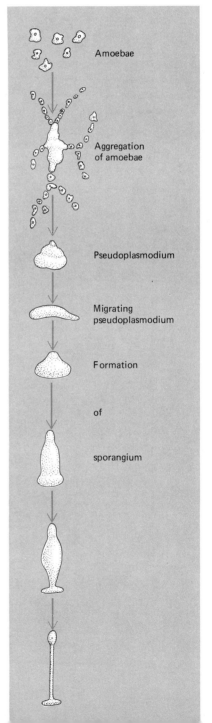

Figure 9.26 Life cycle of a cellular slime mold.

Amoebae

Aggregation of amoebae

Pseudoplasmodium

Migrating pseudoplasmodium

Formation

of

sporangium

230
Viruses; Kingdoms Monera, Protista, and Fungi

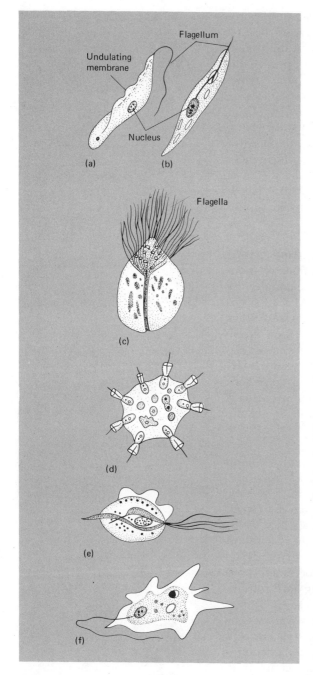

Figure 9.27 Flagellates. (a) *Trypanosoma gambiense,* the cause of a major type of African sleeping sickness. The flagellum is attached to the cell by a thin membranous structure. This undulating membrane, as the flagellum and membrane are called, is used in locomotion. (b) *Heteronema acus,* a free-living euglena-like flagellate. (c) *Calonympha grassii,* a symbiotic flagellate that lives in the gut of termites. These flagellates aid in the digestion of the cellulose eaten by the termites. (d) *Protospongia haeckeli,* a colony of flagellate cells that closely resemble the collar cells of sponges. The individual flagellate cells are embedded in a gelatinous mass. The surface of the collar streams downward toward the cell surface. This enables the collar to act like a continuously moving flypaper that brings trapped food down to the cell surface, where it is engulfed by a food vacuole. (e) *Trichomonas vaginalis,* a flagellate parasitic in the human vagina and male reproductive tract. The organism has four flagella and an undulating membrane. (f) *Mastigamoeba* is particularly interesting because it has a permanent flagellum but utilizes pseudopods for feeding.

Figure 9.28 Diagram of an amoeba engulfing and digesting a small flagellate.

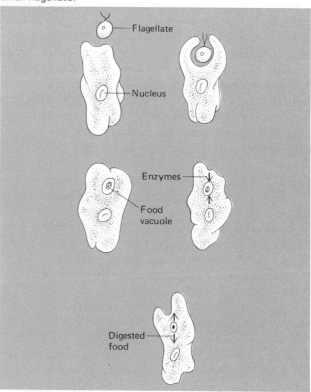

The fossil remains of some of the shelled amoebae date back more than 600 million years.

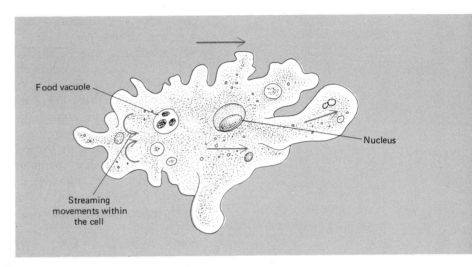

Figure 9.29 *Amoeba*, a sarcodine protozoan.

dysentery. Binary fission, in which the nucleus divides by mitosis, is the usual method of reproduction; however, sexual reproduction also occurs. In this instance, the diploid cell body of the sarcodine undergoes meiosis to form haploid gametes; the gametes from different organisms then fuse to form diploid zygotes, which grow into adult organisms.

Many of the sarcodines, such as *Amoeba,* are naked masses of protoplasm enclosed in a cell membrane (Fig. 9.29); many other amoebae, however, are encased within solid, often brightly colored shells. The snail-like shells of one marine group—the Foraminifera (L. "hole bearers," since their shells are perforated with tiny holes)—form the chalky white cliffs of Dover in England (Fig. 9.30). Such

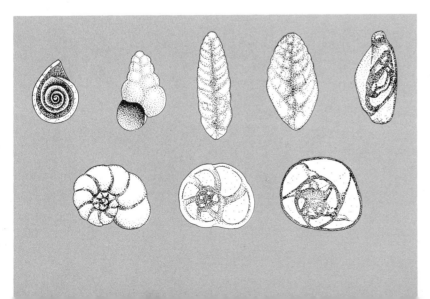

Figure 9.30 Foraminiferans are sarcodines encased within a snail-like shell.

> The macronucleus of a ciliate apparently is necessary for survival. A ciliate can survive without a micronucleus as long as even a small portion of the macronucleus is present.

accumulations of foraminiferan shells are found in many areas of the oceans, where they have been built up over many millions of years. Another group of shelled sarcodines—the Radiolaria (L. *radiolus*, little ray)—are exclusively marine and also have a long evolutionary history; in fact, they are the earliest of the known animal-like organisms on record (Fig. 9.31). Shelled amoebae are also found in the Heliozoa ("sun animals"), which is mainly a freshwater group. Heliozoans are so named because of their numerous needle-like pseudopodia that radiate outward from the body (Fig. 9.32). These pseudopodia, like those of many other shelled sarcodines, are used to trap food and are not used in locomotion. Prey organisms become trapped in a sticky secretion that coats each pseudopod. The pseudopod then retracts into the cell, where the organism is digested.

Class Ciliata. The Ciliata, or ciliates, take their name from the hundreds or thousands of tiny, hairlike cilia (L. *cilium*, eyelid) found on their bodies. Cilia function in locomotion and in obtaining food. Most of the ciliates are free-living and are found in marine waters as well as in fresh water. Unlike other protozoans, ciliates possess two different kinds of nuclei—a large **macronucleus** and one or more smaller **micronuclei**. The macronucleus apparently controls metabolism and growth of the cell, and the micronucleus functions only in reproduction.

Figure 9.31 Radiolarians usually have a shell composed of silica.

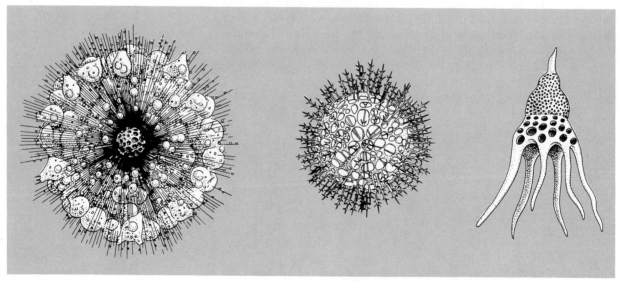

> The class Ciliata is the largest protozoan class and consists of some of the most complex cell types found in the living world.

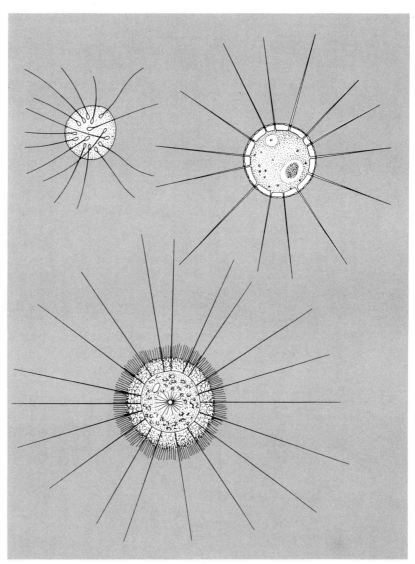

Figure 9.32 Heliozoans, or "sun animals," possess specialized needle-like pseudopodia that can be retracted into the body.

The ciliates reproduce asexually by transverse binary fission, during which the micronuclei divide by mitosis but the macronucleus simply constricts in two without undergoing mitosis. In this manner, each new organism receives a macronucleus and the correct number of micronuclei. Occasionally, many ciliates also undergo a sexual reproductive process called ***conjugation*** (L. *conjugare*, to join

> Sporozoa is the only class of protozoans that is entirely parasitic. Members of this class are most notable as agents of disease.

together) (Fig. 9.33). This involves the side-by-side union of two individual organisms, with the formation of a cytoplasmic bridge between them. The macronucleus in each organism degenerates, and their micronuclei undergo meiosis, with the result that each organism has two haploid micronuclei—a stationary "female" micronucleus and a motile "male" micronucleus. The "male" micronuclei are exchanged across the bridge, after which each fuses with a "female" micronucleus. Fusion of the micronuclei produces new diploid micronuclei in each organism. The conjugating pair separates, and the new micronuclei divide by mitosis to form additional micronuclei and macronuclei. Each organism then undergoes fission, during which the correct number of micro- and macronuclei is apportioned to each individual cell.

Perhaps the best-known freshwater representatives of the ciliates are in the genus *Paramecium,* the somewhat slipper-shaped protozoans commonly found in water containing decaying vegetation (Fig. 9.34). *Paramecium* is an excellent example of single-celled complexity. The organism is covered with thousands of microscopic cilia that allow it to move rapidly through the water, either forward or backward. Cilia are also used to sweep food particles into the *gullet* (L. *gula,* throat), a channel that leads to the interior of the cell. Food particles emptied from the gullet are then enclosed within **food vacuoles** that distribute the food throughout the cytoplasm. Undigested food is eliminated from the organism through an **anal pore** located at the posterior end of the cell. Two contractile vacuoles maintain osmotic equilibrium by continually pumping water to the outside. Embedded in the cell membrane of *Paramecium* and other genera are numerous tiny **trichocysts** (Gk. "hair bladder"), or spearlike capsules, that can be ejected for defensive purposes or for capturing prey (Fig. 9.35).

One of the more unusual ciliates is *Vorticella* (Fig. 9.36). This freshwater organism consists of a bell-shaped body attached to an elongated stalk. The organism anchors itself by the stalk, which then coils up into a tight spiral. Should a tasty morsel happen by, the stalk uncoils rapidly in a "popping" movement, thrusting the body outward. The morsel, with no time to react, is then drawn into the ring of cilia that encircles the open end of the bell.

Class Sporozoa. The Sporozoa ("spore animals") are so named because at one stage in their life cycles they are in the form of small, sporelike cells called ***sporozoites.*** The sporozoites arise from a single zygote that divides by **multiple fission,** i.e., mitotic division that pro-

Kingdom Protista

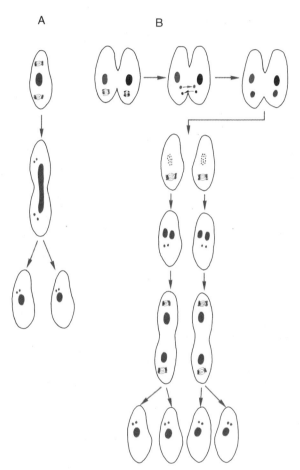

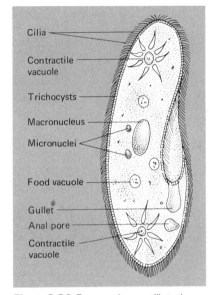

Figure 9.34 *Paramecium*, a ciliated protozoan.

Figure 9.33 Reproduction of *Paramecium*. (a) Asexual reproduction. The micronuclei divide mitotically, but the macronucleus is simply pulled apart. (b) Conjugation. In this simplified scheme, two individuals fuse. One micronucleus degenerates and the other divides mitotically. After mutual exchange, daughter micronuclei fuse together, then divide. The old macronucleus degenerates, and two new ones formed. After division, the new cells finally have the correct numbers of nuclei. (From Clark, M. E.: *Contemporary Biology*. 2nd ed. Philadelphia: Saunders College Publishing, 1979, p. 167.)

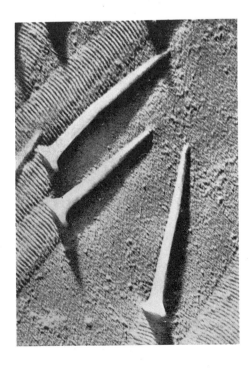

Figure 9.35 Trichocysts are spearlike capsules attached to a long thread. (From Clark, M. E.: *Contemporary Biology*. 2nd ed. Philadelphia: Saunders College Publishing, 1979, p. 166.)

> The fungi, although plantlike in appearance, are heterotrophs, acquiring their nutrients either from dead organic matter or from living hosts.

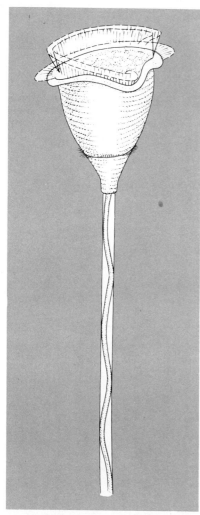

Figure 9.36 *Vorticella*, a stalked ciliate.

duces many individual cells. All sporozoans are internal parasites and are further characterized by the absence of locomotor organelles in the adult stages.

The most infamous of the sporozoans belong to the genus *Plasmodium*, species of which cause malaria. The organism is transmitted from one human being to another by female *Anopheles* (Gk. *anopheles*, harmful) mosquitoes (Fig. 9.37). When a mosquito bites a person infected with malaria, she sucks up blood containing immature *Plasmodium* gametes. In the mosquito's intestine, the gametes mature and unite to form a zygote; the zygote eventually undergoes multiple fission to produce numerous sporozoites, which migrate to the mosquito's salivary glands in the mouth. When the mosquito bites another person, the sporozoites are injected into the skin, where they enter the bloodstream and circulate to the liver. Within the liver cells, the sporozoites divide to produce numerous small cells called **merozoites**. These cells escape from the liver cells into the circulation, where they enter red blood cells and begin to divide asexually. At 48- or 72-hour intervals, depending upon the type of malaria, the infected red cells burst, releasing swarms of merozoites into the bloodstream. The merozoites then enter new red blood cells, undergo division, and burst out 48 or 72 hours later. This process may be repeated many times, with the red cells bursting every 48 or 72 hours. The ruptured red cells release toxins responsible for the chills and fever typical of malaria.

In time, some of the merozoites develop into immature gametes that circulate in the blood. When a person infected with malaria is bitten by an *Anopheles* mosquito, the immature gametes are sucked up in the blood and eventually move into the mosquito's intestine. With the entrance of the immature gametes into the body of the mosquito, the life cycle of *Plasmodium* is completed.

Not only humans but also reptiles, birds, and other mammals are parasitized by various species of *Plasmodium*. Other genera of sporozoans are responsible for *coccidiosis*, a disease particularly devastating to poultry. The disease is also found in other vertebrates, such as rabbits and cattle; coccidiosis even affects humans, in whom the disease produces mild diarrhea and abdominal distress. As with *Plasmodium*, the life cycles of the coccidian sporozoans are quite complex.

KINGDOM FUNGI

Being plantlike in appearance, the fungi were formerly placed in the plant kingdom, but their many unique features warrant classifying

> The evolutionary ancestry of the fungi is essentially unknown, but it has been suggested that they may have arisen from early protozoans or green algae, or both.

them in a kingdom of their own. One unique feature of the fungi is that, in lacking chlorophyll, they cannot make their own food. All the fungi, therefore, are heterotrophs; in fact, most of them are either parasites or saprophytes. Along with certain bacteria, they are important as decomposers in the cycles of living matter. The fungi are quite diverse in their ecological and economic roles and are involved in many activities that affect your life almost every day.

Except for the *yeasts,* which are unicellular, all types of fungi are composed of threadlike filaments called **hyphae** (Gk. "web"). All of the hyphae making up a particular fungus constitute a **mycelium** (Gk. *mykes,* fungus).

Each of the four divisions of fungi we shall consider is named with reference to its method of sexual reproduction. However, all of these groups also reproduce asexually.

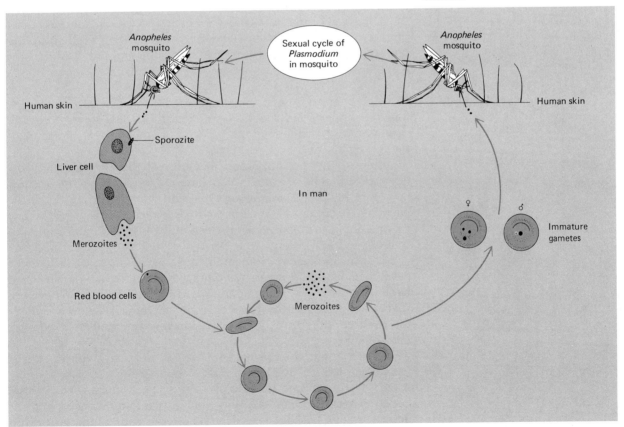

Figure 9.37 Life cycle of the malaria parasite *Plasmodium* (see text for discussion).

Viruses; Kingdoms Monera, Protista, and Fungi

> The Oomycota are the only group of fungi that produce eggs, sperm, and zoospores, i.e., spores equipped with flagella.

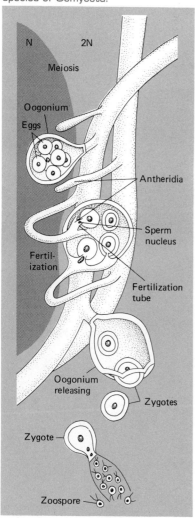

Figure 9.38 Sexual reproduction in a species of Oomycota.

Division Oomycota. The Oomycota, or "egg fungi" (Gk. *oon*, egg), are commonly known as "water molds," although many of them are terrestrial. Like other fungi, most of them are saprophytes, and some are parasites.

During sexual reproduction, some hyphae develop an enlarged *oogonium*—a female reproductive sac—in which haploid eggs are formed by meiosis. Other hyphae give rise to haploid sperm. When the male and female hyphae make contact, the sperm pass through tiny fertilization tubes into the oogonium. Fusion of the sperm and eggs produces diploid zygotes; each zygote then germinates to form a sporangium containing flagellated spores (zoospores). When the spores are released, they travel by wind or water to a suitable place where they grow into a new diploid fungus (Fig 9.38). The production of definitive eggs and sperm, along with flagellated zoospores, is unique to the Oomycota. These structures are unknown in any other group of fungi.

Of great historical significance is a species of Oomycota that caused the potato blights in Ireland during the 1840's. The resulting famine of 1845 to 1847 led to mass starvation and the death of over a million people.

Division Zygomycota. These fungi resemble somewhat the filamentous green algae and are often referred to as algal fungi. They are mainly soil fungi that feed on decaying organic matter, and some of them are parasites.

One of the most common members of the Zygomycota is the black bread mold *Rhizopus*. You have no doubt seen this cottony mass growing on cheese, bread, fruit, and other foods left out in the kitchen. The fungus is anchored to the bread or other food by tiny rootlike hyphae called *rhizoids,* which also function in absorbing nutrients.

During asexual reproduction, the upright hyphae of *Rhizopus* develop globular *sporangia* containing thousands of nonflagellated haploid spores. When the sporangia ripen and burst, the liberated spores are dispersed by wind currents. If they fall onto a suitable food source, the spores germinate, forming the familiar cotton-like mass of hyphae.

Under some circumstances, *Rhizopus* also may reproduce sexually. Two haploid hyphal strands of opposite mating types (designated + and −) line up side by side, each of which then forms a lateral extension. One extension contains haploid + nuclei, and the other contains haploid − nuclei. The extensions continue to grow

Kingdom Fungi

> *Rhizopus* and other saprophytic fungi release enzymes that break down food outside the fungal cells. The digested food then enters the cells by absorption.

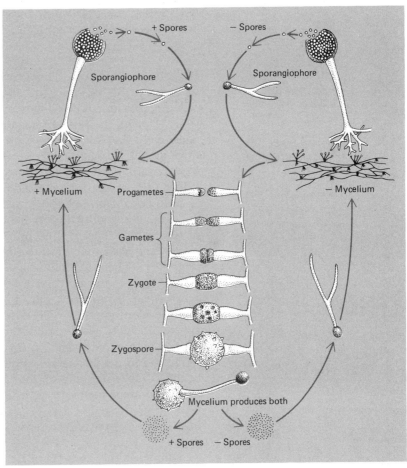

Figure 9.39 The life cycle of the black bread mold *Rhizopus nigricans*. The diagrams at the top indicate the asexual production of mycelia from spores. In the center is a series of stages in sexual reproduction.

laterally until they eventually touch, following which the wall between them disintegrates. This permits the + and − nuclei to fuse and form a diploid zygote. The zygote remains dormant for a time, after which it undergoes meiosis to form a haploid hypha with a sporangium at its tip. When mature, the sporangium bursts, releasing asexual spores that germinate to form a new fungal mass consisting of + and − hyphae (Fig. 9.39).

The method by which the zygote is formed in *Rhizopus* illustrates the derivation of the division name, Zygomycota. The zygote of *Rhizopus* (and other members) is more correctly referred to as a **zygospore**. By definition, a zygospore is a type of spore formed by the fusion of nuclei from two *conjugating* organisms. Zygomycota

Consuming ergot-infected bread can produce constriction of peripheral blood vessels, resulting in a burning sensation in the hands and feet known in the Middle Ages as "St. Anthony's fire." Although this condition is rare today, ergot can be used medically to constrict blood vessels in some cases of internal bleeding.

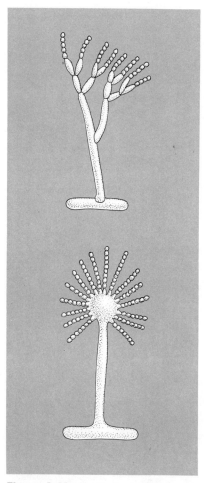

Figure 9.40 Asexual reproduction in some members of the Ascomycota occurs by means of tiny conidia.

(Gk. *zygon*, yoke, + *mykes*, fungus), therefore, refers to fungi that "yoke together," i.e., conjugate, during sexual reproduction.

Division Ascomycota. This group, known as "sac fungi," is characterized by the presence of sporangia in the form of little sacs called **asci** (sing. ascus). The term is from the Greek *askos,* meaning "bag." During the sexual reproductive cycle, each ascus usually contains eight haploid spores. Asexual reproduction usually occurs through the formation of tiny haploid spores called **conidia** (Gk. *konis,* dust), borne at the tips of specialized hyphae (Fig. 9.40). When the conidia are released, each can grow into a new fungus. Asci do not form during the asexual cycle.

Sexual reproduction in the sac fungi involves the germination of two haploid spores; one produces hyphae containing haploid + nuclei, and the other produces hyphae containing haploid − nuclei (Fig. 9.41). These hyphae, along with other hyphae that do not contain + and − nuclei, form a cuplike structure in which many asci are formed. Eventually, one + nucleus and one − nucleus enter each developing ascus, where the two nuclei fuse to form a diploid zygote nucleus. Each diploid nucleus then undergoes meiosis to form four haploid nuclei; in turn, each of these nuclei divides once by mitosis to produce eight haploid nuclei. As the asci mature, each of these nuclei becomes enclosed in a protective wall, with the result that each ascus contains eight haploid spores. When an ascus ruptures at maturity, the spores are spewed into the air. Upon settling in a suitable environment, the spores germinate to begin another cycle.

Some of the more familiar and economically important members of the sac fungi are the **yeasts.** The yeasts are different from other sac fungi, as well as all other fungi, in that they are unicellular. Moreover, yeasts reproduce asexually by **budding,** in which a new yeast cell is formed from an outpocketing of the body wall of the parent cell (Fig. 9.42). From Chapter 6, you will recall that during fermentation, the yeasts produce ethyl alcohol and carbon dioxide. Fermentation, of course, is the basis of the wine and liquor industry, and the baking industry depends upon the carbon dioxide produced for the rising of dough.

The sac fungi are responsible for several plant diseases, which in turn affect human beings. Some of the sac fungi are parasitic, such as the species causing *ergot,* a disease of rye and other cereals. When eaten by man, such contaminated rye may produce serious physical and mental complications. For example, one of the constit-

Kingdom Fungi

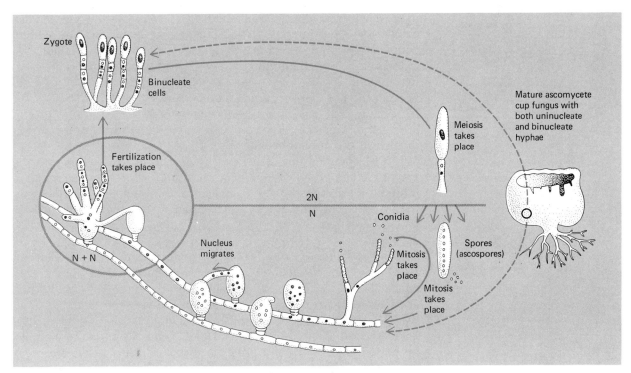

Figure 9.41 Life cycle of a typical sac fungus.

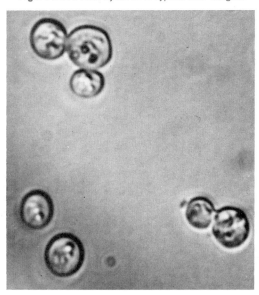

Figure 9.42 Yeast cells reproduce asexually by budding.

242

Viruses; Kingdoms Monera, Protista, and Fungi

> Although the familiar stalk and cap of many mushrooms are above ground, a great portion of the fungus is deeply embedded in the soil.

uents of ergot is lysergic acid, a substance found in the psychedelic drug LSD. Consumers of lysergic acid may experience grotesque hallucinations and other mental delusions. Various other plant diseases, such as Dutch elm disease and chestnut blight, are also caused by sac fungi.

On the positive side, some members of the Ascomycota, such as truffles, are edible delicacies, and members of the *Penicillium* group are used to enhance the flavor of Roquefort cheese and in the production of the antibiotic penicillin.

It was pointed out earlier that the forms of plant life known as *lichens* are composed of an alga and a fungus existing together in a mutualistic relationship. In most lichens, the fungus is a member of the Ascomycota, and the other component is either a green or a blue-green alga.

Division Basidiomycota. The name of this group is derived from *basidium* (Gk. "club"), a specialized club-shaped hypha on which spores are formed (Fig. 9.43). The Basidiomycota, therefore, are referred to as "club fungi." Perhaps the best-known representatives of this division are the mushrooms and toadstools. The difference between the two is that "toadstool" generally refers to inedible or poisonous mushrooms, and "mushrooms" are edible. However, unless one is an expert, there is no reliable way to distinguish between them. Other representatives include bracket fungi (often found growing on rotten logs), puffballs, rusts, and smuts. The last

Figure 9.43 (a) General structure of a mushroom. (b) Spores form on the basidia, which are located on the gills. (c) Close up of basidia and spores on gill.

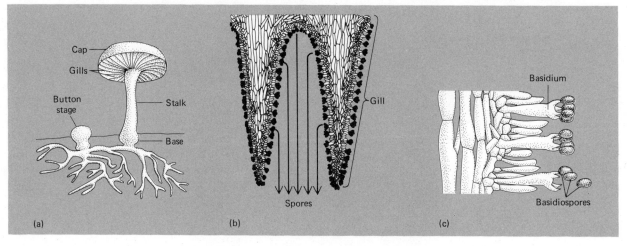

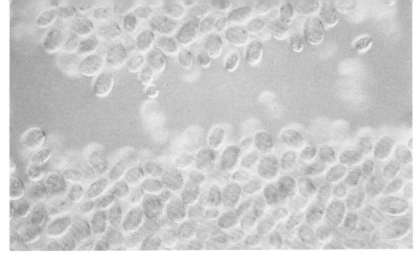

Color Plate 32 *Chlamydomonas,* a unicellular green alga. (Photograph by Bruce Russell, Biomedia Assoc.)

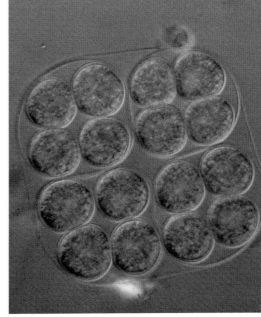

Color Plate 33 Four-celled colonies of *Gonium* immediately following reproduction. (Photograph by Bruce Russell, Biomedia Assoc.)

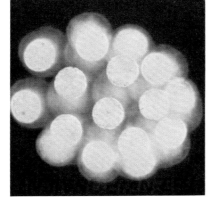

Color Plate 34 *Eudorina,* a spherical colony of *Chlamydomonas*-like cells, which may number 32, 64, or 128. (Photograph by Bruce Russell, Biomedia Assoc.)

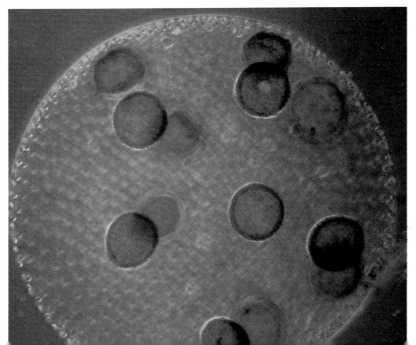

Color Plate 35 *Volvox* with its dark green daughter colonies. Depending upon the species, *Volvox* may consist of 500 to 50,000 individual cells. (Photograph by Bruce Russell, Biomedia Assoc.)

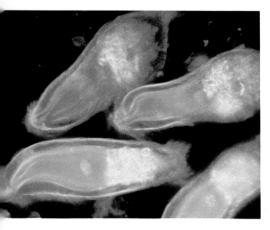

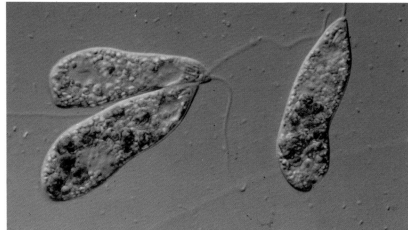

Color Plate 36 Flagellated protozoans from the gut of a termite. (Photograph by Bruce Russell, Biomedia Assoc.)

Color Plate 37 *Peranema* stained to show flagella. (Photograph by Bruce Russell, Biomedia Assoc.)

Color Plate 38 Plasmodium of a slime mold. (Photograph by Bruce Russell, Biomedia Assoc.)

Color Plate 39 Mold fruiting bodies. (Photograph by Bruce Russell, Biomedia Assoc.)

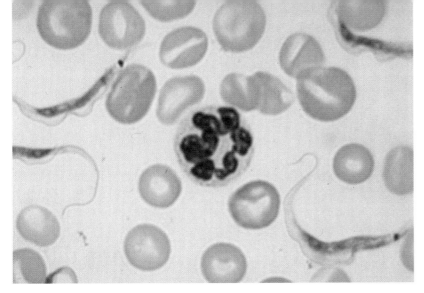

Color Plate 40 *Trypanosoma,* a flagellated protozoan found in the blood of vertebrates. (Photograph by Carolina Biological Supply Co.)

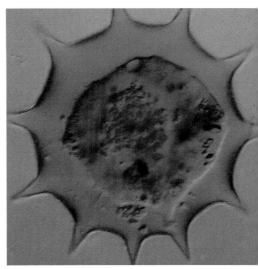

Color Plate 41 *Arcella,* a shelled amoeba. (Photograph by Bruce Russell, Biomedia Assoc.)

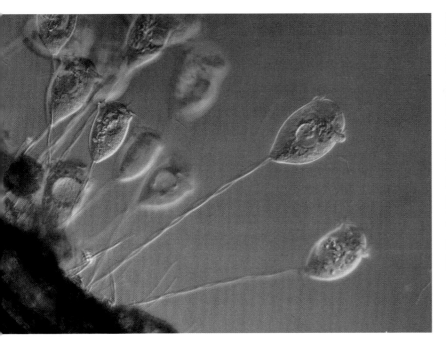

Color Plate 42 A group of the ciliate *Vorticella.* (Photograph by Bruce Russell, Biomedia Assoc.)

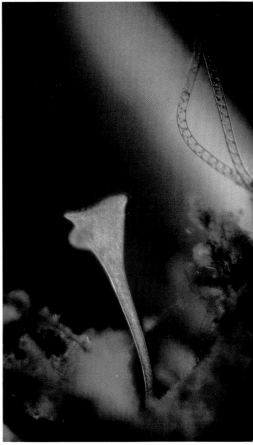

Color Plate 43 *Stentor,* a ciliated protozoan. (Photograph by Bruce Russell, Biomedia Assoc.)

Color Plate 44 Moss sporophytes. (Photograph by Bruce Russell, Biomedia Assoc.)

Color Plate 45 Sori (clusters of sporangia) on the underside of a fern leaf. (Photograph by Bruce Russell, Biomedia Assoc.)

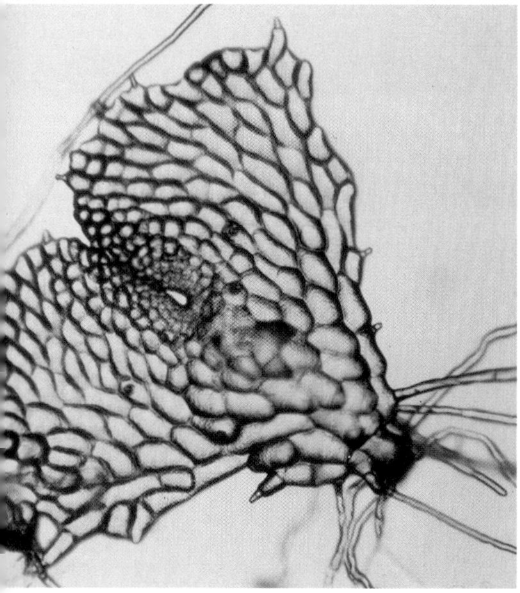

Color Plate 46 A fern gametophyte or prothallus. (Photograph by Bruce Russell, Biomedia Assoc.)

Color Plate 47 The veins of a leaf contain the vascular tissues xylem and phloem. (Photograph by Bruce Russell, Biomedia Assoc.)

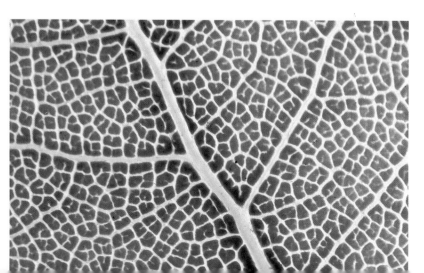

two are parasites that cause extensive economic loss by attacking various cereal grains. If eaten, various species of club fungi can be dangerously hallucinogenic.

In the sexual reproductive cycle, haploid spores germinate to form hyphae; some of these hyphae contain haploid + nuclei, and others contain haploid − nuclei. The two types of hyphae fuse to form a new mycelium, and the + and − nuclei come together to form pairs. After moving into the basidia at the tips of some of the newly formed hyphae, the + and − nuclei fuse to form diploid zygote nuclei. Each of the zygote nuclei then undergoes meiosis to produce four haploid nuclei. All of these nuclei migrate to the edge of a basidium, where each becomes enclosed in a small protuberance of the cell wall that extends outward along the edge. Closure of the cell wall, then, forms haploid spores containing either a + nucleus or a − nucleus. When separated from the basidium, the spores are carried away by the wind and may germinate to form new hyphae.

Fungi Imperfecti

The Fungi Imperfecti constitute a class of fungi in which sexual reproduction is unknown. Consequently, these fungi cannot be classified in any of the four divisions discussed here. The bane of the locker room—"athlete's foot"—is caused by a parasitic species of this group. Other parasitic species are responsible for several diseases of higher plants.

REVIEW OF ESSENTIAL CONCEPTS

1. *Taxonomy* is the biological science concerned with the naming and classification of living things.
2. There are five major kingdoms: *Monera, Protista, Fungi, Plantae,* and *Animalia.* The Monera are prokaryotes; all other organisms are eukaryotes.
3. The seven major taxonomic categories, from the most inclusive to the least inclusive, are the following: *kingdom, phylum* or *division, class, order, family, genus,* and *species.*
4. *Viruses* are noncellular parasites consisting of a nucleic acid surrounded by a protein coat. For their own maintenance and replication, viruses must acquire materials from the host cells.
5. The nucleic acid of most *DNA viruses* enters the nucleus of the host cell, taking control of the genetic machinery of the cell to produce

new viral particles. *RNA viruses* remain in the cytoplasm of the host cell and replicate using *RNA transcriptase* and *reverse transcriptase* enzymes.

6. *Bacteriophages* are DNA viruses that infect only bacteria. Bacteriophages replicate via a *lytic cycle* or *lysogenic cycle.*
7. Some viruses cause the production of *interferon* by infected eukaryotic cells; interferon tends to protect neighboring cells against future viral infection.
8. Viruses cause some cancers in plants and animals, but it is not known if human cancers are virally induced.
9. The kingdom Monera includes the *bacteria* and *blue-green algae;* both groups are *prokaryotes.*
10. Bacteria (Division Schizomycetes) may be *heterotrophic* or *autotrophic.* The forms of bacteria include *cocci, bacilli, spirilla, diplococci,* and *spirochetes.*
11. Heterotrophic bacteria are either *saprophytes* or *parasites.* Autotrophic bacteria are either *photosynthetic* or *chemosynthetic.*
12. Bacteria have beneficial roles, although some are *pathogenic,* causing such diseases as pneumonia, syphilis, botulism, and tetanus.
13. Bacteria are widely used in biological research; one bacterium in particular—*Escherichia coli*—has been commonly employed in *recombinant DNA* research.
14. The *recombinant DNA technique* involves "splicing" segments of bacterial *plasmid* DNA and foreign DNA to form recombinant DNA. Cleaving the plasmid and foreign DNA into segments is accomplished by enzymes called *restriction endonucleases.*
15. Through recombinant DNA technology, it is possible to study genes more thoroughly, synthesize specific proteins coded for by an inserted gene, and perhaps in the future replace defective genes with healthy genes.
16. *Blue-green algae* (Division Cyanophyta) are either unicellular or filamentous photosynthetic prokaryotes.
17. Blue-green algae are important ecologically as *nitrogen fixers, as phytoplankton,* and as *producers of atmospheric oxygen.*
18. Some blue-green algae exist in a mutualistic relationship with a fungus, forming a *lichen.*
19. The kingdom Protista includes the *unicellular algae* (except the blue-green), *slime molds,* and *protozoans.*
20. The three major divisions of unicellular algae are *Euglenophyta, Chrysophyta,* and *Pyrrophyta.*
21. Members of the division Euglenophyta are microscopic, mostly fresh-

water algae, some of which are photosynthetic. A common representative of this division is the green *Euglena*.

22. The division Chrysophyta includes yellow or brown-colored photosynthetic algae, some of which move by flagella. Representatives of this division are the *diatoms*.
23. Members of the Pyrrophyta are called *dinoflagellates*, some of which are photosynthetic, whereas others may be free-living or parasitic heterotrophs.
24. The Division Gymnomycota includes the *plasmodial slime molds* and the *cellular slime molds*. These organisms are nonphotosynthetic, naked masses of protoplasm.
25. The phylum Protozoa includes unicellular or colonial heterotrophs. The four major classes of protozoa are *Mastigophora, Sarcodina, Ciliata,* and *Sporozoa*.
26. Organisms in the class Mastigophora are generally characterized by having *flagella* as locomotor organelles. Most of these protozoans, such as the parasite *Trypanosoma* and the termite flagellates, exist in a symbiotic relationship with plants and animals.
27. Members of the class Sarcodina move by means of *pseudopodia* and may be free-living or parasitic. Sarcodines such as *Amoeba* are naked masses of protoplasm; others—the *Foraminifera, Radiolaria,* and *Heliozoa*—are encased within solid shells.
28. Organisms in the class Ciliata possess hairlike *cilia*, which function in locomotion and feeding. Ciliates have two different kinds of nuclei—a *macronucleus* and one or more *micronuclei*. Representatives of the ciliates include *Paramecium* and *Vorticella*.
29. Organisms in the class Sporozoa are all internal parasites and lack locomotor organelles in their adult stages. Species of *Plasmodium* cause the disease *malaria*.
30. The kingdom Fungi includes nonphotosynthetic heterotrophic organisms, most of which are either parasites or saprophytes. With the exception of the unicellular yeasts, the fungi are composed of *hyphae*, groups of which constitute a *mycelium*. The four divisions of fungi are *Oomycota, Zygomycota, Ascomycota,* and *Basidiomycota;* one group—*Fungi Imperfecti*—constitutes a class in which sexual reproduction is unknown.
31. The division Oomycota, also known as "water molds," includes mostly saprophytic species, although some are parasitic. The production of eggs, sperm, and flagellated zoospores is unique to the Oomycota.
32. Fungi in the division Zygomycota are mainly saprophytic soil fungi, but some are parasites. A common representative of this division is the

black bread mold *Rhizopus*, which reproduces asexually by spores and sexually by *conjugation*.

33. Fungi in the division Ascomycota possess sporangia in the form of *asci*, which are involved in sexual reproduction. Asexual reproduction occurs by *conidia*. Common members of the sac fungi are the unicellular *yeasts*.
34. Parasitic sac fungi are responsible for various plant diseases, such as *ergot*. Some sac fungi are edible; the *Penicillium* group is used to flavor cheese and to produce penicillin. Members of the Ascomycota also are components of lichens.
35. The division Basidiomycota includes mushrooms, toadstools, bracket fungi, puffballs, rusts, and smuts. In this division, spores are formed on club-shaped hyphae called *basidia*.
36. In the Fungi Imperfecti, sexual reproduction is unknown. "Athlete's foot" is caused by a parasitic member of this group.

APPLYING THE CONCEPTS

1. a. Why is taxonomy important in biology?
   b. Why is it necessary to have several major taxonomic categories?
2. Why are viruses considered nonliving "organisms?"
3. Compare the methods of infection and replication of DNA viruses and RNA viruses.
4. How do bacteriophages reproduce?
5. Although bacteria and blue-green algae are both prokaryotic, what are the fundamental differences between the two groups?
6. What are the differences between photosynthetic and chemosynthetic bacteria?
7. Describe in general two diseases caused by bacteria.
8. Outline the procedure for forming recombinant DNA.
9. a. What is a lichen?
   b. Why might such a relationship be necessary?
10. Describe the distinguishing characteristics of *Euglena*.
11. a. What are the basic differences between diatoms and dinoflagellates?
    b. How are these organisms similar?
12. List the characteristics that distinguish each of the four classes of Protozoa.
13. Describe the structure of *Paramecium*.

14. Outline the life cycle of *Plasmodium*.
15. Describe the process of conjugation in *Rhizopus*.
16. What are the distinguishing characteristics of the division Ascomycota?
17. Describe the process of spore formation in the Basidiomycota.
18. Compare the blue-green algae, bacteria, and fungi in terms of: (a) their ecological roles, (b) their harmful effects on plants and animals, and (c) their contributions to the welfare of human beings and other organisms.

10 biology of plants

THE ESSENTIAL OBJECTIVES

You have understood this chapter when you are able to:

1. Name the three major divisions of algae and describe the basic characteristics of each division.
2. Describe the structure of *Chlamydomonas* and explain the significance of the *Chlamydomonas-Volvox* lineage.
3. Describe the structure and reproductive methods of the multicellular green alga *Spirogyra*.
4. Describe the structure of a typical moss and outline its reproductive cycle.
5. Describe the basic differences between nonvascular and vascular plants.
6. Describe the structure of the stems and roots of vascular plants.
7. Explain the processes of water and solute movement in vascular plants.
8. Describe alternation of generations in the ferns, gymnosperms, and angiosperms.
9. Describe the structure and function of a flower and a fruit.
10. Outline the embryonic development of an angiosperm.
11. Explain the physiological action of each of the major plant hormones.

As the major sources of food and oxygen for the living world, green plants are indispensable to the existence of all other forms of life. From the aquatic single-celled algae to the complex land-dwelling angiosperms, green plants display through their common capacity to photosynthesize a basic unity superimposed on immense diversity.

PREVIEW OF ESSENTIAL TERMS

alternation of generations A reproductive cycle in which a haploid (N) gametophyte generation is followed by a diploid (2N) sporophyte generation, which produces the gametophyte, completing the cycle.

zoospore A spore that possesses flagella and is therefore motile.

division of labor The specialization of cells to perform separate specific functions in a colonial or multicellular organism.

antheridium In some plants, the male reproductive structure that produces sperm.

archegonium In some plants, the female reproductive structure that produces eggs.

xylem A specialized vascular plant tissue that conducts water and minerals upward from the roots.

phloem A specialized vascular plant tissue that conducts sugars and other nutrients from the leaves to other parts of the plant.

transpiration The evaporation of water through the stomata of a plant leaf.

apical meristem A region of active mitotic cell division at the tip of a plant stem or root; cell division in the meristem increases the length of the stem or root.

vascular cambium In plant stems, a specialized layer of tissue that produces xylem and phloem; cell division in the vascular cambium results in the increase in diameter of a stem.

translocation The transporting of the products of photosynthesis within the phloem of the roots, stems, and leaves of a plant.

guttation The loss of liquid water from the surface of a plant leaf.

seed A plant organ consisting of an embryo and female storage tissue enclosed in a protective seed coat.

microspore A haploid spore, which in the seed plants develops into the male gametophyte, or pollen grain.

megaspore A haploid spore, which in the seed plants develops into the female gametophyte.

cotyledon An embryonic leaf of a seed plant.

radicle The embryonic root of a seed plant.

hypocotyl The region of a seed plant embryo that forms the lower part of the shoot.

epicotyl The region of a seed plant embryo that forms the upper part of the shoot.

flower The reproductive structure of angiosperms; consists in part of the female organs or the male organs or both.

stamen The male reproductive organ of a flower; consists of a filament and anther.

carpel The female reproductive organ of a flower; consists of the stigma, style, and ovary.

endosperm A triploid (3N) nutritive tissue found only in angiosperms; arises from the fusion of a haploid sperm and two haploid polar nuclei.

fruit The mature plant ovary, along with associated structures; the fruit contains the seeds.

tropism In plants, a turning response resulting from a variety of stimuli; the stimuli influence the activity of one or more plant hormones, which bring about the particular response.

plant hormone A chemical substance, often a growth regulator, that controls a particular plant response. A hormone is produced in one part of an organism and transported to another part, where it exerts a specific biological effect.

KINGDOM PLANTAE

The plant kingdom encompasses a rich and fascinating variety of organisms—some so tiny as to be invisible to the naked eye; others—the giant sequoias—represent the tallest living things on earth. By producing food and oxygen through the process of photosynthesis, plants make possible the existence of virtually every other form of eukaryotic organism on this planet. As sources of food, shelter, clothing, medicines, paper, and a host of other products, plants have been indispensable to the rise and progress of human civilization. In truth, the economic and political value of many plants has influenced dramatically the course of human history. The aftermath of the Boston Tea Party in 1773 is a case in point.

For most people, the term "plant" usually summons images of leafy green structures firmly embedded in the soil. Although this certainly applies to many plants, there are more primitive plant types—the *algae*—that do not have leaves, nor the stems on which to hang them; some of these plants are not green, nor are they adapted to living primarily on land, since they have no roots. The more advanced plants, on the other hand, are terrestrial and highly differentiated, with most groups having distinct roots, stems, and leaves. Moreover, all of the advanced plants are multicellular, whereas the algae include both multicellular and single-celled forms.

The cells of most plants are enclosed within a wall of *cellulose*, a glucose polymer that is tough and flexible (Fig. 10.1). Cell walls of some plants contain additional materials, such as *lignin*, a complex organic substance that makes the walls firm and rigid.

All plants are photosynthetic and contain some type of chlorophyll; in addition, some plants contain other pigments that are responsible for the particular color of the plant.

The life cycle of plants is called *alternation of generations,* in which a sexually reproducing generation alternates with an asexually reproducing generation. In higher plants the zygote is enclosed

> Although almost all plants share the general characteristic of being photosynthetic, there are a few parasitic forms that are nonphotosynthetic.

Figure 10.1 Cellulose fibers in a plant cell wall. (From Camp, P. S., and Arms, K.: *Exploring Biology*. Philadelphia: Saunders College Publishing, 1981, p. 189.)

within female reproductive tissues, where it develops into an embryo; among the algae, however, development of the zygote almost always occurs outside the body of the parent organism.

THE ALGAE

Algae (L. "seaweed") are photosynthetic plants found wherever there is adequate moisture. An entire algal plant is called a ***thallus*** (Gk. *thallos*, young shoot, twig) and does not possess true roots, stems, or leaves. Some of the algae are unicellular, some are multicellular, and some are composed of cells grouped together to form a colony. Although all the algae contain chlorophyll, many species contain additional pigments and may be gold, brown, red, or other colors. In fact, each of the three major divisions of algae is named according to its predominant pigment. These groups include the red algae (Division Rhodophyta), the brown algae (Division Phaeophyta), and the green algae (Division Chlorophyta). The first two divisions are mostly marine, with a few freshwater forms. The green algae are well represented in both marine and fresh waters, in moist soil, and even within the bodies of organisms such as *Paramecium* and *Hydra*. It is believed that some type of primitive green algae gave rise to all the land plants.

Division Rhodophyta

The red algae comprise a group of seaweeds that sometimes attain a length of a yard or more, although most forms are much smaller (Fig. 10.2). They are most commonly found in tropical waters, often several hundred feet below the surface. Red algae usually do not

Biology of Plants

> In the seas of the world, there are more species of red algae than those of all other groups of algae combined.

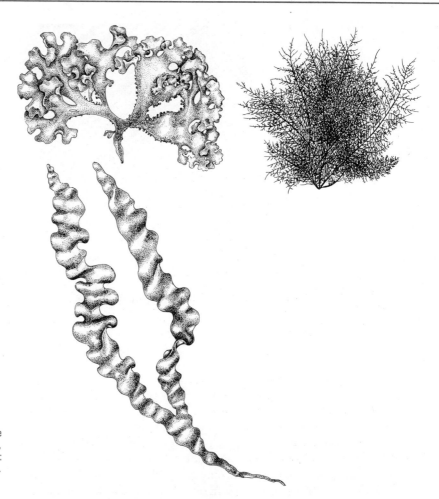

Figure 10.2 Some species of red algae (Division Rhodophyta). (From Villee, C. A.: *Biology*. 7th ed. Philadelphia: Saunders College Publishing, 1977, p. 177.)

float free in the water but are attached to underwater rocks or other aquatic plants.

Some of the red algae are used as food, most notably among the Scots and Japanese. An extract of red algae called **agar** is used to make the culture medium on which bacteria are grown in the laboratory. Another red algal extract called **carragheenin** is a gelatin-like substance used in chocolate milk, ice cream, puddings, toothpaste, and other products.

Various species of red algae are involved with corals in the formation of reefs. Corals are tiny, mostly colonial animals that secrete a calcium carbonate ($CaCO_3$) shell around themselves. When vast numbers of these organisms die, their amassed shells remain as a reef. The red algae accumulate calcium carbonate from the sea water and, by attaching to the reef, contribute to its formation.

> The pigment fucoxanthin is characteristic only of brown algae, and is not found in other living organisms.

Most of the red algae are multicellular, although a few are unicellular. All the Rhodophyta contain chlorophyll, along with the pigments phycocyanin, phycoerythrin, and some carotenoids. Many of these algae are a reddish color because of the predominance of phycoerythrin. However, there are other members of the Rhodophyta that actually are not red but may be yellow, green, brown, and other colors.

Sexual reproduction occurs through the fusion of nonmotile gametes. The Rhodophyta are unique among the major groups of algae in that none of their cells—not even the gametes—possess flagella. Consequently, the sperm is carried to the stationary egg by water currents.

Division Phaeophyta

The brown algae are a group of multicellular, extremely large seaweeds, some of which are over 300 ft in length (Fig. 10.3). They are

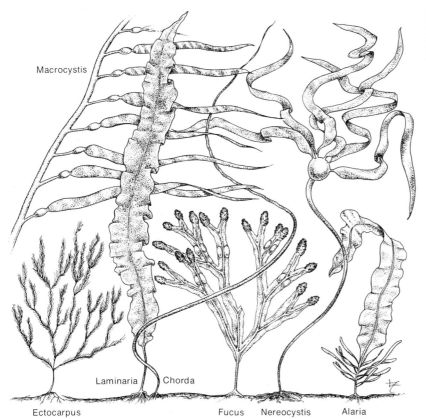

Figure 10.3 Various species of brown algae (Division Phaeophyta). (From Villee, C. A.: *Biology*. 7th ed. Philadelphia: Saunders College Publishing, 1977, p. 175.)

> It is generally believed that the first land plants descended from filamentous green algae some 400 million years ago.

found mainly along the coastlines in temperate and cooler climates. Almost all the brown algae are marine, although there are a few freshwater forms. In addition to chlorophyll, the Phaeophyta contain a carotenoid called *fucoxanthin,* which gives these algae their characteristic brownish color. An extract from brown algae called *algin* is used as a thickening agent in ice cream and is also an ingredient of some candies and toothpastes.

The life cycle of brown algae usually involves alternation of generations, in which a sexually reproducing, or *gametophyte,* generation alternates with an asexually reproducing, or *sporophyte,* generation. The gametophyte ("gamete plant") is so called because it produces gametes; the sporophyte ("spore plant"), on the other hand, produces spores. The gametophyte generation is haploid, and the sporophyte generation is diploid. The gametophyte and sporophyte are both multicellular, and in most brown algae they are structurally quite similar.

In the sexual phase of the life cycle, haploid gametophytes produce gametes by mitosis; the gametes, then, also are haploid. Fusion of gametes from different plants produces a diploid zygote, which germinates to form the diploid sporophyte. The zygote is the first of the structures that make up the asexual sporophyte generation. Within the sporophyte, haploid spores arise by meiosis. After their release into the water, the spores eventually develop into new haploid gametophytes, which completes the cycle.

Although the gametophyte and sporophyte are quite similar in many of the brown algae, this is by no means the situation in all plants. Among the land plants in particular, the gametophyte and sporophyte usually are vastly different, in both size and structure.

Division Chlorophyta

The green algae are more diverse, and include a greater number of living species, than either the red algae or the brown algae (Fig. 10.4). The majority of aquatic green algae are freshwater forms, but there are several species found in marine waters. The Chlorophyta range in size from the microscopic to over 25 ft in length. As indicated earlier, green algae may be found living symbiotically within organisms such as *Paramecium* and *Hydra;* in addition, some species exist in association with fungi to form lichens, and others inhabit moist soil or even melting snow. Along with blue-green algae, diatoms, and dinoflagellates, the aquatic green algae are the major source of photosynthesis in the waters of the world and are the essential source of food for many aquatic animals.

The green algae contain the same chlorophylls and carotenoids as the land plants, but they do not contain the various pigments

Figure 10.4 Green algae. These photographs give some idea of the diversity of form of green algae. These include freshwater forms like (a) the unicellular *Micrasterias* and spherical colonial forms such as (b) *Volvox*. Marine green algae include (c) the filamentous *Enteromorpha* as well as (d) giant unicellular forms such as *Acetabularia*, which have been used extensively in studies of nuclear cytoplasmic interactions. Another impressive marine green algae is (e) *Valonia*. Each *Valonia*, which may be as large as a hen's egg, is a single multinucleate cell and is among the largest cells known. (From Ebert, J. D., Loewy, A. G., Miller, R. S., and Schneiderman, H. A.: *Biology*. New York: Holt, Rinehart and Winston, 1973, p. 578.)

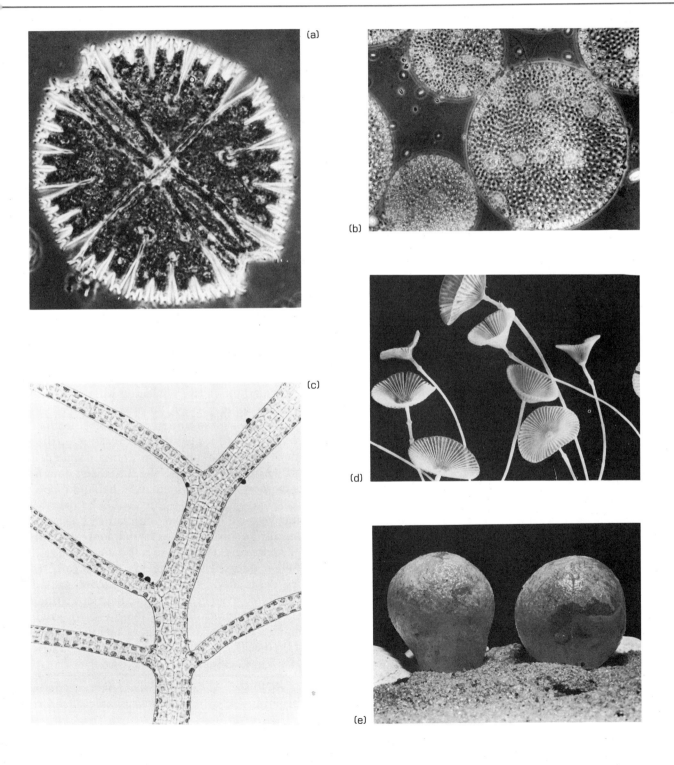

> *Chlamydomonas* and other single-celled forms are among the most primitive of all green algae.

Representative Green Algae

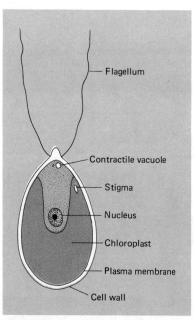

Figure 10.5 *Chlamydomonas*, a unicellular green alga (Division Chlorophyta).

found in the Rhodophyta or Phaeophyta. This is part of the evidence supporting the widely accepted view that Chlorophyta is the group that gave rise to all the land plants.

Chlamydomonas. One of the microscopic forms found in the Chlorophyta is the single-celled freshwater alga *Chlamydomonas* (Fig. 10.5). This protist-like organism is surrounded by a cell wall composed of carbohydrate and protein instead of cellulose. This is in contrast to the cell walls of other green algae, as well as those of higher plants, in which cellulose is the major constituent of the wall. Within the egg-shaped body of *Chlamydomonas*, there is a single large chloroplast containing the same chlorophylls found in higher plants. A pair of anterior flagella allow the alga to swim around like an animal; this movement is facilitated by a tiny red-pigmented stigma, or eyespot, that allows the organism to respond to light. Like many other single-celled freshwater organisms, *Chlamydomonas* possesses small contractile vacuoles that continually discharge water to the outside of the cell.

Reproduction in *Chlamydomonas*—which is itself a haploid cell—may occur both asexually and sexually, although the asexual method is more common (Fig. 10.6). During asexual reproduction, the nucleus divides by mitosis to produce two haploid daughter cells, which remain within the original cell wall. After developing their own cell walls and flagella, the two daughter cells are released when the original wall bursts. However, in some species, the two daughter cells may remain within the original wall and undergo mitosis to produce four haploid daughter cells; these four cells may be released, or they also may divide to produce eight cells, and so on until a number of daughter cells are formed. After their release, the flagellated daughter cells—referred to as **zoospores**—grow into mature haploid *Chlamydomonas* cells.

The sexual reproductive cycle begins when a mature *Chlamydomonas* cell divides mitotically to produce several haploid gametes. The gametes develop flagella and their own cell walls while still within the parent cell. When released from the parent, the gametes of opposite mating types, designated + and −, come together and eventually fuse to form a diploid zygote. The zygote develops a thick protective wall, which is highly resistant to extreme heat or cold. After a period of dormancy, the zygote undergoes meiosis to produce four flagellated haploid cells, which are released into the water, where they develop into mature cells.

From Single Cells to Colonies. A *Chlamydomonas*-like organism probably stood at the beginning of an evolutionary line that led to

> During asexual reproduction in *Volvox*, the cells of the spherical daughter colonies have their flagella oriented inward; before its release from the parent, each daughter colony turns itself inside out so that the flagella are directed outward.

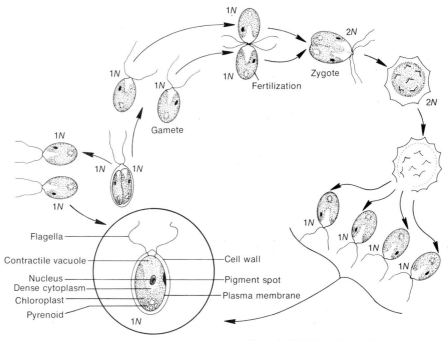

Figure 10.6 The life cycle of the green alga *Chlamydomonas*. Asexual reproduction on the *left*, stages in sexual reproduction on the *right*. *Inset:* Enlarged view of a single individual showing body structures. (From Villee, C. A.: *Biology*. 7th ed. Philadelphia: Saunders College Publishing, 1977, p. 164.)

increasingly more complex colonial forms (Fig. 10.7). There are, for example, living species of colonial algae composed of 4 to 32 individual *Chlamydomonas*-like cells; a somewhat more complex species consists of 8 to 32 cells; another has 16 to 64 cells; and a still more advanced group has 32 to 128 individual cells in the colony. Finally, this line culminates with a large, spherical alga named *Volvox*, in which the *Chlamydomonas*-like cells number from 500 to 50,000. The many cells of *Volvox* are connected by tiny cytoplasmic strands that give the alga the appearance of a single, multicellular organism. All of the cells of *Volvox* have two flagella and thus are involved in locomotion as well as being photosynthetic. However, some of these cells eventually become specialized for reproduction. When conditions warrant, some of the female cells divide by mitosis to form large, dark-green **daughter colonies** within the parent colony. When the parent colony bursts, the daughter colonies are released to begin an independent existence.

At times, some of the cells of the colony become specialized to form gametes, in which case *Volvox* may then reproduce sexually. In some species the colony produces both male and female gametes; in others, only one kind of gamete is produced. The haploid flagellated sperm swim to the egg, which remains within the parent colony. Fer-

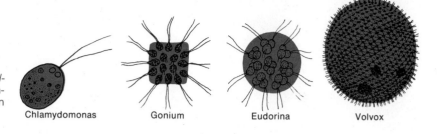

Figure 10.7 The *Chlamydomonas—Volvox* lineage illustrates the principle of increasing complexity and cellular division of labor.

Figure 10.8 *Spirogyra*, a multicellular filamentous green alga. (From Applewhite, P., and Wilson, S.: *Understanding Biology*. New York: Holt, Rinehart and Winston, 1978, p. 114.)

tilization produces a diploid zygote that undergoes meiosis; the resulting haploid cells then divide by mitosis to form a new daughter colony. The daughter colony remains for a time within the parent and is released when the parent colony breaks down.

In considering the lineage from *Chlamydomonas* to *Volvox*, we encounter yet another of the basic biological essentials: namely, that the increasing complexity of living organisms usually accompanies evolutionary advancement. This means that, in the evolutionary process, organisms tend to progress from simpler to more complex forms, which involves a progressive increase in the number of body cells. With more advanced organisms there also is an increasing interdependence among cells. This is a result of what is known as *division of labor,* in which different cells become specialized for specific functions. In the higher animals, for instance, some cells function in movement (muscle cells), some transport oxygen (red blood cells), some form a protective covering (skin cells), and some support the body (bone cells). Thus, the proper functioning of the organism is dependent upon all of these cells working together. As we will discuss, the same principle applies to the functioning of the land plants. The relatively simple lineage leading from *Chlamydomonas* to *Volvox* not only illustrates the principle of increasing complexity and cellular division of labor but further suggests one possible way in which multicellular organisms may have arisen.

A Multicellular Green Alga. An example of a multicellular, and rather unusual, member of the Chlorophyta is *Spirogyra*, familiar to almost everyone as "pond scum" (Fig. 10.8). The alga is filamentous, consisting of rectangular cells arranged like boxcars placed end to end. One of the unusual and distinguishing features of *Spirogyra* is its ribbon-like chloroplast, which forms a spiral along the length of each cell in the filament. Almost equally unusual for green algae is its method of sexual reproduction—a process known as *conjugation*. You will recall our previous encounters with conjugation in organisms such as *Paramecium* and *Rhizopus*.

At the beginning of the sexual process, two filaments, which are haploid in *Spirogyra*, line up side by side, and hollow **conjugation tubes** form between opposite cells of each filament. The entire contents of one cell, the **supplying cell,** then pass through a conjugation tube to fuse with the contents of the **receiving cell** in the other filament. The supplying cell thus acts in the capacity of the male by providing a motile gamete that moves into the receiving cell and combines with the nonmotile female gamete. Fusion of the cellular contents produces a diploid zygote, or zygospore. Later, each zygo-

> In making the transition from sea to land, the first land plants faced the new problems of drying out, transport of eggs and sperm, oxygen delivery, loss of buoyant support, and removal of metabolic wastes.

spore is released from its receiving cell and undergoes meiosis to produce new haploid *Spirogyra* filaments (Fig. 10.9).

Asexual reproduction in *Spirogyra* occurs by fragmentation, in which each broken piece grows into a new filament.

Many other species of multicellular green algae are also filamentous. Like *Spirogyra,* they reproduce asexually by fragmentation and undergo a sexual process involving the formation of haploid gametes and a diploid zygote that produces spores by meiosis. Upon germination, the haploid spores give rise to new filaments, thus completing the sexual cycle.

INTRODUCTION TO THE LAND PLANTS

It began some 400 million years ago—the rise of the autotrophic organisms known as land plants. As noted earlier, land plants are thought to have evolved from some type of green algae that for millions of years floated about in the ancient seas. For organisms to have moved from the sea to land—and to have remained—was a slow and extensive adaptive struggle. The major problem accompanying such a transition concerned the availability and retainment of water. Many modern plants have made full adaptation to a terrestrial existence; but for other plants the transition to land has been

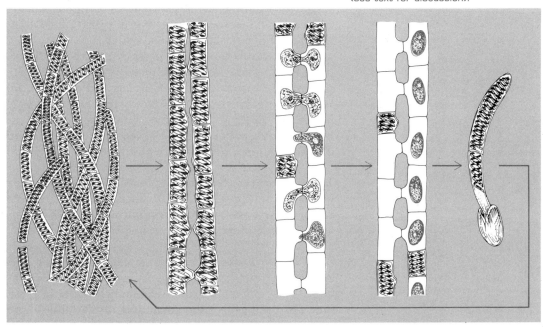

Figure 10.9 Conjugation in *Spirogyra* (see text for discussion).

> The bryophytes are unique in that, although they are land plants, they do not have functional vascular tissues, and they require water to transport sperm to the egg. Furthermore, the gametophyte generation of bryophytes is dominant, and the sporophyte usually is permanently attached to the gametophyte.

only partial, evidenced by the fact that they are almost always found near a direct source of water.

Although many kinds of plants have made the transition to land in the earth's history, most of them are now extinct. Only two major divisions are represented on the earth today—the Bryophyta and the Tracheophyta. Both groups are similar in that there is differentiation of the plant body to form specialized organs and structures; the fertilized egg in both groups develops within female tissues; parts of the plant body are usually covered by a waxy **cuticle** (L. *cutis*, skin), which helps prevent excessive water loss through evaporation; and all undergo a reproductive cycle involving alternation of generations.

There is, however, a most significant difference between the two groups—a difference that has resulted in contrasting abilities to adapt to a terrestrial existence. The tracheophytes have what is called **vascular tissues,** i.e., tissues that form a system of tubelike conducting cells that transport water, minerals, and nutrients from one part of the plant to another. This tissue is mostly lacking in the bryophytes. Tracheophytes, then, are referred to as **vascular plants;** the bryophytes are **nonvascular plants.** Vascular tissue allows the tracheophytes to retain water and to deliver it to all parts of the plant body. The differentiated body parts of the tracheophytes—**roots, stems,** and **leaves**—all contain this type of tissue. The veins in a leaf, for example, are small bundles of vascular tissue. After being taken up by the roots, water is then distributed to the stems and leaves by this internal transport system.

The bryophytes do not possess true roots, stems, or leaves, although the plant body is differentiated into leaflike and stemlike parts, along with some rootlike structures called **rhizoids.** In the absence of vascular tissue, the bryophytes cannot retain water for long periods of time, nor can water be transported up the plant body. Consequently, water must be absorbed directly from the surrounding air or from a nearby source of moisture.

NONVASCULAR PLANTS
Division Bryophyta

The bryophytes (Gk. "moss plants") include the hornworts, liverworts, and mosses (Fig. 10.10). Since the three groups are quite similar in many ways, we will direct our attention to the most numerous of the three in terms of different species—the mosses.

Class Musci. Lacking a vascular system for conducting water, mosses are usually found in moist areas, such as creek banks, swamps, and bogs. There is little reason to take much notice of these plants, since they are ordinarily quite small and display little that might attract one's attention. The fact that mosses are relatively inconspicuous

> In mosses, the first cell of the sporophyte generation is the diploid zygote; the gametophyte generation begins with the haploid spores.

Figure 10.10 Liverworts (a) and a moss (b), two representatives of living bryophytes. (From Arms, K., and Camp, P. S.: *Biology*. 2nd ed. Philadelphia: Saunders College Publishing, 1982.)

suggests something more about vascular tissue. Not only does it conduct materials, but also it serves as a means of mechanical support for a plant. Consequently, without it, mosses and other nonvascular plants never have the opportunity to become very large or tall.

The small leaflike structures of mosses are photosynthetic and possess tiny openings called **stomata** (sing. stoma), through which oxygen and carbon dioxide are exchanged. The upright stemlike parts provide support and also store food. The rootlike structures, or rhizoids, anchor the plant and, like other parts of the plant body, absorb water.

The mosses undergo alternation of generations, which, you will recall, involves two different stages, one sexual (the gametophyte), and the other asexual (the sporophyte), that follow each other in the life cycle of the plant. In lower plants such as mosses, it is usually the gametophyte that is conspicuous, or **dominant,** i.e., the stage in the life cycle you ordinarily see. The sporophyte in these plants is usually quite small, and unless you are a student of botany, you are probably unaware that it even exists.

There are separate sexes in most mosses, a haploid male gametophyte and a haploid female gametophyte (Fig. 10.11). These are the little green plants you actually see on a rock or a tree and recognize as being a moss. On the male plant, haploid sperm are produced by mitosis within reproductive structures called **antheridia;** in the female, haploid eggs develop within **archegonia,** found at the tip of the plant. In the fertilization process, the flagellated sperm require water, ordinarily in the form of rain or dew, to swim to the female plants. Union of sperm and eggs occurs within the archegonia; the resulting diploid zygotes, which will develop within the female plant, are the first cells of the sporophyte generation.

Each zygote divides by mitosis to form an elongated stalk having a capsule, or sporangium, at its tip. Throughout its development the sporophyte (stalk and capsule) remains attached to and nourished

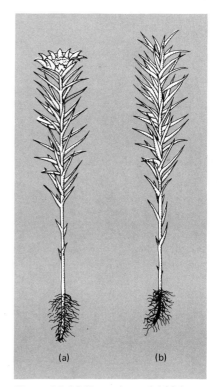

Figure 10.11 Moss plants: (a) Male gametophyte. (b) Female gametophyte.

Figure 10.12 *(Top)* Haploid spore germinating to form a moss protonema. *(Bottom)* The protonema gives rise to the moss gametophyte.

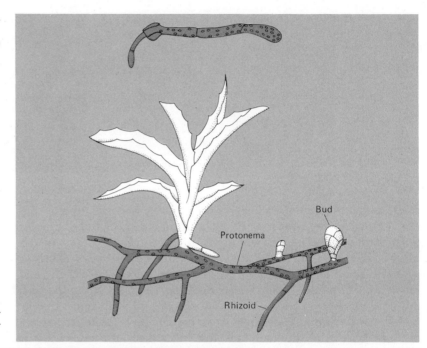

Figure 10.13 Life cycle of a moss. There is a distinct alternation of generations: the sporophyte and the gametophyte.

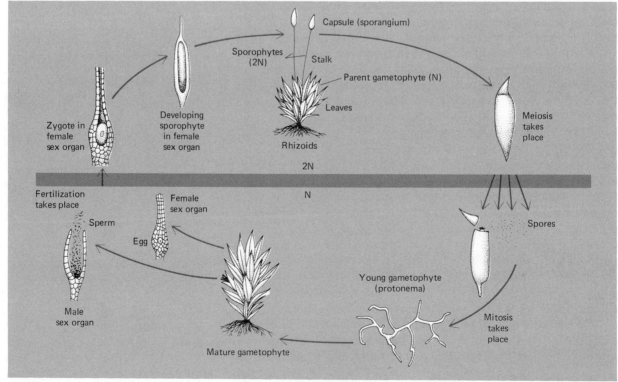

> There are approximately 260,000 known species of tracheophytes, more than 99.9% of which are ferns and seed plants.

by the female gametophyte. Within the capsule, haploid spores are produced asexually by meiosis and are released when the capsule matures. The spores are dispersed by the wind, and if they fall into suitable soil, each germinates into a tiny branched filament called a *protonema* (Gk. "first thread") (Fig. 10.12). These microscopic structures very much resemble filamentous green algae. Each protonema divides by mitosis to give rise to new male and female gametophytes, completing the cycle (Fig. 10.13).

Even though they occupy a lowly position in the plant kingdom, the mosses provide insight into some of the reproductive features that characterize many of the higher forms of life. First, there are often separate, distinguishable sexes in the mosses, each producing unique sex cells, or gametes. Second, fertilization is internal, within the body of the female. Third, development of the zygote also occurs for a time within the body of the female and is dependent upon the female for nourishment. These three general precepts are representative of the majority of the advanced forms of life in both the plant and the animal kingdoms.

VASCULAR PLANTS

Introduction to the Tracheophytes. The tracheophytes are characterized primarily by the presence of a vascular system, which consists of two types of specialized tissues—*xylem* and *phloem.* Xylem (Gk. "wood") functions in conducting water and minerals upward from the roots; phloem (Gk. "bark") transports sugars and other nutrients from the leaves to other parts of the plant. Both xylem and phloem are distributed throughout the roots, stems, and leaves. Vascular tissue also serves as mechanical support for a plant; consequently, some tracheophytes are capable of growing quite tall.

The leaves of tracheophytes are covered with a waxy cuticle that retards excessive water loss. The epidermis of the leaf is perforated with numerous stomata that allow for the exchange of oxygen and carbon dioxide. Each stoma is bordered by a pair of photosynthetic *guard cells* that regulate the size of the stoma (Fig. 10.14). The opening and closing of the stomata ultimately result from the gain or loss of water. As their water content increases, the guard cells become turgid and pull away from each other, opening the stoma. On the other hand, the loss of water causes the guard cells to become flaccid and to move back together, closing the stoma. As you would expect, the stomata are usually open during the day to allow the entrance of carbon dioxide for photosynthesis; at night the stomata normally close. Stomata also are important in allowing water to escape and evaporate through the surface of the leaf, a process called *transpiration.*

Biology of Plants

> Plant stems not only function as organs of conduction and support but also some are photosynthetic and some are edible storage organs.

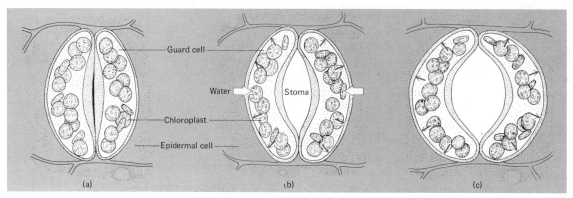

Figure 10.14 Diagrams illustrating the regulation of the size of the stoma by the guard cells. (a) Nearly closed condition. (b) When osmotically active substances such as glucose are produced, water enters the guard cells, turgor pressure increases, and the guard cells buckle so as to increase the size of the stoma. (c) Stoma open.

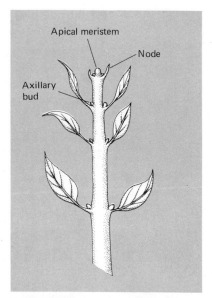

Figure 10.15 Growth areas of a plant stem.

The stems of vascular plants provide strong support and contain xylem and phloem cells continuous with those of the leaves. In many cases, stems also serve as food-storing organs, and green stems are photosynthetic. The tip of the stem has an **apical meristem,** or growth region, where cells actively undergo mitotic cell division. Growth of the meristem is responsible for increasing the length of the stem. Small **nodes** arise on the apical meristem and develop into leaves. In the angle between the developing node and the stem, an **axillary bud** arises that will develop into a lateral stem, or branch (Fig. 10.15).

As the stem lengthens, it also increases in diameter. This is brought about by a specialized layer of tissue called the **vascular cambium.** As the cells of the cambium layer divide, cells on the inner layer differentiate into xylem tissue, and cells on the outer layer differentiate into phloem tissue. As the diameter of the stem increases, the older, outermost phloem cells develop into a layer called the **cork cambium.** This layer continues to be pushed to the outside of the stem as new phloem cells are produced. Cell division within the cork cambium produces **cork cells,** which, together with the phloem tissue, constitute the bark of the stem. Thus, in going from the outside inward, the layers of the stem are cork cells, cork cambium, phloem, vascular cambium, and xylem, which is the woody part of the stem.

Although the foregoing is a general description of stems, there are actually two basic types: **herbaceous** stems and **woody** stems. Herbaceous stems are generally soft and contain bundles of vascular tissue (xylem and phloem) arranged in a ring around the periphery of the stem (Fig. 10.16). On the other hand, the vascular tissues of woody stems are layered throughout the stem (Fig. 10.17). There are some additional minor differences, but basically the two stem types are quite similar.

Vascular Plants

> The functions associated with the roots of vascular plants include anchorage, absorption, and conduction; many roots also function as food storage organs.

Roots, as we have said, anchor the plant and conduct water and minerals upward to the stems and leaves. The vascular tissue of roots is continuous with that of the stem, so that the entire vascular system of a plant is an interconnected network. Large individual

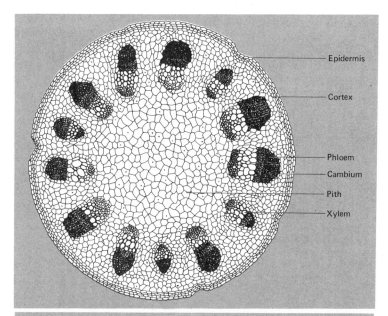

Figure 10.16 Cross section of a herbaceous stem.

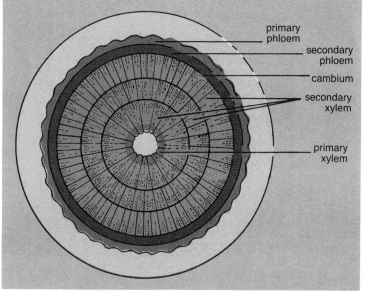

Figure 10.17 Cross section of a woody stem.

> Cells sloughed off the growing root tip secrete a slimy polysaccharide substance that facilitates movement of the root through the soil.

roots known as **taproots,** such as those of carrots and radishes, often serve as storage organs for the products of photosynthesis. A group of slender branching roots of the same size, such as are found in the grasses, are known as *fibrous roots* (Fig. 10.18).

Although there are some variations, the basic structural plan of roots is quite similar for all plants. At the tip of the root is the **apical meristem,** a region of active mitotic cell division. Covering the meristem is a protective **root cap,** which, like all the other tissues of the root, is derived from meristematic cells. Although the root cap cells are worn away as the root tip pushes through the soil, they are continually replaced by new meristem cells. Just above the meristem is the **zone of elongation,** in which the cells lengthen, causing the root to grow longer. Above this region is the **zone of differentiation,** where the elongated cells develop into the conducting cells of xylem and phloem (Fig. 10.19).

The entire root is covered by a thin epidermis; just behind the growing tip, a tiny hairlike extension of each epidermal cell forms a **root hair.** A single root may have billions of root hairs, which enormously increase the surface area for absorption (Fig. 10.20).

Figure 10.18 Types of root systems in plants. (a) The diffuse root system of a grass. (b) The taproot system of a carrot.

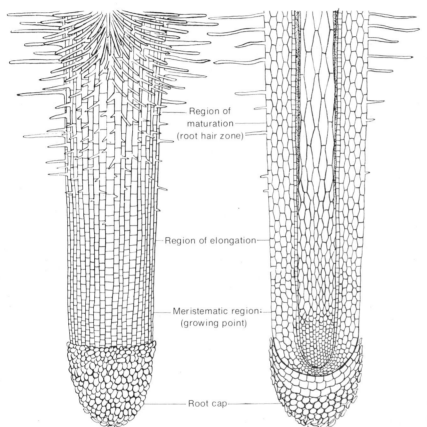

Figure 10.19 An enlarged diagrammatic view of the tip of a young root. *Left*, The surface of the root. *Right*, A longitudinal section of the root showing the internal structure. (From Villee, C. A.: *Biology*. 7th ed. Philadelphia: Saunders College Publishing, 1977, p. 254.)

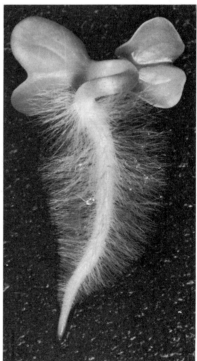

Figure 10.20 Root hairs of the young root of a radish. (From Ebert, J. D., Loewy, A. G., Miller, R. S., and Schneiderman, H. A.: *Biology*. New York: Holt, Rinehart and Winston, 1973, p. 213.)

> The energy for the evaporation of water from the leaf surface—the process necessary for pulling water upward from the roots—is supplied by the sun.

The thick tissue layer beneath the epidermis is the **cortex,** which partly consists of cells that often store starch; other cells of the cortex help support the root. The **endodermis,** a layer one cell thick, lies interior to the cortex. The endodermis surrounds the vascular tissue lying in the center of the root. In some roots the xylem cells form a large "X" in the center of the root; phloem cells are located in the angles of the "X" (Fig. 10.21). Other root types have a mass of cells called **pith** located in the exact center of the root. Bundles of xylem and phloem cells encircle the periphery of the pith.

Transport in Vascular Plants

In vascular plants, water and dissolved materials (Table 10.1) enter the plant through the roots, from where they move upward through the xylem cells into the stems and leaves. Some of the water is used by the leaves in photosynthesis, and the rest eventually evaporates from the surface of the leaves.

The products of photosynthesis are transported from one part of the plant to another—a process called **translocation**—within the phloem cells of the roots, stems, and leaves. Sugars and other organic products, as well as some inorganic solutes, are transported downward from the leaves to the stems and roots. Some of these products are used in cellular metabolism, and some are stored.

Movement of Water. The land plants, as we have said, take in water through their root systems. Water in the soil surrounding the roots has a relatively lower concentration of dissolved solutes than does the cytoplasm of the root cells. This is the same thing as saying there is a higher *solvent* (water) concentration outside the root cells than inside. Consequently, water enters the root, primarily through the

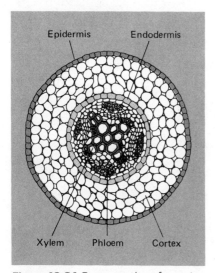

Figure 10.21 Cross section of a root.

Table 10.1 LIST OF MINERALS REQUIRED BY GREEN PLANTS

MACRONUTRIENTS	MICRONUTRIENTS*
Nitrogen	Iron
Phosphorus	Boron
Potassium	Manganese
Sulfur	Zinc
Calcium	Copper
Magnesium	Chlorine
	Molybdenum

*Some micronutrients are required only by certain species of plants and are not included here.

> Water exuded from the leaf during guttation does not escape through the stomates, but through tiny marginal openings called *hydathodes* (Gk. *hydatis* drop of water + *hodos* way, path).

root hairs, by osmosis. Upon entering the root, water continues to move inward, across the cells of the cortex, by osmosis. In this manner, water reaches the xylem cells of the root for transportation upward to the stems and leaves.

The ascent of water for a short distance up the stems of some plants can be explained in terms of osmotic pressure. From the outside of the root inward, all the way to the xylem cells, there is an osmotic gradient along which water moves across the root. Water moving into the xylem creates what is known as **root pressure** (Fig. 10.22). This can be thought of as a type of "push" that forces water up the xylem tissue. In some plants, root pressure is sufficient to force water out of the surface of the leaves, where tiny droplets form. This loss of liquid water resulting from root pressure is called **guttation** (L. *gutta*, drop) (Fig. 10.23).

Although root pressure may account for the upward movement of water in some plants, it is not sufficient alone to raise a column of water to the top of many plants that may be 75, 100, or even 300 ft tall. Root pressure is measured in terms of a unit called an **atmosphere**, which is equivalent to the pressure of air at sea level. One atmosphere of pressure can raise water about 32 ft. Measurements in several species of plants indicate that root pressure is seldom greater than 2 atmospheres. It is obvious, then, that in many plants, the ascent of water cannot be explained solely in terms of root pressure.

The current theory of water transport—known as the **cohesion-tension theory**—holds that water is "pulled" up the plant from above

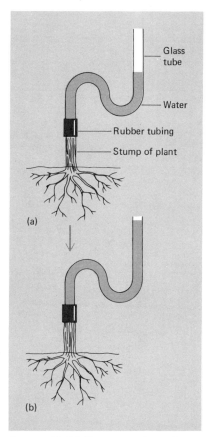

Figure 10.22 Demonstration of root pressure. (a) A curved glass tube is sealed to a freshly decapitated plant and is then filled with water. (b) Sap rising into the tube from the roots raises the water level in the tube until the hydrostatic pressure in the tube equals the root pressure. The pressure can be calculated using the final height.

Figure 10.23 Guttation—the loss of liquid water resulting from root pressure. (From Villee, C. A.: *Biology.* 7th ed. Philadelphia: Saunders College Publishing, 1977, p. 270.)

270
Biology of Plants

> The key to the theory of water movement in plants, i.e., the cohesion-tension theory, is found in the unique properties of water that are a consequence of hydrogen bonding.

Figure 10.24 Experiment measuring the pull of transpiration. A tube filled with water is attached to the cut root of the tree. The tube is placed in a vessel of mercury. As the tree loses water through transpiration, water from the tube, along with the mercury, is drawn upward. The extent of transpiration pull can be determined by measuring how far the mercury is pulled up into the tube. (From Arms, K., and Camp, P. S.: *Biology*. 2nd ed. Philadelphia: Saunders College Publishing, 1982.)

(Fig. 10.24). The basis of the theory involves the unique adhesive, cohesive, and evaporative properties of water (see Chapter 3). You will recall that water molecules are linked together by hydrogen bonds, which join the hydrogen and oxygen atoms of adjacent water molecules. In other words, as a result of hydrogen bonding there is cohesion between the water molecules. As water in the leaf cells is lost by transpiration, the osmotic concentration within the cells increases. Water then moves into the leaf cells from adjacent cells; the osmotic concentration in these cells increases, and they in turn take up water from cells adjoining them. This process continues from cell to cell all the way to the xylem in the veins of the leaf. As their

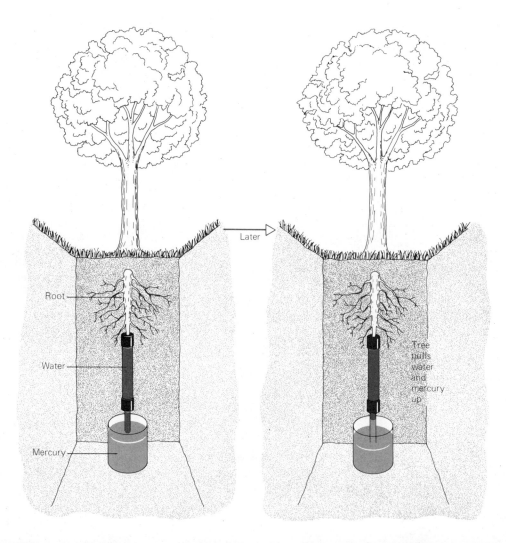

> According to the mass flow hypothesis, the movement of food through the phloem ultimately depends upon the mechanism of osmosis.

osmotic concentration increases, the cells adjacent to the xylem withdraw water from the xylem cells. As water molecules move out of the xylem, they pull other water molecules behind them, creating tension between the molecules; these water molecules in turn pull others upward, and so on all the way down through the stems to the tip of the root. There is, then, an unbroken column of water extending from the leaves downward to the roots.

In addition to cohering to each other, the water molecules also adhere to the inner walls of the xylem. This action tends to add support to the water column. The cohesion-tension theory is also referred to as the *transpiration theory* or, less elegantly, the "push-pull" theory of water transport. It should be emphasized that the key to the upward movement of water in a plant is transpiration. As long as there is a loss of water vapor from the leaf cells, a pulling force is created that allows water to move upward from the roots to the leaves. In turn, the pull of transpiration resides in the strong cohesiveness between individual water molecules.

Movement of Solutes. Translocation of the products of photosynthesis, along with some inorganic solutes, is known to take place within the phloem, although the exact transport mechanism is unclear. The most widely accepted explanation is the *mass flow hypothesis,* which assumes that sugars and other solutes are transported as a result of differences in turgor pressure within the phloem cells. In the photosynthesizing leaves, the phloem cells acquire a high concentration of solutes, particularly sucrose. As a result, water tends to move into the cells, raising their turgor pressure. This high turgor pressure forces the sucrose solution into an adjacent phloem cell having a lower osmotic concentration. The turgor pressure in this cell then rises, and the sucrose solution is forced into the next cell, and so on. Through this mechanism, an entire mass of water and solutes flows from one phloem cell to another along a turgor pressure gradient (Fig. 10.25).

The question arises of why some of the phloem cells have a lower solute concentration than do others. If sugars and other substances are moving downward throughout the plant, one might assume that each phloem cell has about the same osmotic concentration. It must be remembered, however, that in some parts of the plant, particularly in actively growing areas, the sugars and other solutes are removed from the phloem and used by other plant cells for cellular work. Consequently, the phloem cells in these areas lose water as well, since the loss of solutes lowers their osmotic concentration. With the accompanying decrease in turgor pressure, these cells can then take up water and solutes from adjacent phloem cells. It is

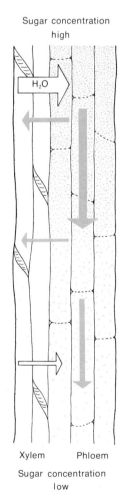

Figure 10.25 Diagram illustrating the mass flow hypothesis. In regions of high sugar concentration, water moves into the phloem cells, raising their turgor pressure. The sugar solution is then forced into regions where the turgor pressure is lower. (From Villee, C. A.: *Biology.* 7th ed. Philadelphia: Saunders College Publishing, 1977, p. 274.)

Biology of Plants

> Ferns vary in size from having leaves only 1/2 inch long to reaching heights of more than 75 feet.

evident, therefore, that mass flow through the phloem cells depends upon differences in cellular osmotic concentration, which in turn regulates the uptake or loss of water by the cells.

Division Tracheophyta

There are three major groups of living tracheophytes, two of which are small and include the club mosses and horsetails. The third group consists of the three major representatives of the land plants: the **ferns** (Class Filicineae), **gymnosperms** (Class Coniferae), and **angiosperms** (Class Angiospermae).

Class Filicineae. Although possessing a well-developed vascular system and true roots, stems, and leaves, the ferns, like the bryophytes, are only partially adapted to land. Accordingly, most species of ferns are found in tropical and subtropical regions, where the supply of water is plentiful (Fig. 10.26). The flagellated sperm require water to swim to the egg, and the gametophyte, which is very much reduced in the ferns, does not contain vascular tissue and must absorb water from the soil.

In comparing alternation of generations in ferns and mosses, we can note a few similarities and some significant differences. The fern plant one would find hanging in a florist shop is the diploid sporophyte generation. In the ferns, then, it is the sporophyte generation that is conspicuous, or dominant, and not the gametophyte, as in the mosses.

On the underside of fern leaves, haploid spores produced by meiosis are contained within clusters of sporangia called *sori* (Gk. "a heap"). When a sorus breaks open, the haploid spores are released and dispersed by the wind. A germinating spore gives rise to a photosynthetic heart-shaped gametophyte called the **prothallus** (Gk. "first shoot") (Fig. 10.27). Male and female reproductive structures—antheridia and archegonia, respectively—develop on the underside of the prothallus. The tiny prothallus is anchored by tiny rootlike structures called rhizoids. Within the antheridia and archegonia, haploid gametes (sperm and eggs) are produced by mitosis. When the mature flagellated sperm are released, and if there is adequate water, they swim to the archegonium and fertilize the egg cell. The resulting diploid zygote grows out of the prothallus and becomes the new diploid sporophyte. When the fern sporophyte develops sori on its leaves, the cycle is completed and may begin anew (Fig. 10.28).

As pointed out earlier, the sporophyte generation is relatively inconspicuous in the mosses and other bryophytes. In the ferns and all higher plants, however, the sporophyte is dominant, and the ga-

Figure 10.26 Ferns. (a) The bracken fern, *Pteridium.* (b) Growth of fern fronds or "fiddleheads" of the Christmas fern, *Polystichum achrostichoides.* (c) A tree fern, *Cyathea,* in a Puerto Rico rain forest. (d) On the undersides of some fern leaves are found clusters of sporangia, called *sori*. This shield fern, *Dryopteris spinalosa,* is covered with sori. (e) This higher magnification view of the sori of a staghorn fern shows clusters of granules, which are the individual sporangia in which spores are formed. (From Ebert, J. D., Loewy, A. G., Miller, R. S., and Schneiderman, H. A.: *Biology*. New York: Holt, Rinehart and Winston, 1973, p. 605.)

> The seed-bearing conifers first appeared on the earth some 290 million years ago.

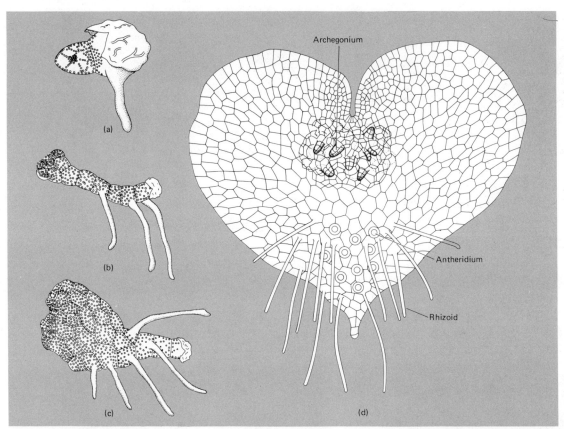

Figure 10.27 The prothallus. (a) Spore germination. (b) and (c) Stages in development of prothallus. (d) Mature prothallus, lower surface.

metophyte, although multicellular, is quite small. In conjunction with this, the sporophyte of ferns and higher vascular plants is not totally dependent upon the gametophyte for its existence. In the mosses, as we have seen, the sporophyte grows out of and is nourished by the female gametophyte. Vascular plants, however, develop independent sporophytes that must absorb and transport water, photosynthesize, and otherwise exist on their own.

Class Coniferae. The Coniferae, or conifers ("cone bearers"), are also known as gymnosperms (Gk. *gymnos,* naked, + *sperma,* seed), since their seeds are not enclosed within female tissues. Instead, the seeds are borne on the surfaces of the female cone scales. By definition, a **seed** is an organ consisting of an embryo and female storage tissue enclosed in a protective seed coat. The only other living representatives of the seed plants are the angiosperms. The conifers are the largest and most familiar members of the gymnosperms and include such trees as cedar, fir, spruce, pine, and the giant redwoods

> The tallest species of trees, the giant redwoods, are conifers that attain heights of up to 380 feet.

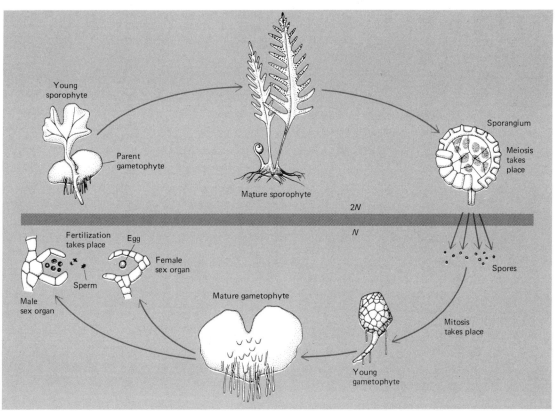

Figure 10.28 Life cycle of a fern. Alternation of generations is shown, but the more conspicuous form is the sporophyte.

of the west coast of the United States. The leaves of most conifers are needle-shaped, although broader leaf types are also found in many species. Conifers are a major source of commercial lumber, the uses of which are legion.

Like all vascular plants, the conifers undergo alternation of generations in which the sporophyte is dominant. We will examine this life cycle as it occurs in a pine tree (Fig. 10.29). The sporophyte (the tree itself) bears small male cones and the larger, more familiar female cones. The male cone consists of scales (which are a type of modified leaves), each of which bears a single sporangium. Within the sporangia, haploid **microspores** arise by meiosis. The microspores eventually develop into millions of microscopic **pollen grains**. The nucleus of the pollen grain undergoes mitosis to produce four haploid cells, two of which degenerate. The two functional cells remaining are known as the **generative cell** and the **tube cell** (Fig. 10.30).

The female pine cone is composed of scales that are usually larger than those of the male cone; each scale of the female cone

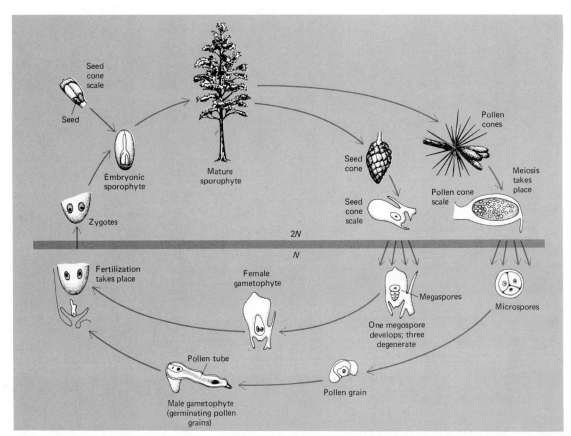

Figure 10.29 Life cycle of a pine tree. A pine tree is the mature sporophyte plant. The gametophytes are much reduced structures and are enclosed within the tissue of the sporophyte. After fertilization, the female gametophyte with its enclosed zygote and a tough seed coat produced by the parent sporophyte becomes the seed that forms on the cone scale.

bears two sporangia. A sporangium produces four haploid *megaspores* by meiosis, but three of these disintegrate. The remaining megaspore undergoes repeated mitotic divisions to produce the multicellular female gametophyte. Within the female gametophyte sev-

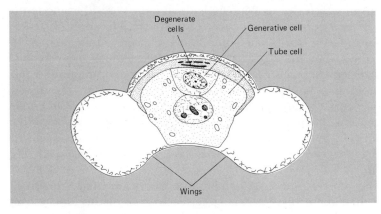

Figure 10.30 Mature pollen grain of pine.

One of the keys to the rise and eventual dominance of angiosperms was their great diversity, and for 80 million years these flowering plants have been the dominant group of vascular plants all over the world.

eral archegonia develop, each containing a haploid egg. The female gametophyte and the surrounding sporangium constitute the *ovule* (L. "small egg"), which, after fertilization, will become the seed (Fig. 10.31).

In the male pine cone, the mature sporangia, or pollen sacs, eventually burst, releasing the pollen grains to be dispersed by the wind. Some of the pollen grains drift down between the scales of the female cone and become caught in a sticky liquid secreted by the ovule. One of the pollen grains is drawn into the ovule as the liquid dries. Within the ovule, the pollen grain continues development by growing an elongated *pollen tube.* The generative cell enters the pollen tube and then divides by mitosis to produce two cells, one of which divides again to produce two haploid sperm cells. In giving rise to gametes, the pollen grain is therefore the male gametophyte. The pollen tube, with its contained sperm cells, continues to grow through the tissues of the ovule until it penetrates one of the archegonia. Both sperm cells are then released, but only one fertilizes the egg to produce the diploid zygote. Note that in the entire process of sexual reproduction—from release of pollen grains from the pollen sacs to fertilization—the pine tree is not directly dependent on water. This is one of the most significant adaptations to land exhibited by the seed plants.

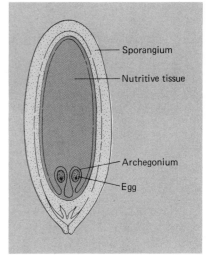

Figure 10.31 Mature female gametophyte of pine.

The zygote, the first cell of the diploid sporophyte generation, divides by mitosis to produce a multicellular embryo. The embryo is surrounded by the tissues of the female gametophyte, which will provide nourishment as the embryonic plant develops. The walls of the ovule soon harden, forming a seed coat around the female gametophyte and embryo. In most conifers, the seed develops a thin wing that aids in dispersing the seed after its release from the cone (Fig. 10.32).

At its upper end, the tiny pine embryo is composed of several *cotyledons,* or embryonic leaves; the lower tip is the *radicle,* which will give rise to the first roots when the seed germinates. The region of the embryo between the cotyledons and radicle is called the *hypocotyl,* which forms a part of the shoot of the young plant. The embryonic pine tree draws on the nutritive tissues of the female gametophyte until it breaks through the soil and is exposed to light. The seed coat and gametophyte tissue then fall away, and the young pine grows into a mature sporophyte (Fig. 10.33).

Class Angiospermae. The most advanced and biologically successful vascular plants are the *angiosperms,* or flowering plants, of which there are over a quarter of a million known species. The term "angiosperm" means "seed vessel" and is a reference to the female tis-

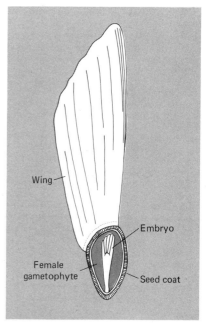

Figure 10.32 A pine seed.

278 Biology of Plants

> The flower, through its reproductive characteristics, protection of the seeds, and pollinator-attracting features, has been crucial to the evolution and biological success of the angiosperms.

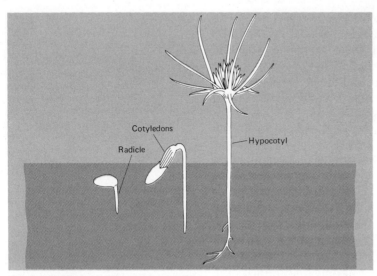

Figure 10.33 Stages in the germination of a pine seed.

sues that cover, or enclose, the seed. The life cycle of angiosperms is similar to that of gymnosperms, a feature one would expect, since the angiosperms are believed to have arisen from an ancient group of gymnosperms. With such an enormous variety of different species, the angiosperms are the dominant forms of plant life on the earth today. Almost any type of vegetable, flower, fruit, nut, cereal grain, grass, or hardwood tree you can think of is an angiosperm. The two most distinguishing features of angiosperms are the *flower* and the *fruit*.

The Flower. The flower functions in reproduction and, depending upon the species, may bear only the male reproductive structures, only the female reproductive structures, or both (Fig. 10.34). The male structure is called the *stamen* (L. "a thread"), which consists of a stalk, or *filament,* having an *anther* at its tip. The anther is actually a group of sporangia in which the haploid pollen grains (male gametophytes) are produced. In the development of the pollen grains, cells within the sporangia undergo meiosis to produce numerous haploid microspores, each of which develops a thick wall. The nucleus of each microspore divides by mitosis to produce a haploid *generative nucleus* and a haploid *tube nucleus.* The mature pollen grain, then, consists of two haploid nuclei surrounded by a thick wall (Fig. 10.35).

In the center of the flower is the female reproductive structure, the *carpel* (Gk. "fruit"). The base of the carpel is swollen to form the *ovary,* which contains the *ovule* or *ovule(s).* The female gametophyte

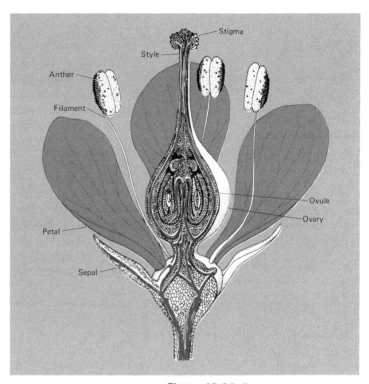

Figure 10.34 General structure of a flower.

develops within the ovule. Extending upward from the ovary is the stalklike *style,* which enlarges at its tip to form the *stigma.* Thus the carpel consists of the stigma, style, and ovary with its contained ovule (or ovules).

Within an ovule, meiosis produces four haploid megaspores, three of which disintegrate. The remaining megaspore undergoes several mitotic divisions to produce seven haploid cells. Three of these cells are arranged at one end of the ovule, and three are arranged at the other end. Each of these cells contains a single haploid nucleus; however, only one of them becomes the functional egg cell. The seventh cell, situated in the center of the ovule, contains *two*

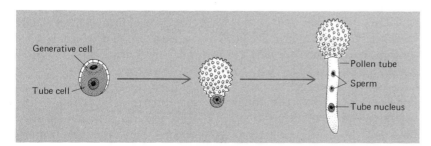

Figure 10.35 The mature angiosperm pollen grain consists of two haploid cells surrounded by a thick wall. The generative cell nucleus later divides by mitosis to produce two haploid sperm.

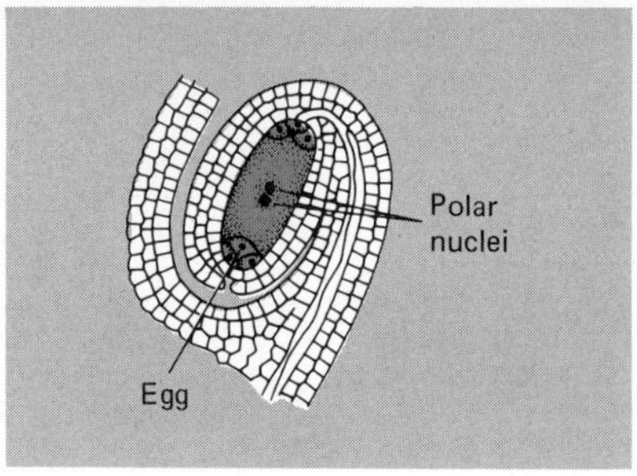

Figure 10.36 The mature female gametophyte, or embryo sac, of angiosperms.

haploid nuclei, referred to as *polar nuclei.* This structure, containing seven cells with eight haploid nuclei, is the female gametophyte, also known as the *embryo sac* (Fig. 10.36).

The other parts of the flower include the *sepals* and *petals,* both of which are types of modified leaves. On many flowers the sepals and petals are brightly colored, an adaptation that attracts insects and other pollinators. The female gametophyte of some plants secretes a sugary liquid called nectar, a highly prized food source of many insects and hummingbirds.

Pollination and Fertilization. When mature, the sporangia (pollen sacs) within the anther rupture, releasing the haploid pollen grains. The pollen grains may be blown about by the wind, transported on the feet of insects, or carried in the fur of animals (Fig. 10.37). Pollen traveling from one flower to another of the same species results in cross-pollination. If pollen remains in the parent flower and simply falls on the stigma, the result is self-pollination.

As pollen settles on the flower, the pollen grains adhere to the sticky surface of the stigma. One pollen grain begins germinating to form a pollen tube, which grows downward through the carpel. The generative nucleus and the tube nucleus move into the pollen tube, and the generative nucleus divides by mitosis to produce two haploid sperm cells. The pollen tube continues its downward growth until it enters the ovary and penetrates an ovule. The two sperm cells are then released into the embryo sac, and one of the sperm fertilizes the egg. The other sperm cell unites with the two polar nuclei, which previously had fused, to form a triploid ($3N$) nucleus (Fig. 10.38). This nucleus undergoes several mitotic divisions to produce a triploid tissue known as *endosperm.* During embryonic development, the surrounding endosperm serves as a source of nourishment. In some plants, the embryo uses up all the endosperm before the seed matures; in other plants the endosperm persists throughout life. Plants such as wheat and rice, for example, are composed predominantly of endosperm tissue. In eating these and other grains, human beings and various other animals are therefore nourished by triploid endosperm tissue.

The reproductive process involving two sperm cells is called *double fertilization.* The accompanying formation of triploid tissue in angiosperms is a phenomenon unknown in any other group of living things.

As we saw in the pine tree, the ovule of the angiosperm develops into a seed, with the nutritive tissue and young sporophyte (developing embryo) enclosed within a seed coat. In the angiosperms, however, the seed is entirely surrounded by the ovary. The ovary is

Figure 10.37 Pollination by insects and other animals. (a) Beetle crawling among goldenrod flowers. (b) Honeybee foraging in a pear flower. Note the filled pollen basket on the left hind leg. (c) Brazilian hermit hummingbird pollinating a flower. (d) "Honey guides" are conspicuous markings that indicate the location of the nectar supply, as in this foxglove. (e) *Coreopsis* has honey guides that are not evident in ultraviolet light *(right)*. Insects see in the UV and can see these honey guides. Flowers normally pollinated by insects have honey guides visible in the UV, whereas flowers pollinated by birds and bats do not. (From Ebert, J. D., Loewy, A. G., Miller, R. S., and Schneiderman, H. A.: *Biology.* New York: Holt, Rinehart and Winston, 1973, p. 617.)

> A fruit is basically a mature ovary. Like the flower, fruits have been instrumental in the biological success of the angiosperms.

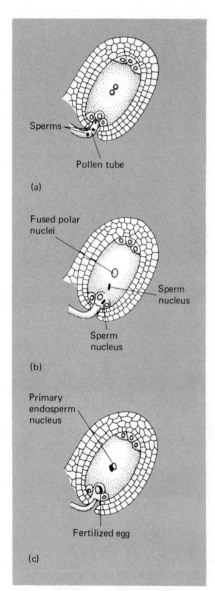

Figure 10.38 Double fertilization. One sperm nucleus fuses with the egg nucleus, the other with the polar nuclei.

the "seed vessel" from which this class of plants derives its name (Fig. 10.39).

The Fruit. The ovary with its contained seeds enlarges to form the *fruit*. In fruits such as apples, the core is the wall of the ovary, which surrounds the apple seeds. The fleshy part of the apple we eat consists mainly of stem. In pea plants, the pod is also the wall of the ovary and is therefore a fruit; the peas within the pod are the seeds. Peaches and plums are composed almost entirely of ovarian tissue. The seeds of these fruits are inside the hard pit.

The fruit serves as protection for the young seeds, particularly during periods of drought. Moreover, dispersal of the seeds is also facilitated by the fruit. Some fruits are distributed over wide areas by wind and water; others may be caught in the fur or feathers of animals and shaken loose many miles away. Animals that feed on fruits may travel to wide-ranging locations and eliminate the seeds in the feces. The fruit, then, is another adaptive feature of the angiosperms and one that has been instrumental in the success and worldwide distribution of these flowering plants.

Embryonic Development and Germination. The tiny zygote, surrounded by endosperm, undergoes mitosis to form a multicellular embryo. At its upper end the embryo develops one or two **cotyledons,** or embryonic leaves. Angiosperms with only one embryonic leaf are called **monocotyledons,** or monocots; those with two embryonic leaves are **dicotyledons,** or dicots. Just above the point of attachment of the cotyledons is the **epicotyl,** the upper part of the shoot. The **hypocotyl,** destined to become the lower part of the shoot, is the region of the embryo below the cotyledons. At the lower tip of the embryo is the **radicle,** which will become the primary root.

When conditions such as water and temperature are favorable, the growing embryo breaks through the seed coat, and the radicle begins to develop a root system. The hypocotyl, epicotyl, and cotyledons then emerge through the soil into the sunlight (Fig. 10.40). The tip of the epicotyl has an apical meristem region, as does the tip of the root. However, the tip of the root develops a **root cap,** which aids in pushing the growing root through the soil. The meristem cells of the epicotyl divide and then elongate, causing the shoot to lengthen. Growth of the shoot and root proceeds as the apical meristem cells continue to divide; however, after the older meristem cells elongate, they begin to differentiate into the various tissues of the plant. As we have discussed, the shoot develops xylem, phloem, vascular cambium, leaf nodes, and so on; the root also develops vascular tissue, along with the cortex, endodermis, and sometimes pith.

> Stimulatory and inhibitory hormones are the major internal factors that control plant growth and development.

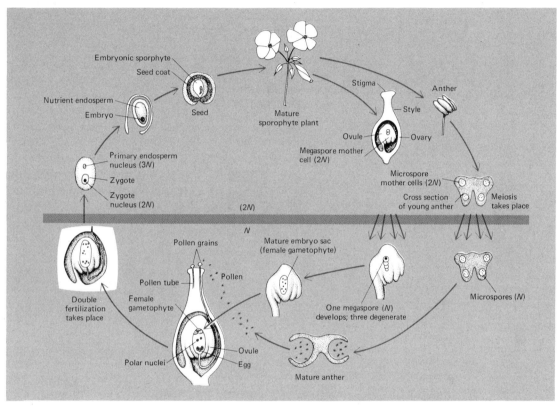

Figure 10.39 Life cycle of an angiosperm. (See text for discussion.)

Having emerged from the soil, the young sporophyte must now survive on its own. The roots absorb water and minerals, and the young leaves begin to photosynthesize. A new angiosperm is thus established, and a new life cycle begins once more. With its unique adaptations for solving the problems of worldwide survival, this young seedling stands with all the other angiosperms at the peak of plant evolution.

PLANT HORMONES

Many of the activities observed in plants are associated with the process of growth. Such activities actually are direct responses to some environmental stimulus. The response is called a *tropism* (Gk. "a turning") and may result from several different stimuli, such as light, touch, certain chemicals, gravity, and so on. In plants the various tropisms are the result of chemical and physical changes within the plant itself. These changes are influenced by one or more plant *hormones,* or growth regulators. By definition, a hormone is a chemical substance produced in one part of the organism and transported

284
Biology of Plants

> Darwin's experiments on grass seedlings led eventually to the discovery of auxins, the plant hormones responsible for virtually all of the plant responses known as tropisms.

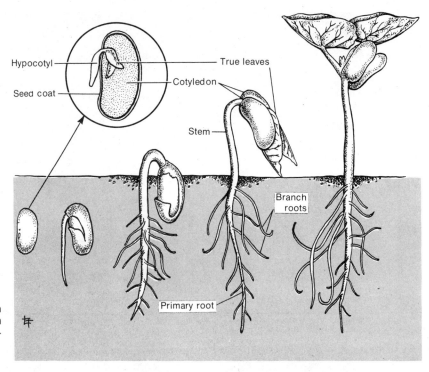

Figure 10.40 Stages in the germination and development of a bean seed. (From Villee, C. A.: *Biology.* 7th ed. Philadelphia: Saunders College Publishing, 1977, p. 228.)

to another part, where it exerts a specific biological effect. A few examples will illustrate the manner in which these chemical messengers work.

One of the most familiar and common plant responses is the bending of the stem toward a light source. In this case, light is the stimulus, and the response of the plant is known as the **phototropism** ("light turning") (Fig. 10.41). Since the response in this instance is *toward* the stimulus, we refer to the response as *positive;* if a response is *away* from the stimulus, it is *negative*. About 1880, the renowned evolutionist Charles Darwin was experimenting with grass seedlings and observed that, when illuminated, the tip of the shoot would bend toward the light source. Since Darwin, experiments have shown that phototropism results from the influence of plant hormones called **auxins.** The word comes from the Greek *auxein,* meaning "to increase." Basically, the major effect of auxins is to cause most plant cells to increase in length. When a plant is exposed to light, the auxins migrate away from the light across to the darker side of the stem. With the auxins on the dark side, the cells in this area elongate more rapidly than cells on the lighted side. As the cells continue to grow faster on the dark side, the stem eventually bends toward the light source. This positive phototropism is, of course, im-

Whether auxins stimulate or inhibit elongation of root cells apparently depends upon the concentration of the hormones.

(a)

(b)

Figure 10.41 Phototropism in a young radish. (a) In the dark or in uniform light, the plants grow straight upward. (b) When exposed to light coming from a single direction, they quickly bend toward it; that is, they are positively phototropic. The photograph on the right was taken half an hour after the one on the left. (From Villee, C. A.: *Biology*. 7th ed. Philadelphia: Saunders College Publishing, 1977, p. 237.)

portant to the process of photosynthesis by exposing the leaves to the sun.

A similar effect of auxins can be seen in what is known as **geotropism** ("earth turning"). If a growing plant is placed on its side and left for a few hours, the *shoot* will begin to bend upward, away from the earth (*negative* geotropism) (Fig. 10.42). In this instance, auxin concentration is greatest on the underside of the shoot, primarily because of the pull of gravity. As a result, auxins stimulate cells on the underside of the shoot to elongate faster than the cells on the upper side, causing the shoot to bend upward.

In the case of the *root* of a plant lying on its side, auxins produce a different response. Although auxins concentrate on the underside of the root, just as they do in the shoot, the root bends downward, toward the earth (*positive* geotropism). Thus, both the shoot and the root of the plant concentrate auxins on their lower sides, but they

(a)

(b)

Figure 10.42 Geotropism in young radish plants. (a) A pot of straight radish seedlings placed on its side and kept in the dark to eliminate phototropism. (b) Within 30 minutes the seedlings have bent. Since they bend away from the direction of the force of gravity, they are negatively geotropic. (From Villee, C. A.: *Biology*. 7th ed. Philadelphia: Saunders College Publishing, 1977, p. 238.)

286
Biology of Plants

> The dropping of fruit and leaves from trees—a process known as abscission—involves the interaction of auxins, ethylene, and, in some plants, abscisic acid.

exhibit opposite responses. Why? It is thought that a high concentration of auxins *inhibits* elongation of root cells, whereas a low auxin concentration stimulates elongation. A root lying horizontally accumulates a high concentration of auxins on its lower side, but there is a relatively low auxin concentration on its upper side. Consequently, cells on the upper side of the root elongate more rapidly than cells on the underside, and the root bends downward.

It should be noted that another plant hormone called **abscisic acid** (ABA) also may influence geotropism in roots. Abscisic acid is produced in mature leaves and is mainly an inhibitory substance. It has been found that the distribution of abscisic acid in a horizontal root is the same as the distribution of auxins. Accordingly, the high concentration of abscisic acid on the underside of the root inhibits the growth of cells in this area, whereas cells on the upper side continue to grow, bending the root downward. Abscisic acid also inhibits cell division in developing leaves and often inhibits the growth of germinating seeds. As long as abscisic acid is active in the branch or twig, fully developed leaves do not form; instead, the young leaf cells form protective scales around the tip of the twig. The inhibition of leaf formation and seed growth is a natural occurrence, characteristic of many plants during the winter.

Auxins are involved with another plant hormone, **ethylene,** in controlling the dropping of leaves and fruit from trees. This process is known as **abscission,** a term from the Latin *abscindere,* meaning "to cut off." The leaves of many plants develop a weak wall of cells called the **abscission layer** at the point where the leaves join the stem (Fig. 10.43). If the leaf is jostled by the wind or otherwise shaken, the cells of the abscission layer break apart and the leaf falls to the ground. Formation of the abscission layer may result from injury, disease, or aging of the leaf, or it may form as autumn approaches. Auxins act as inhibitors of abscission by preventing formation of the abscission layer. On the other hand, ethylene appears to stimulate breakdown of the abscission layer, thereby promoting leaf drop. In many plants, the concentration of auxins decreases in the early months of fall, whereas the concentration of ethylene increases; this permits formation of the abscission layer and promotes falling of the leaves. Similarly, stems of fully ripe fruit separate from the branch in the area of the weak abscission layer, and the fruit drops from the tree.

Ethylene (C_2H_4) is a highly volatile gas, produced naturally by fruits and other plants. Years ago when fruit was ripened in a room with a kerosene stove, it was believed that the heat from the stove

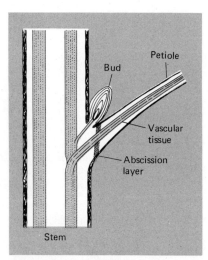

Figure 10.43 Abscission layer in the petiole of a leaf. Several rows of small cells provide a line of weakness from which the leaf falls.

> Gibberellins were first isolated from a fungus and are now believed to occur in all plants. The first of several natural cytokinins was isolated from corn kernels; there also are several synthetic compounds that behave like cytokinins.

was responsible for the ripening. It turned out, however, that it was not the heat but the ethylene gas given off as the kerosene burned. Ethylene has also been suggested as an inhibitor of root growth and is known to interfere with geotropism in some plants. In the presence of ethylene, the shoots do not bend but continue to grow horizontally. Other inhibitory effects regarding plant development have been attributed to ethylene, and as investigations proceed, more specific information concerning its activity should be forthcoming.

Fruit orchards often are sprayed with synthetic auxins to prevent formation of the abscission layer. This allows the fruit to continue to ripen on the trees until it is ready to be picked.

Many aspects of plant growth and development are influenced by two other major growth-regulating substances, the *gibberellins* and *cytokinins*. The most notable effect of gibberellins is their influence on plant growth through stimulation of rapid stem elongation in various plants. For example, certain species of dwarf bean plants when treated with gibberellins grow to the same height as normal tall plants (Fig. 10.44). There are other plants that ordinarily flower

Figure 10.44 An effect of plant hormones on growth. The bean plant on the right was treated with gibberellin, while a similar plant on the left was untreated.

> Plant growth and development apparently involve virtually all the major plant hormones acting in a complex and coordinated interplay.

or undergo stem elongation when subjected to the proper stimuli, such as length of time in the sun and warm temperature. When these plants are treated with gibberellins, both flowering and stem elongation occur *without* these stimuli.

The primary effect of the cytokinins is to induce plant cells to divide, i.e., to undergo mitosis. Experiments have shown that isolated plant cells grown in the laboratory and supplied with the necessary nutrients will enlarge but will not divide. However, with the addition of an adequate concentration of cytokinins, the cells soon begin dividing. Cytokinins also bring about increased stem and root growth, enhance flowering, and stimulate germination of some types of seeds.

In addition to the growth-regulating substances discussed here, there are undoubtedly many others yet to be discovered. It is thought, for example, that there are a variety of growth inhibitor chemicals that play an important role in counteracting the effects of growth-stimulating substances. This interaction between the various growth regulators seems to be the rule in controlling growth activities in most plants. As we have seen, auxins inhibit formation of the abscission layer, whereas ethylene stimulates its breakdown. As another example, the interaction between auxins and cytokinins greatly increases growth in some plants, much more so than when only one of these hormones is present. Control of growth is mediated not only by counteracting hormones but also by the relative concentrations of a single hormone. For example, stem elongation is stimulated by a moderately low concentration of auxins but inhibited by very high concentrations. Thus, it has become evident that control of growth responses in plants is an extremely complex process requiring the coordinated interplay of a variety of growth-regulating substances.

Figure 10.45 (a) The "sensitive plant," *Mimosa pudica*, before being disturbed. (b) The plant 5 seconds after being touched. Note how the leaves have folded and drooped. (From Villee, C. A.: *Biology*. 7th ed. Philadelphia: Saunders College Publishing, 1977, p. 239.)

NONHORMONAL ACTIVITY

A variety of responses in plants are not under hormonal control, nor are they always directly associated with growth of the plant. In general, these are relatively rapid responses, most of which are thought to result from variations in turgor pressure in the stems and leaves.

A number of rapid plant responses are referred to as **nastic** (Gk. *nastos*, close-pressed, compact) **movements**. The opening and closing of flowers, the folding of leaves, and the drooping of stems and leaves are common examples. One of the more familiar nastic movements is that exhibited by a species of *Mimosa*, which is quite sensitive to touch. Specialized supporting cells located at the bases of

Each leaf of the Venus flytrap is equipped with three trigger hairs; the leaves will close only if one hair is touched twice or if two of the hairs are touched in succession.

the leaf and leaflets experience a sudden change in turgor pressure when the plant is touched, causing the leaflets or leaf to droop almost instantly (Fig. 10.45). When normal turgor pressure is restored, these plant structures return to their upright positions. The response of plants to touch is known as *thigmotropism* (Gk. "touch turning").

The opening and closing of carnivorous plants such as the Venus flytrap also result from changes in turgor pressure (Fig. 10.46). The paired leaves of the flytrap resemble somewhat the hinged valves of a clam or mussel. The margins of each leaf are equipped with long, thin spines that interlock when the leaves press together. When an insect walks on the inner surface of a leaf, it brushes against small trigger hairs that initiate the rapid closing of the leaves. Trapped within the interlocking spines, the insect is then slowly digested by enzymes secreted by the leaf cells.

Figure 10.46 A leaf of the Venus flytrap catching and digesting a fly. (From Villee, C. A.: *Biology.* 7th ed. Philadelphia: Saunders College Publishing, 1977, p. 25.)

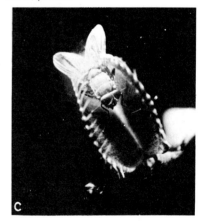

Although there is evidence that the nastic movements of *Mimosa*, the Venus flytrap, and other plants result from changes in turgor pressure, the mechanism for these changes is not fully understood. It is known that plant cells can transmit electrical impulses similar to nerve impulses in higher vertebrates. In some manner, a stimulus, such as touch, may set up impulses that travel along specific cells of the plant. Upon reaching the responding cells, the impulses alter the permeability of the cell membranes, allowing water to move outward. It has also been suggested that nastic movements in some plants may be influenced by special stimulatory chemicals produced within the plants. At present, however, this position is not supported by definitive evidence.

REVIEW OF ESSENTIAL CONCEPTS

1. The plant kingdom includes unicellular and multicellular *algae* and the multicellular *land plants*. All plants are *photosynthetic*, and most undergo *alternation of generations*. Most plant cells are surrounded by a wall of *cellulose*.
2. The algae include the *red algae* (Division Rhodophyta), *brown algae* (Division Phaeophyta), and *green algae* (Division Chlorophyta).
3. The red algae, so-called because of the usual predominance of the pigment *phycoerythrin*, are mostly multicellular, submerged forms found in tropical waters. None of their cells has flagella.
4. The brown algae are large multicellular seaweeds found mostly in marine waters. Their brownish color results from the presence of a carotenoid pigment called *fucoxanthin*.
5. Brown algae undergo alternation of generations, i.e., a life cycle in which a haploid *gametophyte generation* is followed by a diploid *sporophyte generation*. The gametophyte produces haploid gametes by mitosis; fusion of gametes produces the sporophyte, which produces haploid spores by meiosis. The spores develop into new gametophytes.
6. The green algae are mostly freshwater forms and are one of the major photosynthetic groups in the waters of the world. Green algae are considered to be the group from which the land plants arose.
7. A common unicellular freshwater green alga is *Chlamydomonas*; with its light-sensitive eyespot and locomotor flagella, the organism exhibits some animal-like characteristics. *Chlamydomonas* may reproduce asexually through the production of *zoospores* or sexually by the development of gametes.
8. A representative colonial green alga is *Volvox*. The evolutionary lineage exemplified by the *Chlamydomonas-Volvox* line illustrates the development of cellular *division of labor* and the increasing complexity of organisms as a result of evolutionary advancement.

9. A multicellular representative of Chlorophyta is *Spirogyra*, characterized in part by its ribbon-like chloroplast and method of sexual reproduction, called *conjugation*.
10. The *land plants* arose about 400 million years ago; in making the transition from the seas to land, the major problem for these plants was the availability and retainment of water. The two major divisions of living land plants are *Bryophyta* and *Tracheophyta*.
11. Bryophytes are *nonvascular* plants without true roots, stems, or leaves; tracheophytes are *vascular* plants having true roots, stems, and leaves containing vascular tissues.
12. Mosses (Class Musci) are small nonvascular plants, usually found in moist areas. They usually display separate sexes, and their gametophyte generation is dominant. The sporophyte is totally dependent upon the gametophyte for nourishment.
13. The vascular system of tracheophytes consists of *xylem* and *phloem*. Xylem conducts water and minerals; phloem transports sugars and other nutrients.
14. *Stomata* allow for exchange of oxygen and carbon dioxide in the leaf. The escape of water vapor through stomata is called *transpiration*.
15. The tissues of *stems* include *cork cells, cork cambium, phloem, vascular cambium,* and *xylem*. The growing tip of the stem is the *apical meristem*.
16. *Roots* consist of an *apical meristem* covered by a *root cap*; above the meristem is the *zone of elongation* and the *zone of differentiation*. *Root hairs* extend from the *epidermis* covering the root. Beneath the epidermis is the *cortex*, interior to which is the *endodermis*. Mature vascular tissue either lies in the center of the root or is scattered in bundles surrounding the *pith*.
17. *Translocation* is the transport of the products of photosynthesis within the phloem.
18. The loss of liquid water from the plant leaf surface is called *guttation*, which is the result of *root pressure*.
19. The upward movement of water through the xylem is currently explained by the *cohesion-tension theory*. The theory holds that water movement results from *root pressure, transpiration pull,* and *cohesiveness of water molecules*.
20. The mechanism of translocation is thought to occur according to the *mass flow hypothesis*, which assumes that water and solutes flow from cell to cell along a turgor pressure gradient.
21. The three major representatives of the tracheophytes are the *ferns* (Class Filicineae), *gymnosperms* (Class Coniferae), and *angiosperms* (Class Angiospermae).
22. The *ferns* display alternation of generations in which the sporophyte is dominant, and the gametophyte, called the *prothallus*, is very much

reduced. The sporophyte in ferns is only partially dependent upon the gametophyte.

23. *Gymnosperms* are seed plants in which the seed is not enclosed within female tissues. A *seed* is an organ consisting of an embryo and female storage tissue enclosed in a protective seed coat. *Conifers* bear two types of cones: male cones, in which the male gametophytes *(pollen grains)* are produced, and female cones, in which the female gametophytes *(ovule tissues)* are produced. The sporophyte generation is dominant in gymnosperms, whereas the gametophyte generation is greatly reduced.

24. The pine *embryo* consists of *cotyledons* (embryonic leaves), the *radicle* (roots), and *hypocotyl* (shoot).

25. *Angiosperms*, or flowering plants, produce seeds completely enclosed within female tissues. The two most distinguishing features of angiosperms are the *flower* and the *fruit*. The sporophyte generation is dominant in angiosperms, whereas the gametophyte generation is very much reduced.

26. The *flower* is a reproductive structure that bears either the male structure, the *stamen*, or the female structure, the *carpel*, or both. The stamen consists of the *filament* and *anther*, in which haploid *pollen grains* are produced. The carpel consists of the *stigma*, *style*, and *ovary* with its contained *ovule (or ovules)*. The female gametophyte, or *embryo sac*, is formed within an ovule.

27. Fertilization in angiosperms involves two sperm cells, one of which fertilizes the egg; the other unites with two haploid *polar nuclei* to form triploid *(3N) endosperm*, a nutritive tissue.

28. The *fruit*, which contains the seeds, is the mature plant ovary, along with associated structures. The fruit protects the seeds and facilitates their dispersal.

29. The embryo of angiosperms consists of *cotyledons*, *epicotyl*, *hypocotyl*, and *radicle*. Angiosperms with only one cotyledon, or embryonic leaf, are *monocotyledons*; those with two cotyledons are *dicotyledons*.

30. Plant *hormones*, or growth regulators, are chemical substances produced in one part of a plant and transported to another part, where they exert a specific biological effect.

31. In plants, a *tropism* is a turning response resulting from a variety of stimuli. Tropisms may be influenced by *auxins*, which cause most plant cells to increase in length. A tropism may be positive or negative.

32. Geotropism may be influenced by the inhibitory plant hormone *abscisic acid*. This substance also inhibits cell division in developing leaves and often inhibits the growth of germinating seeds.

33. *Ethylene*, along with auxins, is involved in *abscission*, the process responsible for the dropping of leaves and fruit from trees. Ethylene

also is known to interfere with geotropism in some plants and may be an inhibitor of root growth.

34. *Gibberellins* stimulate rapid cell elongation in some plants; *cytokinins* induce plant cells to undergo mitosis, and they also cause increased stem and root growth, enhance flowering, and stimulate germination in some seeds.

35. Nonhormonal activities in plants include *nastic movements*, which result from a change in turgor pressure within the plant. *Thigmotropism*, the response to touch, is a common example.

APPLYING THE CONCEPTS

1. What factors distinguish the algae from the higher land plants?
2. What are the predominant features of each of the three major divisions of algae?
3. Why is *Chlamydomonas* described as both plantlike and animal-like?
4. What is meant by "cellular division of labor"?
5. Compare sexual reproduction in *Chlamydomonas* and *Volvox*.
6. What do you think are the advantages and disadvantages of conjugation in *Spirogyra*?
7. a. In what manner are mosses adapted to life on land?
   b. Why are they considered to be only partially adapted to land?
8. a. Compare the structure of the gametophyte and sporophyte in mosses.
   b. What is the function of each?
9. How is growth in length and diameter accomplished in stems? In roots?
10. a. What is meant by "transpiration pull"?
    b. What is its significance?
11. Explain the concept of a turgor pressure gradient.
12. What is the relationship between the structure of water molecules and the movement of water through the xylem?
13. How does alternation of generations in the ferns differ from that in the mosses?
14. Describe the formation of the male and female gametophytes of gymnosperms.
15. Describe the seed of a gymnosperm.
16. Describe the formation of the male and female gametophytes of angiosperms.
17. a. What is "double fertilization" as applied to angiosperms?
    b. What does it accomplish?
18. How are the flower and the fruit related to the biological success of angiosperms?
19. What is the functional relationship between auxins and phototropism?
20. Explain the relationship between plant hormones and abscission.
21. Compare the effects of cytokinins and gibberellins on plant function.

11 biology of animals

THE ESSENTIAL OBJECTIVES

You have understood this chapter when you are able to:

1. List the major distinguishing characteristics of the Metazoa and describe the function or functions of each.
2. Explain, by example, the difference between homology and analogy.
3. Outline the early embryonic development of the metazoans.
4. List the distinguishing characteristics of the Porifera.
5. List the distinguishing characteristics of the members of the following metazoan groups: Coelenterata, Ctenophora, Platyhelminthes, Aschelminthes, Annelida, Mollusca, Arthropoda, Echinodermata.
6. Trace the life cycles of the Chinese liver fluke, a vertebrate tapeworm, and the nematode *Trichinella*.
7. Explain why the arthropods are considered to be at the peak of invertebrate evolution.
8. Describe the hormonal control of insect molting and metamorphosis.
9. List the distinguishing characteristics of the phylum Chordata.
10. Explain the relationship of the urochordates and cephalochordates to the vertebrates.
11. List the distinguishing characteristics of the seven major classes of vertebrates.
12. Trace the general evolutionary lineage of the vertebrates.

From sponges to mammals, the animals are unified as multicellular, heterotrophic organisms, generally capable of voluntary movement. Through a host of adaptive features, invertebrate and vertebrate animals have become profusely distributed on land, in water, and in the air.

PREVIEW OF ESSENTIAL TERMS

tissue A group of cells similar in structure and function; the four basic animal tissues are *epithelium, connective tissue, muscle,* and *nervous tissue.*

homologous Pertaining to organs having similar structures but different functions; similarity of structure reflects a common embryological and evolutionary origin.

analogous Pertaining to organs having similar functions but different structures, and hence, different embryological and evolutionary origins.

ectoderm The outer embryonic tissue layer; gives rise to skin and the nervous system.

endoderm The inner embryonic tissue layer; gives rise to the lining of the digestive tract, lungs, and other organs.

mesoderm The middle embryonic tissue layer; gives rise to muscle, connective tissue, and blood.

larva In the life cycle of some animals, an immature form quite unlike the adult, into which it develops.

monoecious Having both male and female reproductive structures on the same organism.

dioecious Having male and female reproductive structures on separate organisms.

cephalization Concentration of nerve cells anteriorly, or toward the head, to form a brain.

parthenogenesis Development of an egg without fertilization.

closed circulatory system A system in which blood is confined within blood vessels throughout its circuit through the body.

open circulatory system A system in which blood is not totally confined within blood vessels but also flows into open tissue spaces in its circuit through the body.

trochophore A ciliated, free-swimming larva found in some flatworms, mollusks, and annelids.

molting The process of shedding of the exoskeleton.

metamorphosis A developmental or transformational change in an organism as it passes from a juvenile stage to the adult stage.

notochord In chordates, a dorsal supportive rodlike structure usually extending the length of the animal, just below the dorsal nerve cord.

ostracoderm An ancient jawless, bony-scaled fish, now extinct; considered to be an ancestor of modern agnaths.

placoderm A member of an extinct group of jawed fishes from which the cartilaginous and bony fishes arose; placoderms are the first known vertebrates with jaws.

crossopterygian A member of an ancient group of fishes descended from the original bony fish; crossopterygians are considered to be the line from which amphibians arose.

stegocephalian A member of the original group of amphibians on earth; stegocephalians are thought to have evolved from a group of crossopterygians.

cleidoic Pertaining to a type of egg produced by land vertebrates—reptiles, birds, and monotreme mammals. Within the closed shell, the embryo is surrounded by or associated with the four fetal membranes—the amnion, yolk sac, allantois, and chorion.

cotylosaur An extinct amphibian-like reptile; cotylosaurs are considered to be the immediate ancestors of modern reptiles and are the group from which birds and mammals evolved.

thecodont A member of an extinct reptile group that arose from the cotylosaurs; thecodonts gave rise to dinosaurs, crocodiles, and alligators and are the group from which modern birds descended.

homeothermic Having a constant body temperature, regardless of the temperature of the surrounding environment.

poikilothermic Having a body temperature that varies in accordance with the temperature of the surrounding environment.

therapsid An extinct mammal-like reptile; therapsids are considered to be the immediate ancestors of the mammals.

placenta In mammals, an organ formed from the lining of the uterus and two fetal membranes—the amnion and chorion. The placenta serves to exchange nutrients, oxygen, and wastes between the blood of the mother and fetus.

KINGDOM ANIMALIA

The earliest known fossils of multicellular animal life date back over 600 million years, some 200 million years before the rise of land plants. By the time the land plants had appeared, a few species of animals were already struggling to make the transition from the seas to land. Like most plants, the animals became generally more complex as time progressed.

Animals are conveniently separated into two major groups: the *invertebrates*, so called because they lack a spinal column, and the *vertebrates*, which have a spinal column composed of vertebrae. Nearly 95 percent of all the different kinds of animals are invertebrates; consequently, they are immensely diverse and occupy an extremely wide range of habitats. Although a minority in number, the vertebrates, too, are quite diverse and include among many others one species known as man.

All animals, except for the sponges and one other small group, are classified in a subkingdom called Metazoa (Gk. "later animals"). Although the metazoans are extremely diverse, they exhibit several similar and distinguishing characteristics. All of these animals are multicellular, having groups of similar body cells organized to form

> The four basic types of animal tissues are each composed of individual cells highly specialized structurally and functionally.

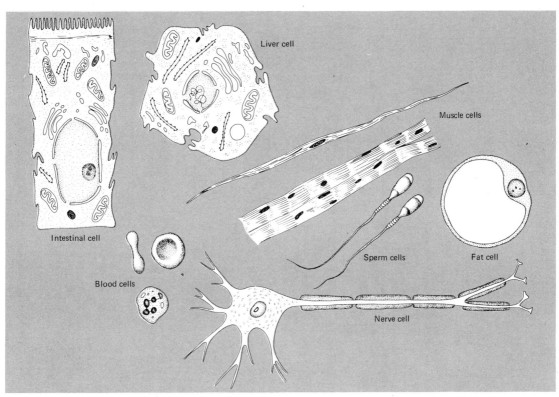

Figure 11.1 Various types of cells, each specialized for a specific task.

tissues. There are four basic types of animal tissues: *epithelial, connective, muscle,* and *nervous* tissues. The diverse functions of these tissues illustrate the concept of division of labor, i.e., the specializing of body cells and structures for specific tasks (Fig. 11.1). It should be emphasized, however, that all animals do not necessarily possess these tissues in as complex a form as described in the following sections.

Epithelial Tissue

Epithelial tissue, or epithelium, consists of tightly joined cells that cover the external and internal surfaces of the body (Fig. 11.2). Epithelium also forms the lining of the tubules and vessels of the body. The skin of animals is partly composed of epithelium, which serves as an outer protective barrier against pathogenic microorganisms, chemicals, and other harmful substances.

Connective Tissue

Connective tissue is quite diverse and includes blood and lymph, connective tissue proper, cartilage, and bone (Fig. 11.3).

Biology of Animals

> Organs are comprised of tissues, and many organs contain a combination of different types of tissues.

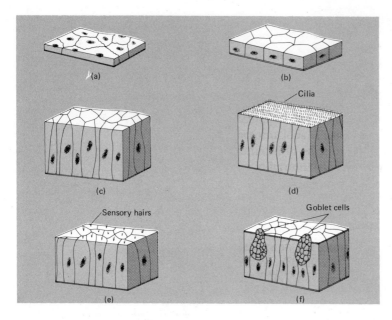

Figure 11.2 Types of epithelial tissue. (a) Squamous epithelium. (b) Cuboidal epithelium. (c) Columnar epithelium. (d) Ciliated columnar epithelium. (e) Sensory epithelium (cells from the lining of the nose). (f) Glandular epithelium: two single-celled glands (goblet cells) in the lining of the intestine.

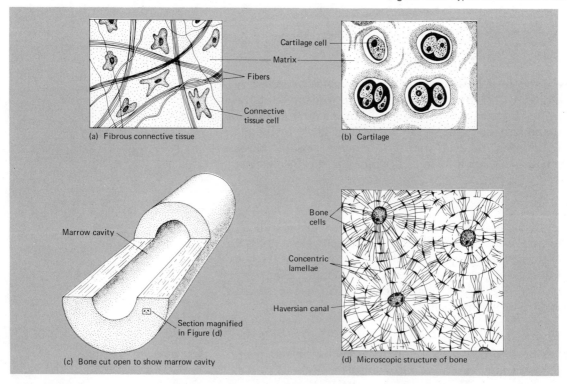

Figure 11.3 Types of connective tissues.

> The intercellular substance of a tissue is called its *matrix*; the matrix of blood tissue, for example, is liquid, whereas that of bone is hard and rigid.

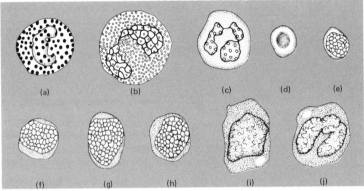

Figure 11.4 Types of blood cells. (a), (b), and (c) are white blood cells having a granulated cytoplasm; (d) is a red blood cell; (e) through (j) are white blood cells having a nongranulated cytoplasm.

Blood and Lymph. Blood and lymph, of course, are partially liquid tissues. The liquid component of vertebrate blood is *plasma*, in which are suspended the formed elements: *red blood cells* (erythrocytes), *white blood cells* (leukocytes), and *platelets* (thrombocytes) (Fig. 11.4). Blood is the major circulatory fluid of the body, transporting the respiratory gases carbon dioxide and oxygen, delivering nutrients to the cells, carrying metabolic waste products to organs of excretion, and distributing hormones throughout the body. Red blood cells function primarily in transporting oxygen, whereas the white blood cells are involved in the defense of the body against disease. Platelets release a substance that is necessary for the normal clotting of blood.

The lymphatic system is involved in the removal of excess fluid and small proteins that leak from the blood into the tissues. Lymph is actually a mixture of these proteins and tissue fluid that has entered the system of lymphatic vessels found throughout the body. Lymph is then delivered to the heart, where it enters the circulating blood.

Connective Tissue Proper. Connective tissue proper includes a variety of fibers, such as those composing ligaments, tendons, and other cementing materials that bind tissues together. In addition, various cells such as large phagocytes and fat-storing cells (adipose tissue) are included as connective tissue proper.

Cartilage. Cartilage is a very strong, flexible elastic tissue that forms the embryonic skeleton of vertebrates. Except for two vertebrate groups, the cartilage is later replaced by bone. However, in most adult vertebrates, cartilage is found between the vertebrae, at the joints, in the nose and ears, and in other areas of the body.

> An example of the interdependence of tissues is seen in muscles and nerves. Muscle tissue depends upon nerves to bring about movement in response to a stimulus, and stimuli must first be detected by nervous tissue.

Bone. Bone is a relatively hard tissue that functions in support, movement, and protection of the vertebrate body. The marrow of some bones produces the blood cells, which subsequently enter the circulation.

Muscle Tissue The most significant characteristic of muscle tissue is its capacity for contraction and relaxation. There are three types of muscle found in vertebrates: *skeletal*, or *striated*, muscle; *smooth* muscle; and *cardiac* muscle (Fig. 11.5). Skeletal muscle is responsible for the voluntary and reflex movements of the body. Involuntary movements of the digestive tract, blood vessels, and internal organs involve the action of smooth muscle. Cardiac muscle is found only in the heart.

Nervous Tissue Nervous tissue makes possible the sensation-response mechanism of the animal body. Specialized types of nerve cells, or *neurons*, receive and transmit stimuli, whereas other neurons bring about a particular response to the stimulus (Fig. 11.6). Some type of nervous system is present in all animal groups except the sponges.

Figure 11.5 Types of muscle tissues.

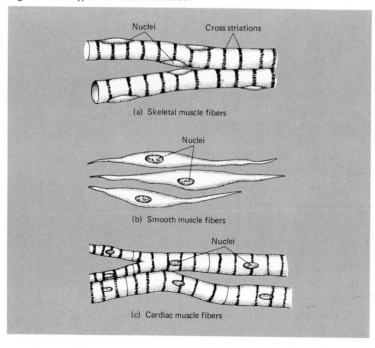

> For systematic biologists, it is important to know if the observed similarities between two species are homologous or analogous.

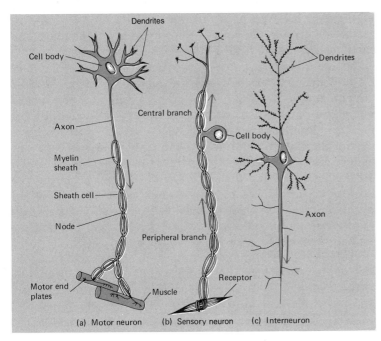

Figure 11.6 Types of neurons. (a) Motor neurons bring about a response in muscles or glands. (b) Sensory neurons receive stimuli and relay impulses to the appropriate area in the brain or spinal cord. (c) Interneurons conduct nerve impulses between sensory and motor neurons.

Animal tissues are joined together to form *organs,* such as the heart, liver, muscles, and kidneys. Various organs of the body make up *systems,* including the circulatory, respiratory, digestive, excretory, and other systems. The sponges have not progressed beyond a crude tissue level of development and therefore lack any distinct organ system. Other animals, however, have definite organs, and most have well-developed systems that are correlated with their particular mode of existence.

Further comparison of the metazoans reveals additional structural or functional similarities. For example, various anatomical features may closely resemble each other, as do the bones in the human arm, the bird's wing, or the flipper of a seal (Fig. 11.7). This similarity of bone structure, even though each appendage has a different function, reflects a common embryological and evolutionary origin. Body parts displaying such a relationship are said to be **homologous** (Gk. *homologos,* agreeing). There are some features, however, that are only similar in function and quite different structurally. The wings of birds and butterflies, for instance, function in flight, but their structural differences are such that they do not reflect a common embryological or evolutionary origin. The wing of a bird and the wing of a butterfly, therefore, are **analogous** (Gk. *analogos,* proportionate) structures. In observing the structural similarities or dif-

Organs and Systems

Figure 11.7 The bones in the forelimbs of birds and man are *homologous* because they reflect a common embryological and evolutionary origin.

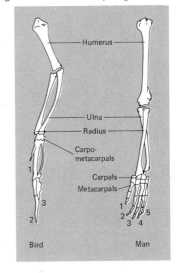

> The science of embryology has provided extensive information concerning evolutionary relationships among the various groups of living animals.

ferences between body parts in different species, investigators can more readily determine the relationships among various living species, as well as their relationships to ancestral forms.

As just mentioned, homologous structures reflect a common embryological origin. **Embryology** is the biological science that attempts to explain the development of an organism from the time of fertilization until birth. In most metazoans, the early stages of embryological development are quite similar.

After fertilization (union of a haploid egg and sperm), the diploid zygote undergoes a series of mitotic divisions called **cleavage** (Fig. 11.8). The zygote first divides into two cells, which in turn divide into four cells, followed by division into eight cells, and so on. Cleavage, then, gives rise to a multicellular **embryo.** Eventually, the embryo develops into a ball of cells called the **blastula.** In many animals the blastula contains a cavity, or **blastocoele.**

Embryological development beyond the blastula stage may vary extensively from one metazoan group to another. Through various inpocketings, foldings, and cell movements, the blastula is transformed into a structure consisting of two layers of cells. The outer cell layer forms one of the embryonic tissue layers, the **ectoderm** (Gk. "outer skin"); the inner layer of cells is the **endoderm** (Gk. "inner skin"). Among those animals in which organs develop, a third embryonic tissue, the **mesoderm** (Gk. "middle skin"), develops between

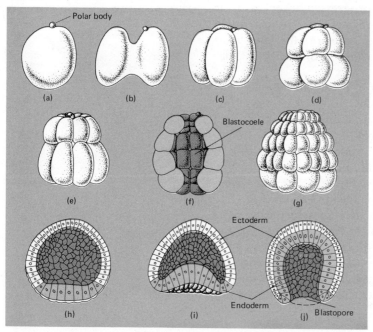

Figure 11.8 Cleavage and gastrula formation in a lower animal. (a) Egg with polar body. (b) through (e) Cleavage showing 2-, 4-, 8-, and 16-cell stages. (f) Thirty-two—cell stage cut open to show blastocoel. Blastula (g) and blastula cut open (h). Early gastrula (i) and late gastrula (j).

> Although they are multicellular and heterotrophic, the sponges have only a few other characteristics usually associated with members of the animal kingdom.

the ectoderm and endoderm. The formation of the embryonic tissues transforms the blastula into a **gastrula**. The gastrula contains a cavity that opens to the outside through the **blastopore;** the cavity will become the cavity of the digestive tract, and depending upon the species, the blastopore becomes either the mouth or the anus.

The embryonic tissues—ectoderm, mesoderm, and endoderm—give rise to the various tissues, organs, and other structures of the animal body. Ectoderm, for example, gives rise to skin and nervous tissue; mesoderm, in those animals that possess it, gives rise to muscle, connective tissue, blood, and so on; and endoderm gives rise to the linings of the digestive tract, lungs, liver, blood vessels, and so forth. Some of the lower invertebrates develop only ectoderm and endoderm, a condition referred to as **diploblastic.** Other animals develop all three embryonic tissues and are described as **triploblastic.**

One other intriguing characteristic of vertebrate embryos is that, in their early stages, most of these embryos bear a striking resemblance to each other (Fig. 11.9). In viewing early human, chicken, and fish embryos, for example, it is extremely difficult to determine which is which. This close parallel in embryological development between different species is indicative of a common evolutionary ancestry.

THE INVERTEBRATES
Subkingdom Parazoa

The Parazoa (Gk. "aside from animals") includes only one major group of animals, the sponges. These organisms have been set "aside from" the metazoans primarily because the evolutionary relationships between the two groups are unclear.

Phylum Porifera. The Porifera (L. "pore bearers") are better known as the sponges. Most of them are marine forms, but a few are found in fresh waters (Fig. 11.10). The sponges are quite simple animals structurally, having attained only a simple multicellular or a crude tissue level of development. They exhibit practically no physical activity and usually remain attached to some underwater object. In fact, sponges are so inactive that the ancient Greeks considered them to be a type of plant.

The body of a sponge is essentially a hollow tube composed of two cell layers (Fig. 11.11). This type of construction allows all the cells to be bathed in water. The body is perforated with numerous tiny pores (hence the name Porifera) that lead into the hollow body cavity, or **spongocoel.** As water passes through the pores into the spongocoel, food particles suspended in the water are filtered out by cells lining the pores; the water is then forced out through the "mouth," or **osculum,** at the upper end of the sponge.

Biology of Animals

Figure 11.10 A sponge. Water enters through pores all over the surface and leaves the animal via the hole in the center. (From Camp, P. S., and Arms, K.: *Exploring Biology.* Philadelphia: Saunders College Publishing, 1981, p. 18.)

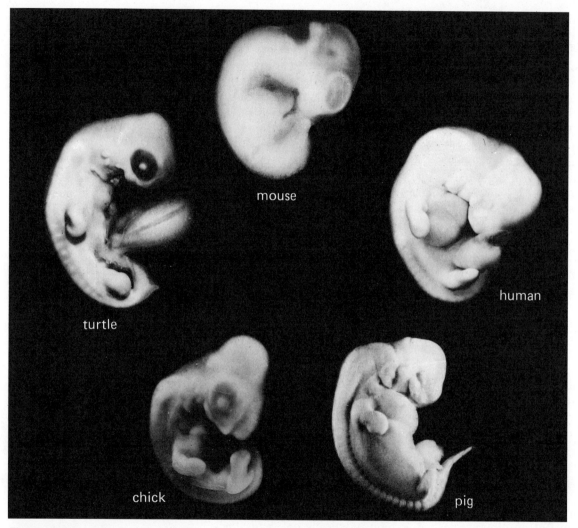

Figure 11.9 Embryos of various vertebrates bear a striking resemblance. (From Ebert, J. D., Loewy, A. G., Miller, R. S., and Schneiderman, H. A.: *Biology.* New York: Holt, Rinehart and Winston, 1973, p. 539.)

The sponges are the only animals without a nervous system and, in fact, have no organs at all. The body is supported by an internal "skeleton" composed either of tiny, needle-like **spicules** of calcium carbonate or silica or of a fibrous protein called *spongin* (Fig. 11.12). The spongin skeletons of some marine species are cleaned and sold commercially for household and industrial use.

The sponges usually reproduce asexually by buds that form on the outer body wall of the parent. The buds may separate from the parent and develop independently, or several buds may remain united in a cluster to form a colony. Among the freshwater sponges in particular, internal buds called **gemmules** (L. *gemma*, bud) are

> Regeneration, the replacement of lost body parts, is characteristic to some degree of almost all animal groups.

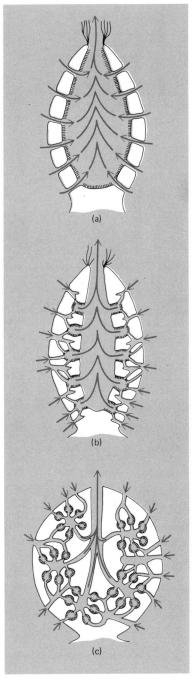

Figure 11.11 Structure of the body wall in sponges. Water enters the sponge through tiny pores in the body wall. After food particles are filtered out, water leaves the sponge through the osculum at the top.

formed during the fall and winter. In the spring, cells within the gemmules emerge and slowly develop into new sponges.

Sexual reproduction involves the production of haploid eggs and flagellated haploid sperm, which unite to form the diploid zygotes. Fertilization may occur in the surrounding water or within the spongocoel. After cleavage, the multicellular embryo develops into a flagellated *larva,* a free-swimming form that eventually settles down and grows into an adult sponge (Fig. 11.13).

Some sponges have both male and female reproductive organs and are therefore **monoecious** (Gk. *monos,* one, + *oikos,* dwelling), or bisexual; in other sponges the sexes are separate, a condition described as **dioecious** (Gk. *di,* double), or unisexual.

Sponges have been used rather extensively in studies on **regeneration**, i.e., the replacement of lost body parts. If a section of the body is cut away or broken off, the sponge will replace, or regenerate, the missing part. One of the classic experiments on these animals involves the passing of the sponge through a silk cloth, separating the body ino tiny pieces and even into individual cells. In the container, the cells and pieces eventually come together to reform the sponge. In fact, if two different species are treated in this manner, and their pieces mixed together in the same container, the parts of one species will aggregate only with parts of that species. By some recognition mechanism, the cells of a species reorganize only with

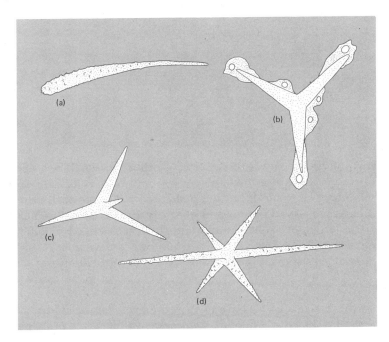

Figure 11.12 Sponge spicules.

306 Biology of Animals

> It has been postulated that metazoans may have arisen from a simple, multicellular form known as a *plakula*—a flattened structure with only two tissue layers and no organs.

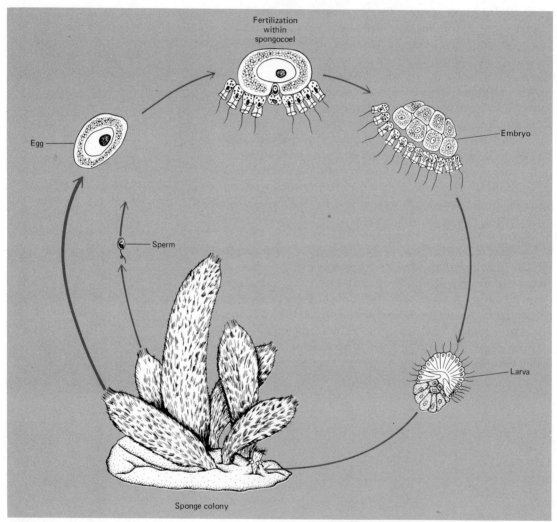

Figure 11.13 Sexual life cycle of a sponge.

each other and not with the cells of other species. It has since been discovered that not only sponge cells but also other cells from the same type of tissue have the capacity to recognize each other. This is observable even in the embryonic tissues ectoderm and mesoderm. If these tissues are disrupted and their cells mixed together, ectodermal cells reorganize only with other ectodermal cells and mesodermal cells similarly reorganize only among themselves.

Subkingdom Metazoa

The metazoans, as we have described, have attained at least a distinct tissue level of development, and most of them have discrete organs

> Having two different types of individuals—polyps and medusae—in their group, the coelenterates display what is known as *dimorphism*, meaning "two forms."

and systems. The complexity of these animals is evident in the increasing degree of cellular specialization, or division of labor.

Phylum Coelenterata. This phylum, also known as Cnidaria, includes the freshwater hydras, along with jellyfish, sea anemones, and corals, almost all of which are marine animals (Fig. 11.14). The body of coelenterates consists of two layers of tissue—an outer ***epidermis*** and an inner ***gastrodermis***—cemented together by a gelatinous ***mesoglea*** (Gk. "middle glue"). The epidermis and gastrodermis are derived from embryonic ectoderm and endoderm, respectively; the mesoglea is a rather poorly developed derivative of embryonic mesoderm.

The term "coelenterate," meaning "hollow intestine," refers to the hollow digestive cavity, called the ***gastrovascular*** ("stomach vessel") ***cavity.*** Either at the anterior end or in the center of the underside of the body is the mouth, which leads into the gastrovascular cavity. The basic hollow tube body plan of the coelenterates is similar to that of the sponges; overall, however, the coelenterates are considerably more complex.

All of the coelenterates have elongated ***tentacles,*** used in capturing food and, among the hydras, also used for locomotion. Tentacles are equipped with tiny specialized cells called ***nematocysts*** (Gk. "thread bladder"), which eject a fine thread when the tentacles are touched. The thread contains a sticky substance or a paralyzing toxin that incapacitates prey organisms. After capture, the prey is pushed into the mouth opening by the tentacles.

As a group, the coelenterates display two different body forms. One type is simply a hollow, cylinder-like form called the ***polyp;*** the other form is somewhat umbrella-shaped and is known as the ***medusa*** (Fig. 11.15). The medusa is familiar to most people as a "jellyfish." Actually, a medusa is essentially a polyp turned upside down and flattened out. Some of the coelenterates, including the hydras, sea anemones, and corals, exist only in the polyp form. Others, such as the jellyfish *Aurelia,* go through both polyp and medusa stages in their life cycles (Fig. 11.16). These life cycles are somewhat analogous to alternation of generations in plants. The medusae, which are of separate sexes, produce haploid eggs and sperm by meiosis. The medusa, then, represents the sexual stage. When released into the water, the egg and sperm unite to form the diploid zygote, which develops into a ciliated larva called a *planula* (L. *planus,* flat). After swimming about for a time and then settling down, the planula grows into a polyp. The polyp soon undergoes mitotic division to produce new medusae asexually. The polyp, therefore, is the asexual stage in the life cycle.

Biology of Animals

> Although the toxins released by nematocysts are not harmful to humans, the sting of some coelenterates, such as the Portuguese man-of-war, can be quite painful.

(a)

(b)

(c)

(d)

Figure 11.14 Various coelenterates. (a) Portuguese man-of-war. (b) Two species of jellyfish attached to red alga. (c) A sea anemone with a cluster of smaller anemone on the left. (d) A colony of corals. (From Ebert, J. D., Loewy, A. G., Miller, R. S., and Schneiderman, H. A.: *Biology*. New York, Holt, Rinehart and Winston, 1973, pp. 642, 643, 644.)

> Most ctenophores are quite small, although one species, known as Venus' girdle, is more than 3 feet in length.

Coelenterates such as *Obelia* exist as colonies of polyps. A single colony consists of feeding polyps and reproductive polyps attached like tree branches to a main stalk. The feeding polyps possess tentacles for capturing prey, and the reproductive polyps produce tiny buds that give rise to medusae asexually. When released, the medusae produce haploid eggs and sperm; after fertilization, the resulting zygote develops into a swimming planula. The planula eventually settles down and develops into a new *Obelia* colony (Fig. 11.17).

The corals, which are tiny, usually colonial, polyps, secrete a protective limestone skeleton around themselves. Massive accumulations of these skeletons create the many coral reefs found in the South Pacific and other parts of the world. Sea anemones, or "sea flowers," are marine polyps that sometimes grow to be several inches in length. Many sea anemones are beautiful animals, having brightly colored tentacles that, when expanding and contracting, give the appearance of the opening and closing of flower petals.

Coelenterates exhibit tremendous powers of regeneration. The freshwater polyp *Hydra* has been used extensively in laboratory regeneration studies for well over 200 years. This organism can be cut into several pieces, and each piece will regenerate into a completely formed *Hydra*.

Phylum Ctenophora. The ctenophores (Gk. *ktenos*, comb, + *phoros*, bearing), also known as "sea walnuts" and "comb jellies," are floating marine animals that are quite similar to the coelenterates (Fig.

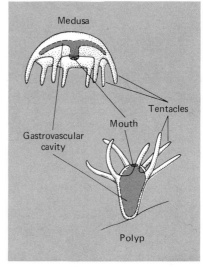

Figure 11.15 The two body forms found in coelenterates. The free-floating medusa is an inverted version of the sessile polyp. Both have tentacles surrounding the mouth, which leads into a blind gastrovascular cavity. Wastes are ejected through the mouth.

Figure 11.16 Life cycle of the jellyfish *Aurelia*.

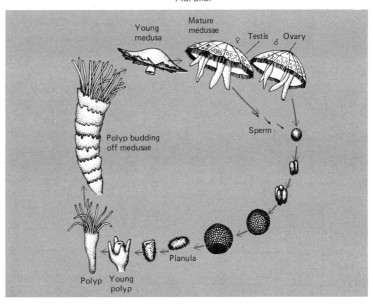

310
Biology of Animals

> The flatworms (and one other small group of worms) are the most primitive animals with *bilateral symmetry*, i.e., having right and left sides that essentially are mirror images of each other.

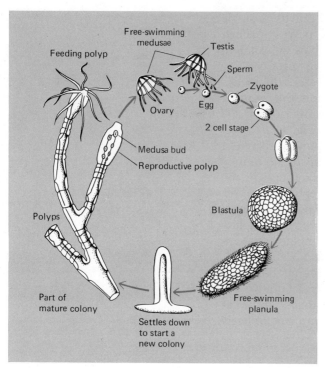

Figure 11.17 Life cycle of the colonial coelenterate *Obelia*.

11.18). They have eight rows of ciliated plates, or combs, that run longitudinally along the surface of the body. Beating of the cilia enables the organisms to swim rather weakly, but ordinarily they are swept along by currents and tides. Although the ctenophores have the same three-layered body plan and have attained the same tissue level of development as the coelenterates, they are different in not having nematocysts, nor do they have a polyp stage in their life cycles. Moreover, ctenophores have only two tentacles instead of several. However, there is a resemblance between adult ctenophores and some coelenterate medusae; consequently, it is believed that the two groups may share some evolutionary kinship.

Ctenophores are monoecious, producing both eggs and sperm in specialized reproductive tissues. The gametes exit through the mouth and enter the water, where fertilization occurs. The zygote develops into an adult, either directly or by first passing through a larval stage. Many of the ctenophores are brilliantly luminescent, often giving off vivid flashes of light that flicker across the waves.

Phylum Platyhelminthes. This group of invertebrates includes the free-living *planarians* and two other types that are entirely parasitic, the *flukes* and *tapeworms*. The phylum name is from the Greek

Figure 11.18 Ctenophores. Note the rows of comb plates and the two long tentacles. (From Ebert, J. D., Loewy, A. G., Miller, R. S., and Schneiderman, H. A.: *Biology*. New York: Holt, Rinehart and Winston, 1973, p. 645.)

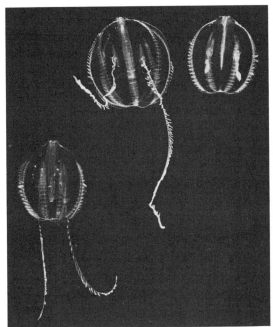

platys, flat, + *helmins*, worm; these flatworms, then, have a body that is flattened dorsally (on the back or upper side) and ventrally (on the front or underside). The flatworms show considerable advancement over the coelenterates in being triploblastic and having attained the organ level of development. Many of the flatworms possess muscular, digestive, excretory, and reproductive systems. It has been suggested that the flatworms may have evolved from a ciliated larval form of some ancient coelenterates.

The Planarians. One of the most common representatives of the free-living flatworms is the freshwater planarian. This organism is covered by a ciliated skin (ectoderm) that secretes a slimy mucus. Using the cilia and rhythmic muscular movements of its body, the planarian is able to slide along in its own slime track.

Planarians have a branching digestive tract that opens to the exterior through the mouth, located on the underside of the body. Planarians are mostly carnivorous ("flesh eaters"), feeding on small worms and insects that are sucked up into the intestine through a protrusible pharynx, which extends through the mouth opening (Fig. 11.19).

In the planarians, we encounter for the first time a well-differentiated excretory system. The system consists of a network of small tubules running the length of the body. Branching off the tubules are *flame cells*, each of which has a small cluster of cilia that undu-

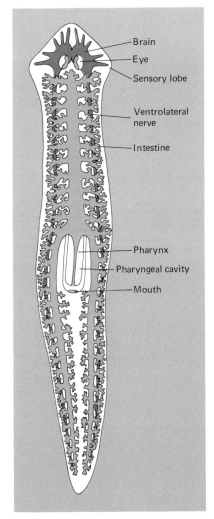

Figure 11.19 A planarian, showing the digestive and nervous systems.

> Cephalization, the concentration of nervous tissues and organs in the head end of the body, is evident mainly in animals with bilateral symmetry.

lates like a flame, sweeping water and waste products along the tubules and out of the body through tiny pores (Fig. 11.20).

Flatworms exhibit further advancement over coelenterates in development of the nervous system. Planarians possess two masses of nerve cell bodies called **ganglia** (sing. ganglion), located at the anterior, or head, end of the organism. Extending posteriorly from each ganglion is an elongated **nerve cord** that runs the length of the body. The nerve cords are connected at intervals by transverse nerves, forming a "ladder type" of nervous system. Also located in the head region is a pair of pigmented **eyespots,** which receive small nerves from the cerebral ganglia. The eyespots enable the planarian to respond to light, although they do not form images. The concentration of nerve cells in the anterior end of the body is called **cephalization.** In more advanced animals, cephalization results in the development of a brain that functions as the major center of nervous control.

Planarians reproduce asexually by constricting transversely and separating into two parts; each part then regenerates into a completely formed planarian.

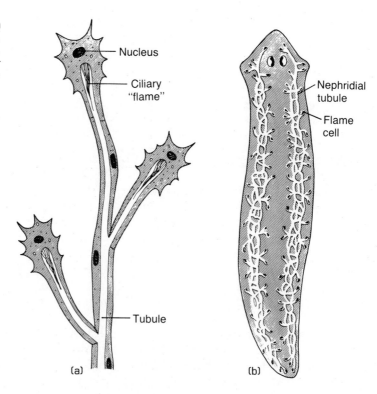

Figure 11.20 Excretory system of a planarian. (a) Flame cells and tubules. (b) Tubule network. (From Villee, C. A., Walker, W. F., and Barnes, R. D.: *General Zoology.* 4th ed. Philadelphia: W. B. Saunders Company, 1973, p. 476.)

> Although most flukes are monoecious, the human blood fluke *Schistosoma* displays separate male and female sexes.

Most flatworms, including the planarians, are monoecious and reproduce sexually by cross-fertilization. Sperm from each partner is deposited within the body of the other, and fertilization of the eggs occurs internally. After the eggs ripen, they are released from the body and develop into mature worms.

The planarian is another animal that has been used extensively in studies on regeneration. Like the hydras, it can be cut into several pieces, each of which regenerates into a complete animal (Fig. 11.21). In addition, planarians have been used in conditioning behavior studies, being trained to navigate mazes in search of food. Flatworms may be the lowest animal group having the capacity for learning through simple conditioning. Such activity indicates considerable advancement in nervous system function compared with that of any lower animal phyla.

The Flukes. The flukes are parasitic flatworms that, like the tapeworms, are most notable as the agents responsible for a variety of diseases affecting higher animals, including humans. Some of the flukes have highly complex life cycles, often involving several hosts. One such example is *Opisthorchis,* the so-called Chinese liver fluke (Fig. 11.22). The adult inhabits the bile duct of human beings, where it lays its eggs. The eggs move into the intestine, from where they are shed to the exterior in the feces. To continue the life cycle, the eggs must first get into water and then must be taken up by a certain species of snail. The eggs hatch in the tissues of the snail, and the resulting forms remain in the snail. They do, however, produce offspring of their own, which leave the snail and swim free in the water for a short time. If these offspring encounter a fish, they burrow into its muscles and encyst. If human beings eat infected fish that have been improperly cooked, the flukes are released in the intestine and migrate to the bile duct, where, after laying their eggs, the flukes may repeat their life cycle.

Not surprisingly, the flukes characteristically produce an enormous number of eggs. With such intricate life cycles, you can well imagine that producing an immense number of eggs is certainly an advantage to a species because a great number of the eggs never make it very far in the cycle, much less complete it. However, if the hosts become more concentrated in number, they are easier for the developing flukes to find, and thus there is an increased likelihood that the life cycle will be completed.

A case in point involved the construction of the Aswan High Dam across the Nile River in Egypt. Before the dam was built, the river flooded in the spring and then subsided during the dry season. With the building of the dam, however, irrigation channels provided

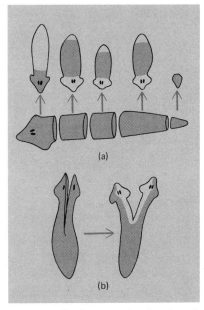

Figure 11.21 Regeneration by planarians. (a) All but the last part of each section will form a new head. (b) This animal, which has been cut along its plane of symmetry, has formed two heads. It will complete the separation and become two planarians.

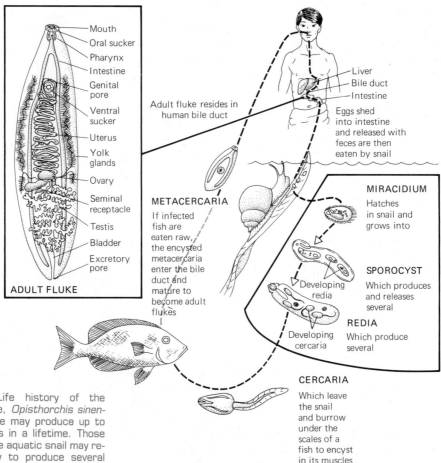

Figure 11.22 Life history of the Chinese liver fluke, *Opisthorchis sinensis*. An adult fluke may produce up to half a million eggs in a lifetime. Those eaten by a suitable aquatic snail may reproduce asexually to produce several hundred offspring each. The motile cercaria discharged from the snail must find a fish host. After burrowing through the fish's skin, the cercaria form cysts in the muscle. Encysted metacercaria hatch when the fish muscle is eaten by a suitable host and make their way to the bile duct, where they mature to the adult form. (From Arms, K., and Camp, P. S.: *Biology.* 2nd ed. Philadelphia: Saunders College Publishing, 1982.)

a continuous source of water for the Nile Valley, which, in effect, eliminated the dry season. As it turned out, the dry seasons were vital in limiting the population of water snails, the hosts of the eggs of a blood fluke called *Schistosoma*. With the increased availability of water, the snails became heavily concentrated in the area, facilitating infestation by the eggs. When the offspring of these eggs come in contact with a human being, they burrow through the skin into a blood vessel and eventually enter the intestine, where they lay their eggs (Fig. 11.23). In the body, these blood flukes cause a weakening, painful, and sometimes fatal disease called *schistosomiasis*.

The flukes characteristically possess two or more suckers for attachment to the host and have a thick body covering that is highly

> The treatment of tapeworm infestation must involve removal of the scolex. As long as the scolex remains embedded in the muscle, it will form new proglottids to replace those that are shed in the feces.

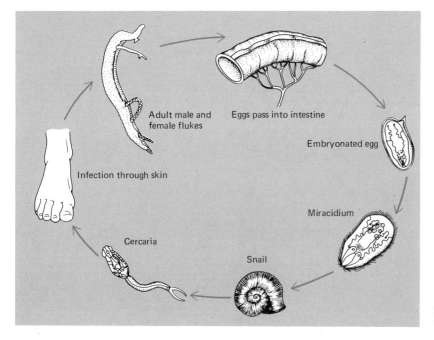

Figure 11.23 Life cycle of the blood fluke *Schistosoma*.

resistant to the intestinal digestive enzymes of the host. Both of these features are significant adaptations to a parasitic mode of existence.

The Tapeworms. The other group of parasitic flatworms comprises the tapeworms, some of which may grow to a length of 75 ft. Like the flukes, they have a resistant body covering and possess several suckers used for attachment to the host. In the tapeworms, however, the suckers are found on an anterior attachment organ called the **scolex.** Posterior to the scolex, the long, flat body is composed of individual segments called **proglottids** (Fig. 11.24).

Each proglottid of a tapeworm contains male and female reproductive structures in various stages of development. The more anterior proglottids behind the scolex are relatively immature, whereas those toward the middle of the body have fully developed male and female organs. Each of the most posterior proglottids contains only the uterus; together, these uteri are engorged with tens of thousands of fertilized eggs. All the other reproductive organs have degenerated. The posterior proglottids eventually break off and are passed to the exterior in the feces of the host. New immature proglottids are produced in the "neck" area near the scolex.

If tapeworm eggs are eaten by a cow, hog, fish, or some other vertebrate, the eggs are carried by the blood to the muscles, where

Biology of Animals

> Nematodes are among the most numerous and widespread organisms on earth, being found in every conceivable habitat where there is adequate moisture.

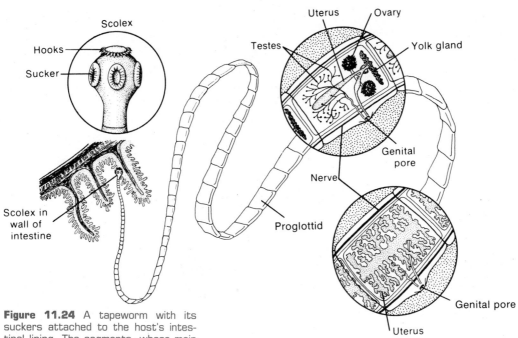

Figure 11.24 A tapeworm with its suckers attached to the host's intestinal lining. The segments, whose main functions are reproductive, are called proglottids. The head, or scolex, and immature and mature proglottids are shown enlarged. (From Clark, M. E.: *Contemporary Biology.* 2nd ed. Philadelphia: Saunders College Publishing, 1979, p. 74.)

they encyst. If humans eat the raw or poorly cooked meat of these infected animals, the eggs pass to the intestine, where they mature into adult tapeworms. With the production of new eggs within the intestine of man, the cycle may begin anew.

Phylum Aschelminthes. The Aschelminthes (Gk. *askos,* bag, + *helmins,* worm) is a group of mostly small, cylinder-shaped worms; some are free-living, whereas others are parasitic. These animals show some advancement over the flatworms in having a complete digestive system with mouth and anus. Moreover, they are usually dioecious, i.e., the sexes are separate. The aschelminths are found in marine and fresh water and in terrestrial habitats. Like the flatworms, they are triploblastic and possess well-developed nervous, excretory, and reproductive systems.

We shall consider two representative classes of the Aschelminthes: Nematoda and Rotifera.

Class Nematoda. The nematodes (Gk. "threadlike") have thin, elongate bodies, usually tapered at both ends. They are found in abundance in soil and water, and some are parasites of both plants and

> A female *Ascaris* worm may release as many as 200,000 eggs a day.

animals. One of the many species parasitic on human beings is *Trichinella spiralis* (Fig. 11.25), the causative agent of the disease **trichinosis**. The intestines of hogs and other mammals harbor adult *Trichinella* worms, the females of which release larval worms into the lymph and blood of the host. The young worms circulate to skeletal muscles, where they grow and become enclosed in a resistant sac, or cyst. Human beings may acquire the encysted worms by eating poorly cooked pork. The worms mature and reproduce in the human intestine, after which the newly produced larval worms circulate in the blood to the skeletal muscles, where they encyst (Fig. 11.26). The resulting disease, trichinosis, is characterized by high fever, general weakness, and extreme muscular pain. Trichinosis can be prevented simply by cooking pork thoroughly enough to kill the encysted worms.

Another parasitic nematode is *Ascaris*, species of which are found in the intestine of humans, horses, and pigs. As in many other roundworms, the female is much larger than the male (Fig. 11.27). Eggs are laid in the intestine of the host and then eliminated in the feces. The eggs, which after two or three weeks contain embryonic worms, may be ingested by eating foods grown in infected soil or by hand-mouth contact. The eggs hatch into embryonic larvae in the intestine and then circulate in the blood to several organs and back to the intestine. The worms mature in the intestine, where, after copulation, the female lays her eggs.

Two other common parasitic nematodes are the **pinworms** and **hookworms**. Pinworms more commonly infect children, who pick up the eggs from contaminated soil. Hookworms are most frequently

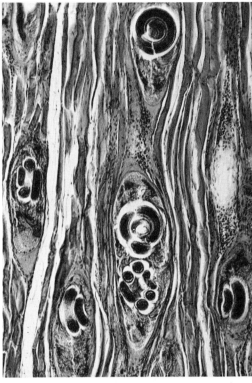

Figure 11.25 Larvae of the nematode *Trichinella* embedded in muscle. (From Clark, M. E.: *Contemporary Biology*. 2nd ed. Philadelphia: Saunders College Publishing, 1979, p. 75.)

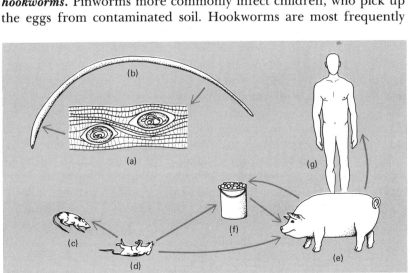

Figure 11.26 *Trichinella spiralis*. Larvae encysted in muscle (a) mature into intestinal worms when eaten (b). These give birth to larvae that burrow into the host, encysting in muscle. The natural reservoir is rodents (c) and similar animals that eat their dead (d). Pigs (e) will also eat dead rodents. Furthermore, killed rodents and pig scraps are fed to pigs in garbage (f). Humans can become infected by eating insufficiently cooked meat containing larvae.

> A feature unique to the rotifers is the *mastax*, a complex muscular pharynx that employs seven jawlike structures to grind food.

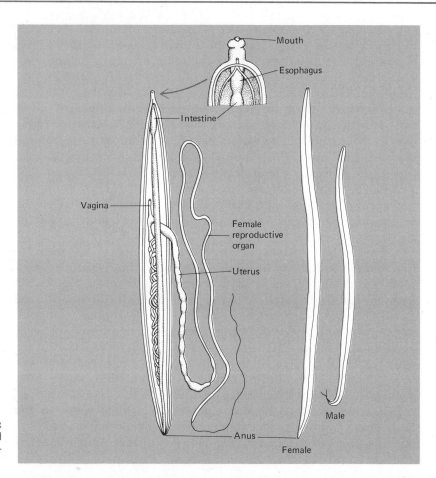

Figure 11.27 *Ascaris*, a parasitic roundworm. Side views of a male and female are shown at the right. The dissected worm at the left is a female.

acquired by going barefoot in soil infested with larvae. Prevention of both pinworm and hookworm infection is best effected through personal cleanliness and sanitation.

A parasitic nematode of tropical and subtropical climates is the **filaria worm,** which is transmitted to the host by mosquitoes (Fig. 11.28). In humans, the worms inhabit the lymph vessels, where they may block the flow of lymph, causing fluid to accumulate in the tissues. The result is **elephantiasis,** a grotesque swelling of the body, particularly of the lower extremities.

Although the nematodes noted in this section are parasitic and harmful, most of the nematodes are harmless and occupy a variety of aquatic and terrestrial habitats.

Class Rotifera. A group of unusual little characters, which look more like protozoans than roundworms, are the **rotifers** (L. "wheel bear-

Parthenogenesis is found not only among rotifers but also in crustaceans, some ants, bees, and plant lice.

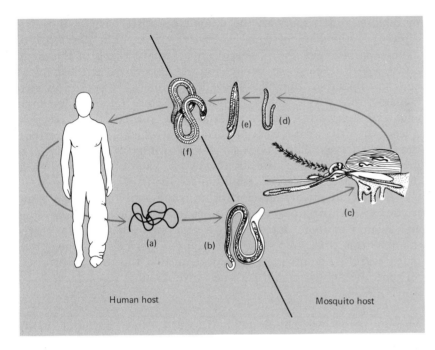

Figure 11.28 Life cycle of a filarial worm. Adult worms in human lymphatic tissues (a) release microscopic larvae into the blood (b). If taken up by a mosquito (c), the larvae migrate to thoracic muscle, where they mature and grow (d - f). The larvae are later introduced into humans when the mosquito feeds.

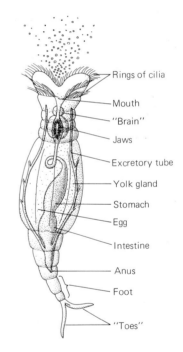

Figure 11.29 Anatomy of a rotifer. Internal organs are easily seen through the transparent body wall of many forms. Cilia around the mouth create a feeding current. Food particles are passed to the jaws in the pharynx, where they are ground up. The "toes" are often provided with glands that secrete glue by which the rotifer can anchor itself to the substratum. Although small males are produced occasionally, most rotifers are females, and their offspring are usually more females (note the single large egg). (From Clark, M. E.: *Contemporary Biology*. 2nd ed. Philadelphia: Saunders College Publishing, 1979, p. 75.)

ers") (Fig. 11.29). These microscopic aquatic animals acquire their name from an anterior ring of motile cilia, which gives the appearance of a revolving wheel. The cilia are used to draw food into the mouth and are also used for locomotion. The body is divided into three sections: head, trunk, and foot. The head is crowned with motile cilia, and the trunk contains most of the internal organs. The pointed foot is used for attachment. In some species, the body sections are retractile, much like a telescope.

One of the unusual and fascinating features of the rotifers is that each member of a given species has exactly the same number of body cells. Furthermore, beyond a certain point in their development, cell division (mitosis) is no longer possible, and the animals can neither grow nor replace broken or missing body parts. Even stranger is the fact that, in most species of rotifers, there are no males. Eggs laid by the females develop parthenogenetically (Gk. *parthenos*, virgin, + *genesis*, origin), i.e., the eggs develop without being fertilized by sperm. In the few species in which they are found, the males live only about 3 days, a seemingly accommodating fate, since they lack a mouth, anus, and most other digestive structures and are unable to feed.

The explanation of the reproductive oddities of rotifers involves two different types of eggs produced by the females. One type arises

> As an earthworm constructs its burrow, much of the soil passes through the digestive tract and eventually is left behind.

by meiosis and is haploid; the other type arises by mitosis and is diploid. If the haploid eggs are not fertilized, they develop into males; diploid eggs always develop into females. Both of these egg types, then, develop parthenogenetically. However, if the haploid eggs are fertilized, they also hatch into females. In species of cold climate rotifers, the male is important because only the fertilized eggs can survive the winter months. Male rotifers, then, are produced when the weather becomes cooler, at which time they fertilize the haploid eggs. These eggs survive the winter and hatch into females in the spring. However, the males, having performed their dutiful task, are no longer around to see the fruits of their labor.

A final adaptive characteristic of some rotifer species is quite remarkable. During periods of severe drought, the body loses water and shrinks to about one-third its normal size. The animals can remain in this dried-out state for extensive periods of time, often many years. When moisture becomes available, the body absorbs water, and the animals again become active within a few minutes or a few hours.

Phylum Annelida. The annelids (L. "little rings") include, among others, the common earthworm, the clam worm, and the leeches (Fig. 11.30). Annelids are found in a variety of habitats, including soil and marine and fresh waters. Many species are free-living, but some, such as various leeches, are parasitic. As their name attests, the annelids have bodies composed of repeating ringed segments. These animals have highly developed organ systems and are the first group we have encountered that possesses a discrete circulatory system.

The Earthworm. The common earthworm, *Lumbricus terrestris* (L. *lumbricus,* worm, + *terra,* earth), not only serves as fish bait but also is otherwise economically and ecologically beneficial. With their underground twisting and churning, earthworms turn over millions of tons of earth every year, thereby aerating the soil and helping maintain its fertility. The outer ventral and lateral walls of the body of the earthworm bear bristle-like structures called **setae.** Setae are used to anchor the worm as it pushes through the soil. Attempts to pull an earthworm intact from its burrow are usually unsuccessful because the setae anchor the body firmly in the soil.

The circulatory system of the earthworms and other annelids is a **closed system.** This means that the circulating blood is confined within blood vessels and does not flow freely into body spaces or sinuses. A closed circulatory system is also found in all the verte-

Figure 11.30 Representatives of the three classes of annelids. (a) Earthworm. (b) Clam worm. (c) Leech. (From Ebert, J. D., Loewy, A. G., Miller, R. S., and Schneiderman, H. A.: *Biology.* New York: Holt, Rinehart and Winston, 1973, p. 656.)

brates. The earthworms have a dorsal blood vessel that functions as a heart in pumping blood throughout the body. Oxygen and carbon dioxide are exchanged across the skin surface, which is richly supplied with blood capillaries.

The nervous system is well developed, consisting of cerebral ganglia, nerve cords running the length of the body, and a ganglion in each body segment. Consequently, earthworms exhibit complex muscular reflexes and well-coordinated movements.

Inside almost every body segment of the earthworm is a pair of primitive excretory organs, the **nephridia** (Gk. *nephros,* kidney). Each nephridium is associated with a network of blood capillaries, through which materials in the blood may enter the nephridium.

> Some leeches are capable of consuming such enormous quantities of blood that they may go without feeding again for several months or even a year.

These materials, along with others from the body fluids, are excreted from the nephridia to the outside through tiny pores in the body wall. The basic internal structure of an earthworm is illustrated in Figure 11.31.

Lumbricus and other large earthworms are monoecious and engage in cross-fertilization. The eggs develop within a mucous **cocoon** (Fr. *coque*, shell) secreted by several segments of the body. In *Lumbricus* only one egg of many in the cocoon develops into a young worm, a process requiring about 2 to 3 weeks.

The Clam Worm. The clam worm, genus *Nereis,* is a marine annelid somewhat resembling the earthworm. However, *Nereis* and similar members have lateral fleshy appendages called **parapodia** (Gk. "side feet"), which bear numerous setae. The parapodia contain a rich network of blood vessels and are used in respiratory gas exchange and in locomotion. Unlike *Lumbricus,* the clam worm has a well-developed head with jaws and various sensory structures. With one major exception, the internal systems of the clam worm are similar to those of the earthworm. The exception is the reproductive system, in that the clam worm and most other similar annelids display separate sexes, i.e., they are dioecious. Fertilization is external, and the embryos of several species pass through a **trochophore** (Gk. *trochos,* wheel, + *phoros,* to bear) larval stage during development. A trochophore larva is shaped something like a child's top and is a form common to many aquatic invertebrates (Fig. 11.32).

Leeches. Leeches are found in marine and fresh waters and on land. Some leeches are parasites on vertebrates and usually possess anterior and posterior suckers used for locomotion and attachment to the host. Sharp, chitinous teeth are present for penetrating the skin, and their salivary glands produce an anticoagulant called **hirudin,** which prevents the clotting of blood. Unlike other annelids, the leeches have no parapodia or setae. They are monoecious and have no larval stage. In bygone days, leeches were used by physicians for bloodletting in the treatment of disease.

Phylum Mollusca. The mollusks (L. *mollis,* soft, a reference to their soft bodies) constitute one of the largest of all animal groups and include such organisms as clams, oysters, scallops, snails, squids, and octopuses. Most of the mollusks have a hard outer shell surrounding their soft bodies, although in the squids the shell is internal, and in the octopus and a few others it is missing altogether. During their development, many of the marine mollusks pass through a trochophore larval stage, similar to that of the annelids.

The Invertebrates

The mollusks are second only to the arthropods in the number of different species. The fossil record of mollusks dates back well over one-half billion years.

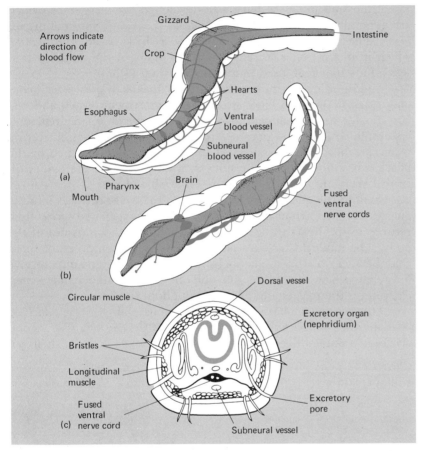

Figure 11.31 The anatomy of an earthworm and the annelid body plan. (a) The digestive tract and the circulatory system. The mouth opens into a highly muscular pharynx that sucks soil containing decaying vegetation into the gut. This material is sorted in the crop and pulverized in the gizzard with the aid of the sand grains in the soil. The rest of the gut consists of a long intestine, where food is digested and absorbed. The circulatory system is composed of three longitudinal blood vessels that run from head to tail, a dorsal vessel, and two ventral vessels. The dorsal vessel receives blood from smaller vessels in each segment. It is extremely muscular and pumps the blood forward. Five muscular pumping vessels, the hearts, propel blood downward to the ventral vessels, from which it flows back to the posterior segments. The dorsal vessels and the hearts have valves that prevent backflow, and as a result the blood flows in a circuit. (b) The nervous system consists of two ventral nerve cords that are fused throughout most of the body. This fused nerve cord divides near the pharynx and fuses again to form the earthworm's brain. (c) Cross section revealing the large coelom in which numerous organs are suspended. The excretory organs, called nephridia, are also visible. Each of these simple kidneys consists of a long tubule lined with cilia. One end is funnel-shaped and opens into the coelom. The other end of the tubule opens to the outside via an excretory pore. Coelomic fluid is drawn into the tubule by the action of the cilia. As the fluid passes down the tubule, various useful materials, such as salts and sugars, are absorbed by the cells of the tubule and returned to the coelom.

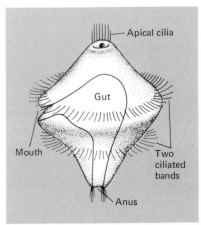

Figure 11.32 Trochophore larva of an annelid. Almost identical trochophore larvae are found in mollusks.

> An unusual feature of the bivalve mollusks is that the intestine passes right through the heart.

Almost all mollusks have well-developed organ systems, which are generally contained within a *visceral* (L. *visceris,* internal organ) *mass.* The visceral mass is draped by a thick tissue, the *mantle,* which in most species secretes the shell. Many of the mollusks possess a muscular *foot,* used in locomotion (Fig. 11.33).

Almost all of the mollusks respire by means of *gills,* which contain a rich network of blood vessels through which oxygen and carbon dioxide are exchanged. Most mollusks have an open circulatory system, in which blood does not remain within blood vessels during its circuit through the tissues. At some point, blood pumped by the heart enters open spaces, or sinuses, where it bathes the tissues directly. From the sinuses, blood re-enters blood vessels that convey the blood to the gills, where oxygen from the water diffuses into the gill vessels and carbon dioxide is given off. Oxygenated blood then returns to the heart to be pumped once again to the tissues of the body. The mollusks have a well-developed heart composed of three chambers: Two upper chambers, the *atria* (sing. atrium), receive oxygenated blood from the gills; the one lower chamber, the *ventricle,* pumps the oxygenated blood throughout the body.

The two major classes of shelled mollusks are Pelecypoda (Gk. "hatchet foot"), also known as *bivalves,* and Gastropoda (Gk. "stomach foot"). Some of the gastropods, however, have lost the shell.

Class Pelecypoda. Included in this class are clams, oysters, mussels, scallops, and a few others (Fig. 11.34). The name of the class refers

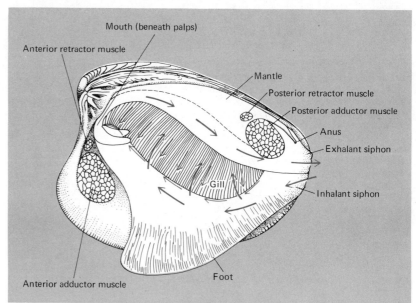

Figure 11.33 Some of the basic structures of a mollusk (clam).

Among the 50,000 named species of gastropod mollusks, some, such as the slugs and nudibranchs, do not have shells.

Figure 11.34 Some members of the Class Pelecypoda (Phylum Mollusca). (a) Scallops, along with an intruding starfish. (b) A mussel anchored to a rock by fine threads that it secretes. (c) A model of a giant clam with coral attached. (From Ebert, J. D., Loewy, A. G., Miller, R. S., and Schneiderman, H. A.: *Biology*. New York: Holt, Rinehart and Winston, 1973, p. 664.)

> The land snails are among the few invertebrate groups that have fully adapted to a terrestrial existence.

to the hatchet-shaped foot that can be extruded from the shell for use in locomotion and burrowing into sand.

The two shells, or valves, of these animals are hinged together by a strong ligament. The valves are lined by the mantle, which surrounds the soft body, or visceral mass. The internal surface of each valve is composed of **nacre,** or mother-of-pearl. Occasionally, a grain of sand or some other irritating material becomes lodged between the mantle and the shell. The mantle then begins to secrete several layers of nacre around the irritating material, producing a pearl. Although an object of pleasure in the human realm, a pearl to a bivalve is but a product of annoyance.

Most adult bivalves creep slowly through sand or mud, using the foot for locomotion; others, however, swim for short distances by clapping their valves together.

Water enters a bivalve through a posterior tubelike siphon and washes over the gills. In the process, tiny food particles adhere to the gills and are swept forward into the mouth. Animals that feed in this manner are called **filter feeders.** After passing over the gills, water is forced out of the body through a second posterior siphon.

Bivalves, of course, are among the choicest seafoods, sought not only by humans but also by various marine predators, such as starfish. In one group of bivalves—the scallops—only the large muscles that hold the valves together are sold for human consumption; the rest of the body is discarded.

Class Gastropoda. This group of mollusks includes snails, slugs, whelks, abalones, and conchs (Fig. 11.35). Most gastropods are marine, but some are found in fresh water and others are terrestrial. Although most gastropods have a single coiled shell, in some species the shell is uncoiled, and in others (slugs) the shell is absent.

Figure 11.35 Gastropods. Snail (a) and slug (b). (From Applewhite, P., and Wilson, S.: *Understanding Biology.* New York: Holt, Rinehart and Winston, 1978, p. 76.)

(a)

(b)

> For many millions of years, the earliest cephalopod mollusks were covered by heavy shells; today, the only shelled cephalopod is the chambered nautilus.

Most terrestrial and freshwater snails do not possess gills. The space between the mantle and visceral mass contains a rich network of blood vessels through which oxygen and carbon dioxide are exchanged. This highly vascularized space, then, serves as a functional lung.

Some of the more unusual gastropods include the oyster borers, which bore through the shells of oysters and suck out their body fluids. Rock borers, using their rasping teeth, can bore through solid rock.

Class Cephalopoda. The cephalopods (Gk. "head foot") are so called because the foot, in the form of arms and tentacles, is located in the head region. All of these animals—squids, octopuses, cuttlefishes, and the chambered nautilus—are marine and are among the most advanced invertebrates (Fig. 11.36). It is often difficult to believe that cephalopods belong to the same phylum as clams, snails, and other mollusks. The squid, for example, is freed of a cumbersome shell and moves in any direction with speed and dexterity. With its ten "arms" the squid can grasp and manipulate its prey. It has a very highly developed brain and a pair of eyes that nearly rival our own. As added attractions, squids also possess a tough beak for tearing their prey and an ink sac that expels a black fluid used for concealment, and they can even change body color as a means of camouflage.

Although most cephalopods are of moderate size, some squids are less than two inches long, while the North Atlantic giant squid—the largest of all living invertebrates—may exceed 50 ft. in length.

Contrary to popular legend, the eight-armed octopuses are neither gigantic nor ferocious. Their bodies usually measure less than a foot in diameter, and they are shy animals, preferring a solitary existence along the ocean shore.

Cephalopods possess gills for respiration and have a closed circulatory system with a well-developed heart. The sexes are separate, and, in the squids, the process of sperm transfer is rather unusual. The male uses a specialized tentacle to remove sperm packets from his own mantle cavity, after which the tentacle deposits the packets in the mantle cavity of the female. After fertilization, the masses of eggs develop in the surrounding water, where they are attached to a rock or some underwater object.

Other Mollusks. There are two other major classes of mollusks: Amphineura and Scaphopoda. The class Amphineura comprises a small group of marine mollusks, some of which are considered to be the most primitive living representatives of the phylum. These animals,

> The trochophore larva is characteristic of some members of several groups of invertebrates including free-living flatworms, annelids, and mollusks.

Figure 11.36 Cephalopods. Common octopus (a) and American squid (b). (From Ebert, J. D., Loewy, A. G., Miller, R. S., and Schneiderman, H. A.: *Biology.* New York: Holt, Rinehart and Winston, 1973, p. 665.)

known as *chitons,* possess a shell divided into eight plates and a broad, flat foot used in adhesion and locomotion. Chitons may vary in size from ½ inch to over a foot in length (Fig. 11.37).

Mollusks in the class Scaphopoda are known as "tooth shells" or "tusk shells." These small marine animals have a shell shaped like a long tooth or an elephant's tusk. In the living organism, both the head and the foot project from the large opening at the anterior end of the shell. The shells of scaphopods were once used as currency by some Indian cultures.

As indicated earlier, many of the marine mollusks, like some of the annelids, have a trochophore larval stage in their development.

The Invertebrates

> The extensive adaptive features of the arthropods have made this group by far the most biologically sucessful of all animals.

Similarities in larval forms are often used by zoologists to determine the evolutionary relationships between various invertebrate groups. Although adult mollusks and annelids are quite different, sharing a similar larval form indicates that both groups probably arose from a common ancestor.

Phylum Arthropoda. It has been estimated that about 80 percent of all known animal species on the earth are arthropods. The word "arthropod" refers to the jointed legs that characterize all members of this group. These animals have developed adaptive features that have placed them at the peak of invertebrate evolution. Most arthropods have a protective, waterproof exoskeleton partially composed of **chitin,** a tough polysaccharide. Periodically, however, in order to grow, the exoskeleton must be shed, a process called **molting.** The arthropods also have an elaborate arrangement of muscles that permit a variety of body movements; a highly developed nervous system; an extensive array of sense organs; and, of course, some of them can fly. The arthropods occupy practically every known habitat on earth, some as free-living forms and some as parasites.

Arthropods have an open circulatory system, which includes a dorsal blood vessel that functions as a heart in pumping the blood forward. Among the terrestrial arthropods, excretory products are formed by organs called **malpighian tubules.** These are elongated, dead-end sacs that branch off the digestive tract. As blood in the body sinuses washes over the tubules, fluid containing waste products is absorbed from the blood and enters the tubules. The fluid then moves toward the open ends of the tubules, and most of the

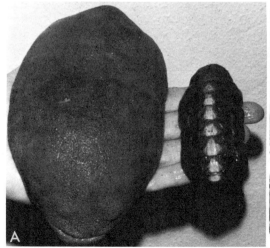

Figure 11.37 Chitons. Class Amphineura. (From Barnes, R. D.: *Invertebrate Zoology.* 4th ed. Philadelphia: Saunders College Publishing, 1980, p. 382.)

> The compound eyes of an arthropod may consist of as many as 30,000 independent visual units called ommatidia.

water is gradually reabsorbed into the blood. The remaining concentrated waste products pass into the digestive tract and are excreted in a dry form through the anus. The arthropod excretory system, like that of many higher animals, is extremely effective in conserving body water. Most aquatic arthropods excrete metabolic wastes through their gills.

Survey of the Arthropods. The diversity of the arthropods is enormous, including such representatives as crabs, crayfish, barnacles, lobsters, mites, ticks, spiders, scorpions, centipedes, millipedes, and the largest group of all the animals, the insects. Although it would be impossible to comment on all the structural and functional variations even in a few of these organisms, we can highlight some of the more distinctive and unusual features that make the arthropods in general such a biologically successful and fascinating group of animals.

One of the most highly developed sense organs found in many arthropods is the **compound eye** (Fig. 11.38). Although it is not as acute as the human eye, the compound eye is better for detecting rapid motion. The arthropods with compound eyes actually see more "slowly" than we do.

Most of the insects undergo successive changes in body form during their development, a process known as *metamorphosis* (Gk. "to transform") (Fig. 11.39). Grasshoppers, crickets, and several other kinds of insects undergo what is called **gradual metamorphosis,** in which the newly hatched insect looks very much like the adult, although it is sexually immature and has no wings. Further development consists simply of growing larger. Insects such as dragonflies and damselflies undergo **incomplete metamorphosis,** in which the eggs are laid in water, and the young first develop into a gill-breathing form. After successive molts (shedding of the exoskeleton), the young become winged, air-breathing adults. About 90 percent of all insects—including beetles, butterflies, mosquitoes, houseflies, fleas, bees, ants, and wasps—undergo **complete metamorphosis,** in which the eggs develop into wormlike *larvae.* The larval stages of insects are commonly referred to as caterpillars, maggots, or grubs. The larva molts several times and then becomes encased within a cocoon, forming the *pupa* stage. Within the cocoon, development proceeds until the adult insect emerges. Oddly enough, in many insects the larval stage may last for years, whereas the adult may exist for only a few days or weeks.

As evidence of their complex nervous systems, the arthropods on occasion demonstrate some unusual and even bizarre behavioral patterns. For instance, imagine a secluded rendezvous between a

Figure 11.38 The large compound eyes of a grasshopper. Each compound eye is composed of hundreds of six-sided lenses (*insert*). (From Applewhite, P., and Wilson, S.: *Understanding Biology.* New York: Holt, Rinehart and Winston, 1978, p. 85.)

331
The Invertebrates

> Metamorphosis is biologically significant in that it permits the immature form and the adult to occupy different habitats, utilize different nutrients, and otherwise follow different life styles.

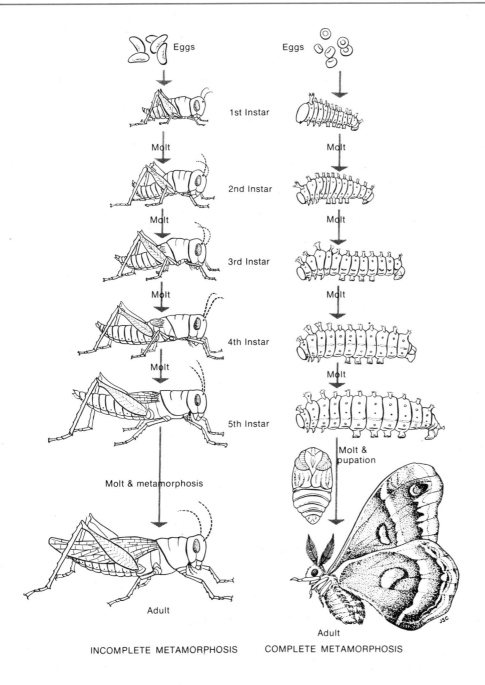

Figure 11.39 Comparison of gradual, or incomplete, metamorphosis of a grasshopper and complete metamorphosis of a cecropia moth. (From Villee, C. A., Walker, W. F., and Barnes, R. D.: *General Zoology*. 5th ed. Philadelphia: Saunders College Publishing, 1978.)

> The most venomous spider is the "black widow" of North and South America. It is the female spider that is poisonous; the much smaller male is harmless.

pair of amorous praying mantises. As the male makes his advances, the female, in sensuous response, tenderly enfolds him in her forearms and proceeds straightway to chew his head off. His lechery unthwarted, the now headless male breaks free of the female and begins to mate with her. Truly, the will shall find a way. Apparently, the loss of certain areas in the anterior portion of the nervous system triggers the mating response in the male. After copulation, the female may devour the entire body of the male.

The vast array of physical adaptations developed by the arthropods includes structures such as claws, fangs, poison glands, silk-secreting glands; receptors for detection of sound, odors, taste, and touch; swimming appendages; and mouth parts modified for biting, piercing, sucking, and grinding. We should add that many arthropods are capable of communicating with each other by means of sight, sound, or chemical substances called **pheromones**. Pheromones are secreted not only by arthropods but by many other animal groups as well. By means of their odors, pheromones function as sex attractants, trail markers, warning signals, and so forth. It is believed that pheromones secreted by a given species can communicate information only to other members of the same species.

Subphylum Chelicerata. There are two subphyla of living arthropods: Chelicerata and Mandibulata. Chelicerates (Gk. *chele,* claw, + *keras,* horn) are so named because they possess a pair of fanglike **chelicerae** used for biting or grasping prey. A second pair of appendages is used for manipulating prey and tearing it apart. These animals also have four pairs of legs, but they have neither antennae nor jaws. The major class of chelicerates is Arachnida, which includes spiders, ticks, mites, and scorpions (Fig. 11.40). Other members of this subphylum are the primitive horseshoe crabs and the marine sea spiders.

Class Arachnida. With their hairy bodies, stealthy movements, and menacing fangs charged with venom, the spiders no doubt have received a bad press, although they are generally quite harmless creatures (Fig. 11.41). Spiders have two body sections, an anterior **cephalothorax** ("head and chest") and a posterior **abdomen.** The cephalothorax often bears several simple eyes, and the tip of the abdomen is supplied with **spinnerets** that release silk. Silk, which is actually a type of protein, not only is used in spinning webs but also serves to bind entrapped victims, house the offspring, and line the nest.

Spiders are carnivorous and use their hollow fangs to inject venom into the body of the prey. The prey is then torn apart and

> Some of the largest spider webs known are nearly 19 feet in circumference.

Figure 11.40 Various arachnids. (a) Black scorpion. (b) Whip scorpion. (c) Dog tick. (d) Harvestman, or "Daddy long legs." (From Ebert, J. D., Loewy, A. G., Miller, R. S., and Schneiderman, H. A.: *Biology*. New York: Holt, Rinehart and Winston, 1973, p. 676.)

covered by digestive enzymes released from the spider's intestine. The enzymes reduce the tissues of the prey to a liquid mass, which the spider sucks up into its stomach.

In some species, two spiders intent on mating often engage in an elaborate ritual in which the male repeatedly circles the female and at times raises up on his back legs, exposing his underside. All of this being too much for the female, she accepts his advances and crawls under him. When mating is completed, passion defers to indifference, and the female proceeds to devour the body of her once amorous suitor.

334
Biology of Animals

> One arachnid, the horseshoe crab, is referred to as a "living fossil," having remained virtually unchanged from its 200 million–year-old ancestors.

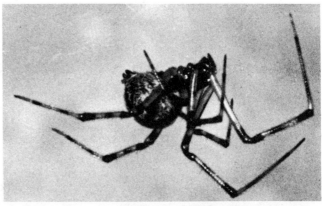

Figure 11.41 Spiders. (a) Wolf spider. (b) House spider. (From Villee, C. A., Walker, W. F., and Barnes, R. D.: *General Zoology*. 4th ed. Philadelphia: W. B. Saunders Company, 1973, p. 586.)

Scorpions have terminal stingers that release a venom affecting the nervous system. Although their sting is quite painful, most scorpions are not dangerous to man. The respiratory structures of scorpions and spiders are called **book lungs,** which consist of small sacs lined with a series of folded plates. The plates increase the surface area for the exchange of oxygen and carbon dioxide (Fig. 11.42).

Subphylum Mandibulata. The mandibulates are arthropods with *mandibles,* or jaws, instead of chelicerae. Mandibles function in biting and grinding and in some species are modified for sucking and piercing. Paired sensory organs called *antennae,* along with a variety of other paired appendages, are located in the head region. Many of these animals have only three pairs of walking legs, whereas others may have four pairs, and still others have dozens of pairs.

> Crustaceans vary in size from giant crabs, whose bodies measure more than a foot across, to water fleas, some of which are less than 1/100 of an inch long.

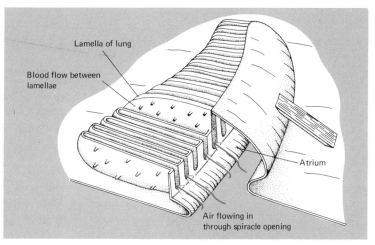

Figure 11.42 Section through a book lung, the respiratory structure of spiders and scorpions.

The major classes in this subphylum are Crustacea, Diplopoda, Chilopoda, and Insecta.

Class Crustacea. The *crustaceans* (L. *crusta*, crust, a reference to the shell) are extremely diverse and include crabs, crayfish, lobsters, shrimp, barnacles, water fleas, and several others (Fig. 11.43). The enormous variety of these animals makes it virtually impossible to portray them by a general description.

As a group, crustaceans are mostly marine, although there are many freshwater forms and a few that are terrestrial. Many of the microscopic crustaceans constitute a part of the zooplankton—the floating life that serves as the food source for many aquatic animals (Fig. 11.44).

The body covering, or exoskeleton, of crustaceans is extremely hard and may consist of segments varying in number from three or four to dozens. For growth to take place, the exoskeleton must be shed periodically. Crustaceans respire by means of soft gills, which are protected by the walls of the exoskeleton.

There are two pairs of sensory antennae and usually one or two compound eyes. The eye is said to be "compound" because it is composed of numerous independent visual units, each of which is called an **ommatidium** (Gk. "little eye"). Each unit has its own lens and retina. Although compound eyes probably do not form a sharp image, they permit the animal to see in practically every direction at once, and they are highly effective in detecting rapid motion.

Many crustaceans are highly valued as food by human beings, although there are also crustaceans that are sometimes a nuisance. Barnacles, for instance, attach themselves in such great numbers to

336
Biology of Animals

> Some millipedes may have as many as 350 pairs of legs; the maximum number of legs for centipedes is almost 175 pairs.

Figure 11.43 Various crustaceans. (a) Brine shrimp. (b) A crab in the process of molting. (c) A "skeleton shrimp." (d) Pill bugs, one of the few kinds of terrestrial crustaceans. (From Ebert, J. D., Loewy, A. G., Miller, R. S., and Schneiderman, H. A.: *Biology.* New York: Holt, Rinehart and Winston, 1973, p. 672.)

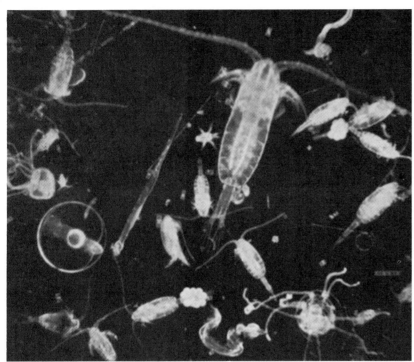

Figure 11.44 Living zooplankton. The shrimplike animals are various small crustaceans; some are carrying eggs. Two tiny jellyfish with long tentacles are also seen. (Magnification X14) (From Clark, M. E.: *Contemporary Biology*. 2nd ed. Philadelphia: Saunders College Publishing, 1979, p. 78.)

the hulls of ships that they have to be scraped off. The barnacles create drag on the ships and also damage the hulls. In addition, some crustaceans are parasitic on food crops, and others are the hosts for pathogenic parasites.

Classes Diplopoda and Chilopoda. The class Diplopoda (Gk. "double foot") includes **millipedes,** the so-called "thousand-legged worms." The body of a millipede may have up to 100 segments, with two pairs of legs on almost every segment. These arthropods are mostly herbivorous and are usually found in dark, damp places. When touched or otherwise disturbed, they often roll up into a ball. Many millipedes have a pair of stink glands on most of the body segments.

The class Chilopoda (Gk. "lip foot") includes the carnivorous **centipedes** (L. "hundred feet"). In these animals each body segment, except the last two, has only one pair of walking legs. The first pair of appendages on the trunk is provided with pincers bearing poison glands. Most centipedes, however, are not harmful to humans, although the bite of some species can be quite painful. Centipedes generally prefer a moist environment and are sometimes found around bathrooms and other damp areas in the home (Fig. 11.45).

Variation among insects is evident not only in the vast number of different species but also in the fact that some are more than a foot in body length and others are smaller than some species of protozoans.

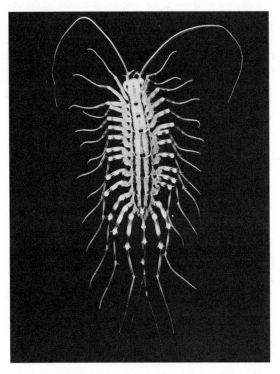

Figure 11.45 The centipede (*above*), *Scutigera*, which is a chilopod, has one pair of legs per segment. The millipede (*right*), which is a diplopod, has two pair of legs per segment. (From Ebert, J. D., Loewy, A. G., Miller, R. S., and Schneiderman, H. A.: *Biology*. New York: Holt, Rinehart and Winston, 1973, p. 671.)

Class Insecta. In terms of numbers, the insects are the indisputable champions of the animal kingdom. Perhaps a million different species are known, about one-third of which are beetles. Insects are practically everywhere, having adapted to every conceivable terrestrial habitat and every aquatic habitat except the sea. In their 300 million years of existence, the insects have been able to exploit every environmental change and to emerge as the most biologically successful terrestrial invertebrates on earth (Fig. 11.46).

The success of the insects is attributable to a number of factors. Since only an encyclopedic entomology (the study of insects) textbook could describe all the classes, orders, and families of insects, we shall employ one representative insect—the grasshopper—as a means of illustrating many of these factors.

The body of a grasshopper, like that of all insects, is divided into three sections: the **head, thorax,** and **abdomen.** As pointed out earlier, the entire insect body covering is composed of chitin, a tough, flexible polysaccharide that affords maximum protection while still allowing a wide range of movements. Chitin is light and waterproof and serves as an attachment for the highly developed insect muscles. The thorax and abdomen of the grasshopper consist

There are more named species of beetles than of any other group of animals on earth.

Figure 11.46 A variety of insects. (a) Giant rhinoceros beetle, the largest living insect. (b) Dragonfly. (c) A flea. (d) The shield bug. (e) An adult mosquito. (From Ebert, J. D., Loewy, A. G., Miller, R. S., and Schneiderman, H. A.: *Biology.* New York: Holt, Rinehart and Winston, 1973, p. 680.)

> The wing beat of insects known as midges may exceed 60,000 beats per minute, the fastest instance of recorded muscular movement in any organism.

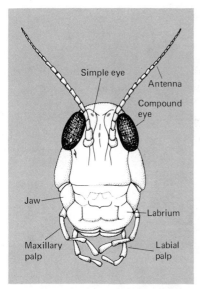

Figure 11.47 Facial view of grasshopper's head.

of several unfused segments, a factor that allows for even more efficient movement.

The head of the grasshopper is admirably supplied with sensory organs (Fig. 11.47). There is one pair of compound eyes, between which are three light-sensitive simple eyes. One pair of antennae, or sensory feelers, extend forward from the head. Quite similar to antennae are two pairs of short palps, which contain taste receptors. For feeding there are a pair of mandibles used for grinding and chewing; a pair of maxillae, or lower jaws, that delicately manipulate food; and an upper lip and lower lip that hold food before it is chewed by the mandibles. Grasshoppers are herbivores, usually feeding on a variety of leaves.

The segmented thorax has three pairs of jointed walking legs, which terminate in short claws (Fig. 11.48). The jointed legs greatly enhance the maneuverability of the grasshopper, as they do for all other arthropods as well. The hind legs of the grasshopper are extremely powerful, and being capable of propelling the animal considerable distances, they illustrate the effectiveness of the insect muscular system. Indeed, the arthropods as a group possess a highly efficient system of striated muscles similar to that of vertebrates. The thorax also bears two pairs of wings. Wings are of obvious adaptive advantage, enabling insects to escape their enemies, widely distribute their species, and obtain food in places inaccessible to many other organisms.

Each lateral surface of the thorax bears a round membrane, or **tympanum** (Gk. *tympanon*, drum), which functions in sound reception. Sound waves cause the tympanum to vibrate in much the same manner as your own eardrum.

Several pairs of **spiracles** (L. *spiraculum*, airhole) are found along either side of the abdomen (Fig. 11.49). Spiracles lead to a system of internal branching tubes called **tracheae**. Air entering the spiracles is pumped by abdominal contractions into the tracheae, al-

Figure 11.48 A female grasshopper.

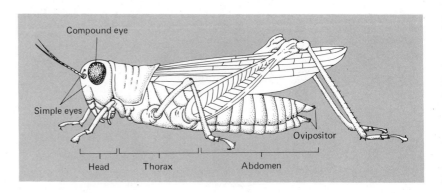

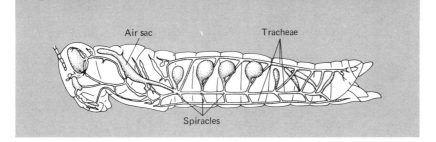

Figure 11.49 Respiratory apparatus of a grasshopper.

lowing oxygen to diffuse into the internal tissues. Carbon dioxide diffusing from the tissues into the tracheae is pumped out of the body through the spiracles (Fig. 11.50). In the female grasshopper, the abdomen terminates in an *ovipositor*, which is involved in egg laying. Grasshoppers and all insects have separate sexes, and fertilization is internal.

The organ systems of the grasshopper are highly developed and include many of the systems found in higher vertebrates. There is a high degree of cephalization, with several pairs of ganglia located in the head. The digestive system releases specific enzymes correlated to the diet of the insect, and excretion via malpighian tubules aids in conserving water. As mentioned, insects and other arthropods have an open circulatory system and a dorsal vessel that pumps blood.

Let us add to this description the facts that protective coloration provides the grasshopper with an effective suit of camouflage, that by rubbing its hind legs over its forewings it can produce mating and warning sounds, and that it probably even sees in color. This is quite an assemblage of adaptive characteristics—and the grasshopper is considered one of the more primitive insects!

Molting and Metamorphosis. As indicated earlier, many arthropods periodically shed their exoskeletons to allow for growth of the body. The process of shedding the exoskeleton, known as molting, is characteristic of almost all the insects and is accompanied in these animals by metamorphosis, i.e., successive changes in body form as the insect matures. From the work of V. B. Wigglesworth in the 1930s, it was determined that molting and metamorphosis are regulated by an interplay of at least three hormones. The insect most extensively

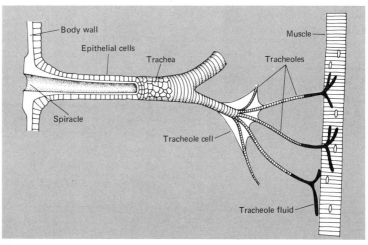

Figure 11.50 Diagram of insect tracheal system.

> With its ability to arrest a developing insect in the larval stage, juvenile hormone could prove effective as a substitute for chemical pesticides.

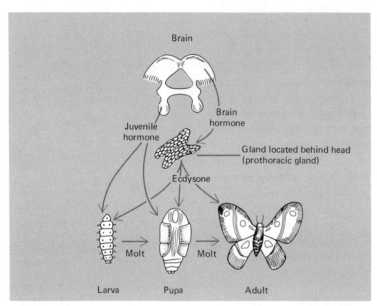

Figure 11.51 Hormonal control of molting in insects (see text for details).

studied in this regard has been the silkworm moth, which undergoes complete metamorphosis.

During the larval stage, a gland in the brain secretes ***juvenile hormone,*** which controls growth and molting of the larva (Fig. 11.51). Each successive molt and accompanying growth of the larva depend upon the amount of juvenile hormone present. A high concentration of juvenile hormone causes the larva to molt into another larval stage. However, when the larva attains a certain stage of growth, the secretion of juvenile hormone begins to decline. This apparently triggers a second gland in the brain to begin secreting ***brain hormone.*** This hormone in turn stimulates glands located just behind the head of the larva; these glands respond by secreting the molting hormone, ***ecdysone.*** With the decline of juvenile hormone and the increasing dominance of ecdysone, the larva molts into the pupal stage. Once the pupa is formed, juvenile hormone is no longer secreted. In the absence of juvenile hormone, along with the continuing effect of ecdysone, the pupa eventually molts into the adult insect. Thus it is juvenile hormone that determines if the insect will progress from the larval stage to the pupal stage. As long as juvenile hormone is in high concentration, the larva will molt only into another larva. However, in the pupal stage, juvenile hormone is absent, and ecdysone becomes the dominating hormone.

> Echinoderms are the only group of animals constructed on a radial, five-part body plan.

Phylum Echinodermata. All the echinoderms (Gk. "spiny skin") are marine and have an internal skeleton bearing small spines that project through the surface skin. Included in this group are starfishes, brittle stars, sea urchins, sand dollars, and sea lilies (Fig. 11.52). The body of an echinoderm is generally constructed on a five-part plan, with each part radiating outward from a central point. These animals have a complete digestive system, but the circulatory system is rather poorly developed. They have no brain, nor are there any excretory organs. The sexes are usually separate (dioecious), and fertilization occurs externally in the surrounding water.

A feature unique in echinoderms is the ***water vascular system,*** an arrangement of internal canals having rows of tiny hollow ***tube feet*** used in attachment and locomotion (Fig. 11.53). Each tube foot has a sucker at one end. The canals radiate outward from the center of the body, and the rows of tube feet extend along the length of each canal. Water enters the canals through a small, porous button on the body surface. When water is forced into the tube feet, they extend, allowing the suckers to attach to some object.

In the starfishes, the water vascular system is employed in an interesting method of feeding. By straddling a clam or oyster, the starfish attaches its tube feet to both shells and begins to pull them apart. Once the shells are slightly separated, the starfish everts part of its stomach through its own mouth and into the opening. Enzymes secreted by the stomach then begin to digest the soft body within the shells. When the meal is finished, the starfish withdraws its stomach into the interior of the body.

The starfishes and brittle stars characteristically have five arms, although some starfishes may have twenty or more. The arms of brittle stars are longer and more slender than those of starfishes.

The sea urchins, sand dollars, and sea cucumbers do not have arms, but they do have five rows of tube feet. The bodies of sea urchins are covered with moveable spines, which, in some species, contain irritants or poisons.

Sea lilies, some of which look more like plants than animals, have stalked bodies with long, branching, feathery arms.

Like some other invertebrates we have mentioned, the echinoderms exhibit remarkable powers of regeneration. A single starfish arm, for example, can regenerate a complete new animal. Equally amazing is the ability of sea cucumbers, after expelling some of their internal organs through the anus, to regenerate all the missing structures.

The sea lilies and feather stars are considered to be the most primitive echinoderms; fossils of these animals date back well over 500 million years.

Figure 11.52 Representatives of the five classes of echinoderms. (a) Starfish, showing its tube feet. (b) Brittle stars. (c) A sea urchin. (d) A sea cucumber. (e) A fossil sea lily. (From Ebert, J. D., Loewy, A. G., Miller, R. S., and Schneiderman, H. A.: *Biology*. New York: Holt, Rinehart and Winston, 1973, p. 691.)

A small phylum of invertebrate chordates, known as Hemichordata, consists of wormlike and seaweed-like marine animals with pharyngeal gill slits, dorsal nerve cord, and a larval stage resembling that of the echinoderms. Accordingly, hemichordates provide evidence for the relationship between the phyla Echinodermata and Chordata.

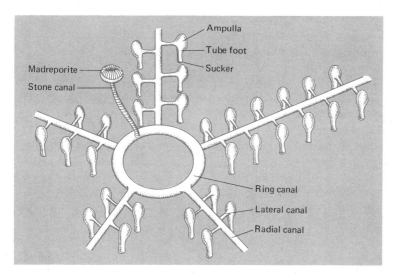

Figure 11.53 Diagram of starfish water vascular system. Water enters the system through the madreporite, a button-like sieve on the body surface. The radial canals extend into each arm of the starfish.

The animals in the phylum Chordata include the so-called invertebrate chordates and a large group that includes all the vertebrates. At some stage in their development—during embryonic life or in the adult stages or both—all chordates share three distinguishing characteristics: a *notochord;* a *dorsal hollow nerve cord;* and *pharyngeal gill slits,* or *pouches.*

The notochord (Gk. *noton,* back, + L. *chorda,* cord) is a supportive, rodlike structure usually extending the length of the animal, just below the dorsal nerve cord. In the vertebrates, the notochord is present during early embryonic development but is later incorporated into a cartilaginous or bony vertebral column (backbone).

The dorsal hollow nerve cord, such as the spinal cord in humans, is enlarged at the anterior end, forming a brain. The nerve cord of vertebrates lies within the vertebrae that make up the backbone. The brain is surrounded by the skull, or cranium.

All chordate embryos, and many adults as well, have slitlike openings leading from the pharyngeal, or throat, cavity to the outside. Adult vertebrates in which the slits persist include fish and some amphibians. In these aquatic animals, the slits eventually develop into gills. Although gill slits (or at least deep grooves) form in human and other mammalian embryos, they do not function in respiration and normally disappear. One gill slit, however, has a specific function in mammals; it forms the *eustachian tube,* the tube connecting the throat with the middle ear.

Other chordate characteristics include a ventral *heart,* a *tail,* which may or may not persist; and a cartilaginous or bony *endoskeleton* found in the vertebrates. Human beings, of course, normally

INTRODUCTION TO THE CHORDATES
Phylum Chordata

Biology of Animals

> Tunicates are most unusual in that they secrete a body covering composed of a cellulose-like substance. Although forming the wall of most plant cells, cellulose is not characteristic of animals.

are not born with tails; there are, however, rare cases in which this has occurred. Fortunately, the tail can be removed surgically. An endoskeleton is highly advantageous in that it grows along with the rest of the body and does not have to be shed. Accordingly, many vertebrates are capable of attaining massive size.

The phylum Chordata includes three subphyla: Urochordata, Cephalochordata, and Vertebrata. The first two subphyla, known as the invertebrate chordates, are considered to be the evolutionary links between invertebrates and vertebrates. The chordates as a group are considered to be closest to the echinoderms in terms of evolutionary kinship. This is not to say that chordates evolved from echinoderms, but that both groups probably arose from a common ancestor. This relationship is evident in the similar embryonic patterns of development in both groups.

Subphylum Urochordata. The most common urochordates are the ***tunicates***, or "sea squirts." The adults are mostly saclike animals, found only in marine waters (Fig. 11.54). These organisms are referred to as tunicates because the entire body is covered by a transparent tunic composed mainly of a carbohydrate similar to cellulose. Urochordates may be sessile (L. *sessilis,* sitting down) or free-swimming, and they exist as solitary individuals or in colonies.

The tadpole-like larva of tunicates possesses all three major chordate characteristics—a notochord, dorsal nerve cord, and pha-

Figure 11.54 An adult tunicate. (From Ebert, J. D., Loewy, A. G., Miller, R. S., and Schneiderman, H. A.: *Biology.* New York: Holt, Rinehart and Winston, 1973, p. 695.)

> The cephalochordate amphioxus burrows in the sand with its tail, using its exposed anterior end to strain food particles from the surrounding water.

Introduction to the Chordates

ryngeal gill slits (Fig. 11.55). In adult tunicates, however, the first two structures disappear, leaving only the gill slits as reminders of their chordate heritage.

Nearly all tunicates are monoecious; they also may reproduce asexually by budding and have great powers of regeneration.

Subphylum Cephalochordata. The common representative of the cephalochordates is amphioxus, the lancelet (Fig. 11.56). The name "amphioxus" is derived from the Greek meaning "pointed at both ends." It is thus fishlike in form and is a burrowing animal found in the shallow regions of the oceans. The adult amphioxus, in having a notochord, dorsal nerve cord, and pharyngeal gill slits, is highly representative of the chordate phylum.

The segmented arrangement of body muscles in amphioxus is characteristic of fishes and, indeed, of all vertebrates. The animal also displays other vertebrate characteristics, including a tail and a rudimentary ventral heart. The sexes are separate in amphioxus, and fertilization is external. The cephalochordates and urochordates share a common primitive characteristic in that both groups are filter feeders.

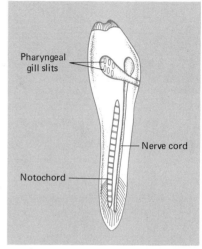

Figure 11.55 A tunicate larvae, showing the three major chordate characteristics.

The Vertebrates

Subphylum Vertebrata. The animals in this large and diverse subphylum are distinguished from all other animals by the presence of a cartilaginous or bony **vertebral column.** The vertebral column replaces the notochord in all adult vertebrates, except the jawless fishes. As we have mentioned, vertebrates have an *internal skeleton* of cartilage or bone, which grows along with the rest of the body. Various parts of the skeleton protect the internal organs and well-developed brain and serve as attachment sites for muscles. The body is covered by **skin** or **derivatives of skin,** such as fur, feathers, or scales. Vertebrates have a pair of usually highly developed, image-forming **eyes,** and a pair of **ears** that function in sound reception and equilibrium, and the blood is pumped through a **closed system** of vessels by a muscular **heart.** In addition, there are a **pair of kidneys** for re-

Figure 11.56 The fishlike *Amphioxus* lives in tropical and temperate coastal waters. It retains its dorsal nerve cord, notochord, and gill slits throughout life. The notochord and gill slits disappear early in the development of the higher chordates. (From Applewhite, P., and Wilson, S.: *Understanding Biology.* New York: Holt, Rinehart and Winston, 1978, p. 85.)

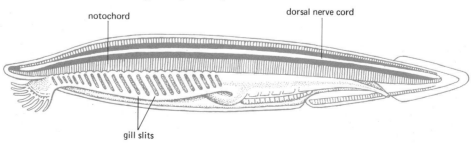

> Although they are considered to be descendants of the extinct ostracoderms, modern lampreys and hagfishes are quite unlike their bony ancestors.

moving wastes from the blood, an **endocrine system** that releases hormones involved in regulating body function, 10 or 12 pairs of **cranial nerves,** and two specialized glands—the **liver** and **pancreas.** Respiration occurs through either **gills** or **lungs.** Almost all adult vertebrates have two pairs of limbs, and the sexes are separate.

In the following overview, we will consider the more distinctive features of the seven major classes of living vertebrates.

Class Agnatha. The agnaths (Gk. *a,* without, + *gnathos,* jaw) include the most primitive and only parasitic vertebrates, the **lamprey eels** and the scavenging **hagfishes.** These animals do not have jaws, scales, or paired fins, and their internal skeletons are composed of cartilage. The agnaths are the only vertebrates in which the notochord persists throughout life; their cartilaginous backbone is only rudimentary.

Lampreys are ectoparasites on fishes, attaching to the host by their round, sucker-like mouth (Fig. 11.57). The horny teeth lining the mouth are used to scrape away the scaly skin of the fish, after which the blood and body fluids are sucked out. Hagfishes usually

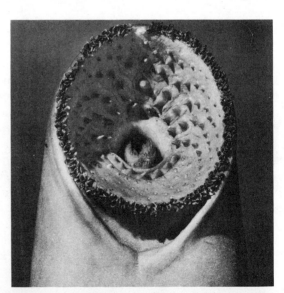

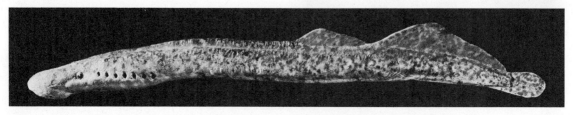

Figure 11.57 A lamprey (*below*). Note the pharyngeal gill slits. On the right is an enlargement of its mouth, which has no jaws. (From Ebert, J. D., Loewy, A. G., Miller, R. S., and Schneiderman, H. A.: *Biology.* New York: Holt, Rinehart and Winston, 1973, p. 699.)

> The jaw of placoderms apparently was derived from the bony structure (gill arches) that support the first pair of gills.

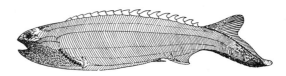

A, Anglaspis

B, Pterolepis

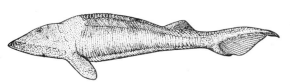

C, Hemicyclaspis

Figure 11.58 Three fossil ostracoderms—primitive, jawless, limbless fishes. (From Villee, C. A.: *Biology*. 7th ed. Philadelphia: Saunders College Publishing, 1977, p. 759.)

attach to the gills of the fish and bore into the flesh using a tongue-like boring organ.

The larval stage of lampreys, called an **ammocoete** (Gk. *ammos*, sand, + *koite*, bed) larva, is strikingly similar to the cephalochordate amphioxus. Although lampreys spawn in fresh water, the adults are either freshwater or marine. Hagfishes, however, are exclusively marine.

The oldest vertebrate fossils on record are a group of jawless, bony-scaled fishes that existed some 350 million to half a billion years ago. These ancient agnaths, known as **ostracoderms** (Gk. *ostrakon*, shell, + *derma*, skin), were probably filter feeders and are the long-extinct ancestors of modern lampreys and hagfishes (Fig. 11.58). The ostracoderms survived for some 150 million years before finally becoming extinct.

Class Chondrichthyes. With the gradual disappearance of the ostracoderms, a second group of bony-plated fishes known as ***placoderms*** (Gk. *plakos*, flat object, + *derma*, skin) made their appearance about 360 million years ago (Fig. 11.59). The placoderms are the first

Figure 11.59 Model of an extinct placoderm. The hinged jaws are noticeable. (From Ebert, J. D., Loewy, A. G., Miller, R. S., and Schneiderman, H. A.: *Biology*. New York: Holt, Rinehart and Winston, 1973, p. 700.)

> The rough skin of sharks and rays consists of scales constructed of the same hard substance that forms their teeth.

known vertebrates with jaws, a feature characteristic of all vertebrates except the agnaths and one that enabled the vertebrates to become highly efficient predators. Moreover, the paired fins of the placoderms greatly enhanced their maneuverability in water. The paired fins of some placoderms later gave rise indirectly to the front and hind limbs of terrestrial vertebrates.

One of the two living classes that arose from the placoderms is Chondrichthyes (Gk. *chondros*, cartilage, + *ichthys*, fish), the cartilaginous fishes. These animals are the only jawed vertebrates with an endoskeleton of cartilage and are found primarily in marine waters. Included in this group are the sharks, skates, and rays (Fig. 11.60). The last two are characterized by flattened bodies and enormous lateral fins. Some rays have an organ that produces an electric shock, and others possess poisonous spines used in defense.

The Chondrichthyes include the largest known species of fishes, the whale shark, specimens of which grow to a length of 55 ft. In fact, sharks are the largest living vertebrates, except for the whales. The whale shark also lays the largest egg of any living animal.

Sharks have poor vision, but their sense of smell is highly developed. Interestingly, sharks must continue to move through the water, or they will "drown." Unlike the bony fishes, sharks have no gill muscles for drawing water across the gills, where gas exchange occurs. By swimming, sharks maintain a constant water current across the gills.

(a)

(b)

(c)

Figure 11.60 Some modern cartilaginous fishes. (a) A stingray. (b) The whale shark. (c) Sand shark. (From Ebert, J. D., Loewy, A. G., Miller, R. S., and Schneiderman, H. A.: *Biology*. New York: Holt, Rinehart and Winston, 1973, p. 701.)

> In the number of species—about 20,000—the bony fishes are the most numerous of all vertebrates.

Sharks and various representatives of all classes of vertebrates have a digestive system that terminates in a *cloaca* (L. "a drain"). The cloaca is a chamber that receives the materials and products of the digestive, urinary, and reproductive systems. The cloaca opens to the exterior through the *vent*. Among the many vertebrates that lack a cloaca, the contents of the digestive, urinary, and reproductive systems do not enter a common chamber and therefore are usually released through separate body openings.

Sharks, as well as all other fishes, have a two-chambered heart, composed of one atrium and one ventricle (Fig. 11.61). Deoxygenated blood from the body tissues enters the heart through the atrium and passes into the muscular ventricle. When the ventricle pumps, blood is sent through arteries to the gills, where the blood is oxygenated. After leaving the gills, blood is circulated throughout the body, where it gives up its oxygen to the cells. This deoxygenated blood then returns through a system of veins to the atrium of the heart. In fishes, then, only deoxygenated blood enters and leaves the heart.

Class Osteichthyes. The second class of vertebrates that arose from the placoderms is Osteichthyes (Gk. *osteon*, bone, + *ichthys*, fish), the bony fishes. Extremely diverse, this group includes salmon, bass, perch, trout, swordfishes, electric eels, seahorses, lungfishes, minnows, and a host of others (Fig. 11.62). Almost all of these fishes have an endoskeleton of bone and are found in marine and fresh waters. The bony fishes display a wide range of sizes and physical

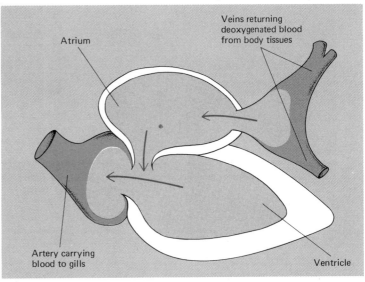

Figure 11.61 The fish heart (see text for description).

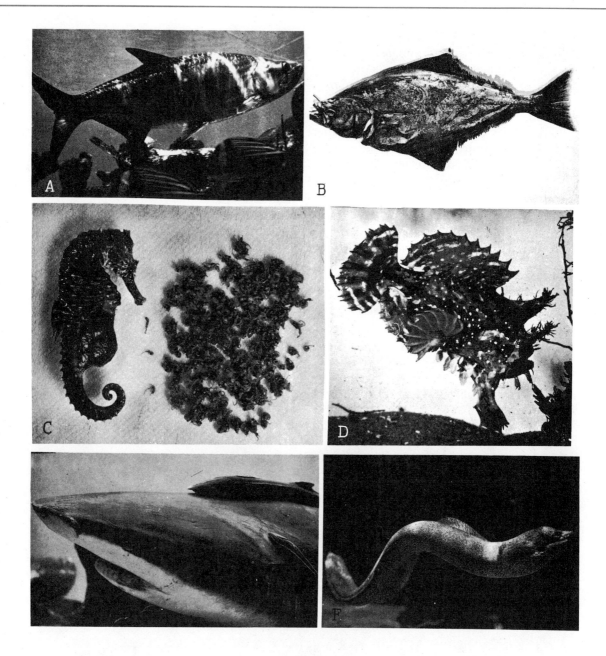

Figure 11.62 A variety of bony fishes. (a) Tarpon. (b) Halibut. (c) Male sea horse. (d) Sargassum fish. (e) Sharksucker. (f) Moray eel. (From Villee, C. A., Walker, W. F., and Barnes, R. D.: *General Zoology.* 5th ed. Philadelphia: Saunders College Publishing, 1978, p. 781.)

> The body of the female deep-sea angler fish may be several thousand times as massive as the body of the male angler.

adaptations. The Russian sturgeon, for example, may exceed 25 ft in length, whereas certain freshwater fishes found in the Philippines are less than 1/3 inch long.

A few of the adaptations found among the bony fish include their ability to change color, which enables them to blend with the background as a means of camouflage (Fig. 11.63); the development of electric organs that emit shocks strong enough to incapacitate prey; the presence of sharp spines, some of which release a powerful toxin used for defense; and the unusual arrangement in the seahorses and pipefishes, in which the males possess a pouch for carrying eggs until they hatch. In addition, flying fishes can glide through the air; a catfish from Southeast Asia introduced into Florida can "walk" on land, using its gills and fin spines for locomotion; a species of perch found in India climbs trees; and in the angler fish, the male is but a tiny parasite attached to the body of the much larger female (Fig. 11.64).

The first bony fishes to evolve from the placoderms are believed to have been a freshwater group that had developed lungs. In almost all present-day bony fishes, the lungs have been modified into a **swim bladder,** an organ that functions in maintaining buoyancy and in detecting sound vibrations. After their rise from the placoderms, the original Osteichthyes soon diverged along two evolutionary lines. One line led to the bony fishes found on the earth today; the other line led to a group of lungfishes, represented today by a few species in Africa, Australia, and South America (Fig. 11.65). This line also includes a single species (genus *Latimeria*) of ancient "lobe-fin" fishes, a diverse group also known as ***crossopterygians*** (Gk.

Figure 11.63 An experiment to show the remarkable ability of the flounder to change its color and pattern to conform with its background. *Top,* a flounder on a uniform, light background; *bottom,* the same fish after being placed on a spotted, darker background. (From Villee, C. A., Walker, W. F., and Barnes, R. D.: *General Zoology.* 5th ed. Philadelphia: Saunders College Publishing, 1978, p. 876.)

Figure 11.64 A deep-sea angler fish. The protrusion above the mouth is a lure; it attracts small fish, the angler's prey. The male is reduced to a tiny parasite, which can be seen attached beneath this female's tail. (From Arms, K., and Camp, P. S.: *Biology.* 2nd ed. Philadelphia: Saunders College Publishing, 1982.)

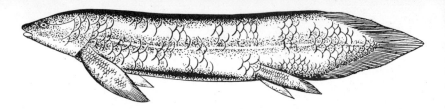

Figure 11.65 The Australian lungfish. (From Villee, C. A., Walker, W. F., and Barnes, R. D.: *General Zoology*. 5th ed. Philadelphia: Saunders College Publishing, 1978, p. 783.)

krossoi, fringe, + *pterygia,* fins). Thought to have been extinct for some 75 million years, a specimen of *Latimeria* was caught off the east coast of Africa in 1939 (Fig. 11.66). There have since been several dozen of these "living fossils" caught in the same area. Although the lungfishes did not invade the land, a group of crossopterygians somewhat similar to *Latimeria* are thought to be the ancestors of amphibians, reptiles, birds, and mammals—the land vertebrates.

The designation "fringe fins" is an appropriate description of the appendages of the ancient relatives of terrestrial vertebrates. Being on the "fringe" between fins and legs, the appendages probably were used by the animals to pull themselves onto a muddy shore or to crawl from one water hole to another. Over millions of years, as the descendants of these ancient fishes slowly adapted to a terrestrial existence, the unique lobed fins evolved into the legs of the first vertebrates to inhabit the land.

Class Amphibia. The name of this class of vertebrates is derived from the Greek words *amphi* ("of two kinds") and *bios* ("life"). Nearly all of these animals, then, lead "two kinds of lives," i.e., part of their lives is spent in water and part on land. The most familiar amphibians are the tail-less frogs and toads, and the salamanders, which have a tail (Fig. 11.67). In addition, there is one group of rare tropical amphibians called *cecilians* that, having no legs, somewhat resemble oversized worms (Fig. 11.68).

The amphibians are thought to have evolved from ancient crossopterygians about 400 million years ago. More precisely, the crossopterygians gave rise to an early amphibian group known as **stegocephalians** (Gk. *stegos,* roof, + *kephale,* head, a reference to their fused skull bones) (Fig. 11.69). The front and hind limbs of these original land vertebrates evolved from the lobed fins of the crossop-

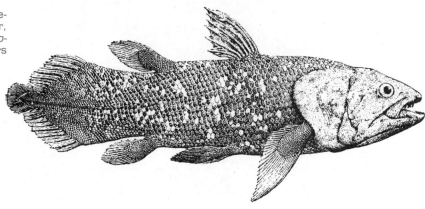

Figure 11.66 *Latimeria,* a living "lobe-fin" fish. (From Villee, C. A., Walker, W. F., and Barnes, R. D.: *General Zoology*. 5th ed. Philadelphia: Saunders College Publishing, 1978, p. 784.)

Figure 11.67 Modern amphibians. (a) The common newt. (b) A frog devouring a grasshopper. (c) The tiger salamander. (d) Bullfrog tadpoles. (From Ebert, J. D., Loewy, A. G., Miller, R. S., and Schneiderman, H. A.: *Biology.* New York: Holt, Rinehart and Winston, 1973, p. 704.)

Figure 11.69 A stegocephalian. (From Villee, C. A.: *Biology*. 7th ed. Philadelphia: Saunders College Publishing, 1977, p. 329.)

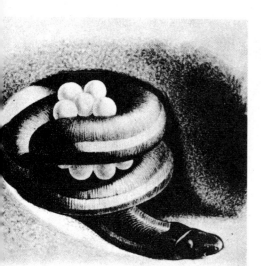

Figure 11.68 A burrowing caecilian with her eggs. (From Villee, C. A., Walker, W. F., and Barnes, R. D.: *General Zoology*. 5th ed. Philadelphia: Saunders College Publishing, 1978, p. 793.)

terygians. Although lobed fins and walking limbs have different functions, there is similarity in their bone structure, which reflects a common evolutionary origin. Thus, the fin bones of crossopterygians and the limb bones of the stegocephalians are homologous.

Amphibians cannot be considered true land vertebrates, since they have never adapted totally to a terrestrial existence. As adults, many amphibians live on land, but most species must return to water to reproduce. The fertilized eggs usually develop into gill-breathing tadpole-like forms that undergo a metamorphosis to produce the adults. Some species, however, have developed methods for laying their eggs that do not require a return to water. Some rather unusual examples include a species of frog in which the male carries the developing young in his mouth; in another frog species, the female carries the young in her stomach.

As a protective measure, some frogs have the ability to change body color to adapt to their background (Fig. 11.70). Salamanders have a remarkable power of regeneration, being able to replace an entire missing leg or tail. As a means of defense against predators, many amphibians secrete a toxic chemical, produced by specialized skin glands. In fact, one of the most deadly venoms known is de-

Figure 11.70 Both color and shape enable certain amphibians like the Surinam toad (*Pipa pipa*) to blend with the background and thereby avoid detection by enemies. (From Orr, R. T.: *Vertebrate Biology*. 5th ed. Philadelphia: Saunders College Publishing, 1982.)

> Unlike the fish heart, the three-chambered amphibian heart receives oxygenated blood as well as deoxygenated blood. The ventricle pumps the oxygenated blood throughout the body, while deoxygenated blood is pumped to the lungs and skin.

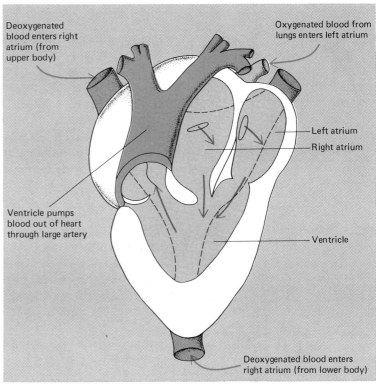

Figure 11.71 Amphibian heart (see text for description).

rived from skin secretions of the so-called arrow-poison frog of western Colombia, South America.

Adult amphibians have a pair of lungs, but they also exchange respiratory gases through their moist, scaleless skin. The amphibian heart has three chambers—a right and left atrium and one ventricle (Fig. 11.71). Blood that has given up its oxygen to the tissues returns to the heart and enters the right atrium. From the right atrium, blood passes into the ventricle. Blood also enters the ventricle from the left atrium, which collects oxygenated blood from the lungs and skin. Thus, in the amphibian heart, specifically in the ventricle, there is a slight mixing of oxygenated and deoxygenated blood. However, muscular folds and flaps within the ventricle prevent the two kinds of blood from mixing excessively. When the ventricle contracts, blood is pumped out through a double-branched aorta. Oxygenated blood goes to the tissues of the body, and deoxygenated blood is delivered to the lungs and skin, where it picks up a fresh supply of oxygen. In amphibians, then, there is a separate *pulmonary* (lung) circuit through which blood passes, in addition to a *systemic* (body) circuit.

Biology of Animals

> The reptiles include some of the longest-lived vertebrates; some tortoises, for instance, have a lifespan often exceeding 100 years.

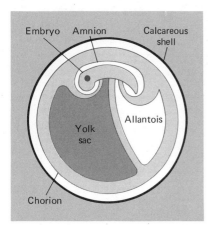

Figure 11.72 The amniotic egg of a reptile or bird. The embryo is nourished by yolk. It floats in a pool of amniotic fluid that protects it from shock. The allantois is a sac that stores the embryo's nitrogenous waste until hatching. Blood vessels grow out from the embryo through the allantoic membranes. Vessels near the chorion pick up oxygen and release carbon dioxide, permitting the embryo to exchange gases with its environment. The chorion is permeable to gases and somewhat permeable to water. A shell, leathery in reptiles and rigid in birds, forms over the outside of the egg shortly before it is laid.

Class Reptilia. The reptiles (L. *reptilis,* crawling) include snakes, lizards, turtles, tortoises, alligators, and crocodiles. These animals have a body covering of waterproof scales, four limbs provided with claws, and highly efficient lungs. Reptiles are true land vertebrates, having made total adaptation to a terrestrial existence. Male reptiles have a penis for depositing sperm, but the cloaca has been retained. Since fertilization is internal, i.e., within the body of the female, reptiles are not dependent upon an external source of water for reproduction. Moreover, the young develop within a fluid-filled **cleidoic** (Gk. *kleistos,* closed) egg that is protected from drying out by a tough, leathery shell (Fig. 11.72). Within the shell, the embryo is suspended in a watery fluid surrounded by a membrane called the **amnion.** As the embryo develops, it draws on food reserves in the form of yolk, which is contained within the **yolk sac.** A third membrane, the **allantois,** absorbs urinary wastes and contains blood vessels across which oxygen and carbon dioxide are exchanged. Finally, the amnion, yolk sac, and allantois are surrounded by an outer membrane, the **chorion.** This type of egg, which is laid on land, often far from an outside source of water, is the primary factor in the success of the reptiles as terrestrial vertebrates.

Most reptiles have a three-chambered heart, but the ventricle is partially divided by a muscular septum, or wall (Fig. 11.73). The septum effectively separates oxygenated and deoxygenated blood.

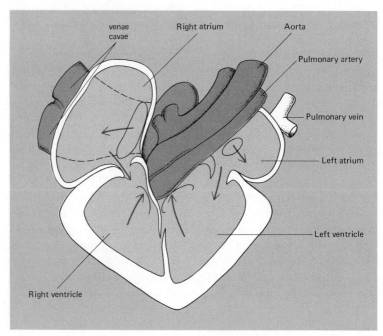

Figure 11.73 Reptilian heart. Note that the ventricles are only partially divided. Deoxygenated blood from the body enters the right atrium through the two venae cavae. After passing into the right ventricle, blood is pumped through the pulmonary artery to the lungs, where oxygenation occurs. Blood returns to the left atrium through the pulmonary veins. In the left ventricle, oxygenated blood is pumped out through the aorta to the body.

Figure 11.74 A cotylosaur, or stem reptile. Cotylosaurs were the earliest reptiles. (From Villee, C. A.: *Biology*. 7th ed. Philadelphia: Saunders College Publishing, 1977, p. 329.)

However, in crocodiles and alligators, the septum completely divides the ventricle, forming a four-chambered heart. This type of construction assures complete separation of oxygenated and deoxygenated blood as it enters and leaves the heart. Consequently, the delivery of oxygen to the cells and the removal of carbon dioxide can be accomplished with great efficiency.

Although the lineage is uncertain, it is believed that the reptiles arose from the stegocephalian group some 350 million years ago. The reptiles on the earth today, with the exception of crocodiles and alligators, are direct descendants of an amphibian-like reptile group called *cotylosaurs,* or "stem reptiles" (Fig. 11.74). The cotylosaurs were extremely important in vertebrate evolution, since they appear to be the "stem" from which all higher vertebrates arose.

The cotylosaurs gave rise to a second reptilian lineage called *thecodonts* ("socket-toothed"), which in turn gave rise to the dinosaurs, pterosaurs ("flying lizards"), aquatic ichthyosaurs ("fish lizards"), crocodiles, and alligators. In addition, thecodonts are the group from which modern birds are descended.

The largest living reptiles are certain species of crocodiles, which average about 14 ft. in length and weigh over 1000 lb. Crocodiles have long, slender snouts, a feature that distinguishes the group from alligators, which have short, broad snouts (Fig. 11.75).

Figure 11.75 The American alligator. (From Orr, R. T.: *Vertebrate Biology*. 5th ed. Philadelphia: Saunders College Publishing, 1982.)

One of the early cotylosaurs, or stem reptiles, was the lizard-like *Seymouria* (its fossil remains were discovered in Seymour, Texas, 1917), which had characteristics of both amphibians and reptiles.

In the United States, crocodiles are found only in Florida, and alligators inhabit the southeastern states and Texas.

The smallest reptiles are believed to be species of wall lizards, or geckos, which measure less than $1\frac{1}{2}$ inch in length.

Turtles and tortoises are shelled reptiles that lack teeth, although many have sharp, horny jaws (Fig. 11.76). Tortoises are usually considered terrestrial forms, and turtles are regarded as aquatic. These animals vary in size from a few inches in diameter to the great 1000 lb. sea turtles.

Probably the most famous, or infamous, reptiles are the snakes (Fig. 11.77). Contrary to general belief, only a few species are dangerous to human beings. Snake venom is produced in salivary glands and released through specialized teeth, or fangs. Most venoms paralyze the nervous system or destroy the red blood cells of the victim. The flicking tongue of snakes is an organ of smell and is not poisonous. Snakes have good vision and can detect sound, and many species can move quite rapidly. The largest of all snakes, the anaconda of South America, sometimes exceeds 30 ft. in length.

One order of reptiles includes only one species, the tuatara, genus *Sphenodon*, of New Zealand (Fig. 11.78). Often referred to as a "living fossil," this lizard-like reptile is the only surviving species of an ancient group of reptiles older than the dinosaurs. A true lizard, the Gila monster of the Southwest, is the only poisonous lizard in the United States (Fig. 11.79).

Figure 11.76 The desert tortoise. (From Orr, R. T.: *Vertebrate Biology*. 5th ed. Philadelphia: Saunders College Publishing, 1982.)

Most of the great herbivorous dinosaurs had extremely tiny brains and an additional brain-like mass of nervous tissue located at the base of the tail.

361
Introduction to the Chordates

Figure 11.77 The king cobra of Asia with its head raised and hood spread, ready to strike. (From Orr, R. T.: *Vertebrate Biology.* 5th ed. Philadelphia: Saunders College Publishing, 1982.)

The most "glamorous" of the reptiles—perhaps in fact and certainly in fable—were the now-extinct dinosaurs, the "great lizards" that first stalked the earth more than 200 million years ago (Fig. 11.80). Perhaps the largest flesh-eating land animal that ever lived was the dinosaur *Tyrannosaurus* ("tyrant lizard"). Over 45 ft. in length, *Tyrannosaurus* weighed in at an estimated 8 tons. However, some of the great plant-eating dinosaurs were even larger. For some 150 million years, not only the dinosaurs but also the reptiles as a group were the dominant animals on earth.

Figure 11.78 *Sphenodon* of New Zealand, a living relict of an ancient group of reptiles. (From Orr, R. T.: *Vertebrate Biology.* 5th ed. Philadelphia: Saunders College Publishing, 1982.)

362 Biology of Animals

> Styles of flight among the birds vary from the 90 wing beats per second of some hummingbirds, to the soaring flight of vultures that may continue for hours without a single wing beat.

Figure 11.79 The gila monster of the American southwest. (From Orr, R. T.: *Vertebrate Biology*. 5th ed. Philadelphia: Saunders College Publishing, 1982.)

Class Aves. The class Aves (L. *avis*, bird) includes all the birds. These animals represent one of the two flying groups that descended from the thecodont reptiles. The other group—the pterosaurs—mentioned earlier, had large, membranous wings and probably were gliders and not capable of true flight (Fig. 11.81). The pterosaurs eventually became extinct, and there is no evidence that the birds arose from them. There is evidence, however, of the reptilian ancestry of the birds. The oldest bird known from the fossil record is the so-called "lizard bird" *Archaeopteryx,* which had features characteristic of both reptiles and modern birds (Fig. 11.82). *Archaeopteryx* had reptilian teeth, a lizard-like tail, and clawed digits, but it also had wings and feathers. Feathers are thought to have evolved from reptilian scales. Modern birds reveal their reptilian heritage in having scales on their legs similar to the scales of reptiles.

The structural adaptations of the birds are related primarily to their capacity for flight. Their bones are hollow, and there are numerous air sacs throughout the body, which make the animals buoyant and light. Birds have massive breastbones, to which the powerful wing muscles are attached. Some birds, of course, such as the penguin and ostrich, are incapable of flight. All birds are characterized by a covering of feathers that aids in decreasing the loss of water and body heat.

Like the reptiles, fertilization is internal in the birds, and their eggs have the same membranes and basic structure as reptilian eggs. Birds may differ from some reptiles in one significant respect, in that they are "warm-blooded," or **homeothermic** (Gk. *homos*, same, + *therme*, heat). Homeothermy refers to the ability of organisms to

Figure 11.80 Giant dinosaurs from the Cretaceous period of western North America. The largest flesh-eating dinosaur known was *Tyrannosaurus*, two of which are shown attacking the herbivorous, horned dinosaur *Triceratops*. *Tyrannosaurus* reached a length of 15 meters and a height of 6 meters. Its head was as much as 2 meters long and was equipped with many sharp teeth. The front legs were small and completely useless; it walked on its powerful hind legs and balanced with its long tail. *Triceratops* was armed with a horn on the nose and a pair of horns over the eyes and was protected by a bony ruff covering the neck and shoulders. The rest of the body was covered with a leathery hide, so that it was vulnerable except when facing its enemy. (From Villee, C. A.: *Biology*. 7th ed. Philadelphia: Saunders College Publishing, 1977, p. 765. Copyright, Chicago Natural History Museum; from the painting by Charles R. Knight.)

Introduction to the Chordates

maintain a constant body temperature regardless of the temperature of their surroundings. Most commonly applied to birds and mammals, homeothermy generally has not been considered a biological feature of any other animal group. Instead, these animals are described as "cold-blooded," or **poikilothermic** (Gk. *poikilos*, various,

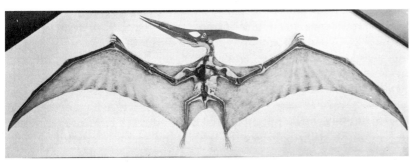

Figure 11.81 Pterosaurs had a wingspread of about 20 feet. A complete Cretaceous fossil is shown with a restoration of the body outline (*top*) and the probable appearance (*bottom*) of a living pterosaur. (From Ebert, J. D., Loewy, A. G., Miller, R. S., and Schneiderman, H. A.: *Biology*. New York: Holt, Rinehart and Winston, 1973, p. 564.)

> Birds exhibit complex behavioral patterns, including such activities as nest building, migration, elaborate courtship rituals, and mimicry.

Figure 11.82 The earliest known bird fossil is of the "lizard bird" *Archaeopteryx*. (From Ebert, J. D., Loewy, A. G., Miller, R. S., and Schneiderman, H. A.: *Biology*. New York: Holt, Rinehart and Winston, 1973, p. 564.)

+ *therme,* heat), i.e., having a body temperature that varies in accordance with the temperature of the surrounding environment. However, the distinction between homeothermic and poikilothermic animals may not be as cut and dried as once believed. Research on such poikilotherms as bony fishes and reptiles indicates that at least some of them also are capable of maintaining a constant body temperature.

In birds, homeothermy is accommodated by a four-chambered heart in which the two ventricles are completely separated. Since birds are quite active animals compared with amphibians or reptiles, they have a high metabolic rate (that is, they expend energy rapidly), which requires highly efficient and rapid delivery of oxygen to the body cells. A four-chambered heart assures that only well-oxygenated blood is delivered. Temperature regulation also is influenced by body feathers, which, as mentioned, decrease the loss of body heat.

Birds vary in size from the African ostrich, which stands 9 ft tall, to the bee hummingbird, which measures a little over 2 inches in length. Most birds have extremely acute vision, many times more

Figure 11.83 A therapsid, a mammal-like reptile considered to be the immediate ancestor of modern mammals. (From Villee, C. A.: *Biology.* 7th ed. Philadelphia: Saunders College Publishing, 1977, p. 762.)

so than that of humans. The fastest animal known is a bird, the spinetailed swift, which has been clocked flying at a speed of over 100 mi. per hour. Some penguins can swim at a speed of over 20 mi. per hour under water.

Class Mammalia. The dominant vertebrates on the earth today are the mammals (L. *mamma,* breast), a group considered to have arisen from a stock of carnivorous, mammal-like reptiles called **therapsids** (Gk. *theraps,* attendant) (Fig. 11.83).

An important adaptation of the therapsids was the orientation of the skeletal system that permitted more efficient use of the limbs. As mammals evolved, they departed from the crawling gait of their reptilian forebears, acquiring limbs that lifted the body high off the ground—a distinct advantage in chasing and being chased.

Mammals are characterized by the presence of hair or fur and mammary glands that secrete milk, and, like the birds, they are homeothermic with a four-chambered heart (Fig. 11.84). The mammalian brain is extremely large in proportion to the rest of the body, and the entire nervous system is highly developed. Mammals have keen senses and demonstrate degrees of intelligence unequaled in the animal kingdom. Although existing along with the dominant reptiles for millions of years, the ancestral reptilian-like mammals

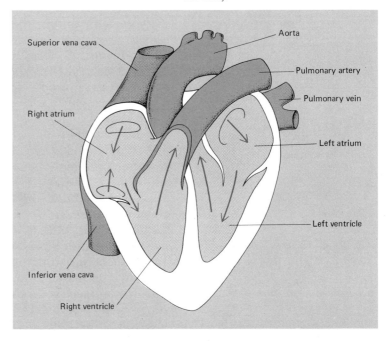

Figure 11.84 Mammalian heart. Note that the ventricles are completely separated. Deoxygenated blood from the body enters the right side of the heart through the two venae cavae. The blood is then pumped out through the pulmonary artery to the lungs, where oxygenation occurs. Oxygenated blood is returned to the left side of the heart via the pulmonary veins. From the left ventricle, blood is pumped out the aorta to the body.

> Although the young marsupial develops for a short time within the uterus, marsupials do not have a true placenta; this organ is characteristic only of the placental mammals.

Figure 11.85 The monotremes (egg-laying mammals) are represented by only two living forms: the duckbill platypus (*left*) and the spiny anteater (*right*). (From Ebert, J. D., Loewy, A. G., Miller, R. S., and Schneiderman, H. A.: *Biology*. New York: Holt, Rinehart and Winston, 1973, p. 567.)

managed to survive, giving rise to the true mammals, which began to spread across the earth some 50 million years ago.

One surviving group of mammals comprises the **monotremes** (Gk. *monos*, one, + *trema*, hole, a reference to the cloaca, a holdover from their reptilian ancestors), the only group of egg-laying mammals (Fig. 11.85). Although prominent millions of years ago, the only present-day monotremes are the duckbill platypus and the spiny anteater, both found mainly in Australia. After hatching from the egg, the young monotreme is nourished by milk secreted by the mammary glands of the female.

A second group, the pouched mammals, or **marsupials** (Gk. *marsypion*, bag), includes the kangaroos, koala bears, wombats, and others of Australia, and the only representative of the Americas, the opossum (Fig. 11.86). When born, the young marsupial is in a very immature stage of development and must crawl up the abdomen to the mother's pouch, where development is completed.

By far the largest group in this class comprises the **placental mammals**, in which the young develop within the uterus for a relatively long period of time and are often well developed at birth. The **placenta,** an organ unique to the mammals, is formed from the lining of the uterus and two fetal membranes—the amnion and chorion—that surround the developing fetus. The placenta serves to exchange nutrients, oxygen, and wastes between the blood of the mother and the fetus. Some placental mammals, such as rabbits and cattle, are able to move around on their own within minutes after birth; others, including humans and squirrels, are virtually helpless at birth. The placental mammal with the longest period of embryonic development is the Asian elephant, with an average period of just over 20 months. The opossum, a marsupial, has a very short gestation period of only 13 days.

Mammals are of significant economic importance as sources of food, clothing, and other products, and as disease carriers and destroyers of crops and trees.

Figure 11.86 Marsupials. (a) An infant kangaroo attached to a nipple in its mother's pouch. The gestation period in marsupials is extremely short and an infant marsupial is born in a very immature state. It continues its further development in the protected pouch. (b) A kangaroo with a well-grown infant called a joey in its pouch. (c) The Virginia opossum is an extremely common marsupial in the United States. It lives mainly in trees and is most active at night. When its young are first born they are barely a half inch long and are kept in the mother's pouch. Later on they are carried on the mother's back. (From Ebert, J. D., Loewy, A. G., Miller, R. S., and Schneiderman, H. A.: *Biology.* New York: Holt, Rinehart and Winston, 1973, p. 714.)

Biology of Animals

> It is generally acknowledged that the evolutionary success of the mammals can be attributed to their high degree of intelligence and to the prolonged close association between parent and offspring.

Figure 11.87 A bat, the only flying mammal. (From Orr, R. T.: *Vertebrate Biology*. 5th ed. Philadelphia: Saunders College Publishing, 1982.)

Placental mammals are quite diverse and are found on land, underground, in trees, in marine and fresh waters, and, in the case of bats, flying through the air (Fig. 11.87). The largest animal on earth, the blue whale, is a placental mammal and may measure 100 ft in length and weigh over 175 tons. By comparison, the smallest mammal known is the pygmy shrew, which measures only 2 to 3 inches in length.

One group of placental mammals—the order Primates—is of particular interest because, along with monkeys, chimpanzees, gorillas, baboons, and several others, it includes a species known as man. All of the primates have highly developed brains, opposable thumbs, eyes in the front, rather than on the sides of the head, stereoscopic vision, and so forth. As we have seen in this survey, every animal group has certain adaptive features for coping with the demands of existence; for human beings, the paramount feature is the brain, the extreme development of which makes man unique among all the animals on earth.

REVIEW OF ESSENTIAL CONCEPTS

1. The earliest known animal fossils date back over 600 million years. Animals are conveniently divided into two major groups: *invertebrates* and *vertebrates*.
2. Almost all animals are included in the subkingdom *Metazoa*. As a group, metazoans are characterized by (1) *heterotrophic* nutrition,

(2) *multicellular* bodies, (3) cells arranged to form *tissues*, (4) tissues joined together to form *organs*, (5) organs combined to form *systems*, and (6) similar early *embryonic development*.

3. Metazoan tissues include *epithelium, connective tissue, muscle,* and *nervous tissue.*
4. Body parts similar in structure but different in function are said to be *homologous*; body parts having the same function but different structures are said to be *analogous*. Homology reflects a common embryological and evolutionary origin; analogy does not.
5. Early embryological development in metazoans involves *cleavage* of the zygote to form a multicellular *embryo*. The embryo develops into a *blastula*, which is followed by the *gastrula*. A gastrula consists of the embryonic tissues *ectoderm, endoderm,* and, in most animals, *mesoderm*. Animals that develop only ectoderm and endoderm are *diploblastic*; animals with ectoderm, endoderm, and mesoderm are *triploblastic*.
6. The only major animal phylum in the subkingdom Parazoa is *Porifera*, the sponges. Sponges are *diploblastic* aquatic invertebrates having developed only to the cellular or crude tissue level. The sponge body is supported by *spicules*, and reproduction may be asexual or sexual. Sponges exhibit great powers of *regeneration*.
7. The subkingdom Metazoa includes all animals except the sponges. Metazoans have attained at least a tissue level of development and exhibit a high degree of cellular *division of labor*.
8. The phylum *Coelenterata* includes hydras, jellyfish, sea anemones, and corals. Coelenterates are *diploblastic* and possess several *tentacles* equipped with *nematocysts*. The phylum displays two different body forms: a sessile *polyp* and a free-swimming *medusa*. Medusae reproduce sexually, and polyps reproduce asexually. Coelenterates exist as solitary individuals or in colonies. They have reached the tissue level of development and have great powers of regeneration.
9. The phylum *Ctenophora* includes diploblastic marine animals known as sea walnuts and comb jellies. Ctenophores have eight rows of ciliated body plates and two tentacles but no nematocysts. These animals are monoecious, and some are luminescent.
10. The phylum *Platyhelminthes* includes three groups of flatworms: the free-living *planarians* and the parasitic *flukes* and *tapeworms*. These invertebrates are *triploblastic* and have attained the organ level of development. Most flatworms are monoecious and reproduce sexually by cross-fertilization. The nervous system is relatively complex, and there is definite *cephalization*.
11. Planarians are carnivorous flatworms that possess well-developed muscular, digestive, excretory, nervous, and reproductive systems.

The excretory system is the *flame cell* type. Planarians and other flatworms have great powers of regeneration.

12. Many *flukes* are pathogenic, having complex life cycles often involving several hosts. Examples of pathogenic flukes include *Opisthorchis* and *Schistosoma*. Flukes possess two or more *suckers* and a highly enzyme-resistant body covering.

13. Many *tapeworms* also are pathogenic, with complex life cycles. Flukes have an anterior attachment organ, the *scolex*, and a segmented body composed of *proglottids*. Humans are infected with tapeworms primarily by eating poorly cooked meat infested with eggs.

14. The phylum *Aschelminthes* includes free-living and parasitic roundworms. They are usually dioecious and are found in marine and fresh waters, and on land. Two major classes of Aschelminthes are *Nematoda* and *Rotifera*.

15. *Nematodes* are generally thin, elongate worms, some of which are human parasites. Two examples include *Trichinella*, the causative agent of *trichinosis*, and *Ascaris*, an intestinal parasite also found in other vertebrates. Other common parasitic nematodes are *pinworms*, *hookworms*, and *filarial worms*.

16. *Rotifers* are protozoan-like roundworms having an anterior ring of motile cilia and a body divided into three parts. Each member of a given species has the same number of body cells, and at a certain point of development, mitosis stops. In most species of rotifers, there are no males. Rotifers produce two different kinds of eggs—haploid eggs that develop parthenogenetically into males and diploid eggs that develop into females. However, if haploid eggs are fertilized, they also develop into females.

17. The phylum *Annelida*, or segmented worms, includes earthworms, clam worms, and leeches. Annelids are found in fresh and marine waters and in the soil; some are free-living, and others are parasites. Annelids have a *closed circulatory system*.

18. The earthworm *Lumbricus* has *setae*, a well-developed *nervous system*, and excretory organs called *nephridia* and is *monoecious*. Earthworm eggs develop within a *cocoon*.

19. The clam worm *Nereis* possesses *parapodia* and numerous setae. The well-developed head has jaws and sensory structures. The clam worm is *dioecious*, and some species pass through a *trochophore larval* stage.

20. Some *leeches* are parasites on vertebrates and possess chitinous teeth and suckers. Leeches have no parapodia or setae and are monoecious.

21. The phylum *Mollusca* includes clams, oysters, scallops, snails, squids, octopuses, and a few others. Most mollusks have an external shell;

Color Plate 48 Sensitive brier plant with its leaves spread. (Photograph by Carolina Biological Supply Co.)

Color Plate 49 Sensitive brier with its leaves closed. (Photograph by Carolina Biological Supply Co.)

Color Plate 50 Poppy plants open in sunlight. (Photograph by Carolina Biological Supply Co.)

Color Plate 51 Poppy plants closed at evening. (Photograph by Carolina Biological Supply Co.)

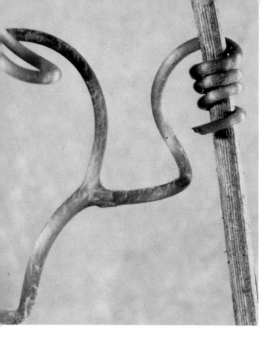

Color Plate 52 Growth response to touch. A grape tendril wraps around a stem. (Photograph by Carolina Biological Supply Co.)

Color Plate 53 Sundew, an insect-eating plant. (Photograph by Carolina Biological Supply Co.)

Color Plate 54 Response of a live oak tree to wind. (Photograph by Carolina Biological Supply Co.)

Color Plate 55 A *Yucca* flower and yucca moth. The plant and the moth are totally interdependent — the plant depends exclusively on the moth for pollination, and the moth deposits its eggs only in the flowers of a *Yucca* plant. (Photograph by Carolina Biological Supply Co.)

Color Plate 56 One of the tallest living things on earth — a giant sequoia. (Photograph by Carolina Biological Supply Co.)

Color Plate 57 Clam worm, a carnivorous marine annelid. (Photograph by Carolina Biological Supply Co.)

Color Plate 58 Snail, a shelled gastropod. (Photograph by Carolina Biological Supply Co.)

Color Plate 59 The horseshoe crab is not really a crab at all but a "living fossil" of an extinct group of chelicerates. (Photograph by Carolina Biological Supply Co.)

Color Plate 60 A centipede is carnivorous and has one pair of walking legs on each body segment except the first, which bears poison claws. (Photograph by Carolina Biological Supply Co.)

Color Plate 61 A millipede is herbivorous and has two pairs of walking legs on most of the body segments. (Photograph by Carolina Biological Supply Co.)

Color Plate 62 About one third of all species of insects are beetles. (Photograph by Carolina Biological Supply Co.)

Color Plate 63 Displaying its protective coloration, a praying mantis dines atop a twig. (Photograph by Carolina Biological Supply Co.)

in others, the shell is internal or absent. Characteristics of the phylum include the *visceral mass, mantle,* and muscular *foot.* Respiration is usually by gills, and except for the cephalopods, the mollusks have an *open circulatory system.* The heart is three-chambered, consisting of two *atria* and one *ventricle.* The major classes of living mollusks are *Pelecypoda, Gastropoda, Cephalopoda, Amphineura,* and *Scaphopoda.*

22. Class *Pelecypoda* includes the *bivalves*—clams, oysters, mussels, scallops, and others. The inner surface of the valves is composed of *nacre.* Bivalves are *filter feeders.*

23. Class *Gastropoda* includes snails, slugs, whelks, abalones, and conchs. The shell may be coiled, uncoiled, or absent. Gastropods are marine, freshwater, and terrestrial. Respiration in terrestrial and freshwater snails is by lungs.

24. Class *Cephalopoda* includes squids, octopuses, cuttlefish, and the chambered nautilus. All are marine and are the most advanced mollusks. Respiration is by gills, and the circulatory system is closed. The contrasting characteristics of the clam and the squid illustrate the great diversity of the phylum.

25. Some of the most primitive mollusks are the *chitons,* class *Amphineura.* The class *Scaphopoda* includes the tooth shells or tusk shells. Many marine mollusks have a trochophore larval stage in their development.

26. Animals in the phylum *Arthropoda* are characterized by *jointed legs* and usually a *chitinous exoskeleton.* Arthropods are immensely diverse and include the largest group of all the animals, the insects. Shedding of the arthropod exoskeleton is called *molting.*

27. Arthropods have an open circulatory system. Terrestrial arthropods possess excretory organs called *malpighian tubules;* aquatic species employ *gills* for excretion. Many arthropods have *compound eyes,* which are effective in detecting rapid motion.

28. Most insects undergo *metamorphosis,* the types of which are *gradual, incomplete,* and *complete.* Arthropods have a highly developed nervous system and a vast array of physical adaptations. Communication among arthropods may occur by sight, sound, or *pheromones.*

29. The two subphyla of living arthropods are *Chelicerata* and *Mandibulata.* Chelicerates possess fanglike *chelicerae* and four pairs of legs, but no antennae or jaws. The major class of chelicerates is *Arachnida,* which includes spiders, ticks, mites, and scorpions.

30. Spiders have two body sections, a *cephalothorax* and an *abdomen.* The abdomen is supplied with *spinnerets* that release silk. Spiders are carnivorous and have fangs through which venom is released. The repiratory structures of spiders and scorpions are *book lungs.*

31. The subphylum *Mandibulata* includes arthropods with *mandibles;* other characteristics include paired *antennae* and three, four, or dozens of pairs of walking legs. The major classes in this subphylum are *Crustacea, Diplopoda, Chilopoda,* and *Insecta.*
32. The class *Crustacea* includes crabs, crayfish, lobsters, shrimp, barnacles, and others. They are mostly marine, and some constitute part of the *zooplankton.* Respiration is by gills, and compound eyes are usually present.
33. The class *Diplopoda* includes the herbivorous *millipedes,* which have two pairs of legs on almost every body segment. The class *Chilopoda* includes the carnivorous *centipedes,* which have one pair of legs on each body segment. The first pair of appendages bears poison glands.
34. The class *Insecta* includes all the insects, the most numerous animals in terms of different species. Insects are found in all habitats except the sea. The biological success of insects can be illustrated by considering the adaptive characteristics of the grasshopper.
35. The insect body is divided into *head, thorax,* and *abdomen.* The head of the grasshoppper bears compound and simple eyes, one pair of antennae, two pairs of *palps,* and mouth parts for manipulating and chewing food. The thorax has three pairs of jointed walking legs, a *tympanum* on either side, and *spiracles* that lead into a system of respiratory *tracheae.* Other adaptive features of the grasshopper include protective coloration, sound production, and color vision.
36. *Molting* and *metamorphosis* in insects are controlled by the interaction of *juvenile hormone, brain hormone,* and *ecdysone.*
37. The phylum *Echinodermata* includes starfishes, brittle stars, sea urchins, sand dollars, and sea lilies, all of which are marine. The skeleton is internal, and the body is constructed on a five-part plan. Echinoderms have no brain or excretory organs; sexes are separate, and fertilization is external. A unique feature of the echinoderms is the *water vascular system.* Echinoderms have remarkable powers of regeneration.
38. The phylum *Chordata* includes the invertebrate chordates and all the vertebrates. Chordates are characterized by having a *notochord, dorsal hollow nerve cord,* and *pharyngeal gill slits* at some stage in their development. Other chordate characteristics include a ventral *heart,* a *tail,* and a cartilaginous or bony *endoskeleton.* The phylum Chordata includes three subphyla, *Urochordata, Cephalochordata,* and *Vertebrata.*
39. The most common members of the subphylum *Urochordata* are the *tunicates,* which are saclike marine animals. The tunicate larva possesses all three chordate characteristics—notochord, dorsal nerve cord, and pharyngeal gill slits. Only the gill slits persist in the adult.

40. The common representative of the subphylum *Cephalochordata* is *amphioxus,* a fishlike, burrowing animal found in shallow marine waters. The adult has a notochord, dorsal nerve cord, and pharyngeal gill slits. The cephalochordates and urochordates are *filter feeders.*
41. Characteristics of the subphylum *Vertebrata* include a cartilaginous or bony *vertebral column;* an *endoskeleton;* body covered by *skin* or derivatives of skin; a pair of image-forming *eyes;* a pair of *ears;* a *closed circulatory system;* a pair of *kidneys;* an *endocrine system;* 10 or 12 pairs of *cranial nerves;* a *liver* and *pancreas; gills* or *lungs;* usually two pairs of *limbs;* and separate sexes. There are seven major classes of living vertebrates.
42. The class *Agnatha* includes the lampreys and hagfishes, characterized by lacking jaws, paired fins, and scales. There is a cartilaginous endoskeleton, and the notochord persists throughout life. Lampreys are ectoparasites, and hagfishes are scavengers. Lampreys pass through an *ammocoete* larval stage. The oldest vertebrate fossils on record—the *ostracoderms*—are the extinct ancestors of modern agnaths.
43. *Placoderms,* the first known vertebrates with jaws, gave rise to the *Chondrichthyes* and *Osteichthyes.* The class Chondrichthyes includes the sharks, skates, and rays—the only jawed vertebrates with a cartilaginous endoskeleton. Sharks have poor vision, a highly developed sense of smell, and a two-chambered heart. The digestive system terminates in a *cloaca.*
44. The class *Osteichthyes* includes the bony fishes, which are extremely diverse and have developed a wide range of adaptations. After their rise from the placoderms, Osteichthyes developed along two evolutionary lines: modern bony fishes and lungfishes. The lungfishes include the *crossopterygians,* an extinct group except for one species, *Latimeria.* A group of crossopterygians are thought to be the ancestors of the land vertebrates.
45. The class *Amphibia* includes frogs, toads, salamanders, and cecilians. Modern amphibians are thought to have arisen from the *stegocephalians.* Amphibians are only partially adapted to land and usually must return to water to reproduce. Adult amphibians have a pair of lungs, a three-chambered heart, and skin without scales.
46. The class *Reptilia* includes snakes, lizards, turtles, tortoises, alligators, and crocodiles. Reptiles are true land vertebrates with internal fertilization. The *cleidoic* egg of reptiles has a leathery shell protecting the embryo. Within the shell, the embryo is associated with the four fetal membranes: the *amnion, yolk sac, allantois,* and *chorion.* Most reptiles have a three-chambered heart, except for the crocodiles and alligators, in which the heart is four-chambered.

47. Stegocephalians gave rise to *cotylosaurs,* from which all modern reptiles, except crocodiles and alligators, are directly descended. Cotylosaurs also gave rise to the *thecodonts,* from which dinosaurs, crocodiles, alligators, and modern birds are descended.
48. The largest living reptiles are crocodiles; the smallest are geckos. Tortoises are terrestrial, and turtles are aquatic. Most snakes are not poisonous, and their tongues are organs of smell.
49. The lizard-like *Sphenodon* is the only surviving species of an ancient group of reptiles. The now-extinct dinosaurs were some of the largest reptiles that ever lived.
50. The birds, class *Aves,* descended from the thecodont reptiles. The reptilian ancestry of the birds is seen in the fossil remains of *Archaeopteryx;* modern birds also have reptilian-like leg scales. Birds are structurally adapted for flight, although a few species cannot fly. Fertilization is internal in birds, and the young develop within a cleidoic egg.
51. Birds have a four-chambered heart and are *homeothermic.* Body feathers aid in temperature regulation by decreasing heat loss. Some animals, except for birds and mammals, are *poikilothermic.* Birds have acute vision and include the fastest animals known.
52. Vertebrates in the class *Mammalia* are characterized by hair or fur, mammary glands, a four-chambered heart, and a large brain. Mammals are thought to have evolved from the *therapsids.* The surviving groups of mammals are the *monotremes, marsupials,* and *placental mammals.*
53. The *placenta* is an organ formed from the lining of the uterus and from the amnion and chorion surrounding the fetus. The placenta serves to exchange nutrients, oxygen, and wastes between the blood of the mother and the fetus.
54. The order *Primates* is of special interest because it includes human beings as one of its species.

APPLYING THE CONCEPTS

1. What is the biological significance of homologous and analogous structures?
2. What is the significance of diploblastic construction compared with triploblastic construction?
3. What basic features distinguish the Parazoa from the Metazoa?
4. What are the relationships between the polyp and medusa stages of coelenterates?
5. What are the essential differences between the phylum Ctenophora and the phylum Coelenterata?

6. Compare the parasitic adaptations of the flukes and tapeworms.
7. What are the basic structural differences between planarians, flukes, and tapeworms?
8. Name five species of parasitic nematodes.
9. Compare the structures, habitats, and activities of the earthworm, clam worm and leech.
10. a. Illustrate diversity in the phylum Mollusca by comparing organisms in any two classes.
    b. List the ways in which these organisms are similar.
11. Explain the difference between an open and a closed circulatory system.
12. Compare the basic characteristics of the subphyla Chelicerata and Mandibulata.
13. List the distinctive features of the class Crustacea.
14. What factors are responsible for the biological success of insects?
15. What is a water vascular system, and what is its function?
16. In what manner do the urochordates and cephalochordates reflect an evolutionary relationship to the vertebrates?
17. Compare the structural differences and similarities in amphioxus and the tunicates.
18. Why are the agnaths considered to be holdovers from primitive vertebrates?
19. What is the significance of the development of jaws in the vertebrates?
20. Explain why amphibians are only partially adapted to land.
21. Compare the circulation of blood in the cartilaginous fish and the amphibians.
22. Describe the cleidoic egg.
23. What are the structural and evolutionary relationships between reptiles and birds?
24. What is the significance of the four-chambered heart in birds and mammals?
25. How is the placenta related to the biological success of the mammals?

12 general ecology

THE ESSENTIAL OBJECTIVES

You have understood this chapter when you are able to:

1. Describe the general composition of the crust, mantle, and core of the earth.
2. Explain what is meant by the biosphere.
3. Describe the major types of terrestrial biomes.
4. Discuss the major life zones associated with the oceans.
5. Compare, generally, oceanic and freshwater ecosystems.
6. Discuss some of the causes and problems of water pollution.
7. Differentiate between species, population, community, and ecosystem.
8. Differentiate between habitat and ecological niche.
9. Explain with examples a food chain and food web.
10. Discuss the relationship between energy and trophic levels in an ecosystem.
11. Describe an energy pyramid and a pyramid of biomass.
12. Explain the 10 Percent Rule and discuss its significance to the production and availability of food.
13. Explain the law of conservation of matter.
14. Outline the carbon, oxygen, nitrogen, and water cycles.
15. Discuss the process of eutrophication.
16. Explain the phenomenon of biological magnification.
17. Describe the factors involved in regulating the size of a natural population.
18. Describe three types of symbiosis.
19. Discuss the stages of ecological succession.

The maintenance of living organisms is dependent upon their interactions with each other and with their nonliving environment. If not upset, these interactions tend to stabilize and perpetuate the ecosystem over the long run.

PREVIEW OF ESSENTIAL TERMS

biosphere The area of the earth capable of supporting life.

ecosystem The highest unit of structure and function in the biosphere; it consists of a community along with its nonliving environment.

biome A major type of terrestrial ecosystem, such as the tundra, tropical rain forest, or desert.

biological oxygen demand (BOD) The amount of oxygen extracted from organically enriched water by metabolizing bacteria.

ecological niche The total interaction an organism has with its environment, i.e., its "occupation" or role in nature.

autotroph An organism that, through photosynthesis or chemosynthesis, is capable of synthesizing organic nutrients from inorganic compounds.

net primary productivity The amount of energy in green plants that is available for use by the consumers.

biotic potential The maximum rate of population growth that would occur if all factors affecting a species' existence were ideal.

carrying capacity The maximum number of organisms of a species that the environment can support.

environmental resistance The total of the factors that inhibit the growth of a population.

law of conservation of matter Matter cannot be created or destroyed; implied in the law is that the total amount of matter on the earth is constant.

biogeochemical cycle The movement of a substance from the nonliving world through the living world in the bodies of living things and back to the nonliving world.

symbiosis An intimate relationship between two or more organisms of different species; includes parasitism, mutualism, and commensalism.

ecological succession A sequence of usually gradual changes in an ecosystem, culminating in a climax community.

climax community The final stage of ecological succession; it is a diverse and highly stable community.

eutrophication Becoming "well nourished." A process by which a body of water becomes rich in nutrients.

> Life on the planet Earth is restricted to the crust and the atmosphere, an area referred to as the *biosphere*.

INTRODUCTION

As a consequence of their existence, living things interact with their physical environment and with each other. The study of these interactions is called **ecology,** a term derived from the Greek *oikos,* meaning "house." This derivative of *oikos* is also found in the word "economics," which literally means "house management." By extension, ecology can be thought of as the science concerned with how the "house" of nature is "managed." Ecology relies upon many other disciplines, such as geology, physics, chemistry, and other biological sciences, in order to explain the interactions that exist in the living world. The principal pursuit of the **ecologist,** then, is to discover and study the factors governing such interactions.

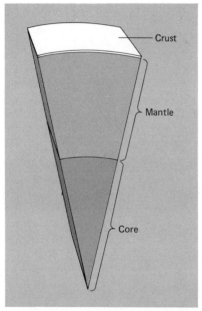

Figure 12.1 The major layers of the earth. Living things exist only on the crust and the atmosphere above it.

THE BIOSPHERE

The mass of the planet Earth is well over 6 sextillion tons (that's 6 followed by 21 zeros) with an estimated volume of nearly 260 billion mi.^{3} Its surface area is almost 197 million mi.^{2} Such immensity is practically impossible for us to visualize; but for all its vastness, only a very small part of the earth is hospitable to living things. Other than a part of the crust, along with the atmosphere extending some 5 mi above its surface, the earth is subjected to such extreme pressures and temperatures that it is unfit for habitation. The **crust,** or surface covering, of the earth averages only about 25 mi in thickness and consists of the land masses, including those covered by salt and fresh waters (Fig. 12.1). Beneath the crust, extending about 1800 mi deep, is an inner layer of nearly molten hot rock called the *mantle.* The center, or core, of the earth is principally iron, nickel, and sulfur at a temperature of 3700°C and at a pressure of over 27,000 tons per square inch.

The crust of the earth and the oxygen-laden atmosphere around it is the home for many hundreds of thousands of species of living organisms. This life-supporting area of the earth is referred to as the *biosphere* (Fig. 12.2). Even within this sphere, though, life is somewhat restricted, primarily by extremes of temperature. With a few exceptions, living things exist only within a temperature range from about 0°C to around 55°C.

On the land masses and in the waters of the crust of the earth, each particular organism or population lives in a particular area called its *habitat.* Any habitat consists of all of those factors that constitute an organism's environment. These factors include temperature, amount of rainfall, soil type, food sources, other organisms,

> Within their respective habitats, living things interact with their environment and with each other. The study of such interactions is known as *ecology*.

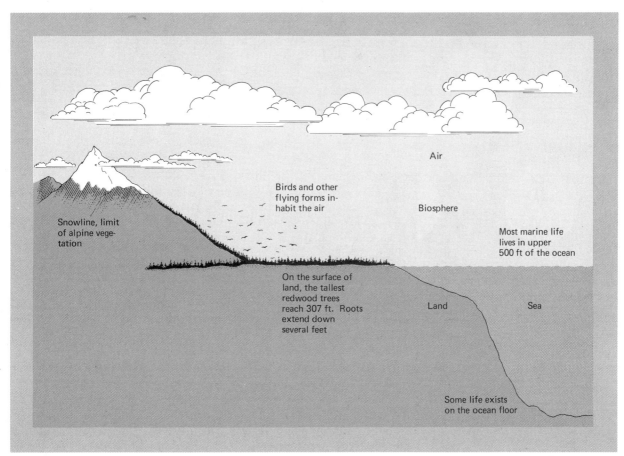

Figure 12.2 Biosphere. Living organisms are present in great numbers and diversity in a small part of the earth and its atmosphere. This has been called "the thin slice of life."

and so forth. There are, of course, myriad habitats to which organisms through eons of evolution have become adapted. Flowers bloom at an altitude of over 20,000 ft in the Himalayas, and starfish exist at depths of nearly 25,000 ft in the Pacific Ocean. Between these extremes, living organisms occupy almost every conceivable habitat within the biosphere. None of these organisms, however, exists alone and apart; each interacts with its physical environment and with other organisms within that environment.

The living plants and animals of a certain area, together with their nonliving surroundings, make up what is called an ***ecosystem,*** the highest level of structure and function within the biosphere. A lake, a thicket, or an open field are familiar examples of natural

General Ecology

> The major northernmost biome is the *tundra*, characterized by sparse vegetation, low temperatures, and permanently frozen subsoil.

Figure 12.3 The Everglades. This river of grass, or Payhay-Okee, as the Seminoles call it, is a strange mix of temperate and tropical zones. Forest and jungle, fresh and salt water ecosystems survive together in the 2000 mi^2 that constitute the national park. The glades are home to rare species of wildlife, such as the crocodile, manatee, and wood stork, which are found nowhere else in the United States. (From Turk, J.: *Introduction to Environmental Studies*. Philadelphia: Saunders College Publishing, 1980, p. 40.)

ecosystems (Fig. 12.3). As you will see, some ecosystems offer a wide variety of habitats, while others provide relatively few. It is often difficult to determine where one ecosystem ends and another begins, since their boundaries usually overlap. In fact, the earth itself is actually a series of ecosystems overlapping to the extent that the entire earth can be referred to as the ***global ecosystem*** (Fig. 12.4).

In a broad sense, there are two principal types of ecosystems—terrestrial and aquatic. The terrestrial ecosystem can be further separated into large major regions called ***biomes.*** The various biomes of the earth are areas characterized by a dominant type of vegetation, which in turn is affected significantly by climate and soil type. A biome, then, may be thought of as a *major* terrestrial ecosystem. Generally, biomes vary according to their location on the globe and, in the case of tall mountains, vary with altitude.

TERRESTRIAL BIOMES

Tundra. Along the northern reaches of Europe, Russia, and North America lies a large land area characterized by a permanently frozen subsoil, low precipitation, and a temperature that seldom rises above 10°C (50°F). This area is a virtually treeless plain called the ***tundra*** (Fig. 12.5). The region has a very short summer growing season, during which the periods of daylight are long. The upper few inches of soil thaw during the summer season but are frozen solid during the long winter. The sparse vegetation of the tundra includes a thin layer of grasses, mosses, sedges, and a few shrubs. In some areas flowering plants are evident, but in the arctic tundra, for example,

> The *taiga* is a region of evergreen forests south of the tundra. The relatively abundant vegetation supports a variety of large and small animals that are well adapted to the cold winters.

the most prevalent plant life is the lowly "reindeer moss." Actually this is not a moss at all, but a type of plant life called a *lichen*, which consists of an alga and a fungus living in close association (Fig. 12.6).

The animals inhabiting this northern wilderness include reindeer, caribou, wolves, lemmings, and polar bears. In the summer, various species of birds are found mainly along the shoreline, but they migrate south for the winter. Although dormant in the winter, teeming hordes of insects appear during the brief summer, providing food for the migratory birds.

Taiga. Moving south of the tundra where the weather is a little warmer, one encounters a region of evergreen forests known as the *taiga* (Fig. 12.7). The summers there are somewhat longer than those of the tundra, and the winters are bitter cold. Lakes and bogs are evident, and vegetation is relatively abundant. The dominant trees include pine, fir, and spruce, which serve as the major sources of commercial lumber. In lesser abundance are trees such as willow, birch, and poplar.

The taiga also is characterized by a variety of species of animals. The most predominant include large animals such as moose, elk, muledeer, black bears, and grizzly bears; smaller animals include

Figure 12.4 Earth. The global ecosystem. (From Arms, K., and Camp, P. S.: *Biology.* New York: Holt, Rinehart and Winston, 1979, p. 7.)

Figure 12.5 The tundra biome in Alaska. (From Johnson, W. H., Delanney, L. E., Cole, T. A., and Brooks, A. E.: *Biology.* 4th ed. New York: Holt, Rinehart and Winston, 1972, p. 854.)

382
General Ecology

> The *deciduous forest* biome, an area of abundant rainfall, covers extensive areas of several continents. The diverse species of trees and other plants support a great variety of animal life.

(a)   (b)   (c)

Figure 12.6 Three main body forms of lichens. (a) Crustose forms may resemble spots of paint on the surface of a rock. (b) Foliose forms are flat but curled at the edges, like leaves. (c) Fruticose forms grow in more complex, shrubby shapes. (From Arms, K., and Camp, P. S.: *Biology*. New York: Holt, Rinehart and Winston, 1979, p. 387.)

Figure 12.7 The taiga biome in Idaho. (From Johnson, W. H., Delanney, L. E. Cole, T. A., and Brooks, A. E.: *Biology*. 4th ed. New York: Holt, Rinehart and Winston, 1972, p. 856.)

> *Grasslands* are found in temperate climates in which the amount of precipitation is relatively small. Large grazing animals feed on the various species of grasses, whereas most of the smaller mammals are burrowers.

wolves, porcupines, snowshoe hares, mice, and several species of birds. In the summer season, there is a veritable swarm of mosquitoes.

Deciduous Forest. South of the taiga lies the large ***deciduous forest*** biome, characterized by a great number of species of trees and an annual rainfall of 30 to 60 inches (Fig. 12.8). "Deciduous" means that the trees shed their leaves periodically. This, of course, is in contrast to the trees of the taiga, which are mainly evergreen. Deciduous forests once covered extensive areas of North America, Australia, Japan, Europe, and South America. However, the dominant trees in each region are different. In the United States, the deciduous forest biome extends from Maine westward to Minnesota and south into Louisiana. The predominant trees throughout the area include oak, maple, beech, white pine, chestnut, birch, and hickory. In addition, deciduous forests support a diverse assortment of smaller plant life. Various shrubs and bushes are relatively abundant, along with several species of grasses, flowers, mosses, and ferns.

With its rich variety of plant life, the deciduous forest provides for a great number of animal species. Large and small mammals such as deer, bears, wolves, foxes, and squirrels range throughout the forest during the spring and summer seasons. In the cold winters of the forest, life is slowed and the animals seek their dens and lairs for comfort and sleep. Most of the numerous species of birds migrate south and return to the forest in the spring.

Grasslands. In temperate climates where the rainfall averages less than 10 to 20 inches a year, the dominant biomes are the ***grasslands,*** or *prairies* (Fig. 12.9). Grasslands cover large areas of North America, Siberia, Argentina, Australia, and South Africa. Various species of grasses, both short and long, grow in abundance, supporting wandering herds of grazing animals, such as bison, antelope, wild horses, and zebras. Most of the small animals, such as prairie dogs and gophers, are burrowers. Representative birds of the grasslands include prairie chickens and hawks. Although the grasses are the dominant plant life, small trees and shrubs can be found dotting the landscape.

Desert. In some parts of North and South America, Africa, Arabia, and Australia, rainfall may average less than 10 inches a year, an amount insufficient even for the survival of grasses. The major biome in these drought-stricken areas is a ***desert*** (Fig. 12.10). Such

Figure 12.8 Deciduous forest biome of the eastern United States. (From Applewhite, P., and Wilson, S.: *Understanding Biology.* New York: Holt, Rinehart and Winston, 1978.)

> The major drought-stricken areas of the earth constitute the *desert* biomes, where vegetation is usually sparse and the air temperature fluctuates widely. Plants and animals of the deserts have developed unique adaptations for conserving water.

arid regions cover nearly an eighth of the earth's land surface. Characteristically, vegetation is sparse in the desert, typically covering less than 20 percent of the area. Low shrubs, sagebrush, and several kinds of cacti are the predominant plants, but some deserts bloom with a carpet of flowers during the brief rainy season. To combat the long drought, some cacti store water, and other plants possess tough, leathery leaves that prevent excessive water loss.

The animals of the desert have developed special and unique measures for coping with their harsh environment. Reptiles such as snakes and lizards, along with a fantastic assortment of insects, have thick outer body coverings that prevent evaporation of water. In addition, most of these animals conserve water by excreting a very highly concentrated waste material that is almost dry. Small desert rodents, such as the kangaroo rat, extract water from seeds and fruits they eat and so can exist without actually drinking water.

Desert areas may reach an air temperature of 110°F (43.3°C) or more during the day and then become freezing cold at night. In the western region of the Sahara Desert in North Africa, the daily temperature may fluctuate more than 80°F (26.7°C).

Tropical Rain Forest. Finally, one of the most complex terrestrial biomes is the ***tropical rain forest*** (Fig. 12.11). Rain forests are complex in the sense that the number of diverse species of organisms is

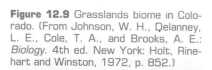

Figure 12.9 Grasslands biome in Colorado. (From Johnson, W. H., Delanney, L. E., Cole, T. A., and Brooks, A. E.: *Biology.* 4th ed. New York: Holt, Rinehart and Winston, 1972, p. 852.)

> Near the equator are found the *tropical rain forests*, areas receiving abundant rainfall and having an extremely diverse assortment of plant and animal species. Vertical stratification in the rain forest provides widely varying habitats for both plants and animals.

enormous. Because of this diversity, no one species of plant or animal life predominates. Rain forests are found near the equator and are characterized by annual rainfalls of 80 to 150 inches or more.

A particularly striking feature is the extensive development of what is known as *vertical stratification*. The trees of the rain forest, mostly evergreen, are quite tall and together form a dense canopy high above the forest floor. Less tall species of plant life growing beneath the canopy include palm trees and a variety of ferns. Because of the thick canopy, much of the sunlight does not reach the forest floor, leaving the ground vegetation rather sparse. The habitats established are possible because the amounts of rainfall, wind, and humidity, as well as sunlight, vary considerably from the top levels to the floor. Temperatures, too, may be quite different at the various levels. In the canopy it may be quite warm, whereas the forest floor is shaded and cool. As a result of these factors, many of the species of plants and animals are separated on the basis of their respective vertical habitats.

Vines and epiphytes (plants supported by and growing on trunks and limbs) hang from the tall trees, accompanied by a diverse assortment of flowering plants, mosses, and other plant life living in

Figure 12.10 Desert biome in Arizona. (From Johnson, W. H., Delanney, L. E., Cole, T. A., and Brooks, A. E.: *Biology*. 4th ed. New York: Holt, Rinehart and Winston, 1972, p. 855.)

General Ecology

> The major terrestrial biomes may be found at different altitudes on tall mountains.

Figure 12.11 Tropical rain forest biome in the Congo. (From Villee, C. A.: *Biology.* 7th ed. Philadelphia: Saunders College Publishing, 1977, p. 841.)

Figure 12.12 The major biomes may be found at varying altitudes as well as at different latitudes.

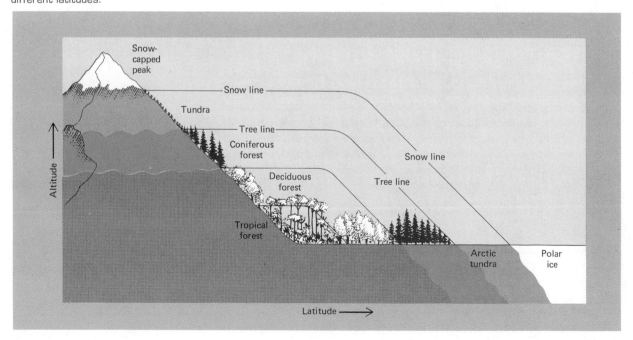

Figure 12.13 The ocean. (From McCormick, J. M., and Thiruvathukal, J. V.: *Elements of Oceanography*. Philadelphia: Saunders College Publishing, 1981, p. 115.)

the canopy. Many species of animals, including various mammals, reptiles, and birds, also inhabit the upper levels.

At the ground level, bare spots are found in many areas, and, as mentioned, vegetation is sparse. The soil is generally infertile, being leached of minerals and nutrients by the abundant rainfall. There is but one season in the rain forest—a warm, eternal summer.

Altitude and Biomes

It was noted earlier that the terrestrial biomes just outlined also may be found at varying altitudes on tall mountains. Below the snow-capped peaks one may find the tundra biome, followed at lower elevations by the taiga, deciduous forest, and tropical rain forest. Here, as with the world biomes just discussed, the altitudinal biomes are separated largely on the basis of climate and soil type (Fig. 12.12).

AQUATIC ECOSYSTEMS

As well as being the structural and functional units of terrestrial regions, ecosystems are units of the marine and fresh waters of the world. The oceans, with their great variety of living things, provide the most varied types of aquatic ecosystems (Fig. 12.13). Physical factors such as water depth, pressure, salinity, light penetration, and temperature determine the types of marine habitats, which in turn determine the kinds of organisms that can be supported

In the open ocean there are two major areas, or zones, where living things are most abundant. The surface waters, called the *pelagic zone,* are inhabited by billions of single-celled microscopic algae, or *phytoplankton* (Gk. "drifting plants"), along with the tiny crustaceans, fish eggs, and invertebrate larvae that make up the *zooplankton* (Fig. 12.14). Both forms of plankton are the major sources of food for most of the swimming sea animals that inhabit the pelagic zone. In addition, the photosynthetic phytoplankton produce a large proportion of the earth's atmospheric oxygen.

The other major area is the *benthic* (Gk. "depths of the sea") *zone* on the bottom of the ocean. This cold region, inhabited by a few species of fish, various invertebrates, and bacteria, is in total darkness (Fig. 12.15). Some light rays from the sun may penetrate to a depth of nearly 2000 ft in the ocean, but below this depth there is perpetual darkness. From the surface waters to a depth of about 650 ft is the *euphotic* (Gk. "well lighted") *zone,* where the amount of light is sufficient for photosynthesis to occur. Below this zone, the light begins to fade gradually until there is nothing but darkness.

General Ecology

> *Phytoplankton* and *zooplankton* are the major sources of food for the swimming animals of the pelagic zone of the ocean.

(a)

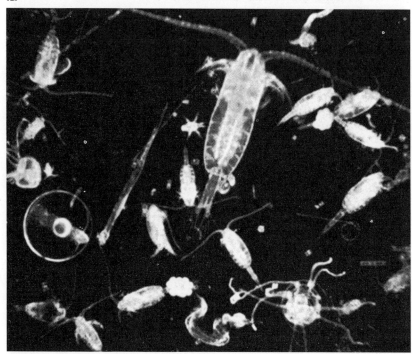
(b)

Figure 12.14 (a) Phytoplankton. These photosynthetic unicellular algae supply much of the earth's atmospheric oxygen and, along with zooplankton (b), provide food for other aquatic animals. (From Clark, M. E.: *Contemporary Biology*. 2nd ed. Philadelphia: Saunders College Publishing, 1981, pp. 33, 78.)

> Nonbiodegradable substances dumped into the waters of the world often accumulate in the bodies of living organisms, sometimes resulting in injury and death.

Since plant life in the oceans is limited to the euphotic zone, the deep-dwelling organisms on the ocean floor depend to a great extent upon nutrients that drift down from the pelagic region above.

Along the coastline, where land and sea merge, the *intertidal zone* provides an environment rich in plant and animal life. In rocky areas, starfish, clams, barnacles, and sea anemones cling to the tide-swept rocks, which are coated with a variety of algae (Fig. 12.16).

On the sandy beach, an assortment of small crustaceans and worms burrow into the sand and emerge to feed on the algae and microscopic organisms washed in by the tide.

The major vertebrates of the intertidal zone are several species of shore birds that feed on the burrowing invertebrates.

The various freshwater systems—ponds, lakes, rivers, and creeks—also have unique physical characteristics that determine the types of organisms that can be maintained (Fig. 12.17). Phytoplankton and zooplankton are ordinarily more abundant in the quiet standing waters of ponds and lakes. Moreover, the shallow waters of ponds permit the growth of bottom-dwelling plants that are important as food sources for aquatic animals. The nutrient content of rivers and creeks depends to a great extent upon the speed of the current. Fast-moving streams usually have fewer nutrients and smaller populations of plant life than do slower-moving streams. Algae and tiny insects cling to the rocks in the shallows of fast-moving streams. A few species of fish feed on the insects swept from the rocks by the current.

Compared with the oceans, freshwater ecosystems do not contain as diverse a variety of aquatic organisms, a result of the more limited availability of habitats.

Figure 12.15 Sea lilies and sea cucumbers in the benthic zone of the ocean. (From McCormick, J. M., and Thiruvathukal, J. V.: *Elements of Oceanography*. 2nd ed. Philadelphia: Saunders College Publishing, 1981, p. 212.)

WATER POLLUTION

For decades many of the freshwater systems of the world have been used as dumping sites for a variety of chemicals, acids, pesticides, and other forms of potentially toxic materials. With the addition of such pollutants, natural ecological interactions can be upset or eliminated entirely. It is known that various synthetic materials such as the PCBs (polychlorinated biphenyls) from plastics, along with the acetates, rayons, and many others, are not *biodegradable,* that is, they cannot be broken down into harmless substances by the natural decaying process of microorganisms (Fig. 12.18). Paper and cardboard, on the other hand, are wood products and are readily biodegradable. When released into lakes and streams, many of the synthetic compounds accumulate in the bodies of living things, leading to injury and death. The insecticide DDT, another nonbiodegradable compound, affects the incorporation of calcium compounds into the

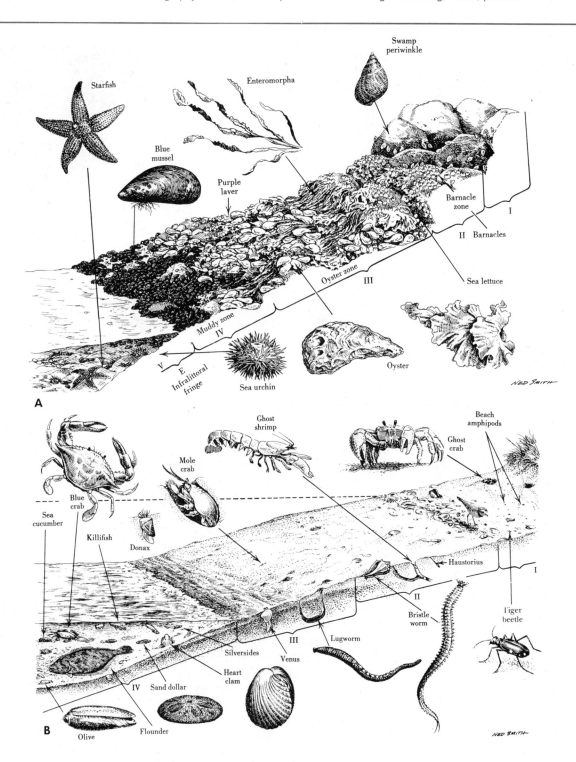

Figure 12.16 The intertidal zone, where land and sea merge, is rich in plant and animal life. (From McCormick, J. M., and Thiruvathukal, J. V.: *Elements of Oceanography*. 2nd ed. Philadelphia: Saunders College Publishing, 1981, p. 209.)

Figure 12.17 Structure of a lake ecosystem. The *limnetic zone,* or zone of open waters, extends downward to the limit of light penetration.

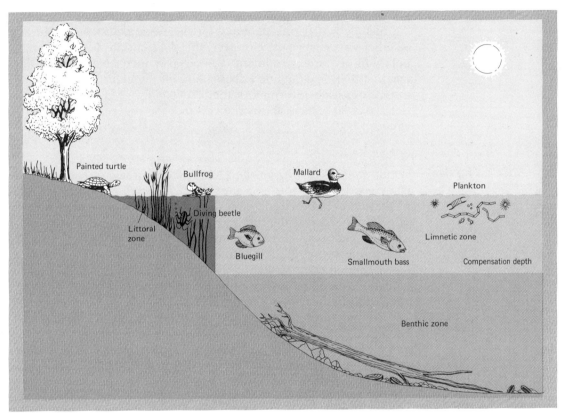

eggshells of birds. As a consequence, the eggshells are thin and usually break before the young have time to develop.

Various organic pollutants added to fresh waters constitute nutrient enrichment and result in a drastic increase in the growth of bacteria. As the bacteria metabolize and multiply, they deplete the amount of oxygen dissolved in the water. Other aquatic organisms, therefore, can be deprived of an adequate oxygen supply. The amount of oxygen required by the bacteria is called the **biological oxygen demand,** or BOD. Since the amount of oxygen used depends upon how much organic matter is present, the BOD is useful as an indicator of the extent of organic pollution.

Another effect that can alter freshwater ecosystems is **thermal pollution.** This results, for example, when heated water from industrial plants is released into lakes and streams. As we have mentioned, most living organisms exist within fairly restricted temperature ranges. Some species, for instance, cannot tolerate temperature changes of more than 2 or 3 degrees. It is possible, therefore, that thermal pollution could eradicate an entire species very quickly.

Figure 12.18 Nonbiodegradable containers. (From Turk, J.: *Introduction to Environmental Studies.* Philadelphia: Saunders College Publishing, 1980, p. 299.)

> Within an ecosystem, each species occupies its own ecological niche; i.e., it has a definite place and function among all other living organisms.

With the world population increasing at a rate of about 60 million people a year, the potential for increased pollution, freshwater or otherwise, is obvious. However, this does not mean that extensive pollution of the environment is necessarily inevitable. What it does mean is that, for pollution abatement, all segments of society need to be realistically informed of the basic biology of ecosystems and of the stakes at hand for themselves and for the generations yet to come.

ECOSYSTEMS: BASIC STRUCTURE AND FUNCTION

Within a given ecosystem there may be only a few or there may be many species of plants and animals, all interacting directly or indirectly with each other and with their environment. Whether we consider the sparsity of the tundra or the luxuriant rain forest teeming with life, there are certain essential ecological relationships that are basic to the existence of all living things. Therefore, every ecosystem has a fundamental structure that determines the possible interactions between living organisms and the interactions these organisms have with their nonliving environment.

We have seen that the various biomes are characterized by certain species of plant life, which to a major extent determine the species of animal life that can be supported. A *species* is a group of organisms with similar structural and functional characteristics that normally breed only with other members of the same species. Together, members of the same species make up a *population,* and populations of different species living in the same area constitute a *community.* A grasslands community, for example, might consist of several species of grasses along with herds of antelope and zebra. A community plus the nonliving physical environment in which it is found constitute an *ecosystem.*

As a part of its respective community, each species of plant or animal occupies or uses what is called its *ecological niche.* This is not the same thing as a habitat, although habitat is certainly involved in the concept. Essentially, ecological niche refers to the unique functions a species may perform within its particular habitat. These functions include how the species obtains its food, what raw materials it utilizes from its environment, its interactions with other species, and so on. Ecological niche is a species' occupation, the total role it plays in nature.

One of the theorems of ecology is that different species in the same area cannot occupy the same ecological niche—one species eventually will dominate or displace the other. This is a statement of what is known as the *principle of competitive exclusion,* or *Gause's principle,* after the Russian biologist G. F. Gause. To keep it simple,

> The interactions and interrelationships between various species result in short-term and long-term effects that tend to stabilize a community.

the principle states that there is only one species in a given ecological niche. Sometimes competitive exclusion is difficult to demonstrate because of the often close similarities between niches.

The maintenance of any community within the biosphere is dependent upon the interactions and interrelationships between the various species of organisms. Such dependency produces both short- and long-term effects that tend, overall, to stabilize the community. The procurement of food, availability of energy, control of population size, reproductive potential, and other factors are normally regulated by many biotic (living) and abiotic (nonliving) processes acting together. These processes constitute the functioning aspect of an ecosystem. How an ecosystem functions is based, of course, upon how it is structured. Our task, then, is to examine more closely the ecosystem in its role as the structural and functional unit of the biosphere. At the same time, we will consider some of the ecological extremes that may influence the physical alteration of an ecosystem.

FOOD CHAINS AND WEBS

In addition to water, the basic commodity vital to every living thing is food. In almost every ecosystem, food is made available initially by the chlorophyll-bearing algae, grass, trees, and so forth through the process of photosynthesis. An exception to this is seen in ecosystems found at great depths on the floor of the ocean. In this eternal darkness, certain chemosynthetic bacteria utilize hydrogen sulfide (H_2S) to provide food for themselves and other benthic species. As we shall discuss, other chemosynthetic bacteria are vital in the cycling and recycling of nitrogen through the ecosystem. In addition, certain photosynthetic bacteria also supply a limited amount of nutrients in various soil and water environments. Bacteria, however, are relatively insignificant in terms of total food production in most ecosystems.

Being able to synthesize their own food, algae, grass, trees, certain bacteria, and so on are referred to as **autotrophs,** i.e., "self-feeders." All other organisms are **heterotrophs,** i.e., "other feeders," and must obtain their nourishment by feeding on the autotrophs or on other heterotrophs.

By capturing and converting the radiant energy of the sun into chemical bond energy, the green plants and algae provide the most important source of useable energy for almost every form of life. The flow of this chemical, or food, energy throughout the ecosystem follows what is called a **food chain**—a sequence of organisms, each of which serves as food for the next organism in the chain (Fig. 12.19). Food chains in both terrestrial and aquatic ecosystems al-

> Food chains always begin with autotrophs and end ultimately with decomposers.

ways begin with the autotrophs. We have mentioned, for example, that the photosynthetic phytoplankton are the major source of food in all marine ecosystems.

The next link in the food chain is the **herbivores** ("plant eaters"), or **primary consumers,** which are heterotrophs that feed directly on the autotrophs. These may include such diverse organisms as insects, birds, or cattle, depending, of course, upon the ecosystem in question.

Animals that feed on the herbivores are called **carnivores** ("flesh eaters"), or **secondary consumers.** The carnivores actually obtain their energy from plants via an indirect route. Fish and frogs, for instance, may devour a plant-eating grasshopper, and a lion may feed on a grazing antelope. Grasshoppers and antelopes, being herbivorous, incorporate nutrients from the plants into their own bodies and, when eaten, pass these nutrients to the carnivores. Still other animals feed on the secondary consumers and thus are referred to as **tertiary consumers.** An owl, for example, may be a tertiary consumer if it eats snake (a secondary consumer) that has devoured a field mouse (a primary consumer). Finally, various animals in the food chain are sometimes **quaternary consumers.** If the owl in the example just cited were eaten by a hawk, the hawk would qualify as a quaternary consumer. These carnivores, then, feed on the tertiary consumers.

The ultimate benefactors at some point in every food chain are the **decomposers,** microscopic bacteria and fungi that break down the remains of plants and animals that have died. All organic matter, in fact, eventually reaches these microorganisms to be decomposed. Even the remains of the decomposers themselves are broken down by other decomposers. The decomposers constitute an irreplaceable link within the ecosystem by returning organic matter to the environment to be used again.

Although the food chain provides a general understanding of the flow of chemical energy within an ecosystem, it is more accurate to speak of such energy flow in terms of a **food web** (Fig. 12.20). Food webs are a number of interconnected food chains that crisscross from one food source to another. Such webs illustrate the fact that a given species may have several available sources of nourishment rather than only one or very few. This is important in that the availability of several food sources tends to stabilize and perpetuate the ecosystem. Obviously, if a species had only limited sources of food, as in a relatively simple ecosystem, such as the tundra, the loss of those sources could be disastrous.

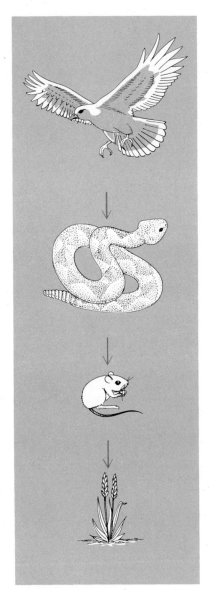

Figure 12.19 Simplified diagram of a food chain.

Through the concept of trophic levels, it is possible to construct an energy pyramid and a pyramid of biomass to demonstrate the relationship between available energy and the biomass that can be supported in a given ecosystem.

Trophic Levels

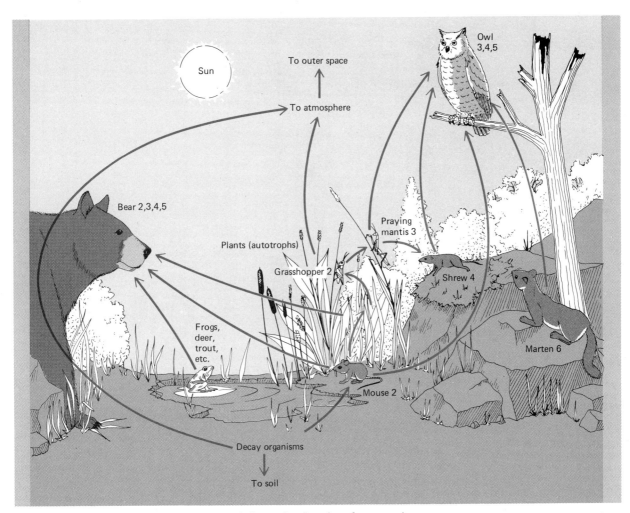

Figure 12.20 A land-based food web. *Arrows* indicate the direction of progressive energy loss at each of the trophic levels (numerals).

TROPHIC LEVELS

The relationships between organisms in a food chain can be explained further in terms of *trophic levels,* or levels of nourishment. The autotrophs, or *producers,* as they are called, constitute the first trophic level; the primary consumers constitute the second trophic level; the secondary consumers constitute the third; and so forth. If we consider the flow of *useable* energy through these trophic levels—that is, the energy available to the organisms—there are some interesting correlations concerning the relationship between energy supply and the maintenance of living things. In moving from the first

General Ecology

> As dictated by the second law of thermodynamics, there is a loss of heat energy at each higher trophic level—a critical factor in limiting the *number* of trophic levels.

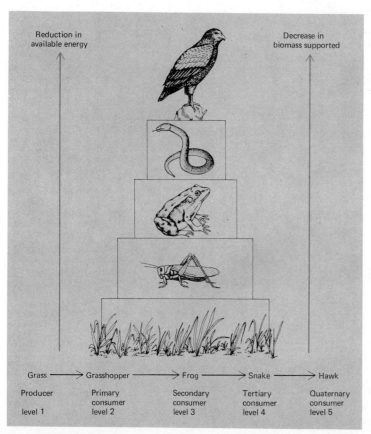

Figure 12.21 Illustration of trophic levels in an ecosystem. The pyramid indicates that, in moving from a lower trophic level to a higher one, there is a corresponding reduction in available energy. Accordingly, less biomass can be supported at a higher trophic level than at a lower level.

trophic level to the second, then to the third, and so on, there is a corresponding *loss* of useable energy from the ecosystem. This means that each trophic level has less energy available to it than the level on which it depends for food. Recalling the second law of thermodynamics (Chapter 4), this reduction in available energy is what you would expect, since some energy is lost from every system as heat.

Ecologists have constructed what is known as an *energy pyramid* to illustrate the distribution of energy between the trophic levels (Fig. 12.21). The producers are found at the base of the pyramid, followed by each successive trophic level on up to the top of the pyramid. The producers, then, have the most energy available to them; the primary consumers have less; the secondary consumers have even less; and so on. As a result of this loss of energy from the ecosystem, it is possible in some cases to construct a *pyramid of biomass* (biomass refers to the total dry weight of all the plants and animals in a given area). This pyramid illustrates the fact that a loss

> The *net primary productivity* represents the energy from green plants that actually is available to all other organisms in the ecosystem.

of energy at each trophic level means that less biomass can be supported. Consequently, the producers usually have the greatest biomass in the ecosystem, followed by the primary consumers, which have a lesser biomass, followed by the secondary consumers, which have still less, and so on.

To illustrate the effect of energy loss, let's consider what actually happens to energy in its course through the ecosystem.

In the process of photosynthesis, green plants utilize the light energy of the sun to form organic products. The amount of *useful* energy derived from photosynthesis, however, is considerably less than the amount of light energy that enters the green plant. According to the second law of thermodynamics, any conversion of energy involves a loss of unavailable heat energy to the environment.

Although an efficient process, photosynthesis uses only about one-tenth of 1 percent of the light energy striking the earth. Of this small fraction, the plants themselves use anywhere from 10 to 50 percent of the energy for their own metabolism. The rest of the useable energy from photosynthesis, called the **net primary productivity,** is available to the various consumers and decomposers. A rough approximation that actually varies considerably in different ecosystems is that, of the energy captured by the organisms at each trophic level, only about 10 percent is captured by the organisms at the next higher trophic level. This is sometimes referred to as the **10 Percent Rule** (Fig. 12.22). Thus, if the net primary productivity of a hypothetical ecosystem amounted to 1000 Kcal, the primary consumers would capture only 100 useable Kcal, the secondary consumers would capture only 10 Kcal, and the tertiary consumers would capture only 1 Kcal. The tertiary consumers, then, would capture only a thousandth of the total available energy stored in the green plants.

At each trophic level about 90 percent of the energy escapes from the ecosystem as heat energy, produced primarily by the metabolic activities of the plants and animals. Obviously, an ecosystem would be short-lived without a continual input of energy from the sun. The availability of energy, then, is a critical as well as a limiting factor in the very existence of all living things. The escape of energy from the ecosystem explains why there is almost never a large number of trophic levels—five is about the usual limit that the availability of energy permits.

It should be noted, however, that the number of trophic levels is also limited by two other major factors. First, some of the organic constituents of organisms at one trophic level are not actually eaten by animals at the next level. These constituents might include fur, bones, pine cones, sharp needles, and various other unpalatable ma-

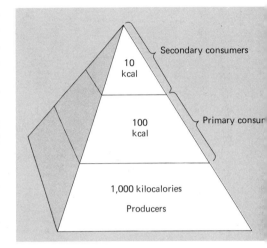

Figure 12.22 Illustration of the 10 Percent Rule. Of a net primary productivity of 1000 Kcal, only 100 Kcal are available to the primary consumers, and just 10 Kcal are available to the secondary consumers. At each trophic level, 90 percent of the energy is dissipated as heat.

> In accordance with the 10 Percent Rule, it requires 10 times the energy expenditure to subsist on meat than on plants.

terials. Second, an animal does not fully utilize all the food it eats. Some of the food is excreted, and although such wastes are available to the decomposers, the energy in this food is no longer directly available to the higher trophic levels.

Knowing that the availability of energy limits the biomass at each trophic level, what is its practical application to human concerns? Human beings, like several other animal species, are both herbivores and carnivores—in a word, **omnivores.** As we have just seen, carnivores have less total food energy immediately available than do herbivores. By feeding on herbivores, the carnivores receive only a hundredth of the energy originally available in the green plants, but the herbivores receive one-tenth of this energy. In terms of energy efficiency, feeding on plants conserves more food energy (calories) than does feeding on animals.

For example, the calories we obtain from steak originally came from green plants. By consuming the plants, cattle utilize some of the calories for their own energy needs and release calories of energy as heat. These calories of heat energy, then, are unavailable to human beings or any other animal that might feed on the steak. In accordance with the 10 Percent Rule, this can be translated into quantitative terms. Generally speaking, it would require about 10 lb of plant food to produce 1 lb of human being, if that person fed directly on the plants. However, if a person became a strict carnivore, feeding on nothing but cattle, for example, this situation changes dramatically. It would take about 100 lb of green plants to produce 10 lb of cow to produce 1 lb of human being. As you can see, it requires 10 times more energy expenditure to subsist on meat than on plants.

What does this mean when applied exclusively to human food consumption? Being an omnivore, man feeds on both producers and consumers, that is, on several trophic levels. However, a more efficient use of food energy would be possible if human beings fed only on the producers. In fact, feeding on plants is the only option for great numbers of the human population. In areas where there is little agricultural land available in the first place, feeding crops to animals and then consuming the animals represents a tremendous loss of food energy, not to mention the loss in the quantity of food. It is apparent, then, that food energy is used more efficiently by human beings if they occupy the second trophic level rather than a higher level.

It is essential to point out, however, that the issue of feeding the human population does not center on food energy (calories) alone. One could obtain the daily requirement of calories by eating nothing but white sugar and still die of malnutrition. The real issue, of

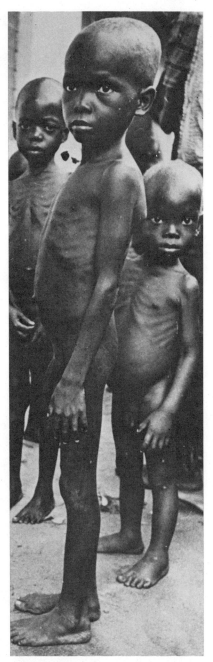

Figure 12.23 These African children are victims of the protein-deficiency disease kwashiorkor. (From Silverstein, A.: *The Biological Sciences.* New York: Holt, Rinehart and Winston, 1974, p. 233.)

> Although energy cannot be reused, matter is continually replenished through many biogeochemical cycles.

course, is the availability of adequate *nourishing* food. A person could suffer from malnutrition on a full stomach if the diet were lacking in vitamins. Similarly, an inadequate intake of protein produces the disease of malnutrition called *kwashiorkor,* so prevalent among African children (Fig. 12.23). On the other hand, even if the diet contained all the essential nutrients, an individual could be undernourished simply from an inadequate *quantity* of food. The tragedy underlying human hunger is that there is sufficient food to satisfy and nourish the human race, but political and economic barriers have often prevented the distribution of food to where it is needed. Thus, the fundamental problem is only partially biological. In fact, over the past 25 years, some countries have dramatically increased food production through the development of new, high-yield grain crops. But even this so-called "green revolution" has had a minimum effect on reducing world hunger. The cultivation of these high-yield grains requires high-energy fertilizers, pesticides, and irrigation systems, often not affordable by poor countries.

In many parts of the world, poverty no doubt is a consequence of overpopulation. Nonetheless, there are political, cultural, or religious concerns that make the prospect of reducing population size unacceptable to many people. Whether food production can keep pace with the relentless growth of world population is uncertain. But this may only beg the essential question: Can we overcome the political and economic barriers that prevent the worldwide distribution of food? For the hungry, an affirmative answer cannot come too soon.

THE CYCLES OF LIFE

With the continual escape of energy, the maintenance of an ecosystem is dependent upon a replenishment of energy from the sun. According to the second law of thermodynamics, energy cannot be recycled and used over again, since all energy conversions result in a loss of nonuseable heat energy that dissipates to the environment. Sooner or later, then, an energy converter such as a living organism would exhaust its energy reserves if no outside source were available.

In addition to energy, living things require a constant supply of matter in the form of elements such as carbon, hydrogen, oxygen, nitrogen, and many others. Like energy, these elements must be readily available for life to continue. However, matter is not continually replenished from some outside source. This is in accordance with another natural law, which states that *matter cannot be created or destroyed.* Known as the **law of conservation of matter,** it refers to the fact that, unlike energy, matter may be used over and over again by

> Photosynthesis incorporates inorganic carbon dioxide into organic compounds used by the plants and by the animals that feed on the plants.

living things. Thus, the earth contains just so much matter—no more and no less.

How is it that matter is never lost? After all, wood is burned, food is consumed, and living organisms die. Do these not represent a loss of matter? Moreover, new forests, foods, and living organisms appear, all requiring an input of matter. Where does this "new" matter come from? The answer involves what are called **biogeochemical cycles,** in which living matter passes from the nonliving world to the living world and back again. In passing through these cycles, matter may assume many forms, but the total amount of matter never changes. Thus, the limited supply of the earth's raw materials is passed around to be used over and over again. These endless, repetitive cycles are the ties that bind all forms of life to the earth from which they came. A brief look at a few of these cycles will illustrate what this means.

The Carbon Cycle. Most of the carbon on the earth is in the form of inorganic calcium carbonate ($CaCO_3$), a salt you would find in such familiar substances as limestone, marble, and chalk. A very tiny percentage is found in fossil fuels (coal and oil) and in often inaccessible dead organic matter deep in the earth or at the bottom of the oceans. The minute, but vitally significant, fraction remaining is circulated through the carbon cycle (Fig. 12.24). We can begin our look at this cycle with carbon dioxide (CO_2), the carbon-containing gas in the atmosphere. If it remained only in its gaseous form, carbon dioxide would be of minimum use to most living things. Conse-

Figure 12.24 The carbon cycle.

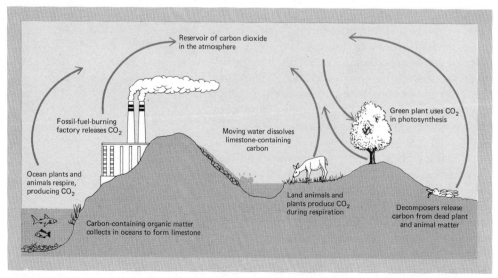

> In one phase of the carbon cycle, carbon dioxide is returned to the atmosphere as a result of the metabolic activities of plants and animals.

quently, before the carbon in CO_2 can be utilized by the living world, it must first be "fixed" into biologically useful compounds. This is initiated by the green plants of the world through the process of photosynthesis.

You will recall that photosynthesis is a series of cellular reactions in which atmospheric CO_2, along with water, is incorporated into the formation of carbohydrates. In the process, the green plants release free oxygen gas (O_2) into the atmosphere. The carbohydrates can be used as an energy source to run the metabolic processes of the plants and as structural components for maintenance, growth, and repair. Any excess carbohydrate not immediately needed is stored within the plants in the form of starch. In addition, carbon also is incorporated into other important organic compounds, such as lipids, proteins, and nucleic acids.

In its continuing journey through the living world, the carbon fixed by green plants may be ingested by herbivores when they feed on the plants. In turn, the carnivores obtain some of the carbon by feeding on the herbivores. Within the bodies of both animal groups, some of the carbon is assimilated to form their essential organic compounds, and some of it is stored.

Within the cells of both plants and animals, some of the organic compounds—carbohydrates, lipids, and proteins—are broken down by glycolysis and cellular respiration. These metabolic processes supply the organisms with energy and, in so doing, produce carbon dioxide as a waste product. Plant cells lose some of this carbon dioxide, and the animals exhale a portion of it back into the atmosphere with every breath. You, of course, perform this latter activity well over 20,000 times a day.

In this one part of the cycle, carbon, in the form of carbon dioxide, has moved from the nonliving world, through the living world in the make-up of living organisms, and back to the nonliving world again as carbon dioxide.

Eventually, the plants and animals die, and their remains are then acted upon by the decomposers. These microorganisms—bacteria and fungi—incorporate the fixed carbon from the dead plant and animal matter. The metabolic processes of the bacteria and fungi also produce carbon dioxide, which is released into the atmosphere. These hidden, seemingly insignificant microorganisms play a crucial role in nature in that the cycling of life's elements simply could not take place without them.

Earlier it was stated that most of the carbon remains in the earth in the form of carbonates. Even in this case, however, some of these salts slowly dissolve in water, where they form carbon dioxide, which escapes into the atmosphere. In addition, dead marine plants and

> Atmospheric carbon dioxide, along with water vapor, acts to shield the surface of the earth against excessive heat loss.

animals often sink to the bottom of the oceans, carrying their fixed carbon with them. At great depths, where there is little oxygen and where the temperature and pressure are suitable, the accumulated carbon of these marine organisms forms deposits of oil, coal, and natural gas. The formation of these energy-rich substances, of course, often requires many millions of years. When man burns these "fossil fuels," a tiny percentage of carbon is returned to the atmosphere in the form of carbon dioxide. In this manner, carbon that has long since been removed is returned to the cycle once again.

In our discussion, there has been repeated reference to carbon dioxide—and with good reason. Although constituting only about 0.03 percent of atmospheric air, carbon dioxide is the single source of carbon for all living organisms. Moreover, it makes yet another vital contribution. Solar energy heats the surface of the earth, and some of this heat tends to radiate back out into space. But atmospheric CO_2, along with water vapor, forms a gaseous blanket or shield around the earth that retards this heat radiation. This aids in preventing excessive heat loss and thus contributes to warming the earth's surface. Without this so-called "greenhouse effect" of carbon dioxide, the average temperature of the earth—about 60°F—would drop some 100 degrees to $-40°F$ (Fig. 12.25).

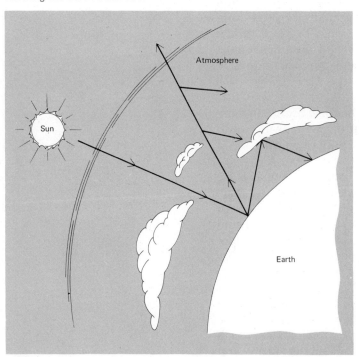

Figure 12.25 The "greenhouse effect." Solar energy heats the earth, and some of the heat energy radiates back out into space. A layer of CO_2 and water vapor in the atmosphere absorbs the heat energy, preventing excessive loss of heat and thereby contributing to warming the earth's surface.

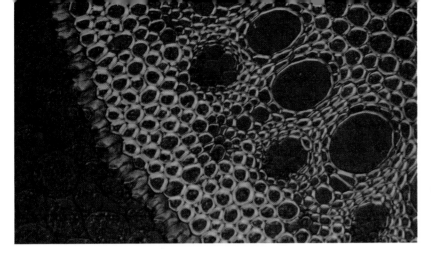

Color Plate 64 Cell layers in a section of a woody stem. (Photograph by Bruce Russell, Biomedia Assoc.)

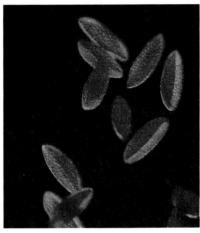

Color Plate 65 Pollen grains from an angiosperm. (Photograph by Bruce Russell, Biomedia Assoc.)

Color Plate 66 Stamens, the male reproductive structures of a flower. Each brightly colored anther (pollen sac) is borne at the tip of a filament. (Photograph by Bruce Russell, Biomedia Assoc.)

Color Plate 67 A pollinating bee on the surface of a flower. (Photograph by Bruce Russell, Biomedia Assoc.)

Color Plate 68 A sunflower, close up. (Photograph by Carolina Biological Supply Co.)

Color Plate 69 St. Johnswort, showing pistil and stamens. (Photograph by Carolina Biological Supply Co.)

Color Plate 70 Cabbage palm, an angiosperm found in Florida. (Photograph by Carolina Biological Supply Co.)

Color Plate 71 A wild fig tree. (Photograph by Carolina Biological Supply Co.)

Color Plate 72 A sponge colony. (Photograph by Bruce Russell, Biomedia Assoc.)

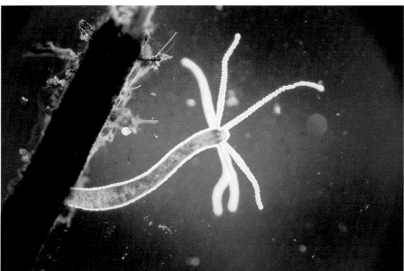

Color Plate 73 *Hydra,* a coelenterate that exists only in the polyp form. (Photograph by Bruce Russell, Biomedia Assoc.)

Color Plate 74 A free-swimming medusa, the "jellyfish" stage found in the life cycles of some coelenterates. (Photograph by Bruce Russell, Biomedia Assoc.)

Color Plate 75 A group of sea anemones. (Photograph by Bruce Russell, Biomedia Assoc.)

Color Plate 76 A coral reef. Corals are polyp coelenterates that secrete a hard limy skeleton. Over the centuries, the skeletons accumulate to form a coral reef, such as this one. (Photograph by Bruce Russell, Biomedia Assoc.)

Color Plate 77 The brown planarian, a nonparasitic flatworm. (Photograph by Bruce Russell, Biomedia Assoc.)

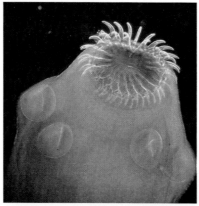

Color Plate 78 A tapeworm scolex with its hooks and suckers. (Photograph by Bruce Russell, Biomedia Assoc.)

Color Plate 79 Vinegar eel, a nonparasitic nematode. (Photograph by Bruce Russell, Biomedia Assoc.)

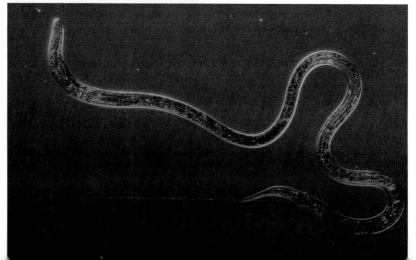

> For some 2 billion years, a reservoir of atmospheric oxygen has been available for use by the earth's living organisms.

Even though its production varies with changing environmental factors, the total content of carbon dioxide in the atmosphere tends to remain fairly constant. There is some indication, however, that with the burning of fossil fuels, the carbon dioxide content of the atmosphere has risen some 0.004 percent in the last 75 years or so.

The Oxygen Cycle. Atmospheric oxygen (O_2) has been present around the earth for about 2 billion years. Its production began with the rise of oxygen-releasing cells in the ancient seas. These early cells are the ancestors of today's green plants, which, through the process of photosynthesis, are the major sources of the earth's atmospheric oxygen.

The oxygen and carbon cycles are closely linked, in part because both involve the processes of photosynthesis and cellular metabolism (Fig. 12.26). Oxygen comprises about 21 percent of atmospheric air at sea level. This reservoir of oxygen, produced mainly by photosynthesizing plants, is available for use by living organisms in cellular metabolism. Within the bodies of plants and animals, oxygen is incorporated into the compounds necessary for life. In animals, oxygen may enter the body through the lungs, gills, skin, or tiny openings in the body wall. Aquatic organisms such as fish, for example, possess gills, which extract oxygen dissolved in the surrounding water.

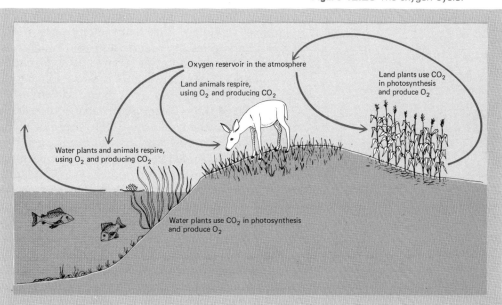

Figure 12.26 The oxygen cycle.

> Ozone, produced by the complexing of oxygen molecules high in the atmosphere, forms a gaseous layer that prevents potentially lethal ultraviolet radiation from reaching the surface of the earth.

The carbon dioxide given off by terrestrial and aquatic organisms is partially composed of oxygen, which is returned eventually to the green plants. In addition, oxygen is a component of inorganic compounds such as water, calcium carbonate, and the phosphates. Oxygen is slowly released from these compounds into the atmosphere, where it is available for use by living things.

Oxygen plays yet another crucial role in the maintenance of life on earth. Although sunlight is the source of life-sustaining energy on earth, some of its component rays—known as **ultraviolet radiation**—are potentially lethal to living organisms. In the upper reaches of the atmosphere, oxygen gas is bombarded by ultraviolet rays, bringing about a complexing of the oxygen molecules to form a layer of ozone, O_3. As the sunlight passes through the ozone layer, most of the ultraviolet rays are absorbed, preventing them from reaching the surface of the earth. Without this protective ozone layer, life on this plant would be impossible. Incidentally, ozone also is generated by the electrical discharge of lightning. You may have detected the pungent odor of ozone immediately after a thunderstorm.

The Nitrogen Cycle. The element nitrogen is a basic constituent of all proteins and nucleic acids. Although comprising about 78 percent of the earth's atmosphere, the gaseous form of nitrogen, N_2, is essentially useless to most living organisms. Nitrogen gas does not tend to form compounds readily with other elements and so must be converted in form, or "fixed," before it can be utilized by most living plants and animals. The fixing of nitrogen is accomplished mainly by certain species of bacteria and to a much lesser extent by some fungi and blue-green algae. Some of the nitrogen-fixing bacteria live free in soil and water, whereas some inhabit the root nodules of plants known as **legumes** (peas, beans, clover, alfalfa, and so on) (Fig. 12.27). All of these bacteria are capable of reducing atmospheric nitrogen (N_2) to the compound NH_3, *ammonia*. Thus, the legumes are immediately supplied with a form of nitrogen they can use, and the free-living nitrogen fixers release ammonia into the soil or water, where it can be taken up by other types of plants.

There is a variety of plants that cannot utilize ammonia and therefore must acquire nitrogen in different forms. These forms may include either **nitrite** (NO_2) or **nitrate** (NO_3). Two different groups of bacteria—called **nitrifiers**—oxidize ammonia to nitrite or nitrate, a process known as **nitrification.**

The various forms of nitrogen are incorporated into organic nitrogen compounds, largely proteins and nucleic acids, within the plant body and within the bodies of animals that feed on the plants.

> Eutrophication may occur artificially through the introduction of fertilizers or sewage into a body of water, or it may occur naturally as nutrients are brought in gradually by wind and water seepage.

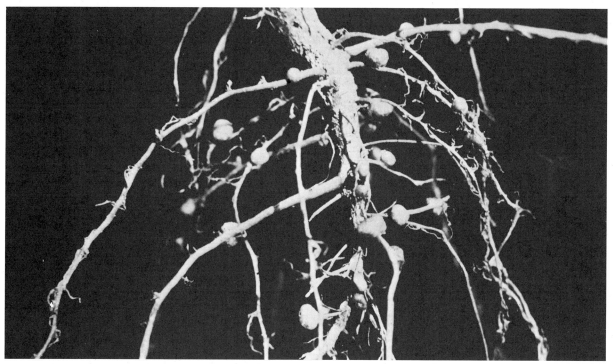

Figure 12.27 Photograph of the roots of a legume (cowpea), showing the nodules inhabited by nitrogen-fixing bacteria. (From Arms, K., and Camp, P. S.: *Biology*. New York: Holt, Rinehart and Winston, 1979, p. 259.)

When the plants or animals die, decomposers break down their nitrogen-containing compounds into ammonia again. Still other bacteria—the ***denitrifiers***—convert the ammonia, and any nitrite or nitrate in the soil, back into atmospheric nitrogen, a process called ***denitrification.*** With the release of N_2 into the atmosphere, the cycle of nitrogen through the living and nonliving worlds is complete (Fig. 12.28).

Ordinarily, the amount of nitrogen fixed by microorganisms is balanced by the amount released into the atmosphere by the denitrifiers. However, with the addition of excessive amounts of fertilizer to the soil and the high concentrations of sewage in certain areas, nitrogen compounds can be washed into lakes and streams, resulting in what is called ***eutrophication,*** a term meaning "well-nourished." A consequence of eutrophication is a drastic overgrowth of algae in a lake or stream. When the algae die, they are acted upon by bacterial decomposers present in the water. In the decomposing process, the respiring bacteria draw on the oxygen dissolved in the water, thereby greatly reducing the amount available to fish and other aquatic organisms. Eventually, these organisms may be wiped out, leaving nothing but satiated bacteria—and stagnant water (Fig. 12.29).

406
General Ecology

> Most plants utilize nitrogen in the form of ammonia (NH_3), or nitrate (NO_3), or nitrite (NO_2).

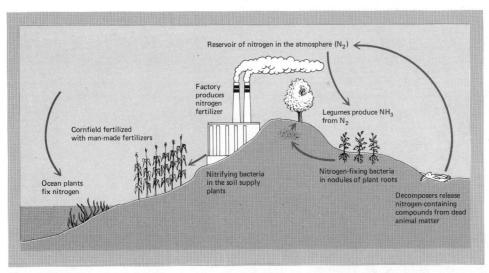

Figure 12.28 The nitrogen cycle.

Figure 12.29 A fish kill such as this one may be the result of eutrophication. (From Turk, J.: *Introduction to Environmental Studies.* Philadelphia: Saunders College Publishing, 1980, p. 273.)

> The water cycle is an endless evaporation—condensation process driven by the energy of the sun.

Although eutrophication may have its harmful effects on an ecosystem through pollution, it is nonetheless a natural process, characteristic of practically all ecosystems. Like all living units, ecosystems age, beginning as relatively unproductive systems and gradually becoming nutrient-rich productive systems. In other words, they become eutrophic. A significant effect of natural eutrophication is the change in the animal and plant life in the area. With the slow but constant influx of nutrients brought in by the wind and by water seepage, the ecosystem is gradually altered both in structure and in function. Although it is possible for natural eutrophication to result in unfavorable changes in an ecosystem, the process ordinarily requires enormous spans of time. However, pollution greatly accelerates the process, often all but destroying an ecosystem in a matter of a few years.

The Water Cycle. Water, like the elements carbon, oxygen, nitrogen, and others, is continually recycled through the living and nonliving worlds (Fig. 12.30). As the most vital liquid on earth, water is found throughout the crust of the earth, in the atmosphere above, and within the bodies of all living things. Supplied with energy from the sun, water continually evaporates from the soil, from the bodies

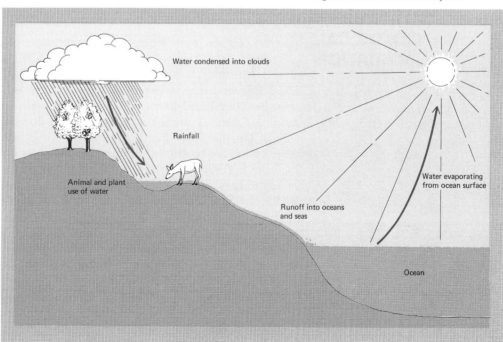

Figure 12.30 The water cycle.

of plants and animals, and from the waters of the world. As the water is heated, it forms water vapor that rises and forms clouds through condensation. When sufficiently cooled, the clouds precipitate the water that falls to the earth as rain or snow.

Some of the water is taken into the bodies of plants and animals, and some trickles down into the soil. Eventually, water deep in the soil finds its way into the lakes, streams, and oceans of the world. From these sources, water evaporates again into the atmosphere.

Many animals release water in the form of urine and as water vapor exhaled with every breath. Land plants lose water vapor through the process of *transpiration,* i.e., the evaporation of water through the stomates.

In the endless evaporation-condensation process of the water cycle, it is possible for various substances dissolved in the water to be redistributed over wide areas of the earth. This has both beneficial and harmful effects. On the one hand, the water cycle serves to distribute some of the nutrient minerals found in the earth; on the other, a variety of chemical contaminants dissolved in or carried by the water can be spread into areas where they may present a hazard to living things. In fact, it is possible for toxic substances to circulate through all the biogeochemical cycles.

BIOLOGICAL MAGNIFICATION

As part of their movement through the various biogeochemical cycles, toxic chemicals must also be passed along many of the food chains. Consider, as an example, a situation in which a certain chemical compound is released into a lake or stream. It is possible that in time the compound may be broken down and incorporated into the natural cycles of matter and thereby rendered harmless to living things. On the other hand, the chemical may not be easily degradable, or it may be altered by natural processes into a different, biologically harmful form. If the chemical is then taken into the bodies of organisms at one trophic level, it may be passed to organisms at higher trophic levels through the food chain (Fig. 12.31). As it is passed along, the chemical tends to become more and more concentrated in the bodies of each successive organism. In effect, the concentration of the chemical becomes magnified as it passes from the bottom of the food chain (the producers, for instance) to the top (such as the tertiary consumers).

If the chemical is a biological hazard, as was the case with mercury in recent years, human beings may experience ill effects from feeding on these top consumers. Organic compounds of mercury, in particular, can be extremely toxic, producing damage to the nervous

> Regulation of the size of a natural population is a confrontation between biotic potential and environmental resistance.

system. Mercury is used industrially in the manufacture of paints, plastics, and pesticides; in polluted waters, mercury is converted to organic compounds by the decomposers.

The increase in concentration of chemicals as they move through the food chain is called *biological magnification.*

REGULATION OF POPULATION SIZE

Being components of all ecosystems, populations of living plants and animals are subject to certain ecological interactions that affect the *size* of these populations. The regulation of population size is a critical factor in the functioning of all ecosystems. In addition to the nutrient and energy relationships just discussed, population size is an essential factor in maintaining and stabilizing the ecosystem.

Almost all species of living organisms have a tremendous *biotic potential.* This refers to the maximum rate of population growth that would occur if all the factors affecting a species' existence were ideal. Such factors include unlimited food resources; freedom from disease, competition, or attacks from other organisms; unlimited breeding opportunities; and so forth. This is a very choice set-up indeed, but unfortunately somewhat outside the realm of reality. There is a collection of environmental and species interactions that result in limiting population growth, the sum of which is known as *environmental resistance.* Every population, then, has a limiting size that represents what is called the *carrying capacity* of the environment. This simply refers to the greatest number of organisms the environment can "carry," or support, for an extended period of time.

Thus, there is a struggle of sorts between the biotic potential of a population and those factors that act against unlimited growth, i.e., the environmental resistance. Such factors are generally separated into *density-independent* and *density-dependent* types ("density," in these instances, refers to the number of organisms living in a given area at any one particular time).

Density-Independent Factors

The primary density-independent factors acting to limit population size can be classified generally as *climatic* and *geological* events. A given physical environment may be altered as a result of famine, fires, storms, and such geological upheavals as volcanic eruptions and earthquakes. These climatic and geological disasters usually affect an entire population regardless of the number of individuals living in the area, i.e., regardless of the density of the population. The result of such density-independent factors acting on a popula-

General Ecology

> The major density-dependent factors regulating population growth are competition, predation, and parasitism.

tion may be that organisms starve or drown or have their habitats destroyed. Extreme changes in temperature also would tend to affect entire populations. You may be aware, for example, of the sharp decline in certain insect populations during the cold winter months. In any of these instances, a small population may be affected as readily as a large one, regardless of the density of either. The result is that over an extended period of time the biotic potential is not realized and population size is held in check.

Density-Dependent Factors

It is practically common knowledge that living organisms cannot exist without adequate nourishment, water, light, or space. As the density of a population increases, *competition* for these resources becomes more and more intense, and some members may simply lose out. Competition for the limited resources of the environment may be confined to members of the *same* population. Such interactions between the same species is referred to as **intraspecific,** i.e., within the species. Competition and other interactions also occur between members of *different* species and thus are **interspecific.** In some forests, for example, trees and other plants of various species may be in competition for light, water, space, and so on in a manner similar to that between members of the same population. In both cases competition results from the increasing density of the populations involved and serves to keep population size in check. Competition, then, is considered to be one of the principal density-dependent limiting factors, both within a species and between different species.

Two other factors that usually are considered along with competition are **predation** and **parasitism.**

Predation. Since all heterotrophs must feed on other organisms, predation is a natural function of all ecosystems. The relationship between predator and prey is often quite interesting and is almost always a delicately balanced interaction. Although it may seem paradoxical, the predator-prey relationship ordinarily is beneficial to both groups. If conditions have been favorable, let us say, the density of a given prey species may increase. As one would expect, the density of the predator species feeding on the prey would also tend to increase. But as a result of increased predation, the density of the prey population may decline. With fewer prey, the density of the predator population after a time also declines. As you will notice, the availability of food (the prey) is a limiting factor on the predator population, and predation tends to limit the density of the prey population.

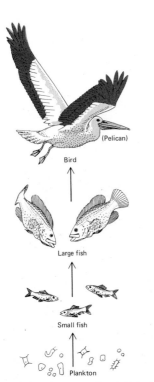

Figure 12.31 An example of biological magnification. A chemical *(colored dots)* may increase in concentration as it moves along the food chain. Often the concentration reaches a fatal level in the top predators of the food chain.

> Parasitism is often a delicately balanced relationship between the demands of the parasite and adaptive resistance of the host.

In the absence of its natural predators, a prey population may reach such a high density that competition for food, space, and other requisites actually may result in a greater loss to the population than if the predators had never been removed.

Such was the case in this country in the state of Arizona during the early part of this century. A population of some 4000 deer living on a plateau near the Grand Canyon increased in number to around 100,000—considerably above the carrying capacity of the area—in less than 20 years. The factor responsible for the 25-fold increase was the intentional killing of thousands of the deer's natural predators—wolves, coyotes, and cougars. There was thus a subsequent increase in the number of deer competing for a limited food supply. Thereafter, nearly 60 percent of the herd eventually died from starvation. This left a population of about half of what could have been supported had the predators not been destroyed.

Parasitism. Parasitism is in many ways similar to the predator-prey relationship. Parasitism, by definition, is an intimate relationship between two different species in which one, the ***host***, is harmed to a greater or lesser extent and the other, the ***parasite***, benefits from the relationship.

Here again, as the host (prey) population increases, it becomes easier for the parasite (predator) to find and victimize members of the host population. Conversely, less victimizing is possible if the density of the host population declines. Eventually, however, the parasite-host relationship must reach a state of balance. For the parasite to survive, it would certainly be counterproductive for it to kill its host. To do so could mean extinction. The host, on the other hand, must be able to exist in reasonable comfort in order to provide the needs of the parasite. What is often seen in nature, then, is that as the host species becomes increasingly resistant to the demands of the parasite, this is balanced by a corresponding adaptive adjustment by the parasite. We know, however, that in some instances parasites do weaken or even kill their host. More often than not, this results from some favorable alteration in the life cycle of the parasite that greatly increases its population size. Consequently, this may lead to massive and thus fatal infestation of the host.

Other Interspecific Relationships

Mutualism. Although not considered to be density-dependent limiting factors, other interspecific relationships, like predation and parasitism, are often essential to the structure and function of a given ecosystem. The intimate association of organisms of different species is referred to as ***symbiosis***, a term meaning "living together."

General Ecology

> Mutualism may occur between two animals, two plants, or an animal and a plant.

One of the more common instances of symbiosis is **mutualism,** defined as a relationship between two different species in which both species benefit. Such is the case, for example, with a type of plant life known as *lichens*. Lichens consist of a green or blue-green alga existing in association with a fungus, usually a cup fungus (Fig. 12.32). The alga, being photosynthetic, provides food for the fungus, and the fungus retains water, which prevents the alga from dehydrating. The fungal mycelium and algal cells form a closely knit mat that grows on tree trunks, bare rocks, and often in a variety of inhospitable regions all over the world. The frozen soil of the tundra, for example, is layered with "reindeer moss," which is actually a type of lichen. In an environment as harsh as the tundra, it is doubtful if either the alga or the fungus could survive on its own.

Another interesting mutualistic relationship exists between the common termite and certain flagellated protozoans that inhabit the termite's intestine. The termite, although capable of chewing and swallowing wood, cannot digest it. The actual digestion of the wood

Figure 12.32 Lichens. (a) A leafy lichen growing on rock. (b) Reindeer "moss," which grows on soil. Lichens are formed by a mutualistic relationship between an alga and a fungus. The photosynthetic alga provides food, and the fungus retains water, preventing dehydration. (From Ebert, J. D., Loewy, A. G., Miller, R. S., and Schneiderman, H. A.: *Biology.* New York: Holt, Rinehart and Winston, 1973, p. 592.)

> Some symbiotic relationships are obligatory for one or both species involved; others are only facultative, since each species could survive on its own if necessary.

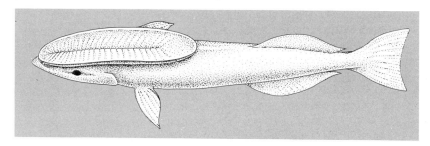

Figure 12.33 The remora fish, or "shark sucker," showing its modified dorsal fin used for attachment to the body of a shark. The remora and the shark share a commensalistic relationship in that the remora benefits by being transported and provided with food, whereas the shark is unaffected.

is accomplished by enzymes released by the protozoans. Thus, the termite benefits from the breakdown of the wood, and the protozoans are provided with food plus a safe habitat.

In human beings, the large intestine normally harbors a large population of mutualistic bacteria. The most common of these bacteria is *Escherichia coli,* or simply *E. coli,* a species widely used in biological research. Intestinal bacteria are vital to human life in that they are involved in the synthesis of vitamin K, a constituent necessary for the normal clotting of blood.

One other common type of symbiosis is called **commensalism.** Roughly translated, the word means "at the table together." More exactly, commensalism is defined as a relationship between two different species in which one member benefits and the other is unaffected. The so-called "shark sucker," or remora fish, attaches itself to the body of a shark and gains a free ride (Fig. 12.33). Moreover, when the shark feeds, the remora dines on the leftovers and returns to the shark to await the next meal. Through it all, the shark apparently is totally indifferent to and unaffected by its freeloading companion.

Commensalism is widespread in the natural world: Barnacles attach themselves to the bodies of whales, a significant means of dispersing their species; various fish and other aquatic organisms inhabit the burrows of larger sea animals; and birds build their nests in trees.

Some symbiotic relationships are absolutely necessary for the survival of one or both of the participants. These relationships are said to be *obligatory,* i.e., one or both organisms are obligated to live together. Certain lichens are classic examples of obligatory symbiosis. Other types of symbiosis, such as the shark-remora relationship, are not mandatory for the survival of either species and are referred to as *facultative.* The involved species have the faculty, or ability, to live together, but they need not necessarily do so.

Some of the symbiotic relationships designated as parasitism, mutualism, or commensalism often are difficult to determine. Just how one of the species in a certain relationship is affected may be

> Ecological succession involves a sequence of community changes that progress from simple to increasingly complex stages, culminating in a climax community.

open to debate. What can be gleaned from the brief survey here, however, is that the interspecific relationships within a community can be quite varied and that they serve their ecological function of maintaining over the long term the stability and overall welfare of the community itself.

ECOLOGICAL SUCCESSION

At first glance, it might appear that a given ecosystem is a rather static, unchanging unit. However, lush meadows or thickets often come to occupy an area that was once only an open, barren field. Equally striking is the appearance of a variety of plant and animal life in a region that was initially a nearly lifeless stretch of sand dunes. Such changes in communities can occur in almost any area of the biosphere. These changes, ordinarily requiring great spans of time, come about gradually and result from both environmental and organismal influences. The process of community change is referred to as *ecological succession* (Fig. 12.34).

Environmental effects, such as changes in climate, shifting winds, volcanic eruptions, and so on, are important to succession in that they alter the physical surroundings and thereby have a direct influence on the types of plant and animal life that will inhabit the area. Succession also involves alterations of the environment caused by incoming organisms or by those already in the area. In time, a species can so alter the environment that it can no longer occupy the area. Such alterations, however, often make the area more suitable for a different species. The new species in turn may repeat the alteration process and eventually be replaced by another species, and so on. Essentially, then, succession involves a modification of habitats brought about to a large extent by the organisms themselves. The result is the replacement of one dominant species by another.

The process of succession eventually ends with the formation of a *climax community.* In contrast to the communities that preceded it, a climax community is stable and self-perpetuating. This is because a climax community does not tend to upset or destroy the conditions necessary for its own survival. The central factor apparently responsible for the stability of climax communities is diversity. Having more ecological niches, a greater number of different species, more intricate food webs, and a greater biomass relative to other communities, a climax community is thus less likely to be disrupted by short-term changes in the environment.

The arctic biome is a classic example of succession. Glacier movement in that area produced at one time a surface of smooth,

Ecological Succession

> The relative stability of a climax community apparently is attributable to its greater diversity as compared with the communities that preceded it.

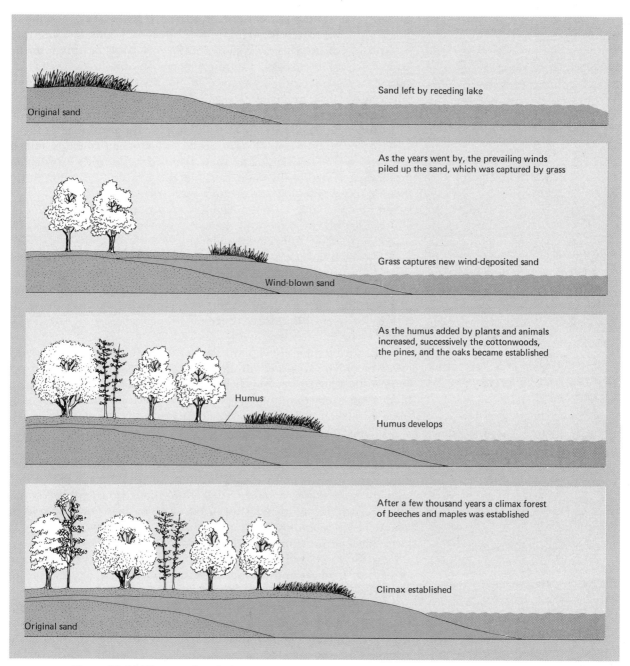

Figure 12.34 Diagram of ecological succession leading to the establishment of a climax community.

> As with individual organisms, climax communities are considered to be homeostatic, since they display a relative constancy or balance.

barren rock. As the temperature rose and fell, the rocks expanded and contracted, often breaking into fragments. Water, seeping into the crevices of the rocks, expanded as it froze, further aiding in the breakdown of the rocks. In time, lichens, blown in by the wind, gained a foothold on the surface of the rocks, establishing themselves as the first, or *pioneer,* stage of succession.

The lichens released corroding acids that further broke the rocks into small particles, forming a base for the production of soil. With the eventual death of some of the lichens, their remains, mixed with the rock particles, continued the soil build-up. Once established, patches of soil made possible the invasion of wind-blown moss spores, initiating the second stage of succession. The established moss plants, through cellular secretions and finally decay, added organic enrichment to the soil. Eventually, the growth of the mosses produced a thick, absorbent mat.

The next invaders—seeds of various herbs—fell into the mat beds and germinated, sending their roots deeper into the soil. Because of the cold climate, more discriminating and complex forms of plant life, such as shrubs and trees, could not survive; the community of herbs, therefore, represents the final stage of succession in the arctic biome.

In more temperate climates, additional stages of succession may occur in which various communities of grasses, shrubs, and trees become established after the lichens, mosses, and herbs have taken hold. With each community replacing the previous one, the environment is substantially altered. Each species adds its waste products and chemical secretions to the soil. Root systems rearrange the soil, affecting aeration and water drainage. The decomposition of the bodies of dead plants and animals adds organic nutrients to the soil, aiding the growth of new and different forms of plant life. Along with the succeeding stages of vegetation, the various species of animal life capable of being supported also change. For example, it would be quite impossible for the lichen community to support a large bird population. Most land birds require seeds or berries or worms for food, and trees in which to nest. With succession, then, come new habitats and the opportunity for varied ecological niches (Fig. 12.35).

With the progressive succession of communities, an area eventually becomes modified to the extent that further stages of succession are unlikely. When this occurs, the climax stage has been reached. If not affected by extreme environmental factors, either geologic or climatic, the climax community may persist for centuries. In a word, it can be said that a climax community is **homeostatic,** i.e., it normally displays a relatively constant ecological balance. The

> The reverse of succession is called *simplification*, in which a community regresses to less complex stages. Climatic changes, forest fires, nutrient depletion, and other factors contribute to simplification.

Figure 12.35 (a) A marsh evolving into a meadow. (b) A meadow evolving into a forest. (From Turk, J.: *Introduction to Environmental Studies*. Philadelphia: Saunders College Publishing, 1980, p. 38.)

world biomes previewed earlier may be considered as major climax formations.

Climax communities, then, may be relatively simple, as in the tundra, or quite complex, as in the case of the tropical rain forest. Whatever the area, the type of climax community is determined by the physical make-up of the environment. There is, for example, a vast difference between temperature extremes, amounts of rainfall, soil types, and so forth in the tundra and rain forest. The total time for succession to progress from the pioneer stage to the climax community may in many instances require hundreds or sometimes thousands of years.

REVIEW OF ESSENTIAL CONCEPTS

1. *Ecology* is the study of the interactions between living organisms and their nonliving environment and between the organisms themselves.
2. The *biosphere* includes that part of the crust of the earth and the atmosphere capable of supporting life.
3. Each of the major terrestrial *biomes* is characterized by a particular type of vegetation, climate, and soil.
4. Oceanic ecosystems include several areas where life is most abundant: the pelagic zone, benthic zone, euphotic zone, and intertidal zone.
5. Freshwater ecosystems are less diverse than oceanic ecosystems because of the more limited availability of habitats.
6. Water pollution may result from the presence of nonbiodegradable compounds, organic nutrients, and heated water.
7. The *biological oxygen demand* (BOD) is an indicator of the extent of organic pollution.
8. *Ecological niche* refers to a species' total role in nature; no two species can occupy the same ecological niche.
9. A *food chain* consists of a sequence of organisms, each of which serves as food for the next organism in the chain; *food webs* are a number of interconnected food chains.
10. The relationships between trophic levels of an ecosystem can be illustrated by an *energy pyramid* and a *pyramid of biomass*.
11. An energy pyramid illustrates the loss of energy that occurs at each trophic level; because of the loss of energy, a pyramid of biomass illustrates that less biomass can be supported at each higher trophic level.
12. Generally, only 10 percent of the total energy available to one trophic level is available to the next higher trophic level. This is an expression of the *10 Percent Rule*.
13. The issue of feeding the world population centers on economic and political factors, in addition to biological factors.
14. The *law of conservation of matter* states that matter cannot be created or destroyed.
15. Among the important biogeochemical cycles are the carbon, oxygen, nitrogen, and water cycles.
16. Biogeochemical cycles conserve matter by circulating nutrients repeatedly through the living and nonliving worlds.
17. Toxic materials tend to accumulate in concentration as they are passed along a food chain, a process known as *biological magnification*.
18. Population size is controlled by environmental and species interac-

tions known as *environmental resistance*. These interactions result from *density-independent* and *density-dependent* factors.
19. The primary density-independent factors are *climatic* and *geologic* events.
20. Density-dependent factors may be intraspecific or interspecific; these include *competition*, *predation*, and *parasitism*.
21. *Symbiosis* is an intimate relationship between different species; examples of symbiosis include *parasitism*, *mutualism*, and *commensalism*.
22. *Ecological succession* is a process of one community replacing another over a certain period of time. The final stage of succession is the *climax community*, characterized by being homeostatic, i.e., in relatively constant ecological balance.

APPLYING THE CONCEPTS

1. Discuss the general characteristics of the biosphere.
2. Comment on the basic characteristics of each of the terrestrial biomes.
3. Describe the oceanic life zones.
4. What would be some essential differences between oceanic and freshwater ecosystems?
5. What are the biological effects of introducing nonbiodegradable materials into the environment?
6. Explain what is meant by biological oxygen demand.
7. Differentiate between habitat and ecological niche.
8. Outline a hypothetical food chain.
9. Explain the biological significance of a pyramid of energy and a pyramid of biomass.
10. What is the application of the 10 Percent Rule to a given ecosystem?
11. Explain how the carbon, oxygen, nitrogen, and water cycles illustrate the law of conservation of matter.
12. What is the "greenhouse effect"? What is its significance to life on earth?
13. Discuss the causes and effects of eutrophication.
14. What is biological magnification? How does it relate to human concerns?
15. Discuss the factors responsible for regulating the size of natural populations.
16. What is symbiosis? Discuss the major types of symbiosis.
17. Discuss the factors contributing to ecological succession. Why is a climax community said to be "homeostatic"?

13 origins and evolution

THE ESSENTIAL OBJECTIVES

You have understood this chapter when you are able to:
1. Describe the theory of the origin of the earth.
2. Name the sources of available energy on the primitive earth.
3. Discuss the major factors involved in chemical evolution.
4. Discuss the significance of the Miller-Urey experiment.
5. Discuss the possible origin of living cells as they relate to proteinoid microspheres.
6. Describe the general characteristics of the original cells on earth.
7. Discuss the rise of autotrophs on the early earth and explain its significance to organic evolution.
8. Discuss the evolutionary significance of the first eukaryotic cells.
9. Outline briefly the developmental history of evolution from the early Greeks to Darwin.
10. Explain what is meant by catastrophism and inheritance of acquired characteristics.
11. Discuss the essential factors involved in the theory of evolution by natural selection.
12. Discuss the application of mutation, recombination, and gene flow to the theory of evolution.
13. Discuss industrial melanism and its application to evolutionary theory.
14. Discuss the significance of the Hardy-Weinberg Law.
15. Determine the gene frequencies and genotype frequencies of a given population using the Hardy-Weinberg formula.
16. Discuss, using examples, the practical application of the Hardy-Weinberg Law.
17. Discuss the general evolutionary history of man from *Dryopithecus* to *Australopithecus*.
18. Discuss the general evolutionary history of the genus *Homo*.

The earth was formed from a whirling mass of gaseous elements, some of which, after the cooling of the earth, became the constituents of living matter. With the eventual rise of living cells, organic evolution began and, over great spans of time, led ultimately to the flourishing of living things.

PREVIEW OF ESSENTIAL TERMS

proteinoid microsphere A hypothetical protocell model consisting of an amino acid polymer (proteinoid) surrounded by a water droplet.

protocell Any primitive, prelife cell type that may have been present on the early earth.

catastrophism The formerly held theory that extinction of species and principal geological changes on the early earth were the result of a series of sudden catastrophes.

adaptation The ability of organisms to adjust to changes in their environment, resulting in their becoming better suited for survival.

natural selection The process whereby certain hereditary traits or variations advantageous to an organism tend to be perpetuated in succeeding generations.

synthetic theory of evolution The combination of evolutionary theory and genetics.

gene flow The movement of genes from one population to another as a result of dispersal of organisms, seeds, pollen, and so on.

gene pool All of the genes or genotypes present in a population.

Hardy-Weinberg Law A statement that the frequency of each member of a pair of alleles remains constant generation after generation in a population that is not evolving.

anthropoid Manlike; referring to the higher primates, as gorillas, monkeys, orangutans, and so forth.

> *Organic evolution* is the process of change in populations of living organisms over a period of time. The diversity and unity of all living things is unified in the concept of organic evolution.

INTRODUCTION

There is no single principle more essential to a unified view of the living world than that of evolution. The word is from the Latin *evolvere,* meaning "to unroll or unfold." Although we sometimes speak of the evolution of the automobile, styles of dress, ideas, and so on, our primary concern in biology is with *organic* evolution—the "unrolling or unfolding" of life. More exactly, organic evolution is basically a process of change in populations of living organisms over a period of time.

Among living things there is great diversity—in size, shape, adaptive characteristics, and so forth—but there is also a fundamental similarity. For example, all living organisms are composed of certain elements and chemicals, all are made up of cells, all use energy in the same basic manner, and so forth. In a phrase, then, life is characterized by both diversity and unity. This might seem like an unlikely phenomenon—unless there were some unifying concept through which this diversity and unity could be reconciled. We find this unifying concept in the theory of organic evolution. Specifically, one of the cornerstones of the theory is that all species of living organisms are descendants of a common form of life. The unity of life, therefore, can be understood in terms of a common ancestry. On the other hand, diversity among living things has resulted from their capacity to adapt to the earth's many environments.

In Chapter 2 it was pointed out that living matter and nonliving matter are fundamentally alike at the chemical level. Through the process of chemical evolution when the earth was very young, interactions between various forms of nonliving matter gave rise to primitive cells, the original forms of life. At this point, organic evolution had begun. There is, then, an evolutionary bond between living things and the earth itself—a bond that was formed millennia ago.

THE ORIGIN OF THE EARTH

Although there is no consensus regarding the origin of the earth, there is impressive evidence to support the view that planetary systems are a by-product of star formation. According to this theory, the sun, the planets, and other bodies in our solar system were formed some 4.6 to 5 billion years ago from an enormous whirling cloud of gas and dust (Fig. 13.1). Evidence indicates that the gases present were primarily hydrogen and helium; the dust particles consisted of metals and various other elements. Apparently, the temperature of this whirling cloud initially was quite cold. Once the cloud reached a critical size, it began to contract. As its density increased, the massive cloud began to contract to form a protostar (which be-

The Origin of the Earth

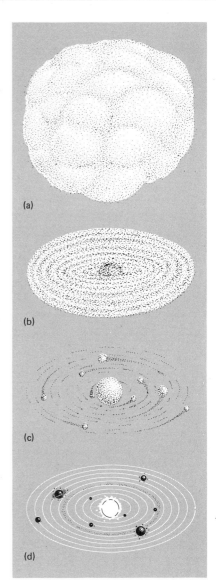

Figure 13.1 Solar system formation. On reaching a critical size and density, a massive cloud of interstellar gas and dust begins to contract (a) forming a protostar (b), and smaller bodies called planetesimals (c). The protostar eventually contracts into a star (our sun), and the planetesimals clump together, forming first protoplanets and then planets and moons (d). The smaller planetesimals from asteroids, meteors, and comets. This whole process probably took place between 4.6 and 5 billion years ago.

came our sun) and a large number of small bodies called planetesimals. As the protosun contracted, it became hotter and its spinning motion increased. This spinning motion threw off more gas and dust, forming more planetesimals, which eventually clumped together to form protoplanets. These protoplanets further contracted, becoming either planets or moons. Leftover dust and debris formed the minor members of the solar system: asteroids, meteors, and comets. The sun eventually contracted to its present size and is the primary source of heat and light for our solar system.

With time, the earth became increasingly hotter, reducing its various metallic elements to a nearly molten state. Deep within the earth, various gases—principally hydrogen, nitrogen, carbon monoxide, and water vapor—were trapped under enormous pressure. Through volcanic eruptions, these gases were spewed outward to become the first constituents of the earth's primitive atmosphere. Water vapor in the atmosphere condensed and returned to earth as torrential rainstorms, forming lakes and seas and muddy pools.

As the temperature of the earth continued to increase, various elements in the interior were reduced to a molten state. Iron, nickel, and sulfur formed the core of the earth, while the mantle largely contained compounds consisting of silicon dioxide (SiO_2), magnesium oxide (MgO), and ferrous oxide (FeO). This is essentially the composition of the mantle and core of the earth today (Fig. 13.2).

During the stage of core and mantle formation, the atmosphere of the primitive earth began to change (Fig. 13.3). Nitrogen gas (N_2) came to predominate, whereas carbon dioxide (CO_2), water vapor, methane (CH_4), and other gases were present in smaller amounts. There was, however, no more than a trace of free oxygen present.

To complete the scenario of activity on the primitive earth, we must consider the probable sources of available energy—energy that

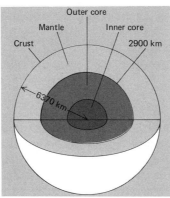

Figure 13.2 The three major layers of the earth: crust, mantle, and core. During the formation of the earth, heavier elements, such as iron, nickel, and some sulfur, became concentrated in the core. Lighter elements, such as silicon and magnesium, constituted part of the mantle. The lightest elements of which living things are composed—carbon nitrogen, phosphorus, and so on—are found predominantly in the crust.

Origins and Evolution

Chemical interactions between the atmospheric gases of the early earth produced organic compounds—some of which became essential to the evolution of living matter. Energy for these interactions came from the sun (in the form of visible light, heat, and ultraviolet radiation), electrical discharge of lightning and geothermal activity (in the form of volcanic eruptions and radioactive decay).

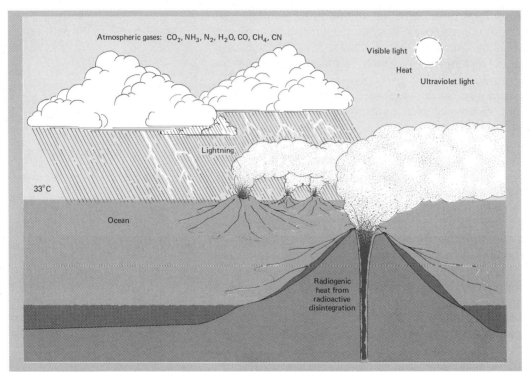

Figure 13.3 A diagram of the earth about 3.5 billion years ago. The atmosphere consisted mainly of CO_2, N_2, and H_2O, with several other gases also present, but no O_2. The surface temperature was hot; shallow oceans formed. Radiant energy from the sun, radiogenic heat from the interior, and lightning all provided potential energy sources for synthesis of organic molecules.

was indispensable to the origin and evolution of life. From the beginning of its formation, the earth was bombarded by the ultraviolet radiation of the sun. For 2.5 billion years this energy was available to drive the reactions of chemical evolution. With the violent rainstorms, the electrical discharge of lightning also provided enormous amounts of useable energy. Heat energy from the outgassing of volcanoes, along with decay of radioactive elements, also contributed as significant sources of energy. Driven by these forms of energy, interactions between the atmospheric gases of the early earth produced a variety of simple organic compounds. Using the same energy sources, these simple compounds reacted with each other to form larger, more complex organic compounds—some of which became the essential constituents of living matter.

CHEMICAL EVOLUTION

As the ancient rains fell, the earth slowly began to cool; as a consequence, there were violent upheavals and bucklings of the earth's crust, forming deep depressions and tall mountains. Rainwater filled the depressions, creating oceans, lakes, and ponds.

As we have noted, a variety of gases were spewing forth from volcanic activity in the earth's interior. At a suitable temperature,

> *Chemical evolution* is the formation of increasingly complex organic compounds over enormous periods of time.

and supplied with energy, these gases interacted to form many other gases. For example, three molecules of hydrogen gas (H_2) react with a molecule of carbon monoxide (CO) to form methane (CH_4) and water (H_2O); also, nitrogen gas (N_2) interacts with hydrogen to form ammonia (NH_3). With these and other atmospheric gases interacting with each other, there is the potential of forming a variety of simple organic compounds. The bonding of carbon, hydrogen, and oxygen atoms, for instance, produces compounds such as acetic acid, formic acid, and formaldehyde; by including nitrogen in the reactions between carbon and hydrogen atoms, compounds such as hydrogen cyanide (HCN) are formed. In turn, reactions between these simple "intermediate" compounds produce more complex organic compounds. These include a variety of amino acids, the building blocks of proteins; fatty acids, the basic constituents of lipids; nitrogen bases, the components of nucleotides; and many other important organic compounds.

Thus, over great spans of time, interactions between simple chemicals led to the formation of increasingly complex organic compounds—a process described as ***chemical evolution*** (Fig. 13.4). Is the case for chemical evolution merely speculative, or is there evidence that these reactions could have occurred on the primitive earth? For more than 80 years, researchers have performed laboratory experiments that provide conclusive evidence that such reactions are in-

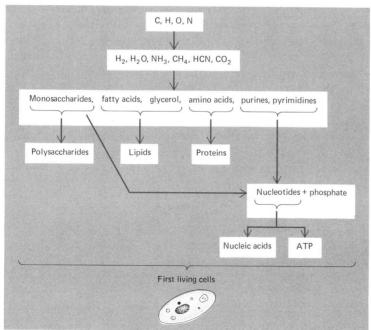

Figure 13.4 An abbreviated sequence of the steps of chemical evolution. Supplied with energy from heat, lightning, ultraviolet radiation, and radioactivity, a few basic elements on the primitive earth interacted to form a variety of simple compounds, which led to the formation of the fundamental organic compounds of life.

Origins and Evolution

> The Miller-Urey experiment proved that small organic compounds could be created from a mixture of simple gases to which energy (in the form of electrical discharges) had been added.

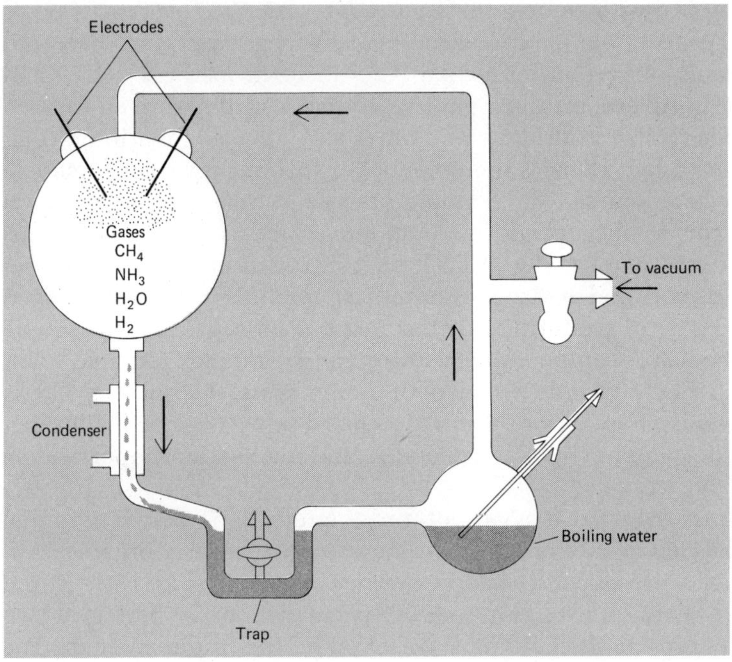

Figure 13.5 Reaction chamber used in the Miller-Urey experiments. Conditions within the chamber simulated those thought to have existed on the primitive earth. Supplied with electrical energy, the circulating gases eventually formed an aqueous mixture containing a variety of simple organic compounds.

deed possible. One of the classic experiments was devised by Stanley Miller and Harold Urey of the University of Chicago in 1953.

In the Miller-Urey experiment, ammonia, hydrogen and methane gases, along with water vapor, were released into a reaction chamber from which air had been removed (Fig. 13.5). The mixture of gases was circulated through the chamber, past electrodes that supplied continuous electrical discharges of energy. After several days, the mixture was removed and analyzed. A variety of small organic compounds, such as amino acids, urea, and an array of organic acids, were recovered from the mixture and identified. Many of these compounds are essential constituents of living matter. Over the years, the Miller-Urey experiment has been refined and repeated, and new chemical evolution experiments have been developed and carried out—all yielding essentially the same results.

Once formed, the products of chemical evolution, along with the various inorganic constituents, were washed into the lakes and seas, creating, as it were, a warm, dilute chemical broth. With abundant sources of energy available through ultraviolet radiation, lightning, and heat, the stage was set.

THE ORIGIN OF LIFE

In the opening sentence of Chapter 7, we quoted the aphorism of Virchow that "All cells come from cells." The fact that all cells arise

> *Proteinoid microspheres exhibit a number of cell-like characteristics but are not "alive" in the biologic sense because they cannot synthesize protein and reproduce.*

from pre-existing cells is one of the cornerstones of the cell theory. The cell theory, however, does not explicitly address the origin of the first cell. As we have seen, chemical evolution led to the formation of a variety of organic compounds, many of which are vital constituents in the make-up of living things. Current theory holds that, given these organic products, along with sufficient energy sources, an oxygen-deficient atmosphere, and enough time, living cells could have arisen on the early earth.

Several *prebiotic* ("before life") structures somewhat resembling living cells have been prepared in the laboratory. One such structure, called a **proteinoid microsphere,** was demonstrated by Sidney W. Fox at Florida State University (Fig. 13.6). Fox heated a dry mixture of amino acids, which resulted in the formation of amino acid polymers (chains) resembling proteins. Fox referred to these protein-like structures as *proteinoids*. Adding water to proteinoids produces the tiny droplet-like proteinoid microspheres. These bodies exhibit several structural and functional characteristics of modern cells. Structurally, they possess a double-layered protein boundary suggestive of a cell membrane, and they also have a heterogeneous internal structure like that of cytoplasm. Functionally, microspheres take up small molecules from the surrounding water, exhibit osmotic properties by swelling or shrinking, and divide in a manner similar to budding in yeast. Moreover, they are capable of a simple metabolism, and some microspheres display a type of movement by sometimes repelling each other and at other times joining together.

Thus, on the basis of experimental evidence, it would have been possible for amino acids adrift in the ancient seas to have been washed onto hot rock, where they polymerized into proteinoids, which, when washed into the sea, became proteinoid microspheres. However, even though this sequence could have occurred, micro-

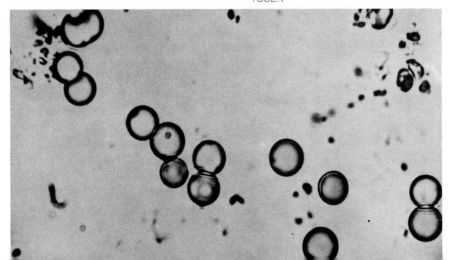

Figure 13.6 Proteinoid microspheres. These hypothetical "protocells" are formed by adding water to prepared amino acid polymers. (From Arms, K., and Camp, P. S.: *Biology*. 2nd ed. Philadelphia: Saunders College Publishing, 1982.)

> The acquisition in early cells of a genetic machinery based on protein synthesis and the self-duplication of DNA marks the beginning of life on earth.

spheres are a long way from being living cells. Essentially, newly formed microspheres lack two fundamental mechanisms displayed by living cells: protein synthesis and reproduction.

As time passed, the microspheres, or *protocells* ("primitive cells") as they are called, probably accumulated various chemicals and organic compounds from the surrounding broth. Perhaps one of the first compounds taken up for energy use was the nucleotide adenosine triphosphate (ATP), which could have been formed through chemical evolution relatively easily. With ATP available, protocells could then carry out increasingly complex energy-requiring reactions. In time, however, the supply of available ATP probably ran low, but it is postulated that by then some protocells could synthesize ATP by breaking down organic nutrients acquired from the surroundings.

Since the first protocells acquired nourishment from an outside source, they were therefore *heterotrophs* ("other feeders"). Metabolism of the acquired foodstuffs most assuredly occurred by *fermentation*, a process discussed in Chapter 6. Remember that on the primitive earth there was no free oxygen, and fermentation is the only way an anaerobic organism can break down foodstuffs to derive energy. Modern cells such as yeast, for example, ferment glucose to yield carbon dioxide, ethyl alcohol, and a relatively small amount of energy. Animal cells, too, ferment sugar; the end-product, however, is lactic acid, not ethyl alcohol. The early protocells, then, had to depend upon nutrient materials that had accumulated in the surrounding chemical broth. These materials were metabolized anaerobically by fermentation to yield carbon dioxide, ethyl alcohol or lactic acid, and ATP.

The nucleic acids constituted one vital group of organic compounds that came to be associated with the proteinoid microspheres. As discussed in Chapter 8, nucleic acids are the carriers of hereditary information and are instrumental in the synthesis of cellular proteins. The available evidence suggests that the various nucleotides that compose nucleic acids were taken up by the protocells from the surrounding broth. Once inside the protocells, the nucleotides became associated with and arranged in sequence by the proteinoids. This is a reversal of what is seen in modern cells, in which the sequence of nucleotides determines the amino acid sequence of the protein.

It has been suggested that the earliest nucleic acids formed in the protocells were probably ribonucleic acid (RNA) molecules and that deoxyribonucleic acid (DNA) appeared at a later time. Eventually, the interactions between proteinoids, RNA, and DNA led to a mechanism of protein synthesis involving the self-duplication of

> *Prokaryotes* are the simplest cellular organisms, lacking a nucleus and other cellular organelles.

> *Autotrophs*, cells capable of synthesizing their own food from carbon dioxide, evolved from the early heterotrophs.

DNA. Thus, through a slow process of accumulation and assimilation of organic compounds, early cells acquired a crude but functional genetic machinery. As you are aware, one of the essential criteria of life is the capacity to pass hereditary information from parent cells to offspring. With the evolution of a genetic machinery, the saga of life had begun.

At this point, then, the picture that emerges of the earliest life form, or "protobiont" ("primitive life"), is one of a microscopic, single-celled, aquatic heterotroph sustaining itself by means of anaerobic fermentation. Lacking a true nucleus and other cellular organelles, it was thus a *prokaryote*, somewhat similar to modern bacteria. In fact, the earliest records of primitive life are fossil bacteria over 3 billion years old (Fig. 13.7).

As with any heterotroph, one of the crucial problems that confronted the protobionts was the need for a continuous supply of food. Had they consumed all the nutrients available in the surrounding broth, they would have soon died, making the story of life only a brief episode. It is intriguing, however, that the metabolic activities of the protobionts themselves were to save the day. You will recall that one of the by-products of fermentation is the gas carbon dioxide. As time progressed, increasing quantities of carbon dioxide were released by the protobionts into the seas and atmosphere. With the continuing depletion of organic nutrients, it would be advantageous for an organism to evolve that could utilize carbon dioxide to synthesize its own nutrients, i.e., an *autotrophic* ("self-nourishing") organism. Evidence indicates that such an organism did indeed evolve from a population of ancient heterotrophs.

EARLY FORMS OF LIFE

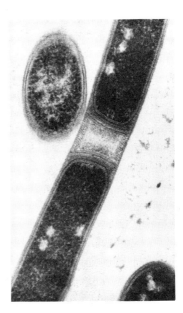

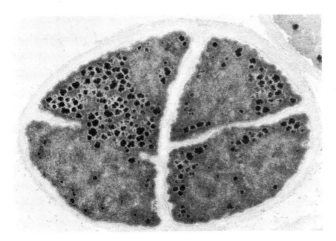

Figure 13.7 Photomicrograph of an archaebacterium, a fossilized 3.5–billion–year–old single–celled organism. (From Pasachoff, J. M.: *Contemporary Astronomy*, 2nd ed. Philadelphia: Saunders College Publishing, 1981).

Origins and Evolution

> Primitive blue-green algae (like their modern-day counterparts) were also autotrophs that used carbon dioxide and water to produce free oxygen through the process of photosynthesis. It is this production of oxygen that changed our atmosphere from a reducing one to an oxidizing one and is an event that has led to the great diversity of life that we know today.

Utilizing carbon dioxide, autotrophic organisms were capable of synthesizing their own organic compounds in a manner similar to that of photosynthesizing plants. The initial autotrophs were probably *chemosynthetic* bacteria, which employed a variety of inorganic compounds as hydrogen donors (see Chapter 9). Later, some bacteria acquired the ability to use water as a source of hydrogen, just as green plants do today. Many present-day autotrophic bacteria have survived virtually unchanged from their original ancestors.

That autotrophic organisms became diversified early in organic evolution has been evidenced by the discovery of fossil blue-green algae, some of which may be nearly as old as fossil bacteria (Fig. 13.8). The primitive blue-green algae, like some of the bacteria, were photosynthetic, utilizing carbon dioxide and water in the process and releasing free oxygen (O_2) into the atmosphere. This latter activity was one of the pivotal events in the story of evolution—an event that would lead to the flourishing of living things.

THE PRIMITIVE ATMOSPHERE

The primitive atmosphere of the earth has been described as a *reducing* atmosphere, meaning that the element hydrogen was in great abundance (recall that the addition of hydrogen to a substance is termed reduction). We have mentioned, for example, that some of the gases present were hydrogen (H_2), methane (CH_4), ammonia (NH_3), and water vapor (H_2O), all of which consist of hydrogen. Oxygen is a highly reactive element. Had it been present in large quantities in the primitive atmosphere, it would have reacted with and destroyed the newly formed organic molecules. This, incidentally, is one reason why new life forms do not arise from nonliving chemicals today.

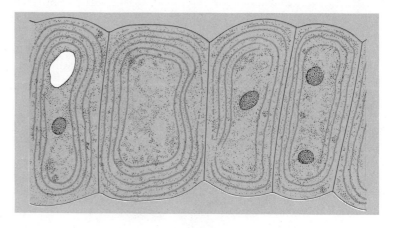

Figure 13.8 Section of a blue-green alga filament. The first photosynthetic prokaryotes on the primitive earth may have resembled this organism.

> Oxygen released into the atmosphere by the primitive autotrophs created the protective ozone layer and provided the opportunity for a more efficient method of energy production in living organisms—*aerobic respiration*.

With the release of oxygen by the primitive autotrophs, the atmosphere of the earth began to change. In the outer reaches of the atmosphere, oxygen molecules, bombarded by the ultraviolet rays of the sun, interacted to form a layer of ozone (O_3). With its formation, the ozone layer absorbed the lethal ultraviolet rays, preventing them from striking the earth. For millions of years, life had been restricted to the seas, where the water surrounding the early life forms protected them from the sun's deadly rays. But the ozone layer made possible the emergence of terrestrial life and new opportunities for organic evolution.

An oxidizing atmosphere had yet another profound effect on the evolution of life—it provided the opportunity for *aerobic* respiration. Anaerobic respiration, i.e., fermentation, is an extremely wasteful process, yielding only two ATPs from the breakdown of one molecule of glucose (see Chapter 6). By comparison, aerobic respiration normally produces 36 ATPs from the complete metabolism of a single glucose molecule. Thus, aerobic respiration is a highly efficient process that provides living organisms with a large and immediate supply of energy. The result is that aerobic organisms can grow and multiply more rapidly, increase in structural and functional complexity, and thereby more effectively exploit their environmental options.

EARLY EUKARYOTIC CELLS

Even with the advent of photosynthesis and aerobic respiration, the early heterotrophs and autotrophs were still prokaryotes, devoid of organelles and unable to undergo either mitotic or meiotic cell division. Today, the two major groups of prokaryotes—bacteria and the blue-green algae—still retain these primitive characteristics.

Fossil evidence indicates that roughly 1 billion years ago, cells arose that possessed a nucleus and perhaps a variety of organelles. These cells, apparently representing types of green algae, are thus the earliest known types of *eukaryotic* cells. How did eukaryotic cells arise? Although a great deal of controversy surrounds the subject, it has been suggested that organelles such as chloroplasts and mitochondria were once independent prokaryotic organisms that established a symbiotic relationship with another prokaryotic cell. Apparently, the chloroplasts and mitochondria were engulfed by the host cell, setting up a mutualistic relationship. Although there is some evidence to support this theory, many scientists are at odds concerning the interpretation of the evidence. A final resolution of the disagreement probably is impossible, but continued investigation should at least narrow the options.

Figure 13.9 Stages in the evolution of life on earth. (a) Formation of simple organic molecules through the complexing of atmospheric gases, using electrical and ultraviolet radiation energy. (b) Formation of organic aggregates (protobionts). (c) Stage of heterotrophic prokaryotes undergoing anaerobic respiration and releasing carbon dioxide into the atmosphere. (d) Rise of photosynthetic autotrophs, producing an oxidizing atmosphere and the development of the ozone (O_3) layer. (e) Appearance of first animals, utilizing aerobic respiration. (f) Appearance of multicellular organisms and the invasion of land.

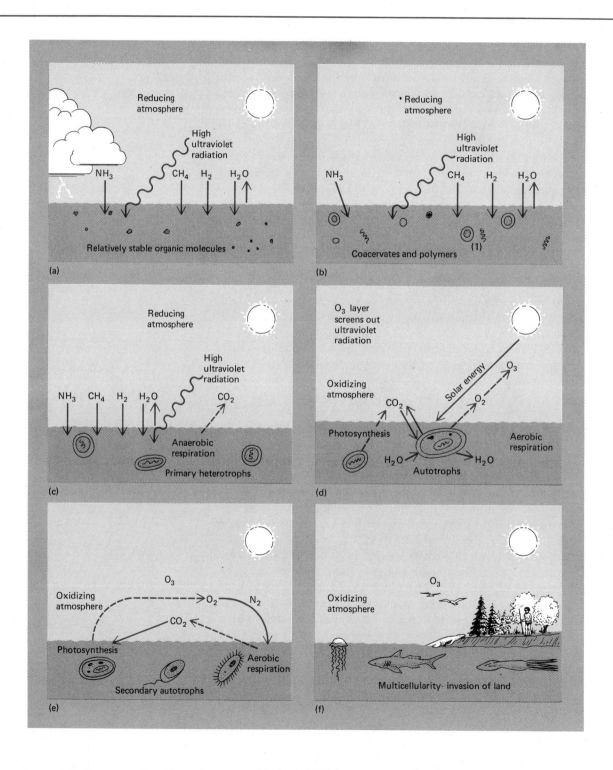

> *Mitotic cell division* involves the passing of the full chromosome complement (diploid number) from parent cell to daughter cells.

> *Meiotic cell division* is the process by which a multicellular organism forms sex cells, or gametes.

The rise of eukaryotic cells on the earth was one of the essential stages in organic evolution (Fig. 13.9). Indeed, these cells held the key to the subsequent evolution of all higher forms of life. That key was *mitotic cell division*. You will recall from Chapter 7 that mitosis involves the passing of the full chromosome complement (diploid number) from parent cell to daughter cells. Some of the primitive eukaryotes employed mitosis in asexual reproduction, producing a genetically identical lineage that persists to the present day. In turn, single-celled organisms capable of mitotic division eventually gave rise to the first *multicellular* organisms a little less than 1 billion years ago.

It was with the appearance of muticellular organisms that organic evolution began to proceed at a rapid pace. Multicellular organisms developed the ability to form sex cells, or gametes, through the process of *meiotic cell division,* which is thought to have evolved from mitosis. Gametes contain only half the number of chromosomes (haploid number) as all other body cells and are formed only by sexually reproducing organisms. When gametes from opposite mating types join at fertilization, the new diploid organism, or zygote, receives heritable characteristics from both parents. Sexual reproduction, then, leads to *genetic variability,* i.e., new and different combinations of genes—and *genetic variability is the essential key to evolutionary advancement,* a point we shall discuss later. Thus, it was the rise of multicellular organisms capable of sexual reproduction that greatly accelerated the rate of evolutionary progress.

The evolutionary "explosion" that followed the appearance of multicellular organisms eventually led to the diversity of life present on the earth today. The time required for this progression—some 1 billion years—is amazingly short in terms of the geological time scale. The first land plants, for example, appeared only 400 million years ago, the dinosaurs arose about 175 million years ago, and pre-humans walked the earth just 5 million years ago. We can establish the times of these developments through the fossil record (Fig. 13.10).

EVOLUTION: AN HISTORICAL BACKGROUND

Most people are probably unaware that evolution is an ancient historical concept, contemplated, however crudely, as far back as the early Egyptians and Chinese. Later, pre-Socratic Greeks such as Anaximander and Empedocles alluded to the possibility of cosmic and biological evolution. Although openly discussed for centuries thereafter, evolution was eventually buried under the intellectual stagna-

Origins and Evolution

> Geological studies of rocks and fossils provided the first evidence to support evolutionary theory.

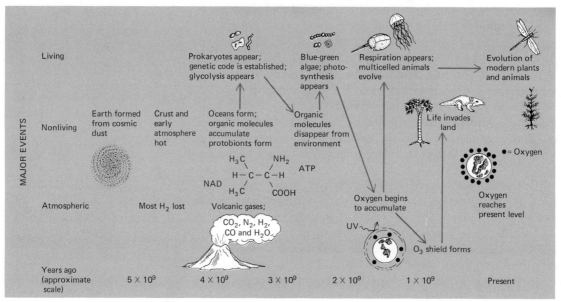

Figure 13.10 The major events hypothesized to explain the origin of life on earth. The time scale is only approximate with respect to some events. *Arrows* indicate what are believed to be cause-and-effect interactions. Many other important events, such as the formation of sedimentary rocks, changes in the shape of the ocean beds, and formation of the continents, are omitted.

tion of the Middle Ages. Its revival was not to begin until the eighteenth century.

Throughout history, opposition to the concept of evolution has been based largely on religious grounds. Such opposition was founded on the belief in the "fixity of species" concept, i.e., the belief that all living things were created in their present form and did not share a common ancestry. An accompanying belief was that the earth had come into its present state through one sudden, explosive act of creation.

During the eighteenth century, the science of geology began to provide evidence that the earth had a much longer history than the 6000 years calculated by the Christian theologians. In addition, fossil discoveries indicated that the history of living organisms was one of diversity and change and, quite importantly, that it accompanied the developmental history of the earth. Much of the fossil evidence was uncovered by the English geologist William Smith (1769–1839). In his study of the earth's surface, Smith found that each different rock layer he examined contained distinctive groups of fossils. His observations attested to the extreme age and formative process of the earth and to the relationship between these factors and the history of life.

Other geologists confirmed Smith's findings, and man's view of the world began to change. In its initial stages, then, evolutionary theory received its greatest impetus from the science of geology. Geology not only revealed the ancient and developmental history of the

> Lamarck's theory of evolution was based on the concept of inheritance of acquired characteristics. Simple observation tells us that acquired characteristics cannot be inherited, and we now know that variations in future generations result from genetic factors.

earth but also provided evidence for the mutability, or changeableness, of living things. Both of these precepts were later to be found indispensable to an integrated theory of evolution.

Even though the geological revelations of the eighteenth century were impressive, there were many scientists of the time who were adamantly opposed to any type of evolutionary theory. One of the staunchest and most influential opponents was the French zoologist Georges Cuvier (1769–1832) (Fig. 13.11). Although an expert on fossil animals, Cuvier was a proponent of the "fixity of species" concept and refused to accept the idea that organisms could change with time. His interpretation of the fossil record was that species had been eliminated by a series of climatic and geological catastrophes. The world was then repopulated by special acts of divine creation. This theory, known as *catastrophism,* dominated the thinking of many scientists throughout the early part of the nineteenth century.

Another Frenchman, Jean Baptiste Lamarck (1744–1829), disagreed with the theory of catastrophism and proposed that all species are descended from pre-existing species (Fig. 13.12). In 1802 Lamarck published his own theory of evolution, the central theme of which centered on the concept of *inheritance of acquired characteristics.* According to this theory, an organ or body structure becomes larger, more functional, stronger, and so on with use and smaller, less functional, weaker, and so forth through disuse. Whether obtained through use or disuse, these acquired characteristics are hereditary and will be transmitted to the offspring. The classic illustration of Lamarck's theory involves the "explanation" of how the giraffe acquired its long neck (Fig. 13.13). The theory proposed that short-necked ancestors needed to stretch their necks to feed on leaves high in a tree. This inner "need" resulted in longer necks that were then passed to the offspring, which stretched their necks, and so on until all descendants had very long necks. Thus, by the use of a structure (the neck), the giraffe acquired a characteristic (a longer neck) that could be inherited by the offspring.

At first glance, Lamarck's theory may seem reasonable, but simple observation alone indicates that acquired characteristics cannot be inherited. For example, a 97-lb "weakling" who later acquires a "Mr. Universe" physique through weightlifting does not pass his acquired muscles to his offspring. Similarly, should one accidentally lose a finger, there is no evidence to indicate that succeeding generations will also be fingerless. Today we know that only the information coded in DNA can be passed from parent to offspring. Therefore, variations observed in future generations result not from acquired characteristics but from genetic factors. What all living things inherit, then, is a particular set of genes (genotype) and not

Figure 13.11 Georges Cuvier (1769–1832), French zoologist and proponent of the theory of catastrophism. (From Stansfield, W.: *The Science of Evolution.* New York: Macmillan, 1977.

Figure 13.12 Jean Baptiste Lamarck (1744–1829). Although Lamarck recognized the fact of evolution, his theory of evolution based on the inheritance of acquired characteristics cannot be supported scientifically. (From Arms, K., and Camp, P. S.: *Biology.* 2nd ed. Philadelphia: Saunders College Publishing, 1982.)

> Darwin's concept of *natural selection* describes a process whereby certain hereditary traits or variations advantageous to an organism tend to be perpetuated in succeeding generations.

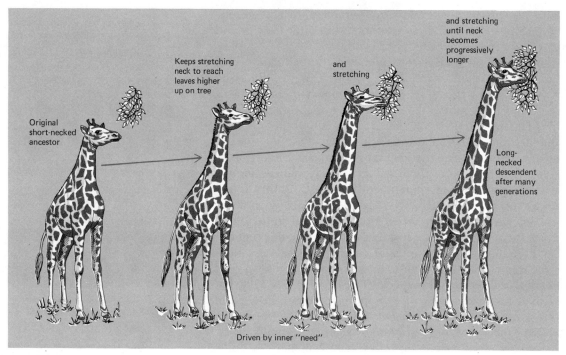

Figure 13.13 Illustration of Lamarck's concept of inheritance of acquired characteristics.

necessarily any set of physical appearances (phenotype). One could inherit genes for dark skin, but one could not inherit a suntan.

DARWINIAN EVOLUTION

The most notable figure ever associated with evolutionary theory was the English naturalist Charles Robert Darwin (1809–1882) (Fig. 13.14). Darwin's major contribution to the theory of organic evolution was his explanation and application of the concept of **natural selection.** Basically, natural selection is the process whereby certain hereditary traits or variations advantageous to an organism tend to be perpetuated in succeeding generations. In a general sense, natural selection can be described by the phrase "survival of the fittest."

Darwin based his theory of evolution by natural selection on observations he had made while on a 5-year voyage aboard the scientific research ship H.M.S. *Beagle*. Signing on as the ship's naturalist, Darwin spent his time collecting animal and plant life and taking notes on everything he saw.

After his return to England in 1836, Darwin continued to accumulate data on evolution by natural selection, an endeavor he was to pursue for more than 20 years. In 1858, Darwin received a letter from Alfred Russell Wallace (1823–1913), a fellow naturalist who had also formulated a theory of evolution by natural selection. So

The theory of evolution by natural selection is often referred to as the Darwin-Wallace Theory.

similar were the theories of the two men that they agreed to publish their findings simultaneously. In 1858, their papers appeared in the same issue of a British journal. Thus, the theory of evolution by natural selection is often referred to as the Darwin-Wallace Theory.

After 23 years of exacting observation and research, Darwin presented his work publicly in his book *On The Origin of Species by Means of Natural Selection, or the Preservation of Favored Races in the Struggle for Life*. Published on November 24, 1859, all available copies were sold the first day. Although an immediate "success," the book created a storm of controversy; there were impassioned confrontations, not only between scientists and the public but also between the scientists themselves. Fundamentalists were outraged; catastrophists scorned the book; and the Church attacked the theory with vengeance. Darwin was even accused of asserting that man had descended from the apes, a proposition that does not even appear in his work. Indeed, Darwinian evolution had provoked an uproar that has subsided only a little in over 120 years.

Figure 13.14 Charles Robert Darwin (1809–1882). In the Darwinian theory of evolution, the prime force is natural selection. (From Silverstein, A.: *The Biological Sciences*. New York: Holt, Rinehart, and Winston, 1974, p. 501.)

THE THEORY OF NATURAL SELECTION

One of the fundamental tenets of evolutionary theory is that great spans of time are required in the formation of new species. In Darwin's day, it was widely believed that the earth, according to the genealogy of Adam in the Bible, was about 6000 years old. For Darwin, as well as his opponents, this was much too brief a span for any major evolutionary change to occur. On his around-the-world voyage aboard the *Beagle*, Darwin read a newly published book entitled *Principles of Geology*, written by Charles Lyell (1797–1875) (Fig. 13.15). Lyell did not agree with the theory of catastrophism but contended that the earth, through the effects of natural causes, had undergone a process of long, continuous change. Darwin seized upon the idea, reasoning that living organisms might also have undergone a similar change. Armed with Lyell's theory, Darwin could proceed with the conviction that there was, indeed, time enough.

Given that there was adequate time for organisms to evolve, what natural process might effect their evolution? For Darwin, that process was *natural selection*. Although the concept did not originate with Darwin, he showed that natural selection was the essential force behind organic evolution. In arriving at this conclusion, Darwin considered the following:

1. There are many more organisms born in any population than will survive or reach reproductive age.

2. Over long periods of time, however, the size of most natural populations tends to remain fairly constant. This is because there

Figure 13.15 Charles Lyell (1797–1875). Lyell's book *Principles of Geology* had a profound effect on Darwin in developing the theory of evolution by natural selection. (From Arms, K., and Camp, P. S.: *Biology*. New York: Holt, Rinehart and Winston, 1979, p. 800.)

> Environmental factors drive the process of natural selection.

are limited environmental resources, and many individuals die before reaching reproductive age.

3. Individual organisms in a given population have variable characteristics, i.e., they are not all exactly the same in their anatomical, physiological, and behavioral make-up.

4. Because of certain characteristics, some organisms have a better chance of surviving and reproducing than others, i.e., some organisms are better adapted to their environment than are others.

5. Favorable characteristics tend to be transmitted to the offspring in greater proportions than characteristics of the less adapted organisms.

Over great spans of time, the cumulative effects of these factors could produce new species of organisms. If, as we have stated, natural selection can be described as "survival of the fittest," what determined who is "fit" and who is not? In other words, what is it that "selects" some organisms and not others? The answer is the *environment*. The environment is all of the forces that affect an organism, such as climate, availability of food, disease, enemies, and so on. Depending upon the variable characteristics of organisms, environmental conditions favor the perpetuation of some of these characteristics (that is, they are selected for) and tend to eliminate the others.

This can be illustrated by using again the example of the evolution of the long neck in giraffes (Fig. 13.16). According to Darwinian evolution, the ancestral population of giraffes probably exhibited inherited variations in neck length. Some necks were a little longer than others, and some were a little shorter. Once the leaves and fruit on the lower levels of a tree had been consumed, it was only the longer-necked individuals that could reach food at the higher levels. Consequently, it would be more likely that the longer-necked individuals could survive. These organisms would then tend to perpetuate their kind in greater numbers than would the shorter-necked

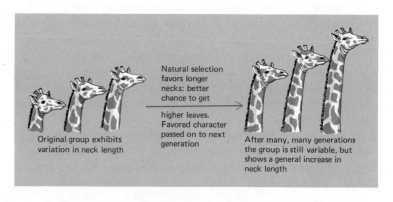

Figure 13.16 Illustration of the Darwinian concept of evolution by natural selection. Compare with Figure 13.13.

> Natural selection does not *produce* variation in organisms but acts in a manner more favorable to some variations than to others.

individuals. Through repetition of this process over many generations, the descendant giraffe population would have longer necks than the ancestral population.

It is important to point out that natural selection does not *produce* variation in organisms; rather, it acts in a manner more favorable to some variations than to others. The result is a high level of adaptability of a population to its environment.

Although the theory of organic evolution by natural selection confirmed that change was the rule in nature, that new species could evolve from old ones, that the effect of the environment on variation was the essential key, Darwin nonetheless could not answer with convincing evidence the most crucial question: What is the *source* of the variations observed in natural populations? It was to be decades before the answer emerged.

THE MODERN THEORY OF EVOLUTION

That organisms within a species are variable is usually obvious; we need only observe a group of people, a herd of cattle, or a forest of pines to confirm the fact. Although Darwin considered variation within a population to be of vital significance in evolution, he knew practically nothing of the mechanism of heredity responsible for variation.

In 1866, 7 years after the publication of Darwin's *Origin of Species*, the genetic discoveries of Gregor Mendel (see Chapter 8) appeared in the Proceedings of the Natural History Society in the town of what was then Brunn, Austria. Apparently, Darwin was unaware of Mendel's work, and even those who read of it were almost totally unimpressed. Thereafter, Mendel and the basic laws of heredity were all but forgotten for 35 years.

During the last quarter of the nineteenth century, advancements in cell biology led to the discovery of chromosomes and the processes of mitosis and meiosis. By the turn of the century, Mendel's basic laws were "rediscovered" and subsequently were incorporated into the *Chromosomal Theory of Heredity*, i.e., the theory that hereditary traits are determined by genes, which are located on the chromosomes. The Dutch botanist Hugo de Vries (1848–1935), one of the discoverers of Mendel's work, contributed the concept of **mutations** to the chromosomal theory (Fig. 13.17). An increasing number of biologists were delving into the mysteries of heredity, confirming Mendel's principles and establishing new ones. In 1908, G. H. Hardy and William Weinberg applied hereditary principles to populations, formulating what is now known as the **Hardy-Weinberg Law**. We shall discuss this important concept later in this section. In his studies on the fruit fly *Drosophila* in 1910, Thomas Hunt Morgan

Figure 13.17 Hugo de Vries (1848–1935), Dutch botanist who contributed the concept of mutations to the science of genetics. (From Stansfield, W.: *The Science of Evolution*. New York: Macmillan, 1977.)

> The *synthetic theory of evolution* involves the interaction of natural selection and heredity.

Figure 13.18 Thomas Hunt Morgan (1866–1945), American geneticist who introduced the concept of sex-linked traits. (From Stansfield, W.: *The Science of Evolution*. New York: Macmillan, 1977.)

(1866–1945) introduced the concept of sex-linked traits and provided further evidence that genes are located on chromosomes (Fig. 13.18). The infant science of genetics was indeed on its way.

Although genetics began to make tremendous strides in the early twentieth century, the study of evolution was largely ignored until 1935. However, in the decades that followed, advances in genetic and biochemical research revealed the structure and function of DNA; the deciphered genetic code attested to the fundamental unity of all forms of life; and new dating techniques confirmed the antiquity of the earth and its first living inhabitants. As a consequence, evolution again came to the forefront, but this time as the essential principle reconciling the unity and diversity of life. Darwinian evolution coupled with the science of genetics provided an integrated, more comprehensive interpretation of the history of life on earth. This modern view of evolution, then, involves the interaction of natural selection and heredity, which together constitute the so-called **synthetic theory of evolution.**

THE BASIS OF THE SYNTHETIC THEORY OF EVOLUTION

Charles Darwin was unable to account for variation within a population because the principles of inheritance were unknown to him. He was convinced that evolution—the process of change in populations from generation to generation—resulted from the action of natural selection on heritable variations within a population. But he did not know the *source* of these variations nor its relationship to the

The Basis of the Synthetic Theory of Evolution

> A *gene mutation* is an alteration in the nucleotides of DNA; a *chromosome mutation* is similar but involves extended segments of the DNA molecule.

reproductive process. He knew nothing about chromosomes, genes, DNA, or mutations.

Today, we know that every living thing possesses some type of chromosomal material, that genes—the units of heredity—are located on the chromosomes, that genes are actually certain sequences of nucleotides in a segment of DNA, and that a mutation is some inheritable change in the DNA molecule. With this knowledge, we can define evolution in a general sense as the process of *genetic* change in populations from generation to generation. Thus, variation among organisms in a population results initially from particular genetic changes. Still, the essential question remains: What is the source of genetic, i.e., inheritable, variations? Basically, there are three sources: (1) mutation, (2) recombination, and (3) gene flow. We shall consider briefly the evolutionary application of each of these.

Mutation

As we saw in Chapter 8, mutations are genetic errors involving a change in a gene or a change in a chromosome. Mutations may result from exposure to radiation, various chemicals, and so on, but most mutations occur spontaneously for reasons not as yet known.

A ***gene mutation*** is an alteration in the nucleotides of DNA. Such an alteration may involve the *addition* of a nucleotide, the *loss* of a nucleotide, or a *substitution* of one nucleotide for another. As a consequence, a mutant DNA consists of a sequence of nucleotides different from that of the original DNA.

The two types of DNA will direct the synthesis of different enzymes and other proteins. The inheritance of the mutant DNA is important in the evolutionary process *if its effect brings about some change in the offspring.* In other words, a mutation must alter the phenotype (physical expression or appearance) of an organism before it can be determined if the mutation is of adaptive value in a given environment. If the mutation results in an advantageous variation, natural selection would dictate the perpetuation of the variation in succeeding generations. If the variation conferred a disadvantage on the organism, making it less able to adapt to its environment, the organism may not survive long enough to reproduce, and its DNA will be eliminated from the population.

Chromosome mutations are essentially similar to gene mutations except that extended segments of the DNA molecule are involved rather than individual nucleotides (Fig. 13.19). The segments are actually broken pieces of chromosomes that have been lost, or they may reattach to the same chromosome in an inverted position, or they may become attached to another chromosome. In some cases,

> *Recombination* is the method by which genes are brought together in new combinations with other genes.

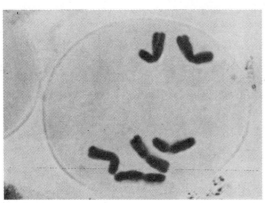
(a)

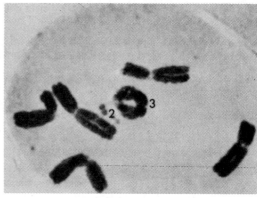
(b)

Figure 13.19 Chromosomes of the microspores of the spiderwort, *Tradescantia*. (a) Normal chromosomes. (b) Abnormal chromosomes, after irradiation. Note abnormal ring (3) and fragments (2). (From Clark, M. E.: *Contemporary Biology*. 2nd ed. Philadelphia: Saunders College Publishing, 1979, p. 429.)

entire chromosomes or groups of chromosomes may be lost or gained. In any of these instances, either the number or the arrangement of the genes is affected. The genetic consequences of chromosome mutations are basically the same as for gene mutations.

Recombination

Recombination refers to the methods by which genes are brought together in new combinations with other genes. In meiosis, crossing over—i.e., the exchange of genetic material between homologous chromosomes—results in new combinations of genes on each homologue. In addition, when homologous chromosomes separate from each other at anaphase I, their distribution into the gametes is independent of any other pair of homologues. For example, an organism having the genotype *AaBb* could produce gametes containing *AB, Ab, aB,* or *ab* allele combinations, each of which has an equal chance of being formed. Thus, the separation of homologues "shuffles" the genes to produce a variety of combinations in the gametes.

Finally, genes are recombined at fertilization when two gametes (the egg and sperm) unite. With genes contributed by each parent, the zygote may possess a new combination of genes and thus a new phenotype, which may or may not be of adaptive advantage.

We mentioned earlier that genetic variability is the key to evolutionary advancement. In the final analysis, it is the interplay of mutation and recombination that contributes most significantly to the variation essential to evolutionary change.

Gene Flow

The third source of variability in populations is *gene flow*, also referred to as *immigration*. If natural populations of the same species are not isolated, there is often movement or dispersal of individuals, seeds, pollen, and so on from one population to another (Fig. 13.20).

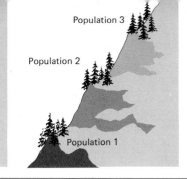

Figure 13.20 Diagrammatical representation of the kind of gene flow that would occur between populations at different heights on a mountain. Some pollen grains will blow away and carry their genes from one population to the next, although more pollen grains will fertilize other flowers within their own populations. The intensity of color indicates the density of pollen grains at particular locations. (In animal populations, genes will be carried between populations by the movement of individuals or of sperm and eggs in water.)

Interbreeding introduces new genes into populations, providing the opportunity for increased variability.

In Chapter 12 we defined a population as a group of organisms of the same species living in a certain area. Genetically, every population is characterized by its **gene pool,** i.e., the total of all the genes of all the organisms in the population. The concept of populations is central to the study of evolution because *it is the population that evolves, not the individual organism.* A population may vary in its genetic make-up from one generation to the next, but an individual organism has the same genetic make-up all its life. Thus, mutations, recombination, and gene flow act to alter the gene pool of a natural population, thereby providing the opportunity for increased variability and adaptive advantage. This is essentially what evolution is all about.

Having discussed the sources of variation in populations, we can now restate Darwin's theory in modern terms as *the change in the gene pool of a population brought about by the action of natural selection on heritable variation; the sources of variation include mutations, genetic recombination, and gene flow.*

> It is important to remember that it is the population that evolves, not the individual organism.

EVOLUTION IN ACTION

For populations, the long-term effect of evolution is *adaptation.* As we stated in Chapter 1, evolution can be thought of essentially as long-term adaptation. In the evolutionary process, either populations become better adapted to changing environments, or, if they cannot adapt, they become extinct. In the long history of life, most of the species that ever existed apparently could not adapt and eventually disappeared from the earth.

One of the classic examples illustrating adaptation and the mechanism of natural selection involves what is known as **industrial melanism** (Fig. 13.21). This is the change of an organism to a darker color associated with the release of industrial smoke and soot. During the first half of the nineteenth century, a pale gray moth, *Biston betularia,* was found throughout most of England. Other specimens infrequently found were almost black. Ordinarily, the black variety of the moth was quite rare, but toward the end of the nineteenth century, the black form came to predominate in the industrial areas. In the unpolluted regions of England, the pale gray moth remained the more common form.

Prior to pollution of the industrial areas, the tree trunks in the vicinity were encrusted with pale lichens. For the pale gray moths that rested on the trees, the lichens provided a camouflaged background as protection against predatory birds. However, as pollution blackened the tree trunks, the pale gray moths could be seen more

Figure 13.21 Industrial melanism. (a) The dark form of the moth *Biston betularia* contrasts sharply against the lichen-covered background but is practically invisible against the blackened tree (b). The light-colored moth, however, is quite conspicuous against the dark background. (From Clark, M. E.: *Contemporary Biology.* 2nd ed. Philadelphia: Saunders College Publishing, 1979, p. 620.)

easily by the birds, and their numbers began to dwindle. With the dark background as protection, the population of black moths began to increase until eventually almost all of the moth population in industrial areas was black. What had caused this change in the moth population, and how did a black moth appear in the population in the first place?

To answer the second question first, it is now known that the black color in *Biston* is the result of a rare mutation; therefore, as we have discussed, it is a genetically determined variation. The change in color of the moth population was an adaptation to a new environment. Against a dark background, the black moths were difficult for the predatory birds to find. The pale gray moths, however, lost their protective camouflage and became easy prey. Hence, more black moths than gray ones could survive long enough to reproduce. Over succeeding generations, the population of black moths would tend

> The Hardy-Weinberg Law applies to a population in genetic equilibrium.

to increase, whereas the population of gray moths would decrease. Thus, as natural selection dictates, an advantageous variation tends to be perpetuated in succeeding generations. Given the particular soot-covered environment and its relationship to the predatory capabilities of the birds, the black moths were more "fit," i.e., better adapted, for survival than were the gray moths. Conversely, with the reduction of pollution in England during the 1950s, the population of gray moths began to increase dramatically in proportion to the black variety.

In the example of industrial melanism we can see in microcosm the general trend of the evolutionary process. With changing environments, populations must either adapt or perish. If favorable genetic variations arise, natural selection—i.e., the impact of the environment on certain phenotypes—tends to perpetuate those variations in succeeding generations. In the process, the gene pool of the population is changed, and over the long term the population as a whole becomes better adapted to its ecological niche.

THE HARDY-WEINBERG LAW

Recall that in 1908, one of the disciplines that emerged from the study of evolution and heredity was the science of *population genetics*. Since populations, not individuals, evolve, it is necessary to understand the genetic structure of populations and how this structure might be altered by variation. The *Hardy-Weinberg Law* serves as a starting point for the study of population genetics because it describes a hypothetical situation in which there is no change in the gene pool from generation to generation and, hence, no evolution.

Since the Hardy-Weinberg Law applies to a static, i.e., nonevolving, population, there are certain restrictions that must be met: (1) the population must be large (to lessen the probability that chance alone will affect changes in the gene frequencies), (2) mating is random, (3) no mutations occur, (4) there is no gene flow, and (5) natural selection does not occur. A population meeting these conditions is said to be in *genetic equilibrium* (Fig. 13.22). This means that the frequencies of the alleles, i.e., the proportion of one allele relative to another allele of the same gene, will remain the same in the population generation after generation. In other words, the allele A and its counterpart a will remain in the same proportion to each other forever, as long as the population is in genetic equilibrium. The consequence, as mentioned, is that evolution does not occur.

The Hardy-Weinberg Law can be expressed mathematically as follows:

$$p^2 + 2pq + q^2 = 1$$

> In the equation describing the Hardy-Weinberg Law ($p^2 + 2pq + q^2 = 1$),
>
> p^2 = the frequency of the homozygous dominant genotype
> $2pq$ = the frequency of the heterozygous genotype
> q^2 = the frequency of the homozygous recessive genotype

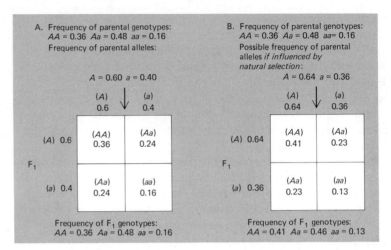

Figure 13.22 (a) According to the Hardy-Weinberg Law, gene frequencies would stay the same generation after generation *if a population could remain in genetic equilibrium.* (b) If certain alleles are selected for in a population, gene frequencies in succeeding generations will be different from those of the parental generation. Thus, the population is evolving.

where p is the frequency of the *dominant* allele in the population and q is the frequency of the corresponding *recessive* allele. Therefore, $p + q = 1$, which means that the frequencies of the two alleles added together equal 100 percent (all the alleles in the population).

In the expanded formula, then, p^2 = the frequency of the *homozygous dominant* genotype. Why? If p = the frequency of the dominant allele, then p^2, i.e. ($p \times p$), represents the frequency of a genotype having two dominant alleles (such as *AA*, *BB*, and so on). Similarly, $2pq$, i.e. ($2 \times p \times q$) = the frequency of the heterozygous genotype, since $2pq$ represents the frequency of a genotype having one dominant and one recessive allele (such as *Aa*, *Bb*, and so forth). The heterozygote is multiplied by 2 because in matings involving two alleles (p and q), the heterozygote appears *twice* as often as the other genotypes. You can verify this by drawing a Punnett square and crossing the gametes from two heterozygous individuals (such as *Aa* × *Aa*). Finally, q^2, i.e. ($q \times q$) = the frequency of the *homozygous recessive* genotype, i.e., the frequency of a genotype having two recessive alleles (such as *aa*, *bb*, and so on). The sum of the three genotypes equals 1, or 100 percent, which is all the genotypes in the population.

By employing the Hardy-Weinberg formula, we can determine the expected frequencies of all the genotypes and the frequencies of both alleles in a given population. For example, consider a population of 500 organisms containing the alleles *A* and *a* in its gene pool. There are, therefore, three possible genotypes in the population: *AA*, *Aa*, and *aa*. However, we do not know the *frequencies* of these genotypes. The initial task, then, is to find the frequencies of the dominant (*A*) and recessive (*a*) alleles. This is easy to do because we know simply *by looking* at the organisms which ones are homozygous recessives (*aa*). For instance, if *A* is the allele for dark-colored fur and *a* is the allele for light-colored fur, then all the organisms in the population with light-colored fur would have the genotype *aa*. Obviously, an organism having the dominant allele *A* in its genotype (*AA* or *Aa*) would have dark-colored fur.

Let us say, then, that we count 80 homozygous recessive (*aa*) individuals out of a population of 500. These individuals therefore make up 16 percent of the population (80/500 = 0.16), which is their frequency, or relative proportion, in the population. Since q^2 = the frequency of the homozygous recessive genotype, we now know that q^2 = 0.16, or 16 percent. To find q, the frequency of the recessive allele, we simply take the square root of q^2(0.16), which is 0.4, or 40 percent. If q = 40 percent, then the frequency of the dominant allele p = 60 percent, because $p + q$ = 1 (100 percent). Thus, in our population of 500 individuals, the frequency of the dominant allele *A* = 60 percent, and the frequency of the recessive allele *a* = 40 percent. But in what proportion are they distributed among the genotypes? We already know that 16 percent of the population are homozygous recessives (*aa*), but how are the *A* and *a* alleles distributed among the remaining 84 percent of the population? In other words, what are the frequencies of the *AA* and *Aa* genotypes?

Finding the answer is simply a matter of inserting the values of p and q in the Hardy-Weinberg formula. Thus, $p^2(0.60)^2$ + $2p(0.60)q(0.40)$ + $q^2(0.40)^2$ = 1, or 0.36 + 0.48 + 0.16 = 1. In our population, then, p^2 = 36 percent (the frequency of the homozygous dominant individuals, *AA*); $2pq$ = 48 percent (the frequency of the heterozygotes, *Aa*); and q^2 = 16 percent (the frequency of the homozygous recessive individuals, *aa*). Of 500 organisms, 180 (0.36 × 500) have the *AA* genotype, 240 (0.48 × 500) have the *Aa* genotype, and as previously observed, 80 (0.16 × 500) have the *aa* genotype.

We have stated that the Hardy-Weinberg Law applies only to populations that are *not* evolving, i.e., populations in which the gene (allele) frequencies remain the same generation after generation. In the example just presented, the frequency of the *A* allele = 0.60, and the frequency of the *a* allele = 0.40. Since these numbers represent the frequencies of the alleles of all the males and females in the population, "crossing" the alleles will show the frequencies of the *AA*, *Aa*, and *aa* genotypes in their offspring. As shown in Figure 13.22, the F_1 genotypes and their frequencies are as follows: 36 percent, *AA*; 48 percent, *Aa*; and 16 percent, *aa*—exactly the same as those of the parent generation. Moreover, the frequencies of the *A* and *a* alleles have not changed. The *AA* individuals of the F_1 con-

> The Hardy-Weinberg Law can also be used to determine the proportion of carriers of human hereditary diseases.

tribute 36 percent of the *A* alleles, and the *Aa* individuals contribute 24 percent of the *A* alleles, for a total of 60 percent. In addition, the *Aa* individuals contribute 24 percent of the *a* alleles, and the *aa* individuals contribute 16 percent, a total of 40 percent. Therefore, the frequencies of the two alleles have not changed from one generation to the next. This would still be the case no matter how many successive generations might be considered. However, in reality, mating is *not* always random within a population, mutations *do* occur, gene flow *is* possible, and natural selection *does* exert its effects. Consequently, gene frequencies change, and the population evolves. The Hardy-Weinberg Law illustrates the conditions under which evolution will *not* occur, while at the same time showing that evolution probably *will* occur in the absence of these conditions.

In addition, the law has a most significant practical application. Many human diseases, for example, result from the effects of either dominant or recessive alleles. The human hereditary disease *galactosemia*, for instance, occurs in individuals who are homozygous for the recessive allele *g*. Normal individuals have the genotypes *GG* or *Gg*; the heterozygote (*Gg*), however, is a carrier of the allele for galactosemia. Infants who have inherited the homozygous recessive genotype (*gg*) possess a defective enzyme and therefore cannot metabolize the monosaccharide *galactose*. You will recall that the disaccharide lactose (milk sugar) can be split into the simple sugars glucose and galactose. While galactosemic individuals can metabolize glucose, they cannot metabolize galactose. As a result, galactose and other substances accumulate in the body, and the child falls victim to mental retardation, cirrhosis of the liver, lens cataracts, and other abnormalities.

The incidence of galactosemia in the United States is about 6 of every 100,000 births. Knowing the number of afflicted individuals, we can use the Hardy-Weinberg formula to determine the number of individuals who are carriers of the allele, i.e., who have the genotype *Gg*. The frequency of recessive genotype $q^2 = 6/100,000$, or 0.00006. The frequency of the recessive allele q is thus the square root of 0.00006, or approximately 0.0077. Therefore, the frequency of the dominant allele $p = 1.00 - 0.0077$, or 0.9923. The frequency of the heterozygote ($2pq$) in the population is then 2(0.9923)(0.0077), or about 0.015. This means that about 1 in every 67 persons, 1.5 percent of the United States population, is a carrier of the allele for galactosemia.

Such information is particularly valuable to genetic counselors, who deal with prospective parents. For example, a person who is a known carrier of a harmful recessive allele may wish to know the

> Anthropoid apes from the genus *Dryopithecus* are considered the forerunners from which modern apes and humans evolved.

probability of marrying another person who is also a carrier. In the case of galactosemia, that probability is 1 in 67.

THE EVOLUTION OF MAN

Perhaps for most people, the greatest interest (and controversy) concerning evolution centers on the developmental history of *Homo sapiens*, the "wise man." Intriguing, indeed, is the question of where we came from. Unfortunately, the fossil record of human evolution is sketchy at present, and accordingly, investigators admit to considerable disagreement regarding its interpretation. Nonetheless, it is all but certain that human-like species have inhabited the earth for at least 3 million years and probably much longer. Compared with other organisms, however, we are quite recent arrivals on the biological scene.

Human beings belong to a group of mammals known as *primates* (Gk. "first" or "highest"). Our closest primate relatives are the *anthropoid* (manlike) apes: the chimpanzee, gorilla, orangutan, and gibbon (Fig. 13.23). All primates have several distinguishing characteristics, such as hands or feet adapted for grasping; three-dimensional vision; a shortened nose; upright posture; wide, flattened nails; and a highly developed brain.

Anthropoid apes became generally widespread over a large part of the earth some 25 million years ago. One large group of these animals has been placed in the genus *Dryopithecus*, meaning "oak ape," a reference to the trunks of oak trees found in the same deposit where the fossil apes were discovered. Apparently, the dryopithecines were about the size of a present-day chimpanzee, although they may have been somewhat lighter in build (Fig. 13.24). Some of them had faces slightly more flattened than those of modern apes, and their teeth were partially human-like.

Although the dryopithecines roamed the earth for some 20 million years, there is no definitive fossil evidence bridging the gap between these early apes and later forms. Although their own fossil evidence is fragmentary—consisting mostly of jaws and teeth—it has been strongly suggested that *Dryopithecus* is the forerunner from which modern apes and humans evolved (Fig. 13.25).

In 1932, a fossil jaw fragment along with several teeth was discovered in an area north of New Delhi, India. They have since been dated to be around 10 to 15 million years old. Their discoverer, G. Edward Lewis, observed that some of the features of the jaw and teeth were definitely man-like. Lewis placed the find in a new genus *Ramapithecus* ("Rama's ape"), after the name of a mythical Indian prince. During the 1960s and 1970s additional *Ramapithecus* jaws

Figure 13.23 The anthropoid apes. (a) Gibbon. (Courtesy of New York Zoological Society.) (b) Chimpanzee mother and baby. (Courtesy of San Diego Zoo. Photo by Ron Garrison.) (c) Orangutans. Larger individual is a male; smaller one, a female. (Courtesy of New York Zoological Society.) (d) Gorilla. (From Johnson, W. H., Delanney, L. E., Cole, T. A., and Brooks, A. E.: *Biology.* 4th ed. New York: Holt, Rinehart and Winston, 1972, pp. 782–783.)

(a)

(b)

(c)

(d)

Figure 13.24 Artist's reconstruction of East Africa as it probably appeared 20 million years ago. *Dryopithecus,* the unspecialized common ancestor of both great apes and humans, lived in the tropical forests that existed then. (From Clark, M. E.: *Contemporary Biology.* 2nd ed. Philadelphia: Saunders College Publishing, 1979, p. 634.)

> *Ramapithecus*, one of the descendent lines of *Dryopithecus*, shows considerable adaptations to environmental changes.

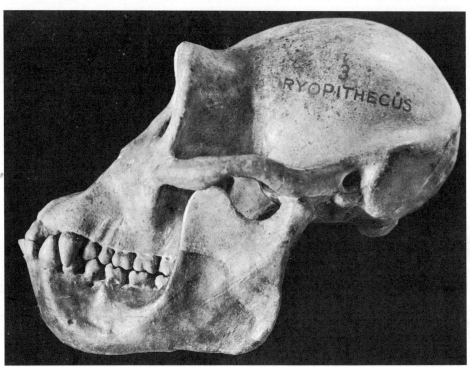

Figure 13.25 Restored skull of *Dryopithecus*, the fossil "oak ape," possibly the ancestor common to gorillas, chimpanzees, and humans. (From Villee, C. A.: *Biology*. 7th ed. Philadelphia: Saunders College Publishing, 1977, p. 791.)

and teeth were discovered and described. The latest of these fossils indicate that *Ramapithecus* was one of the descendant lines of a group of *Dryopithecus*. There is considerable evidence, too, that *Ramapithecus* subsisted on a diet different from that of his forebears and had adapted to a different way of life. As we have discussed, adaptations arise through the selective pressures of the environment. Those organisms whose variations offer an adaptive advantage in a certain environment have the greatest potential for survival.

Some of the dryopithecines were forest-dwellers that fed on the abundant leaves and soft fruit in the tree tops. Fossils of these animals show that the lower jaw was relatively thin, light, and narrow and that their molar teeth were small with a thin covering of enamel. These structures, then, were fully adequate for an animal with a soft diet.

Over many millions of years, the forest habitat of *Dryopithecus* began to change from one of continuous vegetation to one of seasonal vegetation. Leaves and fruit thus became scarce, and the animals were forced to forage elsewhere for food. On the ground and in the open woodland beyond the forest, the available foods, such as nuts and roots, would be of a tough consistency. With their particular dental system, it is unlikely that the forest-dwelling dryopithe-

> The link between *Ramapithecus* and *Australopithecus* is unclear because of a gap in our knowledge about man's evolutionary history between the times these two creatures lived.

cines could have survived this dietary shift. However, the fossils of *Ramapithecus* show a relatively wide, thickened lower jaw, with broad, flattened molars. Such a dental system would have been a distinct advantage to organisms having to grind tough, harsh food. Natural selection would favor the survival of organisms with such an advantage, making it likely that they would perpetuate their kind in increasing numbers.

Thus, *Ramapithecus* is considered to have diverged from its dryopithecine relatives essentially as a result of an adaptation to changing demands of the environment. Keep in mind, however, that evolution is a long-term adaptive process; in the lineage from *Dryopithecus* to *Ramapithecus,* the process consumed some 15 million years. Although the fossil record is far from complete, what we have learned of *Ramapithecus* indicates that it was distinctly manlike, so much so that it may be a close ancestor to modern human beings.

After the rise of *Ramapithecus,* the next 10 to 12 million years of man's evolutionary history are essentially unknown. The next significant find, a group of fossils dated between 1 and 4 million years old, was recovered from various sites in southern and eastern Africa. The original specimen, discovered in 1924, was described by Raymond A. Dart as the skull of a juvenile female. He named the group to which the skull belonged *Australopithecus* ("southern ape") (Fig. 13.26). The various fossils of *Australopithecus* indicate that these human-like creatures were about 4 to 5 ft tall and apparently walked upright. They had large brains for their body size, and there is some evidence that they may have been tool-makers.

There has been speculation that a population of *Ramapithecus* living in Africa may have given rise to *Australopithecus*. However, the fossil records of both groups are so fragmentary that many of the necessary pieces are missing. For example, more recent discoveries

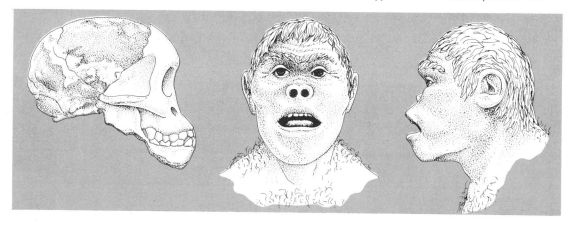

Figure 13.26 From fossil skulls (a), anthropologists have reconstructed the appearance of *Australopithecus* (b).

> Some investigators believe that there were two species of *Australopithecus*: *A. bosei* and *A. robustus*.

suggest to some investigators that there were two species of *Australopithecus*, one a tool-maker and one not. Moreover, L.S.B. Leakey discovered a group of fossils he called *Homo habilis* ("skillful man"), but considered by many to be another form of *Australopithecus*. Until further evidence is forthcoming, the exact status of *Ramapithecus* and *Australopithecus* in the history of man remains uncertain (Fig. 13.27).

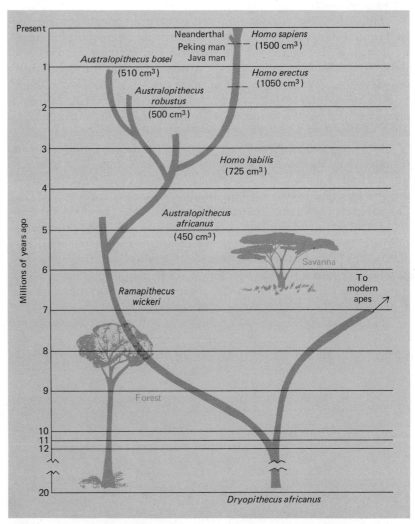

Figure 13.27 A tentative evolutionary tree of *Homo sapiens* based on current fossil evidence. Major disagreement exists over the details of the branch points shown here as occurring 3.5 million years ago. Numbers in parentheses are approximate brain volumes (in cubic centimeters) estimated from sizes of cranial vault. (From Clark, M. E.: *Contemporary Biology*. 2nd ed. Philadelphia: Saunders College Publishing, 1979, p. 633.)

> *Homo erectus* was the first true man and is considered the direct ancestor of modern man.

THE GENUS *HOMO*

The first fossil specimens clearly belonging to the genus *Homo* were discovered in Java and China. These early men have been designated *Homo erectus* ("upright man") and arose approximately 1 million years ago (Fig. 13.28). They apparently lived in caves and had fashioned a variety of stone tools. Charred bones found in the caves attest to their use of fire, and they probably had developed clothing for protection against the cold. The origin of *H. erectus* seems to be Africa, where he may have arisen from a group of australopithecines. What happened to him is unknown, but it has been postulated that he is the immediate ancestor of modern man.

The final stage in the evolutionary history of man is the appearance of *Homo sapiens*. The earliest records of *H. sapiens* are fossil skulls from England and Germany; these skulls have been dated at about 250,000 years old. One of the more familiar members of *H. sapiens* was "Neanderthal man," who appears in the record about 80,000 years ago. The fossil record between the earliest discovered *H. sapiens* and Neanderthal man is virtually blank. It appears that Neanderthals were a little over 5 ft tall, with large brains, thick, jutting brows, and teeth quite similar to those of modern man. Although widely distributed throughout Europe, Neanderthal man suddenly disappeared some 40,000 years ago. The cause of his disappearance remains a mystery.

After the extinction of Neanderthals, populations closely resembling modern man arose in Europe and parts of the Old World. These men, called Cro-Magnon, replaced the Neanderthals about 40,000 to 50,000 years ago. Cro-Magnon man is perhaps best known for his paintings on the walls of caves; some of these paintings are

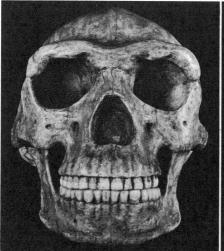
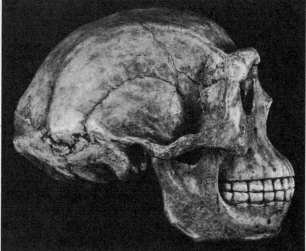

Figure 13.28 Front and side views of a reconstructed skull of Peking man, *Homo erectus*. Note the massive bony ridges over the eyes, the low, retreating forehead, the protruding jaws, and the absence of a chin. (From Villee, C. A.: *Biology*. 7th ed. Philadelphia: Saunders College Publishing, 1977, p. 794.)

Origins and Evolution

Figure 13.29 The art of Cro-Magnon Man. These paintings were discovered on the walls of caves in southern France. (From Villee, C. A.: *Biology*. 7th ed. Philadelphia: Saunders College Publishing, 1977, p. 805.)

10,000 to 25,000 years old (Fig. 13.29). As with species of early man, the origin of modern man is unknown, but he survived to populate the earth while all others had failed. For some 40,000 years, *H. sapiens* has been the only species of man on earth (Fig. 13.30).

Figure 13.30 (a) Skulls of prehistoric men (left to right: *Homo erectus*, Neanderthal, and Cro-Magnon). (b) Restorations of these three prehistoric men.

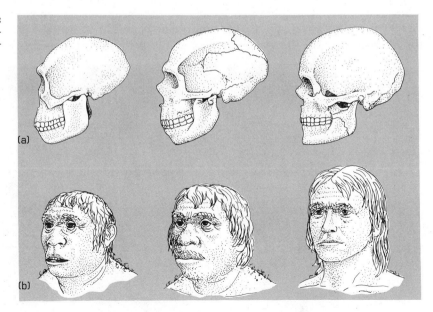

> Though man appears to have separated in an evolutionary sense from the anthropoid apes 25 million years ago via the dryopithecines, there are numerous anatomical and biochemical similarities that indicate man and apes are closely related.

The evolutionary line leading to the anthropoid apes and that leading to modern man appear to have separated from a dryopithecine stock about 25 million years ago. That man and the anthropoid apes are closely related is supported not only by anatomical similarities but by biochemical evidence as well. Studies indicate that humans are most closely related to African chimpanzees and gorillas. These apes each have 48 chromosomes, compared with 46 for man. The pigment hemoglobin of red blood cells is identical in adult humans and chimpanzees. In fact, most human and chimpanzee proteins are alike. These proteins, of course, are produced by genes, and it is the extent of genetic relationships that determines most specifically the evolutionary similarities (or differences) between living things.

REVIEW OF ESSENTIAL CONCEPTS

1. The *theory of organic evolution* is the concept reconciling unity and diversity in the living world.
2. Evidence indicates that our solar system is a by-product of star formation. The earth was formed from condensation of gaseous elements and dust particles.
3. The earliest atmosphere of the earth is thought to have consisted of hydrogen, nitrogen, carbon monoxide, and water vapor, which were spewed from the earth by volcanic eruptions.
4. With the formation of the core and mantle of the earth, the gas nitrogen came to predominate in the atmosphere, whereas carbon dioxide, water vapor, and methane were present in smaller amounts.
5. The forms of available energy on the primitive earth included ultraviolet radiation, lightning, heat, and radioactivity.
6. Supplied with energy, atmospheric gases complexed to form organic compounds, which in turn interacted to form even more complex compounds. This sequence is known as *chemical evolution*.
7. Evidence for the possibility of chemical evolution has been indicated by the Miller-Urey and other experiments.
8. A possible prebiotic cell-like structure is represented by *proteinoid microspheres*. These droplet-like structures are amino acid polymers surrounded by water.
9. Protocells accumulated nutrients from the surrounding water and were therefore *heterotrophs*. Metabolism of nutrients occurred by *fermentation*.
10. Protocells devised a genetic machinery by incorporating nucleotides that were sequenced by the proteinoids.
11. The earliest protocells or protobionts were heterotrophic *prokaryotes*, which eventually gave rise to the first *autotrophs*.

12. The first autotrophs were probably *chemosynthetic bacteria* and *blue-green algae*.
13. With the rise of oxygen-releasing autotrophs, the early atmosphere of the earth began to change from a *reducing* to an *oxidizing* state. This resulted in the formation of the protective *ozone layer* and the introduction of *aerobic respiration*, both of which made possible the advancement of organic evolution.
14. The earliest *eukaryotic* cells were probably types of unicellular green algae. The origin of eukaryotic cells is uncertain, but they may have arisen as a result of free-living organelles establishing a symbiotic relationship with prokaryotic cells.
15. Eukaryotic cells developed the capacity for *mitotic cell division*, which in turn led to the rise of multicellular organisms. Mitosis was followed by the evolution of *meiosis* and the beginning of sexual reproduction.
16. Historically, the theory of evolution has been opposed on religious grounds, to a large extent on the basis of the "fixity of species" concept.
17. The science of geology provided the greatest impetus to evolutionary theory in the eighteenth century; however, the theory was opposed by many, including the proponents of *catastrophism*.
18. Lamarck explained the theory of biological evolution in terms of *inheritance of acquired characteristics*, a concept that has since been shown to be scientifically indefensible.
19. Charles Darwin explained biological evolution in terms of the effect of *natural selection* on variations within a population. Natural selection is the process whereby favorable inherited variations tend to be perpetuated in succeeding generations.
20. The theory of evolution by natural selection coupled with the science of genetics constitutes the *synthetic theory of evolution*.
21. The sources of genetic variation in natural populations are *mutation*, *recombination*, and *gene flow*.
22. An example of evolution in action is *industrial melanism*, which, through soot pollution, alters the environment in favor of dark-colored moths.
23. Population genetics involves a basic understanding of the *Hardy-Weinberg Law*, which demonstrates that gene frequencies in a population will remain the same unless influenced by *nonrandom mating*, *mutations*, *gene flow*, and *natural selection*. Since these factors normally do occur, gene frequencies change and the population evolves.
24. The Hardy-Weinberg Law can be used in genetic counseling to determine the frequency of various potentially harmful genotypes in the human population.

25. The evolutionary forerunner of modern apes and humans may have been a group of anthropoid apes designated *Dryopithecus*.
26. A descendant of *Dryopithecus* called *Ramapithecus* may have been a close ancestor to man.
27. *Ramapithecus* may have given rise to the human-like *Australopithecus*, believed to have walked upright and used tools.
28. *Homo erectus*, the original representative of the genus *Homo*, appears to be the immediate ancestor of modern man.
29. The earliest records of *Homo sapiens* are about 250,000 years old; representatives of *Homo sapiens* include Neanderthal man and Cro-Magnon man.
30. The evolutionary relationship between man and the anthropoid apes is supported by anatomical and biochemical evidence.

APPLYING THE CONCEPTS

1. Discuss the formation of the earth and the development of its original atmosphere.
2. What is the relationship between chemical evolution and biological evolution?
3. Discuss the possible origin of the first living cell.
4. What is the significance of the Miller-Urey experiment?
5. What was the significance of the rise of autotrophic cells on the early earth?
6. In what manner did the rise of eukaryotic cells affect the evolutionary history of life?
7. What is catastrophism? Do you think it is a reasonable theory? Why?
8. Explain why the theory of inheritance of acquired characteristics is not scientifically defensible.
9. What factors did Darwin have to consider in formulating his theory of evolution by natural selection?
10. What is the role of the environment in the evolutionary process?
11. What is the synthetic theory of evolution?
12. Discuss the sources of genetic variation in natural populations.
13. How does industrial melanism illustrate evolution by natural selection?
14. What is the significance of the Hardy-Weinberg Law?
15. Assume that of a population of 200 individuals, there are 25 homozygous recessives. Determine the frequencies of the dominant and recessive alleles, the frequencies of the possible genotypes, and the number of each genotype in the population.
16. Comment on the evolutionary history of man leading to the genus *Homo*.
17. Trace the general development of the genus *Homo* leading to modern man.

the nervous system

THE ESSENTIAL OBJECTIVES

You have understood this chapter when you are able to:

1. Describe the structure and function of the three types of neurons.
2. Discuss, using an example, the components and physiological action of a reflex arc.
3. Explain how a nerve impulse is conducted along a nerve fiber and across the synapse.
4. Differentiate between excitatory and inhibitory neurons.
5. Describe the major structures and functions of the forebrain, midbrain, and hindbrain of man.
6. Describe the general structure and functions of the following: spinal cord, somatic nervous system, and autonomic nervous system.
7. Contrast the nervous systems of selected invertebrates with that of man.
8. Describe the organs of special senses in terms of general structure and function.

The human nervous sytem is a vast network of communicating nerve cells. Through the sensation-response mechanism, we are informed of and react to the countless stimuli constantly impinging on the body.

PREVIEW OF ESSENTIAL TERMS

homeostasis The maintenance of an internal physiological balance by an organism; constant internal conditions, such as body temperature, water balance, and so on.

neuron A nerve cell; the unit of structure and function in the nervous system.

axon The cytoplasmic extension of a neuron that conveys the nerve impulse *away* from the cell body of the neuron.

dendrite A cytoplasmic extension of a neuron that conveys the nerve impulse *toward* the cell body of the neuron.

reflex An automatic response to a stimulus performed without conscious thought; involves at least a sensory neuron and a motor neuron.

synapse The junction between the axon of one neuron and the dendrite or cell body of another neuron.

central nervous system (CNS) The division of the vertebrate nervous system that includes the brain and spinal cord.

peripheral nervous system (PNS) The division of the vertebrate nervous system that includes the somatic (spinal and cranial nerves) system and the autonomic (involuntary) system.

neurotransmitter A chemical substance released by the axon of a neuron that travels across the synaptic junction to excite or inhibit another neuron.

deactivating enzyme A chemical released by a dendrite or neuron cell body that destroys a neurotransmitter at the synapse.

The Nervous System

> Anatomy is the study of the *structure* of the body; physiology is the study of the *function* of the body.

Preview of Anatomy and Physiology

Over 26 centuries ago, in the fertile and venturous minds of the early Greeks, science was born; but it was another 1000 years before science would apply its critical methods to the structure and function of the human body. During the Dark Ages, dissection of the human body was a sacrilegious offense, and science in general was forced to labor under the burden of millennia of bigotry and ignorance. With the Renaissance, however, came a gradual rebirth of the exacting observational approach to science instituted by Aristotle centuries before. Knowledge of anatomy was advanced through the artistic genius of Leonardo da Vinci, and the great turning point came with the publication of the book *On the Fabric of the Human Body* by Andreas Vesalius in 1543 (Fig. 14.1). The later discovery of the circulation of blood by William Harvey laid the foundation of human physiology, and a new understanding of man had begun (Fig. 14.2).

Anatomy and physiology are two of the major biological sciences and are closely interrelated. Anatomy is the science of the *structure* of the body and includes both gross anatomy and microscopic anatomy. Physiology is the science concerned with the *function* of the body and thus attempts to explain the workings of the various body organs and systems. Since it is virtually impossible to separate the function of a body part from its structure, the sciences of anatomy and physiology together help us to view the body as an integrated whole.

Throughout the following chapters, there is one vital concept that stands as the integrating theme and focal point of your study of

Figure 14.1 Andreas Vesalius (1514–1564). His book *On The Fabric of the Human Body* established the basis for modern human anatomy. (From Villee, C. A., Walker, W. F., and Barnes, R. D.: *General Zoology*. 4th ed. Philadelphia: Saunders College Publishing, 1973, p. 10.)

Figure 14.2 William Harvey (1578–1657). His proof of the circulation of blood was a major contribution to the science of physiology. (From Villee, C. A., Walker, W. F., and Barnes, R. D.: *General Zoology*. 4th ed. Philadelphia: Saunders College Publishing, 1973, p. 10.)

> *Homeostasis* is the physiological balance that must be maintained within the body of every living thing in order for the organism to remain healthy.

the human body. The concept is **homeostasis.** In essence, homeostasis means the condition of "staying the same." It refers to the physiological balance that must be maintained within the body of every living thing. For health or life to continue, the conditions within the organism must be kept relatively constant. A familiar example of this is the maintenance of a constant body temperature irrespective of the temperature outside the body. Human beings, for instance, have a normal oral temperature of 98.6° F, or 37° C. If this temperature fluctuates a few degrees in either direction, serious complications may result. If, for example, the body temperature rises above 106° F, many of the cells of the body are damaged; or if the body temperature falls a few degrees, the heart rate is greatly decreased and so on. As you can see from this example, the upsetting of one homeostatic mechanism (temperature regulation) affects the functioning of other body systems.

It is when all of the systems are functioning properly that we say that the body is in homeostatic balance. The need for homeostatic adjustment is determined by information received from within the organism itself and from its external environment. Using such information, the body systems act in concert to ensure that this adjustment is made and that the essential balance is restored.

In considering the various systems responsible for homeostatic control, we shall also cite further examples of what happens when this control system is upset.

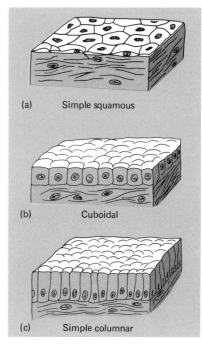

Figure 14.3 The three basic cell types found in epithelial tissue. (a) Flattened, or squamous, (b) Cuboidal, (c) Columnar.

INTRODUCTION TO THE HUMAN BODY
Cells and Tissues

Like every other living thing, the human body is composed of microscopic **cells,** the structural and functional units of life. Man, of course, is a multicellular organism, a composite of trillions of individual cells. The cells of the body are organized into different types of **tissues,** defined as groups of cells that are similar in structure and function. Although we often speak of "reproductive" tissue, "lung" tissue, "skin" tissue, and so forth, there actually are only four basic types of tissues that constitute all the organs and structures of the body. These are designated (1) **epithelial** tissue, or **epithelium;** (2) **connective** tissue; (3) **muscle** tissue; and (4) **nervous** tissue.

Epithelium is composed of flattened, cuboidal, or columnar cells that cover the surface of the body (skin) as well as the surface of the internal organs (Fig. 14.3). Epithelium also lines the inside of the body cavities, blood vessels, and other tubular structures and is the predominant tissue composing the glands of the body.

Connective tissue is a general term for those tissues that function in support, protection, binding together of body structures, and production of blood cells. Connective tissues include *loose* connective

464
The Nervous System

Connective tissue supports, protects, and binds together body structures and produces blood cells. Some of the various types are loose connective tissue, adipose tissue, tendons, ligaments, cartilage, bone, blood, and lymph.

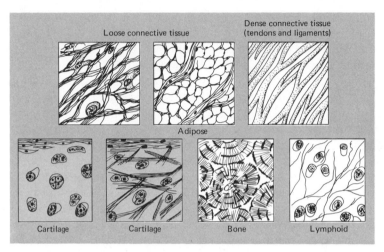

Figure 14.4 Some representative types of connective tissue.

tissue, *fat* (adipose) tissue, *tendons, ligaments, cartilage, bone, blood,* and *lymph* (Fig. 14.4).

The significant feature of muscle tissue is its *contractility,* i.e., it is capable of shortening and then returning to its original shape. The three types of muscle tissue are **striated,** or **skeletal,** muscle; **smooth** muscle; and **cardiac** muscle (Figs. 14.5 and 14.6). Striated muscle, so called because of its striped appearance, is attached to bones and is thus responsible for movement of the body. For the most part, striated muscle can be controlled voluntarily. Smooth muscle is not striped and is found in the walls of the internal organs and blood vessels. Usually, we cannot consciously control the action of smooth muscle. Cardiac muscle is found only in the heart and, like smooth muscle, is normally not under voluntary control. Cardiac muscle fibers branch and unite to form an interconnecting network that makes possible a rhythmic heartbeat.

Nervous tissue comprises mainly **neurons,** or nerve cells, which are responsible for the sensation-response mechanism of the body (Fig. 14.7). Neurons transmit nerve impulses to muscles, glands, and all parts of the nervous system itself. Through the interplay of bil-

Figure 14.5 Types of muscle tissue. (From Villee, C. A.: *Biology.* 7th ed. Philadelphia: Saunders College Publishing, 1977, p. 72.)

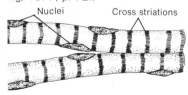

A, SKELETAL MUSCLE FIBERS

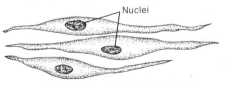

B, SMOOTH MUSCLE FIBERS

C, CARDIAC MUSCLE FIBERS

465
Introduction to the Human Body

> *Organs*, composed of several different types of tissues, perform specific functions in the body.

Figure 14.6 The three types of muscle in vertebrates. (a) Smooth muscle (from the gut wall). (b) Cardiac (heart) muscle. Each cell is lightly striped; (c) Striated (striped) muscle (from a leg) is unique in that many cells fuse to form a fiber, each of which is lined by many nuclei. (From Camp, P. S., and Arms, K.: *Exploring Biology,* Philadelphia: Saunders College Publishing, 1981, p. 429.)

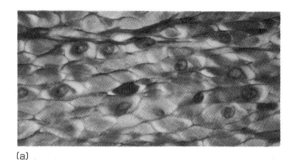
(a)

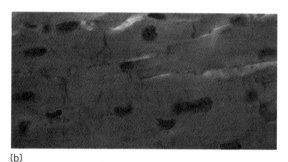
(b)

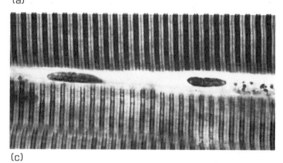
(c)

lions of neurons, the nervous system directly or indirectly regulates and coordinates most of the activities of the human body.

The tissues listed in the previous section are organized to form the various *organs* of the body. An organ is usually composed of several

Organ Systems

Figure 14.7 Some of the types of neurons found in human beings and other vertebrates.

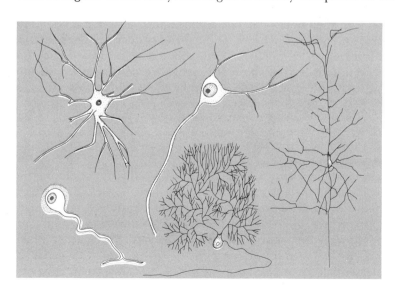

> Neurons consist of a *cell body* and two or more extensions called *nerve fibers*. The nerve fibers are of two types: *dendrites* and an *axon*.

different types of tissues and is a structure that performs a specific function in the body. A group of related organs make up each of the various body **systems**. There are eight major systems that constitute the human body: **nervous, musculoskeletal, respiratory, circulatory, digestive, urinary, reproductive,** and **endocrine**. In the chapters to follow, there are other systems that we shall include as being a part of one or more of the eight listed.

All of the various systems of the body interact to produce the specific physiological effects that would usually be impossible for one system acting alone. This interaction is the underlying mechanism of homeostatic control. For proper functioning of the body as a whole, all the systems interact with each other, so much so that they are almost totally interdependent.

INTRODUCTION TO THE NERVOUS SYSTEM

One of the basic principles applicable to any living organism is its capacity to sense changes in its environment and then to make an appropriate response to those changes. In this manner, the organism adjusts to its surroundings, an activity known as adaptation. Depending upon the environmental change, or **stimulus,** the response among higher animals may be movement of the body, quickening of the heart rate, increased glandular secretions, or any number of physical and mental adjustments. Such responses are usually beneficial, serving to promote the general welfare of the organism.

The overall control, integration, and coordination of the activities of the human body are managed by the nervous system in conjunction with the endocrine system. The nervous system acts rapidly to inform and adjust the body, whereas the endocrine system carries out its functions in a much slower manner. The rapid action of the nervous system is attributable to the structure and function of some of the most highly specialized cells in the body—the **neurons.**

NEURONS

The structural and functional units of the nervous system are the *neurons,* or nerve cells (Fig. 14.8). Neurons are of varied shapes and sizes, but typically, each neuron consists of a **cell body** containing the nucleus, mitochondria, lysosomes, and other organelles. In addition, most neurons have two or more extensions of the cell body called *nerve fibers*. These fibers are of two types: (1) **dendrites,** which conduct nerve impulses *toward* the cell body, and (2) an **axon,** which conducts the nerve impulse *away* from the cell body. A **nerve,** in fact, is essentially a bundle of axon fibers. A neuron may have one

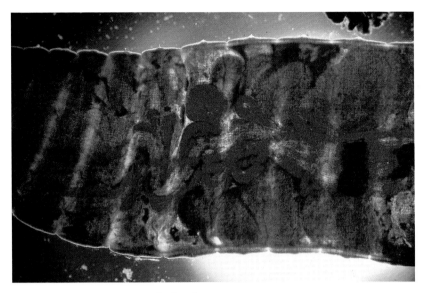

Color Plate 80 The "hearts" of a live earthworm. (Photograph by Bruce Russell, Biomedia Assoc.)

Color Plate 81 A marine annelid. (Photograph by Bruce Russell, Biomedia Assoc.)

Color Plate 82 This nudibranch is a member of a group of gastropods in which the shell is absent. (Photograph by Bruce Russell, Biomedia Assoc.)

Color Plate 83 *Daphnia,* a tiny freshwater crustacean. (Photograph by Bruce Russell, Biomedia Assoc.)

Color Plate 84 Copepods, free-swimming crustaceans, with their eggs. (Photograph by Bruce Russell, Biomedia Assoc.)

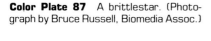

Color Plate 85 The praying mantis, a carnivorous insect, has powerful forelegs used to grasp its prey. (Photograph by Bruce Russell, Biomedia Assoc.)

Color Plate 86 A starfish. (Photograph by Bruce Russell, Biomedia Assoc.)

Color Plate 87 A brittlestar. (Photograph by Bruce Russell, Biomedia Assoc.)

Color Plate 88 Tunicate, a urochordate. (Photograph by Carolina Biological Supply Co.)

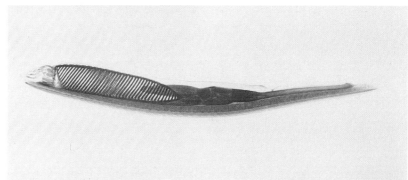

Color Plate 89 Amphioxus, a cephalochordate. (Photograph by Carolina Biological Supply Co.)

Color Plate 90 Lamprey eel, a jawless vertebrate. (Photograph by Carolina Biological Supply Co.)

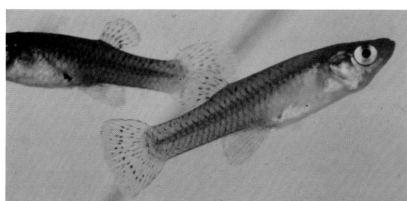

Color Plate 91 Mosquito fish, one of many species of bony fishes. (Photograph by Carolina Biological Supply Co.)

Color Plate 92 Grass frog, a tailless amphibian. (Photograph by Carolina Biological Supply Co.)

Color Plate 93 Newt, a small salamander. (Photograph by Carolina Biological Supply Co.)

Color Plate 94 Pygmy rattlesnake. (Photograph by Carolina Biological Supply Co.)

Color Plate 95 A male mallard. (Photograph by Carolina Biological Supply Co.)

> Neurons are of three different types: *sensory* neurons, *motor* neurons, and *interneurons*.

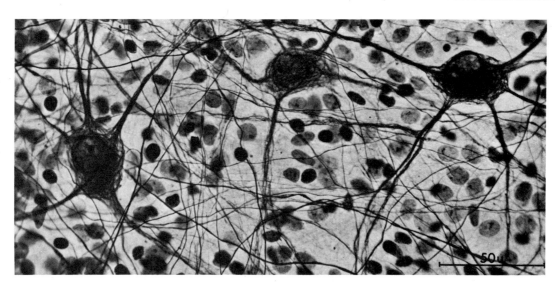

Figure 14.8 Nervous tissue is composed mainly of neurons, or nerve cells. These neurons are from the brain of a rat.

or many dendrites, but it has only one axon, although the axon may have many branches.

The functions of neurons may be described generally as *reception, conduction,* and *response;* that is, they receive or detect stimuli, conduct the nerve impulse, and bring about a response in some body part. On the basis of these functions, neurons can be separated into three different types: **sensory** neurons, **motor** neurons, and **interneurons** (Fig. 14.9). A sensory neuron, as the name implies, receives sensations, or stimuli, from a receptor organ such as the skin, eyes, ears, and so forth and conducts a nerve impulse *toward* the central nervous system (brain and spinal cord).

A motor neuron conveys the nerve impulse *away* from the central nervous system and terminates in what is called an **effector organ.** Effectors are usually either muscles or certain glands. The motor neuron, then, brings about movement in the case of a muscle, or if terminating in glandular tissue, the motor neuron would cause the gland to secrete or release, its contents.

Between the sensory neuron and motor neuron is the interneuron, which is located in the spinal cord. Nerve impulses coming into the central nervous system via a sensory nerve go to the spinal cord, where they are usually passed to an interneuron. The interneuron then passes the impulse to a motor nerve, which carries the impulse on to the effector. Thus the interneuron is a sort of "connection" or relay station between sensory and motor neurons. Interneurons also lie within the brain, where they function in relaying impulses to various areas of the entire central nervous system.

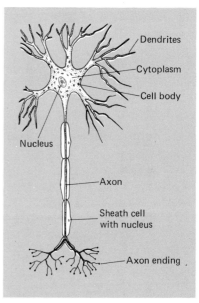

Figure 14.9 A motor neuron, showing the basic structural parts.

THE REFLEX ARC

The fundamental physiological activity of the nervous system is the *reflex,* a response to a stimulus performed automatically without conscious thought. The pathway followed by a nerve impulse from the point of stimulation to the responding organ is called a **reflex arc** (Fig. 14.10). Thus, reflex arcs involve sensory neurons, motor neurons, and, in most instances, interneurons. The simple knee jerk and the withdrawal of the hand from a painful stimulus are familiar reflexes, but many other vital physiological processes of the body also are under reflex control (Fig. 14.11). For example, the heartbeat, breathing rate, movements of the digestive tract, coughing, and many others are involuntary responses to internal and external stimuli. We shall comment on the nervous control of these reflexes later in this section.

To illustrate a simple three-neuron reflex arc, let's use the common *withdrawal reflex* as an example. Everyone is familiar with the pain of touching a hot stove and with the fact that the finger does not remain on the stove for long. The physiological activity that occurs from the time you sense the heat until you remove your finger constitutes the reflex arc. A receptor organ in the fingertip detects the stimulus (heat), and a nerve impulse is sent up the arm along the dendrite of a sensory neuron. The impulse passes to the cell body of the sensory neuron, which is located in a **ganglion** just outside the spinal cord. A ganglion is simply a mass of neuron cell bodies that usually lies outside the central nervous system. From the cell body, the impulse continues along the axon of the sensory neuron, which terminates in the spinal cord. In the cord the axon synapses with the dendrite of an interneuron. Moving through the cell body of the interneuron, the impulse continues along its axon, which synapses with a dendrite of a motor neuron. The impulse is conveyed through the cell body and out of the spinal cord along the axon of

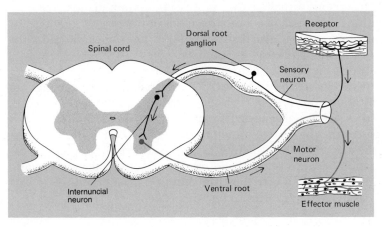

Figure 14.10 Components of a simple reflex arc.

> The normal plantar reflex includes a downward movement (flexion) of the foot and toes. An abnormal plantar reflex, known as the Babinski reflex, results in an upward movement (extension) of the foot and toes.

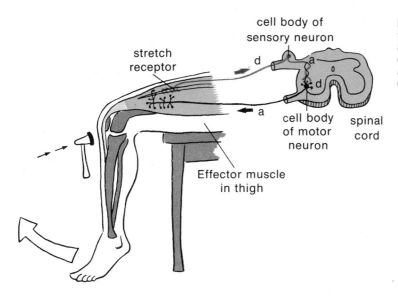

Figure 14.11 The simple knee jerk reflex involves only a sensory neuron and a motor neuron. a = axon; d = dendrite. (From Clark, M. E.: *Contemporary Biology*. 2nd ed. Philadelphia: Saunders College Publishing, 1979, p. 286.)

the dendrite, which terminates in an effector, in this instance, a muscle. The action of the motor neuron causes arm muscles to contract, and the finger is removed from the stove—all in less than a second.

In the spinal cord the impulse is also sent to other interneurons that convey the impulse up the cord to the brain. It is in this fashion that you are informed of the nature of the stimulus, after which your reaction may be to kick the stove, cry, or engage in some other personally appropriate activity.

For the most part, reflexes serve to promote the general welfare of an organism, protecting it from excessive injury and ensuring proper physiological adjustments. Moreover, reflexes are often important in determining the general condition of the nervous system. A normal reflex is a good indication that the part of the nervous system associated with the reflex is functioning as it should. On the other hand, if a particular part of the nervous system has been injured, an abnormal or absent reflex may aid in locating the injured site. For example, stroking the sole of the foot elicits a *plantar reflex,* which normally includes a *downward* movement, or flexion, of the foot and toes. However, if there has been injury to certain nerve pathways in the spinal cord, the response is *upward* movement, or extension, of the big toe and a fanning out of the smaller toes. This abnormal plantar reflex is known clinically as the **Babinski reflex;** it is, however, considered normal in infants under a year old, in whom the spinal cord is not yet fully matured.

In our discussion of the reflex arc, we have mentioned two concepts that need to be examined in order to understand the physiol-

The Nervous System

> The *nerve impulse* results from successive ionic reversals traveling along a nerve fiber. The two ions that set up opposite each other across the nerve fiber membrane are sodium ions and potassium ions.

ogy of the nervous sytem as a whole. The first of these concerns the movement, or conduction, of the nerve impulse along the nerve fiber, and the second involves the passage of the impulse from one neuron to another across the synapse. We shall consider each of these in turn.

THE NERVE IMPULSE

Like all living cells of the body, neurons contain and are surrounded by a watery solution consisting of a variety of ions. As far as the neuron is concerned, there are two different ions specifically involved in the nerve impulse: *sodium ions* (Na^+) and *potassium ions* (K^+). When the membrane covering a nerve fiber (axon or dendrite) is in a *resting state*, i.e., when no impulse is being transmitted, the *inside* of the nerve fiber has a high concentration of potassium ions and a low concentration of sodium ions. The fluid on the *outside* of the nerve fiber is just the reverse, with a high sodium ion concentration and a low potassium ion concentration.

The differences in the concentrations of the two ions across the nerve fiber membrane are such that the interior of the fiber is electrically negative *relative* to the outside. Before it is stimulated, the nerve fiber membrane is mostly impermeable to the sodium ions. But when a stimulus is applied, the permeability of the membrane is altered, and the sodium ions on the outside rush into the fiber. For a fraction of a second, the inrushing sodium ions cause a reversal of the electrical charges of the membrane, i.e., the *inside* of the fiber now becomes positively charged relative to the outside of the fiber, and the *outside* becomes relatively negative. The reversal of electrical charges at one point on the fiber membrane stimulates an adjacent point on the membrane. The electrical charges become reversed at this point, and this reversal in turn stimulates the next adjacent point on the membrane, and so on in a "domino effect" along the length of the fiber. These successive ionic reversals traveling along a nerve fiber constitute the **nerve impulse** (Fig. 14.12).

In the next fraction of a second after the influx of sodium ions, the fiber membrane becomes permeable to the potassium ions, and they rush through the membrane to the outside of the fiber. With the potassium ions now on the outside, the original electrical negativity inside the fiber is restored, and the impulse stops.

Now, how are the original concentrations of sodium and potassium ions eventually restored outside and inside the fiber? This is accomplished through the action of *ion pumps* located in the nerve fiber membrane. Although the exact nature of the pumps is unclear, they are known to transport the ions against a concentration gra-

> The *threshold* is the minimum stimulus required to actuate a nerve impulse. Once the threshold is reached, intensifying the stimulus will not increase the strength of the impulse. Thus a nerve fiber responds completely or not at all; this phenomenon is known as the *all-or-none law.*

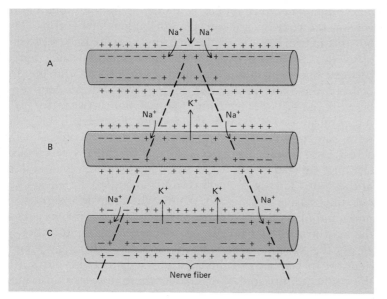

Figure 14.12 Conduction of a nerve impulse. (a) Stimulation of the nerve fiber (arrow) causes Na^+ ions to rush to the interior of the nerve fiber; this reverses the electrical charge on the fiber, so that the outside becomes negative relative to the inside of the fiber. The reversal of charges stimulates adjacent points along the fiber, thereby conducting the nerve impulse. (b) Within a fraction of a second after the influx of Na^+ ions, some of the K^+ ions rush to the outside of the fiber. (c) The outward movement of the K^+ ions continues as the impulse travels along the fiber. In the next fraction of a second, the ion pumps restore the original resting state of the fiber.

dient. Sodium ions, then, are pumped back to the outside of the fiber, and potassium ions are pumped back in. This restores the original resting state of the membrane, and the nerve is ready to "fire" again. The energy to run the ion pumps is supplied by adenosine triphosphate (ATP).

The speed at which an impulse moves along a nerve fiber is proportional to the diameter of the fiber—the greater the diameter, the faster the impulse. In addition, impulses are transmitted more rapidly by those nerve fibers that are covered by a thick, fatty tissue called the **myelin sheath.** The sensory and motor nerves of the withdrawal reflex arc are typical examples of nerves having this sheath. In a period of 1 second, some of the largest nerve fibers can transmit an impulse a distance just a little over the length of a football field—a velocity of nearly 270 mi per hour. At the other extreme, some of the smallest fibers transmit impulses at a rate of only about 1 mi per hour.

A certain minimum, or **threshold,** stimulus is required to actuate a nerve impulse, but a more intense stimulus does not increase the strength of the impulse. In other words, as long as there is a threshold stimulus, the nerve fiber responds completely; it cannot respond to any greater degree even if the stimulus is much more intense than threshold. Thus a nerve fiber responds completely, or it does not respond at all. This phenomenon is known as the **all-or-none law.**

> Nerve impulses move from neuron to neuron across the synapse by way of chemical substances known as *neurotransmitters*.

THE SYNAPSE

The human nervous system is a vast communications network involving the coordinated interplay of billions of neurons. Throughout this network, none of the neurons actually makes physical contact with another neuron. Instead, there is an extremely narrow gap between one neuron and the next. More exactly, this gap is found between the axon terminal of one neuron and the dendrite or cell body of the adjacent neuron. The junction between the axon and the dendrite or cell body is the *synapse* (Fig. 14.13). The nerve impulse can continue from one neuron to another only if it can pass across the synapse. Evidence indicates that this is accomplished by chemical substances known as **neurotransmitters,** released at the tips of the axon. The tip of each axon (or axon branch) encloses numerous tiny saclike vesicles containing the neurotransmitter. As the nerve impulse reaches the tip of the axon, the vesicles rupture and the neurotransmitter is released. The neurotransmitter then diffuses across the synaptic junction and stimulates the membrane of the dendrite or cell body of the adjacent neuron. This neuron transmits the nerve impulse out to its axon, where a neurotransmitter is released that stimulates the next neuron, and so on. In this manner, the nerve impulse travels from one neuron to another.

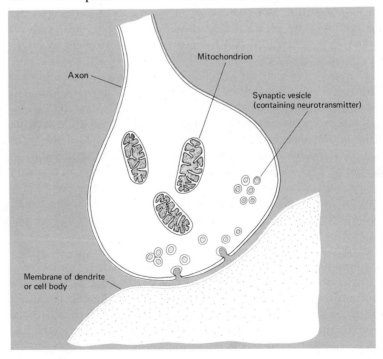

Figure 14.13 Diagram of a synapse. As the nerve impulse reaches the tip of the axon, synaptic vesicles release neurotransmitter into the gap. The neurotransmitter diffuses across the gap and stimulates the dendrite or cell body. After the nerve impulse passes, deactivating enzymes destroy the neurotransmitter.

> Proper nervous control depends on regulating the amount of neurotransmitter present at the synapse. This is accomplished through a *deactivating enzyme*.

Immediately after the neurotransmitter stimulates a neuron, it is quickly destroyed by a type of **deactivating enzyme**. These enzymes are released from a dendrite or cell body where they synapse with an axon. The destruction of the neurotransmitter is necessary because its continued presence at the synapse would bring about constant stimulation, and nervous control would be impossible. Apparently, not all of the available neurotransmitter is released by a single nerve impulse, whereas repeated impulses traveling through the axon at rapid rates may exhaust the supply of neurotransmitter. When this occurs, the nerve impulse cannot travel from one neuron to the next until more neurotransmitter is synthesized in the axon.

There are several known types of neurotransmitters in the human nervous system, the most prevalent of which appears to be **acetylcholine**. Acetylcholine and a number of other neurotransmitters act as *excitatory* chemicals, i.e., they *increase* the permeability of the nerve cell membrane to sodium ions and thus facilitate the passage of the nerve impulse from one neuron to another. However, there are other chemicals released from axons that *decrease* membrane permeability to sodium ions and thus have an *inhibitory* effect on the passage of the nerve impulse. In other words, inhibitory chemicals tend to prevent a nerve impulse from passing across the synapse. There are, then, **excitatory neurons** that release excitatory neurotransmitters and **inhibitory neurons** that release inhibitory substances. Both types of neurons are found within the central nervous system, where together they sort out, integrate, coordinate, and modify the myriad stimuli that impinge upon the body every second.

Various chemicals and drugs affect the activity of the neurotransmitters and deactivating enzymes by causing either excessive stimulation or a slowing of the nervous response. Some of the *nerve gases* and *phosphate pesticides* are known to block deactivating enzymes. The result is that the neurons continue transmission of the nerve impulse, causing strong contraction and paralysis of the skeletal muscles. The *amphetamines* stimulate the release of neurotransmitter, resulting in excessive transmission of the nerve impulse. The effect here is that the nervous system becomes extremely sensitive and more easily excited than normal. The various *tranquilizers* exert their effects by inhibiting the nerve impulse at the synaptic junction. This, of course, results in the well-known calming effect of these drugs. Other substances, including depressants (such as alcohol and anesthetics), stimulants (caffeine, nicotine, and so forth), and toxins (botulin, snake venom, and so on), also influence the transmission and regulation of the nerve impulse.

> The two major divisions of the human nervous system are the *central nervous system* (CNS) and the peripheral *nervous system* (PNS).

DIVISIONS OF THE NERVOUS SYSTEM

Human beings (and most other vertebrates) have two major divisions of the nervous system. The first of these is the the **central nervous system** (abbreviated CNS), which includes the *brain* and the *spinal cord* (Fig. 14.14). Second, there is the **peripheral nervous system,** or PNS, which is divided into the *somatic system* and the *autonomic system*. The somatic system consists of 12 pairs of **cranial** nerves and 31 pairs of **spinal** nerves. The cranial nerves are specialized sensory or motor nerves (or both) that arise from areas in or near the brain. The spinal nerves are connected to the spinal cord and are involved in glandular secretion and control of voluntary muscle movement. The other division of the PNS is the **autonomic nervous system,** or ANS, which controls the involuntary activities of the body, such as breathing, heart rate, and so on. The nerves of the ANS arise from areas in the brain and spinal cord and pass out to the internal organs, blood vessels, and other structures. The ANS is actually two nervous systems, as we shall see.

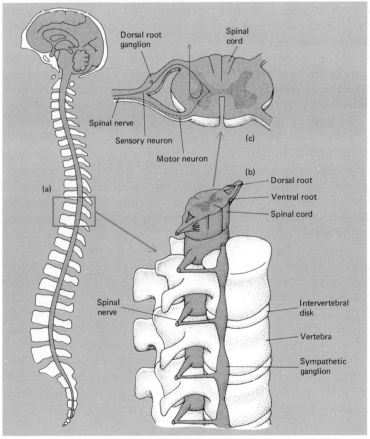

Figure 14.14 (a) The human spinal cord and brain (central nervous system) are housed in the vertebrae and skull, respectively. Nerves running from the spinal cord in spaces between the vertebrae connect the central nervous system with all parts of the body. (b) However, the spinal cord is protected by vertebrae. Adjacent vertebrae are separated by an intervertebral disk, a cartilaginous structure that cushions the vertebrae and further protects the spinal cord. (c) Cross section of the spinal cord.

> The cerebrum, covered by the cerebral cortex, stores and processes vast amounts of information.

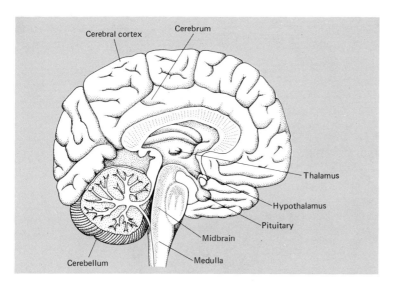

Figure 14.15 Major areas of the human brain (right side removed).

The Central Nervous System: The Brain

In man, *cephalization,* the development and specializing of the anterior or head end of an animal, reaches its highest stage of development. The brain is quite large in relation to the rest of the body and is capable of exerting considerable influence over the functioning of other body organs. The brain of human beings (and other higher vertebrates) is composed of three major divisions: the ***forebrain,*** the ***midbrain,*** and the ***hindbrain*** (Fig. 14.15). It is primarily the increased development of the forebrain that has separated man from his vertebrate relatives. Each of the major divisions of the brain exerts control over various physiological activities. These divisions for the most part do not function separately but interact in a coordinated fashion to influence each other as well as other parts of the nervous system (Fig. 14.16).

The forebrain has three major functional areas: the ***cerebrum,*** the ***thalamus,*** and the ***hypothalamus.*** The cerebrum is the highest brain center, and its surface, the ***cerebral cortex,*** is primarily a storage area for an incredible amount of information. In this area, memories are stored, abstract thinking is performed, and previous motor responses are remembered that can be called forth at almost any time to aid in control of the movements of the body. The cerebrum is capable not only of storing information but of processing information as well. The various physiological activities of speech, vision, hearing, and thought, for example, are analyzed and processed in certain areas of the cerebrum. Much of the control of motor function by the cerebral cortex is voluntary, whereas the midbrain and hindbrain exert control subconsciously.

> The thalamus acts as a relay for nerve impulses passing from the sensory organs to the cerebrum.

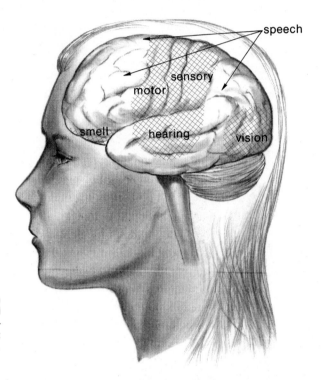

Figure 14.16 Areas of the human brain in which various senses are determined and controlled. (From Clark, M. E.: *Contemporary Biology.* 2nd ed. Philadelphia: Saunders College Publishing, 1979, p. 292.)

The thalamus is located below the cerebrum and is connected by nerve tracts to the cerebral cortex. It functions primarily as a relay area for nerve tracts passing from the sensory organs to the cerebrum. This means that when a stimulus is perceived by a sensory neuron, the resulting impulse is relayed to the thalamus and then to the cerebrum for processing.

The hypothalamus, located just below the thalamus, is vital in controlling many of the involuntary functions of the body. These functions are referred to as *vegetative* functions, meaning that they are necessary for life. Experimental research with various vertebrates has shown that stimulating the hypothalamus indicates that it contains nerve centers that regulate such factors as blood pressure, body temperature, thirst, excretion of water, appetite, pleasure, pain, sleep and wakefulness, and so on. When an electric current is sent through small electrodes implanted in the hypothalamus of an experimental animal, the current acts as a nerve impulse, and the animal can be made to eat, drink water, cringe in fear, become violent, and so forth. The fact that much of our behavior is influenced by our reaction to pleasure or pain imparts considerable significance to the importance of the hypothalamus. Obviously, if we derive plea-

> The midbrain integrates reflex movements of the head in response to sudden visual and audio stimuli.

sure from doing something, we continue doing it; if an action we take is painful, we try to stop.

The hypothalamus manufactures various hormones that are secreted by specialized neurons. From the hypothalamus, the hormones pass to the pituitary gland, which is located just below the hypothalamic area. The pituitary gland then releases the hormones into the bloodstream. In addition, the hypothalamus produces *releasing factors,* which are chemical substances that control the release of other hormones from the anterior portion of the pituitary gland.

By receiving information from the blood and sensory neurons, the hypothalamus is constantly informed of the physiological state of the body. Through the secretion of its hormones and releasing factors, the hypothalamus exerts control over most of the body systems and thus acts to maintain homeostasis.

The second major division of the brain is the midbrain. The midbrain plays a role in the familiar visual reflex and subsequent reflex movements of the head known to vaudevillian enthusiasts as the "double take." The midbrain also apparently integrates reflex movements of the head in response to sudden loud sounds such as firecrackers, sirens, screams of biology students, and so on. It also serves as a relay center between the cerebrum and parts of the hindbrain.

The remainder of the subconscious activities of the body are controlled mostly by the hindbrain. Two of the most prominent structures of the hindbrain are the **medulla oblongata** and the **cerebellum.** The medulla oblongata is the posterior portion of the hindbrain and is continuous with the upper portion of the spinal cord. It functions as a conduction pathway for sensory and motor impulses between the spinal cord and the higher brain centers. It is also the center for several vital reflexes that regulate heartbeat, breathing, blood vessel diameter, coughing, sneezing, and so on. Many of these reflexes are necessary for normal body functioning and even for life itself. You can easily see why a karate chop to the base of the brain is not only unpleasant but actually dangerous.

Of the structures of the brain, the cerebellum is second in size only to the cerebrum. The cerebellum is a coordinating area for the maintenance of muscle tone, balance of the body (equilibrium), posture, and the normal fluid motion of voluntary muscles. Damage to the cerebellum can result in uncoordinated movements such as the inability to touch your nose with your finger, a staggering walk, jumbled mouth movements when attempting to talk, and other outward signs of impaired muscle control.

> The spinal cord, the other major structure of the CNS, is involved in almost all of the reflexes of the body.

The Central Nervous System: The Spinal Cord

The other major structure of the CNS is the spinal cord (Fig. 14.17). The cord continues posteriorly from the medulla oblongata and is surrounded by the bony vertebral column. The spinal cord was mentioned previously in our discussion of the simple reflex arc and, in fact, is involved in almost all of the reflexes of the body. In addition, it conducts sensory and motor impulses between the brain and other body structures. For example, nerve impulses coming into the cord through sensory neurons may be relayed to various areas of the brain. The impulses are relayed along nerve tracts, which often extend the entire length of the cord. From the brain, other nerve tracts carry motor impulses back down the cord. These impulses are then sent out from the cord through motor nerves to the appropriate effector.

An interesting phenomenon occurs at the junction of the spinal cord and medulla oblongata at the base of the brain. Most of the nerve fibers in an area on the left side of the medulla cross over and run down the right side of the spinal cord; similarly, most of the fibers in an area on the right side of the medulla cross over and run down the left side of the cord. This crossing over of nerve fibers is

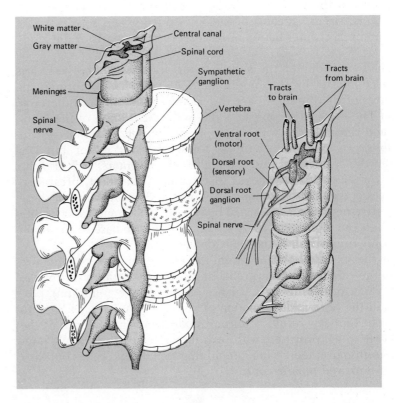

Figure 14.17 The human spinal cord. The cord is surrounded and protected by the meninges and by the vertebral column. The paired spinal nerves protrude through spaces between vertebrae. Gray matter is composed of cell bodies, white matter is made up of bundles of myelinated axons.

> *Decussation* is the crossing over of nerve fibers from the medulla to the spinal cord.

called *decussation*. Since the spinal cord, medulla oblongata, and cerebral cortex are interconnected by nerve tracts, it turns out that the right cerebral cortex is involved in control of voluntary muscle movement on the left side of the body, and the left cerebral cortex has similar control of the right side of the body. With this arrangement, a blow to one side of the head can often cause paralysis or other complications on the opposite side of the body.

Both the brain and the spinal cord are covered by three separate layers of tissues called **meninges** (Gk. *meninx*, membrane). The spaces between the meninges are filled with **cerebrospinal fluid,** a derivative of blood plasma. The meninges and the cerebrospinal fluid function in cushioning and protecting the central nervous system against hard blows and jarring.

Infection of the meninges by certain bacteria or viruses causes an inflammatory condition called *meningitis*. This is a potentially serious condition because the infection can spread to the vital tissues of the central nervous system. Another unfortunate condition—the infamous alcoholic "hangover"—most likely results from irritation of the meninges of the brain.

The Peripheral Nervous System (PNS): Somatic System

The somatic division of the PNS is composed of all the sensory nerve fibers leading from the skin and skeletal muscles, which go to the central nervous system, and all the motor nerve fibers that run from the central nervous system back to the skin and skeletal muscles. These sensory and motor fibers constitute the 31 pairs of **spinal nerves.** The basic reflex arc involves one sensory neuron and one motor neuron of a spinal nerve. Having both sensory and motor components, all the spinal nerves are *mixed* nerves. Each nerve is formed from two short *roots* that emerge from the spinal cord. The *dorsal root* on the back surface of the cord contains the sensory neuron. You will recall that the cell body of a sensory neuron lies in a ganglion just outside the cord. The ganglion appears as an enlargement of the dorsal root.

The *ventral root* of a spinal nerve emerges from the underside of the spinal cord and contains the axon of a motor nerve.

The dorsal and ventral roots unite just outside the spinal cord to form a spinal nerve, each of which passes outward to supply the skeletal muscles and some of the glands of the body (Fig. 14.18).

The somatic system also includes 12 pairs of **cranial nerves,** which originate in or near the brain and extend into the head, neck, and internal organs. Some of the physiological activities in which the cranial nerves are involved include smelling, seeing, tasting, hearing, equilibrium, chewing, and movements of the eyes, facial muscles,

> The ANS is composed almost entirely of motor fibers.

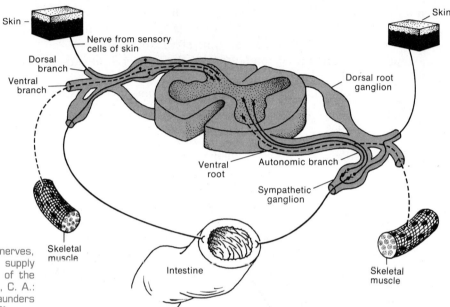

Figure 14.18 Paired spinal nerves, emerging from the spinal cord, supply the skeletal muscles and some of the glands of the body. (From Villee, C. A.: *Biology.* 7th ed. Philadelphia: Saunders College Publishing, 1977, p. 486)

and tongue. One cranial nerve, the *vagus*, which is also a vital component of the autonomic nervous system, supplies nerve fibers to many of the internal organs. Stimulation of the vagus influences the physiological activities of the heart, lungs, intestinal tract, liver, pancreas, and other organs. Some of the other cranial nerves will be mentioned later in association with specific body organs.

The Peripheral Nervous System: Autonomic System (ANS)

Earlier, we stated that the ANS governs the involuntary activities of the body, i.e., those activities over which we have no conscious control—and it does this automatically. Everyone is familiar with the pounding of the heart brought on by the sound of a footstep in the dark. We are familiar, too, with the onset of "goose pimples," a rush of blood to the face, sweating palms, or that surge of energy that comes from "somewhere" when we are tired or frightened. These and many other involuntary reactions of the body are called forth automatically without prior thought. The division of the nervous system responsible for such activities is the autonomic nervous system.

The central nervous system and the somatic nervous system are composed of both sensory and motor fibers; the autonomic nervous system, however, is almost entirely *motor*. This means, of course, that the ANS functions in causing certain responses in various effector organs. The ANS supplies nerves to cardiac muscle, found in the

> *Enkephalins* and *endorphins* are part of a group of brain chemicals known as neuropeptides that act, it is thought, as pain inhibitors.

heart; to smooth muscle, found in the internal organs and blood vessels; and to most of the glands of the body.

Structurally, the ANS actually consists of two sets of nerves, one set belonging to what is termed the ***sympathetic division*** and the other set belonging to the ***parasympathetic division*** (Figs. 14.19 and 14.20). The organs, smooth muscle, and most of the glands of the body receive two nerves, one from each division. This pair of nerves works antagonistically, i.e., one nerve has a stimulating effect, and the other has an inhibitory effect. For instance, consider the example of a pounding heart. When you are apprehensive, the sympathetic nerve to the heart is stimulated, resulting in an increase in the rate and the force of the heartbeat. On the other hand, parasympathetic stimulation (actually, the vagus nerve) slows the heart rate and decreases the force of the heartbeat. The sympathetic system, however, does not always produce a stimulating effect, nor does the parasympathetic system always inhibit. In the digestive tract, stimulation of the smooth muscle walls by the parasympathetic nerves causes increased muscular contraction, whereas sympathetic stimulation relaxes the muscle. This is just the opposite of the effects on the heart. Basically, then, the sympathetic and parasympathetic systems act in concert to help regulate and maintain homeostatic control, or balance. Although the two divisions complement each other, the sympathetic system is usually called forth during periods of stress, and the parasympathetic system is more active in influencing the more routine functions of the body.

ENKEPHALINS AND ENDORPHINS

One of the most fascinating discoveries related to the nervous system concerns a group of brain chemicals known as ***neuropeptides.*** These are short-chain peptides, ranging from 2 to 39 amino acids in length. Many of these have been known for years and include some of the hormones found in the pituitary gland and hypothalamus (these will be discussed in Chapter 19). However, the discovery of two types of neuropeptides—the ***enkephalins*** and ***endorphins***—is relatively recent, and their significance to overall brain function is just beginning to be understood.

Both of these brain chemicals closely resemble the narcotic morphine. It is known that certain regions of the brain contain specific receptors that bind morphine and other similar drugs; moreover, these same receptors also bind enkephalins and endorphins. Research has indicated that these receptors are located in regions of the brain (and spinal cord) concerned with pain and emotions. Since morphine suppresses painful stimuli, it is possible that the enkephalins and endorphins may act in a similar manner. For instance,

482
The Nervous System

Figure 14.19 The sympathetic division of the autonomic nervous system. Axons of sympathetic nerves extend outward from the spinal cord (a) and synapse with neurons lying in ganglia within the *sympathetic trunk* (b), a chain of ganglia lying on either side of the spinal cord (one of the chains is not shown here). Long axons from the neurons within the ganglia pass outward and enter various internal organs. Numbers 1 to 12 on the spinal cord denote the twelve thoracic vertebrae; numbers 1 to 5 below the last thoracic vertebrae denote the lumbar vertebrae, and the last five numbers denote the sacral vertebrae.

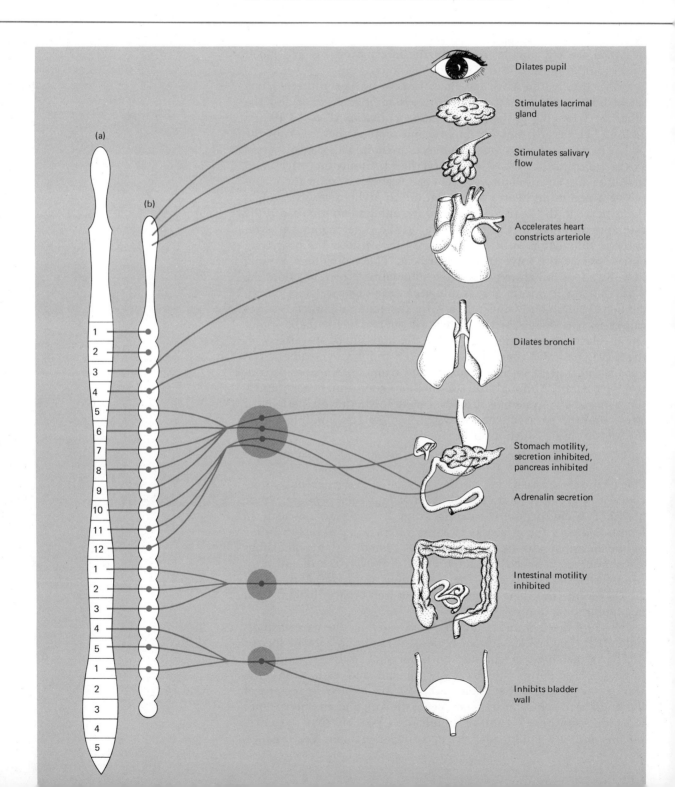

Figure 14.20 The parasympathetic division of the autonomic nervous system. The long axons of the parasympathetic division extend from the brain and sacral vertebrae. When the axons reach the internal organs, they synapse with other neurons, whose axons then extend deep into the organ.

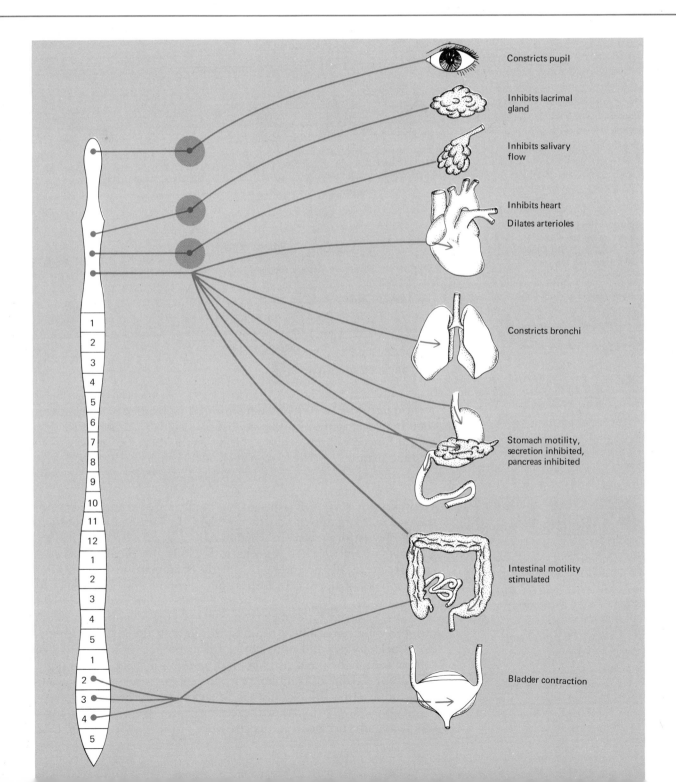

484
The Nervous System

> Coelenterates have a nerve net type of nervous system.

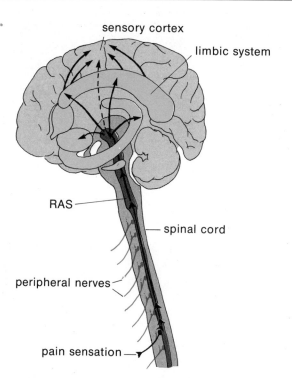

Figure 14.21 The pathway of diffuse pain sensations. Impulse enters the spinal cord, where several synapses occur (arrows). It then travels to the brain, entering the RAS. Further synapses send signals directly to the sensory cortex (dashed line) or into the limbic system, where further relay to the sensory cortex occurs. Enkephalin or opiate receptor sites occur in the colored regions and presumably could inhibit incoming pain impulses at synapses in these areas. The RAS (reticular activating system) stimulates the cortex, causing conscious awareness and wakefulness. The limbic system is a major area of the cerebral cortex and is concerned generally with involuntary aspects of behavior. (From Clark, M. E.: *Contemporary Biology.* 2nd ed. Philadelphia: Saunders College Publishing, 1979, p. 303.)

it is thought that the enkephalins may act as inhibitory transmitters, thereby blocking the transmission of pain signals. Although the biological function of endorphins is less certain, they, too, appear to act in some manner to relieve pain (Fig. 14.21).

NERVOUS SYSTEMS IN LOWER ANIMALS

We find some sort of nervous system in all animal groups except for the acellular organisms and the sponges. In fact, the coelenterates are the lowest organisms in which a true nervous system is present. It is not certain whether neurons with axons and dendrites as such are present in all coelenterates, but these animals do possess some type or types of nerve cells for reception, conduction, and response. Classically, the freshwater coelenterate *Hydra* is used to illustrate the primitive *nerve net* type of nervous sytem. This is simply a network of interlacing nerve fibers running throughout the body and tentacles. Coelenterates and other lower forms do not possess a brain or control center for the coordination of nervous activity. Accordingly, they are incapable of complex reactions characteristic of higher animals and are thus limited in their behavioral patterns.

As we move "up" from the coelenterates, we encounter some significant advancements in the structure of the nervous system (Fig.

Cephalization is the development and specializing of the anterior or head end of an organism.

14.22). You will no doubt recall the free-living planarian, one of the flatworms. This organism typifies the beginning of *cephalization,* the development and specializing of the anterior or head end of an animal. Cephalization has resulted in the concentration of nervous tissue anteriorly to form a "brain." This is important when you consider that the head end of the body usually has first contact with the environment. In planaria, the "brain" is hardly more than a couple of tiny, swollen masses of nervous tissue at the anterior end of the organism. This "brain," however, no doubt has very limited influence over the rest of the nervous system. It is still not a brain in the sense of being a dominant control center. The nerve fibers of the planarian form more definite pathways than those of *Hydra,* as evidenced by the presence of two distinct nerve cords extending from the brain posteriorly to the tail end. Also, these two longitudinal cords are connected by several transverse nerve fibers to form a "ladder type" of nervous system. Tiny nerve segments also connect the brain with the eyespots. Although it is pretty certain that the planarian cannot "see," the specialized nerves in the head end associated with the eyespots permit the organism to respond to light.

A still further advancement is seen in the nervous system of the common earthworm. It, too, possesses a pair of longitudinal nerve cords, but each cord has small, swollen masses called *ganglia* (sing. ganglion) located along its length. A ganglion, as mentioned, is sim-

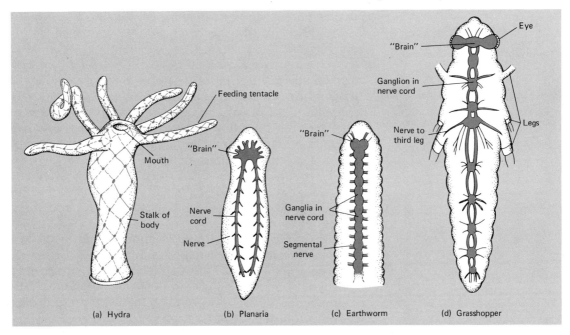

Figure 14.22 Increasing complexity of nervous systems. (a) The simple nerve net of the coelenterate *Hydra* has no ganglia. Synapses conduct in both directions. (b, c, and d) Central nervous systems of the flatworm (*Planaria*), the earthworm, and the grasshopper show increasing organization into ganglia. Cephalization is apparent in the increasing size and importance of the brain.

> *Compound eyes*, made up of many thousands of simple eyes, enable their possessors (usually arthropods) to detect rapidly moving objects well.

ply a mass of neuron cell bodies. Small, specialized nerve fibers run out of each ganglion to the various body structures. Ganglia at the anterior end form the brain, from which the two longitudinal nerve cords extend posteriorly. Overall, the nervous system of the earthworm is considerably more complex than that of the planarian, but much less so than the nervous systems of vertebrates. Nervous systems in other invertebrates—mollusks, echinoderms, insects, and so on—vary somewhat in their degree of complexity. You should keep in mind that it is the type of nervous system that determines the capacity of an organism to communicate with and respond to its surrounding world.

THE SPECIAL SENSES

In this section we shall consider what are called the *special senses* of the body: seeing, hearing, equilibrium, taste, and smell. For the most part, these senses require highly specialized organs for detecting stimuli. The sensory neurons associated with these organs relay impulses to the brain through specific cranial nerves.

Seeing

In some lower animals, such as the planarian, small masses of cells form what are often termed photoreceptors, or light-sensitive organs. These are considered to be very primitive "eyes." The photoreceptors, however, merely enable an organism to sense light and actually do not form a visual image. On the other hand, the eyes of some other invertebrates are quite well developed and do form visual images.

Some arthropods, for example, have **compound eyes,** which in certain species may be made up of many thousands of *simple eyes* (Fig. 14.23). Each simple eye has a nerve fiber that carries the light stimulus to the brain. The vision of arthropods is less acute than that of vertebrates, but their compound eyes are better for detecting motion. They actually see more "slowly" than do vertebrates. A rapidly moving object that appears blurred to us would be seen quite distinctly by a bumblebee. Or, if the bumblebee is flying at a high rate of speed, a stationary object would still be seen distinctly. Obviously, this is a great advantage to any speeding insect flying over the object of its desire.

Eyes quite similar to our own are found in the octopus. The octopus eye, however, apparently is incapable of *stereoscopic vision*, i.e., viewing an object in three dimensions. Consequently, the octopus is very poor at estimating depth or distance. A relative of the octopus, the squid, also possesses highly developed eyes and similarly acute vision.

The Special Senses

> The human eye provides stereoscopic and color vision.

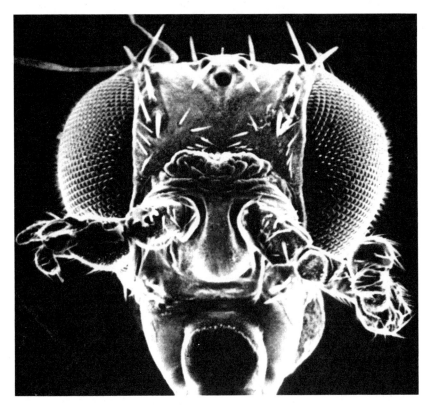

Figure 14.23 Electron micrograph of the compound eyes of an insect (*Drosophila*, the fruit fly). Each compound eye is composed of numerous simple eyes. (From Ebert, J. D., Loewy, A. G., Miller, R. S., and Schneiderman, H. A.: *Biology*. New York: Holt, Rinehart and Winston, 1973, p. 436).

The Human Eye. The human eye is a remarkable organ, providing stereoscopic and color vision (Fig. 14.24). Although we do not see as "slowly" as arthropods, our vision is much more acute. The eyeball of human beings is a spherical mass of gelatin-like substance, the

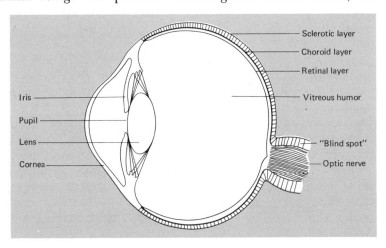

Figure 14.24 Basic structure of the human eye.

> The dilation and constriction of the *pupil* is controlled by the autonomic nervous system.

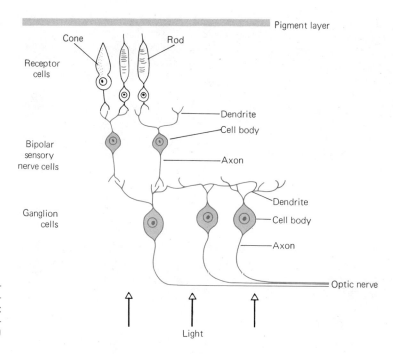

Figure 14.25 Cross section of the retina of a vertebrate. Nerve cells are colored. (From Camp, P. S., and Arms, K.: *Exploring Biology*. Philadelphia: Saunders College Publishing, 1981, p. 418.)

vitreous humor, partially surrounded by three layers of specialized tissues. The outer layer, called the **sclera** or "white of the eye," is a tough, protective covering. The **cornea** is the anterior part of the sclera and is transparent, permitting light to enter the eye. The middle tissue layer, the **choroid,** contains a rich network of tiny blood vessels. The third, inner layer of tissue is the **retina.** The retina is a complex structure and contains two types of highly specialized photoreceptor cells—the **rods** and **cones.** We depend upon the rods for seeing in dim light, whereas the cones are responsible for color vision (Fig. 14.25).

Immediately behind the transparent cornea is the **iris.** When we speak of brown eyes, blue eyes, green eyes, or whatever, we are describing the color of the iris. In the center of the iris is the **pupil,** a small opening through which light passes to the inside of the eyeball. The dilation and constriction of the pupil is controlled by the autonomic nervous system. The size of the pupil depends upon the degree of constriction of tiny muscles in the iris. When we get excited or angry, the sympathetic nerves cause our pupils to dilate. Bright light entering the eye stimulates parasympathetic nerves, which cause the pupil to constrict; in the dark the parasympathetic nerves are less stimulated, and the pupil dilates. This occurrence is familiar

> *Nearsightedness*, the ability to see distinctly at short distances only, and *farsightedness*, the ability to see distinctly at long distances only, result from an abnormally shaped eyeball or a malfunctioning lens.

to anyone who has walked out of a darkened theater into bright sunlight, or vice versa.

Behind the iris is the **lens.** This structure refracts (bends) incoming light rays so that they all come to the same point on the retina. This is important in perceiving a clear image. Axons from special nerve cells on the back of the retina all converge to form the **optic nerve** (one of the cranial nerves), which passes into the brain. The point at which the optic nerve passes out of the eyeball is a "blind spot." This area is so named because of its lack of photoreceptor cells.

Functionally, the eye is much like a camera. Light rays from an object pass through the transparent cornea, into the pupil, and through the lens. The lens focuses the image of the object on the retina. The image is inverted, but the brain interprets it as being right side up. Thus, our visual interpretations of what we see are performed not by the eye but by the brain.

Sometimes, incoming light rays do not come together to focus exactly on the retina. This may be the result of an abnormally shaped eyeball or some malfunction of the lens. If, for example, the eyeball is too long, the light rays focus on a point in *front* of the retina. The lens may be functioning properly, but it cannot focus the image of the object directly on the retina. An elongated eyeball results in what we refer to as **nearsightedness,** i.e., the ability to see distinctly at short distances only (Fig. 14.26). If the eyball is too short, in which case the image is focused *behind* the retina, we have the condition of **farsightedness** (Fig. 14.27). Both of these conditions may be corrected with glasses that bend the light rays so that they focus the image exactly on the retina. *Cataracts* are a very common abnormality of the lens. These are cloudy or opaque areas caused by the breakdown of the lens proteins. Cataracts, which occur more often as we grow older, obscure or prevent normal transmission of light to the retina.

Glaucoma is another serious eye condition in which fluid pressure within the eye rises to extremely high levels. Such pressure can

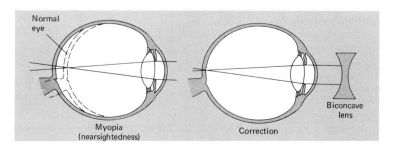

Figure 14.26 Myopia, or nearsightedness; note how the image focuses in front of the retina. A biconcave lens is used as a corrective device for this condition.

490
The Nervous System

> Sound vibrations are passed from the eardrum, through the bones of the middle ear, to the *cochlea*, a fluid-filled structure of the inner ear. Fluid motion in the cochlea stimulates surrounding neurons to send out nerve impulses that are passed to the *auditory nerve* and then on to the brain for interpretation.

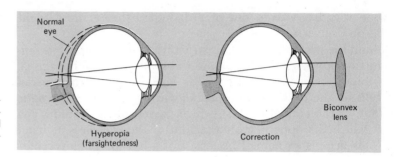

Figure 14.27 Hyperopia, or farsightedness: note how the image focuses behind the retina. A biconvex lens is used as a corrective device for this condition.

damage the optic nerve and may also shut off the supply of blood to the eye by compressing the tiny surrounding blood vessels. If the pressure is not relieved, permanent blindness may result. Glaucoma can be treated surgically to facilitate drainage of the fluid, or certain drugs that promote drainage, such as pilocarpine, can be used.

Hearing

Hearing is a nervous response to the vibration of sound waves. Only the vertebrates and some of the arthropods have true receptor organs for detecting sound. The antennae of the mosquito, for example, contain tiny hairs that detect vibrations (Fig. 14.28). Other insects such as crickets and grasshoppers possess an eardrum-like membrane that vibrates in response to sound waves. Depending upon the type of insect, the membrane may be located on the foreleg, thorax, or abdomen. Most aquatic vertebrates have sound receptors located near the surface of the skin. Vibrations in the water stimulate tiny hair cells located in the receptors. Movement of the hair cells sets up nerve impulses that are transmitted to the ear.

In human beings, the ear is divided into three main areas: the **outer ear, middle ear,** and **inner ear** (Fig. 14.29). The familiar external flap of the outer ear collects sound waves, which then pass down the **ear canal** to the **eardrum.** As sound waves strike the eardrum, it begins to vibrate. The vibrations of the eardrum are relayed to three very tiny interconnected bones in the middle ear. These tiny bones have been named according to their shapes and are called the **hammer, anvil,** and **stirrup.** The hammer is attached directly to the eardrum, and the stirrup is attached to a fluid-filled structure in the inner ear called the **cochlea.** The anvil is situated between the hammer and stirrup.

Leading from the cochlea is the **auditory nerve** (one of the cranial nerves), which goes to the cerebral cortex. When the eardrum is stimulated by sound waves, its vibrations are picked up by the hammer, which also begins to vibrate. From the hammer, the vibrations pass to the anvil, which transmits the vibrations to the stirrup. As the stirrup vibrates, it sets up movements of the fluid in the cochlea.

> The *eustachian tube* connects the middle ear with the throat, allowing air to enter or leave the middle ear so that the pressure inside and outside the ear is equalized.

Tiny cells within the cochlea move in response to the fluid, and, by stimulating surrounding neurons, these cells set up nerve impulses that are passed to the auditory nerve. Through the auditory nerve the impulses enter the brain, where they are interpreted.

Located on the cochlea is a small membrane-covered opening called the *round window*. As the fluid in the cochlea moves, there is a certain amount of pressure exerted against the walls inside the cochlea. Some of this pressure can be relieved by the bulging in and out of the round window membrane. If such were not the case, your ears would "ring" constantly.

The cavity of the middle ear has a small opening that leads into the *eustachian tube.* This tube connects the middle ear with the pharynx (throat). Changes in air pressure, such as occur while one is flying in an airplane at 30,000 ft, could cause damage to the eardrum. The eustachian tube, however, permits air to enter or leave the middle ear so that pressure inside and outside the ear is equalized. By swallowing or yawning, you actually change the pressure in the middle ear to correspond with the air pressure around you. Unfortunately, the eustachian tube is also one passageway through which microorganisms cause infections in the middle ear.

Figure 14.28 Antennae of an immature male mosquito *(upper)* and a sexually mature male. *(lower)*. The hairs on the antennae of the mature male mosquito stand out and are very sensitive to the sound vibrations produced by the wings of a female.

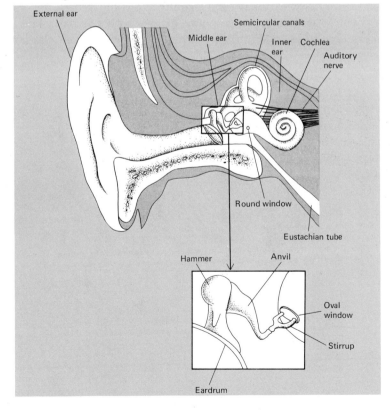

Figure 14.29 Basic structure of the human ear.

> The three fluid-filled *semicircular canals* maintain the position of the body when the head is suddenly rotated by informing the brain via nerve fibers of the body's orientation in three dimensions.

Equilibrium Normally, you know "up" from "down" and whether or not your head is rotating. The sense organs responsible for orienting your body with respect to gravity and for detecting rotation of the head are located in the inner ear (Fig. 14.30).

Near the cochlea are three fluid-filled canals called **semicircular canals.** These function in maintaining the position of the body when you suddenly rotate the head. The canals are oriented at right angles to each other so that they represent the three dimensions of height, width, and depth. Inside the canals there are small crests of tissue to which are attached tiny hair cells. Leading from each hair cell are sensory nerve fibers that pass to a branch of one of the cranial nerves. This nerve ultimately leads to the cerebellum (remember our earlier reference to the equilibrium function of the cerebellum). When you turn your head, the hair cells are stimulated by the movement of the fluid within the canals. The resulting nerve impulses inform the cerebellum of the movement of the head, and if necessary, to adjust the position of the body accordingly. This adjustment is necessary, for example, if you are running and suddenly change direction. Another well-known experience is the feeling of dizziness after being spun around a few times. Once you stop spinning, the fluid in the semicircular canals continues to move back and forth in a manner similar to water being carried in a flat pan. This movement continues to stimulate the hair cells, which relay the impulses

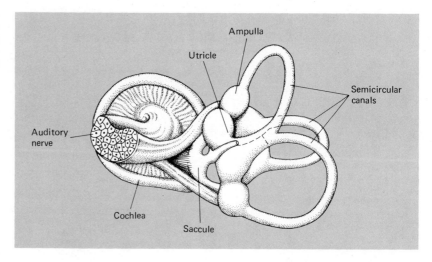

Figure 14.30 The sense organs of equilibrium in the inner ear. The semicircular canals detect rotation of the head, and the utricle informs the brain of the position of the head with respect to the pull of gravity.

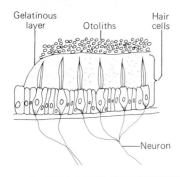

Figure 14.31 The position of the head with respect to gravity is detected by hair cells within the utricle. Tiny calcium carbonate crystals called *otoliths* stimulate the hair cells in response to the pull of gravity. (From Arms, K., and Camp, P. S.: *Biology.* 2nd ed. Philadelphia: Saunders College Publishing, 1982.)

to the brain. You may have stopped spinning, but your brain doesn't think so. The dizziness usually subsides when the fluid comes to rest.

The other sense organ of equilibrium is the **utricle** ("small bag"), which is located near the semicircular canals in the inner ear. The function of the utricle is to inform the brain of the position of the head with respect to the pull of gravity. More simply put, it tells you if you are upside down, right side up, or whatever. Inside the utricle there are hair cells coated with a sticky substance in which tiny crystals of calcium carbonate are embedded. These crystals are called **otoliths** ("ear stones") (Fig. 14.31). Here again, sensory nerve fibers from the hair cells enter a nerve branch leading to the cerebellum. The weight of the otoliths causes stimulation of the hair cells in the following manner: If you are standing erect with your head upright, the otoliths compress *downward* on the hair cells. If you bend forward at the waist, as if to touch your toes, the weight of the otoliths bends the hair cells *forward*. When you throw your head *backward*, the hair cells bend in that direction. Finally, if you were standing on your head, the weight of the otoliths would *pull* on the hair cells. In each one of these instances, stimulation of the hair cells sends nerve impulses that inform the brain of the position of the head.

Taste and Smell

The special senses of taste and smell are called chemoreceptive senses. **Chemoreception** simply means the receiving of chemical stimuli (Fig. 14.32). Many of the lower animals have chemical receptors that are used to locate food or to move away from some noxious substance. Such behavior in response to chemical stimuli is called **chemotaxis.** Some insects have extremely sensitive chemoreceptors on their antennae and mouth parts. A species of hawk moths, for example, is some 200 times more sensitive than we are in detecting the taste of sucrose (table sugar). Taste receptors of fish and amphibians may be widely distributed in the skin. Many fish, by tasting the water, can detect food or the presence of an enemy fish in the area.

In man, the receptors for taste are located in the **taste buds** found mostly on the tongue. There are four basic tastes that we can detect: sweet, sour, salty, and bitter. Sweet and salty are tasted mostly on the tip of the tongue, sour along the sides, and bitter at the back of the tongue. The various substances that we taste must be in solution. This is one of the functions of saliva in the mouth. Taste impulses are transmitted from the taste buds along several cranial

494
The Nervous System

> *Taste buds* on the tongue are the sensory receptors for taste; *olfactory cells* found in epithelial tissue lining the upper portion of the nasal cavities are the sensory receptors of odors.

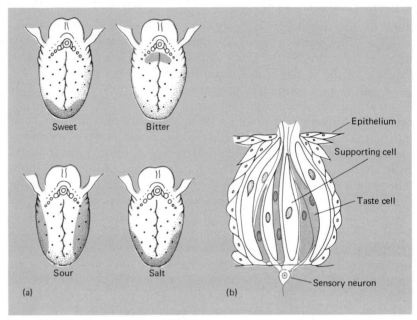

Figure 14.32 Chemoreceptors involved in the sense of taste in humans (a) The distribution of taste buds sensitive to sweet, bitter, sour, and salt on the tongue. (b) A taste bud in the tongue.

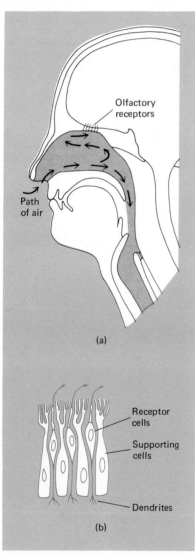

Figure 14.33 (a) Olfactory receptors are found in the top of the nasal cavity in a region of spongy tissue. (b) Chemoreceptors in the nose are free dendrites of neurons in the olfactory epithelium.

nerves to the brain. As we grow older, the taste buds tend to degenerate, and we become less sensitive to various tastes or flavors.

The sense of smell varies considerably among animals. Most mammals have a highly developed sense of smell, whereas this sense is very poor in most birds. Some whales are believed to have no sense of smell at all. Various forms of animal behavior are influenced by smell. Smell may be used by fish such as salmon to return to their spawning grounds. The mother seal smells her newborn infant and thus is able to recognize her pup from the others.

Our sense organs of smell are found in epithelial tissue lining the upper portion of the nasal cavities. Neurons called *olfactory cells* lie within the epithelium and are the actual sensory receptors of odors (Fig. 14.33). Nerve impulses from the olfactory cells are transmitted along the olfactory nerve (one of the cranial nerves) to the cerebral cortex, where they are interpreted as odor. We have millions of these olfactory cells, which are capable in many instances of detecting thousands of different odors. It is interesting that we remember odors quite well and recognize them if we smell them at a later time. This applies with equal emphasis to body odors ("B.O.") or Chanel No. 5.

NERVOUS SYSTEM DISORDERS AND DISEASES

Malfunction of the human nervous system encompasses a great variety of disorders and diseases, many of which affect individuals of all ages. Nervous system disorders not only are the direct cause of many deaths but also contribute indirectly to the cause of death related to other body systems.

Epilepsy. Epilepsy, a term derived from a Greek word meaning "a condition of seizure," is one of the most common nervous system disorders. Epilepsy results from uncontrolled firing of neurons in the cerebrum. It has been suggested that some brain neurons may be hypersensitive to normal stimuli and consequently send out impulses repeatedly. There are several types of epileptic seizures, such as *petit mal* (Fr. "minor attack"), in which there is a lapse of consciousness for only a few seconds, and a type called *grand mal* (Fr. "major attack"), characterized by prolonged unconsciousness and strong muscle contractions. Petit mal occurs primarily in children, whereas grand mal is much more prevalent in adults. Many conditions, including diseased blood vessels in the brain, meningitis, alcoholism, and so on, are known to cause epileptic seizures; however, seizures often arise for reasons as yet unknown.

Parkinson's Disease. Parkinson's disease, named for the English physician James Parkinson (1755–1824), is perhaps the most common nervous system disorder of the elderly. The disease is characterized by rigidity of the muscles and usually tremor of the body when at rest. It results from destruction of a mass of neurons and ganglia near the base of the brain.

Multiple Sclerosis (MS). This disease attacks the central nervous system, particularly among young people. Either the brain or the spinal cord, or both, develops hard patches that destroy the myelin sheath surrounding a nerve. Although the cause of MS is unknown, it is suspected of being virally induced. Some of the symptoms of MS include weakness of the limbs, numbness, blurred vision, and mental changes, including loss of memory.

Descriptive Terms. Many abnormal conditions can be attributed directly to brain lesions. A **lesion** is a harmful change in the structure of an organ, resulting from injury or disease. Some of the more common disorders associated with brain lesions are presented in Table 14.1

Finally, the human mind is sometimes subject to influences that lead to the development of a **mania** or a **phobia**. A mania is an extraordinary craving, attraction, interest, and so on; a phobia is an exaggerated fear or aversion. Listed in Table 14.2 are a few of the more common manias and phobias, along with some rather unusual examples. The word or words to the right describe the mania or phobia.

Table 14.1 BRAIN LESION DISORDERS

Aphasia: Loss of ability to speak, write, or understand words.
Alexia: Inability to read; *dyslexia* refers to impaired reading ability.
Apraxia: Loss of coordinated movement.
Dysbasia: Difficulty in walking.
Paralogia: Impaired ability to reason.
Paraphrasia: Disordered arrangement of words when speaking.

Table 14.2 MANIAS AND PHOBIAS

MANIAS	*PHOBIAS*
Pyromania: setting fires	*Acrophobia:* heights
Monomania: one idea or subject	*Algophobia:* pain
Megalomania: delusions of grandeur	*Astraphobia:* lightning; thunderstorms
Dipsomania: alcohol	*Cynophobia:* dogs
Kleptomania: stealing	*Eremophobia:* being alone
Logomania: talking too much	*Hemophobia:* blood
Phagomania: abnormal craving for food	*Musophobia:* mice
Hypnomania: sleep	*Nyctophobia:* night; darkness
Arithromania: counting things	*Ponophobia:* work
Nostomania: one's home; extreme homesickness	*Tonitrophobia:* thunder
Hippomania: horses	*Vermiphobia:* worms
	Xenophobia: strangers

And in the unlikely event you are afraid of *everything*, you then suffer from *panaphobia*.

REVIEW OF ESSENTIAL CONCEPTS

1. The *nervous system*, in conjunction with the endocrine system, controls, integrates, and coordinates the activities of the human body.
2. The *neuron* is the structural and functional unit of the nervous system. A neuron consists of dendrites, an axon, and the cell body.
3. There are three basic types of neurons: *sensory*, *motor*, and *interneurons*.
4. The fundamental physiological activity of the nervous system is the *reflex*. All reflexes involve a sensory neuron and motor neuron and usually an interneuron.
5. The *nerve impulse* is basically a succession of sodium and potassium ion reversals along the length of a nerve fiber.
6. The junction between the axon terminal of one neuron and the dendrite or cell body of an adjacent neuron is called a *synapse*.
7. The nerve impulse is conveyed across the synapse by *neurotransmitters*, chemicals released at the axon terminal.
8. Neurotransmitters are destroyed at the synapse by *deactivating enzymes* released by the dendrites or cell body.
9. Neurons may be either *excitatory* or *inhibitory*, depending upon the type of chemical released at the axon terminal.
10. Various drugs and chemicals affect the transmission of the nerve impulse across the synapse; the result is excessive stimulation or slowing of the nervous response.
11. The *central nervous system* (CNS) consists of the brain and spinal cord. The *peripheral nervous system* (PNS) consists of the *somatic system* and the *autonomic system* (ANS).
12. The major functional areas of the human forebrain are the *cerebrum*, *thalamus*, and *hypothalamus*.
13. The midbrain is a major reflex center and relays impulses between the cerebrum and hindbrain.
14. The major functional areas of the hindbrain are the *medulla oblongata* and the *cerebellum*.
15. *Enkephalins* and *endorphins* are brain chemicals thought to be involved in the suppression of pain stimuli.
16. The *spinal cord* functions as a reflex center and conducts sensory and motor impulses between the brain and other body structures.
17. The somatic nervous system consists of 31 pairs of *spinal nerves* and 12 pairs of *cranial nerves*.
18. Spinal nerves are *mixed* nerves, having both sensory and motor components.
19. The autonomic nervous system governs the involuntary activities of the body. The ANS consists of the *sympathetic* and *parasympathetic* divisions, which act antagonistically.

20. The nervous systems of lower organisms, such as coelenterates, flatworms, roundworms, and so on, exhibit varying degrees of complexity. The most complex nervous systems are found in the vertebrates, in which cephalization is highly advanced.
21. The *special senses* of man include seeing, hearing, equilibrium, taste, and smell.

APPLYING THE CONCEPTS

1. Cite an example of homeostasis and explain why homeostasis is crucial.
2. Outline a basic reflex arc involving an interneuron. What is the total role of the spinal cord in a reflex arc?
3. Describe the process by which a nerve impulse travels along an axon.
4. What essential physiological activities occur at the synapse? What do these activities accomplish?
5. What might be some possible results of damage to the hypothalamus?
6. How does the cerebellum contribute to normal body functioning?
7. In what manner are the sensory and motor components separated in a spinal nerve?
8. Cite a specific example of the antagonistic effects of the sympathetic and parasympathetic divisions of the ANS.
9. What is the significance of cephalization? How is it involved in the nervous behavior of an organism?
10. Give the anatomical locations of the major structures of the human eye. How is the human eye functionally like a camera?
11. How are sound vibrations conveyed from the outer ear to the brain?
12. What are the functional differences between the semicircular canals and the utricle?
13. Why are taste and smell referred to as chemoreceptive senses?

skin, skeleton, and muscles

THE ESSENTIAL OBJECTIVES

You have understood this chapter when you are able to:

1. Describe the general structure and functions of human skin.
2. Compare the functions of cartilage and bone.
3. Describe the structure and function of an Haversian system.
4. Describe some of the disorders and diseases associated with the human skeletal system.
5. Describe the general structure of skeletal muscle and relate it to the sliding filament theory of muscle contraction.
6. Describe the chemical mechanism of muscle contraction.
7. Describe the general structure and function of a motor unit.
8. Differentiate between "slow-twitch" and "fast-twitch" muscle fibers.
9. Distinguish between isotonic and isometric muscle contraction.

The skin is the body's outer protection against invading microorganisms and chemicals and is also an organ of widely diversified functions. The skeletal system, while providing support and protection, also acts in conjunction with the skeletal muscles to move the body.

PREVIEW OF ESSENTIAL TERMS

epidermis The epithelial cells that form the thin outer layer of skin.
dermis The connective tissue that forms the thick inner layer of skin beneath the epidermis.
subcutaneous layer A layer of fatty tissue lying beneath the dermis.
melanin A brownish-black pigment present in the hair and skin of many animals.
sebaceous gland A skin gland that secretes an oily substance (sebum).
ligament A tough connective tissue that binds one bone to another or anchors a body organ.
tendon A fibrous connective tissue that binds a muscle to a bone.
Haversian system A structural unit of bone composed of concentric layers of bone proteins, in which bone cells and blood vessels are embedded.
myofibril A tiny contractile structure of skeletal muscle composed of thick (myosin) and thin (actin) protein filaments.
motor unit A motor neuron plus all the skeletal muscle fibers it stimulates.
synovial fluid A thick lubricating fluid secreted by connective tissue membranes located at the moveable joints of the body.
slow-twitch fiber A type of skeletal muscle fiber that contracts slowly and is resistant to fatigue.
fast-twitch fiber A type of skeletal muscle fiber that contracts rapidly and fatigues quickly.
neuromuscular junction The site at which a motor neuron axon and muscle fiber come in close contact.
creatine phosphate A high-energy compound that stores energy in muscle cells; the energy in creatine phosphate can be transferred to ADP to form ATP when necessary.
atrophy A wasting away or degeneration of a muscle, nerve, or other body part.
myoglobin A hemoglobin-like pigment in muscle cells that binds strongly with oxygen.

Skin, Skeleton, and Muscles

> All vertebrates have an *endoskeleton* of bone or cartilage that grows along with the rest of the body. Invertebrates, such as insects and crustaceans, have an *exoskeleton* that must be shed periodically to allow for growth of the animal.

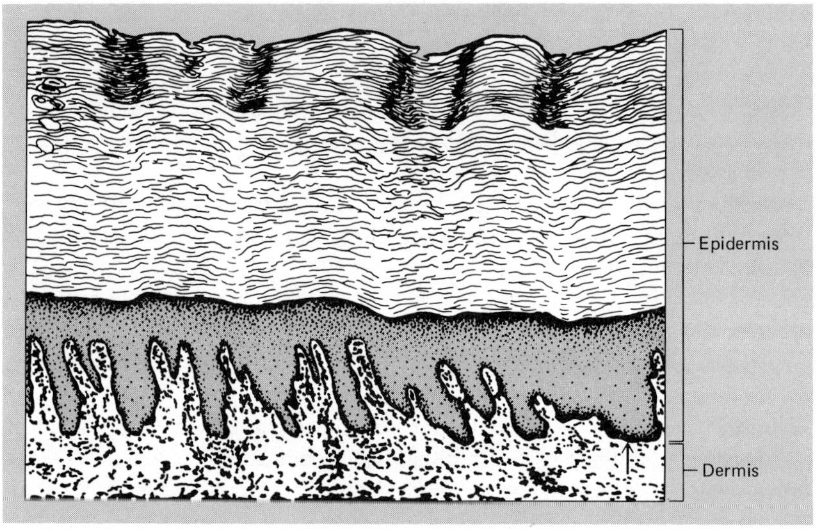

Figure 15.1 A section of human skin taken from a fingertip. The outermost layer is composed of dead epithelial cells. The deepest layer *(arrow)* consists of cells that undergo mitosis to replace dead cells that are sloughed off.

— Epidermis

— Dermis

INTRODUCTION

The skin is a remarkable organ, forming the tough, outer protective covering of the body. It consists of layers of epithelial and connective tissues in which are embedded glands, hair, nails, nerve fibers, and blood vessels (Fig. 15.1). The fur, feathers, and scales of other animals are actually derivatives of skin. Not only is skin a protective covering, preventing the entrance of bacteria and many harmful chemicals, but it also aids in regulating body temperature; produces vitamin D; secretes milk, ear wax, and oil; responds to a variety of sensory stimuli; and functions as an organ of excretion.

The skeleton of man and other animals functions in support and protection and serves to anchor the skeletal muscles used in movement of the body (Fig. 15.2). The skeleton also stores inorganic salts of calcium, phosphorus, magnesium, and sodium. In addition, the *marrow* of certain bones produces most of the body's blood cells. All of the vertebrates, with two exceptions, have an **endoskeleton** composed of bone and cartilage. The two exceptions are the sharks and their relatives and the jawless vertebrates—lampreys and hagfishes—which have an endoskeleton of cartilage only. The invertebrates, such as insects and crustaceans, characteristically have an outer skeleton, or **exoskeleton,** that must be shed periodically to allow for growth of the animal (Fig. 15.3). In contrast, the endoskeleton of vertebrates grows along with the rest of the body.

Skeletal muscles give shape to the body and act in conjunction with the skeletal system to bring about movement (Fig. 15.4). Skeletal muscles are supplied by the spinal nerves and are involved in the

Introduction

> *Skeletal muscles* give shape to the body and act with the skeletal system to bring about voluntary movements of the body.

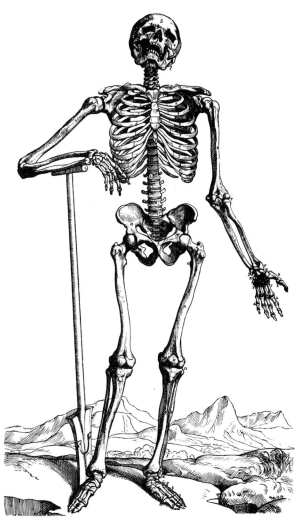

Figure 15.2 The human skeleton, drawn by the anatomist Andreas Vesalius in the sixteenth century. (From Clark, M. E.: *Contemporary Biology.* 2nd ed. Philadelphia: Saunders College Publishing, 1979, p. 310.)

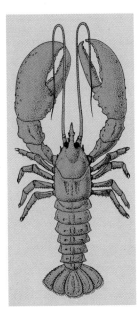

Figure 15.3 The exoskeleton of the lobster must be shed periodically to permit growth of the animal.

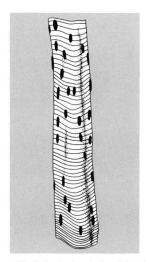

Figure 15.4 A single skeletal (striated) muscle fiber from a human calf muscle. This type of muscle is attached to bones and is responsible for movement of the body.

voluntary movements of the body and in many reflexes. Movement of the body is brought about by muscles acting as antagonistic pairs, that is, while one muscle is contracting, its "partner" is relaxing. For example, bending the arm at the elbow requires contraction of the biceps muscle while its antagonist, the triceps muscle, is relaxing. To straighten the arm, the reverse occurs, and the triceps contracts

Skin, Skeleton, and Muscles

> *Ligaments* attach bones to each other at the joints; *tendons* attach muscles to bones.

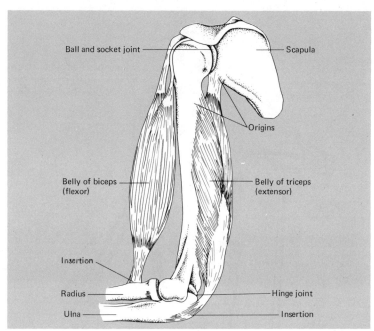

Figure 15.5 The arrangement of skeletal muscles across a joint (elbow). The *origin* of a muscle refers to its fixed, or immoveable, end; the moveable end is the *insertion*. When the biceps contracts, the insertion is pulled toward the origin. Note that the muscles originate and insert on the bones by means of tendons.

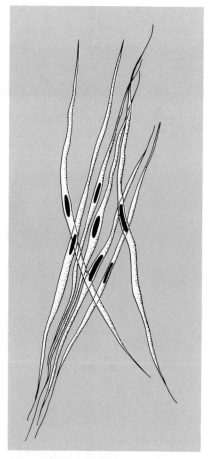

Figure 15.6 Individual smooth muscle cells. Smooth muscle composes part of the walls of most of the tubular structures in the body.

while the biceps relaxes. Movement of the body is made possible because the opposite ends of skeletal muscles are attached to different bones across a *joint* (Fig. 15.5).

A joint is a junction where two bones meet, or *articulate*. With the muscle situated across a joint, the bones may move toward each other, as when the biceps contracts, or the bones move away from each other, as when the triceps contracts. Some joints, of course, such as those between the bones of the skull, are immoveable, whereas others display varying degrees of movement. At the joints the bones are attached to each other by **ligaments,** which are composed of tough, fibrous connective tissue. Muscles are attached to bones by similar types of connective tissue known as **tendons.**

In contrast to skeletal muscle, smooth muscle is characterized by its slow, rhythmic contraction and relaxation (Fig. 15.6). Smooth muscle action is regulated by the autonomic nervous system and thus is not under conscious control. The walls of almost all tubular structures in the body are composed in part of smooth muscle (Fig. 15.7). The digestive tract, urinary bladder and ducts, lymph and blood vessels, and so on contract and relax in a wavelike motion that aids in moving their contents onward. A type of smooth muscle is found in the iris of the eye. Contraction of the iris muscles regulates the constriction and dilation of the pupil. This action is controlled by the

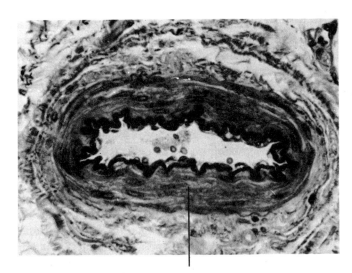

Smooth muscle wall

Figure 15.7 Cross section of an artery showing location of smooth muscle. (From Leeson, C. R., and Leeson, T. S.: *Histology.* 2nd ed. Philadelphia: W. B. Saunders Co., 1970, p. 222.)

autonomic nervous system in response to light intensity or one's emotional state.

Cardiac muscle is found only in the heart and is structured somewhat like skeletal muscle (Fig. 15.8). However, it contracts rhythmically like smooth muscle, although the contractions last

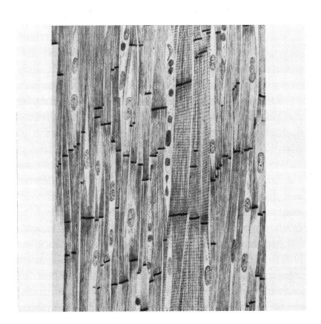

Figure 15.8 Longitudinal section of human cardiac muscle, showing intercalated disks (From Bloom and Fawcett: Textbook of Hystology. 9th Edition. Philadelphia: W.B. Saunders Company, 1968, page 289.

> Human skin is composed of an outer layer of epithelial cells—the *epidermis*—and a deep inner layer—the *dermis*.

longer. The contraction phase of cardiac muscle is called *systole* (Gk. "contraction"), and the relaxation phase is called *diastole* (Gk. "lengthening"). The rate and force of the heartbeat is controlled by the autonomic nervous system in response to a variety of stimuli. We shall consider the heart and its action in our study of the circulatory system in Chapter 17.

THE SKIN

Human skin is composed of an outer layer of epithelial cells, which form the **epidermis**. Epithelial cells forming the top layer on the surface of the body are dead and eventually are sloughed off. These cells are replaced from beneath by the deepest layer, which lies just above the thick inner skin layer, the **dermis** (Fig. 15.9).

In addition to functioning as a barrier to harmful microorganisms and chemicals, the epidermis contains specialized cells that produce the pigment **melanin.** Each of us inherits his or her skin color, which is lighter or darker depending upon the amount of melanin produced. Dark-skinned individuals have cells that produce larger quantities of melanin than do the cells of fair-skinned individuals. An *albino* is a person whose epidermal cells do not produce melanin at all. The production of melanin is increased by exposure to the sun, resulting in the darkening of the skin known as a suntan. In the process, melanin absorbs much of the ultraviolet light of the

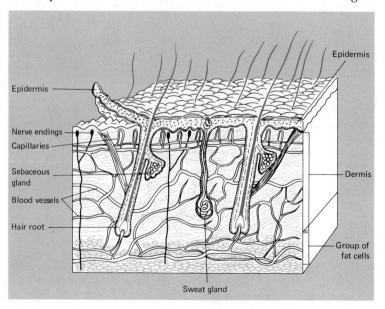

Figure 15.9 Drawing of a section of skin showing some of the major structures.

> Sweat glands regulate body temperature through the release and evaporation of water; they also excrete waste products in the form of dissolved salts and urea.

sun and thereby protects the skin from becoming sunburned or blistered. However, if the skin is overexposed to the sun, melanin cannot absorb all of the ultraviolet rays, and the skin becomes inflamed. Skin cells also use ultraviolet rays to produce vitamin D, a substance essential to the normal growth and maintenance of bone. The skin cells synthesize a derivative of the steroid *cholesterol,* which is converted into vitamin D upon exposure to the sun.

The dermis consists primarily of tough connective tissue, which is supplied with a rich network of small blood vessels. Under the control of the autonomic nervous system, the vessels constrict and dilate, an activity that aids in regulating body temperature. When the body temperature is dropping below normal, the blood vessels constrict, keeping the warm blood deeper in the tissues away from the surface of the body. As the body temperature rises, the blood vessels dilate, and the warm blood, being closer to the surface, can release heat to the outside of the body.

The dermis also contains a variety of sensory nerve receptors that are sensitive to touch, pressure, heat, cold, and pain. Hair is a specialized structure of skin that extends from the surface down into the dermis. Like the skin itself, hair derives its color from the relative abundance of melanin. Red hair, however, contains a pigment not found in hair of any other color. Each hair on the body arises from a structure called a **hair follicle** (L. *folliculus,* bag). Attached to each follicle is a small muscle that, when it contracts, causes the hair to "stand on end." The familiar "goose pimples" result from very tight contraction of the follicle muscles.

Fingernails and toenails are derived from specialized epithelial cells of the epidermis. The most actively growing area is the whitish, half-moon–shaped *lunula* at the base of the nail.

Two major types of glands are found in the skin: **sebaceous,** or oil, glands and **sweat** glands. Sebaceous glands secrete an oily substance called *sebum,* which helps keep the hair and skin soft and waterproof.

There are over 2.5 million sweat glands in the body; these function in the regulation of body temperature and in the excretion of waste. An increase in body temperature, as during exercise, stimulates the release of sweat. As the sweat evaporates, it absorbs heat from the skin, and the body becomes cooler. Sweat is mostly water in which there are dissolved salts and a nitrogen waste product called **urea.** To some extent, then, skin functions as an organ of excretion by releasing the salts and urea from the body through perspiration. However, profuse sweating may deplete the body of needed salts, which must be replaced. This is why someone engaged

Figure 15.10 Sites in men and women where fat tissue tends to accumulate.

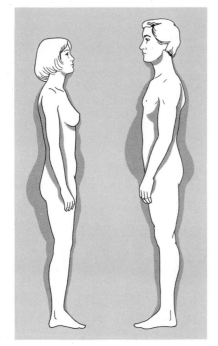

THE SKELETAL SYSTEM

Figure 15.11 A section of hyaline ("resembling glass") cartilage. This cartilage is the most common type found in the body and is located in the nose, in parts of the respiratory passageways, and in many joints. (From Bloom, W., and Fawcett, D. W.: *A Textbook of Histology*. 2nd ed. Philadelphia: W. B. Saunders Co., 1970, p. 213.)

in strenuous exercise occasionally takes salt tablets or a salt-fortified liquid.

The skin around the ear canal is provided with special wax glands that secrete *cerumen,* more commonly known as ear wax. Since the epithelial cells lining the ear canal are sloughed off periodically, the ear wax is removed along with them and normally does not tend to accumulate. Ear wax functions in keeping foreign substances from entering the ear canal.

Beneath the dermis is a layer of fatty (adipose) tissue called the **subcutaneous layer.** The fat cells serve to insulate the body from excessive heat loss and constitute a source of reserve energy. Moreover, subcutaneous tissue anchors the skin to the muscles and bones below and protects underlying organs from potentially damaging shocks and blows. On an esthetic note, the contours of the female body result from the particular distribution of subcutaneous fat. In fact, the average thickness of subcutaneous fat in women is about twice that of normal men. Most of the fat in women is concentrated in the breasts, hips, and legs; in men, fat accumulates primarily around the abdomen (Fig. 15.10).

The skeletal system of man is composed of two types of connective tissues: *cartilage* and **bone**. There are several kinds of cartilage that perform various functions in the body (Fig. 15.11). For example,

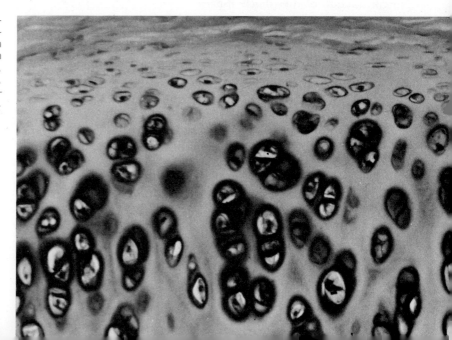

> As osteoblasts (bone cells) develop into mature osteocytes, concentric rings of bone proteins are deposited around a central blood vessel, forming the *Haversian system*. Once formed, calcium salts and other minerals infiltrate the bone tissue, causing it to harden.

one type of cartilage is found at the ends of bones where they meet at a joint. This same cartilage is also laid down in the embryo and is later replaced by bone. The ears contain a type of elastic cartilage, and still another type functions as a cushion between bones that would otherwise press against each other. For instance, most of the vertebrae in the spinal column are separated by discs of cartilage that act as shock absorbers by preventing the vertebrae from jamming together when one walks or runs (Fig. 15.12).

Although cartilage is a relatively hard connective tissue, bone, of course, is even more so. Much of the hardness of bone is due to its high concentration of inorganic salts, such as calcium carbonate and calcium phosphate.

Bone tissue itself is composed of two kinds of proteins that are produced by the bone cells, or *osteoblasts*. These cells replace the cartilage cells after the embryo has developed for about 2 months. The osteoblasts deposit the bone proteins in concentric rings around a central blood vessel. As the bone is deposited, the osteoblasts become trapped in little spaces called **lacunae** between the layers. The osteoblasts develop into mature *osteocytes*, which have long cellular processes that extend through the bone tissue. The concentric layers of bone protein, along with the bone cells and blood vessels, make up what is called the **Haversian system** (Fig. 15.13). In the center of the system, the blood vessel is contained within the **Haversian canal**. The extensions of the bone cells pass through tiny canals, or **canaliculi**, that permit nutrients to pass from the blood vessel to the bone tissue. After the formation of each Haversian system, the calcium salts and other minerals infiltrate the bone tissue, which then takes on its characteristic hardness.

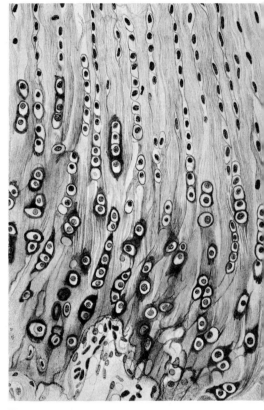

Figure 15.12 A section of fibrocartilage, a very tough tissue found in the discs between the spinal vertebrae. (From Bloom, W., and Fawcett, D. W.: *A Textbook of Histology.* 2nd ed. Philadelphia: W. B. Saunders Co., 1970, p. 219.)

THE DIVISIONS OF THE SKELETON

There are reportedly 206 bones in the human body, although the number may vary slightly from one person to another. As mentioned, the skeletal system functions in support, protection, movement, storage of inorganic salts, and production of various blood cells.

The human skeleton is conveniently divided into (1) an ***axial skeleton,*** consisting of the *skull,* the *vertebral column,* and the *rib cage;* and (2) the ***appendicular skeleton,*** which includes the *upper and lower limbs,* the *shoulders,* and the *pelvis* (Fig. 15.14).

The Axial Skeleton

The skull consists of the bones of the face along with the ***cranium,*** which houses the brain. On the undersurface of the cranium there

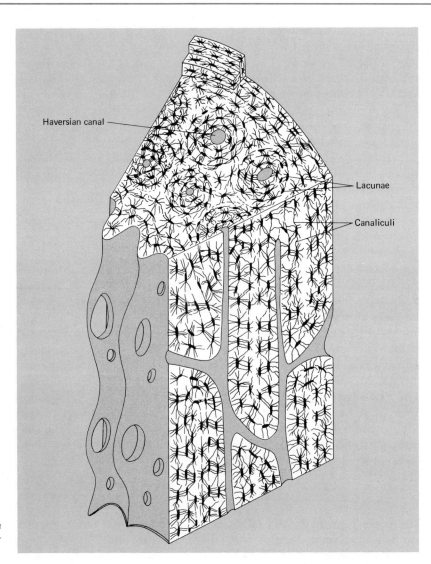

Figure 15.13 A portion of bone tissue showing the circular Haversian systems.

is a large opening through which the spinal cord emerges posteriorly. The teeth are set in sockets in the bones of the upper jaw (maxilla) and the lower jaw (mandible). The joint between the upper and lower jaws is quite remarkably constructed and permits mobility in four directions—up and down and laterally to both sides. This feature is of great significance in the complex motions of speaking and chewing.

Figure 15.14 Diagrams of the human body showing (a) the bones of the axial skeleton and (b) the bones of the appendicular skeleton.

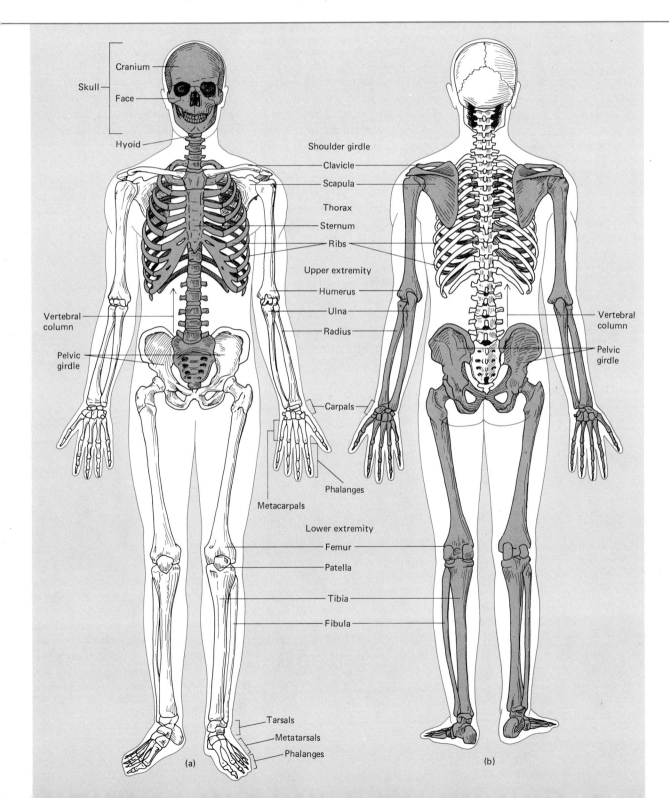

Skin, Skeleton, and Muscles

The ***vertebral column,*** or spine, is the vertical axis of the body and supports about 80 percent of the body's weight (Fig. 15.15). The individual bones of the column are the ***vertebrae,*** most of which are separated from each other by pads of tough cartilage known as ***intervertebral discs.*** There are actually 33 separate vertebrae that make up the vertebral column, but in the adult some of the vertebrae fuse, resulting in a total of 26 bones.

The spine extends from the base of the skull to the hip bones, or ***pelvis,*** and is quite flexible, permitting a variety of bending, turn-

Figure 15.15 Human vertebral column. Most of the vertebrae are separated by pads of fibrocartilage called intervertebral discs, which function as shock absorbers.

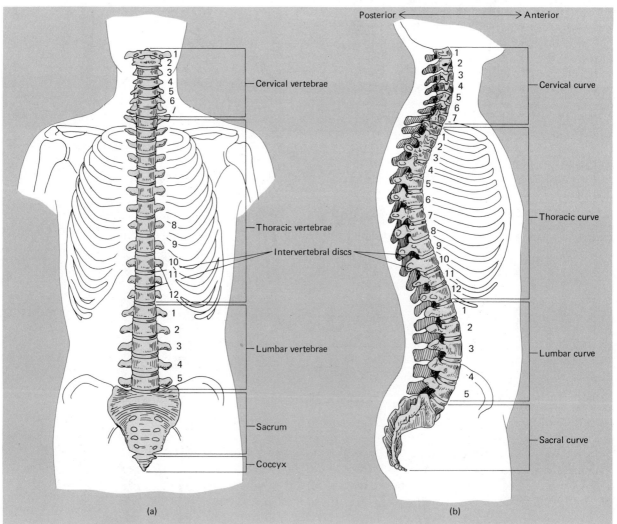

> The five distinct groups of vertebrae from top to bottom are the *cervical* (7), the *thoracic* (12), the *lumbar* (5), the *sacral* (5 fused to form sacrum), and the *coccygeal* (4 fused to form coccyx).

The Divisions of the Skeleton

ing, and rotating movements. A major function of the vertebral column is protection of the delicate **spinal cord,** which passes along most of the length of the column through openings, or canals, in the vertebrae.

Adjacent vertebrae of the spine are so arranged that there are spaces provided for the exit of the 31 pairs of spinal nerves. Back injuries sometimes cause the intervertebral discs to herniate, or protrude, and press against the root of a spinal nerve, causing pain or even paralysis of a part of the body. Contrary to popular usage, a "slipped disc" really does not exist; instead, the disc bulges out, as mentioned, and presses against a nerve.

There are five distinct groups of vertebrae in the spine, each of which has a specific role to play. The first group consists of seven **cervical,** or neck, vertebrae, which permit a wide range of head movements. The next 12 **thoracic,** or chest, vertebrae serve as attachment sites for ribs. The five **lumbar** vertebrae form the small of the back and support most of the body's weight. The next five **sacral** vertebrae are fused together to form the **sacrum.** The sacrum is connected to the pelvis. Finally, there are usually four small *coccygeal* vertebrae that are fused to form the **coccyx,** or "tailbone." Each of these groups of vertebrae further serves to anchor a variety of muscles and ligaments.

The rib cage consists of the **sternum,** or breastbone, 12 pairs of **ribs,** and the 12 thoracic vertebrae to which the ribs are attached dorsally.

The sternum is the elongated, centrally located bone on the front of the chest. It lies as a protective barrier over the heart and anchors the ribs. The marrow of the sternum is an important site of blood cell formation.

All 12 pairs of ribs attach posteriorly to the thoracic vertebrae. The first seven pairs, or *true ribs,* attach anteriorly directly to the sternum by means of flexible cartilage. The next three pairs, or *false ribs,* are attached to the cartilage of the ribs above them and not directly to the sternum. The last two pairs are called *floating ribs* because they are completely unattached anteriorly. The flexible nature of both the ribs and the cartilage allows for the movements of breathing and provides a resilient shield against blows to the chest.

The Appendicular Skeleton

The arrangement of the bones of the upper and lower limbs follows a similar pattern. The upper arm bone, extending from the shoulder to the elbow joint, is the **humerus.** There are two forearm bones, the **radius** and the **ulna.** The radius is located on the thumb side of

> The leg bones include the *femur*, the *tibia*, the *fibula*, and the *patella*.

the forearm and is somewhat shorter than the ulna. The hand is connected to the radius, so that when the radius rotates, the hand rotates with it.

The wrist bones are the **carpals,** of which there are eight. The palm of the hand consists of five **metacarpals,** and the **phalanges** (sing. phalanx) are the bones of the fingers. There are a total of 14 bones in the fingers of each hand, 3 in each finger and 2 in each thumb.

The upper leg bone, the **femur,** extends from the pelvis to the knee joint. The femur, also called the thigh bone, is the longest bone in the human body.

There are two bones in the lower leg, the **tibia,** or shinbone, and the smaller **fibula,** located laterally (to the outside) of the tibia. At the junction of the femur and tibia at the knee joint, there is a small, flat bone called the **patella,** or kneecap.

The ankle consists of seven **tarsals,** the largest of which is the **calcaneus,** or heel bone. Five **metatarsals** form the foot, and like the hands, 14 **phalanges** make up the toes of each foot.

The shoulders, or **pectoral girdle,** consist of a right and left **clavicle,** or collarbone, and a right and left **scapula,** or shoulder blade. The lateral ends of the clavicles actually join with the scapulae, whereas the other ends join with the sternum in front.

The pelvis, or **pelvic girdle,** includes two large bones (hip bones), each of which is composed of three fused smaller bones. Included in these bones are the right and left **pubic bones,** which join together at the bottom of the pelvis. The joint between the pubic bones is tough cartilage like that found in the intervertebral discs. Prior to childbirth the cartilage in women softens, allowing the pubic bones to spread apart. The back of the pelvis is formed by the sacrum and coccyx.

Usually, but not always, there is a distinct difference between the pelvises of males and females. Generally, the female pelvis is wider, with more flaring hip bones. Like many other bones of the body, the pelvic bones of females are lighter and smoother than those of the male (Fig. 15.16).

There are a variety of abnormalities and disorders that affect the bones and joints. Among the more common and severe are fractures, dislocations, and sprains. A **fracture** is simply a broken bone. A **dislocation,** of which there are several kinds, is essentially a displacement of a bone from its proper position. For example, a severe blow may displace the humerus from its socket in the shoulder, or the bones of the leg may be separated at the knee joint. If the ligaments around a joint are wrenched or torn, but the bones are not dislocated, the condition is called a **sprain.**

In *rheumatoid arthritis* connective tissue is inflamed, synovial fluid in the joint thickens, surrounding cartilage is damaged, and bone tissue may erode; ultimately the joint may become immobile.

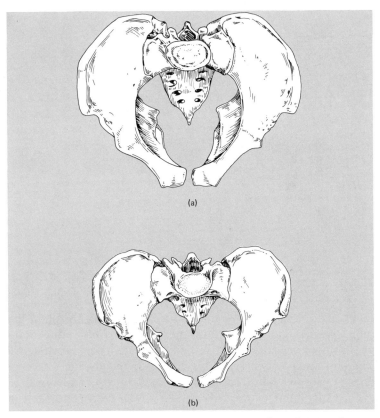

Figure 15.16 Comparison of the (a) male pelvis and (b) female pelvis (top view).

A chronic disease of the joints is ***rheumatoid arthritis,*** characterized by inflammation of connective tissue. Where bones meet at a joint, there are connective tissue membranes that secrete ***synovial fluid,*** which functions as a lubricant. When these membranes become inflamed, the synovial fluid becomes thickened and loses its normal lubricating capacity. In addition, the surrounding cartilage is damaged, and there may be erosion of bone tissue. Eventually, the bones across the joint may fuse together, making movement impossible.

THE MUSCULAR SYSTEM

In this section we shall focus our attention on the structure, function, and types of skeletal muscles, the voluntary muscles that move the body (Fig. 15.17). You will recall that the skeletal muscles are under the control of spinal nerves from the somatic division of the peripheral nervous system.

A skeletal muscle is composed of individual muscle ***fibers,*** which in turn are composed of numerous ***myofibrils*** lying parallel to each

> How muscles contract is presently explained by the *sliding filament theory.*

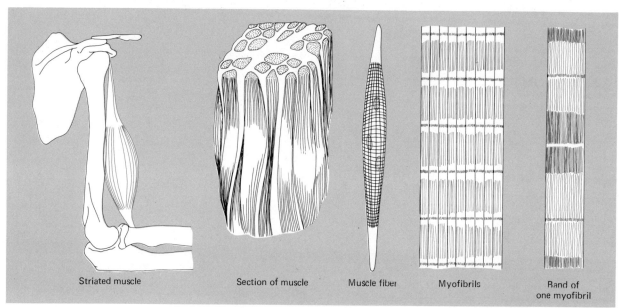

Figure 15.17 General structure of a skeletal (striated) muscle.

other. The myofibrils are made up of large, thick protein filaments called **myosin** and smaller thin ones called **actin.** In the composition of the myofibril, the large and small proteins are arranged lengthwise so that they overlap. This arrangement produces the light and dark striations characteristic of skeletal muscle. More importantly, the arrangement of the two proteins with respect to each other is the key factor in the mechanism of muscle *contraction*. Within the myofibril, the thick protein filaments and the thin protein filaments are connected along their lengths by tiny *cross bridges*. When a muscle contracts, the thick and thin filaments are pulled past one another by the cross bridges. This action of muscle is called the ***sliding filament theory*** of muscle contraction (Fig. 15.19). Contraction of skeletal muscle is controlled by the somatic nervous system. The spinal nerves that supply the skeletal muscle fibers contain the axons of motor neurons. Outside a muscle the axon divides, sending its branches to individual muscle fibers. Thus, one axon may supply many muscle fibers; together, the motor neuron and its group of muscle fibers constitute a ***motor unit.*** An impulse from the motor neuron causes all of the fibers in the motor unit to contract at the same time. Some of the larger muscles in the body contain hundreds or thousands of motor units. Whether a muscle contraction is strong or weak depends upon the number of motor units that are stimulated at one time.

Actually, the number of motor units supplying a skeletal muscle depends upon the type of muscle in question. For example, the tiny

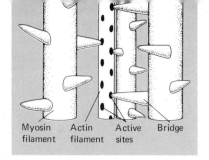

Figure 15.18 Enlarged view of myosin and actin arrangement in skeletal muscle.

muscles that move the eye have a high ratio of motor neurons to the number of muscle cells. This means that the axon branches of a single motor neuron supply only a few muscle cells—for the eye muscles, the ratio is one motor neuron for about every three muscle cells. This arrangement is necessary because movement of the eye may be rapid and yet minute, calling for precision control.

In large muscles, such as those of the thigh, the ratio of motor neurons to the number of muscle cells is very low. A single motor neuron may send its axon branches to hundreds of muscle cells. In this case, stimulation of the motor neuron causes coarse, rather than precision, movement of the leg.

The fibers of skeletal muscle are classified as either *slow-twitch* fibers or *fast-twitch* fibers. Both types of fibers are usually found within a single muscle. Slow-twitch muscle fibers contract slowly and are highly resistant to fatigue; fast-twitch fibers contract rapidly and fatigue quickly. The thigh muscles, which contract slowly, usually have a greater proportion of slow-twitch fibers, whereas the rapidly acting eye muscles contain predominantly fast-twitch fibers.

Muscles in which there are more slow-twitch fibers are sometimes referred to as *red muscle*. This is a reference to the reddish tint caused by the presence of **myoglobin,** a pigment similar to hemoglobin. Myoglobin combines with oxygen inside the muscle and releases the oxygen when needed by the cells. Red muscle is also surrounded by an extensive network of blood capillaries.

Muscles in which fast-twitch fibers predominate are called *white muscle.* These contain little myoglobin and have a much less extensive supply of capillaries than does red muscle.

The tip of the motor nerve axon and a muscle fiber actually do not make physical contact but are separated from each other at the **neuromuscular junction** (Fig. 15.20). The axon and muscle fiber make contact through a **synapse,** as described in Chapter 14. When a nerve impulse reaches the tip of the axon, a neurotransmitter is released that diffuses across the synapse and stimulates the muscle fiber to contract. The contraction of muscle fibers involves the same basic process of sodium ion and potassium ion movement that causes the nerve impulse (Chapter 14).

In the "resting state" the interior of the muscle fiber is electrically negative relative to the outside. When the fiber is stimulated by the neurotransmitter, the sodium ions rush to the inside of the fiber. In response, calcium ions (Ca^{++}) are released from vesicles within the fiber, and the myofibrils contract. Within a fraction of a second, the potassium ions diffuse out of the fiber, restoring the positive electrical charge on the outside. The original distribution of the sodium ions on the outside of the fiber and potassium ions on the

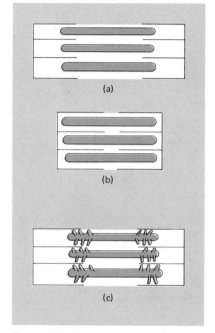

Figure 15.19 Arrangement of myosin (thick horizontal bands) filaments in a myofibril of skeletal muscle. (a) The appearance of the filaments when the muscle is relaxed and (b) when the muscle is contracted. (c) According to the sliding-filament theory, myosin possesses tiny cross bridges that hook onto the actin filaments at specific sites. By bending inward, the cross bridges pull the actin filaments past the myosin filaments. The cross bridges then release, bend outward and reattach to the actin, and pull again.

Skin, Skeleton, and Muscles

> The energy for muscle contraction comes from ATP and creatine phosphate.

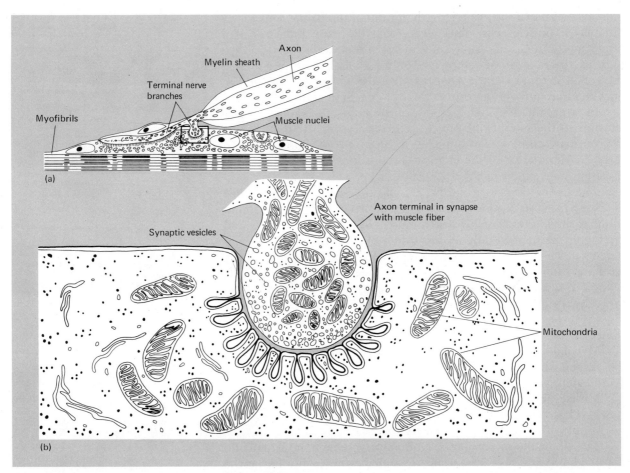

Figure 15.20 Diagram of (a) a motor nerve axon making contact with a muscle fiber at a neuromuscular junction. (b) An enlargement of the rectangle in (a), showing the synaptic region between one axon terminal and the muscle fiber.

inside is subsequently established by the ion pumps. This restores the original "resting state" of the muscle fiber. In the meantime, a deactivating enzyme at the synapse destroys the neurotransmitter so that the muscle fiber is not continually stimulated. Finally, the calcium ions are transported back into the vesicles, and the muscle relaxes.

The energy for muscle contraction comes from adenosine triphosphate (ATP) and another high-energy compound found in

> Heavy exercise creates a lactic acid buildup in the muscles.

muscles called **creatine phosphate**. This substance stores energy that can be used to generate ATP from ADP during periods of high energy use (exercise, for example) when ATP is being broken down rapidly.

In our discussion of glycolysis and cellular respiration in Chapter 6, it was pointed out that in the absence of sufficient oxygen, muscle cells produce *lactic acid* and only a small amount of ATP. Such anaerobic respiration accompanies strenuous exercise and incurs an "oxygen debt" to the body that must be "paid back." This is accomplished when exercise stops and deep breathing delivers oxygen to the muscles. With sufficient oxygen, the lactic acid is eventually converted to pyruvic acid, and the supply of ATP and creatine phosphate is restored.

The *size* of muscle fibers increases with exercise, but the *number* of fibers does not. Enlargement of muscle fibers occurs through **isotonic** exercise or **isometric** exercise. During isotonic exercise, such as weight lifting, the muscles shorten, and the tension on the muscles remains fairly constant throughout the contraction. In isometric contraction, which results when one pushes against an immoveable object, for instance, the muscles do not shorten, but the tension increases. In our everyday activities, the muscles of the body contract both isotonically and isometrically. Walking or running primarily involves isotonic contractions, whereas standing still requires prolonged tensing of the leg muscles, an instance of isometric contraction.

Repeated, *forceful* use of skeletal muscles is required to increase their size and strength. As the muscle fibers increase in size, they are capable of contracting more strongly. On the other hand, muscles that are not used tend to *atrophy* ("waste away"); consequently, they decrease in size and strength. Muscles also atrophy if their motor neurons are severed or are otherwise nonfunctional.

Skeletal muscles continually receive a few impulses from the nervous system, which maintains a slight tension or contraction of the muscles. This is referred to as **muscle tone** and keeps the muscles from becoming flabby or completely relaxed, should they need to respond quickly. However, during sleep the impulses subside, and the muscles become totally relaxed.

As we grow older, the muscle fibers in all of our muscles decrease in number, and once they die, they cannot be replaced by new fibers. Instead, the muscle fibers are replaced by a tough connective tissue. Soon after death, the body becomes stiff owing to par-

> The human body contains approximately 650 muscles, all of which account for nearly one-half the total body weight.

tial contraction of the skeletal muscles, a condition called *rigor mortis*. The condition persists until the dead body begins to decompose.

Some Major Skeletal Muscles

There are approximately 650 muscles in the human body, accounting for nearly one-half of the total body weight. Skeletal muscles vary in size from the massive muscles of the thigh to the tiny and delicate muscles that move the bones of the middle ear (Fig. 15.21).

The muscles of the head and neck include those responsible for facial expressions, speaking and chewing, and moving the head. One of the most powerful muscles of the body is the **masseter** (Gk. "chewer"), used to raise the jaw and clench the teeth. The ability to wink or close the eyes is a function of the **orbicularis oculi** (L. "eye circle") muscle, one of which surrounds each eye. The **orbicularis oris** (L. "mouth circle"), or "kissing muscle," encircles the mouth, causing the lips to close and pucker.

By name, one of the favorite muscles of anatomy students is the **sternocleidomastoid,** the powerful muscle on either side of the neck. It turns the head from one side to the other or pulls the head downward toward the chest. The name of the muscle is taken from the sternum, collarbone, and mastoid process, a bony projection behind the ear.

On the back, at the base of the neck, a triangular muscle, the **trapezius,** extends across the shoulders and down the middle of the back. The trapezius raises and lowers the shoulder blades and aids in maintaining posture. When the shoulders are in a strained position for long, as in writing or painting, massaging the trapezius usually brings welcome relaxation.

To either side, and extending below the trapezius, is the broadest muscle of the back, the **latissimus dorsi.** It pulls the shoulders downward and back and moves the humerus. Capping each shoulder is the **deltoid** (shaped like the Greek letter delta, Δ) muscle, which is active in raising and lowering the arms.

The large muscle on either side of the upper chest is the **pectoralis major.** It functions in moving the arm up and down and across the chest.

The only muscle on the back of the upper arm, and the only one that extends the elbow, is the **triceps** muscle. On the front of the upper arm is the **biceps** muscle, which flexes the elbow joint and rotates the hand outward.

The muscles of the forearm are responsible for the movements of the wrist, hand, and fingers. Mucles on the front of the forearm

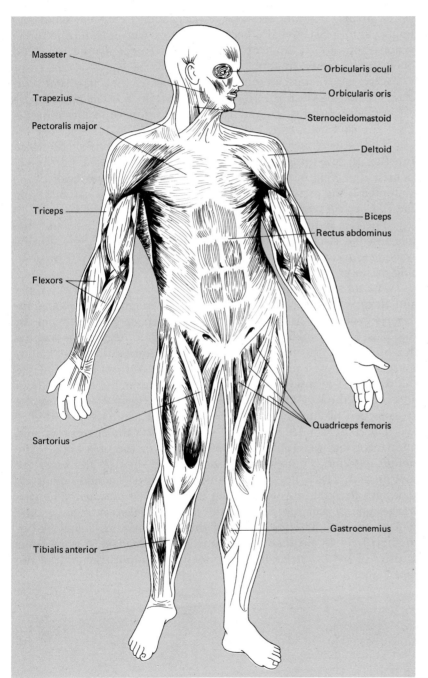

Figure 15.21 Some of the major skeletal muscles of the human body.

> Muscles acting as *flexors* bend body parts inward or forward at the joint, as in closing the hand and fingers into a fist. Muscles acting as *extensors* bend body parts outward or backward at the joint, as in opening the hand and fingers. This movement is opposite to that caused by the flexors.

act as **flexors,** bending the wrist inward and closing the hand and fingers into a fist. On the back of the forearm, **extensors** perform actions opposite to those of the flexors; they extend the wrist and open the hand and fingers.

Two sets of muscles vital to breathing are the **intercostal** muscles, situated between the ribs. The *external* intercostals raise the rib cage during inspiration, and the *internal* intercostals lower or depress the rib cage during exhalation.

There are four major muscles of the abdomen, all of which function in breathing, bending the body, and compressing the abdomen. The long muscle running vertically along the exterior of the abdomen is the **rectus abdominis** (L. *rectus,* straight). It extends from the ribs and sternum down to the pubic bones of the pelvis.

The major hip or buttocks muscle is the **gluteus maximus,** the heaviest muscle in the body. It causes the leg to straighten during walking, running, or climbing stairs and raises the body from a sitting or stooping position.

The muscles of the thigh include two major groups of extensors and flexors. On the front and sides of the thigh is the **quadriceps femoris** group, consisting of four muscles that extend the knee. On the back of the thigh is the "hamstring" group, three muscles that flex and rotate the leg. In addition, the **sartorius** (L. *sartor,* tailor) muscle, the longest muscle in the body, extends from the hip across the thigh to the inside of the knee. It also flexes the leg and thigh and rotates the leg. The sartorius muscle is so named because of the cross-legged position assumed by a tailor while sitting and sewing.

The muscles of the lower leg serve to move the ankle, foot, and toes. Lying on the front of the lower leg over the tibia bone is the **tibialis anterior,** a muscle that flexes the foot upward and turns the foot inward. Two other muscles, one in the lower leg and one in the foot, also flex the foot upward but turn the foot outward. The **gastrocnemius,** or calf muscle, and the adjacent **soleus** muscle flex the foot downward. The lower end of the gastrocnemius is attached to the heel bone by means of the strong **Achilles tendon.** Various other muscles of the lower leg move the toes and support the arch of the foot.

DISEASES AND DISORDERS OF THE SKIN, SKELETON, AND MUSCLES

SKIN

Dermatitis. Dermatitis, meaning "inflammation of the skin," occurs in a variety of forms. Causes of dermatitis include ultraviolet rays, drugs, chemicals, and allergies. Although the reaction varies with the causative agent, the skin generally becomes reddened and scaly, accompanied by various lesions, itching, and burning.

Carbuncle. A carbuncle is an infection that extends deep into the subcutaneous tissue. It is caused by a species of the bacterium *Staphylococcus*.

Shingles. Shingles is a viral disease characterized by inflammation of spinal nerve ganglia. Blisters arise on the skin in the area supplied by the nerve.

Seborrhea. Seborrhea results from oversecretion of the sebaceous, or oil, glands, resulting in the formation usually of brownish-gray scales and dandruff.

Psoriasis. Psoriasis, as in "heartbreak of," literally means "itching condition" (Gk. *psora*, an itch). It is characterized by reddish patches and silvery scales on the skin.

Other common skin disorders include *athlete's foot*, a fungus infection; *acne*, a skin condition of the face and back caused by inflammation of the sebaceous glands; and *warts*, a virus-caused condition producing small bumps on the skin.

SKELETON

Osteoarthritis. This is a degenerative disease of the joints, the cause of which is unknown. The term literally means "inflammation of bone and joint" and is a disease mainly of people past middle age. Osteoarthritis is characterized by degeneration of joint cartilage and enlargement of the bones where they articulate. The joints often become swollen and painful.

Bursitis. Many of the joints of the body, such as the knee and elbow joints, are associated with small fluid-filled sacs called *bursae* (sing. bursa). Overstressing these joints sometimes causes inflammation of the bursae, making movement difficult and painful. A common example of bursitis is "tennis elbow."

Syndactylia. Syndactylia is a hereditary disease in which the fingers or toes are joined together, or "webbed." The extent of fusion varies from the union of only two fingers to the extreme condition in which all of the fingers are joined, giving the hand the appearance of a lobster claw.

Rickets. Rickets is primarily a disease of children and is caused by a deficiency of vitamin D. The bones become softened and are easily bent and distorted.

Several skeletal disorders affect the spinal column. The condition known as "humpback" is called *kyphosis* (backward curvature of the spine). An abnormal hollow in the back caused by an inward curvature of the spine is called *lordosis*. Finally, a lateral curvature of the spine is called *scoliosis*. In most cases, the cause of these abnormalities is unknown; in other instances, however, they are attributable to polio or tuberculosis of the spine.

MUSCLES

Trichinosis. This is an inflammatory muscle condition caused by the parasitic nematode *Trichinella*. Humans acquire the parasite by eating infected, poorly cooked pork. Symptoms of trichinosis include fever, severe muscle pains, anemia, and general weakness. The simple preventive measure is to cook pork thoroughly, thereby killing the encysted worms.

Myasthenia Gravis. Myasthenia gravis, meaning "grave muscle weakness," results from impairment of the neuromuscular junction. Impulses from the motor neurons are too weak to stimulate the muscle fibers to respond. Consequently, the muscular system generally is quite weak and fatigues easily.

Muscular Dystrophy. Muscular dystrophy is a degenerative disease resulting from an X-linked condition. It is characterized by progressive muscular atrophy and weakness. The affected individual usually dies before reaching maturity.

Muscle Cramp. This familiar condition can result from overexercising, reduced blood flow to a muscle, and other factors. Whatever the irritation, nerve impulses are transmitted from the muscle to the spinal cord, which sends impulses back to the muscle, causing increased contraction and cramping.

Flutter and Fibrillation. These terms are ordinarily applied to rapid muscular contractions of the heart. Contractions at a rate of 200 to 300 beats per minute are called flutter, whereas more rapid contractions are called fibrillation. Both flutter and fibrillation result from abnormal electrical impulses within the heart muscle.

REVIEW OF ESSENTIAL CONCEPTS

1. Human skin consists of an upper layer of cells called the *epidermis* and a lower layer called the *dermis*. Beneath the dermis is the fatty *subcutaneous layer*.
2. The epidermis is a protective layer and contains the pigment *melanin*; the epidermis also produces fingernails and toenails. The dermis contains blood vessels, sensory receptors, hair follicles, and oil and sweat glands.
3. The subcutaneous layer stores fat, which insulates the body and serves as a reserve store of energy.
4. The tissues of the skeletal system include *cartilage* and *bone*. Bone is constructed of units called *Haversian systems*.
5. The human skeleton is divided into an *axial* skeleton and an *appendicular* skeleton.
6. Common skeletal disorders include fractures, dislocations, and sprains.
7. Skeletal muscle is composed of numerous fibers that are made up of *myofibrils*. Myofibrils consist of thick protein filaments of *myosin* and thin filaments of *actin*.
8. In the myofibril, the thick and thin filaments are connected by tiny *cross bridges*, which pull the two filaments past each other. This action is the basis for the *sliding filament theory* of muscle contraction.
9. A group of muscle fibers and the neuron supplying it is called a *motor unit*. Rapidly acting muscles have a high ratio of motor neurons to the number of muscle cells; slow-acting muscles have a low ratio.
10. *Slow-twitch* muscle fibers (red muscle) contract slowly and are resistant to fatigue; *fast-twitch* muscle fibers (white muscle) contract rapidly and fatigue quickly. Red muscle contains more of the pigment *myoglobin* than does white muscle.
11. A motor nerve axon communicates with a muscle fiber across the *neuromuscular junction*.
12. The *neurotransmitter* released at the neuromuscular junction stimulates the muscle fibers to contract. Contraction involves the movement of sodium ions into the fiber and the outward movement of potassium ions. This is accompanied by release of calcium ions from within the fiber. Relaxation occurs when this process is reversed, and calcium ions move out of the fibers. The energy for skeletal muscle contraction is supplied by ATP and creatine phosphate.
13. Muscle fibers enlarge through *isotonic* or *isometric* exercise. Muscles not exercised tend to atrophy.
14. *Muscle tone* results from continual nerve impulses supplied to the skeletal muscles by the nervous system.
15. There are approximately 650 skeletal muscles in the body, each of which is involved in the movement of a bodily structure.

APPLYING THE CONCEPTS

1. Discuss the various ways in which the skin serves to protect the body.
2. Describe the fundamental differences between skeletal, smooth, and cardiac muscle.
3. Describe the formation and function of an Haversian system.
4. Name three functions of cartilage.
5. How is the structure of skeletal muscle related to its function?
6. What is the significance of the ratio of motor neurons to the number of fibers in skeletal muscle?
7. Distinguish between slow-twitch and fast-twitch muscle fibers.
8. Explain the role of ions in skeletal muscle contraction.
9. Using examples, explain the difference between isotonic and isometric muscle contraction.

16 respiration

THE ESSENTIAL OBJECTIVES

You have understood this chapter when you are able to:

1. Describe generally the respiratory processes of selected lower animals.
2. Identify and give the functions of the major structures of the human respiratory system.
3. Explain how air mechanically enters and leaves the lungs.
4. Explain the exchange of the respiratory gases between the lungs and the blood and between the blood and the body cells.
5. Discuss the basic factors involved in oxygen and carbon dioxide transport and release.
6. Explain how breathing is controlled under normal conditions and during exercise.
7. Discuss the causes and general effects of the major respiratory disorders and diseases.

In all aerobic organisms, there is a respiratory process or mechanism that serves to supply cells with oxygen and to remove carbon dioxide. In man the delivery of oxygen and removal of carbon dioxide depend upon the combined functions of the respiratory and circulatory systems.

PREVIEW OF ESSENTIAL TERMS

respiration In higher animals, the process involving the inhalation and exhalation of air through the lungs.
alveoli Tiny thin-walled sacs that function as the sites of respiratory gas exchange between the lungs and the blood.
oxyhemoglobin Hemoglobin that has combined chemically with oxygen.
bicarbonate ion HCO_3^-; the form in which most of the carbon dioxide is transported in blood plasma.
respiratory center An area near the medulla oblongata of the brain responsible for the control of inhalation and exhalation.
carbaminohemoglobin The compound formed when carbon dioxide complexes with hemoglobin in red blood cells.
stretch receptor Sensory tissue in the lungs that stimulates the respiratory center to prevent overinflation or excessive deflation of the lungs.
chemoreceptors Tiny nerve structures in the walls of the aorta and carotid arteries; they relay nerve impulses to the respiratory center when the blood oxygen level is low and thereby cause the breathing rate to increase.

Respiration

> Insects use small tubelike *tracheae* to transport oxygen from the atmosphere into their bodies. Mollusks and fish use *gills* to extract oxygen from the water surrounding them.

INTRODUCTION

Except for a few anaerobic microorganisms (mainly bacteria and yeasts), every form of life on earth requires an almost constant supply of oxygen. As we discussed in Chapter 6, the major source of atmospheric oxygen (O_2) is photosynthesis. This is the gaseous oxygen that you and all other aerobic organisms utilize every day. In many of the lower forms of life, such as protozoans, sponges, flatworms, and so on, oxygen simply diffuses from the surrounding water or air into the cells. Some worms, and to some extent even frogs, take in oxygen through the surface skin. Oxygen then diffuses into networks of blood-filled capillaries that extend throughout the skin and into the body. The blood distributes the oxygen via the capillaries to the body cells. Insects and other arthropods possess tiny tubelike structures called *tracheae* (Fig. 16.1). Oxygen from the atmosphere enters the body through tiny pores and is then delivered throughout the body by a network of tracheae. Structures called *gills* are found in many organisms, such as mollusks and fish. Gills contain a rich network of blood vessels that take up oxygen from the surrounding water. After entering the gills, oxygen circulates in the blood to the body cells; on its return to the heart, the blood is again pumped to the gills to pick up fresh oxygen.

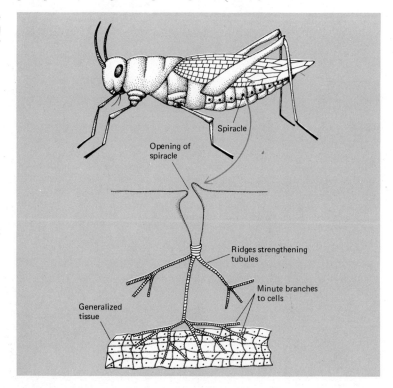

Figure 16.1 Grasshopper tracheal system. Air enters the spiracle opening and travels through the tracheal tubules to the cells.

> In *cellular respiration* foodstuffs are metabolized in conjunction with oxygen to produce cellular energy and the by-product carbon dioxide.

Every cell in the body is dependent upon oxygen for the metabolism of foodstuffs and the accompanying production of cellular energy. This metabolic activity is known as *cellular respiration* and involves the Krebs cycle and electron transport system. While utilizing oxygen, metabolizing cells release the by-product carbon dioxide. In many lower organisms, carbon dioxide is eliminated directly by diffusion from the interior of the body to the outside. In man and higher animals, carbon dioxide is removed from the body via the lungs (Fig. 16.2). In a reciprocal manner, green plants utilize this carbon dioxide in photosynthesis to produce and replenish the oxygen in the atmosphere.

As we shall use it in this chapter, the term "respiration" (L. *respirare*, to breathe again) refers to the processes of inhalation (intake of air) and exhalation (release of air) through the lungs. Associated with these processes is the exchange of the respiratory gases—oxygen and carbon dioxide—between the lungs and the blood and between the blood and the body cells. Upward of 20,000 times a day, human beings inhale and exhale, moving 350 ml of air into and out of the lungs with each breath. Inhaled air contains a high concentration of atmospheric oxygen, which is delivered by the lungs to the blood and then through blood vessels to the cells of the body. About one-third of the inhaled oxygen is given up to the blood, and the

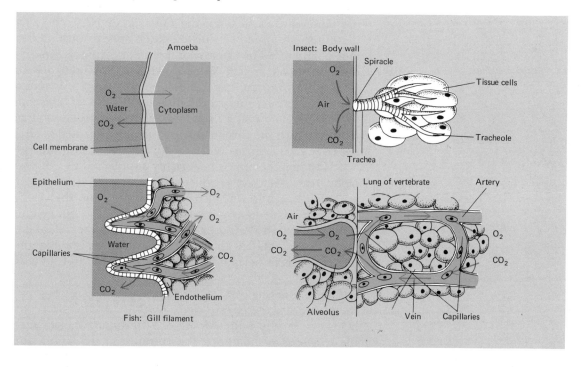

Figure 16.2 A diagram of some of the types of respiratory organs present in animals.

> The *epiglottis* serves to cover the glottis when one swallows food.

remainder is exhaled. However, exhaled air also contains a high concentration of carbon dioxide, the gaseous by-product produced by respiring cells. As carbon dioxide increases in concentration within the cells, it passes into the blood to be transported to the lungs. At the same time, oxygen, which is in high concentration in the blood, moves into the cells.

The structures that conduct air into and out of the body are collectively known as the ***respiratory system.*** Its essential function is to deliver oxygen to the blood and to remove excess carbon dioxide from the body. As we shall discuss, the respiratory system performs its function in close association with the circulatory system. This, of course, is another example of systems acting in concert to maintain homeostasis.

THE HUMAN RESPIRATORY SYSTEM

The major structures of the respiratory system include the ***nose*** and the associated ***nasal cavities;*** the ***pharynx,*** or throat; the ***larynx,*** or "voice box"; the ***trachea,*** or "windpipe"; and the ***lungs, bronchi, bronchioles,*** and ***alveoli.*** In addition, certain muscles, principally the ***intercostals*** and the ***diaphragm,*** are essential to normal respiratory function (Fig. 16.3).

Atmospheric air normally enters the body through the nose, although it often enters through the mouth as well. Upon entering the nasal cavities through the nostrils, air passes over coarse hairs that filter out large particulate matter, such as dust, pollen, and various pollutants. The interior of the nasal cavities is lined with ciliated epithelial tissue, which contains glands that secrete a watery mucus. In passing over the wet epithelium, inhaled air is moistened as it moves posteriorly toward the pharynx, or throat. The constant beating of the cilia assists in the air filtering process and propels the mucus toward the throat. During the day, repeated swallowing empties over a pint of this mucus from the throat into the esophagus.

Nasal epithelium is also richly supplied with a network of tiny blood vessels—a fact of which we are uncomfortably reminded with a nosebleed. However, blood circulating within the vessels serves to warm the inhaled air as it passes over the nasal epithelium. Thus, in passing through the nose, air is filtered, moistened, and warmed as it comes in contact with the nasal cavities. Moistening and warming inhaled air is important because dry, cool air can lead to infection of the lungs.

From the pharynx, inhaled air passes through an opening called the ***glottis*** into the *larynx* ("voice box" or "Adam's apple"). Extending below the larynx is the *trachea,* or "windpipe." A flap of tissue called the ***epiglottis*** is situated at the entrance of the larynx and serves to

> The *vocal cords* are folds in the walls of the larynx that vibrate from expired air, producing the voice.

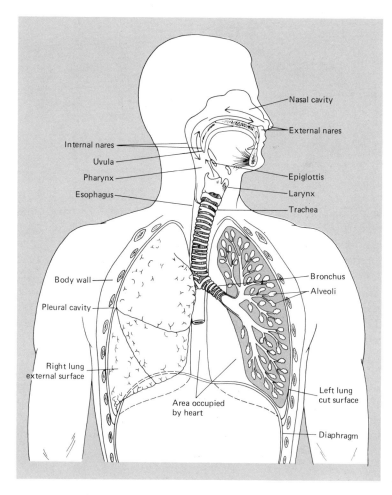

Figure 16.3 Respiratory system of man; the connection between the pharynx and the larynx is broken in the figure.

cover the glottis when one swallows (Fig. 16.4). By placing a finger on your larynx, you can feel it move upward and backward when you swallow. The movement of the larynx closes the epiglottis over the glottis automatically. In this manner, food is kept out of the respiratory passageways, unless we attempt to force feed ourselves with both hands. The choking that often ensues is a reflex effort to expel food that has slipped past the epiglottis into the larynx.

The larynx houses the *vocal cords,* which are actually folds in the membrane lining the lateral walls of the larynx (Fig. 16.5). Air being expired causes the vocal cords to vibrate, producing the voice. Our speech patterns are composed of vibrations from the larynx in association with the structures of the mouth (lips, tongue, and soft palate).

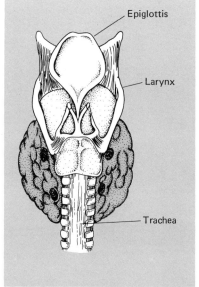

Figure 16.4 Posterior view of the larynx showing the location of the epiglottis. When a person swallows, upward movement of the larynx raises the epiglottis, which is compressed downward by the base of the tongue.

> Within the lung, bronchi branch and divide into smaller and smaller *bronchioles*. These bronchioles eventually form a large network of tiny respiratory ducts that convey air into and out of the body.

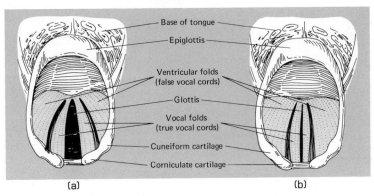

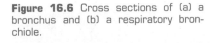

Figure 16.5 Larynx, (a) Anterior view. (b) Posterior view. In this figure on the left, the true vocal cords are relaxed. In the figure on the right, the true vocal cords are taut.

The trachea is a tubelike structure, partially surrounded at intervals by incomplete rings of C-shaped cartilage. The rings prevent the trachea from collapsing. Like the nasal cavities, the trachea is lined with a mucus-coated, ciliated epithelium. The cilia propel the mucus and small inhaled foreign particles toward the pharynx, where they are coughed up or swallowed. The lower end of the trachea divides into the two main *bronchi,* each of which enters a lung. The bronchi are structured much like the trachea and are also lined with mucus-secreting epithelium and cilia (Fig. 16.6*A*).

Within the tissues of the lungs, the bronchi divide into smaller bronchi called *bronchioles* (Fig. 16.6*B*). These in turn divide and subdivide to form still smaller and smaller bronchioles. In effect, then, the lungs are supplied with a profuse network of tiny respiratory ducts that convey air into and out of the body. Each fully developed human lung is a spongy, elastic organ, consisting of several large sections, or *lobes*. The lungs are housed within the thoracic (chest)

Figure 16.6 Cross sections of (a) a bronchus and (b) a respiratory bronchiole.

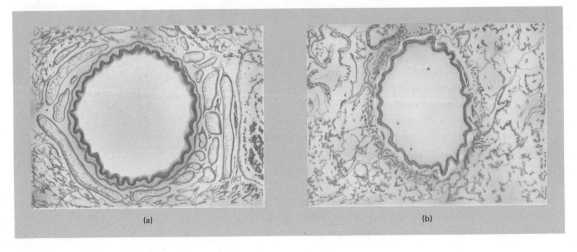

Color Plate 96 The arctic tundra in mid-summer. (Photograph by Carolina Biological Supply Co.)

Color Plate 97 An evergreen forest. (Photograph by Carolina Biological Supply Co.)

Color Plate 98 A deciduous forest. (Photograph by Bruce Russell, Biomedia Assoc.)

Color Plate 99 Sage prairie ecosystem. (Photograph by Bruce Russell, Biomedia Assoc.)

Color Plate 100 A desert ecosystem of the American southwest. (Photograph by Bruce Russell, Biomedia Assoc.)

Color Plate 101 Rocky coast ecosystem. (Photograph by Bruce Russell, Biomedia Assoc.)

Color Plate 102 A freshwater pond environment. (Photograph by Bruce Russell, Biomedia Assoc.)

Color Plate 103 Marine biome. (Photograph by Bruce Russell, Biomedia Assoc.)

Color Plate 104 Four or five billion years ago a newly formed planet revolves around an average-sized yellow star. (Photograph by Bruce Russell, Biomedia Assoc.)

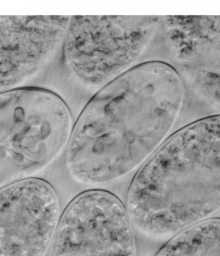

Color Plate 105 *Glaucocystis,* a primitive eukaryote, containing symbiotic blue-green algae. (Photograph by Bruce Russell, Biomedia Assoc.)

Color Plate 106 In the evolutionary process, variation is the key. (Photograph by Bruce Russell, Biomedia Assoc.)

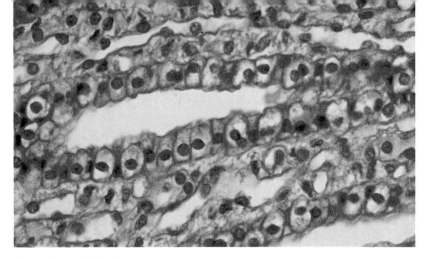

Color Plate 107 Simple cuboidal epithelium. (Photograph by Carolina Biological Supply Co.)

Color Plate 108 Adipose tissue (fat cells). (Photograph by Carolina Biological Supply Co.)

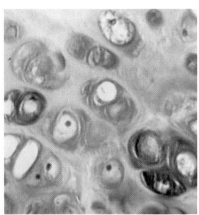

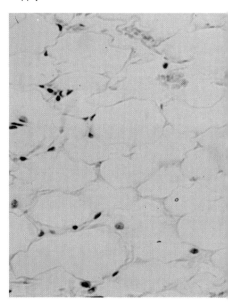

Color Plate 109 Hyaline cartilage, a type found in the nose and trachea of many adult vertebrates. (Photograph by Carolina Biological Supply Co.)

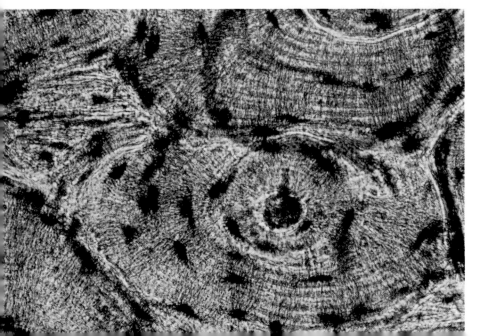

Color Plate 110 A section of human bone. (Photograph by Carolina Biological Supply Co.)

> It is in the capillary-rich *alveoli* that the exchange of respiratory gases (oxygen and carbon dioxide) occurs.

cavity and are covered with a thin membrane called the **pleura**. A painful condition called *pleurisy* results from inflammation of the pleura.

Throughout the lungs, the tiniest bronchioles eventually terminate to form numerous thin-walled sacs called *alveoli* (L. "a hollow") (Fig. 16.7). There are reportedly about 150 million alveoli in each lung. Each alveolus is virtually covered by a network of *capillaries*, the exceedingly thinwalled blood vessels found throughout the body. It is this intimate association of the alveoli and the pulmonary (lung) capillaries that provides for exchange of the respiratory gases (oxygen and carbon dioxide) in the lungs (Fig. 16.8). There is no gas exchange through the bronchi or most of the bronchioles because their walls are simply too thick. After a brief look at how air gets into and out of the lungs, we shall consider the exchange of respiratory gases in the lungs and correlate this with gas exchange in the cells of the body.

THE MECHANICS OF BREATHING

The process of breathing consists of two basic phases—inhalation and exhalation (Fig. 16.9). Both of these phases involve the action of the *intercostal muscles* and the *diaphragm*. The intercostal muscles are situated between the ribs, and the broad, flat diaphragm is located between the thoracic cavity and the abdominal cavity.

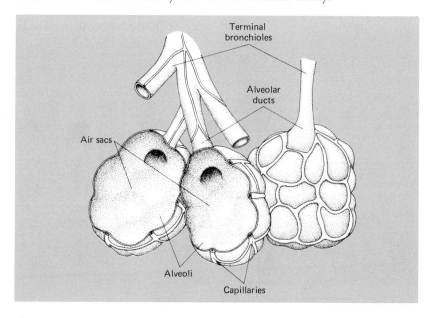

Figure 16.7 Diagram showing the intimate association between alveoli and the pulmonary (lung) capillaries.

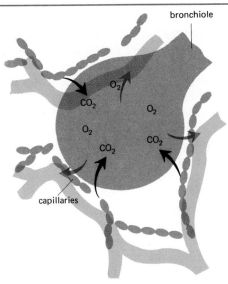

Figure 16.8 Gas exchange between an air sac (alveolus) and capillaries. (From Applewhite, P., and Wilson, S.: *Understanding Biology.* New York: Holt, Rinehart and Winston, 1978, p. 183.)

When we inhale, the chest cage is elevated and expanded by some of the intercostal muscles. At the same time, the diaphragm contracts downward. The overall effect, then, is enlargement of the chest cavity. Since the lungs are elastic, they distend on inhalation, and the air pressure within the chest cavity falls *below* the atmospheric pressure outside the body. Air then rushes into the lungs

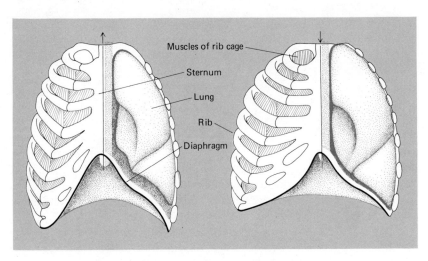

Figure 16.9 How breathing occurs. When the diaphragm, a large sheet of muscle stretching across the bottom of the chest cavity, relaxes, it becomes somewhat dome-shaped. As a result, the volume of the chest cavity decreases, and air is forced out of the lungs. This is exhalation or expiration. When the diaphragm contracts, it flattens, the volume of the chest cavity increases, lung pressure decreases, and air is sucked into the lungs. This is inhalation, or inspiration. Although one can control inhalation and exhalation to a certain extent, they are normally under involuntary control.

> Gas exchange between the alveoli and the blood occurs through the process of *diffusion*.

until the pressures inside and outside the body are equal. You can demonstrate this equalization of pressures by forcefully inhaling; beyond a certain point, you cannot force any more air into your lungs.

During exhalation, the intercostal muscles relax, pulling the chest cage downward; the diaphragm also relaxes and moves upward. In this manner, the size of the chest cavity is reduced. As a result, the lungs are compressed, and the air pressure within the chest cavity rises *above* atmospheric pressure. Consequently, air moves out of the lungs until the two pressures are equal.

Oddly enough, the lungs have a natural tendency to collapse. This results not only from the elasticity of the lungs but also from the surface tension (see Chapter 3) of the watery fluid adhering to the walls of the alveoli. The tension created by the water molecules tends to pull on the alveolar walls, causing them to collapse. However, a substance called **surfactant,** secreted by special alveolar cells, counteracts the effect of surface tension and thereby normally prevents collapse of the lungs. Surfactant acts much like soap and detergents, which lower the surface tension of water so that it can more readily penetrate a fabric and loosen the dirt. Surfactant actually lines the inner moist surface of each alveolus, where it continuously maintains a lowered surface tension of the fluid. Newborn babies who do not secrete sufficient quantities of surfactant suffer from *hyaline membrane disease*. This condition makes expansion of the lungs difficult and is usually fatal either soon after birth or in early infancy.

RESPIRATORY GAS EXCHANGE

The parts of the respiratory system extending from the nose to the bronchioles are essentially only passageways through which air enters and leaves the body. None of these structures is directly involved in the vital process of gas exchange between the lungs and the blood. That function is performed almost exclusively by the 300 million alveoli. It has been estimated that the total surface area (see Chapter 5) of the alveoli in the lungs of a normal adult is in excess of 750 ft^2. Such a large surface area ensures the efficient exchange of gases between the alveoli and the blood. In addition, the walls of the alveoli are extremely thin, as are the walls of the many capillaries that make contact with the alveoli. Consequently, oxygen and carbon dioxide move across these walls with relative ease.

What causes the movement of these gases? Basically, the answer is *diffusion*. As we discussed in Chapter 5, diffusion involves the

> Diffusion takes place because of pressure differences (i.e., concentration) of oxygen or carbon dioxide in the alveoli and blood.

movement of a substance from an area where it is in high concentration to an area where it is in lower concentration, until equilibrium is reached. In the same manner, gases such as oxygen and carbon dioxide move from areas of higher *pressure* to regions of lower *pressure*. This applies not only to gas exchange in the lungs but also to the exchange of gases between the blood and the cells of the body (Fig. 16.10).

The atmospheric air we inhale contains oxygen that exerts pressure within the alveoli. Since the concentration, or amount, of oxygen inhaled is quite high, its pressure within the alveoli is also high. However, the concentration and pressure of oxygen in the capillary blood are very low. This is because the capillary blood gave up most of its oxygen to the cells as it circulated through the body. You will recall from Chapter 6 that oxygen is required by the body for the complete metabolism of foodstuffs. In the metabolic process, oxygen is used almost as rapidly as it enters the cells. The immediate source of this oxygen is the blood that continually flows past every living cell in the body. Thus, having given up its oxygen, the now *deoxygenated* blood must return to the lungs to be recharged with fresh oxygen. How does this recharge occur? Since the oxygen pressure is higher in the lungs than in the returning capillary blood, oxygen diffuses out of the alveoli and into the blood. This newly *oxygenated* blood is then conveyed to the heart to be pumped once again throughout the body.

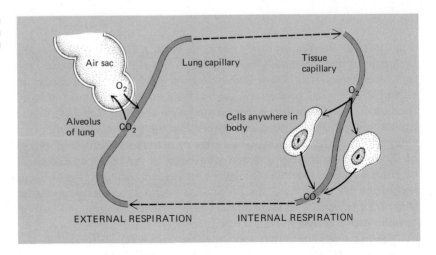

Figure 16.10 Diagram illustrating the exchange of respiratory gases between the lungs and the blood, and between the body cells and the blood.

Figure 16.11 Human red blood cells photographed with a scanning electron microscope. (From Applewhite, P., and Wilson, S.: *Understanding Biology.* New York: Holt, Rinehart and Winston, 1978, p. 183.)

Although the concentration and pressure of oxygen in the lung capillary blood are low, the concentration and pressure of *carbon dioxide* are high. Why is this so? As we have stated, in its course through the body, blood gives up most of its oxygen to the cells. However, at the same time it is releasing its oxygen, the blood takes up carbon dioxide from the cells. You will again recall from Chapter 6 that carbon dioxide is a major waste product of cellular metabolism. As its concentration and pressure rise within the cells, carbon dioxide diffuses outward and enters the circulating blood. On arriving at the lungs, this blood has a higher carbon dioxide concentration and pressure than that of the air in the alveoli. Accordingly, carbon dioxide diffuses out of the blood and enters the alveoli. This is the carbon dioxide we exhale with every breath.

Oxygen Transport. Oxygen is delivered to all parts of the body by the circulating blood. Upon entering the bloodstream, almost all the oxygen passes into the red blood cells (Fig. 16.11), where it combines chemically with the pigment **hemoglobin** (Fig. 16.12). More specifically, oxygen combines with the iron atoms of hemoglobin to form a substance called **oxyhemoglobin** (Fig. 16.13). If we fail to include an adequate supply of iron in our diet, the oxygen-carrying capacity of the blood is reduced, and we suffer from *anemia,* usually characterized by that "tired blood" feeling. However, anemia also results from the loss of blood, from a decrease in the production of new red cells by the bone marrow, or from an increased rate in the destruction of red cells. This last event occurs, for example, in sickle-cell anemia. The sickled red cells become so fragile that they are destroyed in great numbers.

Oxygen is bonded rather weakly to hemoglobin so that as the oxygen pressure decreases, oxygen is released (Fig. 16.14). As we have just seen, oxygen can then diffuse from the blood into the cells. An increasingly greater amount of oxygen is released as the carbon dioxide pressure, acidity, and temperature of the blood increase. These factors explain why more oxygen is released from the blood to the muscles during exercise. As one exercises, the cells of the body metabolize more rapidly, producing greater amounts of carbon dioxide, which diffuses into the blood. Active muscles also produce lactic acid, which lowers the pH of the blood, i.e., makes it more acidic. Finally, cells that are actively metabolizing generate heat throughout the body, thereby increasing the temperature of the

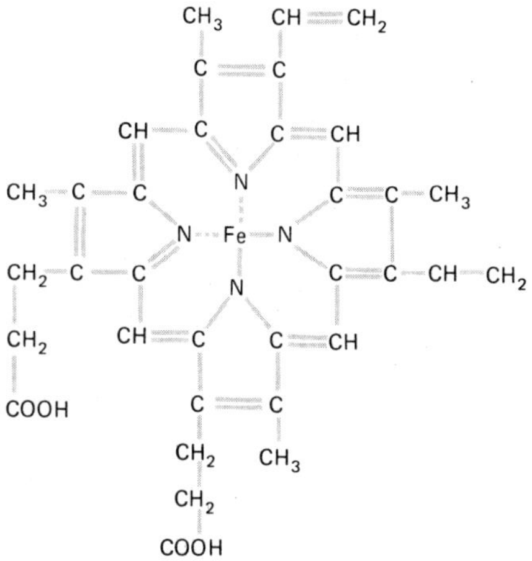

TRANSPORT OF THE RESPIRATORY GASES

Figure 16.12 Structure of a heme group. Each hemoglobin molecule is a combination of protein and four heme groups. The oxygen binds to the Fe (iron) part of the molecule. (From Applewhite, P., and Wilson, S.: *Understanding Biology.* New York: Rinehart and Winston, 1978, p. 183.)

> Carbon dioxide is transported in the plasma in the form of the *bicarbonate ion*.

blood. Not only do the working muscle cells receive more oxygen as a result of these factors but also less active cells receive proportionately less.

Carbon Dioxide Transport. Carbon dioxide is also transported in the blood, but most of it is found in the plasma (the liquid component of blood), with only a small percentage bound to hemoglobin. The complexing of carbon dioxide with hemoglobin forms **carbaminohemoglobin.** By far the largest amount of carbon dioxide is transported in the plasma in the form of the **bicarbonate ion** (HCO_3^-), and a very small amount simply dissolves in the plasma (Fig. 16.15A). When carbon dioxide reacts with water in the plasma, carbonic acid (H_2CO_3) is formed: $CO_2 + H_2O \rightarrow H_2CO_3$. The carbonic acid dissociates (breaks apart) almost immediately into bicarbonate ions and hydrogen ions: $H_2CO_3 \rightarrow HCO_3^- + H^+$. The free hydrogen ions bond almost instantly with hemoglobin in the red cells. This is extremely important because a high concentration of free hydrogen ions would cause a significant change in the pH of the blood, making it much more acidic. The carbon dioxide, now in the form of the bicarbonate ion, is left in the plasma to be transported to the lungs. When the blood reaches the lungs, the process just described is reversed, that is, the hydrogen ions in the red cells recombine with the

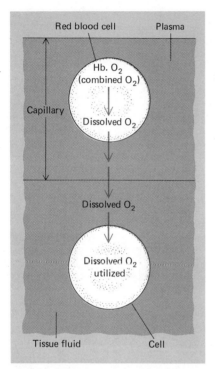

Figure 16.13 Oxyhemoglobin (HbO_2) as a reservoir for the replenishment of dissolved oxygen. Oxygen is taken up and utilized by cells, causing a fall in the concentration of dissolved oxygen in the tissue fluid and plasma. Dissolved oxygen is then replenished from the store of oxygen that is loosely bound to hemoglobin in red blood cells. *Arrows* indicate the direction of diffusion of oxygen.

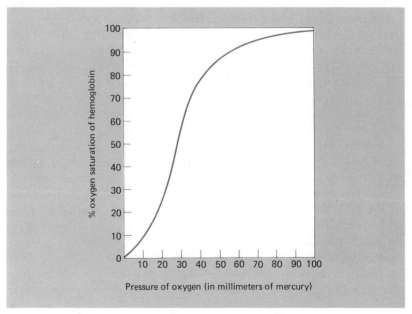

Figure 16.14 Oxyhemoglobin dissociation curve at body temperature. The curve indicates that increased amounts of oxygen are released from hemoglobin as the pressure of oxygen decreases.

bicarbonate ions to form carbonic acid, which separates into carbon dioxide and water: $H^+ + HCO_3^- \longrightarrow H_2CO_3 \longrightarrow CO_2 + H_2O$. The carbon dioxide then diffuses into the alveoli and is exhaled from the body (Fig. 16.15B). At the same time, the carbon dioxide bound to hemoglobin and the carbon dioxide dissolved in the plasma also diffuse into the alveoli to be exhaled.

THE CONTROL OF BREATHING

Breathing is an automatic process in that we do not have to remind ourselves to inhale and exhale, whether we are awake or asleep. At the base of the brain around the medulla oblongata, there is an area

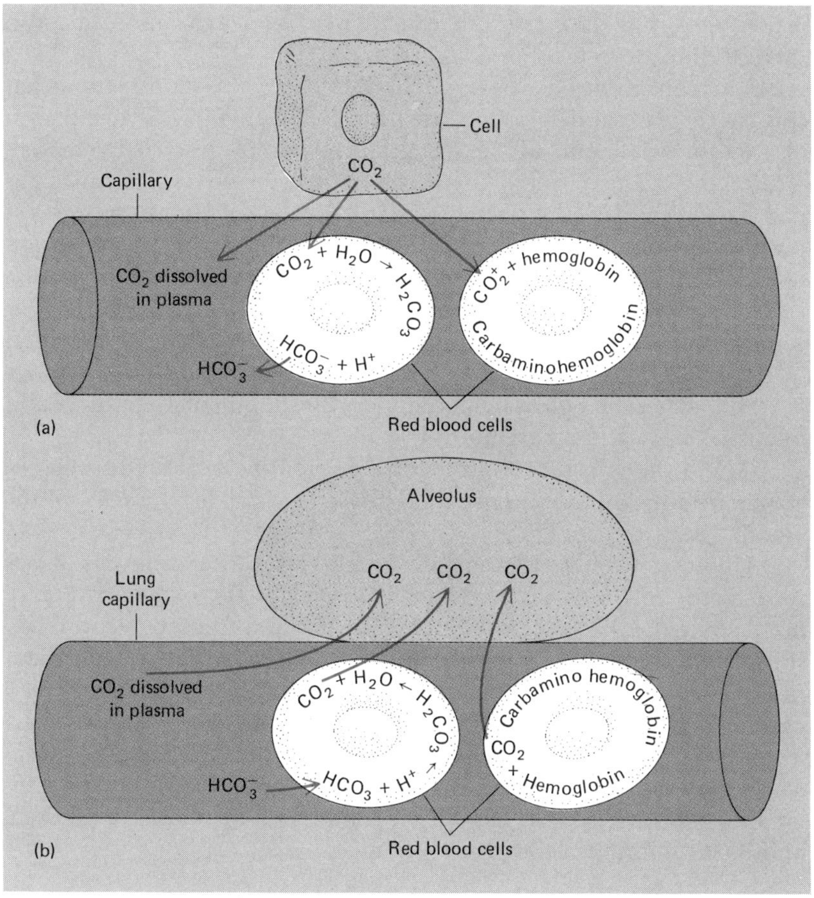

Figure 16.15 Transport of carbon dioxide. (a) After diffusing from metabolizing cells into the capillary blood, carbon dioxide may be transported in the dissolved state in the plasma or complexed with hemoglobin (to form carbaminohemoglobin) or in the form of bicarbonate ions HCO_3^-). (b) When the blood transporting the carbon dioxide enters the lungs, the processes in (a) are reversed, and the carbon dioxide diffuses from the lung capillaries into the alveoli.

> If oxygen pressure in the blood falls low enough, *chemoreceptors* signal the respiratory center and the breathing rate increases, resulting in a more rapid intake of oxygen.

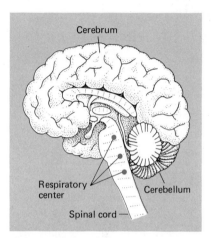

Figure 16.16 Diagram of location of the respiratory center at the base of the brain.

known as the **respiratory center** (Fig. 16.16). Specialized neurons in this area are responsible for initiating nerve impulses that govern inhalation and exhalation. The nerve impulses are conducted along various nerves that lead to the respiratory muscles. When the neurons that cause inhalation are stimulated, they send nerve impulses to the respiratory muscles, which then contract, and we inhale. At the same time, *inhibitory* impulses are sent to the neurons of exhalation so that we cannot exhale at the same time we inhale. In about 2 seconds the neurons of inhalation fatigue and stop sending impulses. This frees the neurons of exhalation, which begin to send out impulses to the respiratory muscles, and we exhale. At the same time, inhibitory impulses are sent to the neurons of inhalation, preventing us from inhaling. When the neurons of exhalation are fatigued, the cycle is repeated. Thus, the neurons of the respiratory center not only act to stimulate inhalation and exhalation but also serve to inhibit each other.

The respiratory center is also influenced by several factors that control the rate and depth of breathing. For example, overinflation or excessive deflation of the lungs is prevented by **stretch receptors** located in various parts of the lungs. When we inhale, the lungs expand and stimulate the stretch receptors to send impulses to the respiratory center. This causes inhibition of the neurons of inhalation so that we do not overinflate the lungs. During exhalation the stretch receptors relax, and the neurons of exhalation in the respiratory center are inhibited. This prevents excessive exhalation or deflation of the lungs. These respiratory activities, known as the *inflation reflex* and *deflation reflex,* help regulate the depth of breathing and maintain the normal rhythm of breathing.

The other factors that control respiration are the relative concentrations of *oxygen, hydrogen ions,* and *carbon dioxide* in the blood.

Under normal conditions, oxygen apparently is unimportant as a regulator of the breathing rate. However, if the oxygen pressure in the blood falls *low* enough, the respiratory center is indirectly stimulated, and the rate of breathing increases. What actually happens is that low blood oxygen is sensed by tiny structures called *chemoreceptors,* located in the walls of the aorta and the carotid arteries (Fig. 16.17). The chemoreceptors then transmit impulses along nerves to the respiratory center, and the breathing rate increases. This, of course, results in a more rapid intake of oxygen, and its level in the blood rises.

> Exercise is the most common way in which the blood levels of oxygen, hydrogen ions, and carbon dioxide change.

Both hydrogen ions and carbon dioxide in *high* concentrations also cause an increase in the rate of breathing. Although both substances may affect breathing indirectly by stimulating the chemoreceptors, their most significant effect results from direct stimulation of the respiratory center. As blood high in carbon dioxide (in the form of bicarbonate ions) and hydrogen ions circulates through the brain, the respiratory center is stimulated and sends out nerve impulses that increase the breathing rate. The result in this case is that excess carbon dioxide is more rapidly exhaled, and the hydrogen ion concentration is prevented from rising too high. This latter point is important because, as mentioned, a higher hydrogen ion concentration increases the acidity of the blood.

Why would the levels of oxygen, hydrogen ions, and carbon dioxide in the blood change in the first place? The simplest and most common reason is exercise. Heightened muscular activity requires a greater supply of oxygen and produces increased amounts of carbon dioxide. As the oxygen is being utilized by the working cells, the blood oxygen level falls. This stimulates the chemoreceptors, which send impulses to the respiratory center, and the breathing rate increases. In addition, as the muscle cells metabolize, they produce increasing amounts of carbon dioxide and lactic acid, which also raise the hydrogen ion concentration of the blood. These activities also stimulate the respiratory center to increase the rate of breathing. Thus, one breathes deeply and rapidly in an effort to supply oxygen to the cells and to rid the body of excess carbon dioxide. When exercise ceases, the demand for cellular oxygen declines, the rate of cellular metabolism slows, and less carbon dioxide is produced. The respiratory center is then less stimulated, and the rate of breathing eventually returns to normal.

During extremely strenuous exercise, when the breathing rate is quite intense, it is unlikely that the factors just outlined could alone account for such a great change in respiration. It has been postulated, therefore, that the brain and special receptors in the limb joints are the factors primarily responsible for this change. Since the cerebral cortex of the brain sends nerve impulses that control the skeletal muscles, it is believed that the cortex transmits impulses to the respiratory center at the same time to accelerate breathing. When skeletal muscles contract during exercise, they stimulate receptors associated with the joints of the arms and legs. The receptors then transmit impulses to the respiratory center, and the breathing rate increases.

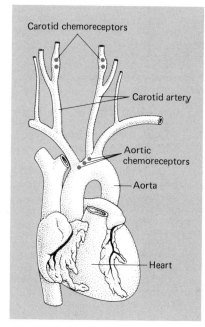

Figure 16.17 Diagram of location of respiratory chemoreceptors in the aorta and carotid arteries.

RESPIRATORY PROBLEMS AND DISEASE

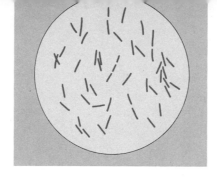

Figure 16.18 *Mycobacterium tuberculosis*, the causative agent of tuberculosis.

The respiratory system is subject to an extensive variety of disorders, many of which affect the delivery and removal of respiratory gases in other body systems.

Hyperventilation. Taking repeated rapid and deep breaths has the effect of lowering the level of carbon dioxide in the blood. Since this reduces stimulation of the respiratory center, one can resist breathing for several minutes. This is of obvious importance to divers and underwater swimmers. Holding the breath for too long, however, leads to dizziness and sometimes unconsciousness. Breath-holding is a favorite ploy among overcoddled children who have discovered its terrifying effect on their elders. However, should unconsciousness ensue, the high carbon dioxide level in the blood stimulates the respiratory center, and breathing resumes automatically. It is impossible to commit suicide simply by holding one's breath.

Another effect of hyperventilation is a sometimes drastic fall in blood pressure. When large amounts of carbon dioxide are blown off, as when attempting to start a fire by blowing on hot coals, there is dizziness and often a feeling of faintness. Not only does the blood pressure fall, but also oxygen tends to remain in combination with hemoglobin in the red cells. As a result, the brain cells are deprived of adequate oxygen, and dizziness or fainting usually follows.

Cyanosis. Literally, cyanosis means "blue condition" and refers to a bluish coloration of the skin and nails. Cyanosis results from insufficiently oxygenated hemoglobin present in the blood vessels of the skin. Respiration may be insufficient to oxygenate hemoglobin adequately in the lungs, or a reduction in blood flow may lower the oxygen content in venous blood. This latter instance may occur when one is exposed to very cold temperatures. Whatever the case, hemoglobin low in oxygen takes on a dark blue color, which is responsible for the typical cyanotic appearance of the skin or nails.

Pneumonia. Pneumonia is a common respiratory disease in human beings and is caused by a variety of microorganisms, including viruses, fungi, and bacteria. These agents cause inflammation of the lung tissue, accompanied by headache, fever, coughing, and fatigue. In the course of the disease, the alveoli swell and become highly porous. This allows fluids and blood cells to enter the lungs, which greatly reduces the surface area available for gas exchange. Although pneumonia can affect people of all ages, young children and the elderly, especially those with chronic diseases, are particularly susceptible.

Tuberculosis. Tuberculosis, or "TB," as it is popularly known, is an infectious disease caused by the tubercle (hence, the name of the disease) bacillus, *Mycobacterium tuberculosis* (Fig. 16.18). After infection of the lungs, tough connective tissue begins to develop and eventually forms walls or capsules around the bacteria. These capsules, called *tubercles*, help to prevent further spread of the bacteria in the lungs. Later, however, secondary infection by other bacteria causes progressive destruction of lung tissue. This, in turn, results in a decrease in the surface area of the alveoli and a thickening of their walls. Consequently, breathing is laborious, and oxygenation of the blood is diminished.

Lung Cancer. A cancer is a malignant mass of cells that divides uncontrollably to the extent that normal cells are deprived of nutrients. The great majority of lung cancers arise in the tissues that line the bronchi (Fig. 16.19). The cause of lung cancer is unknown, but statistical evidence indicates that atmospheric pollutants and especially heavy cigarette smoking are contributing factors. Cigarette smoke is known to paralyze the cilia in the bronchial tubes, resulting in the accumulation of mucus and the familiar "smoker's cough." Eventually, the alveoli are destroyed, greatly reducing the surface area available for gas exchange. Lung cancer also appears to be related to some occupations, as evidenced, for example, by the much higher lung cancer rate in asbestos workers than in the general population.

Asthma. Asthma, from the Greek word for "panting," is usually the result of an allergic reaction to some foreign substance, such as pollen, lint, dust, or sometimes even food. The allergic reaction causes the cells lining the bronchioles to secrete great amounts of mucus and causes the bronchioles to constrict. With constricted bronchioles full of thick mucus, an asthmatic has great difficulty in breathing. Inhalation is less of a problem because the lungs expand, opening the bronchioles somewhat. On exhalation, however, the bronchioles are compressed, making it difficult to move air out of the lungs.

Other respiratory disorders include *emphysema*, which is characterized by destruction of the alveolar walls and a subsequent reduction in the surface area of the lungs (Fig. 16.20). *Bronchitis* and *laryngitis* result from inhalation of various pollutants leading to inflammation of the bronchi and larynx, respectively. And in case you have ever wondered, a *hiccup* results when the diaphragm suddenly contracts, causing an inrush of air that strikes the vocal cords. Why hiccups occur and their functional significance, if any, are unknown.

541
Respiratory Problems
and Disease

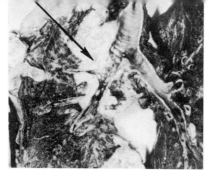

Figure 16.19 Lung cancer. (a) Section of normal human lung. (b) Section of lung of a chronic smoker who died of lung cancer. *Arrow* indicates an area of cancer. (From Ebert, J. D., Loewy, A. G., Miller, R. S., and Schneiderman, H. A.: *Biology.* New York: Holt, Rinehart and Winston, 1973, p. 369.)

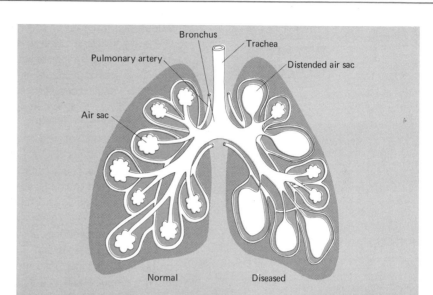

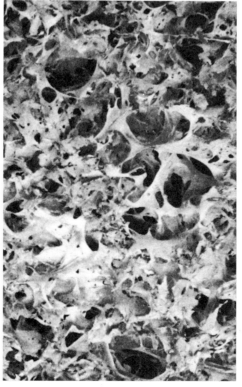
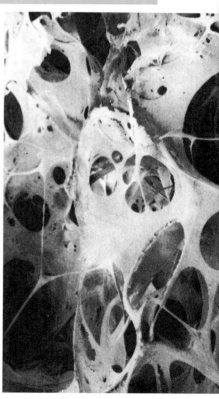

Figure 16.20 Effects of emphysema on the lungs. Normally, there are many tiny alveoli or air sacs. But in emphysema the membranes of several alveoli break down, forming one large sac. Since the combined surface area of several small alveoli is greater than that of one large sac, the area over which oxygen and carbon dioxide can be exchanged between the alveoli and the blood is reduced. In addition to a decrease in the number and the total surface area of the alveoli, their walls become less elastic than before. Furthermore, the diameter of the bronchi decreases, and air movement is restricted. Emphysema is currently the leading respiratory disease. However, many of its victims die of heart disease. Why? Contrast of the emphysematous lung (right) with the normal lung (left), showing extensive alveoli destruction. (From Guyton, A. C.: *Textbook of Medical Physiology.* 5th ed. Philadelphia: W. B. Saunders Co., 1976.)

REVIEW OF ESSENTIAL CONCEPTS

1. The essential function of the *respiratory system* is to supply oxygen to the body cells and to remove carbon dioxide.
2. Metabolizing cells utilize oxygen for respiration and release carbon dioxide as a waste product.
3. The human respiratory structures include the *nose, nasal cavities, pharynx, larynx, trachea, lungs, bronchi, bronchioles, alveoli*, and the chief respiratory muscles—the *intercostals* and *diaphragm*.
4. The major function of the nose is to *filter, moisten,* and *warm* inhaled air.
5. Swallowing forces the larynx upward, thereby closing the epiglottis over the entrance to the larynx; this prevents food from entering the respiratory passageways. The larynx also houses the *vocal cords*.
6. The trachea divides into two *bronchi*, each of which enters a lung. The trachea and bronchi contain cilia that propel mucus and foreign particles toward the throat.
7. Each bronchus divides into many small *bronchioles* that form a network throughout the lungs. The smallest bronchioles terminate in tiny sacs called *alveoli*, which are surrounded by lung capillaries. Alveoli are the sites of respiratory gas exchange in the lungs.
8. Mechanically, breathing consists of *inhalation* and *exhalation*, which involve the action of the respiratory muscles.
9. The tendency of the lungs to collapse is counteracted by *surfactant*, which lowers the surface tension of the fluid in the alveoli.
10. Exchange of respiratory gases between the blood and the lungs and between the blood and the body cells occurs by *diffusion*. Oxygen diffuses from the alveoli into the blood; at the cells, oxygen diffuses from the blood into the cells. Carbon dioxide diffuses out of the body cells into the blood and from the blood into the alveoli, where it is exhaled.
11. In the blood, oxygen combines chemically with hemoglobin to form *oxyhemoglobin*. Oxygen is released from hemoglobin as the oxygen pressure decreases. Increased carbon dioxide pressure, acidity, and temperature of the blood cause more oxygen to be released to the cells.
12. Most of the carbon dioxide is transported in the plasma in the form of the *bicarbonate ion*, which is produced when carbonic acid dissociates.
13. Some carbon dioxide is transported in chemical combination with hemoglobin, a complex called *carbaminohemoglobin*. A very small percentage of carbon dioxide is transported dissolved in the plasma.
14. Breathing is controlled by the *respiratory center* in the medulla oblongata. The neurons of the respiratory center stimulate inhalation

and exhalation and inhibit each other to prevent inhaling and exhaling from occurring at the same time.
15. The rate and depth of breathing are also influenced by the *inflation* and *deflation reflexes, oxygen pressure,* and the concentrations of *hydrogen ions* and *carbon dioxide*. In addition, *joint receptors* increase the breathing rate during exercise.

APPLYING THE CONCEPTS

1. What is the relationship between oxygen, carbon dioxide, and cellular respiration?
2. In what basic manner is respiratory gas exchange different in man and lower forms of life?
3. How do the nasal cavities, larynx, and trachea contribute to the respiratory process?
4. Explain the role of the intercostal muscles and diaphragm in breathing.
5. Explain why the blood in the lung capillaries is high in oxygen but low in carbon dioxide.
6. Discuss the physical movement of the respiratory gases between the lungs and the blood and between the blood and the body cells.
7. How is oxygen transported in the blood? What factors could affect the amount of oxygen transported?
8. How is carbon dioxide transported in the blood? How is carbonic acid involved in this transport?
9. Discuss the factors that influence the control of respiration by the respiratory center.

17 circulation and immunity

THE ESSENTIAL OBJECTIVES

You have understood this chapter when you are able to:

1. Describe the general functions of the components of blood.
2. Explain how red blood cells are produced and how this production is controlled.
3. Explain, using an example, the concept of negative feedback.
4. Describe the process of inflammation.
5. Discuss the formation of a platelet plug and a blood clot.
6. Discuss the human blood groups and explain the compatibility or incompatibility resulting from transfusing each type.
7. Explain the role of the Rh factor in blood transfusions and in the disease erythroblastosis fetalis.
8. Describe the basic structure of the human heart and trace the circulation of blood through the heart.
9. Explain why it is necessary for blood to recirculate continually through the heart.
10. Discuss the concept of blood pressure.
11. Describe the structural and functional differences between arteries, veins, and capillaries.
12. Explain how the blood is involved in regulating body temperature.
13. Discuss the regulation of the heartbeat.
14. Describe the causes and effects of various cardiovascular disorders.
15. Describe the formation of lymph and explain the role of the lymphatic system.
16. Describe two mechanisms involved in nonspecific immune responses.
17. Discuss the possible causes and effects of autoimmunity.

The human circulatory system, along with the lymphatic system, is involved in the transport of vital materials and wastes, in the regulation of body temperature and fluid balance, and in the protection of the body against disease.

PREVIEW OF ESSENTIAL TERMS

erythropoietin A hormone that stimulates bone marrow tissue to produce red blood cells.
negative feedback A body control mechanism that restores normal physiological conditions, should those conditions attempt to change.
antigen A substance capable of stimulating the formation of antibodies.
antibody A protein that interacts with the antigen, or a similar antigen, that caused its formation.
open circulatory system A system common in many lower animals in which the blood vessels open into the tissues so that blood comes in direct contact with the cells.
closed circulatory system A system in which the blood is confined to the blood vessels as it circulates from the heart to the tissues and back again; found in some invertebrates and all vertebrates.
systole The contracting, or pumping, phase of the heart cycle.
diastole The dilating, or relaxation, phase of the heart cycle.
lipoprotein A substance composed of a lipid in combination with a protein.
lymph Tissue fluid (plasma water and small proteins) that enters the lymph vessels.
lymphocyte A type of white blood cell involved in protecting the body against invading antigens.
autoimmunity The development of immunity against one's own body tissues.

Circulation and Immunity

INTRODUCTION

In all living things, it is essential that the body cells receive an almost continuous supply of oxygen, foodstuffs, water, and other vital materials. In addition, metabolic wastes must be removed from the cells and transported to excretory structures or organs for release from the body. In unicellular organisms, various substances dissolved or suspended in the cell fluid may be distributed throughout the cell by movement of the cytoplasm. This is the case in *Paramecium*, for example, in which food digested within the food vacuoles is circulated to all parts of the cell (Fig. 17.1*A*). Undigested residue remains in the vacuoles and is excreted through pores to the outside.

In multicellular organisms there is usually some type of passageway or a system of tubes through which nutrients and wastes are circulated. Sponges obtain nutrients through tiny body pores that lead into the hollow body cavity, or spongocoel (Fig. 17.1*B*). The surrounding water in which the nutrients are suspended also passes into the spongocoel. Within the sponge, water serves as a circulatory fluid to distribute nutrients and remove wastes. Circulation within *Paramecium* and the sponge, then, occurs in the absence of a separate and distinct circulatory system. Many invertebrates are relatively small and inactive and do not require rapid and direct transport of substances to and from the cells. Other more active invertebrates, however, such as earthworms, squids, insects, and so on, as well as all the vertebrates, have definite circulatory systems consisting of pumping organs called *hearts* and a network of circulatory vessels containing fluid. In most animals this fluid is *blood*.

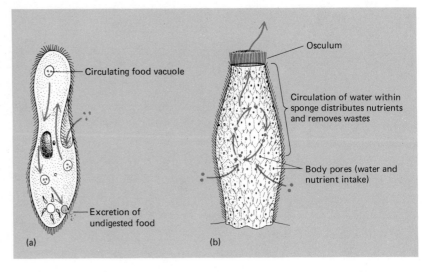

Figure 17.1 Examples of circulation in lower organisms. (a) In *Paramecium* food taken in from the surrounding medium is enclosed within food vacuoles. Movement of the cytoplasm distributes the vacuoles throughout the interior of the organism. Undigested food is released from the vacuoles through pores to the outside. (b) In sponges water and nutrients enter the spongocoel through tiny body pores; within the sponge, water serves as a circulatory fluid to distribute the nutrients and to remove wastes through the "mouth", or osculum. (Organisms are not drawn to scale.)

> Human blood, a type of connective tissue, consists of *solids* (red blood cells, white blood cells, platelets) suspended in the liquid *plasma* (water plus some proteins, hormones, ions, and wastes).

THE HUMAN CIRCULATORY SYSTEM

In man, the circulatory system consists of the **blood,** the **heart,** and a network of **blood vessels.** Taken as a group, these three components also are referred to as the **cardiovascular** ("heart and vessels") **system.** In addition, an accessory circulatory system, the **lymphatic system,** serves to clear the body tissues of excess fluid and return it to the blood.

Not only is it the major transportation system of the body, but the circulatory system also is involved in regulating body temperature and fluid balance and in protecting the body against disease.

BLOOD

Human blood is a type of connective tissue, consisting of various *solids* suspended in the liquid *plasma.* Although it varies with the size and weight of an individual, the volume of blood in the body averages about 5 to 6 qt. The blood solids include the **red blood cells, white blood cells,** and **platelets** (Fig. 17.2). Plasma is about 92 percent water; the remaining constituents include plasma proteins, foodstuffs, hormones, various ions, and metabolic wastes. Some plasma proteins are important as *antibodies,* which protect the body against disease; others are involved in the blood clotting process and in the transport of hormones, fatty acids, and other lipids. Blood has a normal pH range between 7.35 and 7.45 and is thus slightly alkaline.

Blood Solids

Red Blood Cells, or Erythrocytes. Mature red blood cells are tiny biconcave discs that function in the transport of oxygen (Fig. 17.3). In adults, red cells are formed primarily in the bone marrow of the vertebrae, ribs, and sternum. During their formation in the marrow, the red cells have a nucleus; but prior to their release into the cir-

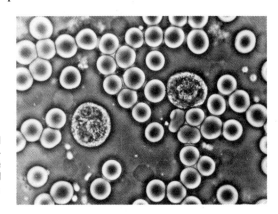

Figure 17.2 A blood smear. The small cells are red blood cells, the big, granular ones are white blood cells, and the faint, transparent circles are red blood cell ghosts—empty plasma membranes. (Biophoto Associates)

> Oxygen deficiency causes the release of the hormone *erythropoietin*, which in turn stimulates the production of red blood cells in bone marrow.

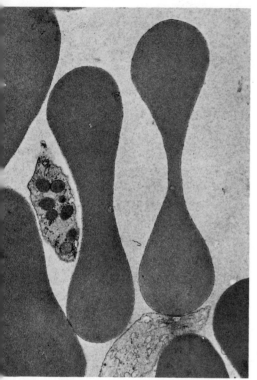

Figure 17.3 Electron micrograph of mature red blood cells showing their biconcave shape. The smaller cells are platelets. (From Bloom, W., and Fawcett, D. W.: *A Textbook of Histology.* 2nd ed. Philadelphia: W. B. Saunders Co., 1970, p. 129.)

culating blood, all normal cells lose their nuclei. Consequently, the life span of these cells in the bloodstream is only about 120 days. However, nearly 2.5 million new red cells are produced every second to replace those that are lost. As the red cells age and wear out, they are destroyed by large phagocytic cells in the liver and spleen.

Every red blood cell contains the pigment **hemoglobin,** which gives blood its characteristic color. Within the red cell, oxygen is transported in association with the hemoglobin molecules. Actually, oxygen combines with *iron* (Fe) atoms, which are part of the structure of hemoglobin. Insufficient iron in the diet can result in *iron deficiency anemia,* a condition characterized by persistent tiredness, lack of energy, and shortness of breath, even on mild exertion.

Under normal conditions, the number of red blood cells in the circulation remains fairly constant. A healthy adult male has about 5.4 million red cells per cubic millimeter (mm^3) of blood, or approximately 30 trillion in the entire circulation. Adult females have slightly fewer, averaging about 4.8 million red cells per cubic millimeter. Given their vital function of transporting oxygen, a significant decrease in the number of red cells, through either blood loss or bone marrow disease, can have serious consequences.

What stimulates the production of red blood cells in the first place, and how is this production controlled? The initial stimulus for red cell production appears to be a deficiency of oxygen in the body. As the oxygen level falls, a hormone called **erythropoietin** is released from the kidneys and perhaps the liver. Erythropoietin circulates in the blood to the bone marrow, where it stimulates tissues in the marrow to produce more red cells. When the number of new red cells produced is adequate to supply the body's oxygen need, erythropoietin is no longer released, and the bone marrow reduces its production of red cells. This process occurs, for example, at high altitudes where the oxygen pressure is reduced. Breathing, then, is difficult until the production of red cells increases sufficiently to supply the cells with oxygen.

The mechanism described here for the control of red blood cell production is an example of what is called **negative feedback** (Fig. 17.4). Essentially, this means that if the level of a substance in the body is too high or too low, various mechanisms act to correct the level of the substance. In the example just given, the low level of oxygen in the blood was elevated by the mechanism of erythropoietin stimulating the bone marrow to produce more red cells. This is called "negative" feedback because the result (a higher level of oxygen) is opposite to the initial stimulus (a low level of oxygen).

We encountered this same mechanism in the previous chapter; excess carbon dioxide stimulates the respiratory center to increase

> Most homeostatic control mechanisms of the body operate on the principle of negative feedback.

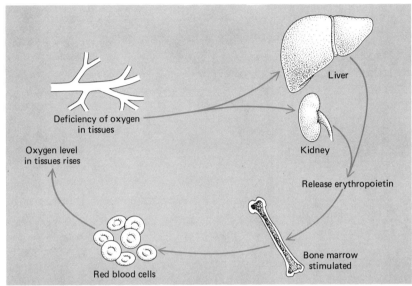

Figure 17.4 Control of red blood cell production by negative feedback. A deficiency of oxygen in the body tissues stimulates release of the hormone erythropoietin by the liver and kidney. Erythropoietin causes an increase in red blood cell production by the bone marrow. With the release of more red cells into the circulating blood, there is increased oxygen transport and a rise in the level of oxygen in the tissues. The increased supply of oxygen inhibits further release of erythropoietin, and red cell production by the bone marrow declines.

the rate of breathing. As more carbon dioxide is blown off, its level in the blood begins to fall. In this case, the stimulus (a high level of carbon dioxide) is opposite, or negative, to the effect (a reduced level of carbon dioxide).

Probably the most familiar example of negative feedback, and one that will help you understand and remember the principle, is that involving the thermostat on a home air conditioner. If the thermostat is set at 75°F, the air conditioner will click on if the room temperature rises above 75°F. Once the room is cooled and the temperature drops down to 75°F, the air conditioner clicks off. In this analogy, the stimulus (a rise in temperature) is negative to the effect (a drop in temperature).

It is important to understand the process of negative feedback because virtually all of the control mechanisms of the body operate on this principle to maintain homeostasis.

White Blood Cells, or Leukocytes. There are several different types of white blood cells in the body, some of which are about twice the size of a red blood cell, whereas others are nearly three times larger (Fig. 17.5). The primary function of white cells is to protect the body against potentially harmful microorganisms that invade the blood and other tissues. Most of the white cells accomplish this by *phagocytosis*—surrounding and engulfing the microorganism and then destroying it with digestive enzymes (Fig. 17.6). Other white cells are able to form antibodies, which, as we will discuss later, are part of the body's immunity system. All white blood cells are nu-

550
Circulation and Immunity

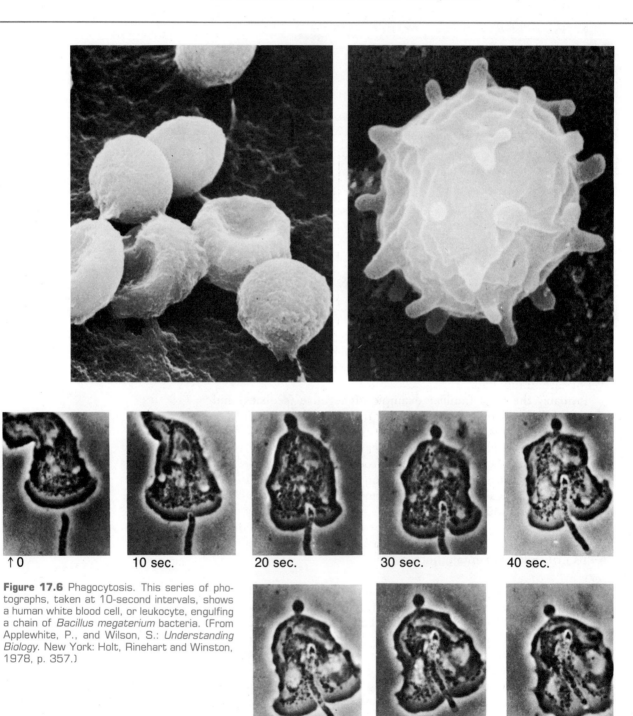

Figure 17.5 Scanning electron micrographs of human red blood cells (left) and a human white blood cell (right). Note the many pseudopods of the white blood cell. (From Ebert, J. D., Loewy, A. G., Miller, R. S., and Schneiderman, H. A.: *Biology*. New York: Holt, Rinehart and Winston, 1973, p. 380.)

Figure 17.6 Phagocytosis. This series of photographs, taken at 10-second intervals, shows a human white blood cell, or leukocyte, engulfing a chain of *Bacillus megaterium* bacteria. (From Applewhite, P., and Wilson, S.: *Understanding Biology*. New York: Holt, Rinehart and Winston, 1978, p. 357.)

> *Platelets* stop the flow of blood from a cut or ruptured vessel.

cleated and are formed in the red bone marrow and in the lymph nodes.

The life span of some white blood cells is quite short, ranging from a few hours to several days. The reason is that, in performing their protective function, they are usually destroyed in the process. The substance indelicately referred to as *pus* is largely an accumulation of white cells that died after their engulfment of invading bacteria. Other white cells, however, may exist in the body for months or perhaps years.

The number of white cells in the circulation is the same for both males and females, averaging about 7000 per cubic millimeter of blood. An abnormal increase in this number—a condition called **leukocytosis**—can result from infection, heavy exercise, and other physical states. An uncontrolled production of white cells, many of them immature, is called **leukemia,** or "cancer of the blood." Even though the white cell count is drastically elevated, cancerous white cells are mostly nonfunctional, thereby lowering the body's resistance to infection. Leukemic cells multiply so rapidly that they utilize the foodstuffs normally intended for healthy cells. As a result, there is usually weight loss, extreme fatigue, and an increasing susceptibility to even mild infections.

There are several viral diseases, such as influenza, measles, and chicken pox, in which there is a decline in the number of white cells in the blood—a condition known as **leukopenia**.

Inflammation. What actually happens to cause the redness and swelling associated with infection or injury? No matter what the cause, injured tissue releases various substances, principally **histamine,** that increase local blood flow by dilating the capillaries. In addition, the walls of the capillaries become more permeable, and fluid leaks from the blood into the surrounding tissues. The increased blood flow causes the area to redden, and the leak of fluids causes the swelling. Once histamine and the other substances are released by the injured tissue, white cells begin to migrate toward the damaged area, a phenomenon called **chemotaxis** (an attraction caused by chemicals). In time, then, the injured tissue becomes well supplied with white blood cells that protect the area against invading bacteria.

Platelets, or Thrombocytes. In the red bone marrow, extremely large cells are formed, which, at maturity, break apart into small cell fragments called platelets. Platelets do not have nuclei and are less than half the size of red blood cells. The essential function of platelets is to stop the flow of blood from a cut or ruptured vessel.

> A *thrombus* is a stationary clot within a vessel; an *embolus* is a loose clot circulating in the bloodstream.

Healthy adults of both sexes have, on the average, about 300,000 platelets per cubic millimeter of blood. Through continual production and destruction, the body replaces the platelets in the blood about once every 4 days.

A serious decline in the number of circulating platelets is called ***thrombocytopenia***—a condition that produces an increased tendency to bleed. This should not be confused with the hereditary disease ***hemophilia,*** which results from the absence of some clotting factor in the blood.

Blood Clotting. In many lower animals, the prevention of blood or fluid loss is accomplished simply by contraction of the muscles and circulatory vessels. This alone is adequate to stop the flow of blood or fluid from the wound. In most of the higher animals and man, various substances in the blood, together with the blood cells, form a plug or clot that stops further blood loss.

When a blood vessel is cut, there is first a constriction of the wall of the vessel, which reduces the flow of blood. As the blood oozes from the wound, the platelets begin to accumulate at the site of the cut. They also become sticky and begin to release adenosine diphosphate (ADP), which tends to attract other platelets. These platelets in turn release ADP, attracting still more platelets, and so on until several layers of platelets form a ***plug.*** This mechanism is adequate for stopping blood loss from very small holes, but larger openings require a blood ***clot*** in addition to the platelet plug.

A blood clot is initiated by an activator, or "starter" chemical, released from the platelets and from the injured wall of the blood vessel. This chemical, along with calcium ions (Ca^{++}), converts a plasma protein, ***prothrombin,*** into an enzyme called ***thrombin.*** Thrombin acts on a second plasma protein, ***fibrinogen,*** converting it into a threadlike form called ***fibrin.*** Threads of fibrin form a network that traps the blood cells and platelets, resulting in a ***blood clot*** (Fig. 17.7). Soon thereafter, the clot begins to contract and shrink, pulling the walls of the cut vessel together and stopping the bleeding.

In cases in which the lining of a blood vessel becomes roughened through aging or disease, or if the blood is moving sluggishly through a vessel, a stationary clot may form *within* the vessel. In these instances, the clot is referred to as a ***thrombus.*** If the thrombus breaks loose and circulates in the bloodstream, it is then called an ***embolus.*** The obvious danger of internal clotting is that either a thrombus or an embolus can obstruct the flow of blood to vital organs, depriving them of oxygen and nutrients.

Figure 17.7 Scanning electron micrograph of a red blood cell caught in threads of fibrin (magnification × 10,250). (From Biological Sciences Curriculum Study, Inc., Postlethwait, S. N., Director: *Animal Structure and Function.* Philadelphia: W. B. Saunders Co., 1976, p. 23.)

> Blood group *antigens*, which determine a person's blood type, are associated with an individual's red blood cells.

What prevents the blood from clotting in the circulatory system under normal conditions? First, the lining of healthy blood vessels is extremely smooth; this prevents the initiation of the clotting process described previously for injured vessels. Second, the smooth lining of blood vessels carries a negative electrical charge that repels the platelets and clotting factors in the blood, again preventing the initiation of the clotting process. In addition, the blood contains an anticoagulant called **antithrombin**, which prevents the enzyme thrombin from converting fibrinogen into fibrin. Another powerful anticoagulant is **heparin**, a substance thought to be produced by many different cells of the body.

Blood Groups. Although it has the same physical appearance, the blood of all human beings is not exactly alike. You are probably aware, for instance, that the transfusion of blood from one person to another is not always safe. Before blood can be given or received, it must be determined if the blood of the donor and the recipient are compatible. If they are not, the red blood cells of the donor will clump together, or *agglutinate*, when mixed with the blood of the recipient. This agglutination reaction is the result of substances present in the red blood cells and plasma.

The red cells of some individuals contain substances called ***antigens***, which are inherited (see Chapter 8). There are two types of blood group antigens, designated *A* and *B*. Persons with type *A* antigens in their red blood cells have type *A* blood; those with type *B* antigens have type *B* blood. If *both* A and B antigens are present in the red cells, the blood is type *AB*. Finally, individuals with *neither* of these antigens have type *O* blood.

Although the *A* and *B* antigens are associated with the red cells, other substances called ***antibodies*** may be present in the *plasma*. Blood group antibodies are designated either *a* (also referred to as anti-A) or *b* (anti-B). The presence or absence of these antibodies is determined by the type of antigen in the red blood cells. By definition, an antigen is any substance that is capable of causing the formation of antibodies. An antibody is a type of protein produced by the body that interacts with the antigen that caused its formation or with a closely related antigen.

In the antigen-antibody relationships, individuals with type *A* blood have *b* antibodies in their plasma, and those with type *B* blood have *a* antibodies in their plasma. People with type *AB* blood have *neither* of these antibodies, whereas the plasma of type *O* persons contains *both* of the antibodies.

Circulation and Immunity

> Phagocytic attack on agglutinated red cells releases free hemoglobin into the circulatory system where it is converted into *bilirubin*. Excessive amounts of bilirubin cause *jaundice*.

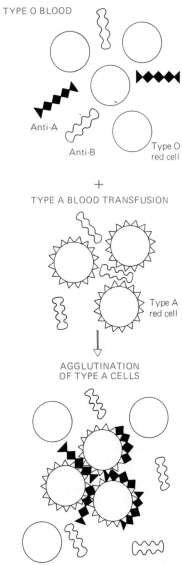

Figure 17.8 Agglutination of type A blood cells given in transfusion to a type O person. The recipient's anti-A antibodies attach to the glycoprotein A on the surface of the type A red cells. Clumps of type A red cells and anti-A antibodies precipitate out of the bloodstream and may clog small vessels, cutting off circulation. (From Arms, K., and Camp, P. S.: *Biology.* New York: Holt, Rinehart and Winston, 1979, p. 455.)

Blood Transfusions. The agglutination of red blood cells that occurs when incompatible blood types are mixed results from specific antigen-antibody reactions (Fig. 17.8). What, then, makes certain blood types incompatible? Basically, the answer involves two interrelated concepts. First, in transfusing blood from one person to another, one must *consider the effect of the donor's red cells on the plasma of the recipient*. In other words, it is the effect of the *donor's antigens* on the *antibodies* of the recipient that is important. This leads us to the second concept. If the donor's antigens and the recipient's antibodies are of the same kind, the red blood cells of the donor will agglutinate. For example, if the donated blood is type *A* and the recipient's blood is type *B*—indicating that the plasma contains *a* antibodies—the red cells will agglutinate. The same applies to *B* antigens donated to a person with *b* antibodies (type *A* blood, for example). As a memory device, remember that the same two letters—a donated *A* and a recipient's *a* or a donated *B* and a recipient's *b*—cannot safely be together when blood is transfused.

Normally, type *O* blood may be *donated* to anyone. This is because there are no antigens in type *O* blood to react with the antibodies (if any) in the plasma of the recipient. On the other hand, persons with type *AB* blood normally may *receive* transfusions of any of the other blood groups. Since type *AB* blood has no antibodies in its plasma, the antigens (if any) in the donated blood cannot be involved in an agglutination reaction.

Could type *AB* blood be *donated*, to, say, a person with type *B* blood? No, because the *A* antigens of the donor and the *a* antibodies of the recipient would result in agglutination. What about transfusing type *A* blood to a person with type *O* blood? This, too, would result in agglutination, since the *A* antigen of the donor's blood will react with the *a* antibodies in the plasma of the recipient. As you might expect, the preferred procedure in transfusing blood is to use donor blood of the same type as the blood of the recipient. The characteristics of the blood groups and the types of transfusions possible are summarized in Tables 17.1 and 17.2.

As mentioned, it is the *donor's* red blood cells that agglutinate within the blood of the recipient. As a result of the agglutination reaction, the clumped red cells often plug up small blood vessels. Eventually, phagocytic cells attack the clumps and in the process release hemoglobin from the red cells into the circulatory system. Some of the free hemoglobin is broken down by other wandering cells and is converted into a substance called **bilirubin.** If enough bilirubin is present in the system, the skin and other tissues take on a yellowish tint, a condition known as **jaundice.**

> The presence or absence of *Rh factor* antigen in blood cells also affects blood compatibility between donor and recipient. If Rh positive and Rh negative blood are mixed, agglutination can result, leading to jaundice.

Table 17.1 SUMMARY OF BLOOD GROUP ANTIGENS AND ANTIBODIES

BLOOD TYPE	RED CELL ANTIGEN	PLASMA ANTIBODY
A	A	b (anti-B)
B	B	a (anti-A)
AB	A and B	None
O	None	a and b

Rh Factor. In addition to the A and B antigens, there is another major antigen that may be present in the blood. This antigen is called the **Rh factor,** so named because it was first discovered in the Rhesus monkey. Individuals who have the factor in their blood cells are **Rh positive;** those who do not have the factor are **Rh negative.** The blood types of human beings, therefore, are designated not only as *A, B, AB,* and *O* but also as *Rh positive* and *Rh negative.*

The problem arising from mixing Rh positive and Rh negative blood is somewhat similar to that of mixing *A, B, AB,* and *O* groups. When Rh positive blood is given to an individual with Rh negative blood, the Rh negative blood begins to form *anti-Rh* antibodies against the Rh factor (antigen) of the donor's blood. If the Rh negative individual is exposed to enough of the Rh antigen (as through multiple transfusions), high concentrations of anti-Rh antibodies develop in the blood. The antibodies attach to the Rh positive red blood cells of the donor, causing agglutination. As before, these clumped, or agglutinated, red cells can then plug up small blood vessels. The clumps are later attacked by phagocytes, and free hemoglobin is released into the circulation. The hemoglobin is converted to bilirubin, leading to the condition of jaundice.

Table 17.2 COMPATIBILITY OF BLOOD TYPES

BLOOD TYPE	CAN DONATE BLOOD TO	CAN RECEIVE BLOOD FROM
A	A, AB	A, O
B	B, AB	B, O
AB	AB	A, B, AB, O
O	A, B, AB, O	O

Circulation and Immunity

> Rh negative blood mixed with Rh positive blood causes no agglutination reaction because Rh negative blood contains no Rh antigen.

Figure 17.9 Diagram of the sequence of events leading to erythroblastosis fetalis, a condition in which the red cells of the fetus clump within the uterus. (a) Red cells pass from an Rh⁺ fetus to its Rh⁻ mother through some defect in the placenta. (b) Maternal white cells produce anti-Rh⁺ antibodies. (c) During a subsequent pregnancy anti-Rh⁺ antibodies pass from the maternal blood to the fetal blood stream. (d) The reaction of Rh⁺ cells with anti-Rh⁺ antibodies causes clumping of the fetal red cells.

If Rh negative blood is donated to an Rh positive individual, there is no agglutination reaction. You will recall that it is the *antigen* in the donor's blood that reacts with the antibodies of the recipient. Since there is no Rh antigen in Rh negative blood, agglutination does not occur.

A potentially dangerous agglutination reaction can occur in some newborn infants. The result of such a reaction is a disease called **erythroblastosis fetalis** (Fig. 17.9). The disease most often occurs in an Rh positive infant whose mother is Rh negative and whose father is Rh positive. Before birth, the blood of the fetus releases its Rh antigens, which diffuse across the placenta into the mother's

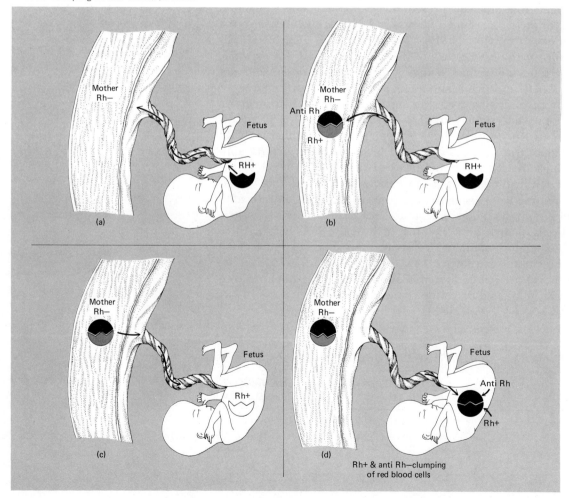

> A child with erythroblastosis fetalis can become severely anemic.

blood. Anti-Rh antibodies begin to develop in the mother's blood and later seep back across the placenta into the blood of the fetus. With the mixing of the antigens and antibodies in the fetal blood, the red blood cells agglutinate. The clumped cells plug up small blood vessels and later release free hemoglobin into the fetal circulation. The subsequent conversion of hemoglobin into bilirubin results in jaundice, and the child becomes severely anemic.

From the preceding discussion, it should be apparent that the mother cannot develop anti-Rh antibodies if the child is Rh *negative*. The blood of an Rh negative child would have no Rh antigen and thus cannot stimulate the production of antibodies in the mother's blood. Usually, the first Rh positive child of an Rh negative mother does not develop erythroblastosis fetalis. This is because the mother ordinarily does not develop a high enough concentration of antibodies to affect the first child. She does, however, become more sensitized to the Rh antigen during the first pregnancy. In the second and subsequent pregnancies with Rh positive children, antibodies generally are produced more rapidly and in greater amounts. The Rh positive children in these instances are progressively more likely to contract the disease.

One treatment given to prevent erythroblastosis fetalis is to inject anti-Rh antibodies into the mother immediately after the birth of her first Rh positive child. These antibodies, obtained from another Rh negative person, destroy the child's Rh antigens and thereby prevent the mother from developing her *own* antibodies. Thus, in the second pregnancy, she is not sensitized to the Rh antigen, should she bear another Rh positive child.

A more recent treatment is the injection of a substance called RhoGAM, which contains antibodies that destroy the anti-Rh antibodies of the mother. RhoGAM is also given after the birth of the first Rh positive child.

If erythroblastosis fetalis has already developed, the newborn child usually is given a transfusion of Rh negative blood while its contaminated Rh positive blood is being removed. Since the life span of a red blood cell is about 120 days, the body eventually destroys all the Rh negative red cells and replaces them with the child's own Rh positive cells.

THE CARDIOVASCULAR SYSTEM
Introduction

Circulatory systems are generally designated as either *open* or *closed*. An open system is one in which the blood vessels open into the tissues so that the blood comes in direct contact with the cells. After circulating among the cells, the blood re-enters open blood vessels to be returned to the heart. All of the arthropods have open circulatory

> Arthropods and many mollusks have *open* circulatory systems. Some invertebrates and all vertebrates have *closed* circulatory systems.

systems, as do many of the mollusks. In a closed system the blood is confined to the blood vessels as it circulates from the heart to the tissues and back again. The only exit for blood in a closed circulatory system is *through* the walls of the blood vessels. Some invertebrates and all vertebrates, including man, have a closed circulatory system (Fig. 17.10).

Among the major groups of vertebrates—fish, amphibians, reptiles, birds, and mammals—we find significant variations in the anatomy of the circulatory system, particularly in the case of the *heart* (Fig. 17.11). In fishes, the heart is basically two-chambered. One thin-walled chamber, the *atrium*, receives blood after it has circulated through the body. From the atrium, the blood passes into the second chamber, the thick-walled *ventricle,* which pumps the blood to the gills. In the gills, the blood is *oxygenated* and continues on to circulate through the body. Upon reaching the atrium, the blood has given up its oxygen to the tissues and so must once again enter the ventricle to be pumped to the gills. Note that in the two-chambered fish heart only *deoxygenated* blood goes to the heart.

Amphibians have a three-chambered heart—a right and left atrium and one ventricle. Deoxygenated blood from the body enters the right atrium and passes to the ventricle, where it is pumped to the lungs. In the lungs the blood picks up oxygen and returns to the left atrium of the heart. This oxygenated blood then enters the ven-

Figure 17.10 Circulatory system of the grasshopper: an open circulatory system. Circulatory system of the fish: a closed circulatory system (with a two-chambered heart and arteries shown in color.)

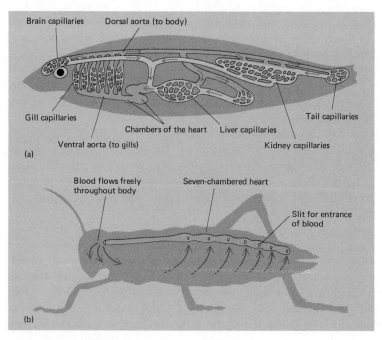

Fish have two-chambered hearts; amphibians have three-chambered hearts. Reptiles, birds, and mammls have four-chambered hearts, though it is only in birds and mammals that the ventricles are entirely separate.

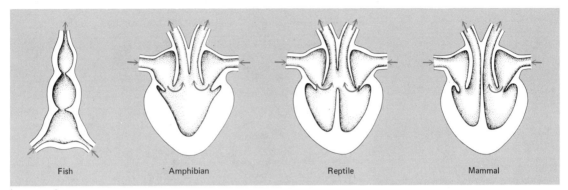

Figure 17.11 Structure of the heart in different vertebrates. The fish heart has only two chambers, an atrium and a ventricle. However, there is no mixing of oxygenated and nonoxygenated blood; oxygenated blood leaving the gills does not return to the heart but circulates throughout the body. The heart simply pumps nonoxygenated blood returning from the body back to the gills. Amphibians and many reptiles have three chambers, two atria and a single ventricle. This ventricle is partially divided but the two halves are not completely separated. As a result, some oxygenated and nonoxygenated blood is mixed. In crocodiles, birds, and mammals, the ventricles are completely separated, and the heart has four chambers. Thus no mixing of oxygenated and nonoxygenated blood occurs. This is one reason why circulation in birds and mammals—animals with a high metabolic rate and a fast-paced way of life—is more efficient than in amphibia and most reptiles.

tricle to be pumped out of the body. As blood is entering the ventricle from the right atrium, it is also entering from the left atrium. The heart is never empty. In the amphibians, then, there is some mixing of deoxygenated and oxygenated blood in the ventricle. As a result, some of the deoxygenated blood is pumped to the cells along with the oxygenated blood. Amphibians are "cold-blooded," with low metabolic rates and somewhat limited physical activity. Consequently, their need for totally oxygenated blood is less crucial than that of the "warm-blooded" birds and mammals.

The reptile heart is essentially four-chambered, having two atria and two ventricles. The ventricles, however, are not entirely separated from each other except in crocodiles and alligators. For the most part, then, deoxygenated blood from the body mixes very little with oxygenated blood from the lungs.

In the birds and mammals, the heart is four-chambered with the ventricles entirely separated. Since there is no mixing of blood in the ventricles, the delivery of oxygenated blood to the body cells is much more efficient than in other vertebrates.

In addition to the heart, the circulatory system consists of several types of conducting vessels. The blood vessel, or **vascular,** system includes the **arteries, veins,** and **capillaries** (Fig. 17.12). Arteries are thick-walled, elastic vessels that carry blood out of or *away* from the heart. Veins have somewhat thinner walls and carry blood *toward* the heart. Veins are also quite elastic, which, on occasion, permits them to temporarily store blood that is moving sluggishly through the system. The tiny capillaries are extremely thin-walled, porous vessels that form extensive networks throughout the body. Anyone who has shaved a face or a leg has probably drawn blood from a capillary. The fact that capillaries are thin-walled and porous is the key to their function. The thin walls permit the entrance and exit of diffusible substances, and the pores restrict the passage of larger molecules while permitting smaller molecules to pass through.

560
Circulation and Immunity

> The human heart has four separate chambers: two upper atria and two lower ventricles.

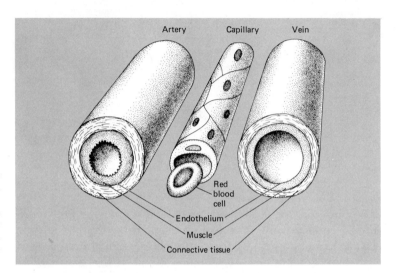

Figure 17.12 Structure of the walls of three types of blood vessels. Both arteries and veins have three-layered muscular walls (note the greater thickness of the arterial walls). Capillary walls consist of a single layer of epithelium and are permeable to gases and liquids.

Moreover, as we shall see, capillaries also serve as links between the arteries and veins.

The Human Heart

Contrary to popular opinion, the human heart is not the center for our emotions of love or hate or for any other emotions. We love or hate not from the heart but from the mind. Essentially, the heart is a muscular organ for pumping blood. It is about the size of your fist and is composed of cardiac muscle, which is capable of rhythmic contraction and relaxation. On the average, the heart beats about 70 times per minute. The heart is situated between the lungs in the middle of the chest cavity, protected in front by the sternum, or breastbone, and surrounded by the rib cage (Fig. 17.13). As with all mammalian hearts, the human heart has four separate chambers—two upper atria and two lower ventricles—through which blood circulates before being returned to the body. We will first look at how blood gets into and out of the heart and then consider why this trip is necessary.

The basic scheme of circulation is relatively simple. Deoxygenated blood from the body cells enters the right side of the heart. The blood is then pumped to the lungs, where it picks up oxygen, and returns to the left side of the heart. From here the newly oxygenated blood is pumped out to the body (Fig. 17.14). Obviously, this brief overview gives us very little insight into the actual workings of the heart. With the addition of some necessary details, we can gain insight that is also essential to a basic understanding of the entire circulatory system.

> A contraction of the heart is called *systole*; the relaxation phase is called *diastole*.

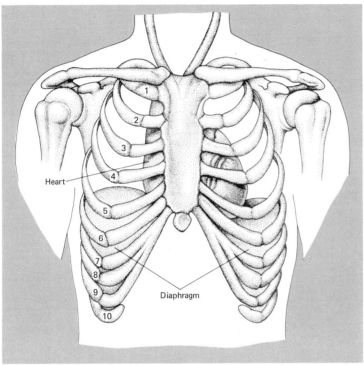

Figure 17.13 The heart is protected in front by the sternum and surrounded by the rib cage.

Blood that has circulated to the body cells, giving up its oxygen, returns to the heart through the venous system. All the venous blood from *above* the heart (from the head, neck, shoulders, and so forth) eventually collects into one large vein, the ***superior vena cava,*** which empties into the right atrium of the heart (Fig. 17.15). From *below* the heart (feet, legs, abdomen, and so on), venous blood collects into the ***inferior vena cava,*** which also empties its blood into the right atrium. Remember that the blood being emptied into the right atrium has returned from metabolizing body cells and is thus high in carbon dioxide but low in oxygen. From the right atrium, blood passes down through an opening, the ***tricuspid valve,*** into the right ventricle. As the right ventricle pumps, or contracts, the tricuspid valve closes, preventing blood from flowing backward into the right atrium. If a heart valve fails to close completely, allowing the backflow of blood, a hissing sound called a **heart murmur** is produced. Any contraction of the heart muscle is called ***systole*** ("contraction"); the relaxation phase is called ***diastole*** ("dilation"; "expansion"). We may speak of atrial or ventricular systole and atrial or ventricular diastole.

From the right ventricle, blood is pumped through another opening, the ***pulmonary semilunar valve,*** into the ***pulmonary artery.***

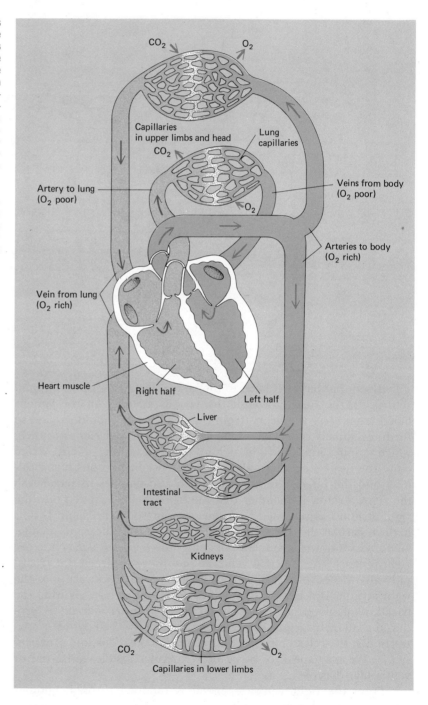

Figure 17.14 Major circulatory routes in man. A drop of blood going from the heart to the big toe, for example, is pumped from the left ventricle into the aorta. From the aorta it moves to the arteries of the leg, foot, and big toe. On its return trip, the blood leaves the capillaries of the toe and enters the venules, and from there it goes to the veins of the legs. Blood in these veins is collected, along with blood from other parts of the lower body, in the inferior vena cava vein, which empties it into the right atrium. From the right atrium, blood flows into the right ventricle, which pumps blood into the lungs via the pulmonary arteries. Oxygenated blood leaves the lungs and enters the left ventricle via the pulmonary veins. Blood flows from the left atrium into the left ventricle, thus completing one round trip.

> *Pulmonary veins* carry oxygenated blood, whereas the *pulmonary artery* carries deoxygenated blood. These are the only exceptions to the rule that arteries carry oxygenated blood and veins carry deoxygenated blood.

563
The Cardiovascular System

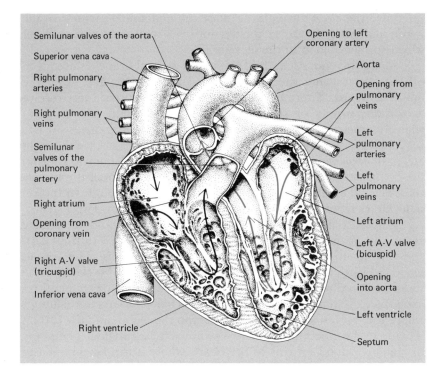

Figure 17.15 The human heart. Note the location of the valves to the heart chambers and blood vessels. The *arrows* indicate the direction of blood flow.

This valve prevents the backflow of blood into the right ventricle. The pulmonary artery exits from the right ventricle and divides into two branches—one going to the left lung and one going to the right lung. Blood pumped through these pulmonary branches goes to each lung, where it picks up oxygen and releases its carbon dioxide. The blood is now ready to return to the left side of the heart.

From each lung, two **pulmonary veins** carry the oxygenated blood to the left atrium. (Note that the pulmonary artery carries *deoxygenated* blood and that the pulmonary veins carry *oxygenated* blood. These are the only exceptions to the rule that arteries carry oxygenated blood and veins carry deoxygenated blood.) Blood passes from the left atrium through the **bicuspid,** or **mitral, valve** into the left ventricle. The bicuspid valve prevents blood from flowing back into the left atrium.

As the left ventricle contracts, the blood is pumped through the **aortic semilunar valve** into the large **aorta.** The aortic semilunar valve prevents the backflow of blood into the left ventricle. The aorta emerges from the left ventricle, arches to the left, and contin-

> *Systolic pressure* measurements indicate arterial pressure when the ventricles contract; *diastolic pressure* measurements indicate arterial pressure when the ventricles relax.

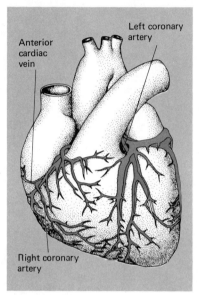

Figure 17.16 Coronary circulation, the routes that convey blood to the heart muscle. The arteries are colored and the veins are gray.

ues down the body through the chest and abdominal cavities. In the lower abdomen the aorta divides into two main arteries, each of which supplies blood to the legs and feet.

Along the entire length of the aorta, arterial branches are given off that extend into the neck, head, arms, muscles, and all of the internal organs. There is one organ in particular that must receive a constant supply of fresh blood—the heart itself. At the point where it emerges from the top of the heart, the aorta gives off two small vessels that enter the heart muscle. These vessels are the right and left *coronary arteries.* The coronary arteries subdivide within the heart to form tiny branches that deliver blood to all the muscle cells. Several small veins, including the **cardiac veins,** return the venous blood from the heart muscle to the right atrium (Fig. 17.16).

Why is it necessary for the blood to circulate and recirculate again and again through the heart to the lungs, back to the heart, and out again? There are two basic reasons for this. The venous blood, which has given up its oxygen to the cells and picked up the waste carbon dioxide, must continually acquire fresh oxygen and release the carbon dioxide. This gaseous exchange occurs in the lungs, where the oxygen you inhale diffuses into the blood, and the carbon dioxide in the blood diffuses into the lungs to be exhaled. The newly oxygenated blood returns to the heart, where it can be pumped to the body cells. It would be quite useless for the heart to continue to pump venous blood to the cells.

The other reason for the repeated circulation of the blood is that the *pressure* of the blood has to be maintained at certain levels. With the pumping force of the left ventricle behind it, oxygenated blood in the arteries is under relatively high pressure and can be delivered effectively to all parts of the body. **Blood pressure** is simply the force of the blood pushing against the wall of the blood vessel. This is the pressure responsible for the **pulse,** which is caused by the expansion and recoil of the arteries as blood is pumped through them.

Blood pressure is usually measured in millimeters of mercury (mmHg). For example, the average blood pressure among healthy young adults is about 120/80. The top number, called the **systolic pressure,** indicates that when the ventricles contract, they exert a pressure within the aorta and pulmonary arteries that can raise a column of mercury to a height of 120 mm. The bottom number, called the **diastolic pressure,** represents the pressure in the arteries when the ventricles relax. This pressure, then, can raise a column of mercury to a height of 80 mm.

> One-way valves within the veins prevent the backflow of venous blood, which is caused by the relatively low venous pressure and the pull of gravity.

If pressure in the arteries of the body falls low enough, the vessels tend to collapse. This can result in the blood vessels becoming so narrow that neither the cells nor the plasma can pass through. The effects of the stoppage of blood flow to the brain, liver, kidneys, and other organs, even for a short period of time, could be fatal.

The pressure of the venous blood returning to the right atrium is quite low compared with the pressure in the arteries. Not having as much force behind it as arterial blood, venous blood, particularly below the heart, often tends to slow, or back up and "pool." To prevent the backflow of blood, there are small *one-way valves* situated at intervals along the inner wall of the veins. As the blood moves toward the heart, the valves open and then are forced shut if the blood flow reverses (Fig. 17.17). Much of the reverse flow in many veins is caused simply by the pull of gravity. You can demonstrate this quite easily by allowing your arms to hang limp at your sides for a few seconds. Venous blood will reverse its flow and collect in the surface veins on the back of the hand. By watching your hand as you raise it, you can see that the veins will empty, once your arm is high enough. The most efficient method in assisting the flow of venous blood is muscular contraction. Anytime you tense or move your muscles, venous blood is forced toward the heart. One of the benefits of exercise is that the kneading action of the skeletal muscles aids in moving the blood through the veins.

Blood moving through the circulatory system performs yet another vital function. If the body begins to overheat, as from vigorous exercise, for example, the hypothalamus in the brain is stimulated,

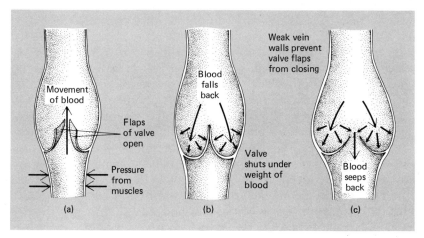

Figure 17.17 Valves. (a) When blood flows toward the heart, the valve opens and allows it to pass. (b) If blood moves in the reverse direction, it fills the cuplike flaps of the valve and presses the edges together, preventing backflow (c) The walls of a varicose vein are weak and allow blood to collect and distend them so that the edges of the valve flaps cannot meet; blood may then return through the valve, and circulation is impaired.

> Cells exchange carbon dioxide and other wastes for oxygen and nutrients through the *capillaries*.

and nerve impulses go out that cause blood vessels in the skin to dilate. This brings blood in the vessels closer to the surface of the body, allowing heat to radiate from the blood through the skin. The hypothalamus also stimulates the sweat glands to increase the rate of perspiration. Together, the radiation of heat from the blood and increased sweating act to cool the body. Should the body temperature fall below normal, this process is essentially reversed. The hypothalamus signals the blood vessels in the skin to constrict and shuts down the activity of the sweat glands. In this manner, warm blood remains deeper in the tissues, and there is little perspiration released. Thus, body heat is conserved, and the temperature of the body is brought up to its normal range.

Exchange of Materials: The Capillaries

The circulation of blood through the heart and blood vessels is of little significance unless at some point materials are exchanged between the blood and the cells. The blood vessels through which this exchange occurs are the *capillaries*. When they reach the muscles and organs of the body, the arteries divide and subdivide into smaller and smaller arteries. Subdividing continues until an extensive network of microscopic capillaries is formed. These tiny, thin-walled vessels are distributed throughout the body tissues and therefore lie in close proximity to the individual cells. Some capillaries are so small that red blood cells must squeeze through them in single file.

As the pulsating blood in the arteries moves into the capillary bed, its rate of flow is slowed, and oxygen and nutrients are released through the capillary walls into the cells. Carbon dioxide and other wastes from the cells pass in the opposite direction into the blood (Fig. 17.18). The blood, which is now deoxygenated, continues its journey through the capillaries until it encounters somewhat larger vessels, which are the small veins. These vessels lead to still larger veins until the blood collects into one large vein leading out of the muscle group or organ. The large vein, or another vein to which it is connected, conveys the blood to one of the venae cavae, which carries the blood to the right atrium of the heart.

As you can see, the capillaries form a link between the arterial and venous systems. As blood moves through the capillaries, its pressure is greatly reduced. By the time it enters the veins, there is very little force of the heartbeat behind the blood, and, as we have discussed, the pressure drops considerably below what it was in the arteries. Because of their thickness, there is no exchange of materials

> The *sinoatrial (SA) node* (also known as the pacemaker) initiates the heartbeat by sending out electrical impulses to the surrounding atrial muscles, causing them to contract. The *atrioventricular (AV) node* conveys the impulses from the atria to the ventricles along the Purkinje fibers, causing the ventricles to contract.

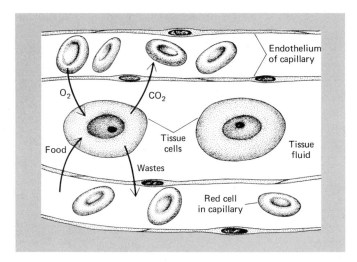

Figure 17.18 The slow movement of blood through the capillaries allows oxygen and food to diffuse into the tissue cells while carbon dioxide diffuses from the cells into the blood.

across the walls of the arteries and veins. The significance, then, of capillary exchange is that it is only through these tiny, porous vessels that all of the cells of the body receive vital nutrients and give up unwanted waste products.

Regulation of the Heartbeat

A heart removed from the body will continue to beat for several hours if supplied with the proper nutrients. Although the heart is regulated by its own conduction system in association with several nerves, the conduction system will continue to function even if the nerves are severed.

In the right atrium there is a small mass of specialized muscle tissue called the ***sinoatrial,*** or SA, ***node*** that initiates the heartbeat. This node is also generally known as the *pacemaker*. The SA node sends out electrical impulses to the surrounding atrial muscles, causing the atria to contract (systole). From the atria, the impulse is conducted to the ***atrioventricular,*** or AV, ***node,*** located in the lower part of the right atrium. When the impulse leaves the atria, they relax (diastole). The AV node conveys the impulse to the ventricles along a group of muscle fibers called the **Purkinje fibers.** These fibers branch into the cardiac muscle of the ventricles, causing the ventricles to contract (systole). Once the impulse passes through the ventricles, they relax (diastole). Then, the SA node "fires" again, and the heartbeat cycle is repeated (Fig. 17.19).

Sympathetic stimulation along accelerator nerves increases the heartrate; parasympathetic stimulation along the vagus nerve decreases the heartrate.

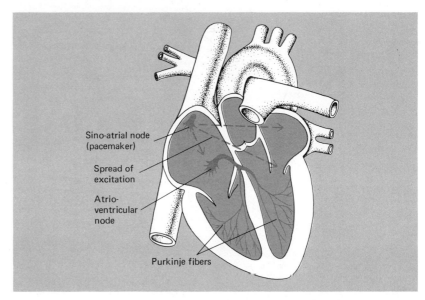

Figure 17.19 Spread of excitation through the heart of a mammal. Impulses start spontaneously in the specialized muscle cells of the AA node. From there the impulses spread through the atria, which contract simultaneously. The impulses finally reach the AV node. The fibers of the AV node conduct slowly, and the atria contract before impulses reach the ventricles. From the AV node the impulses spread throughout the ventricle through specialized muscle cells called the Purkinje fibers. This rapid spread of impulses enables the ventricles to contract as a whole.

In the living body, nerves from the sympathetic and parasympathetic divisions of the autonomic nervous system (ANS) supply the SA and AV nodes and the heart muscle itself. Sympathetic stimulation along *accelerator nerves* causes an *increase* in the rate and force of the heartbeat. A rapidly beating heart is familiar to anyone who has ever been afraid, apprehensive, or in love. Parasympathetic stimulation along the *vagus* nerve primarily acts on the SA and AV nodes to *decrease* the rate and force of the heartbeat. Under normal conditions, it is mainly the parasympathetic system that influences nervous control of the heart; in times of emotional or physical stress, control shifts predominantly to the sympathetic system.

The electrical activities that occur within the atria and ventricles of the heart can be recorded on an instrument called the *electrocardiograph.* The record produced by this instrument is the *electrocardiogram,* or EKG. As shown in Figure 17.20, the EKG consists of three major components: the ***P wave,*** the ***QRS complex,*** and the ***T wave.*** The P wave represents electrical activity in the atria just before they contract. The QRS complex is caused by electrical activity in the ventricles just prior to their contraction. Finally, the T wave results from electrical activity that occurs when the ventricles recover. The EKG is immensely valuable in determining the general condition of

> A *heart murmur* occurs when the heart valves fail to close properly and leak blood backward, producing a hissing sound detectable with a stethoscope.

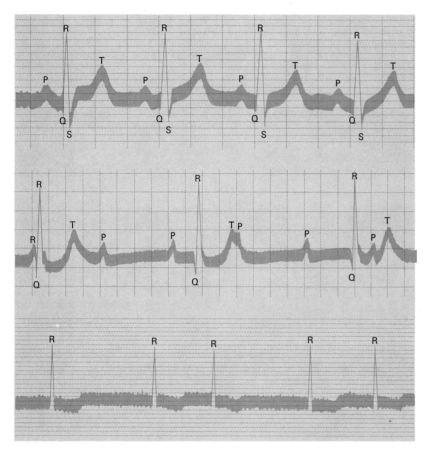

Figure 17.20 Electrocardiograms. (Top) Tracing from a normal heart. Compare this with the middle tracing, which shows the P waves (atrial contraction) and the QRS complex (ventricular contraction) occurring at regular intervals, but note that there is no relation between the P wave and QRS complex. This indicates that the atria and ventricles are beating independently, which occurs when the impulse is blocked at the AV node. (Bottom) EKG of rapid and irregular atrial contraction (fibrillation). There is no P wave, and the spacing of the QRS complexes indicates irregular beating of the ventricles.

the heart. Such conditions as abnormal heart rhythms, slow or rapid heart rates, and evidence of "heart attacks" are easily observable in the EKG.

Another invaluable instrument in cardiac physiology is the **stethoscope,** which amplifies the **heart sounds.** Generally described as "lub-dup," the heart sounds are produced by the closing of the heart valves. The "lub," or first heart sound, is caused by the closure of the tricuspid and bicuspid valves. The "dup," or second heart sound, results when the aortic and pulmonary semilunar valves close. As mentioned, if the heart valves fail to close properly, blood will leak backward, producing a hissing sound called a *heart murmur*. This sound, too, is readily detectable with a stethoscope.

CARDIOVASCULAR DISORDERS AND DISEASE

Diseases of the heart and blood vessels are presently responsible for the majority of deaths in the United States. However, the incidence of heart disease has been declining since 1967. There is no definite agreement on all the factors that contribute to some cardiovascular diseases, but circumstantial evidence points to such factors as heavy cigarette smoking, diet, lack of exercise, obesity, and heredity.

Arteriosclerosis. This condition, generally known as "hardening of the arteries," is characterized by a thickening of the walls of the arteries, which subsequently lose their elasticity. In the development of arteriosclerosis, various materials such as cholesterol, proteins, and calcium accumulate in the walls of the arteries, forming **plaques.** These deposits tend to roughen the inner lining of arteries and thereby contribute to the formation of blood clots, or thrombi. In addition, as the plaques increase in size, they impede or block the flow of blood in an artery (Fig. 17.21).

Arteriosclerosis contributes to a number of complications, including **hypertension,** or high blood pressure; **coronary occlusion** (blockage of the coronary arteries); and **strokes** (blockage of an artery supplying the brain).

Hypertension. High blood pressure can result from nervous disorders, excessive intake of salt, pain, and other factors. However, the origin of hypertension in most cases is unknown. It is interesting that arteriosclerosis contributes to hypertension and that hypertension can contribute to arteriosclerosis. One of the two major effects of hypertension is damage to the arterial walls caused by the high pressure of the blood. This effect contributes to arteriosclerosis in many arteries of the body. The second major effect is damage to the heart resulting from an increased work load. With the elevated pressure in the arteries, the heart must work harder to pump blood. This overload on the heart is particularly damaging to the coronary arteries, which rapidly develop arteriosclerosis.

Coronary Occlusion. Blockage of the coronary arteries is usually the result of arteriosclerosis. Small clots, or emboli, may circulate through the coronary system until they encounter a vessel too small to pass through. At that point the occlusion prevents blood from reaching a part of the heart muscle, and the muscle cells eventually die. This is one of several examples of a "heart attack."

Strokes. A stroke results when there is a loss of blood supply to the brain. Diabetes and especially hypertension appear to be the most serious risk factors associated with strokes. High salt intake, overweight, and cigarette smoking are also contributing factors. With the reduced flow of blood to the brain, stroke victims experience a variety of symptoms, such as dizziness, weakness in the arms or legs, loss of memory, and numbness of the face.

Varicose Veins. Veins near the surface of the body, particularly in the legs, sometimes become stretched and dilated. This occurs most commonly in people who must stand or sit for long periods of time. As a result of the pull of gravity, blood "pools" in the veins, causing them to stretch and dilate. In the process, the valves are pulled apart, which allows blood to flow backward and accumulate (Fig. 17.22). Varicose veins in the area of the rectum are called *hemorrhoids,* which may occur, for example, in prolonged constipation and during pregnancy (Fig. 17.21).

CHOLESTEROL AND TRIGLYCERIDES IN HEART DISEASE

In addition to the risk factors just listed, there are two lipid substances in the body that recent research indicates may play a key role in the incidence of heart disease. These substances are **cholesterol** and the **triglycerides** (fats and oils). Both of these lipids are important to normal body functioning and are transported to the cells by the blood. Since they do not dissolve in the plasma, they become attached to proteins, forming particles called **lipoproteins.** Cholesterol and triglycerides, then, are actually transported in the blood in association with a protein.

There are three different forms of lipoproteins, designated as **very low density lipoproteins,** abbreviated VLDL; **low density lipoproteins,** or LDL; and **high density lipoproteins,** or HDL. Most of the triglycerides are carried by the VLDL. Although the health effects of high or low levels of VLDL in the blood are uncertain, we know a little more about the effects of LDL and HDL.

The LDL contain only a few triglycerides but carry the highest percentage of cholesterol. HDL, on the other hand, are mostly protein with only small amounts of lipids attached. High levels of LDL in the blood have been associated with arteriosclerosis. Apparently, LDL deposit their cholesterol in the walls of the arteries, which, as mentioned previously, contributes to the formation of plaques. HDL, however, apparently function as "scrapers" to remove cholesterol deposits, which are then metabolized by the body and eventually excreted in the feces. It appears, then, that from the standpoint of health, it is preferable to have low levels of LDL in the blood and high levels of HDL.

So how do we raise or lower the levels of lipoproteins in the blood? There is evidence to suggest that diet is a factor in raising or lowering the levels of LDL, i.e., increasing the intake of fatty foods tends to raise the level of LDL, whereas avoiding these foods tends to lower the level. In the case of HDL, however, diet seems to have little effect. The factor that may contribute most significantly to HDL levels is exercise. Vigorous, sustained exercise, such as running, appears to raise the levels of HDL considerably above those of sedentary individuals.

Although all the answers are not yet available, the evidence suggests that persons with low LDL levels and high HDL levels have a relatively lower, or "less than average," risk of developing heart disease.

> The *lymphatic system* acts as an accessory to the cardiovascular system, returning fluid lost from the blood to the heart and serving to defend the body from disease.

571
Cardiovascular Disorders and Disease

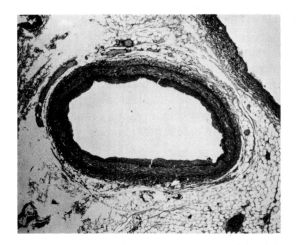

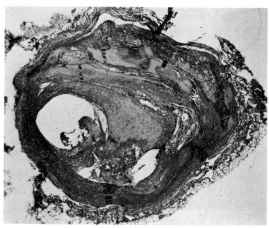

Figure 17.21 (Top) Cross section of a normal artery. (Bottom) Cross section of an artery partially occluded by arteriosclerosis. (From Ebert, J. D., Loewy, A. G., Miller, R. S., and Schneiderman, H. A.: *Biology.* New York: Holt, Rinehart and Winston, 1973, p. 385.)

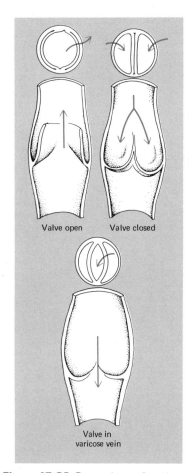

Figure 17.22 Comparison of action of valves in normal vein and in varicose vein. Dilation or stretching of the varicose vein prevents normal closure of the valve, allowing blood to flow backward.

The lymphatic system is actually an accessory to the cardiovascular system, in that it returns fluid lost from the blood back to the heart. In addition, the lymphatic system is involved in the defense of the body against disease, a subject we will discuss in the *Immunity* section of this chapter. Finally, most of the fats broken down in the digestive system are absorbed in the intestine and transported to the circulating blood with the assistance of the lymphatic system. The role of the lymphatic system in digestion will be discussed in Chapter 18.

THE LYMPHATIC SYSTEM

Circulation and Immunity

Lymph Formation and Flow. As blood circulates through the vast network of body capillaries, a small amount of plasma water and small blood proteins leaks through the capillary walls, out into the surrounding tissues. Here the plasma water and small proteins constitute *tissue fluid,* or extracellular fluid, which surrounds the cells. Once the tissue fluid enters the lymph vessels, it is called *lymph*.

The lymphatic system is somewhat similar to the venous system of the body (Fig. 17.23). Tiny lymph capillaries, extending through-

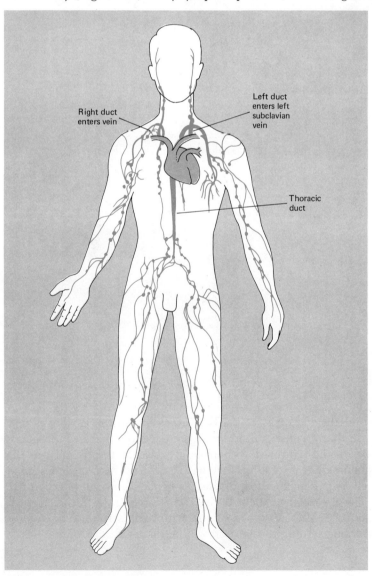

Figure 17.23 The lymphatic system has a one-way flow, returning excess tissue fluid to the heart. The major collecting vessel is the thoracic duct, which empties into the left subclavian vein shortly before it enters the heart. A smaller lymph duct also empties on the right side.

> The lymphatic system aids the blood vascular system in maintaining fluid balance in the body.

out the tissues, lead into large lymph vessels, which eventually lead into still larger **lymph ducts.** These ducts carry the lymph to veins in the neck, which empty into the right atrium of the heart. In this manner, any excess fluid or protein escaping into the tissues is returned to the circulation. Thus, the lymphatic system, in conjunction with the blood vascular system, aids in maintaining fluid balance in the body.

The large lymph vessels are structurally similar to veins in having one-way valves to prevent the backflow of lymph. In addition, lymph is forced through the vessels primarily by the kneading action of skeletal muscles. Obstruction of the lymph vessels results in a type of **edema,** or swelling (Fig. 17.24). Edema is simply the result of an excess of fluid in the tissues. Since the lymphatic system is the *only* means by which excess fluid and protein are removed from the tissues, clogging of the lymph vessels prevents normal drainage. In the tropics especially, infection by small roundworms, called **filarial worms,** is a major cause of lymph vessel obstruction. The larvae of these worms are transmitted to humans by mosquitoes. As the worms mature in the body, there is increasing inflammation and obstruction of the lymphatic channels, resulting in severe edema. Often, certain areas of the body, principally the legs and genitals, become grotesquely swollen, a condition known as **elephantiasis** (Fig. 17.25).

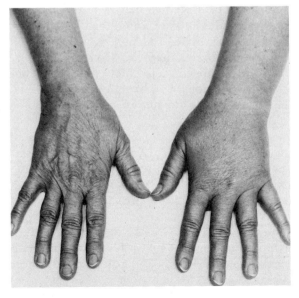

Figure 17.24 Lymphedema, or swelling, due to inadequate lymphatic drainage from the arm. The lymph nodes in the left armpit were destroyed during surgery. Proteins and fluids have accumulated in the tissues. (From Arms, K., and Camp, P. S.: *Biology.* New York: Holt, Rinehart and Winston, 1979, p. 457.)

> *Lymph nodes* filter foreign matter from the lymph and produce *lymphocytes*, which function as part of the body's immune system.

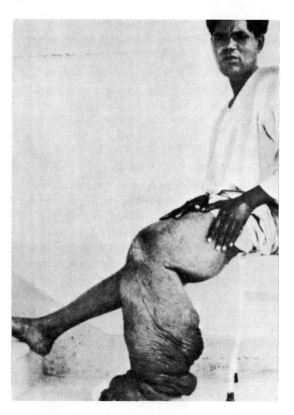

Figure 17.25 A case of elephantiasis caused by a roundworm that blocks the lymph passages, resulting in fluid accumulation in the affected limb. (From Clark, M. E.: *Contemporary Biology.* 2nd ed. Philadelphia: Saunders College Publishing, 1979, p. 75.)

Lymph Nodes and Nodules. Located at intervals along the lymph vessels are bean-shaped masses of tissue called **lymph nodes.** Groups of these nodes also are found mainly in the armpits, neck, and groin area. Lymph nodes function as filters to remove bacteria, viruses, and other foreign particles from the lymph as it passes through. Various phagocytic cells then engulf and digest the trapped particles. In addition, lymph nodes produce a large number of white blood cells called **lymphocytes,** which function as part of the immune system of the body.

Scattered throughout the passageways of the body are small masses of tissue called **lymph nodules.** These, too, contain lymphocytes, which protect the passageways against entrance of harmful microorganisms. The familiar, and sometimes irritating, **tonsils** on either side of the throat are masses of lymph nodules.

The Spleen. The **spleen** is an organ that functions very much like lymph nodes except that it filters blood rather than lymph. Located behind and to the left of the stomach, the spleen is the largest organ

> The *thymus gland*, which produces the hormone *thymosin*, is responsible for developing or differentiating *T lymphocytes*.

of the lymphatic system. As blood passes through the interior of the spleen, old red cells are broken down and finally destroyed by large phagocytic cells. These large cells also serve to cleanse the blood of invading microorganisms. In addition, the spleen contains a large quantity of lymphocytes, which have essentially the same function as those of the lymph nodes.

The Thymus. Located just above the heart and between the lungs is the **thymus gland.** Before birth and in young children, the gland is large and quite conspicuous, but it begins to degenerate after puberty. In the adult it has practically disappeared. The thymus apparently is necessary for the development or differentiation of specific types of white cells, called **T lymphocytes** (Figs. 17.26 and 17.27). These cells are vital in attacking specific foreign antigens that invade the body. The thymus secretes a hormone, or hormone-like substance, called **thymosin.** Apparently, thymosin is necessary for the differentiation and effectiveness of the T lymphocytes.

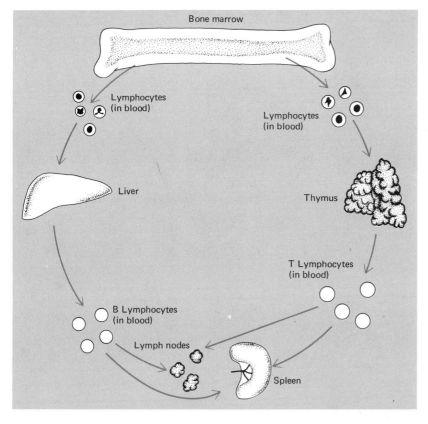

Figure 17.26 Formation of B. lymphocytes and T lymphocytes. Lymphocytes produced in the bone marrow enter the circulating blood and travel to the liver (or perhaps some other area of the body); another group of lymphocytes produced in the bone marrow travel to the thymus. Cells entering the liver differentiate into B lymphocytes; those entering the thymus become T lymphocytes. Both the B lymphocytes and the T lymphocytes enter the circulating blood and are carried to the lymph nodes and spleen. This entire process normally occurs during fetal life and for a few months after birth.

Circulation and Immunity

> Phagocytic cell action and the production of antibodies are responses of the body's *immunity,* or ability to recognize and eliminate harmful substances.

IMMUNITY: BIOLOGICAL DEFENSE

At various points throughout this chapter we have referred to physiological mechanisms that defend the body against the effects of invading microorganisms and other substances. Consider, for example, the role of the white blood cells in response to inflammation and the cleansing of the blood by phagocytic cells in the spleen. These and other mechanisms that protect the body against infection and disease constitute the **immune system**.

Many tissues of the body, including those of the liver, spleen, lymph nodes, lungs, digestive tract, and bone marrow, contain phagocytic cells that engulf and digest foreign invaders that enter the body. In addition, some of these tissues are activated to produce antibodies against such invaders. These responses are the essence of **immunity,** the ability of the body to recognize and eliminate harmful substances.

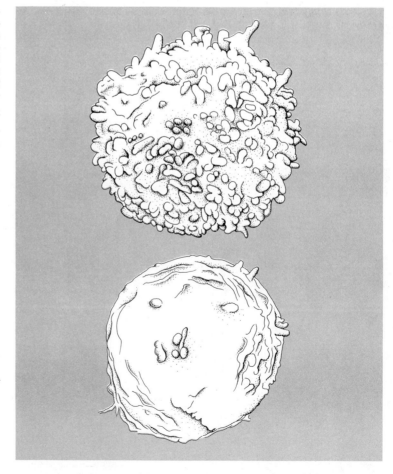

Figure 17.27 *A,* Scanning electron micrograph of a typical circulating lymphocyte with its fimbriated surface. This surface is frequently associated with B-derived lymphocytes. *B,* Another cell with a relatively smooth surface with few microvilli; this surface is frequently found in thymic lymphocytes. (From Polliack, A., Lampen, N., Clarkson, B. D., et al.: Identification of human B and T lymphocytes by scanning electron microscopy. J. Exp. Med., *138*:607, 1973.)

> Invading substances, or antigens, can stimulate the formation of antibodies.

The responses of the immune system; i.e., phagocytosis or antibody formation, may be **nonspecific**. This means that the cells or their antibodies attack a foreign substance in the body without having to "know" the *specific* identity of that substance. Any foreign microorganism, be it bacterium, fungus, or whatever, can be attacked by these cells or antibodies the very first time it enters the body. Thus, we say that the body has a **natural immunity** against many infectious or disease-causing substances.

We also have a system classified as **specific** immune responses. These responses are not immediate when the body is invaded but require several days to a few weeks before immunity is developed. The invading substance, whether it triggers a nonspecific or a specific immune response, is called an *antigen* because it is capable of stimulating the formation of antibodies.

When a foreign antigen enters the body, there are two ways in which the specific immune response may occur. Before birth and for a few months thereafter, some of the lymphocytes produced in the bone marrow migrate to the thymus gland, where they differentiate and are "activated" (apparently through the influence of thymosin) to become *T lymphocytes*. These cells then move into the circulating blood and are carried mainly to the lymph nodes and spleen, where they are found throughout life. When an invading antigen enters the body, some of the T lymphocytes become stimulated and begin to multiply. Masses of these cells then leave the lymph nodes or spleen and move in on the invading antigen, destroying it with enzymes and with the help of phagocytic cells attracted to the area. Although great numbers of T lymphocytes leave the lymph nodes or spleen to attack the antigen, some of the cells remain in these tissues as **memory cells**. Since these cells have already been exposed, or "sensitized," to the antigen, they will attack it even more quickly, should it invade the body a second time. Thus, the memory cells constitute an exceptionally strong immune defense that may last for many years. The specific immune response described here for the T lymphocytes is known as **cellular immunity**.

Other lymphocytes, also formed in the bone marrow, do not migrate to the thymus but are "activated" in some other tissue, possibly the liver, parts of the bone marrow, or some other location in the body. Whatever the case, these lymphocytes differentiate to become **B lymphocytes**. These cells, too, circulate in the blood and then infiltrate the lymph nodes and spleen. The B lymphocytes do not leave the nodes or spleen when a foreign antigen enters the body; instead, they produce specific types of antibodies that move out to attack the invader. More exactly, the presence of the antigen "sensitizes" some of the B lymphocytes, which divide and differentiate into

> Immunity to many diseases can be conferred through *vaccination*.

large *plasma cells* (Fig. 17.28). It is actually the plasma cells that produce the antibodies. Other activated B lymphocytes do not develop into plasma cells but remain in the lymph nodes and spleen as memory cells. The specific immune response resulting from the activities of the B lymphocytes is called **humoral immunity**.

Through the two types of specific immune responses—cellular immunity and humoral immunity—we develop what is called **acquired immunity**. In other words, we must "acquire" these types of immunity in response to an invading antigen. In the case of natural immunity (nonspecific), we already have the necessary antibodies, blood cells, and so on *before* the antigen enters the body.

Vaccines. Many of the diseases that once plagued man and other animals have been controlled by conferring immunity through the use of vaccines. The practice of **vaccination,** or immunization, be-

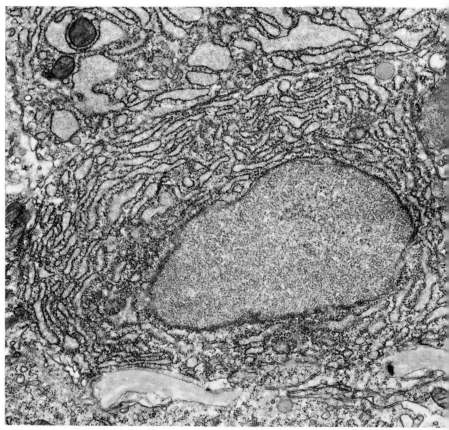

Figure 17.28 Electron micrograph of a plasma cell in mouse spleen. Plasma cells arise from B lymphocytes that have been "sensitized" by an invading antigen. (From Leeson, C. R., and Leeson, T. S.: *Histology*. 10th ed. Philadelphia: W. B. Saunders Co., 1976, p. 101.)

> Long-term *active immunity* results from the introduction into the body of killed or weakened antigens that induce the body to produce its own antibodies. Short-term *passive immunity* results from the introduction into the body of an *antiserum* containing antibodies obtained by injecting another animal with a specific antigen.

gan with the English physician Edward Jenner (1749–1823), whose pioneer work eventually led to the conquering of smallpox.

Basically, vaccines are of two types. Most vaccines contain "killed" or weakened antigens that, when administered, cause the body to form its own antibodies. The Salk vaccine for polio contains a "killed" form of the polio virus, while the Sabin vaccine contains a weakened form of this virus. The virus (antigen) stimulates the production of antibodies, which are then available to attack a live polio virus, should it enter the body. Vaccines containing the appropriate antigen are also used to confer immunity against such diseases as diphtheria, measles, typhoid fever, and whooping cough. Because the antigen induces the body to produce its *own* antibodies, this type of immunity, along with cellular and humoral immunity, is known as **active immunity**. Active immunity is usually long-term, often lasting for a lifetime.

The other type of vaccine is prepared with antibodies instead of antigens. The antibodies, obtained from an animal such as a horse, confer a short-term **passive immunity**. When injected with the antigen, the animal forms its own antibodies that circulate in the blood. The blood is then withdrawn from the animal, and an **antiserum** is prepared. Injections of the antiserum into humans provide antibodies that the body could not produce, or did not have time to produce, for itself. In the case of snakebite, for example, one cannot wait several days for the body to build up antibodies against the venom.

As mentioned, passive immunity is short-term, usually lasting only a few weeks. This is because the body has not had the opportunity to "recognize" the antigen and form antibodies and memory cells against it. Consequently, after the antiserum antibodies are disposed of by the body, there is no remaining defense against the antigen, should it reenter the body. Since both active and passive immunity must be developed, they, too, come under the general heading of acquired immunity.

Autoimmunity. Unfortunately, all of the immune defenses of the body are not always beneficial. Sometimes situations arise in which the body cannot identify "self" from "nonself," i.e., cannot distinguish between its own antigens and foreign antigens. For example, an antigen may enter the body that is similar to one of the body's own proteins. Not only does the immune system form antibodies against the invading antigen but also these antibodies attack the similar body protein. In addition, the body may contain antigens that have been hidden since childhood. The release and exposure of these antigens can cause an immune response in which antibodies

> Transplantation of tissues or organs from one person to another is difficult because of the immune system responses.

are produced against the antigens. Thus, the body is literally attacking itself. The development of immunity against one's own body tissues is known as **autoimmunity**. Various conditions such as rheumatic fever, thyroiditis (inflammation of the thyroid gland), connective tissue disorders (e.g., arthritis), certain allergies, and others are considered to result from autoimmune responses.

ORGAN AND TISSUE TRANSPLANTS

In most instances, the transplanting of tissues or organs from one human being to another, or from an animal to a human being, is only temporarily successful. The immune system of the recipient reacts against the transplanted tissue or organ in the same manner as it would against some other foreign antigen. Within days, the T lymphocytes migrate to the graft and begin to destroy the transplanted tissue. Exceptions to this are found in identical twins, who, having identical antigens, can donate their organs or tissues safely to each other. In addition, grafting tissue from one part of a person's body to another part is almost always successful. This may be necessary, for example, in the case of a burn, in which skin tissue from the leg might be grafted to the affected arm.

Great strides have been taken in the science of **immunology** to overcome the immune response in transplanted tissues and organs. For example, a serum can be prepared that destroys most of the T lymphocytes that attack grafted tissue. In addition, X-rays, gamma rays, and various drugs can be used in some cases to suppress antibody formation. So far, however, most of these measures only extend the life of the transplanted tissue and usually do not prevent the immune response over long periods of time. Consequently, the transplant eventually is rejected by the host's immune system.

REVIEW OF ESSENTIAL CONCEPTS

1. In man, the circulatory system consists of the *blood*, the *heart*, and a network of *blood vessels*; in addition, the lymphatic system is an accessory circulatory system.
2. Human blood consists of various *solids* and the liquid *plasma*. The solids include red blood cells, white blood cells, and platelets. Plasma is 92 percent water and contains proteins, foodstuffs, hormones, ions, and wastes. Blood has a normal pH range between 7.35 and 7.45.

3. In man, mature *red blood cells* are enucleated, biconcave discs that function in transporting oxygen.
4. Production of red blood cells is controlled by negative feedback involving the hormone *erythropoietin*, released from the kidneys. In response to a decrease of oxygen, erythropoietin circulates in the blood to the bone marrow, where it stimulates the production of red cells.
5. *White blood cells* function primarily to protect the body against disease.
6. *Inflammation* of tissues involves the release of histamine and other substances that dilate and increase the permeability of capillaries. This increases local blood flow and causes swelling of the tissues.
7. The function of *platelets* is to stop the flow of blood from a cut or ruptured vessel.
8. Blood clotting involves the formation of a platelet *plug* or *clot*. Platelets accumulate at the site of the cut to form a plug. A blood clot involves the conversion of the plasma protein *fibrinogen* into *fibrin* by the action of the enzyme *thrombin*.
9. A stationary blood clot within a vessel is called a *thrombus*; a moving clot is an *embolus*. Blood normally does not clot within a vessel because the walls of the vessel are smooth and carry a negative electrical charge that repels the blood clotting factors. In addition, the blood contains powerful anticoagulants.
10. Human blood groups are designated A, B, O, and AB according to the type of antigen in the red cells. Blood group antibodies are found in the plasma; the type of antibody is determined by the type of red cell antigen.
11. Safe blood transfusions involve a compatibility between the antigens of the donor and the antibodies of the recipient.
12. Another blood factor is the *Rh antigen*, present on the red cells of Rh positive individuals; persons lacking the antigen are designated Rh negative. Rh negative individuals who receive transfusions of Rh positive blood develop anti-Rh antibodies against the Rh antigen, and the red cells agglutinate.
13. The infant disease *erythroblastosis fetalis* results when the Rh antigens of the fetus cause the development of anti-Rh antibodies in the blood of the mother, who is Rh negative. The antibodies seep across the placenta and agglutinate the red cells of the fetus.
14. Circulatory systems may be *open*, or *closed*. Human beings have a closed circulatory system, consisting of a heart, arteries, veins, and capillaries.
15. The human heart is a four-chambered muscle that pumps blood. The

right side of the heart pumps blood to the lungs; the left side pumps blood to the rest of the body.
16. Blood must continually circulate through the heart so that it can be pumped to the lungs to acquire oxygen and release carbon dioxide. In addition, the heart maintains the *pressure* of the blood.
17. Arteries are vessels that conduct blood *away* from the heart; veins conduct blood *toward* the heart. Backflow of blood in the veins is prevented by one-way valves.
18. Blood functions in association with the hypothalamus and sweat glands to regulate the temperature of the body.
19. *Capillaries* are extremely thin-walled vessels through which respiratory gases, nutrients, and wastes are exchanged between the blood and the body cells.
20. Regulation of the heartbeat involves the interaction of an *intrinsic conduction system* and several *nerves* of the ANS. The electrical activities of the heart are observable on the EKG; heart sounds are detectable with a stethoscope.
21. Cholesterol and triglycerides are transported in the blood in the form of VLDL, LDL, and HDL. High HDL and low LDL levels are associated with a reduced risk of heart disease.
22. The lymphatic system functions in removing fluid from the tissues and returning it to the circulating blood; in addition, the lymphatic system is involved in immunity and digestion of lipids.
23. The lymphatic system consists of *vessels, capillaries*, and *nodes*. Lymph nodes filter out foreign particles from the lymph and produce lymphocytes.
24. The *spleen* filters the blood and contains phagocytes that destroy red blood cells.
25. The *thymus gland* secretes thymosin, which apparently stimulates the development or differentiation of T lymphocytes.
26. The immune system protects the body against foreign antigens. Immunity is generally classified as *natural* or *acquired*.
27. Vaccines contain killed or weakened antigens, or they contain antibodies. Antigen vaccines confer a long-term active immunity; antibody vaccines confer a usually short-term passive immunity.
28. The development of antibodies against one's own body tissues is called *autoimmunity*.
29. Organ and tissue transplants between persons other than identical twins are usually unsuccessful over the long term because of rejection of the transplanted tissue by the host's immune system.

APPLYING THE CONCEPTS

1. a. Why is it necessary to maintain a relatively constant number of red blood cells in the body?
   b. Describe the mechanism by which this number is maintained.
2. What is the role of platelets in the blood clotting process?
3. Why does blood clot more readily outside the body than within a blood vessel?
4. Determine and give your reasons for the compatibility or incompatibility of the following blood type transfusions:
   Donor: Type AB Positive—Recipient: Type B Positive
   Donor: Type O Negative—Recipient: Type A Positive
   Donor: Type B Negative—Recipient: Type O Negative
   Donor: Type A Negative—Recipient: Type AB Positive
5. a. What are the conditions for the development of erythroblastosis fetalis?
   b. What are the physical effects of this disease?
6. Trace the flow of blood in the human circulatory system from the superior vena cava to the arch of the aorta.
7. Why is it necessary for the blood to circulate from the heart to the lungs, back to the heart, and out again repeatedly?
8. What is the functional significance of systole and diastole of the heart? How are these regulated?
9. What are the functions of the lymphatic system? What is the relationship between the lymphatic system and the development of edema?
10. Explain the function of T lymphocytes and B lymphocytes.
11. What is the essential difference between active immunity and passive immunity?
12. Explain what is meant by autoimmunity.

18 digestion and excretion

THE ESSENTIAL OBJECTIVES

You have understood this chapter when you are able to:
1. Explain the process of hydrolysis of foodstuffs.
2. List the major foodstuffs and their constituent units.
3. Trace the digestion of each major foodstuff through the digestive system.
4. Describe the process of absorption for each of the major foodstuffs.
5. Describe the general functions of the large intestine.
6. Discuss the basic energy needs of the human body.
7. List the major structures of the human urinary system and give the functions of each.
8. Describe the basic structure of a nephron.
9. Using examples, describe generally the mechanisms of filtration, reabsorption, and secretion by the nephron.
10. Explain the relationship between aldosterone and potassium ion balance in the body.
11. Explain the function of angiotensin in regulating blood pressure.
12. Describe the disorders associated with the digestive and urinary systems.

The human digestive system breaks down large food molecules mechanically and chemically into smaller molecules that can be absorbed into the circulating blood.

Excretion is the process of ridding the body of metabolic wastes. The urinary system functions to clear blood plasma of toxic materials and to maintain water and ionic balance in the body.

PREVIEW OF ESSENTIAL TERMS

peristalsis Rhythmic contraction and relaxation of the muscular wall of the digestive tract and other organs.

hydrolysis The splitting of a large compound into smaller components by the addition of water.

sphincter A ringlike band of muscle fibers that constricts a body opening or passage.

chyme A semiliquid mass consisting of a mixture of partially digested food and gastric juice.

emulsification In the digestive system, the breaking up of fat globules into smaller particles by the action of bile.

villus A small, finger-like protrusion, many of which line the small intestine. Villi greatly increase the absorptive area of the small intestine.

microvilli Microscopic projections found along the surface of intestinal cells. Microvilli further increase the surface area of the small intestine lining.

glomerulus A tiny ball or cluster of capillaries through which plasma is filtered in the nephron of the kidney.

nephron The anatomical and functional filtering unit of the kidney. A nephron consists of a glomerulus, Bowman's capsule, proximal tubule, loop of Henle, distal tubule, and collecting duct.

vitamin An organic compound necessary for normal functioning of the body; a vitamin cannot be synthesized by the body but must be obtained in the diet.

mineral An inorganic compound required by the body in small amounts.

aldosterone A hormone secreted by the cortex of the adrenal glands; aldosterone is involved in regulating the level of potassium ions in the body.

586
Digestion and Excretion

> *Digestion*, the process by which food is broken down mechanically and chemically into usable forms, takes place in a step-by-step manner in the *digestive tract*.

INTRODUCTION TO DIGESTION

Digestion is the process by which food is broken down mechanically and chemically into forms that can be used by the body. All higher animals, and many lower ones as well, have some type of **digestive system** that reduces large food molecules to smaller molecules in a step-by-step manner. Food is digested within the tubelike **digestive tract**, also called the **gastrointestinal** (GI) **tract** or **alimentary canal**. In man and the higher animals, food enters the **mouth** and then passes through the **pharynx, esophagus, stomach, small intestine,** and **large intestine** during the digestive process (Fig 18.1). In addi-

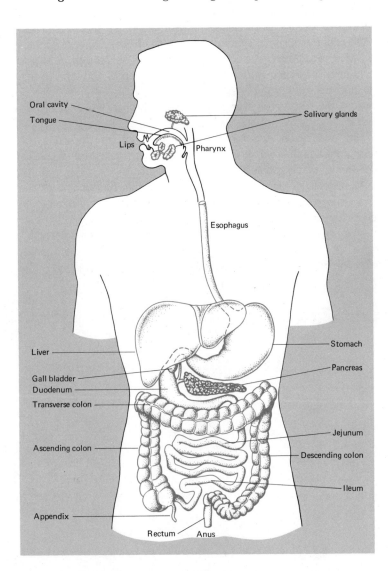

Figure 18.1 The human digestive tract.

Digestion is partially dependent on secretions from the *salivary glands*, the *liver*, and the *pancreas*.

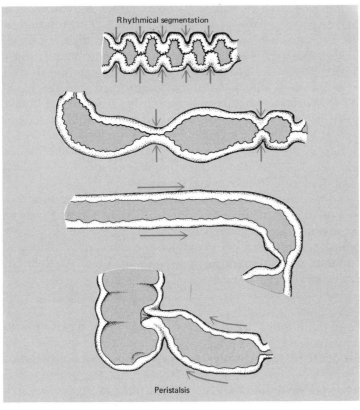

Figure 18.2 Diagrams to illustrate the churning action of rhythmical segmentation in the intestine and the movement of food through the digestive tract by peristalsis.

tion, digestion is dependent upon the secretions of three types of accessory glands—the **salivary glands,** the **liver,** and the **pancreas.**

Food is slowly propelled through the digestive tract by rhythmic waves of contraction and relaxation of the muscular walls. This action, known as **peristalsis,** also serves to break down food mechanically, i.e., breaks down large food particles into physically smaller particles (Fig. 18.2). Peristalsis aids in mixing food with the secretions of the digestive tract so that chemical digestion is more efficient (Fig 18.3). The essential substances involved in the chemical breakdown of food are the digestive **enzymes.** Secreted throughout most of the digestive tract, the enzymes reduce large food molecules to smaller molecules by **hydrolysis.** The term literally means "water splitting" and refers to the addition of a water molecule (H_2O) when a large organic molecule is split into simpler molecules. For example, the hydrolysis of sucrose, ordinary table sugar, can be shown as follows:

$$C_{12}H_{22}O_{11} + H_2O \xrightarrow{\text{enzyme}} C_6H_{12}O_6 + C_6H_{12}O_6$$
$$\text{(sucrose)} \quad \text{(water)} \quad\quad\quad \text{(glucose)} \quad \text{(fructose)}$$

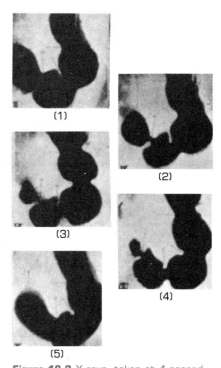

Figure 18.3 X-rays, taken at 4-second intervals, of peristaltic contractions during grinding of food in the human stomach. (From Clark, M.E.: *Contemporary Biology.* 2nd ed. Philadelphia: Saunders College Publishing, 1979, p. 188.)

> Peristalsis and enzyme secretion are largely regulated by the autonomic nervous system.

In this instance, sucrose, a disaccharide, is hydrolyzed by a specific enzyme to form two monosaccharides. This reaction illustrates the fundamental process that occurs in the digestion of all major foodstuffs. Why is hydrolysis necessary? Quite simply, the carbohydrates, lipids, and proteins we consume daily are mostly in the form of large molecules that cannot be absorbed into the bloodstream. What hydrolysis (digestion) accomplishes, then, is the breakdown of these large molecules into forms that are small enough to be absorbed into the blood. Once absorbed, the food must then be transported in the blood to the cells of the body. It would be impossible for most large food molecules to gain entrance to the cells, whereas small molecules can be taken up easily.

The activities of peristalsis and enzyme secretion are regulated to a great extent by the autonomic nervous system. Parasympathetic stimulation results in increased activity of the digestive system, and sympathetic stimulation generally decreases, or inhibits, activity. In addition, regulation of the digestive system is also influenced by several hormones.

THE PRODUCTS OF DIGESTION

Keeping in mind that the primary function of digestion is to reduce large molecules to smaller ones, we need to review briefly the major biological macromolecules and their constituent units (also see Chapter 3).

Carbohydrates are generally classified as sugars or starches. The most common form of long-chain carbohydrates, or *polysaccharides*, taken into the body is *starch*. On hydrolysis, starch is initially broken down primarily into *maltose*, a disaccharide. Two other disaccharides normally included in the diet are *sucrose* and *lactose* (milk sugar). Each of these disaccharides—maltose, sucrose, and lactose—can be hydrolyzed to their constituent *monosaccharides*. Maltose, then, is reduced to two *glucose* molecules; sucrose is reduced to *glucose* and *fructose;* and lactose is reduced to *glucose* and *galactose*. Thus, almost all the carbohydrates are absorbed into the blood in the form of three monosaccharides: glucose, fructose, and galactose.

Lipids in the diet are predominantly fats and oils. These compounds, also called *triglycerides,* consist of three *fatty acid* molecules bonded to a molecule of *glycerol*. On hydrolysis, then, the fats and oils are reduced to individual fatty acids, glycerol, and *monoglycerides*—glycerol molecules with only *one* fatty acid attached. In these forms, the fats and oils can then be absorbed.

The final group of foodstuffs—the proteins—are found in both meats and vegetables. Proteins are broken down initially into large *polypeptides,* and these in turn are reduced to small polypeptides.

> Proteins are hydrolyzed first to *polypeptides* and then to absorbable *amino acids.*

Hydrolysis is completed when the small polypeptides are split into their constituent *amino acids.* The amino acids are then absorbed into the circulating blood.

As you can see, the digestion of each of the major foodstuffs occurs in a step-by-step manner until they are broken down into their smallest components. Each of these steps is catalyzed by a specific enzyme and occurs in a particular area of the digestive tract.

THE PHYSIOLOGY OF DIGESTION
The Mouth

The process of digestion begins in the mouth, where food is torn and chewed by the teeth. Adults normally have 32 *permanent* teeth, which replace the original 20 *deciduous,* or "baby," teeth. As food is being chewed, it is mixed with secretions of glands located in and around the mouth. There are three pairs of *salivary glands*—the **parotid, submaxillary,** and **sublingual glands**—that release saliva directly into the mouth (Fig. 18.4). A parotid gland lies just below, toward the front of each ear. The submaxillary glands lie on either side of the neck at the angle of the jaw, and the sublingual glands are located under the tongue. In addition, the lining of the mouth cavity is supplied with numerous small glands that secrete a watery **mucus.**

Saliva consists of mucus and the digestive enzyme **salivary amylase,** or *ptyalin.* When mixed with saliva, food molecules go into solution and can then be tasted. The taste buds, located on the tongue, are not stimulated unless the food we eat is in solution. In addition, saliva lubricates the mouth and throat, making the food easier to chew and swallow.

Salivary amylase initiates the chemical digestion of carbohydrates by breaking down starch into maltose units. If you chew a cracker long enough, you begin to detect the sweet taste of maltose

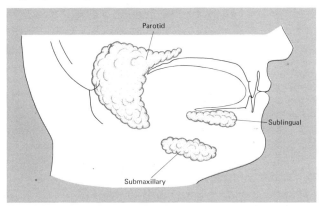

Figure 18.4 Location of the salivary glands.

> Food in the pharynx stimulates the swallowing reflex, which propels the bolus into the *esophagus* and downward into the stomach.

as the starch is being broken down. Although lipids and proteins are also mixed with saliva, there is no chemical breakdown of these foodstuffs in the mouth. The pH of saliva ranges between 6.0 and 7.4, a necessary condition for the effectiveness of salivary amylase.

Secretion by the salivary glands is regulated by branches of sympathetic and parasympathetic nerves that lead from the base of the brain. Parasympathetic stimulation greatly increases the secretion of saliva in response to the sight, smell, taste, or even the thought of a good meal. On the other hand, if these stimuli are unpleasant, parasympathetic stimulation is inhibited, and salivation decreases. Stimulation of the sympathetic nerves produces a small amount of rather thick saliva.

The Pharynx and Esophagus

After food is chewed, moistened, and mixed in the mouth, it is in the form of a semisolid mass, or **bolus** (Gk. "clod"). From the mouth, the bolus is forced into the pharynx by the tongue. Food in the pharynx elicits the swallowing reflex that propels the bolus into the esophagus (Fig. 18.5). The act of swallowing forces the larynx upward, so that its opening is covered by the epiglottis. This prevents food from entering the respiratory passageways.

From the pharynx, the esophagus descends behind the trachea and extends downward to the stomach. The inner lining of the esophagus is supplied with glands that secrete a thick, lubricating mucus. The film of mucus along the inner wall of the esophagus helps prevent irritation or damage to the cells of the lining as food passes through. Mucus-secreting cells are especially abundant at the lower end of the esophagus near its junction with the stomach. Coated with a thick layer of mucus, the lower region of the esoph-

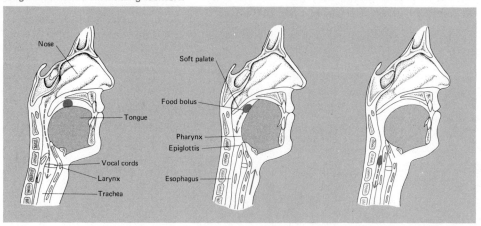

Figure 18.5 Diagram of the position of the tongue and epiglottis during breathing (left) and swallowing (center and right). Note how a food bolus is pushed from the mouth into the pharynx by the tongue to initiate swallowing (center).

> Food in the stomach causes release of the hormone *gastrin*, which then stimulates the production of *gastric juice*—a mixture of mucus, digestive enzymes, and HCl.

agus is usually protected from the highly acid contents that might be regurgitated from the stomach. At times, however, the regurgitated acids irritate the lining of the esophagus, producing the condition commonly known as "heartburn." If this condition persists, the acids may digest a portion of the muscular wall, leaving a raw, open sore, known as an esophageal *ulcer*.

Within the esophagus, the bolus is forced along toward the stomach by a series of peristaltic waves. Although peristalsis aids in the mechanical breakdown of food, there is no chemical digestion in the esophagus, since its secretions do not contain digestive enzymes.

Just above the junction of the esophagus and stomach, there is a ringlike band of muscle forming a constriction that closes off the opening to the stomach. This type of muscle, called a ***sphincter***, keeps food from being dumped into the stomach and prevents regurgitation of the stomach contents into the esophagus (Fig. 18.6). As a peristaltic wave passes to the lower end of the esophagus, the sphincter muscle relaxes momentarily, allowing the bolus to ease into the stomach.

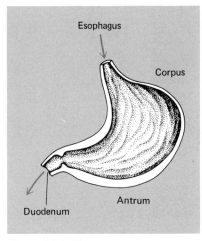

Figure 18.6 The stomach. In the corpus, or body of the stomach, food is stored; in the more muscular antrum, food is mixed with digestive juices and pumped into the duodenum.

The Stomach

The stomach is a pouchlike organ that lies next to the left lobe of the liver, just under the diaphragm. Peristalsis in the stomach continues the mechanical breakdown of food, and aids in mixing food with the secretions released from the stomach cells.

A variety of substances are secreted in the stomach, including mucus, enzymes, hydrochloric acid (HCl), a hormone, and a substance called ***intrinsic factor***. Together, mucus, digestive enzymes, and hydrochloric acid form a mixture referred to as ***gastric juice***. Distention of the stomach by food causes release of the hormone ***gastrin*** from the stomach wall. Gastrin is absorbed into neighboring blood vessels and carried to stomach glands, where it stimulates release of gastric juice. Peristaltic action thoroughly mixes food with gastric juice, forming a semiliquid mass called ***chyme***.

The major enzyme present in gastric juice is ***pepsin***, which begins the chemical breakdown of proteins. Pepsin hydrolyzes almost all food proteins into large polypeptides. Pepsin is released from the stomach cells in an inactive form called ***pepsinogen***. When mixed with hydrochloric acid, pepsinogen is converted into the active enzyme pepsin. The hydrochloric acid secreted by the stomach has an extremely acid pH of about 0.8. But when hydrochloric acid is mixed with chyme, its pH rises to about 2.0, which is the optimum pH for the activity of pepsin. However, the acidic environment of the stomach is unsuitable for the activity of salivary amylase. Thus, salivary amylase loses its enzymatic capacity, and there is no further

> Secretion of *intrinsic factor* is necessary for adequate absorption of vitamin B_{12} from the small intestine. A lack of B_{12} can result in *pernicious anemia*.

breakdown of carbohydrates in the stomach. Moreover, there is only an insignificant amount of lipid digestion in the stomach; nearly 99 percent of lipid digestion occurs in the small intestine.

Gastric juice also contains a substance called *intrinsic factor,* which is secreted by the same cells that produce hydrochloric acid. Intrinsic factor is necessary for adequate absorption of vitamin B_{12} from the small intestine. Vitamin B_{12} is essential for normal growth of the body and for the maturing of red blood cells. A deficiency of vitamin B_{12} often results in **pernicious anemia,** a condition in which the red blood cells do not mature in the bone marrow. Treatment of this type of anemia involves periodic injections of vitamin B_{12}.

With the high acidity in the stomach, along with the presence of protein-digesting enzymes, one might wonder why the stomach itself is not digested by its own secretions. One protective measure is the production of a thick alkaline mucus that coats the inner lining of the stomach. The mucus acts to neutralize the hydrochloric acid in the area around the stomach wall and forms a protective barrier against erosion of the wall by gastric juice. In addition, hundreds of millions of cells from the stomach lining are shed and replaced every day. In this manner, the stomach lining is continually renewed as damaged cells are sloughed off. However, for reasons not fully understood, these protective mechanisms sometimes fail, and a small area of the stomach wall is digested. This leaves an open sore, or **gastric ulcer,** in the lining of the wall. (Fig. 18.7). If an ulcer extends deep into the wall, it may rupture large blood vessels, resulting in severe hemorrhage.

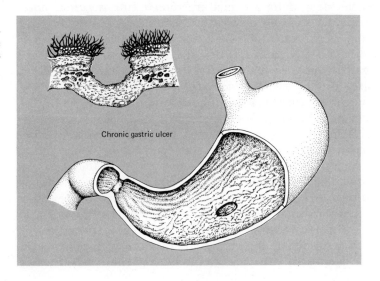

Figure 18.7 A chronic gastric ulcer. The insert on the upper left shows diagrammatically the microscopic damage to the stomach lining as viewed in cross section.

> The release of chyme into the *duodenum* is regulated by the *enterogastric reflex.*

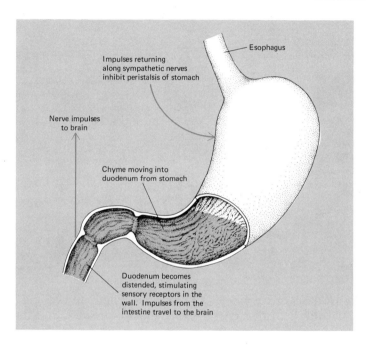

Figure 18.8 Diagram of the enterogastric reflex. This reflex helps to prevent overloading the small intestine with food.

As peristalsis continues, chyme is moved along toward the lower end of the stomach. Near its junction with the small intestine, the stomach narrows to form a sphincter muscle that closes off the entrance to the **duodenum,** or upper section of the small intestine. Peristaltic waves allow the sphincter to relax periodically, permitting a small amount of chyme at a time to enter the duodenum. This action is important in preventing the small intestine from being overloaded with food. In fact, the release of chyme is actually regulated by what is called the **enterogastric** ("intestine-stomach") **reflex** (Fig. 18.8). As chyme enters the intestine, it distends the walls of the duodenum, which in turn stimulates sensory receptors in the wall. This sets up nerve impulses that travel from the intestine along nerves to the brain and back to the stomach. The impulses inhibit peristalsis of the stomach, and the movement of chyme is slowed. As food moves on through the duodenum, the nerve impulses subside, the sphincter opens, and another portion of chyme enters the intestine.

The Small Intestine

The small intestine lies coiled in the abdominal cavity and extends from the stomach to the large intestine. As mentioned, the upper section of the small intestine (the first 10 inches) is called the **duodenum;** the remainder of the small intestine consists of the **jejunum** (about 3 ft in length), and the **ileum** (about 7 ft in length), which

> *Villi* and *microvilli* are finger-like projections that greatly increase the surface area within the small intestine, thereby enhancing the absorption of nutrients.

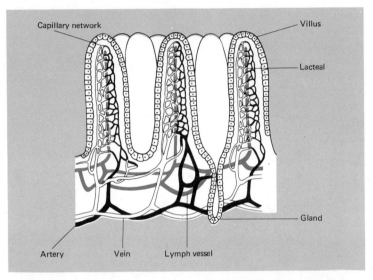

Figure 18.9 Villi in small intestine. These numerous microscopic finger-like projections greatly increase the absorptive surface in the small intestine.

joins the large intestine. The extreme length and coiling of the small intestine provide an enormous surface area for the digestion and absorption of nutrients. The surface area is further increased by millions of tiny finger-like projections called *villi* (L. "tufts of hair") that form a velvety lining along the inner wall (Fig. 18.9). In addition, the intestinal cells possess microscopic projections called **microvilli,** which also increase the surface area and greatly enhance the absorption of nutrients (Fig. 18.10).

Within each villus there are small blood capillaries and a lymphatic vessel called a **lacteal,** which functions in absorption of lipids. Mucus-secreting glands are located throughout the length of the small intestine, including a group of special mucus glands found in the upper portion of the duodenum. These glands secrete a thick

Figure 18.10 Electron micrograph of microvilli projecting from the surface of intestinal cells.

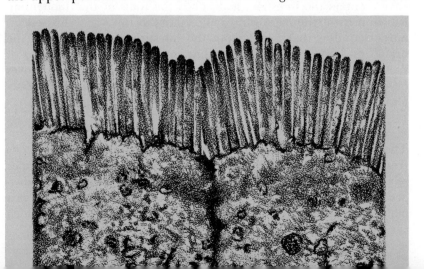

> Secretions of the pancreas include digestive enzymes, sodium bicarbonate, and hormones.

mucus that protects the duodenum from the acid contents released from the stomach.

The intestinal cells produce several different enzymes that act on all of the major foodstuffs. Before considering these, we first need to examine the digestive roles of the other two accessory glands—the pancreas and the liver.

The Pancreas

The pancreas is an elongated gland extending from the curve in the duodenum backward behind the stomach (Fig. 18.11). The pancreas is often referred to as a "double gland," meaning that it performs two different functions. Some pancreatic cells release digestive enzymes that pass through a duct into the duodenum; other cells secrete hormones as a part of the endocrine system, a subject we will discuss in the following chapter.

As mentioned, the pancreas releases enzymes that act on all three major foodstuffs; in addition, enzymes are also released that break down nucleic acids into their component nucleotides. Another important secretion of the pancreas is **sodium bicarbonate** ($NaHCO_3$), which aids in neutralizing the acidic chyme as it enters the duodenum from the stomach. This can be illustrated by the following reaction:

$$HCl + NaHCO_3 \longrightarrow NaCl + H_2CO_3$$
(hydrochloric acid) (sodium bicarbonate) (sodium chloride) (carbonic acid)

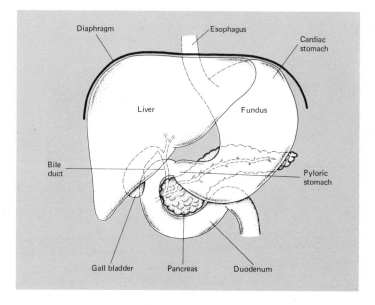

Figure 18.11 Diagram of the relationship of the stomach, liver, pancreas, and duodenum.

> The hormone *secretin* stimulates the pancreas to release sodium bicarbonate; the hormone *cholecystokinin* stimulates the release of digestive enzymes by the pancreas.

Thus, the reaction between an acid (HCl) and a base ($NaHCO_3$) forms a neutral salt (NaCl) solution that remains in the duodenum. The carbonic acid breaks up into carbon dioxide and water, and the carbon dioxide is absorbed by the body. If this process is ineffective, gastric juice may digest a part of the wall of the duodenum, resulting in the development of **duodenal ulcers.** You will recall that the pH of hydrochloric acid in chyme is about 2.0; the neutralizing effect of sodium bicarbonate on hydrochloric acid raises the pH in the duodenum to about 8.0, providing a suitable environment for the activity of the pancreatic enzymes.

What causes the pancreas to release its secretions into the duodenum? For the most part, this is regulated by two hormones—*secretin* and *cholecystokinin*—present in the intestinal wall. As chyme enters the duodenum, it causes the release of both hormones, which are absorbed into the blood and carried to the pancreas. Secretin causes the pancreas to release sodium bicarbonate, whereas cholecystokinin stimulates the release of digestive enzymes and constricts the gallbladder, which empties bile into the duodenum (Fig. 18.12).

The carbohydrate-digesting enzyme in the pancreas is **pancreatic amylase,** which breaks down the polysaccharides starch and glycogen into disaccharides. By the time some of the polysaccharides enter the small intestine, they have already been hydrolyzed to disaccharides by salivary amylase in the mouth. Pancreatic amylase breaks down the remaining polysaccharides into disaccharides in the duodenum.

Figure 18.12 The release of sodium bicarbonate and digestive enzymes from the pancreas is controlled by the hormones secretin and cholecystokinin.

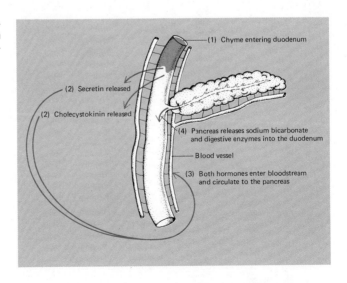

> The enzyme *pancreatic lipase* enters the duodenum, where it acts to digest lipids.

Lipids are split into monoglycerides, fatty acids, and glycerol by the enzyme **pancreatic lipase.** Thus, the significant digestion of lipids begins in the duodenum as pancreatic lipase enters from the pancreas. However, as pointed out in the next section, the digestion of lipids also involves the action of bile secreted by the liver.

There are three major protein-digesting enzymes released from the pancreas: **trypsin, chymotrypsin,** and **carboxypeptidase.** These are secreted in an inactive form, which ensures that the pancreas will not be digested by its own enzymes. If the pancreas is damaged or if its duct into the duodenum is blocked, these enzymes may digest the pancreas itself, resulting in a condition called **acute pancreatitis.** Under normal conditions, the protein-digesting enzymes become active after they enter the duodenum.

The Liver

The liver, which lies just under the diaphragm, is one of the most complex structures in the body. While carrying out hundreds of other activities, the liver also plays an important role in digestion. Specifically, the liver manufactures a yellowish-green liquid called **bile,** which is involved in the digestion of lipids. Bile is temporarily stored in the saclike **gallbladder,** located on the underside of the right lobe of the liver. A duct from the gallbladder joins a duct from the liver to form the **common bile duct,** which opens into the duodenum (Fig. 18.13).

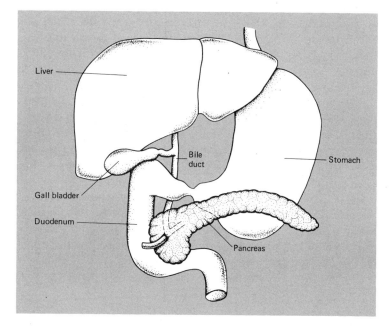

Figure 18.13 The accessory digestive organs—the liver and the pancreas—release their contents through ducts into the duodenum. However, before entering the duodenum, bile produced by the liver is temporarily stored in the gallbladder.

> In the small intestine, *maltase* converts maltose to glucose units; *sucrase* hydrolyzes sucrose to glucose and fructose; *lactase* hydrolyzes lactose to glucose and galactose.

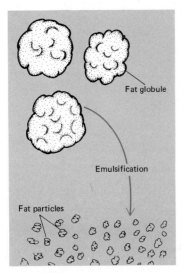

Figure 18.14 Emulsification results from the action of bile on large fat globules. Bile lowers the surface tension of the fat globules, making it easier for the churning movements of the small intestine to break the globules into small particles.

Bile is composed mostly of water, bile salts, bilirubin, cholesterol, and several ions, such as Na^+ (sodium), K^+ (potassium), and Cl^- (chloride). The most important constituents in relation to digestion are the bile salts, which aid in breaking fat globules into small particles and assist in the absorption of lipids from the small intestine. Bilirubin, you will recall, is a product of the breakdown of hemoglobin. Liver cells are responsible for the breakdown of hemoglobin, and the bilirubin formed is excreted in the bile.

Normally, the cholesterol in bile remains in solution, but if the gallbladder is inflamed, causing too much water to be absorbed from bile, cholesterol comes out of solution and forms crystals that develop into **gallstones**. Gallstones may move into the bile duct, obstructing the flow of bile into the duodenum and thereby preventing normal lipid digestion and absorption.

It is important to point out that bile is not an enzyme and therefore is not involved in the *chemical* digestion of lipids. Instead, bile enhances digestion by coating the fat globules, which lowers their surface tension. This allows the churning movements of the intestine to break the globules into small particles. The small size of the particles facilitates their chemical digestion by pancreatic lipase. The breaking of fat globules through the action of bile is called **emulsification** (Fig. 18.14).

Digestion In the Small Intestine

In passing from the stomach, chyme enters the alkaline environment of the duodenum. Digestive enzymes from the pancreas, along with bile from the liver, begin their breakdown of the various foodstuffs. Polysaccharides not digested in the mouth are hydrolyzed to maltose units by pancreatic amylase. After being emulsified by bile, the lipids are broken down into monoglycerides, fatty acids, and glycerol by the enzyme pancreatic lipase. The proteins, initially reduced to large polypeptides by pepsin in the stomach, are further hydrolyzed to small polypeptides and a few amino acids by the enzymes trypsin, chymotrypsin, and carboxypeptidase from the pancreas.

As the foodstuffs move through the small intestine, their digestion is completed by enzymes secreted from the intestinal cells. Each maltose unit is split into two molecules of glucose by the enzyme **maltase**. If the disaccharides sucrose and lactose have been included in the meal, they too are split into their component monosaccharides in the small intestine. Sucrose is hydrolyzed to glucose and fructose by the enzyme **sucrase**, and lactose is hydrolyzed to glucose and galactose by the enzyme **lactase**. Thus, the final products of carbohy-

> Liver and muscle cells store excess glucose in the form of *glycogen*.

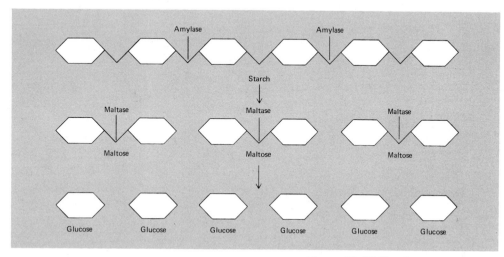

Figure 18.15 The digestion of starch. In the mouth, salivary amylase splits starch into individual maltose units. Digestion is completed in the duodenum, where maltase splits each maltose unit into its glucose constituents.

drate digestion are the monosaccharides glucose, fructose, and galactose (Fig. 18.15).

For the most part, lipid digestion is completed by the action of bile and pancreatic lipase in the duodenum. As mentioned, most of the final products of lipid digestion are monoglycerides, fatty acids, and glycerol.

Several protein-digesting enzymes secreted by the intestinal cells hydrolyze small polypeptides into amino acids. The digestion of all proteins, then, finally releases free amino acids as the final products.

Absorption. The substances absorbed from the small intestine include water, ions, vitamins, bile salts, and the products of carbohydrate, lipid, and protein breakdown. These substances move into the villi by diffusion or active transport (see Chapter 5) and then eventually enter the circulating blood (Fig. 18.16).

The absorption of monosaccharides and amino acids occurs by active transport through the cells of the villi. From the villi, these nutrients pass into the blood and are transported to the liver, where they are processed for storage or excretion. For example, there may be an excess of glucose in the blood after a high carbohydrate meal. Liver cells remove the excess glucose and store it in the form of the polysaccharide **glycogen.** In addition, muscle cells also store a substantial amount of glycogen (Fig. 18.17). When the blood glucose level begins to fall, as it does between meals, the liver cells release glucose back into the blood. The body is extremely dependent upon a constant level of blood glucose for two reasons: First, brain cells

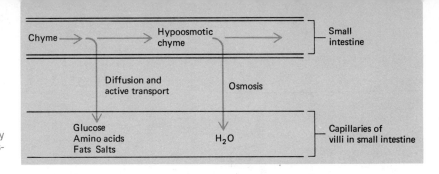

Figure 18.16 Absorption of water by osmosis in the villi of the small intestine.

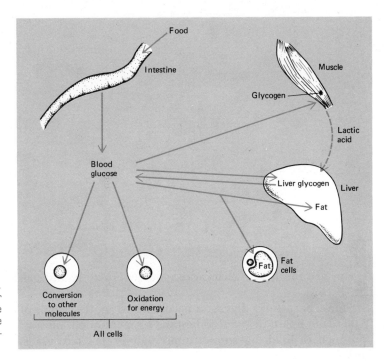

Figure 18.17 Utilization of glucose. Glucose may be stored as glycogen or fat, it may be metabolized to provide other organic molecules, or it may be oxidized to CO_2 and H_2O to provide energy for the synthesis of ATP.

normally cannot use any other fuel as a source of energy. Deprived of glucose for even a few minutes, the brain cells can no longer function. Second, glucose, you will recall, is the principal energy source utilized by the body for cellular respiration, the process that synthesizes adenosine triphosphate (ATP).

Excess amino acids are not stored in the cells in large quantities. Instead, they enter the liver, where their nitrogen groups are removed. These groups are converted into ammonia, which is then converted into *urea* and excreted by the kidneys. The remaining part of each amino acid is retained in the liver, where it may be converted into carbohydrate or lipid.

The absorption of monoglycerides, fatty acids, and glycerol depends upon their association with bile salts. These lipid end-products dissolve in the bile salts and thereby are absorbed much more efficiently by the cells of the villi. The bile salts actually transport the dissolved lipids to the cells, where the monoglycerides, fatty acids, and glycerol diffuse into the cells, leaving the bile salts behind (Fig 18.18). The bile salts are then free to pick up more digested lipids and transport them to the cells. It is apparent, then, that bile from the liver is essential to adequate digestion and absorption of lipids.

> Proper fat absorption is necessary for the absorption of fat-soluble substances known as *vitamins A, D, E,* and *K*.

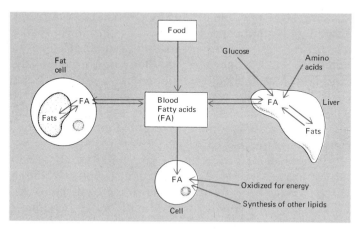

Figure 18.18 Flow of fatty acids (FA) in the body. Fat is transported in the body mainly in the form of fatty acids.

Moreover, proper fat absorption is necessary for the absorption of some essential substances known as *vitamins A, D, E,* and *K*. These vitamins are fat-soluble and therefore can be absorbed from the intestine only in the presence of lipids.

After the monoglycerides, fatty acids, and glycerol diffuse into the villi cells, they are reassembled into triglycerides, which coalesce into small globules. These globules are then surrounded by a coat of protein and subsequently move out of the villi cells and enter the lacteals. From the lacteals the globules move through the lymph capillaries into a large lymph vessel that leads to the veins of the neck. In this manner, the globules reach the circulating blood leading to the heart. After leaving the heart, the blood distributes the globules to the body's fat tissues, where they are stored. When necessary, these stored lipids can be converted into fatty acids and used as fuel for energy. Thus, the stored lipids represent the body's reserve supply of energy.

As chyme approaches the end of the small intestine, almost all of the digested nutrients, along with the water you drink, have been absorbed into the circulation. Eventually, most of the bile salts also are absorbed and are returned to the liver to be used again.

The Large Intestine

The large intestine, also known as the **colon,** arises in the right lower abdomen and extends upward (ascending colon) just below the liver. It then passes across the abdominal cavity (transverse colon) and descends on the left side of the body (descending colon). The lower few inches of the descending colon form the **rectum,** which terminates in the **anus.**

> The upper half of the large intestine absorbs water and ions and harbors mutualistic bacteria that produce vitamins such as K, B_1 (thiamine), and B_{12}; the lower half of the large intestine functions as a storage area for feces.

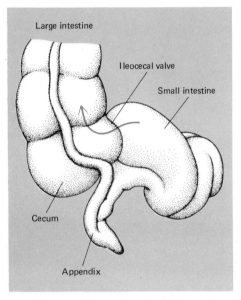

Figure 18.19 Junction of small and large intestines.

The large intestine originates as a blind pouch called the **cecum.** Extending from the end of the cecum is the worm-shaped **appendix,** a structure with no known function in modern humans. Inflammation of the appendix, called **appendicitis,** is a potentially dangerous condition in that, should the appendix rupture, the contents of the colon may enter the abdominal cavity, causing widespread infection. This condition, known as **peritonitis,** results from inflammation of the connective tissue lining of the abdominal cavity as well as the connective tissue that covers the abdominal organs.

The small intestine joins the large intestine about 3 inches above the cecum (Fig. 18.19). At the junction, a sphincter muscle forms the **ileocecal valve,** which prevents the contents of the colon from moving backward into the ileum. Peristaltic waves periodically relax the ileocecal valve, allowing the undigested residue from chyme to enter the large intestine.

The first half of the large intestine functions mainly to absorb water and various ions. There is no digestion of nutrients in the large intestine, since it contains no digestive enzymes. There are, however, numerous glands that secrete a thick mucus, which protects the intestinal lining. This part of the colon also contains numerous bacteria, which have established a mutualistic relationship with their human host. The bacteria produce various vitamins, such as *vitamin K,* which is essential in blood clotting, *thiamine* (vitamin B_1), a necessary factor in normal nervous system, cardiovascular, and gastrointestinal function; and *vitamin B_{12},* discussed previously.

The lower half of the large intestine functions principally as a storage area for feces. Normal feces consist mainly of water, undigested roughage, bacteria, epithelial cells, and bile pigments. The brownish color of feces results from derivatives of bile pigments, formed by bacterial action.

Feces are propelled through the colon by propulsive movements of the intestinal wall. However, when approaching the rectum, feces are forced through a weak sphincter muscle that normally closes the entrance to the rectum. When the rectum is distended, nerve impulses are transmitted to the spinal cord and back to the colon, causing a **defecation reflex** characterized by strong peristaltic waves. Although this is sufficient to empty a large portion of the colon contents, it is not the only factor involved in defecation.

The anal opening at the end of the colon is constricted by two different sphincter muscles—an **internal anal sphincter** that relaxes involuntarily when the rectum is distended and an **external anal sphincter** that must be relaxed voluntarily. Thus, with the defecation reflex and voluntary relaxation of the external anal sphincter, the feces are eliminated from the body.

DIGESTIVE SYSTEM DISORDERS

We have discussed several of the disorders associated with the digestive system, such as ulcers, pernicious anemia, gallstones, and appendicitis; in addition to these, there are a few others that commonly affect human beings.

Constipation. Constipation refers to slow movement of the feces through the large intestine, accompanied by difficult elimination. These conditions usually arise when feces have remained too long in the colon. As a result, more and more water is absorbed from the feces, and they become hard and dry. Constipation may result from irregular bowel habits, insufficient bulk in the diet, and emotional factors. Commercial laxatives are often sought to relieve constipation, but it is more often preferable to alter the diet by including more plant foods that provide bulk and act as natural laxatives.

Diarrhea. Diarrhea literally means "flowing through" and is essentially the opposite of constipation. As fecal matter moves rapidly through the colon, only a slight amount of fluid is absorbed, and the stool remains soft and watery. Diarrhea is usually the result of an infection in the digestive tract that irritates the intestinal lining, causing increased secretion of fluid and strong propulsive movements of the colon. However, various foods and certain emotional states also may lead to diarrhea. Diarrhea is often beneficial in flushing the infectious organism out of the colon; however, prolonged diarrhea results in excessive loss of body fluids and various ions. As a result, the body becomes dehydrated, and its ionic balance is upset. The treatment called for is to replace the water and ions as quickly as possible.

Cirrhosis of the Liver. This disease, most commonly associated with alcoholism, is characterized by the development of bands of connective tissue in the liver that destroy normal liver cells and interfere with blood circulation. If cirrhosis persists, the liver eventually ceases to function. Since there is no known medication that will eliminate the developing connective tissue, treatment for cirrhosis must first center on the cause, usually alcoholism, and then deal with the complications, such as internal bleeding and free fluid in the abdominal cavity.

Many other disorders of the gastrointestinal system are inflammatory conditions resulting from infective viruses, bacteria, or protozoans. The disease *cholera*, for example, is caused by a bacterium that inhabits the lining of the digestive tract. The bacterium releases toxins that induce severe diarrhea and vomiting. *Colitis*, meaning "inflammation of the colon," also results from bacterial infection. Colitis is characterized in part by almost constant propulsive movements of the large intestine.

A digestive system disease resulting from viral infection is *hepatitis*, or "inflammation of the liver." There are several forms of viral hepatitis, each of which may be transmitted from one person to another. This may occur, for example, in drug addicts who share the same needle, or transmission sometimes occurs through routine blood transfusions.

NUTRITION AND DIET

Proper nutrition involves the intake of a sufficient quantity and balanced proportion of nutrients so that the body's metabolic needs are met. Carbohydrates, lipids, and proteins are generally referred to as *macronutrients*, i.e., nutrients required by the body in large quantities. Nutrients required only in relatively small amounts are called *micronutrients*. The micronutrients necessary for normal metabolism are classified as *vitamins* and *minerals*. A vitamin is an organic compound that cannot be synthesized by the body but must be obtained in the diet. In humans the necessary water-soluble vitamins include the large group of B complex vitamins (B_1, B_2, B_6, B_{12}, and so on) and vitamin C; as mentioned previously, the fat-soluble vitamins are A, D, E, and K. Table 18.1 lists some of the vitamins and the symptoms resulting from their deficiency in the body.

Minerals are inorganic compounds, usually in the form of ions, that the body requires in relatively small amounts. These include sodium (Na^+), chloride (Cl^-), potassium (K^+), calcium (Ca^{++}), magnesium (Mg^{++}), iodine (I^-), zinc (Zn^{++}), and several others. Each of

> Carbohydrates, lipids, and proteins release chemical energy during cellular metabolism.

Table 18.1 THE MAJOR WATER-SOLUBLE AND FAT-SOLUBLE VITAMINS AND SOME OF THE SYMPTOMS RESULTING FROM THEIR DEFICIENCY IN THE BODY

	VITAMIN	SYMPTOMS OF DEFICIENCY
Water-soluble:	Vitamin B_1 (thiamine)	Impaired function of CNS; weakened heart muscle; gastrointestinal disorders
	Vitamin B_2 (riboflavin)	Scaly skin; cracking of skin at the angles of the mouth
	Vitamin B_6 (pyridoxine)	Inflammation of the skin; gastrointestinal disturbances; convulsions
	Vitamin B_{12} (cobalamin)	Pernicious anemia; occasional loss of nervous sensations and development of paralysis
	Vitamin C (ascorbic acid)	Prolonged wound healing; cessation of bone growth; fragile capillaries (scurvy)
	Niacin	Inflammation of the skin; gastrointestinal disturbances; mental disorders
	Folic acid	Impaired growth; anemia
Fat-soluble:	Vitamin A	Failure of growth; scaly skin; impaired function of reproductive system; nightblindness
	Vitamin D	Bone deformities; rickets in children
	Vitamin E	Impairment of growth; male sterility
	Vitamin K	Prolonged blood clotting time

these is required in varying amounts. Some of the minerals, along with their functional roles, are given in Table 18.2.

Energy Needs. Each of the major macronutrients—carbohydrates, lipids, and proteins—contains chemical energy that is released during cellular metabolism. Each gram of carbohydrate or protein contains approximately 4.0 Cal (Kcal), and each gram of lipid contains 9.0 Cal. The amount of energy needed by the body varies with size, sex, and extent of physical activity. The maintenance of good nutrition for a 70-kg (154-lb) adult male requires about 2500 to 3000 Cal per day; a 54-kg (120-lb) female requires about 2000 to 2200 Cal per day. During pregnancy and nursing, the daily requirement for females includes an additional 300 to 500 Cal. These caloric requirements, of course, may be much higher if someone is engaged in a vigorous daily exercise program.

In terms of good health, one of the major concerns is the effect of caloric intake on body weight. If the number of calories taken

> Normal health requires 10 *essential amino acids* that the human body cannot synthesize but that must be included in the diet.

Table 18.2 SOME OF THE MINERALS AND THEIR FUNCTIONAL ROLES IN THE HUMAN BODY

MINERAL	FUNCTIONAL ROLE
Sodium (Na)	Maintenance of osmotic pressure of body fluids; necessary for muscle contraction and conduction of nerve impulses
Chlorine (Cl)	Maintenance of osmotic pressure of body fluids; important in acid-base balance
Potassium (K)	Aids in maintaining osmotic pressure within cells; necessary for muscle contraction and conduction of nerve impulses
Calcium (Ca)	Constituent of bones and teeth; necessary for blood coagulation, muscle contraction, and conduction of nerve impulses
Magnesium (Mg)	Necessary for normal muscle and nerve function; associated with production of ATP
Phosphorus (P)	Constituent of bones and teeth; component of DNA, RNA, and ATP
Iodine (I)	Necessary for the synthesis of thyroid hormones
Iron (Fe)	Constituent of hemoglobin, myoglobin, and respiratory enzymes
Copper (Cu)	Constituent of many enzymes; necessary for hemoglobin synthesis and production of melanin
Zinc (Zn)	Constituent of many enzymes; necessary for normal wound healing

into the body is equal to the number expended, the body weight remains constant. When more calories are consumed than given off, the excess food is stored as fat, and the individual *gains* weight. Since each gram of lipid contains 9.0 Cal, the body stores 1 gram of lipid for every 9.0 Cal of *excess* food energy consumed. On the other hand, one *loses* weight when the number of calories expended is greater than the number taken into the body. This is because the stored fat is "burned" and is not replaced by an equivalent amount of food energy.

Malnutrition. The issue of malnutrition has two extremes—an inadequacy of proper food and an excessive intake of food. Although they may get as many calories as they need, most undernourished people do not acquire the essential amino acids or vitamins in their diet. Almost all animals, including man, cannot synthesize some of the amino acids and must get them in their diets. There are 10 of these **essential amino acids** that the human body cannot synthesize; if these are not supplied in the diet, many of the body's vital proteins, such as enzymes, hormones, and tissue constituents, cannot be produced. A lack of the proper kinds of proteins results in brain damage, kwashiorkor, and retarded growth. The remaining amino

> *Excretion* is the process of metabolic waste removal; *elimination* is the removal of undigested wastes from the digestive tract.

acids that are synthesized by the body, and therefore not required in the diet, are called **nonessential amino acids.** The term "nonessential" does not mean that these amino acids are any less important to the body than the "essential" amino acids. Rather, "nonessential" means that these amino acids are not required in the diet.

When there is a lack of any kind of food, malnutrition takes the form of starvation. The human body normally relies almost entirely upon carbohydrates and lipids for energy sources. When these are depleted, the body begins to draw first on the blood amino acids and then the tissue proteins. At this point, the body is literally feeding on itself; when about half of the body proteins have been depleted, cellular function is impaired and death follows.

At the other extreme, an excessive intake of food results in a form of malnutrition called **obesity.** Although fat cells in the body can increase their fat storage about 50 percent above normal, it is believed that obesity more likely results from an excess *number* of fat cells. Overfeeding small children and adolescents apparently increases the formation of new fat cells, but this formation stops once adulthood is reached. Overfed children, then, may have several times the normal number of fat cells, a condition that persists throughout life.

It is thought that persons with excess fat cells can eat more before the so-called **satiety** ("fullness") **control center** in the hypothalamus signals them to stop. The hypothalamus also contains a hunger center, which, when stimulated, causes an organism to look for food and eat it. Once the desire for food is satisfied, the satiety center *inhibits* the hunger center, resulting in the cessation of eating. If the satiety control mechanism has a high setting, which is thought to be the case in obese individuals, it tends to permit a greater intake of food before inhibiting the hunger center. To put it simply, obese persons must eat more before the brain signals them to stop.

As pointed out, weight gain results when the number of calories taken into the body is greater than the number given off. Thus, the only way to lose weight is to reverse this energy relationship so that more calories are expended than are consumed. The only way to do this is to eat less or exercise more, preferably both. In either case, the body has to draw on its excess fat stores for energy, and body weight declines.

INTRODUCTION TO EXCRETION

Excretion is the process of ridding the body of metabolic wastes. It is not the same thing as elimination, which refers to the removal of undigested wastes from the digestive tract.

> Nitrogen, a toxic metabolic waste component, is excreted from the body as *urea* in the urine.

All animals are confronted with the problem of removing waste materials produced by cellular metabolism. One of the most toxic components of metabolic wastes is nitrogen. You will recall that nitrogen is one of the constituent elements of all proteins. Consequently, when proteins are metabolized, nitrogen is liberated, but it cannot be allowed to accumulate in the body. In most aquatic animals, nitrogen waste is converted into the compound ammonia (NH_3), which is excreted into the surrounding water. Amphibians and mammals, including man, convert ammonia into the relatively nontoxic compound **urea** ($CO(NH_2)_2$), which is excreted in the urine. The formation of urea is yet another of the many functions of the liver.

THE HUMAN URINARY SYSTEM

The primary organs of the urinary system are the **kidneys,** located just below the diaphragm at the back of the abdominal cavity (Fig. 18.20). The kidney is generally described as bean-shaped and is

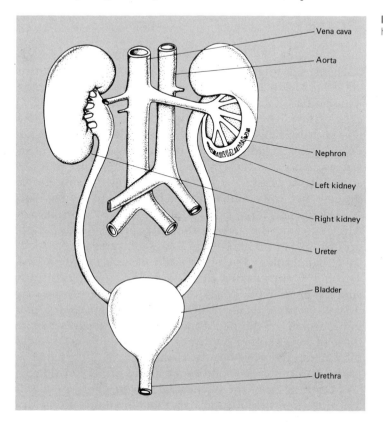

Figure 18.20 General structure of the human urinary system.

> *Nephrons* are microscopic filtering units contained in the kidneys.

about the size of your fist. Two branches of the aorta, the right and left **renal arteries,** supply oxygenated blood to each kidney. Blood exits from each kidney via a **renal vein,** which leads into the inferior vena cava.

The other major structures of the urinary system include the **ureters,** one of which extends from each kidney downward to the **urinary bladder.** Urine formed in the kidneys passes down through the ureters and is temporarily stored in the bladder. At the point where each ureter enters the bladder there is a small flap of tissue that functions as a one-way valve. When the bladder is distended with urine, the valves prevent backflow of urine into the ureters and kidneys.

From the urinary bladder, urine passes through the **urethra** to the outside of the body. In males the urethra is relatively long and passes through the penis; in females the urethra is short and opens to the outside just above the opening of the vagina. Because of the short urethra, females are considerably more susceptible to bacterial infection of the urinary bladder than are males.

PHYSIOLOGY OF THE KIDNEYS

The Nephrons. Each human kidney contains well over a million microscopic filtering units called **nephrons.** The basic function of the nephron is to remove unwanted substances from the blood plasma. This process is called cleaning, or "clearing," of plasma. In general, clearing involves ridding the plasma of urea and other metabolic wastes and removing excess ions such as sodium, potassium, and hydrogen ions. In addition, the nephrons are involved in maintaining water balance and thereby aid in regulating blood volume. They also adjust blood pressure and, by controlling the concentration of hydrogen ions, contribute to maintaining the pH of the blood.

The general structure of the nephron is shown in Figure 18.21. After entering the kidney, the renal artery divides and subdivides until it forms numerous tiny balls of capillaries, called **glomeruli** (sing. glomerulus). Each glomerulus is a part of an individual nephron and is surrounded by a double-layered structure called **Bowman's capsule.** Extending from the outer layer of the capsule is the twisted **proximal tubule,** which joins the long **loop of Henle.** The terminal section of the nephron is the **distal tubule,** which, along with those of several other nephrons, empties into a **collecting duct.** From the numerous collecting ducts throughout each kidney, the urine passes into the ureters and down to the urinary bladder.

Leading *out* of the glomerulus is a small artery that divides into a profuse network of capillaries that surrounds the nephron. Blood

> The clearing of blood plasma and the formation of urine involve three mechanisms: *filtration* of blood, *reabsorption* of water, *secretion* of ions.

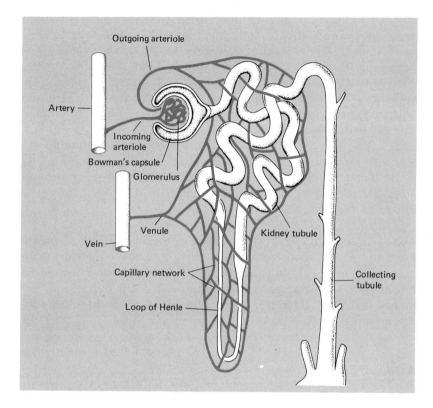

Figure 18.21 A nephron. Over a million of these microscopic filtering units are found in each kidney.

that has been filtered, or cleared, flows from the capillary network into a large vein that empties into the renal vein. This network serves to *reabsorb* some of the water and other substances from the nephron tubules and return them to the blood. The excess water and unwanted wastes and ions that are not reabsorbed remain in the tubules to become urine.

Plasma Clearance and Formation of Urine. The clearing of blood plasma and the accompanying formation of urine involve three basic mechanisms: (1) **filtration** of the blood in the glomerulus, (2) **reabsorption** of water and other vital materials removed during the filtration process, and (3) **secretion** of various substances, such as hydrogen and potassium ions, from the blood into the nephron tubules. Let's look generally at each of these mechanisms.

Filtration. You will recall from Chapter 17 that blood moving through an artery is under high pressure. This is also true of the

> Through a *countercurrent mechanism*, the loop of Henle establishes a region of high sodium chloride concentration in the tissue fluids around the loop and collecting duct.

blood that enters the glomerulus. The capillaries that constitute the glomerulus are extremely thin-walled, as is the inner layer of Bowman's capsule (Fig. 18.22). The high pressure of the blood within the glomerulus forces the plasma water, along with other small blood constituents—such as ions, salts, glucose, amino acids, urea, and so on—through the capillary walls and into Bowman's capsule. Large blood constituents, such as blood cells and some of the plasma proteins, cannot be filtered out and so remain to circulate within the glomerulus and out into the capillary network surrounding the nephron. Essentially, then, most of the water and small constituents of the blood are filtered out and pass into Bowman's capsule; the larger blood constituents that cannot pass through the capillary walls remain in the blood.

Reabsorption. After entering Bowman's capsule, the filtered plasma water and small blood constituents move into the proximal tubule. As the fluid moves through the tubule, most of the water, along with the amino acids, glucose, some of the salts, and other vital substances, is reabsorbed as needed into the blood circulating within the surrounding capillary network. Most of the urea and other unwanted substances remain in the tubule to continue their journey through the loop of Henle, distal tubule, and into the collecting duct.

The essential function of the loop of Henle is to establish a region of high concentration in the tissue fluids around the loop and the collecting duct. This is accomplished by a "countercurrent" mechanism involving the salt sodium chloride (NaCl). The *descending* limb of the loop of Henle is highly permeable to water and sodium chloride, whereas the *ascending* limb is impermeable to water but actively pumps sodium chloride out into the surrounding tissues (Fig. 18.23). Some of the sodium chloride pumped out of the ascending limb diffuses across into the descending limb, which also contains sodium chloride filtered from the blood in the glomerulus. The sodium chloride flows from the descending limb, around the tip of the loop, and up the ascending limb. Most of this sodium chloride is then pumped out again into the tissues, and some of it re-enters the descending limb to repeat the process. By this continuous cycling, or "countercurrent," mechanism, the tissue fluid around the tubules contains a high concentration of sodium chloride.

What is the significance of this high salt concentration in the tissues? As we have seen, in moving through the descending limb to

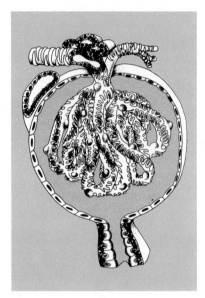

Figure 18.22 Diagram of Bowman's capsule cut away to show the glomerulus inside.

> The high sodium chloride concentration in the tissue fluid establishes an osmotic gradient that draws water out of the collecting ducts; a concentrated urine forms that conserves water and prevents dehydration.

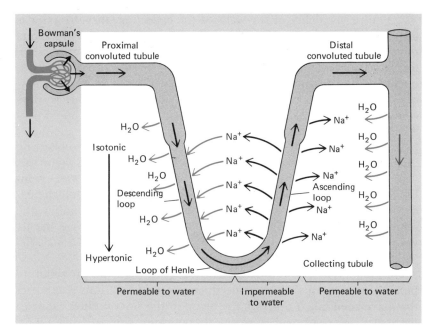

Figure 18.23 Some mechanisms of urine formation in the nephron. Filtration of blood in Bowman's capsule yields filtrate isotonic to the blood. As the filtrate passes down the proximal convoluted toward the loop of Henle, water diffuses out into the surrounding tissue fluids, and sodium ions diffuse in, resulting in a hypertonic fluid in the tubule. Passing through the ascending loop of the distal convoluted tubule, the filtrate loses Na^+ by diffusion and becomes hypotonic; during the flow down the collecting tubule, water diffuses out into the salty surrounding fluids, resulting in the final hypertonic filtrate, urine. Processes of active transport also contribute to urine formation, adding certain substances to the filtrate against the concentration gradient and removing others.

the ascending limb of the loop of Henle, the tubular fluid has a high concentration of sodium chloride. However, after some of the sodium chloride is pumped out by the ascending limb, the fluid becomes more dilute. In fact, by the time the fluid reaches the distal tubule, it is hypotonic to the tissue fluid outside the tubules. Given this situation, you would expect water to move out of the distal tubule into the tissues, as the principles of osmosis would dictate. This does not occur because the distal tubule is highly *impermeable* to water.

On the other hand, the collecting duct is normally *permeable* to water. Consequently, as the dilute fluid moves into the collecting duct, water is lost by osmosis into the surrounding, highly concentrated tissue fluid. Thus, the significance of the high sodium chloride concentration in the tissue fluid is to establish an osmotic gradient that draws water out of the collecting ducts. The result is that the urine formed within the collecting duct is fairly concentrated. Why is this important? Basically, excretion of a concentrated urine *conserves water*. In other words, instead of remaining in the collecting ducts to be excreted, water moves out into the surrounding tissues to re-enter the blood and other body fluids. In this manner, the body does not become dehydrated.

> Hydrogen ion secretion into the filtered plasma aids in maintaining blood pH at the proper level.

The permeability of the walls of the collecting ducts to water is actually regulated by **antidiuretic hormone,** or ADH. (Diuresis refers to an increased flow of urine.) ADH (also called *vasopressin*) is released by the pituitary gland in response to the concentration of water in the body. More ADH is secreted when the body water is low, and less is secreted when there is excessive water intake. If you become dehydrated, as when exercising, for instance, the amount of ADH in the blood rises, and the collecting ducts become more permeable to water. Consequently, water moves out of the collecting ducts, as discussed, instead of being lost from the body in the urine. Conversely, if you drink too much water, the release of ADH is inhibited, the collecting ducts become more impermeable, and the excess water remains in the ducts to be excreted. In this manner, the kidneys aid in regulating the volume of the blood (which is mostly water) and other body fluids.

Secretion. As filtered plasma moves along through the proximal tubule all the way to the collecting duct, it receives various substances that have been taken up from the blood in the surrounding capillary network. These substances, primarily potassium and hydrogen ions, along with certain drugs such as penicillin, are taken up from the blood by the cells of the tubules and then secreted into the filtered plasma. These substances eventually move into the collecting duct and are excreted in the urine. It is absolutely essential that excess potassium ions be removed because an elevated potassium ion concentration usually causes abnormal heart rhythms.

The concentration of potassium in the body—and to some extent the concentration of sodium—is controlled by the hormone **aldosterone,** which is secreted by the **adrenal glands.** In humans, one adrenal gland sits atop each kidney. As the concentration of potassium ions in the body increases, the secretion of aldosterone also increases. This causes more potassium to be taken up by the nephron and excreted in the urine. Since the concentration of potassium in the body is now lowered, the adrenal glands reduce the secretion of aldosterone. The potassium-aldosterone mechanism is another example of negative feedback and functions to maintain homeostasis of the body fluids (Fig. 18.24).

The secretion of hydrogen ions into the filtered plasma aids in maintaining the pH of the blood. However, if the blood becomes too basic, i.e., has too low a concentration of hydrogen ions, fewer hydrogen ions are secreted into the tubules, and the body conserves the needed hydrogen ions, lowering the pH of the blood.

Blood pressure is also influenced by the hormone *angiotensin*, a powerful vasoconstrictor formed when the enzyme renin splits the plasma protein angiotensinogen.

613
The Kidneys and Regulation of Blood Pressure

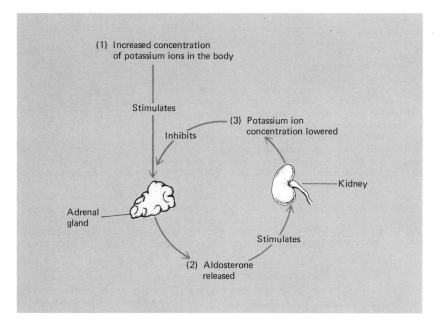

Figure 18.24 Negative feedback control of potassium ion concentration. An increase in potassium ion concentration stimulates the adrenal glands to release the hormone aldosterone. Aldosterone in turn causes the kidney to excrete potassium. As the potassium level falls, the adrenal glands reduce the secretion of aldosterone.

Blood pressure, you will recall, is the force of the blood pushing against the wall of a blood vessel. The control of blood pressure is influenced not only by the nervous system but by hormonal mechanisms as well. One of these mechanisms involves the hormone *angiotensin*, derived from one of the plasma proteins. Should the blood pressure in the arteries fall too low, the kidney responds by secreting an enzyme called **renin**. In the blood, renin splits one of the plasma proteins (angiotensinogen) to form angiotensin. This hormone is an extremely powerful *vasoconstrictor,* i.e., it causes blood vessels to contrict. In particular, angiotensin constricts small arteries and thereby raises the arterial pressure back toward normal. In addition, angiotensin also stimulates the release of aldosterone from the adrenal glands. Aldosterone causes the kidney to decrease its excretion of water, which elevates the blood volume. The higher blood volume also aids in raising the blood pressure back to normal.

This same mechanism occurs if blood flow to the kidney is reduced through arterial obstruction or disease. The result in this case, however, is that if the obstruction is not removed or the disease corrected, the blood pressure remains too high, and the individual develops hypertension.

THE KIDNEYS AND REGULATION OF BLOOD PRESSURE

DISEASES AND DISORDERS OF THE URINARY SYSTEM

There are a considerable number of diseases and disorders of the urinary system, many of which are often indicated by examination of the urine. This procedure, known as **urinalysis**, is one of the most common and highly indicative diagnostic tests. Urinalysis involves observing the color, clarity, and odor of urine, as well as examining it microscopically for the presence of blood cells, bacteria, crystals, and other substances.

Diabetes insipidus. Through disease or injury, nerve cells in the brain may cease or reduce the production of ADH. As we have seen, this condition decreases the permeability of the walls of the collecting ducts, preventing water from being reabsorbed into the body tissues. Consequently, water remains in the collecting ducts, forming a very dilute urine. Since so much water is lost in the urine, individuals with diabetes insipidus must consume enormous quantities of water daily to compensate for the loss. If the water is not replaced, the body becomes dangerously dehydrated, and normal ionic balance is upset.

Diabetes insipidus should not be confused with so-called "sugar diabetes," or **diabetes mellitus**, which is a disease of the pancreas. Diabetes mellitus will be discussed in the following chapter.

Gout. One of the waste products of body metabolism is *uric acid* ($C_5H_4N_4O_3$), derived from the breakdown of the purines adenine and guanine (see Chapter 3). Normally, uric acid is excreted by the kidneys; however, if too much uric acid is produced or if too little is excreted, the excess uric acid crystallizes and is deposited in the joints or urinary tract. In various joints, such as those of the toes, fingers, knees, and elbows, uric acid crystals cause inflammation, pain, and swelling, a condition known as gout. Deposits of uric acid crystals in the kidneys or urinary tract may lead to serious renal disease.

Gout may result from a prolonged diet of foods such as liver, brains, and sardines, which are high in purines. Excessive consumption of alcohol also may be a contributing factor.

Uremic Syndrome. As we have discussed, toxic substances in the blood are normally filtered out by the nephron and excreted in the urine. However, if the filtration mechanism fails, these toxic substances accumulate in the blood, a condition called **uremia**. The presence of toxic wastes in the blood produces a vast array of symptoms affecting virtually every system in the body. Some of the symptoms include pH and fluid imbalance, hypertension, edema, anemia, nausea, diarrhea, muscle twitching, bruises, and so on. This complex of symptoms constitutes the uremic syndrome, another way of describing terminal kidney failure.

Treatment for the uremic syndrome includes dietary regulation that controls the intake of fluids, ions, and food. If the kidneys continue to fail, the only additional means to preserve life is **dialysis** or a kidney **transplant**. Dialysis is based on the principle of diffusion (see Chapter 5) and is the process utilized in the "artificial kidney" machine (Fig. 18.25). Basically, the blood of the patient passes through a tube that has selectively permeable walls, i.e., the wall of the tube allows some substances to move into or out of the blood but restricts the passage of larger substances (mainly the plasma proteins). The tube is in contact with a dialysis "bath," which contains the correct concentrations of most of the blood plasma constituents. As the blood moves through the tube, substances move into or out of the blood from regions of high concentration to regions of lower concentration.

For example, if the concentration of potassium ions in the blood is too high, the net movement of potassium will be *out* of the blood, through the porous tube, and into the dialysis bath. On the other hand, if the concentration of a blood constituent is too low, such as calcium ions, for example, the net movement of calcium ions will occur from the dialysis bath *into* the blood. Movement of substances out of or into the blood occurs until the concentrations of the diffusing substances in the blood and dialysis bath are equal. As the blood enters the dialysis bath, heparin is added to prevent coagulation of the blood. After dialysis, the blood is returned to the patient via a vein. During this procedure, an antiheparin substance is added to the blood to prevent bleeding in the patient.

In the preceding chapter it was pointed out that organ transplants between individuals other than identical twins are usually rejected by the host's immune system. Since identical twins are relatively rare, the more common and next best procedure is to transplant a kidney from a blood relative, preferably a brother or sister. This reduces the chances that the grafted organ will be rejected immediately, as is often the case with transplants between individuals who are not related.

Having only one kidney seems to present no particular problem. A single kidney increases in size and apparently functions about as effectively as two.

Kidney Stones. The normal pH of urine in healthy adults averages 6.0, with a range throughout the day between 4.5 and 8.0. However, a variety of physiological disorders and certain diets produce urine that is either too acidic or too alkaline. For example, fever may result in an acidic urine, and some urinary tract infections produce an alkaline urine. In addition, foods such as meat and bread tend to produce an acidic urine, whereas milk and vegetables tend to make the urine more alkaline.

Oversecretion of parathyroid hormone (see the following chapter) causes large amounts of calcium to be released into the blood. As the blood is filtered by the kidneys, excessive levels of calcium appear in the urine. Some of this calcium forms insoluble crystals that precipitate and collect into kidney stones. This is the case in both acidic and alkaline urine. It is interesting, however, that the composition of kidney stones formed in acidic and alkaline urine is not exactly the same. Although calcium is usually present in all kidney stones, there are other metabolic products, found in either acidic urine or alkaline urine, that combine with calcium to form the stones. Once formed, kidney stones may move into the ureters or urethra, blocking the flow of urine. As it accumulates, urine is backed up into the kidney, causing infection and degeneration of kidney tissues.

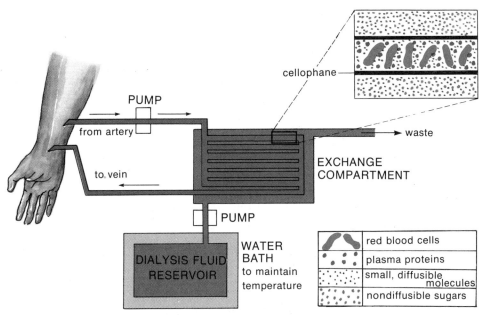

Figure 18.25 Diagram of an artificial kidney machine. Blood flows between sheets of cellophane (supported on plastic—not shown here). Dialysis fluid, warmed to body temperature, contains normal blood nutrients plus nondiffusible table sugar. The latter helps draw excess water from the blood. Na^+, urea ammonia, H^-, and other wastes diffuse from blood to dialysis fluid. (From Clark, M. E.: *Contemporary Biology*. 2nd ed. Philadelphia: Saunders College Publishing, 1979, p. 270.)

REVIEW OF ESSENTIAL CONCEPTS

1. The digestive system breaks down foodstuffs mechanically and chemically into forms that can be used by the body.
2. *Peristalsis* serves to break down foodstuffs mechanically, whereas the digestive *enzymes* break down food chemically by *hydrolysis*.
3. Peristalsis and enzyme secretion in the digestive tract are regulated by the *ANS* and several *hormones*.
4. In the digestive tract, carbohydrates are hydrolyzed ultimately to *monosaccharides;* lipids are hydrolyzed to *fatty acids, glycerol,* and *monoglycerides;* and proteins are hydrolyzed to *amino acids*.
5. Digestion begins in the mouth, where carbohydrates are broken down into *maltose* units by the enzyme *salivary amylase*.
6. Digestion of proteins begins in the stomach, where the enzyme *pepsin* hydrolyzes proteins into *large polypeptides*. Secretions of the stomach also include *mucus, HCl,* the hormone *gastrin,* and *intrinsic factor*.
7. The movement of food from the stomach into the duodenum is controlled by the *enterogastric reflex*.

8. The small intestine consists of the *duodenum*, *jejunum*, and *ileum*. The inner lining of the small intestine is provided with *villi*, which increase the surface area for digestion and absorption. *Microvilli* of intestinal cells also increase the surface area.
9. The *pancreas* releases *sodium bicarbonate*, which neutralizes the acid contents entering the duodenum from the stomach. The digestive enzymes secreted by the pancreas include *pancreatic amylase*, a carbohydrate-digesting enzyme; *pancreatic lipase*, which breaks down lipids; and the protein-digesting enzymes *trypsin, chymotrypsin*, and *carboxypeptidase*.
10. Release of sodium bicarbonate from the pancreas is regulated by the intestinal hormone *secretin*, whereas enzyme secretion, along with the release of bile from the gallbladder, is regulated by the hormone *cholecystokinin*.
11. The liver produces *bile*, which is stored in the gallbladder. Bile functions in emulsifying lipids.
12. Digestion is completed in the intestine by the action of intestinal enzymes. Maltose is split into glucose molecules by the enzyme *maltase*, sucrose is hydrolyzed to glucose and fructose by *sucrase*, and lactose is hydrolyzed to glucose and galactose by *lactase*. Lipids are broken down into monoglycerides, fatty acids, and glycerol by *bile* and *pancreatic lipase*. Polypeptides are split into *amino acids* by several intestinal enzymes.
13. *Absorption* of foodstuffs into the blood occurs in the small intestine. Monosaccharides and amino acids are actively transported through the *villi*. Monoglycerides, fatty acids, and glycerol first dissolve in bile salts and then enter the villi. In the villi, the lipid products are reassembled into triglycerides, coated with protein, and absorbed into the *lacteals*. The lipids then enter lymph capillaries and are transported to the heart, where they enter the circulating blood.

    Proper fat absorption is necessary for the absorption of vitamins A, D, E, and K.
14. The small intestine joins the large intestine, or colon, about 3 inches above the *cecum*. The *appendix* extends from the end of the cecum. At the junction of the small and large intestines, the *ileocecal valve* prevents the contents of the colon from backing up into the small intestine.
15. The upper half of the colon absorbs water and ions; the lower half is a storage area for feces. The upper colon also contains bacteria that produce various vitamins. The colon is emptied by the *defecation reflex* and by relaxation of the external and internal anal sphincters.

16. Disorders of the digestive system include ulcers, gallstones, constipation, diarrhea, cirrhosis of the liver, cholera, colitis, and hepatitis.
17. *Macronutrients*, which include carbohydrates, lipids, and proteins, are required by the body in large quantities. *Micronutrients*, which include a variety of *vitamins* and *minerals*, are required by the body in relatively small amounts.
18. Proper nutrition involves the intake of a sufficient quantity and balanced proportion of carbohydrates, lipids, and proteins. *Malnutrition* results from inadequate intake of proper food or from an intake of too much food.
19. *Obesity* most likely results from an excess number of fat cells in the body, a consequence of overfeeding young children.
20. The desire for food is controlled by the "hunger center" in the hypothalamus; when the hunger is satisfied, the *satiety control center* in the hypothalamus inhibits the hunger center, and the desire for food declines.
21. *Excretion* is the process of ridding the body of metabolic wastes. In humans, nitrogen is excreted in the form of *urea*.
22. The primary organs of the human urinary system are the *kidneys*; other major urinary structures include the *ureters*, *urinary bladder*, and *urethra*.
23. Clearing of blood plasma is accomplished by *nephrons*, the functional units of the kidneys.
24. A nephron consists of the *glomerulus*, *Bowman's capsule*, the *proximal tubule*, *loop of Henle*, the *distal tubule*, and a *collecting duct*.
25. Clearing of plasma and urine formation involve *filtration*, *reabsorption*, and *secretion*.
26. Filtration removes most of the water and small blood constituents, which pass through the glomerulus into Bowman's capsule. Reabsorption returns needed water and blood constituents into the blood circulating within the capillaries surrounding the nephron; urea and other unwanted substances remain in the kidney tubules to be excreted.
27. The loop of Henle aids in maintaining a region of high NaCl concentration around the loop and collecting duct. This region is established by a "countercurrent" mechanism by which NaCl is continually cycled from the ascending limb of the loop of Henle, out into the tissues, and into the descending limb.
28. The region of high NaCl concentration in the kidneys conserves water by drawing water out of the collecting ducts. Water then enters the blood or body fluids, preventing dehydration.

29. The permeability of the walls of the collecting ducts is regulated by *antidiuretic hormone* (ADH). ADH is released by the posterior pituitary gland in response to the concentration of water in the body.
30. Secretion by the nephron tubules is a process by which substances in the capillary blood, primarily potassium and hydrogen ions, enter the tubular cells, which then secrete the ions into the filtered plasma to be excreted. The kidney tubules also secrete hydrogen ions, which aid in maintaining the pH of the blood.
31. The concentrations of potassium and sodium ions are controlled by the hormone *aldosterone*, which is secreted by the adrenal glands.
32. The kidneys aid in regulating blood pressure through the release of the enzyme *renin*. Renin splits the plasma protein *angiotensinogen* into *angiotensin*, which raises the blood pressure primarily by causing constriction of small arteries.
33. Diseases of the urinary system include *diabetes insipidus*, *gout*, and *uremic syndrome*. Treatment for uremic syndrome may require *dialysis* or a kidney transplant.
34. *Kidney stones* are formed when excess calcium, along with various metabolic products, precipitates out of the urine, forming insoluble crystals.

APPLYING THE CONCEPTS

1. Why is hydrolysis important in the digestive process? How is hydrolysis accomplished in the human digestive tract?
2. What are the roles of peristalsis in digestion?
3. What are the final products of the digestion of the major foodstuffs?
4. Cite three examples of the significance of pH in the digestion of foodstuffs.
5. In what manner do the hormones gastrin, secretin, and cholecystokinin affect the digestive process?
6. What is the functional role of hydrochloric acid in the stomach?
7. What are the digestive functions of the salivary glands, pancreas, and liver?
8. What is the mechanism for neutralizing the acidic stomach contents as they enter the small intestine?
9. What is an ulcer? What factors normally prevent development of ulcers in the digestive tract?
10. What are the functions of villi?
11. What are the roles of bile and bile salts in digestion?
12. How is absorption of each of the major foodstuffs accomplished?
13. What are the general functions of the large intestine?

14. Give two examples illustrating the influence of the nervous system on the digestive process.
15. What is the relationship between caloric intake, caloric output, and body weight?
16. What is malnutrition?
17. What is a glomerulus? What is the functional role of the glomerulus in excretion?
18. What is the fate of the nitrogen derived from the metabolism of proteins in the body?
19. What is the functional relationship between the nephron tubules and the capillary network that surrounds them?
20. What is the importance of reabsorption of substances from the nephron tubules?
21. What is meant by "filtration" as applied to nephron function?
22. In relation to the kidney, what is the "countercurrent" mechanism? What does this mechanism accomplish, and why is it important?
23. In what manner does antidiuretic hormone affect kidney function? How is the secretion of ADH regulated?
24. How are the adrenal glands involved in kidney function?
25. Explain the general mechanism of dialysis as applied in the artificial kidney machine.

the endocrine system

THE ESSENTIAL OBJECTIVES

You have understood this chapter when you are able to:

1. Name and give the location of the endocrine organs of the human body.
2. Explain the general role of the hypothalamus in endocrine function.
3. Describe two mechanisms by which hormones work.
4. Give the general functions of growth hormone and prolactin and explain how the secretion of each is controlled.
5. Describe the effects of abnormal secretion of growth hormone.
6. Give the basic functions of antidiuretic hormone and oxytocin and explain how the secretion of each is controlled.
7. Explain the relationship between adrenocorticotropic hormone (ACTH) and the hormones of the adrenal cortex.
8. Give the functions of aldosterone and cortisol.
9. Describe the secretion and physiological effects of epinephrine and norepinephrine.
10. Describe the action of thyroxine and explain how its secretion is controlled.
11. Give the effects of parathyroid hormone and calcitonin and explain how the secretion of each is controlled.
12. Discuss the major effects of insulin and glucagon and explain how the secretion of each is controlled.

Acting in concert with the nervous system, the endocrine system aids in maintaining homeostasis through the release of chemical messengers called hormones. Hormones exert their effects by altering the metabolic activities of their target tissues.

PREVIEW OF ESSENTIAL TERMS

hormone A chemical substance produced in one part of the body and transported in the blood to another part, where it exerts a specific biological effect.

target organ or tissue The body structure on which a hormone acts or exerts its influence.

endocrine gland A gland that releases its secretion (hormone) directly into the bloodstream.

tropic hormone A hormone that acts directly on an endocrine gland.

cyclic adenosine monophosphate (cAMP) A compound formed from adenosine triphosphate (ATP) by the action of the enzyme adenyl cyclase. Cyclic AMP is involved in the mechanism of action of many hormones.

inhibitory and releasing factors Chemical substances that, when released by the hypothalamus, control the secretion of hormones by the pituitary gland.

The Endocrine System

> Endocrine glands release *hormones*, chemical messengers that are transported in the blood to their target organs.

INTRODUCTION The regulation and maintenance of homeostasis in the body are controlled predominantly by two major systems. One of these—the nervous system—exerts control in a sudden, rapid-acting manner that results in almost immediate physiological adjustments. The second major control system is the **endocrine system,** which performs its physiological activities usually at a much slower rate than does the nervous system. Keep in mind, however, that the two systems supplement each other and almost always act together.

The endocrine system is composed of several specialized glands that release chemical "messengers" called **hormones.** The word "hormone" is from the Greek, meaning "to stir up" or "set in motion." A hormone is defined as a chemical substance produced in one part of the body and transported to another part, where it exerts a specific biological effect. The structures of the body on which hormones act are referred to as **target tissues** or **target organs.** The activities of single organs, several organs, or all of the cells of the body may be influenced by hormonal action.

The glands that release hormones are called **endocrine glands** and are located in several areas of the body (Fig. 19.1). In contrast to salivary glands, sweat glands, oil glands, and so forth, the endocrine glands do not release their contents through ducts. Instead the hormones are secreted into the surrounding tissues and diffuse into the blood through the walls of capillaries.

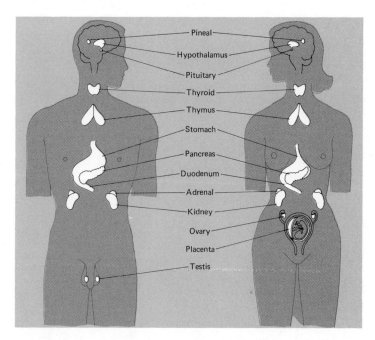

Figure 19.1 Location of the major endocrine organs of human beings.

> The hypothalamus, in response to a variety of internal and external stimuli, secretes releasing and inhibitory factors that regulate the secretion of hormones by the pituitary gland.

Basically, hormones are either *steroids* (see Chapter 3) or *proteins*. The body synthesizes steroids partly from cholesterol, whereas proteins are constructed of amino acids bonded together. Each different hormone, however, has a structure uniquely its own.

Role of the Hypothalamus. A general knowledge of the role of the hypothalamus is essential to understanding endocrine physiology. As discussed in Chapter 14, the hypothalamus is located in the forebrain and is a receiving center for information coming into the brain from the blood and all parts of the nervous system. Through this sensory input, the hypothalamus is constantly informed of the physiological state of the body and is thereby prepared to make any necessary adjustments. This is crucial because the hypothalamus controls most of the vegetative functions of the body, i.e., the involuntary functions, such as heart rate, body temperature, hunger, blood pressure, and so on, that are necessary for life. One of the most vital functions of the hypothalamus is its control of the secretions of the pituitary gland. This is accomplished through the secretion of **releasing** and **inhibitory factors** by the hypothalamus in response to the sensory stimuli it receives from the blood and nervous system. As we shall discuss, the releasing and inhibitory factors in turn control the secretion of hormones by the pituitary gland.

The endocrine glands of the human body include the **pituitary, adrenals, ovaries, testes, thyroid, parathyroids, pancreas, thymus,** and **pineal** (Fig. 19.2). In addition, hormones such as acetylcholine, gastrin, erythropoietin, and so forth are released from other body

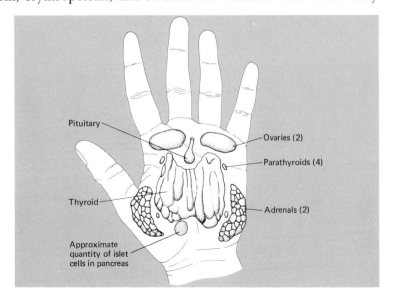

Figure 19.2 The relative size and amounts of the body's endocrine tissue.

> The functional activity of protein hormones involves the formation of cAMP from ATP. Known as the "second messenger" (the stimulating hormone is the "first messenger"), cAMP initiates a particular biological effect in the target cell.

organs, as we have discussed in preceding chapters. These and other hormones *not* released by specific endocrine glands are called **local hormones,** so named because they have local effects. On the other hand, hormones released by the endocrine glands are known as **general hormones.**

The endocrine functions of the ovaries and testes will be discussed in Chapter 20 and therefore will not be covered here. The functions of the thymus gland were discussed in Chapter 17.

HOW HORMONES WORK

The mechanisms by which hormones exert control in the body are rather complex and not totally understood in all cases. In controlling what goes on within the body cells, hormones actually alter the way the cells function. For example, hormones may increase or decrease the permeability of cell membranes, may activate protein synthesis, or may initiate muscle contraction, and so on. How do hormones accomplish these physiological tasks?

One control mechanism involves the messenger substance **cyclic adenosine monophosphate,** or cAMP (Fig. 19.3). This substance is formed by the removal of two phosphate groups from adenosine triphosphate (ATP). When a protein hormone circulating in the blood arrives at its target organ, it binds with a specific receptor site

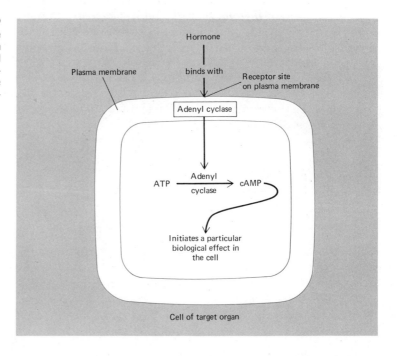

Figure 19.3 Hormonal-cyclic AMP mechanism. Binding of the hormone with a receptor site on the plasma membrane activates the enzyme adenyl cyclase. In the cytoplasm, adenyl cyclase converts ATP into cAMP. The cAMP then brings about the effect dictated by the hormonal stimulation.

The pituitary gland functions in close association with the hypothalamus.

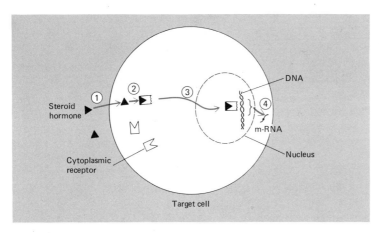

Figure 19.4 Theory of steroid hormone action. Target cells for a given hormone bear unique, highly specific cytoplasmic receptor proteins. When a steroid hormone molecule diffuses into the cell (1), it binds with the receptor (2), causing it to change shape. The hormone-receptor complex then diffuses to the nucleus (3), and binds with a specific part of the DNA, causing it to synthesize mRNA (4).

on a cell membrane. This binding action activates an enzyme called *adenyl cyclase,* which is present in the membrane. Adenyl cyclase moves into the cytoplasm, where it converts ATP into cAMP. The cAMP then brings about the biological effect characteristic of the particular target organ. As mentioned previously, this effect may be to alter membrane permeability, to activate protein synthesis, to initiate muscle contraction, or to influence a variety of other cellular activities. When cAMP acts on an endocrine gland, it regulates the release of hormones by that gland. Thus, it is cAMP that actually triggers the response in many target organs. The cyclic AMP mechanism was first demonstrated by Earl W. Sutherland, who received the 1971 Nobel prize for his discovery.

Another major mechanism of hormonal action involves the steroid hormones (Fig. 19.4). These hormones are released by the ovaries, the testes, and the outer layer, or cortex, of the adrenal glands. Steroid hormones pass through the cell membrane and enter the cytoplasm of the cell, where they bind with specific receptor molecules. The receptors with their attached hormone move into the nucleus of the cell. In some manner, this initiates protein synthesis, which introduces new or additional proteins into the cellular machinery.

PITUITARY GLAND

The pituitary gland is located on the underside of the brain and is about the size of a pea. The gland is attached to the hypothalamus above by a tiny stalk. The pituitary gland is separated into two distinct lobes—the *anterior pituitary* and the *posterior pituitary*— each of which functions differently (Fig. 19.5). The hormones of each and their physiological effects are presented in Table 19.1.

> A *tropic hormone*, such as thyroid-stimulating hormone, acts on an endocrine gland.

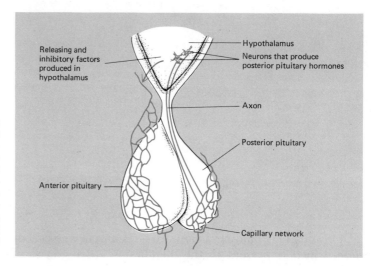

Figure 19.5 The pituitary gland lies at the base of the brain. Secretion of the anterior pituitary hormones is controlled by releasing and inhibitory factors in the hypothalamus. The factors enter the capillaries and are transported in the blood to the anterior pituitary. Hormones of the posterior pituitary are produced by neurons in the hypothalamus. The hormones travel along the axons of the neurons and are absorbed into the capillary blood.

Anterior Pituitary Gland

As pointed out in the introduction to this chapter, control of the secretions of the pituitary gland is regulated by releasing and inhibitory factors produced in the hypothalamus. When released, these factors are absorbed into a blood capillary system in the brain and transported to the anterior pituitary.

The anterior pituitary gland secretes six major hormones, four of which are called *tropic* hormones (Fig. 19.6). A tropic hormone is one that stimulates another endocrine gland. The hormones of the anterior pituitary are **growth hormone** (GH), **prolactin,** and the four tropic hormones, **adrenocorticotropic hormone** (ACTH), **thyroid-stimulating hormone** (TSH), **follicle-stimulating hormone** (FSH), and **luteinizing hormone** (LH). The last two hormones control the activities of the ovaries and testes and will be discussed in the following chapter on reproduction.

Growth Hormone (GH). Growth hormone affects the body in several ways. Mainly, it causes cells to grow and multiply. Under the influence of GH, amino acids enter the cells faster, resulting in increased protein synthesis and a rapid build-up of cellular proteins. This is believed to be the central factor responsible for the increase in the growth rate of the body. In addition, GH causes glucose to enter the liver and muscle cells, where it is converted into glycogen and stored. However, if the cells become saturated with glycogen, no more glucose can be stored, and it remains in the blood. At the same time glucose is being converted into glycogen, GH causes fats to be released from the body's store of adipose tissue. The fats are then used to supply most of the energy for the body cells.

> *Growth hormone* causes cells to grow and multiply by increasing the rate at which amino acids enter the cells.

Table 19.1 THE HORMONES OF THE HUMAN PITUITARY GLAND AND THEIR MAJOR PHYSIOLOGICAL EFFECTS

ANTERIOR PITUITARY HORMONES	EFFECT
Growth hormone (GH)	Causes cells to grow and multiply; enhances uptake of glucose by liver and muscle cells while blocking uptake of glucose by other body cells; causes fats to be released from adipose tissue.
Prolactin	Stimulates cells in the breasts to secrete milk.
Adrenocorticotropic hormone (ACTH)	Regulates secretion of some of the hormones of the adrenal cortex.
Thyroid-stimulating hormone (TSH)	Regulates secretion of hormones from the thyroid gland.
Follicle-stimulating hormone (FSH)	In the female, stimulates enlargement and maturation of the follicle and stimulates enlargement of the primary oocyte; in the male, stimulates the production of sperm.
Luteinizing hormone (LH)	Involved in rupture of the mature follicle and release of the secondary oocyte (ovulation).
Interstitial cell–stimulating hormone (ICSH)	Stimulates cells in the testes to produce testosterone.

POSTERIOR PITUITARY HORMONES	EFFECT
Antidiuretic hormone (ADH)	Causes increased permeability of the collecting ducts of the kidneys, thereby reducing the excretion of water; causes an increase in blood pressure.
Oxytocin	Causes contraction of the walls of the pregnant uterus; stimulates release of milk from the breasts.

But what about all that glucose in the blood? Another effect of GH is to block the uptake and utilization of glucose by the cells. In other words, even though the blood glucose level is high, the body cells cannot use the glucose for energy. Thus, the released fats are used for energy in preference to glucose. This is extremely important during periods of protein or glucose deficiency.

If the process just described is prolonged because of the release of too much growth hormone, the high blood glucose level affects the pancreas, resulting in diabetes. On the other hand, a high concentration of growth hormone has a temporarily positive effect if a person is fasting or starving. Poor nutrition causes the release of large quantities of GH. As we have discussed, the body then cannot utilize glucose for energy but switches to its released store of fats. If

The Endocrine System

> Undernourishment causes increased secretion of growth hormone, which results in the use of fats for energy; this tends to prevent the body from using its proteins as an energy source.

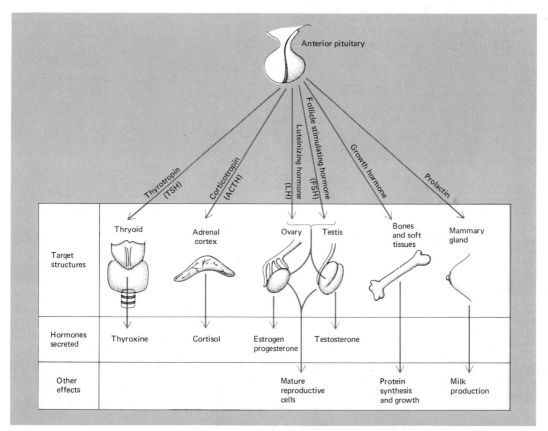

Figure 19.6 Hormones of the anterior pituitary. The diagram indicates the target structures and functions of each hormone.

it were not for the fats, the body cells would have to draw on the only energy sources remaining—the proteins. This means that eventually the body would have to draw on its own tissues, i.e., it would be digesting itself. This, of course, is what happens in the final stages of starvation. The use of fats for energy, therefore, tends to "spare" the body proteins in times of severe undernourishment.

How is the release of growth hormone regulated? As we shall discuss throughout this chapter, the mechanism for regulation of hormonal secretion is negative feedback (see Chapter 17). When the levels of protein or blood glucose begin to fall, the hypothalamus is stimulated to release *growth hormone releasing factor* (GRF). In turn, GRF stimulates the anterior pituitary to secrete growth hormone. As we have seen, GH raises the levels of protein and glucose in the body. As these levels rise, signals are sent to the hypothalamus, which shuts down its release of GRF. Thus, there is no further secretion of growth hormone until the levels of protein or glucose fall again.

> The effects of the undersecretion or oversecretion of growth hormone are evident in the conditions of dwarfism, gigantism, and acromegaly.

Abnormal Secretion of Growth Hormone. If the anterior pituitary fails to secrete adequate amounts of GH during childhood, the result is a condition known as *dwarfism*. The body structures of the affected individual are correctly proportioned to each other, but the growth of the entire body is limited. In most cases, pituitary dwarfs do not mature sexually, but in others sexual development is normal, and reproduction sometimes occurs.

At the other extreme, an excessive secretion of growth hormone during childhood results in *gigantism* (Fig. 19.7). Giants may reach a height of 8 or 9 ft.

If there is oversecretion of growth hormone during adulthood rather than adolescence, the individual cannot grow taller. Instead, many of the bones increase in thickness, and there is general enlargement of the hands, feet, face, and some internal organs. This condition is known as *acromegaly*, or "enlargement of the extremities" (Fig. 19.8).

Prolactin. The hormone prolactin is released from the anterior pituitary of females immediately after childbirth. Prolactin stimulates cells in the breasts to secrete milk. Normally, a substance called *prolactin inhibitory factor* (PIF) produced in the hypothalamus prevents release of prolactin from the anterior pituitary. However, when the

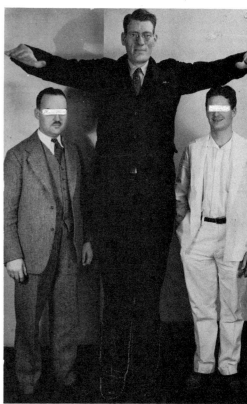

Figure 19.7 A hyperpituitary giant. The giant, approximately 7 ft, 3 inches tall, had a large pituitary tumor and exhibited some of the characteristics of acromegaly—enlarged lower jaw and hands. (From Villee, C. A.: *Biology*. 7th ed. Philadelphia: Saunders College Publishing, 1977, p. 564.)

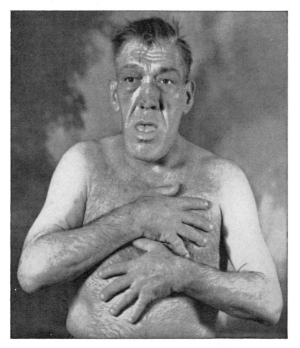

Figure 19.8 A case of acromegaly. The oversecretion of growth hormone of the pituitary in the adult results in an overgrowth of those parts of the skeleton that are still able to respond. Note the enlargement of the lower jaw and hands and the thickening of the nose and ridges above the eyes. (From Turner, C. D.: *General Endocrinology*. Philadelphia: W. B. Saunders Co., 1948.)

> The primary function of *antidiuretic hormone* is conservation of body water.

baby sucks at the breast, it is thought that impulses that suppress the release of PIF are sent to the hypothalamus. Prolactin can then be released, and milk production continues. It is apparent, then, that milk production can continue only if the baby continues to suckle. Otherwise, milk production ceases within 1 or 2 weeks.

Posterior Pituitary Gland

The hormones released by the posterior pituitary gland are produced by specialized neurons in the hypothalamus. The hormones travel down the axons of the neurons to the posterior pituitary and then are absorbed into blood capillaries.

There are two hormones released from the posterior pituitary gland: **antidiuretic hormone** (ADH) and **oxytocin**.

Antidiuretic Hormone (ADH). Antidiuretic hormone, also called *vasopressin*, was discussed briefly in the preceding chapter in the section on excretion. The major effect of ADH is to conserve body water by increasing the permeability of the collecting ducts in the kidneys. Instead of being lost in the urine, water moves out of the collecting ducts and is reabsorbed into the blood. On the other hand, an excessive amount of water in the body inhibits the release of ADH, and the collecting ducts retain water, which is eventually excreted in the urine.

The control of ADH release by the posterior pituitary is regulated by the volume and concentration of the blood. If the concentration of the blood increases or the blood volume falls because of too little water, the hypothalamus is stimulated to produce more ADH. This prevents water from being lost through the kidneys and restores the correct concentration and volume of the blood. When the reverse occurs and the blood is too dilute and blood volume is rising, ADH release is inhibited, and the excess water is excreted in the urine.

ADH also aids in regulating blood pressure. A severe loss of blood volume through hemorrhage causes a marked increase in ADH secretion. ADH then causes constriction of small arteries, which raises the arterial pressure. (This is why ADH is also called vasopressin.) A factor thought to contribute to ADH release when the blood volume falls is the relaxation of stretch receptors in the atria of the heart. The receptors relax because there is lower blood volume and hence lower pressure in the atria. As they relax, the stretch receptors send impulses to the hypothalamus to increase the secretion of ADH. This prevents any further loss of water and, as mentioned, causes the constriction of small arteries. which raises the blood pressure.

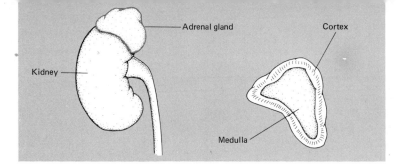

Figure 19.9 Location and general structure of a human adrenal gland.

You may have wondered why frequent trips to the bathroom are necessary when you are drinking alcoholic beverages. Alcohol inhibits the secretion of ADH, which causes the kidneys to retain water and thus increase the output of urine. With a considerable loss of water from the body, you can also understand why extreme thirst often follows an episode of alcoholic indulgence.

Oxytocin. The other hormone released by the posterior pituitary is oxytocin, which causes contraction of the pregnant uterus and acts in conjunction with prolactin to stimulate release of milk from the breasts.

As the uterus stretches during the end of pregnancy, nerve impulses that cause the release of oxytocin from the posterior pituitary are sent to the hypothalamus. Oxytocin then circulates in the blood until it arrives at the uterus, where it stimulates strong uterine contractions. This, of course, assists in expelling the baby through the birth canal.

Oxytocin is also released from the posterior pituitary when a baby suckles the breast. The hormone circulates to the breasts, where it causes contraction of cells that surround the milk glands. This action ejects milk from the glands down into the ducts, so that it is available to the nursing infant.

It is interesting that prolactin and oxytocin are also present in males; however, the functions of these hormones in men are unknown.

THE ADRENAL GLAND

Human beings have two adrenal glands, one sitting atop each kidney. Each gland is separated into two functional areas—an outer layer of tissue called the **adrenal cortex** and an inner tissue layer called the **adrenal medulla**. Both areas of the adrenal glands secrete hormones, but their release is controlled by different mechanisms (Fig. 19.9). Release of some of the adrenal cortex hormones is controlled by adrenocorticotropic hormone (ACTH), secreted by the anterior pituitary gland. Release of hormones from the adrenal medulla results from stimulation of the sympathetic nervous system.

The Adrenal Cortex

Although more than 30 steroid hormones have been found in the adrenal cortex, only two—*aldosterone* and *cortisol*—are of major biological importance.

Aldosterone. The major effect of *aldosterone* is to increase the rate of reabsorption of sodium from the nephron tubules. This ensures that

> The hormone *aldosterone* acts primarily to regulate the concentration of sodium and potassium in the body.

adequate amounts of sodium are returned to the blood and not excreted in the urine. As discussed in the preceding chapter, aldosterone also promotes secretion of potassium from the blood into the nephron tubules. Aldosterone also indirectly aids in maintaining the pH of the blood, since hydrogen ions are periodically secreted into the nephron tubules in place of potassium. While promoting the reabsorption of sodium and the excretion of potassium, aldosterone also aids in maintaining blood and tissue fluid volume by reducing the output of urine.

There appear to be several factors that control the release of aldosterone from the adrenal cortex. Two of the most significant are the concentrations of sodium ions and potassium ions in the body. If the concentration of sodium ions in the body falls, the adrenal cortex is stimulated to increase aldosterone secretion. As discussed, aldosterone enhances the reabsorption of sodium from the nephron tubules into the blood. As the level of sodium rises toward normal, the adrenal cortex is inhibited, and aldosterone secretion declines.

A feedback mechanism also applies to the control of aldosterone secretion by potassium ions. However, an *increase* in the concentration of potassium ions causes increased secretion of aldosterone. Aldosterone then acts on the nephron tubules, causing more potassium to be excreted. When the potassium ion concentration returns to normal, the secretion of aldosterone declines.

Thus, regulated by separate feedback mechanisms, aldosterone acts on the kidneys to maintain homeostasis of sodium and potassium ions in the body.

Cortisol. The second major hormone secreted by the adrenal cortex is *cortisol*. This hormone influences several metabolic activities of the body by *increasing* the levels of amino acids, fatty acids, and glucose in the blood. However, cortisol also causes a slight decrease in the *use* of glucose by the body cells. Consequently, the body must also utilize some fatty acids for energy. The essential result of these actions is to increase the formation of glucose, both in the blood and in the liver, where the glucose is stored as glycogen.

From the blood, the amino acids can be taken up by the liver, where they are converted into glucose or glycogen. Although cortisol causes the *release* of amino acids from almost all body cells, it enhances the *uptake* of amino acids by liver cells. In addition, the high level of amino acids in the blood serves another vital function. Within the liver, the amino acids also can be used for the synthesis of new proteins, as well as for the formation of new glucose or glycogen.

> The major effect of the hormone *cortisol* is to increase the formation of glucose in the body.

The control of cortisol secretion is regulated by negative feedback involving *adrenocorticotropic hormone* (ACTH), released from the anterior pituitary gland. In turn, the release of ACTH is controlled by **corticotropin releasing factor** (CRF) from the hypothalamus (Fig. 19.10).

Given the metabolic effects promoted by cortisol, how do they all interact and what is their physiological significance? The key to the answer—the major factor that initiates release of cortisol—is *stress*. This may be mental stress, such as fear, anxiety, or worry; or it may be physical stress, such as disease, injury, infection, and so on.

When a person experiences a stressful situation, nerve signals are relayed to the hypothalamus, which begins to secrete CRF. After its release, CRF travels in the blood to the anterior pituitary gland, where it causes the release of ACTH. ACTH enters the circulating blood and is carried to the adrenal glands, where the hormone stimulates the adrenal cortex to release cortisol. This hormone then brings about the metabolic effects discussed previously.

The combined action of these effects is directed toward providing defenses against the stressful condition. The decreased use of glucose for energy in preference to fatty acids conserves glucose and glycogen and thereby ensures an adequate supply of cellular fuel if the stress persists. If body tissues are damaged or become depleted of vital proteins by a stressful condition, amino acids in the blood can be taken up by the liver and synthesized into new proteins.

As the level of cortisol in the blood rises, signals are sent to the hypothalamus to decrease the release of CRF. In the absence of CRF, the anterior pituitary shuts down its secretion of ACTH. This, of course, reduces the output of cortisol from the adrenal cortex. Through this negative feedback mechanism, the concentration of cortisol in the blood is maintained at a normal level when the body is not under stress.

In Chapter 17 we referred to the process of *inflammation*, which results when body tissues are injured or damaged. The redness and swelling associated with tissue damage are caused by an increase in local blood flow and the escape of fluid through the capillary walls into the surrounding tissues. It is known that large amounts of cortisol counteract the inflammatory process, presumably by decreasing the permeability of the capillary membranes so that fluid does not leak into the tissues. Cortisol may also block the release of cellular enzymes, which usually destroy some of the tissues in the injured area. Therefore, cortisol and some of the other compounds of the adrenal cortex are *anti-inflammatory agents*. These compounds are used in medicine to treat inflammatory diseases such as arthritis, al-

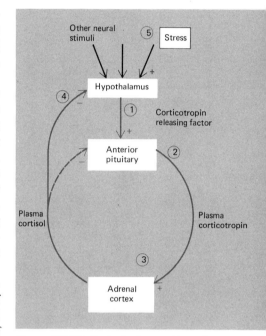

Figure 19.10 Control of secretion of cortisol. (1) Secretion of corticotropin releasing factor by the hypothalamus stimulates the anterior pituitary (2) to release corticotropin, which stimulates the adrenal cortex (3) to secrete cortisol. The secretion of releasing factor is determined in part by negative feedback of cortisol (4) and in part by neural stimuli, particularly those arising from stress (5).

> The medulla of the adrenal glands releases the hormones *epinephrine* and *norepinephrine.*

lergies, and various infectious conditions. It should be pointed out that cortisol and related compounds do not eliminate the cause of a disease but rather help prevent further tissue damage caused by inflammation.

If the adrenal cortex is destroyed through injury or disease, the *lack* of aldosterone and cortisol secretion results in **Addison's disease.** A deficiency of aldosterone causes great amounts of sodium and water to be lost in the urine. In addition, the potassium concentration in the body rises significantly, causing abnormal heart action. If this condition remains untreated, the patient usually dies of shock within a few days.

The lack of cortisol removes the ability of the individual to resist even mild forms of stress. An ordinarily minor infection can often cause death.

Overuse of cortisol medication or disease of the pituitary or adrenal glands may lead to an abnormally *high* level of cortisol in the blood. This usually results in a condition called **Cushing's disease.** As you know, cortisol raises the levels of fatty acids, glucose, and amino acids in the blood. With the upset in fat metabolism, persons with Cushing's disease develop a large deposit of fat around the chest region. The abnormally high level of blood glucose causes "adrenal diabetes," which can result in permanent diabetes. With the continued loss of amino acids from the body cells, protein synthesis is diminished, which leads to muscle and bone weakness and a greater susceptibility to infection.

The Adrenal Medulla

The inner tissue layer of the adrenal glands—the *medulla*—is essentially a mass of neurons of the sympathetic nervous system (see Chapter 14). When stimulated, the adrenal medulla releases two similar-acting hormones, **epinephrine,** commonly known as *adrenalin* (this latter term, with slight spelling variation—adrenaline—is used in Great Britain rather than in the United States), and **norepinephrine,** or *noradrenalin.* As with cortisol, these hormones are released in times of physical or emotional stress. However, secretion of epinephrine and norepinephrine greatly increases the ability of the body to respond to instant emergencies.

Any time you encounter a threatening situation, the sympathetic nervous system is activated, causing epinephrine and norepinephrine to be released into the blood. Instantly, your heart rate increases, and blood vessels in the muscles, brain, and heart dilate—all geared to increasing the oxygen supply to these organs. This is accompanied by a decreased blood flow to less active organs. Your respiratory rate quickens, and more energy is made available as glyco-

> *Thyroxine*, released by the thyroid gland, causes an increase in the metabolic rate of the body.

gen is broken down into glucose in the liver, thereby raising the blood glucose level. Your blood pressure rises, and you become keenly alert. Overall, the rate of cellular metabolism increases. You are now prepared for what is called the "fight or flight" response. Assuming, for example, that you heard the sound of approaching footsteps in a dark alley, your response might be to stand and fight or to depart in haste. Whichever action discretion dictates—fighting or taking flight—is undertaken with an intensity far greater than normal. You can think more quickly, fight more furiously, or run much faster than would ordinarily be the case.

Once the stressful stimulus is removed, most of the epinephrine and norepinephrine is absorbed from the blood, and the body eventually returns to its normal physiological state.

THE THYROID GLAND

The thyroid gland is situated around the trachea (windpipe) just below the larynx (voice box). The gland secretes several iodine-containing hormones, the most abundant and important of which is **thyroxine**. It also secretes the hormone **calcitonin**, which will be considered later in the discussion of the parathyroid glands (Fig. 19.11).

Thyroxine. The principal effect of thyroxine on the body is to increase the rate of metabolism. Thus, under the influence of thyroxine, bone growth increases, glucose is metabolized more quickly, oxygen is consumed more rapidly, the heart rate increases, food is absorbed more readily from the intestine, and so on. Thyroxine also affects overall growth of the body and appears to be necessary for the normal functioning of growth hormone.

To illustrate the metabolic effect and feedback control of thyroxine secretion, consider the following example. Assume that you have been transported via jet from a tropical island to the Arctic Circle. As you step out of the plane into the subzero temperature, receptors in the skin send nerve impulses to the brain, informing the body of the extreme cold. The impulses stimulate the hypothalamus, which begins to secrete **thyroid-stimulating hormone releasing factor** (TRF). In turn, TRF stimulates the anterior pituitary to increase its output of **thyroid-stimulating hormone** (TSH). This hormone circulates in the blood to the thyroid gland, which responds by releasing thyroxine into the circulation. As pointed out, thyroxine acts on body cells to increase their rate of metabolism. You will recall that metabolizing cells release *heat energy*, which is necessary to maintain the temperature of the body. Cells metabolizing more rapidly release more heat energy. In the frozen Arctic, this increased heat energy is required to *keep* the body temperature within its nor-

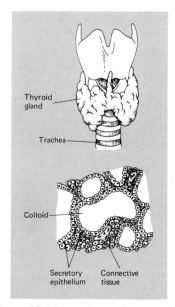

Figure 19.11 Location and microscopic anatomy of the thyroid gland. The thyroid gland appears as a "bow tie" across the trachea. The gland is made up of follicles, and each follicle is lined with secretory epithelium. The colloid in the follicle is inactivated thyroid hormone; it becomes activated before it enters the blood.

The Endocrine System

> Goiter, an enlargement of the thyroid gland, may be associated with an oversecretion or an undersecretion of thyroxine.

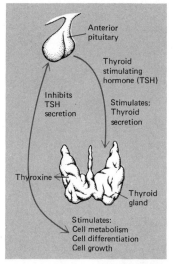

Figure 19.12 Negative feedback of thyroxine in TSH secretion. Thyroid-stimulating hormone (TSH) stimulates thyroxine secretion, and thyroxine inhibits TSH secretion. A delicate balance between the hormones is maintained.

Figure 19.13 Simple goiter, an enlargement of the thyroid gland, can be prevented by the addition of iodine to the diet. Certain brands of table salt (NaCl) contain added iodine. (From Applewhite, P., and Wilson, S.: *Understanding Biology*. New York: Holt, Rinehart and Winston, 1978, p. 322.)

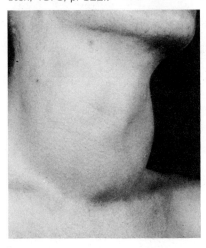

mal range. Thus, people who move to cold regions normally develop a higher metabolic rate.

Once the appropriate metabolic rate is set, further secretion of thyroxine is inhibited by negative feedback. The elevated level of thyroxine in the blood decreases the secretion of TRF and TSH, and the thyroid gland is no longer stimulated. In this manner, the level of thyroxine in the blood can be adjusted to maintain homeostasis (Fig. 19.12).

Hypothyroidism. An insufficient amount of thyroid hormone secretion is known as hypothyroidism. An infant affected with hypothyroidism develops a condition called *cretinism,* which is characterized primarily by stunted growth. In addition, the child is sluggish, has an abnormally low metabolic rate, and is mentally retarded. However, early treatment with thyroid hormones usually prevents cretinism.

A fully grown adult with hypothyroidism develops a condition called **myxedema** ("mucus swelling"). Myxedema results from an accumulation of mucus-containing substances in the body tissues, which causes swelling of the face and body. The affected person also has an abnormally low metabolic rate, is mentally sluggish, and often develops severe arteriosclerosis as a result of high blood cholesterol levels. Treatment of hypothyroidism involves daily use of tablets containing measured amounts of thyroid hormones.

Hyperthyroidism. An oversecretion of thyroid hormones is called hyperthyroidism. The elevated metabolic rate resulting from hyperthyroidism increases hunger, and the individual eats more. However, there is usually weight loss because the food intake is not sufficient to accommodate the high metabolic rate. Other characteristics of hyperthyroidism include muscular weakness, nervousness, extreme fatigue, and emotional instability.

The cause of hyperthyroidism is uncertain, although an antibody found in the blood of hyperthyroid patients appears to be a significant factor. The antibody is thought to exert a prolonged stimulating effect on the thyroid gland, causing abnormal amounts of thyroxine to be released. This overactivity results in enlargement of the thyroid gland, a condition known as **goiter** (L. *guttur*, throat) (Fig. 19.13).

Goiter is also associated with hypothyroidism. In this instance, goiter results from a deficiency of dietary *iodine* (Fig. 19.14). Since iodine is a constituent element of the thyroid hormones, the thyroid gland cannot manufacture the hormones without it. As we have discussed, a certain level of thyroxine in the blood triggers a feedback

> *Parathyroid hormone* controls the balance of calcium and phosphate ions in the body.

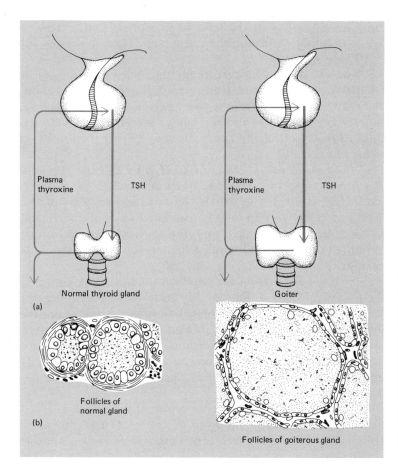

Figure 19.14 Goiter due to iodine deficiency. (a) Left, the normal condition, in which a high level of thyroxine in the blood inhibits the secretion of TSH Right, The intake of iodine is inadequate, and the secretion of thyroxine is therefore reduced. The low level of thyroxine in the blood is insufficient to inhibit the secretion of TSH by the anterior pituitary. The thyroid tissue is therefore constantly stimulated to grow and produces a large amount of thyroxine-deficient colloid, which distends the follicles, forming a goiter. (b) Cross section of the follicles of a normal gland and of a goiterous gland.

mechanism, causing the brain to inhibit the release of TRF and TSH. If thyroxine is absent in the blood, TRF and TSH are not inhibited but are released almost continually. Thus the thyroid gland is overstimulated and becomes enlarged.

This type of goiter is easily treated by including iodine in the diet. Today, many countries have solved the problem by the use of iodized salt.

THE PARATHYROID GLANDS

The parathyroid (meaning "near the thyroid") glands are actually embedded in the thyroid tissues (Fig. 19.15). The parathyroid glands secrete *parathyroid hormone* (PTH), which controls the balance of calcium and phosphate ions in the body. More exactly, PTH causes an *increase* in the level of calcium ions in the blood and tissue

> A decreased secretion of parathyroid hormone can result in tetany; an oversecretion can lead to weakened and brittle bones.

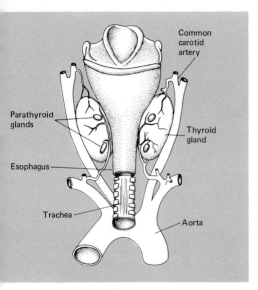

Figure 19.15 Posterior view of the thyroid showing the location of the parathyroid glands. The parathyroid glands —usually four—secrete parathyroid hormone (PTH), which controls the balance of calcium and phosphate ions in the body.

fluid and a *decrease* in the level of phosphate ions. These activities are accomplished through the effect of PTH on the bones, intestine, and kidneys.

You are probably aware of the importance of calcium in maintaining strong bones and teeth. Practically all of the calcium in the body is found in bone tissue, but there is a continual exchange of calcium between bone and the blood. Secretion of PTH causes calcium to move out of the bones, thereby raising the level of calcium in the blood.

The second major effect of PTH is to enhance the absorption of calcium from the intestine, which also raises the blood calcium level. PTH controls this activity by its action on vitamin D. In the liver and kidneys, vitamin D is converted to a substance that, when activated by PTH, actually causes the absorption of calcium from the intestine. You will recall from Chapter 15 that the skin cells form vitamin D upon exposure to the sun.

Finally, PTH acts on the nephron tubules of the kidney to cause the loss of phosphate ions in the urine. This occurs because PTH prevents phosphate ions in the tubules from being reabsorbed into the blood. At the same time, PTH causes the opposite effect in the case of calcium ions. Thus calcium ions are reabsorbed from the nephron tubules into the blood instead of being excreted in the urine. This mechanism ensures against excessive loss of calcium from the bones.

Maintaining the correct concentration of calcium in the blood is essential because, as pointed out in previous chapters, calcium ions are vital in the blood clotting process and for normal muscle contraction.

The secretion of PTH is regulated by the level of calcium in the blood. A decrease in calcium ion concentration causes the parathyroid glands to secrete more PTH. On the other hand, as the blood calcium level rises, the parathyroid glands secrete less PTH.

Calcitonin. Calcitonin, a hormone secreted by the thyroid gland, acts antagonistically to PTH, i.e., it *lowers* the level of calcium ions in the blood. It does this by preventing the loss of calcium from bones. Calcitonin is normally released in large amounts only if the blood calcium level is quite high. The hormone then acts very rapidly to prevent further loss of calcium from the bones. Ordinarily, this is a short-term effect that plays a minor role in the lifelong regulation of calcium ion levels in the body. The major role belongs to the parathyroid glands and PTH.

As mentioned, an increase in the blood calcium level causes increased calcitonin secretion by the thyroid gland. This in turn re-

> The pancreas is both an endocrine organ and an accessory organ to the digestive system.

duces the level of calcium in the blood and inhibits further secretion of calcitonin. While serving as another means of regulating the blood calcium level, notice, however, that the calcitonin feedback mechanism is opposite to that involving the parathyroid glands.

Hypoparathyroidism. A significant reduction in the secretion of PTH lowers the level of calcium in the blood. As a consequence, muscles go into spasms or even severe contractions, a condition known as **tetany**. This is the only major physiological effect of hypoparathyroidism, since tetany usually spreads rapidly to the respiratory muscles, causing death. Hypoparathyroidism is usually treated by administration of calcium and large quantities of vitamin D compounds.

Hyperparathyroidism. Oversecretion of PTH most often is caused by small tumors of the parathyroid glands. With the loss of great amounts of calcium, the bones become weakened and break easily. With the excessively high level of calcium in the blood, the kidneys are taxed in their effort to excrete it in the urine. Accordingly, the urine contains so much calcium that some of it precipitates out, forming kidney stones.

THE PANCREAS

In Chapter 18 we encountered the pancreas in its role as an accessory organ to the digestive system (Fig. 19.16). The pancreas also is partly an endocrine organ that secretes two important hormones—

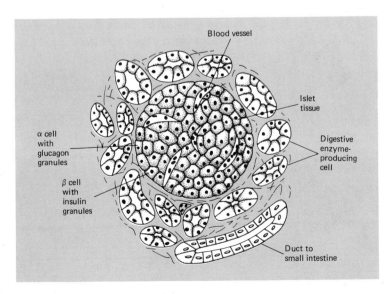

Figure 19.16 Drawing of a section of the human pancreas showing an islet with alpha (α) cells, which make glucagon, and beta (β) cells, which produce insulin. The surrounding cells synthesize digestive enzymes, which pass, via a system of ducts, to the small intestine.

The Endocrine System

> The hormone *insulin* increases the uptake of glucose by the cells and thereby prevents development of hyperglycemia.

insulin and *glucagon*. Scattered throughout the pancreas are patches of cells called **islets of Langerhans**—after Paul Langerhans (1847–1888), a German biologist. Some of the cells of the islets—called **beta cells**—secrete insulin; others known as **alpha cells** secrete glucagon (Fig. 19.17).

Insulin. The hormone insulin acts on the cell membrane to enhance the uptake of glucose by the cells. Glucose enters most body cells by the process of *facilitated diffusion* (see Chapter 5); if insulin is not present, this process is severely impaired. The essential function of insulin, therefore, is to *lower* the blood glucose level by causing the body cells to take up glucose from the blood (Fig. 19.18). Insulin also stimulates the conversion of glucose into glycogen in the liver and muscles and increases the synthesis of proteins and fats (Fig. 19.19).

Glucagon. The major effects of the hormone glucagon are opposite those of insulin, i.e., glucagon *raises* the blood glucose level by stimulating the liver and muscles to convert glycogen into glucose. In addition, glucagon enhances the breakdown of fats, thereby increasing the levels of fatty acids and glycerol in the blood.

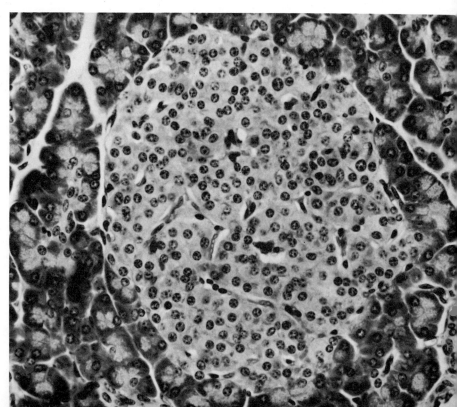

Figure 19.17 A single islet of Langerhans (large, clear area) in the pancreas. Alpha and beta cells are scattered throughout the islet. The dark-staining cells surrounding the islet secrete digestive enzymes. (From Bloom, W., and Fawcett, D. W.: *A Textbook of Histology.* 2nd ed. Philadelphia: W. B. Saunders Co., 1970, p. 615.)

> The hormone *glucagon* increases the level of blood glucose and thus aids in preventing hypoglycemia.

Regulation of Insulin and Glucagon Secretion. Secretion of insulin and glucagon from the pancreas is regulated by the level of glucose in the blood. When the blood glucose level rises, as it does soon after a meal, the beta cells are stimulated to release insulin. As insulin works, the blood glucose level falls, and the secretion of insulin declines.

Between meals the blood glucose level begins to fall, and the alpha cells are stimulated to release glucagon. As the blood glucose level rises, further secretion of glucagon is inhibited.

You will recognize these mechanisms as examples of negative feedback. In working opposite to each other, insulin and glucagon maintain a relatively constant blood glucose level. By lowering the level of glucose in the blood, insulin prevents the development of **hyperglycemia,** an abnormally high blood glucose level; on the other hand, glucagon prevents **hypoglycemia,** an abnormally low blood glucose level, by raising the level of glucose in the blood.

As previously discussed, other hormones also act to prevent hypoglycemia. If the blood glucose level falls, as it does during exercise or reduced carbohydrate intake, the sympathetic nervous system is stimulated, causing the release of epinephrine and norepinephrine from the adrenal medulla. Like glucagon, these hormones act fairly rapidly to raise the blood glucose level. If the blood glucose level falls drastically, the anterior pituitary is stimulated to release growth hormone, and cortisol is released from the adrenal cortex. Although the mechanism of growth hormone and cortisol release is slow-act-

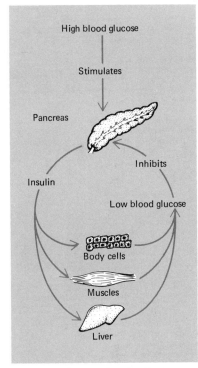

Figure 19.18 Negative feedback control of insulin secretion by the pancreas. Insulin causes the body cells to take up excess glucose from the blood and stimulates the conversion of glucose into glycogen in the liver and muscles. As the blood glucose level falls, the pancreas reduces its secretion of insulin.

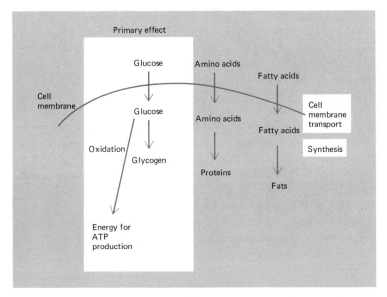

Figure 19.19 Metabolic effects of insulin. The primary effect of insulin is to increase the utilization of glucose. Its general metabolic effect is to increase the rate of transport of glucose, amino acids, and fatty acids into cells and to promote their synthesis into larger molecules.

> *Diabetes mellitus* results from destruction of the insulin-secreting cells of the pancreas.

ing, usually requiring several hours, these hormones are nonetheless vital in preventing hypoglycemia.

Diabetes Mellitus

Diabetes mellitus is usually a hereditary disease, which is transmitted as a recessive genetic trait. It results from destruction of the insulin-secreting beta cells of the pancreas. The deficiency of insulin in the blood leads to excess glucose in the blood and urine, constant thirst, progressive weight loss, and acidity of the blood.

With destruction of the beta cells, the pancreas does not release insulin, and the blood glucose level may rise as high as 10 to 12 times its normal value. Although the kidneys normally remove glucose from filtered plasma and return it to the blood, the amount of glucose in the blood of diabetics is so high the kidneys cannot handle it. Consequently, some of the glucose remains in the nephron tubules and spills out into the urine. The high concentration of glucose within the kidney tubules tends to draw water out of the surrounding tissues by osmosis, resulting in dehydration of the body. Thus, the diabetic experiences almost constant thirst.

In the absence of insulin, the body cells cannot take up glucose from the blood and so must rely on another source of energy. Lack of insulin promotes the release of stored fat, resulting in an extremely high level of fatty acids in the blood. Some of the fats are deposited in the walls of blood vessels, leading to development of arteriosclerosis. Although they can be used by the body cells for energy, fats metabolized in the cells release several acidic by-products into the blood. As these acid compounds accumulate, the pH of the blood may fall to a dangerous level. If the blood pH, which is normally 7.4, falls to 6.9, the central nervous system becomes depressed, and the individual usually passes into a coma. As with the surplus glucose in the blood, some of the excess acid compounds are excreted in the urine.

As mentioned, one of the effects of insulin on the body is to increase the synthesis of proteins. Without insulin, amino acids are continually lost from the cells through the breakdown of body proteins. In other words, a lack of insulin causes depletion of proteins in the body. The diabetic, then, gradually loses weight and becomes emaciated.

The well-known treatment for diabetes mellitus is daily administration of enough insulin to maintain a normal blood glucose level. In addition, the diet must be carefully controlled, particularly the amounts of carbohydrates.

> The pineal gland secretes *melatonin* and perhaps other substances; however, the physiological effects of these secretions on humans are uncertain.

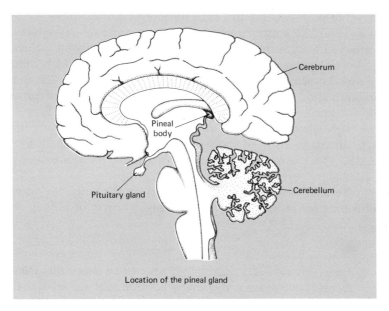

Figure 19.20 Location of the pineal gland.

THE PINEAL GLAND

The pineal is a small endocrine gland located in the forebrain near the area of the thalamus (Fig. 19.20). Knowledge concerning the function of the pineal gland is extremely limited, but it is known to secrete a hormone called ***melatonin,*** which inhibits the development and functioning of the ovaries in rats. Whether or not this also applies to human females is presently unknown. The regulation of melatonin secretion seems to depend upon the amount of light stimulating the brain through the visual pathway. More light causes a decrease in melatonin secretion, whereas less light increases its secretion.

In some lower vertebrates, melatonin is thought to influence pigmentation of the skin. Depending upon the amount of melatonin released, the skin color varies, allowing the organism to assume the protective coloration of its surroundings.

There is some evidence that the pineal gland secretes other hormones or other chemical substances, but at this time concrete evidence concerning their effects on the human body awaits further study.

PROSTAGLANDINS

In the last 25 years, at least 14 related substances called ***prostaglandins*** have been found in many different tissues of the body. These substances are derivatives of fatty acids and exert effects similar to

> *Prostaglandins*, hormone-like substances derived from fatty acids, are produced in many tissues throughout the body.

those of both general and local hormones. For example, prostaglandins stimulate contraction of the uterus, raise or lower blood pressure, reduce gastric secretions, affect the transmission of nerve impulses at the synapse, and inhibit release of female sex hormones. It is now known that the ability of aspirin to reduce fever, pain, and inflammation results from its blocking the synthesis of prostaglandins.

It has been suggested that prostaglandins, depending upon the type of tissue involved, may stimulate or inhibit formation of cyclic AMP. They apparently do this by activating or repressing the enzyme adenyl cyclase.

Because of their numerous effects on the body, the prostaglandins may prove to be useful in treating a variety of human diseases.

REVIEW OF ESSENTIAL CONCEPTS

1. The *endocrine system* consists of specialized glands that release *hormones*. A hormone is a chemical substance produced in one part of the body and transported to another part, where it exerts a specific biological effect.
2. The body structures on which hormones act are called *target tissues* or *target organs*.
3. Hormones are either *steroids* or *proteins*.
4. The *hypothalamus*, located in the forebrain, secretes *releasing* and *inhibitory factors* in response to sensory stimuli. These factors control the secretion of hormones by the pituitary gland.
5. The glands of the endocrine system include the *pituitary, adrenals, ovaries, testes, thyroid, parathyroids, pancreas, thymus,* and *pineal*.
6. Protein hormones and steroid hormones exert control within the cell by different mechanisms. Protein hormones stimulate the formation of *cyclic adenosine monophosphate (cAMP)*, which brings about the biological effect characteristic of a given target organ. Steroid hormones form a complex with a cytoplasmic receptor molecule; the complex enters the nucleus of the target organ cell, where it initiates protein synthesis.
7. The *pituitary gland* consists of two lobes—the *anterior pituitary* and the *posterior pituitary*. Hormones secreted by the anterior pituitary are *growth hormone* (GH), *prolactin, adrenocorticotropic hormone* (ACTH), *thyroid-stimulating hormone* (TSH), *follicle-stimulating hormone* (FSH), and *luteinizing hormone* (LH). Hormones secreted by the posterior pituitary are *antidiuretic hormone* (ADH) and *oxytocin*.
8. Growth hormone causes body cells to grow and multiply, and it blocks the uptake of glucose by the cells. The body then draws on its re-

serve supply of fats for energy, a vital factor during periods of glucose or protein deficiency.
9. Release of growth hormone is regulated by *growth hormone releasing factor* (GRF), secreted by the hypothalamus in response to low levels of protein or blood glucose. As the levels of protein or glucose rise under the influence of GH, the hypothalamus reduces its secretion of GRF.
10. An inadequate secretion of growth hormone during childhood results in *dwarfism;* an excessive secretion causes *gigantism*. An oversecretion of growth hormone during adulthood results in *acromegaly*.
11. The hormone *prolactin* stimulates milk secretion by the breasts. The release of prolactin is normally inhibited by *prolactin inhibitory factor* (PIF), produced in the hypothalamus. Release of PIF is inhibited when the baby sucks at the breast; this allows the secretion of prolactin and the release of milk from the breasts.
12. Antidiuretic hormone (ADH) causes increased permeability of the collecting ducts of the kidneys. This action helps conserve body water. ADH secretion is controlled by the volume and concentration of the blood—low blood volume or high concentration stimulates ADH secretion; the reverse inhibits ADH secretion.
13. ADH also raises arterial blood pressure by causing constriction of small arteries; for this reason, ADH is also called *vasopressin*.
14. The hormone *oxytocin* causes contraction of the pregnant uterus and acts with prolactin to stimulate release of milk from the breasts.
15. Oxytocin is released in response to stretching of the uterus and when a baby suckles the breast.
16. In humans, an *adrenal gland* sits atop each kidney. The outer tissue layer of an adrenal gland is the *cortex;* the inner layer is the *medulla*.
17. The major hormones of the adrenal cortex are *aldosterone* and *cortisol*. Aldosterone increases the rate of sodium reabsorption from the kidney tubules, promotes secretion of potassium from the blood into the tubules, and aids in maintaining blood and tissue fluid volume by reducing urine output.
18. Aldosterone is released from the adrenal cortex when the concentration of sodium ions in the body falls and when the concentration of potassium ions increases.
19. Cortisol causes an increase in the formation of glucose in the liver and raises the blood glucose level. Cortisol also promotes the use of fats for energy.
20. The secretion of cortisol is regulated by *adrenocorticotropic hormone* (ACTH), released from the anterior pituitary. The release of ACTH is controlled by *corticotropin releasing factor* (CRF) from the hypothalamus.

21. Cortisol helps the body adapt to *stress* by conserving glucose and by mobilizing amino acids for protein synthesis.
22. Cortisol and other compounds of the adrenal cortex are *anti-inflammatory agents*.
23. The lack of aldosterone and cortisol results in *Addison's disease*; a high level of cortisol usually results in *Cushing's disease*.
24. The *adrenal medulla* is composed of neurons of the sympathetic nervous system. The hormones of the adrenal medulla are *epinephrine* and *norepinephrine*.
25. Epinephrine and norepinephrine increase the ability of the body to respond to instant emergencies. This involves an increase in the heart rate, elevation of blood pressure, increased rate of cellular metabolism, and so on. These physiological activities prepare the body for the "fight or flight" response.
26. The *thyroid gland* secretes the hormones *thyroxine* and *calcitonin*. Thyroxine causes an increase in the rate of body metabolism and affects overall growth of the body.
27. The secretion of thyroxine is controlled by *thyroid-stimulating hormone* (TSH) from the anterior pituitary; release of TSH is regulated by *thyroid-stimulating hormone releasing factor* (TRF), secreted by the hypothalamus in response to various stressful situations.
28. *Hypothyroidism* in infants causes *cretinism*; adults with hypothyroidism develop *myxedema*.
29. *Hyperthyroidism* may result in *goiter*, an enlargement of the thyroid gland caused by abnormally high amounts of thyroxine in the blood. Goiter may also result from hypothyroidism caused by a deficiency of *iodine* in the diet.
30. The *parathyroid glands* secrete *parathyroid hormone* (PTH), which causes an increase in the level of calcium ions in the body and a decrease in the level of phosphate ions. These actions are accomplished by the effect of PTH on the bones, intestine, and kidneys.
31. *Calcitonin*, released by the thyroid gland, lowers the level of calcium in the blood by preventing the loss of calcium from bones.
32. *Hypoparathyroidism* results in low blood calcium levels that cause *tetany* of the muscles; *hyperparathyroidism* results in the loss of calcium from bones, which become weakened and break easily.
33. As an endocrine organ the *pancreas* secretes the hormones *insulin* and *glucagon*. In the *islets of Langerhans*, the *alpha cells* secrete glucagon, and the *beta cells* secrete insulin.
34. Insulin functions in lowering the blood glucose level; glucagon raises the blood glucose level. Secretion of these hormones is regulated by the level of glucose in the blood—high levels of glucose stimulate the

release of insulin; low levels stimulate the release of glucagon. Accordingly, insulin prevents *hyperglycemia* and glucagon prevents *hypoglycemia.*

35. *Diabetes mellitus,* a hereditary disease, results from destruction of the insulin-secreting cells of the pancreas. Diabetes mellitus is characterized by excess glucose in the blood and urine, constant thirst, weight loss, and acidity of the blood. Treatment involves daily administration of insulin.
36. The *pineal gland* secretes the hormone *melatonin* in response to the amount of light stimulating the brain. The effects of melatonin on human physiology are at present unclear.
37. *Prostaglandins* are derivatives of fatty acids and exert hormone-like effects. Prostaglandins may exert these effects by stimulating or inhibiting the formation of cyclic AMP.

APPLYING THE CONCEPTS

1. Select any protein-type hormone and explain how cAMP could be involved in its mechanism of action.
2. What is the general role of the hypothalamus in endocrine function?
3. Explain the physiological effects of growth hormone during periods of fasting or starvation.
4. Describe the mechanism by which antidiuretic hormone aids in preventing dehydration.
5. What is the relationship between adrenal gland function and physical or mental stress?
6. What factors are involved in the "fight or flight" response?
7. Diagram the negative feedback mechanism involved in regulating the activities of the thyroid gland.
8. Describe the causes and effects of disorders of the thyroid gland.
9. How do the parathyroid glands regulate the levels of calcium and phosphate ions in the body?
10. Explain how insulin prevents hyperglycemia and how glucagon prevents hypoglycemia.
11. What are the symptoms of diabetes mellitus? What factors are responsible for these symptoms?
12. Describe the mechanisms for regulating the blood glucose level.

reproduction

THE ESSENTIAL OBJECTIVES

You have understood this chapter when you are able to:

1. Name and give the functions of the reproductive structures of the human male.
2. Describe the functions of testosterone and explain how its secretion is controlled.
3. Name and give the functions of the reproductive structures of the human female.
4. Outline the human menstrual cycle and explain the regulation of the female sex hormones.
5. Describe the process of fertilization and the factors involved in preparation of the uterus for pregnancy.
6. Describe the available methods of contraception, including sterilization and abortion.

Reproduction involves the passing of genetic information from parent to offspring. In human beings, the reproductive process is a complex interplay of mental, nervous, and hormonal factors that includes the formation of gametes, sexual intercourse, fertilization, and pregnancy.

PREVIEW OF ESSENTIAL TERMS

primary sexual characteristics Those structures of the male and female directly concerned with reproduction, such as the penis and testes of the male and the vagina and ovaries of the female.

secondary sexual characteristics Those characteristics representative of maleness and femaleness, but not directly concerned with reproduction, such as beard growth in the male and breast development in the female.

fertilization Union of the male and female gametes, followed by fusion of their nuclei.

zygote The cell resulting from the union of the male and female gametes; the fertilized ovum.

testosterone The sex hormone of the human male; it is produced in the testes and is responsible for the male sexual characteristics and for maturation of sperm.

estrogens A group of female sex hormones produced primarily in the ovaries and placenta; they are responsible for female sexual characteristics and for preparation of the uterus for pregnancy.

progesterone A female sex hormone produced in the corpus luteum and placenta; its principal function is to prepare the uterus for implantation of the embryo.

human chorionic gonadotropin (HCG) A hormone released by the developing embryo early in pregnancy; HCG stimulates the corpus luteum to continue its secretion of estrogens and progesterone.

corpus luteum A yellowish structure that forms after the ovum is discharged from the follicle; it secretes progesterone and the estrogens.

pronucleus The nucleus of either the male or the female gamete; at fertilization, the two pronuclei fuse to form the nucleus of the zygote.

> *Asexual reproduction* does not require a mate and results in offspring that are genetically identical to the parent.

INTRODUCTION

According to the cell theory, the cells of all living things contain genetic information that is passed to future cells. Succeeding generations of offspring receive this information through the process of ***reproduction.***

The perpetuation of any living species is dependent upon its capacity to reproduce others of its own kind. Indeed, evolutionary success is ultimately a measure of species survival. Throughout the history of life, an essential feature of survival has been the development of an adaptive reproductive method. Although they may vary from one species to another, all reproductive methods come under the general heading of either ***asexual*** reproduction or ***sexual*** reproduction.

Asexual reproduction results in offspring that are genetically identical to the parent. As discussed in Chapter 13, this lack of genetic variability affords a species little adaptive advantage in a changing environment. However, asexual reproduction is a rather rapid process and is convenient in that it does not require finding a mate. Thus, asexual reproduction offers some advantages and in a stable environment produces offspring equally well adapted as the parent.

Lower organisms such as protozoa and algae may reproduce asexually by ***fission,*** i.e., by splitting into two or more separate individuals (Fig. 20.1*A*). Species of yeasts, *Hydra,* and sponges may undergo ***budding,*** in which the young buds develop from an outpocketing of the body wall of the parent (Fig. 20.1*B*). A more unusual form of asexual reproduction—often considered a modification of sexual reproduction—is ***parthenogenesis*** (Gk. "virgin origin"), which refers to the development of a new individual from an unfertilized egg. Among honeybees, for example, unfertilized eggs develop into males, or drones. Many of the worms, echinoderms, and other invertebrates reproduce asexually by ***fragmentation.*** If large pieces of these animals are broken off, each piece will regenerate a complete new organism.

In the land plants and fungi, asexual reproduction occurs by ***spore formation.*** Part of the life cycle of both nonvascular and vascular plants depends upon the formation of spores that arise asexually from tissues of the sporophyte generation. After release from the parent, fungal spores, such as those of *Rhizopus,* germinate to form the adult organism (Fig. 20.1*C*).

In contrast to the methods just outlined, sexual reproduction involves the union of genetically different male and female sex cells, or *gametes.* The offspring, then, are not genetically identical to the parents but constitute a combination of parental traits. However, in many lower organisms, distinct sexes cannot always be distinguished. For example, *Paramecium* and *Spirogyra* reproduce sexually by ***con-***

> Some lower animals reproduce sexually by *conjugation*.

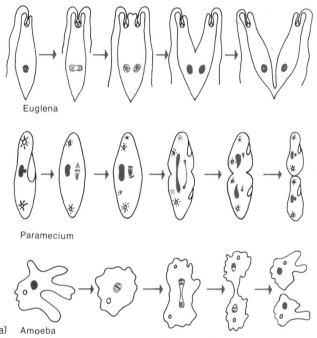

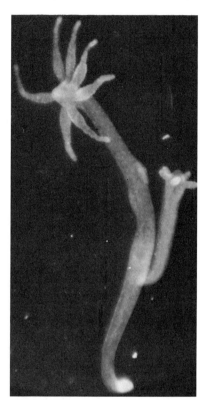

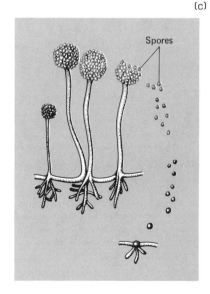

Figure 20.1 (a) Asexual reproduction in *Euglena, Paramecium,* and *Amoeba.* In each, one cell divides mitotically to give rise to two cells. (b) Budding in animals. Budding of a new individual from the stalk of the freshwater coelenterate *Hydra.* In a few days the bud will develop tentacles like its parent. Eventually it will break free and settle independently. (c) Asexual reproduction in *Rhizopus.* Sporangia form atop aerial hyphae, and the spores can be carried away by air currents. ((a) from Villee, C. A.: *Biology.* 7th ed. Philadelphia: Saunders College Publishing, 1977, p. 587; (b) from Clark, M. E.: *Contemporary Biology.* 2nd ed. Philadelphia: Saunders College Publishing, 1979, p. 375.)

jugation (see Chapter 9); in these instances, the conjugating pairs are simply referred to as opposite mating types (Fig. 20.2). Nonetheless, there is fusion of nuclear material, producing offspring genetically different from the parents. Many of the flatworms, tapeworms, and a few other invertebrate groups are **hermaphroditic.** Hermaphrodites are organisms that possess both male and female sex organs and therefore can produce both sperm and eggs. Although these organisms are capable of self-fertilization, this usually does not occur if a mate is available.

The most advanced and complex reproductive methods are found among the higher animals, especially the placental mammals. Sexual reproduction in these animals almost always involves the production of a large, nonmotile egg, or *ovum,* which is fertilized by a small, highly motile *sperm.* While containing genetic information in

Reproduction

Figure 20.2 Sexual reproduction in *Paramecium*. Two individuals with diploid micronuclei undergo conjugation (first figure). After meiosis (second and third figures), three of the nuclei degenerate, and the fourth nucleus divides by mitosis (fourth figure). The organisms undergo mutual fertilization: One nucleus passes from each organism to the other (fifth figure). This is followed by fusion of the haploid nuclei to form a new diploid nucleus in each of the two organisms (last figure). The original macronuclei disappear. The two new diploid nuclei divide several times by mitosis and eventually establish both the new micronuclei and the new macronuclei. (From Villee, C. A.: *Biology*. 7th ed. Philadelphia: Saunders College Publishing, 1977, p. 608.)

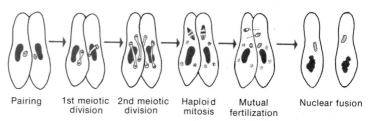

the form of deoxyribonucleic acid (DNA), the ovum is also a reservoir of nutrients that is used by the embryo during its early development. Sperm, on the other hand, are essentially packets of DNA, propelled to their target by tiny flagella.

The union of sperm and ovum is called *fertilization*, a process that produces the single-celled *zygote* (Fig. 20.3). The activities required of both sexes in producing the zygote constitute the science of reproductive physiology. We shall examine these activities as they occur in human beings.

Figure 20.3 The union of egg and sperm. (From Ville, C. A.: Biology. 7th ed. Philadelphia: Saunders College Publishing, 1977, p. 610.)

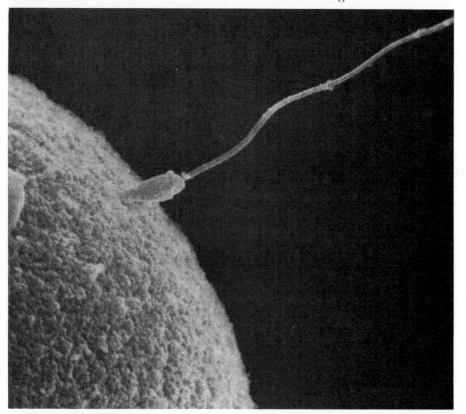

The Male Reproduction System

> *Fertilization*, the union of sperm and ovum, produces the single-celled *zygote*.

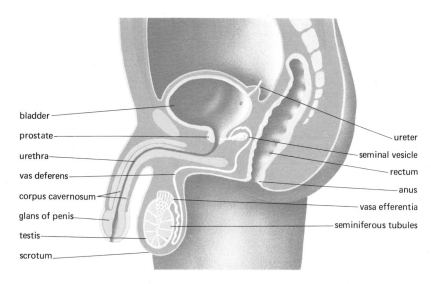

Figure 20.4 Reproductive organs of the human male. (From Ebert, J. D. Loewy, A. G., Miller, R. S., and Schneiderman, H. A.: *Biology.* New York: Holt, Rinehart and Winston, 1973; p. 463.)

Reproductive function in the human male involves the production of sperm, or *spermatogenesis* (refer to Chapter 7, if necessary), and fertilization of the ovum through sexual intercourse.

The primary sex organs (Fig. 20.4) of the male are the *testes* (sing. testis), which consist of the **seminiferous** ("sperm bearing") *tubules* (Fig. 20.5). Beginning at puberty and continuing throughout life, these tubules produce haploid sperm by meiotic cell division (spermatogenesis). From the seminiferous tubules, the sperm move out of the testes into the highly coiled **epididymis,** one of which lies on the outer surface of each testis. Here the sperm mature and are temporarily stored. Each epididymis widens into a **vas deferens,** which conducts the sperm to the paired **seminal vesicles** (Fig. 20.6). These saclike structures release a mucus-like fluid containing sugars, amino acids, prostaglandins, and other organic compounds. Some of these compounds provide nutrition for sperm after they enter the female reproductive tract. From the seminal vesicles, sperm pass into the **urethra,** which extends downward through the penis (Fig. 20.7). Other glands, including the **prostate** and **Cowper's glands,** also release secretions into the urethra.

The prostate gland, located next to the seminal vesicles, releases an alkaline fluid that mixes with the sperm in the urethra. The alkaline pH of the prostate fluid is thought to be necessary to neutralize the highly acid secretions of the female reproductive tract. This is important because functionally active sperm require a surrounding pH of about 6.5.

THE MALE REPRODUCTIVE SYSTEM

Figure 20.5 Cross section of human testis showing seminiferous tubules. Interstitial cells are located in the tissue between the tubules. (From Leeson, C. R., and Leeson, T. S.: *Histology.* 10th ed. Philadelphia: W. B. Saunders Co., 1976, p. 447.)

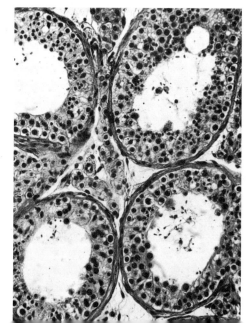

> *Cowper's glands* release a thick mucus in response to sexual stimulation.

REPRODUCTION

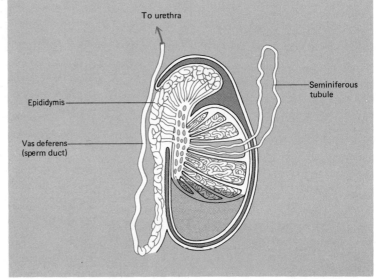

Figure 20.6 General structure of a testis. Sperm produced in the seminiferous tubules pass into a network of small tubules that leads into the epididymis. After a brief maturation period, the sperm enter the vas deferens and ultimately are ejaculated through the urethra.

The small, pea-shaped Cowper's glands—after William Cowper (1666–1709), English anatomist—are located just below the prostate gland. Cowper's glands release a rather thick mucus in response to sexual stimulation. The fluids secreted by the seminal vesicles, prostate gland, and Cowper's glands, along with the sperm, constitute *semen*.

Figure 20.7 Structure of the human penis.

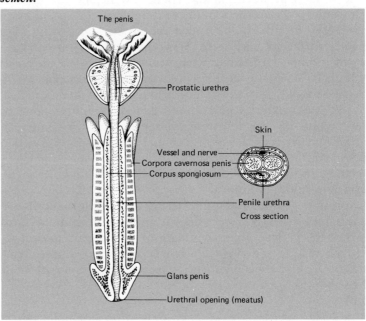

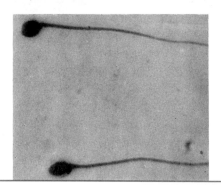

Figure 20.8 Photomicrograph of human sperm. (From Silverstein, A.: *The Biological Sciences.* New York: Holt, Rinehart and Winston, 1974, p. 445)

There are approximately 3.5 ml of semen ejaculated from the urethra during intercourse; normally, this quantity of semen contains an average of 400 million sperm (Fig. 20.8). However, when the number falls below about 70 million, the man is usually sterile. The sperm contain various enzymes that apparently are necessary for breaking down the cell layers and mucous coat surrounding the ovum. If the number of sperm is insufficient for this task, fertilization cannot occur, and the male is considered to be sterile.

The testes are suspended within the saclike *scrotum* at the base of the penis (Fig. 20.9). About a month prior to birth, the testes of the developing fetus normally descend from the lower abdominal cavity into the scrotum. Failure of the testes to descend from the abdominal cavity is called *cryptorchidism* (Gk. "hidden testes"). As a consequence, cells within the seminiferous tubules degenerate, and spermatogenesis cannot occur.

The location of the testes outside the body is necessary because the normal internal temperature of the body prevents the production of active sperm. The temperature within the scrotum is several degrees cooler than the normal 37° C body temperature, a condition that must be maintained for the testes to function properly. To ensure a relatively constant temperature of the testes, the skin of the scrotum relaxes when the temperature is warm, and the testes are suspended away from the body. In cold temperatures the scrotum contracts, pulling the testes closer to the body.

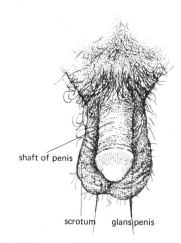

Figure 20.9 Male genitalia.

MALE SEX HORMONES

The attributes associated with maleness, such as a deep voice, heavy muscular development, beard growth, and so on, are referred to as *secondary sexual characteristics*. The external sexual organs—penis, testes, and scrotum—are the *primary sexual characteristics*. Development of both the primary and the secondary characteristics is influenced predominantly by the male sex hormone *testosterone,* which is produced by the specialized *interstitial cells* of the testes.

The production and secretion of testosterone are controlled by *interstitial cell–stimulating hormone* (ICSH), released from the anterior pituitary gland. (This hormone is also known as *luteinizing hormone,* or LH, named originally for its function in the female.) When released into the bloodstream, ICSH is carried to the interstitial cells of the testes, where it stimulates the production of testosterone. After its release into the blood, testosterone is carried away from the testes and out to the body cells, where it maintains the male sexual characteristics. A deficiency of testosterone impairs the functioning of the internal reproductive structures and affects the male secondary sexual characteristics. It is known, for example, that if *castration* (removal of the testes) is performed before puberty, the secondary sexual characteristics fail to develop. Even after puberty, castration causes partial disappearance of adult sexual characteristics. In addition to its masculinizing effects, testosterone is necessary in spermatogenesis. After spermatogenesis is initiated, a process to be discussed further on, testosterone is required in small amounts to complete the process.

Regulation of ICSH Secretion. The secretion of ICSH from the anterior pituitary is controlled by the hypothalamus. In response to nervous stimuli, the hypothalamus produces *luteinizing hormone releasing factor* (LRF), which diffuses into the blood and is carried down to the anterior pituitary gland. In turn, LRF stimulates the release of ICSH, which enters the circulating blood and is transported to the testes. The interstitial cells of the testes are then stimulated to secrete testosterone. As the level of testosterone in the blood rises, the hypothalamus is inhibited and reduces its secretion of LRF. Consequently, less ICSH is released from the anterior pituitary, and the level of testosterone in the blood is stabilized. Should the level of testosterone in the blood begin to fall, the hypothalamus is again stimulated, LRF and ICSH are released, and the level of testosterone is brought back to normal. Thus, through the same negative feedback mechanism characteristic of other hormonal systems, the body is protected against an inadequate or excessive production of testosterone.

Regulation of Spermatogenesis. In the human male, the production of sperm is a constant, uninterrupted process. This is not true in all vertebrate groups, however, since animals such as deer, frogs, freshwater fish, and many others produce sperm only during mating season.

You will recall from Chapter 7 that the seminiferous tubules contain large diploid cells, the spermatogonia, that undergo meiosis to produce the haploid sperm. The initiation of this process is controlled by *follicle-stimulating hormone* (FSH), secreted by the anterior pituitary gland. (This hormone, like LH, is named for its function in the female.) When released into the blood, FSH is carried to the testes, where it stimulates the seminiferous tubules to produce sperm.

The secretion of FSH is controlled by the hypothalamus in essentially the same manner as the secretion of ICSH. When stimulated, the hypothalamus produces *follicle-stimulating hormone releasing factor* (FRF), which passes to the anterior pituitary gland, causing the secretion of FSH. In some unknown manner, spermato-

> Male sterility can be caused by the venereal diseases syphilis and gonorrhea or by the viral disease mumps.

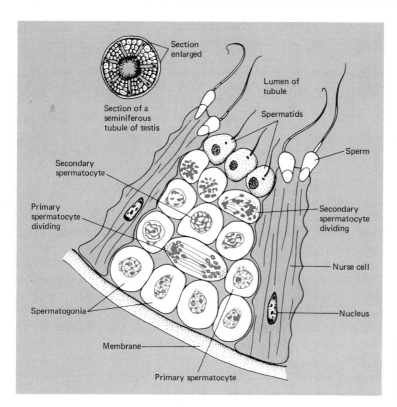

Figure 20.10 Diagram of part of a section of a human seminiferous tubule to show the stages in spermatogenesis and in the transformation of a spermatid into a mature sperm.

genesis exerts feedback control on the secretion of FSH. That is, if the rate of spermatogenesis is adequate, the secretion of FSH is inhibited; conversely, if the rate slows, the secretion of FSH increases.

Although FSH initiates spermatogenesis, mature sperm are not formed unless small amounts of testosterone are present in the seminiferous tubules. Thus, spermatogenesis is dependent upon the combined effects of FSH and ICSH acting on the seminiferous tubules (Fig. 20.10).

MALE SEXUAL DISORDERS

Disorders of sexual function in the human male may result from either physical or psychological factors.

As we have mentioned, sterility results from an inadequate production of sperm. This may occur in males who are otherwise normal in every way. However, sterility may also result from the venereal diseases *syphilis* and *gonorrhea* or from the viral disease *mumps*. Both syphilis and gonorrhea are caused by bacteria, which are almost always transmitted through sexual intercourse.

> *Oogenesis* in the human female results in the production of ova in the primary follicles of the ovaries.

A disorder more often associated with psychological problems is *impotence,* the inability to attain erection of the penis.

In some older men, there is degeneration of the prostate gland, accompanied by a decline in the production of testosterone by the testes. In other cases, the prostate gland enlarges, causing constriction of the urethra and painful urination.

The prostate gland is also a rather common site of cancer in men over 50 years of age. However, cancer of the prostate gland is responsible for only about 2 or 3 percent of all male deaths.

THE FEMALE REPRODUCTIVE SYSTEM

Reproductive function in the human female involves the production of ova, or *oogenesis;* sexual intercourse, by which the sperm are received; preparation of the uterus for nourishment of the embryo during pregnancy; and giving birth.

The primary sex organs of the female are the **ovaries,** which are located near the lateral walls of the pelvic cavity (Fig. 20.11). Just before birth each ovary contains an average of about one-half million **primary follicles** (L. "little bag"), masses of tissue in which the ova are produced (Fig. 20.12). The original cells that develop within the follicles are the diploid **oogonia,** which will undergo meiotic cell division to produce the haploid ova (oogenesis). However, these cells enlarge before birth to become the **primary oocytes** ("egg cells"). A primary oocyte will not undergo the first meiotic division until shortly before it is released from the ovary. Normally, only one follicle with its primary oocyte will develop to this stage of maturity each month. Thus, in her entire reproductive lifetime, from the age of about 12 until the age of about 45, a human female will produce only approximately 450 mature follicles out of the original 1 million. At menopause, when the monthly cycles cease, the remaining follicles slowly degenerate.

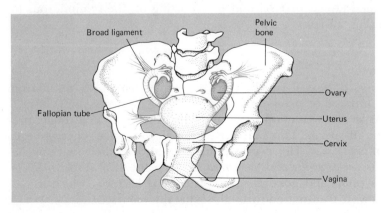

Figure 20.11 The female reproductive organs as they are situated within the pelvic bone.

Color Plate 111 Skeletal muscle. (Photograph by Carolina Biological Supply Co.)

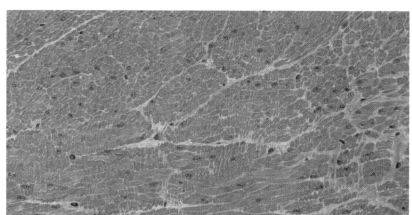

Color Plate 112 Smooth muscle. (Photograph by Carolina Biological Supply Co.)

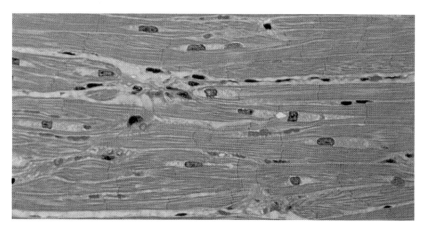

Color Plate 113 Cardiac muscle. (Photograph by Carolina Biological Supply Co.)

Color Plate 114 Innervation of skeletal muscle fibers by the terminal branches of a motor neuron. (Photograph by Carolina Biological Supply Co.)

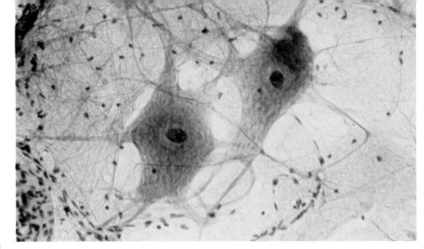

Color Plate 115 Vertebrate neurons. (Photograph by Carolina Biological Supply Co.)

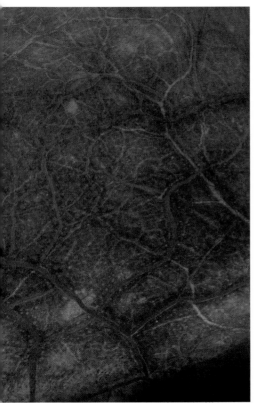

Color Plate 116 Capillary beds in human skin. (Photograph by Bruce Russell, Biomedia Assoc.)

Color Plate 117 Human blood smear. Three white blood cells are scattered among the pink-stained red blood cells. (Photograph by Carolina Biological Supply Co.)

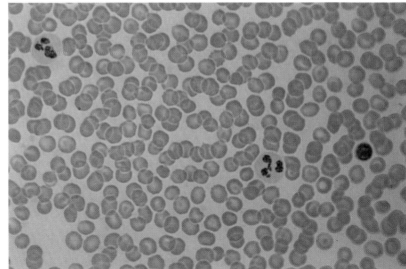

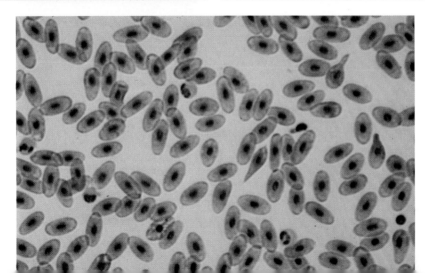

Color Plate 118 Frog red blood cells. The mature red blood cells of frogs and other vertebrates (except the mammals) retain their nuclei. (Photograph by Carolina Biological Supply Co.)

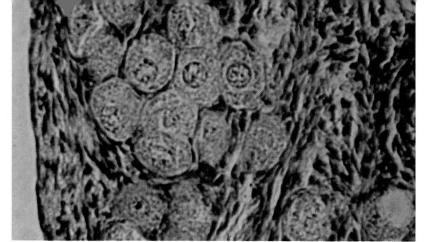

Color Plate 119 Primary follicles in the ovary. (Photograph by Carolina Biological Supply Co.)

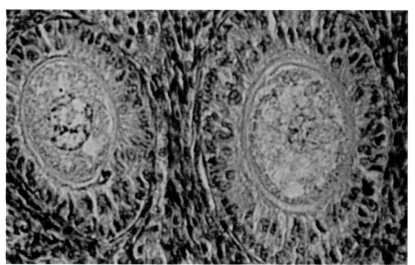

Color Plate 120 Growing follicles. (Photograph by Carolina Biological Supply Co.)

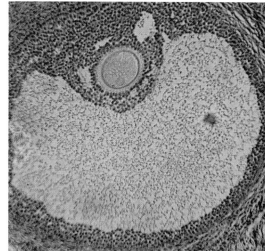

Color Plate 121 Mature follicle with oocyte. (Photograph by Carolina Biological Supply Co.)

Color Plate 122 Corpus luteum, a degenerating follicle that forms after ovulation. (Photograph by Carolina Biological Supply Co.)

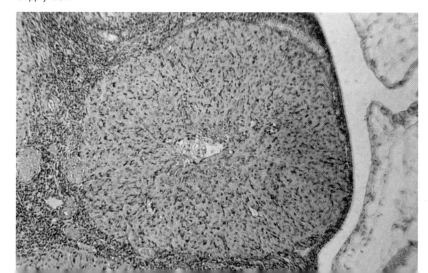

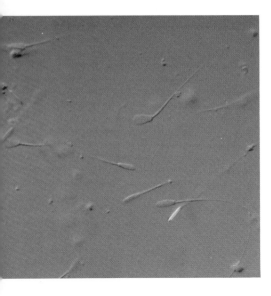

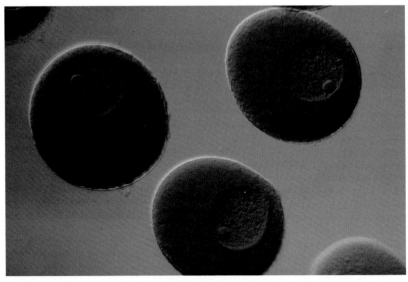

Color Plate 123 Bull sperm. (Photograph by Carolina Biological Supply Co.)

Color Plate 124 Sea urchin eggs. One of the fundamental questions in developmental biology is: How does a fertilized egg give rise to hundreds of kinds of cells? (Photograph by Bruce Russell, Biomedia Assoc.)

Color Plate 125 Zygotes from conjugating *Spirogyra* filaments. (Photograph by Bruce Russell, Biomedia Assoc.)

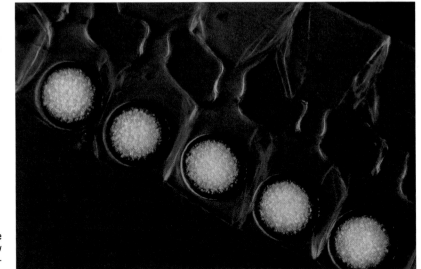

Color Plate 126 A mutation, or change in the genetic material, can produce new phenotypes. (Photograph by Bruce Russell, Biomedia Assoc.)

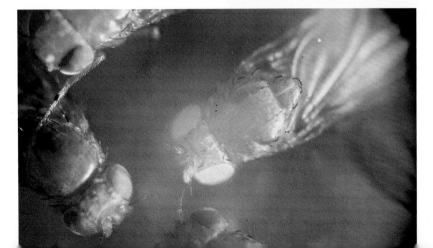

The Female Reproductive System

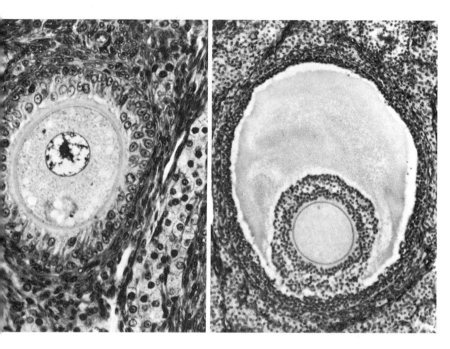

Figure 20.12 A growing follicle containing an immature ovum. A more mature follicle, showing the still immature ovum surrounded by follicle cells. Growth of the follicle is influenced by follicle-stimulating hormone released by the anterior pituitary gland. (From Leeson, C. R., and Leeson, T. S.: *Histology*. 10th ed. Philadelphia: W. B. Saunders Co., 1976, p. 419.)

In addition to the ovaries, the other principal female reproductive organs include the *vagina,* the *fallopian tubes,* and the *uterus* (Fig. 20.13).

The vagina receives the penis during sexual intercourse and serves as the birth canal (Fig. 20.14). The opening into the vagina is partially blocked by a thin membrane called the **hymen.** Although

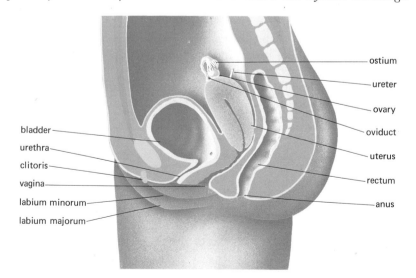

Figure 20.13 Reproductive organs of the human female.

660
Reproduction

> The *estrogens* and *progesterone* are hormones responsible for the development of the primary and secondary sexual characteristics in the human female.

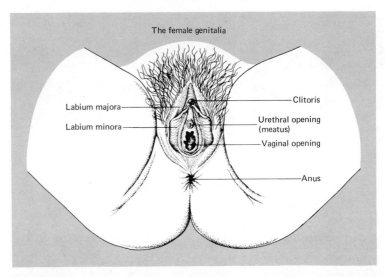

Figure 20.14 The female genitalia.

once considered the anatomical badge of virginity, the hymen may be ruptured by means other than sexual intercourse. The most common cause of rupture is physical exercise. The tissue lining the vagina secretes a thin lubricating mucus, which has an acid pH of about 4 or 5.

The fallopian tubes, or oviducts, extend from the ovaries to the uterus. After an ovum has been expelled from the ovary (a process called ovulation), it moves into the upper portion of the tube, where fertilization normally occurs. Cilia lining the wall of the fallopian tube, aided by peristaltic waves, help propel the ovum toward the uterus. However, if fertilization does not occur, the ovum usually degenerates somewhere within the tube.

The uterus is a thickly muscled, pear-shaped organ located above the vagina. The lower, more narrow section of the uterus, the **cervix,** communicates with the vagina through a small opening. The fallopian tubes enter the uterus on both sides of its upper section. The principal function of the uterus is to sustain the developing embryo throughout pregnancy.

THE FEMALE SEX HORMONES AND MENSTRUAL CYCLE

The development of the primary and secondary sexual characteristics of the human female is brought about by two types of hormones, the **estrogens** and **progesterone.** The estrogens are responsible for enlargement of the sexual organs after puberty, the same function served by testosterone in the male. The influence of estrogens on the secondary sexual characteristics is evidenced by breast develop-

> The reproductive *menstrual cycle* of the human female occurs on the average of once every 28 days.

ment; broadening of the pelvis; deposition of fat in the breasts, hips, and thighs; and distribution of body hair.

The most important function of progesterone is to prepare the lining of the uterus for implantation of the embryo. For this reason, progesterone is sometimes referred to as the "hormone of pregnancy." In addition, progesterone causes the breasts to enlarge, primarily as a result of increased development of the milk-secreting cells.

The Menstrual Cycle. On the average of once every 28 days (the number may vary widely), the human female normally releases a single egg cell, an activity accompanied by changes in the uterus that prepare it for implantation of the embryo. These reproductive *menstrual cycles* begin around the age of 12 and continue until age 45. Associated with the menstrual cycle are drastic changes in the amounts of female hormones released (Figs. 20.15 and 20.16). As in the male, secretion of the female sex hormones is controlled by releasing factors from the hypothalamus and tropic hormones from

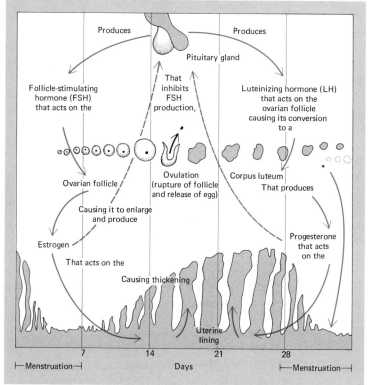

Figure 20.15 The interactions of the pituitary hormones and female sex hormones in the menstrual cycle.

*Disintegration of corpus luteum results in breakdown of uterine lining.

> During maturation of the follicle, the primary oocyte divides into the *secondary oocyte* and *polar body*.

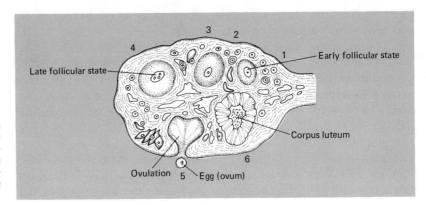

Figure 20.16 A representation of a section through the human ovary. In stages 1 to 4, the ovum is shown maturing. In stage 5, ovulation is shown taking place. In stage 6, the corpus luteum is shown. Mature follicles may be several millimeters in diameter.

the anterior pituitary gland. These secretions and their specific actions will be considered in the following discussion of the menstrual cycle.

During early childhood, the primary follicles in the ovaries do not mature because of a lack of sufficient hormonal stimulation. However, beginning at about the age of 12, the hypothalamus increases its production of *follicle-stimulating hormone releasing factor* (FRF), which causes the anterior pituitary gland to secrete *follicle-stimulating hormone* (FSH). In addition, the hypothalamus also produces *luteinizing hormone releasing factor* (LRF), which causes the anterior pituitary to secrete a small amount of *luteinizing hormone* (LH). As its name indicates, FSH stimulates the follicle, resulting not only in enlargement and maturation of the follicle but in enlargement of the primary oocyte as well. Normally, as mentioned, only one follicle matures during the cycle. This same follicle-stimulating process occurs every month in the nonpregnant female until the end of her reproductive years.

As the follicle continues to grow and mature, it is also secreting increasing amounts of estrogens. After their release into the circulating blood, the estrogens act to develop and maintain the female sexual characteristics. In addition, the estrogens bring about a thickening of the epithelial and muscular lining of the uterus. The high level of estrogens in the blood triggers the feedback mechanism, causing the hypothalamus to slow its production of FRF. This reduces the secretion of FSH, which prevents the maturation of a new follicle. However, FSH secretion increases again near the fourteenth day of the cycle, by which time the developing follicle has completely matured. During the maturation process, the primary oocyte completes its first meiotic division to form two cells: the large *secondary oocyte* and a tiny *polar body*. The polar body may later divide to form two polar bodies, both of which eventually disintegrate.

> *Ovulation* occurs when the follicle ruptures and the secondary oocyte is released.

With maturation of the follicle and elevation of the level of estrogens, the anterior pituitary begins to secrete luteinizing hormone (LH) in large quantities. When LH reaches the ovary, it acts in combination with FSH to cause rupture of the follicle and release of the secondary oocyte, a process known as ***ovulation***. In a 28-day menstrual cycle, ovulation normally occurs on the fourteenth day of the cycle.

After expulsion from the ovary, the secondary oocyte is drawn into the fallopian tube by currents set up by beating cilia and perhaps by chemical means. As mentioned, if fertilization does not occur, the oocyte will degenerate within the tube. However, if the oocyte is fertilized within about 24 hours, it will then undergo the second meiotic division to produce the haploid ovum. Twenty-four hours after ovulation, the secondary oocyte is no longer viable and begins to degenerate.

After ovulation, the empty follicle in the ovary is transformed physically and chemically into the ***corpus luteum*** ("yellow body"). This transformation is influenced predominantly by LH. As the corpus luteum develops, it secretes large amounts of estrogens and progesterone. Progesterone, the "hormone of pregnancy," acts on the lining of the uterus, preparing it for possible implantation of the embryo. Under the influence of progesterone, the uterus becomes thick and spongy, developing a rich network of blood capillaries and nutrient-secreting glands. Moreover, progesterone acts with the estrogens to inhibit release of FSH from the pituitary gland.

If fertilization does *not* occur, the corpus luteum soon begins to degenerate. As it does, the amounts of estrogens and progesterone in the blood rapidly decline. As a consequence, the lining of the uterus can no longer be maintained, and its tissues soon disintegrate. These tissues, along with blood and fluids from ruptured capillaries, pass downward and out through the vagina as the ***menstrual flow***. Menstruation usually occurs on the twenty-eighth day of the cycle and continues for 3 to 7 days before bleeding stops.

With the rapid decline in the levels of estrogens and progesterone in the blood, there is no longer feedback inhibition of the hypothalamus, and the anterior pituitary begins to secrete FSH and LH. As before, these hormones stimulate the maturation of a new follicle, and another cycle is underway.

THE SEXUAL ACT

Sexual intercourse is normally preceded by physical and psychic stimulation that results in changes in the reproductive structures of both sexes. In the male, sexual excitement first causes erection of the penis. Erection results from nerve impulses that travel along par-

> Sperm remain viable in the fallopian tubes for only about 48 hours.

asympathetic nerves leading from the spinal cord to the penis. These impulses dilate arteries in the penis, allowing them to fill with blood. In addition, Cowper's glands are stimulated to release mucus, which lubricates the lining of the urethra.

In the female, sexual excitement sets up parasympathetic nerve impulses that travel from the spinal cord to the vaginal area. A small, sexually sensitive organ, the *clitoris,* located just above the urethral opening, becomes swollen as its arteries fill with blood. Surrounding tissues also swell, and there is a greatly increased secretion of mucus from the vaginal lining. In both sexes, muscle tension increases and blood pressure rises; the heart rate accelerates, and breathing becomes shallow and rapid.

During intercourse, friction between the penis and vagina produces intense stimulation that causes reflex centers in the spinal cord to send out sympathetic impulses to the genital organs. In the male, the impulses cause contraction of the epididymis, vas deferens, seminal vesicles, and prostate gland, all of which expel their secretions into the urethra. As the secretions mix with sperm in the urethra, nerve impulses are sent to the skeletal muscles in the penis, causing rhythmic contractions that ejaculate the semen from the urethra to the outside. This is the male orgasm, or climax.

In the female, continued stimulation increases the flow of blood to the vaginal area, which becomes even more swollen. Although the clitoris retracts near the height of sexual excitement, it is thought to be important in causing that excitement. Once the sexual structures receive sufficient stimulation, sympathetic impulses from the spinal cord bring about rhythmic contractions of skeletal muscles in the vaginal walls, and in the muscular walls of the uterus and fallopian tubes. This is the female orgasm, or climax.

Thus, in both sexes, orgasm is a reflex action that results in muscular response. Soon after orgasm, blood flows out of the sexual structures, muscles relax, and the body returns to its normal physiological state.

FERTILIZATION AND PREPARATION OF THE UTERUS

After sexual intercourse, sperm pass through the opening at the cervix and then move through the uterus into the fallopian tubes. Aided by rhythmic contractions of the uterus and fallopian tubes, as well as by their own flagella, sperm arrive at the upper ends of the fallopian tubes within about 5 minutes after their release. Perhaps no more than one-half of 1 percent—about 2 million—of the sperm actually reach the site of fertilization. Within the fallopian tubes, sperm remain viable for only about 48 hours. As pointed out earlier, the secondary oocyte remains viable for only about 24 hours after

Fertilization is complete when the *male pronucleus* fuses with the *female pronucleus*, producing the zygote.

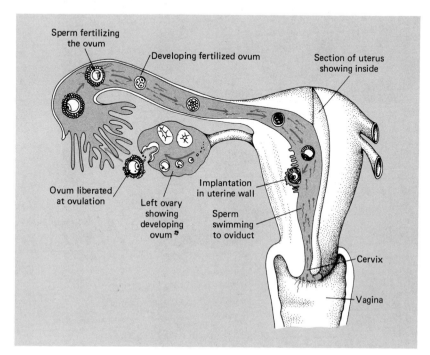

Figure 20.17 Events during conception. The released ovum, picked up by the ciliated cells on the fingers of the oviduct, is soon fertilized by a sperm. As it travels toward the uterus, the zygote divides several times. In a few days, it reaches the receptive uterus, where it implants, forming a hormone-secreting placenta.

ovulation. Therefore, assuming that ovulation occurred on the fourteenth day of the menstrual cycle, intercourse must occur either on day 13, 14, or 15 of the cycle if fertilization is to take place (Fig. 20.17).

Upon reaching the secondary oocyte, the sperm release enzymes that break down the mucous coat and layers of follicle cells surrounding the oocyte. One of the sperm finally penetrates to the surface of the oocyte; this excludes all other sperm, which, though still attempting to penetrate, cannot do so and eventually die. Once a single sperm reaches the oocyte, a "fertilization membrane," impenetrable by other sperm, forms around the oocyte. After it enters the secondary oocyte, the sperm usually loses its flagellum, and its head expands to form the **male pronucleus.** Entry of the sperm stimulates the oocyte to undergo the second meiotic division, thereby completing oogenesis. The secondary oocyte divides to produce the large haploid ovum and a small polar body, which eventually disintegrates. The ovum contains the **female pronucleus,** which fuses with the male pronucleus, completing fertilization. The male and female pronuclei each contain 23 chromosomes; thus, at fertilization, the newly formed *zygote* is a 46-chromosome composite of paternal and maternal DNA, the first cell of a new and unique human being (Fig. 20.18).

> Human chorionic gonadotropin (HCG), released by embryonic cells, prevents the degeneration of the corpus luteum, which secretes the large amounts of estrogens and progesterone needed to maintain the uterine lining.

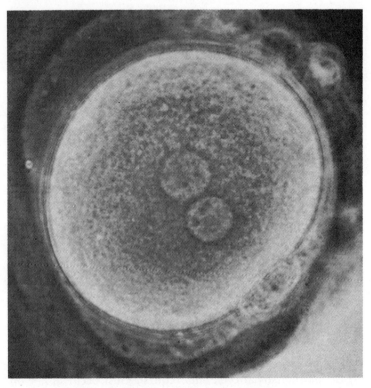

Figure 20.18 Fertilization of the human egg. The sperm head enters the egg, swells, and is approaching the egg nucleus in the photograph. (From Edwards and Fowler: Human embryos in the laboratory. Scientific American, December, 1970. Courtesy of R. G. Edwards.)

During the second half of the menstrual cycle, the corpus luteum, you will recall, secretes large amounts of the estrogens and progesterone. These hormones act to develop the lining of the uterus in preparation for implantation of the embryo, and they suppress the release of FSH and LH from the anterior pituitary gland. If fertilization does not occur, the corpus luteum degenerates, and the levels of estrogens and progesterone in the blood decline rapidly. Consequently, the lining of the uterus is sloughed off, FSH and LH secretion resumes, and a new cycle begins.

However, if the ovum is fertilized, the lining of the uterus must be maintained for pregnancy to continue. How is this maintenance accomplished? Part of the answer involves the developing embryo. After the embryo implants in the wall of the uterus, some of the embryonic cells begin to secrete a hormone called **human chorionic gonadotropin** (HCG). When released into the blood HCG acts on the corpus luteum to prevent its degeneration. The corpus luteum, then, continues to secrete large amounts of estrogens and progesterone, which stimulate continued growth and development of the uterine lining. During the first 2 months of pregnancy, the embryonic cells secrete large amounts of HCG, but secretion declines con-

> From about 3 months after fertilization until birth, the *placenta* secretes the estrogens and progesterone needed to maintain the uterine lining.

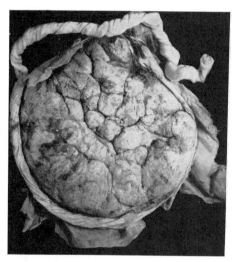 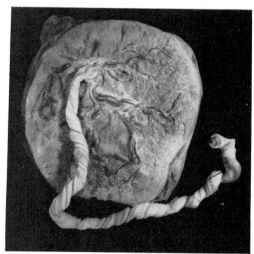

Figure 20.19 Photographs of the maternal (left) and fetal (right) surfaces of a human placenta at the end of a normal pregnancy. (From Greenhill, J. P.: *Obstretrics*. 13th ed. Philadelphia: W. B. Saunders Co., 1965.)

siderably by the end of 3 or 4 months. This results in a reduction in the secretion of estrogens and progesterone by the corpus luteum. By this time, however, the lining of the uterus and the tissues surrounding the embryo have formed the **placenta,** which secretes its own estrogens and progesterone (Fig. 20.19). Thus, from about 3 months after fertilization until birth, the uterine lining is maintained primarily by estrogens and progesterone secreted by the placenta.

We have mentioned that fertilization normally takes place high in the fallopian tube and that the embryo moves down the tube to the uterus, where it implants. Instances arise, however, in which the embryo becomes implanted elsewhere, resulting in an *ectopic* (Gk. "out of place") *pregnancy*. For example, the embryo may implant in the tissues of the ovary, in the abdominal cavity, or, more commonly, in the fallopian tube. A fallopian tube is only about ¼ inch in diameter; as the embryo grows, the tube usually ruptures, causing massive hemorrhaging. The embryo and the fallopian tube must then be removed surgically. Ectopic pregnancies rarely come to term because the embryo does not receive adequate nutrients that normally would be obtained via the placenta.

CONTRACEPTION

Contraception literally means "against conception" and refers to any method that will *prevent* pregnancy. Although there are many contraceptive methods available, all of them have certain drawbacks, both physical and psychological.

The most widely used contraceptives for women include "the pill," intrauterine devices (IUDs), and the diaphragm.

> The diaphragm, a small rubber cap that fits over the cervix, prevents sperm from entering the uterus.

The oral birth control pill is composed of synthetic estrogens and progesterone. These pills are usually taken for 21 days, beginning on the fifth day of the menstrual cycle. By maintaining a high level of estrogens and progesterone in the blood, the birth control pills normally prevent ovulation. You will recall that these hormones, in high concentrations, trigger feedback to the brain, preventing the secretion of the hypothalamic releasing factors. Without ovulation, of course, there is no oocyte available, and fertilization cannot occur.

If taken according to the prescribed schedule, the oral pill is almost 100 percent effective in preventing pregnancy. Women who become pregnant while taking the pill usually do so because of their failure to follow the schedule. In some women, the pill produces unpleasant side effects, such as nervousness, irritability, weight gain, nausea, or darkening of the skin. A very small percentage of women (about 3 or 4 of every 100,000) taking the pill develop blood clots in various parts of the body. As a health hazard, however, this is a much lower risk than the 25 deaths per 100,000 women that result from the complications of pregnancy and childbirth.

The IUDs are small plastic or metal coils that must be inserted into the uterus by a physician (Fig. 20.20). Although it is not known how they function, the hypothesis is that IUDs prevent implantation of the embryo in the lining of the uterus. They are about 99 percent effective, although pregnancy can occur if the IUD is displaced from its proper position. The common side effects of using the IUD include increased menstrual bleeding, cramps, and inflammation or infection of the uterus. However, in the absence of these effects, IUDs are preferred by many women because of the convenience of not having to follow a regimen of pill-taking or to prepare otherwise for intercourse.

A diaphragm is a small rubber cap that fits over the cervix, thereby preventing sperm from entering the uterus. Used in conjunction with the diaphragm are spermicidal ("sperm killing") creams or jellies. In fact, various spermicidal creams, jellies, and foams are used without the diaphragm. When used alone, however, these agents have a failure rate of 20 to 30 percent. Although a properly fitted diaphragm used with a spermicide is about 95 percent effective in preventing pregnancy, their use is highly inconvenient in that both should be inserted just before intercourse. Therefore, even though the diaphragm causes no harmful side effects, it is somewhat less popular than the oral contraceptive or the IUD.

Another contraceptive practice is the *rhythm method,* which involves avoiding intercourse during the "fertile" days of the female cycle. As we have seen, these are generally the thirteenth, four-

The *rhythm method*, in which intercourse is avoided during the period of female ovulation, is unreliable since the time of ovulation can vary from month to month.

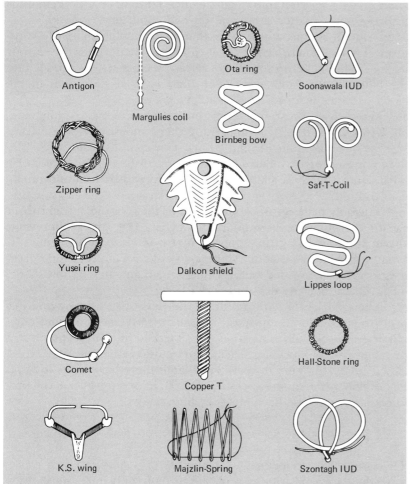

Figure 20.20 Types of intrauterine devices. All are flexible for insertion through the cervical canal by a doctor or trained assistant. Most usually have a thread or chain projecting into the vagina to detect for retention and for removal if necessary.

teenth, and fifteenth days of the cycle, i.e., about a day before to about a day after ovulation. However, the rhythm method is highly unreliable because of the unpredictability of the time of ovulation. On the day of ovulation the body temperature rises about one-half degree; if a record of daily body temperatures is kept for a few months, it is sometimes possible to predict the day on which ovulation will occur. Nonetheless, this method is only moderately dependable, having a failure rate of around 35 percent. The other disadvantage of the rhythm method is that it requires abstinence from sexual intercourse during several days of the female cycle.

Other female contraceptives have been under study in recent years, particularly those that can be administered infrequently but have long-acting effects. For example, compounds have been devel-

> *Vasectomy* is the sterilization procedure in men.

oped that are injected into a muscle once a month or every 3 months to prevent pregnancy. Both the once-a-month and 3-month injections contain synthetic hormones that act to inhibit follicle development. There are also devices that can be planted under the skin or in the vagina, where they slowly release female sex hormones into the bloodstream. Here again, the high level of hormones in the blood prevents maturation of new follicles.

Research on these and other contraceptive methods is ongoing, but problems still persist concerning their undesirable side effects, such as bleeding, nausea, and so on. There is, then, no "ideal" female contraceptive available at present; accordingly, the choice of a birth control method should be considered with reason and intelligent caution.

The only male contraceptive device is the condom, a thin rubber sheath that fits closely around the penis. The condom prevents sperm from entering the vagina or uterus following ejaculation. Although the condom is about 90 percent effective in preventing pregnancy, it must be fitted before intercourse, an interruption that is inconvenient and usually undesirable.

The practice of withdrawal of the penis from the vagina just prior to ejaculation (known as *coitus interruptus*) has its obvious emotional drawbacks and is a rather poor method of contraception. It is possible that, just prior to ejaculation, sufficient sperm may be released for fertilization to occur. Although there have been attempts to develop male chemical contraceptives that suppress sperm production, so far these have met with little success.

STERILIZATION

The most effective method of birth control is sterilization, a procedure that prevents transport of sperm or ova. In men sterilization involves cutting and tying off the vasa deferentia (sperm ducts) so that sperm cannot leave the testes. This simple operation, known as a **vasectomy**, can be done with local anesthesia in a physician's office (Fig. 20.21A). A vasectomy does not prevent the production of sperm or the secretion of testosterone. Consequently, masculinity is not affected. The sperm produced are destroyed by phagocytic cells in the testes, leaving only the fluid content of semen to be ejaculated. Sterilization by vasectomy is usually irreversible, although in some cases, surgically reuniting the cut ends of the vasa deferentia has restored fertility. An alternative method to vasectomy is the insertion of a tiny valve in the vas deferens. Although still an experimental device, the valve can be opened or closed to allow or restrict the passage of sperm.

Tubal ligation, in which the fallopian tubes are cut, is the sterilization procedure in women.

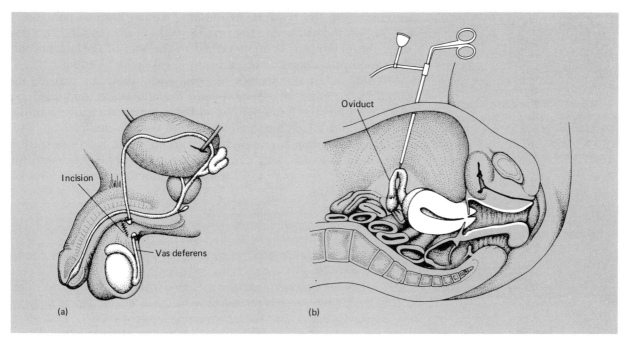

Figure 20.21 Sterilization operations (a) Male: A vasectomy is a relatively simple procedure in which each vas deferens (sperm duct) is cut. (b) Female: Tubal ligation involves cutting or pinching off the fallopian tubes (oviducts).

Sterilization of the female usually involves a procedure known as **tubal ligation,** in which the fallopian tubes are cut and their ends tied off (Fig. 20.21*B*). The ovaries continue to function normally, but the sperm cannot reach the ova. Tubal ligation does not alter the normal secretion of hormones and has no adverse effect on femininity. Reversal of sterilization in women has a very low rate of success; consequently, tubal ligation should generally be considered a permanent condition.

ABORTION

Abortion refers to the premature expulsion of a fetus from the uterus. The term is from the Latin *abortus* "to miscarry." A large percentage of abortions are "natural," or spontaneous, and often occur very early in pregnancy. This type of abortion, more commonly called a miscarriage, occurs when the embryo is in some way defective or abnormal.

Abortions performed as a means of birth control are called *induced abortions.* One type of induced abortion involves dilation of the cervix and gently drawing the embryo from the uterus using a suction device. Another method is called *dilation and curettage* (D & C), in which the cervix is dilated and the lining of the uterus is scraped to remove the embryo and surrounding tissues. Both suction and D & C are performed within the first 12 weeks of pregnancy; after this time, the placenta is firmly attached to the wall of the uterus, and such attempted abortions often result in extensive bleeding.

After the fifteenth week of pregnancy, abortions may be induced by injecting salt solutions into the amnion, a fluid-filled sac that surrounds the fetus. Within an hour or two after injection, the fetus dies and is later expelled along with the placenta.

The death rate for women undergoing legal abortions during the first 12 weeks of pregnancy is about 3 per 100,000 women; for abortions undergone after 12 weeks, the number rises to about 25 per 100,000 women.

REVIEW OF ESSENTIAL CONCEPTS

1. *Reproduction* is the process by which a living organism gives rise to another of its own kind.
2. *Asexual reproduction* results in offspring that are genetically identical to the parent. Asexual reproductive methods include *fission, budding, parthenogenesis, fragmentation,* and *spore formation.*
3. *Sexual reproduction,* which involves the union of genetically different gametes, produces offspring that are not genetically identical to the parents.
4. A relatively simple form of sexual reproduction in lower organisms is *conjugation.*
5. Individual organisms that produce both sperm and eggs are called *hermaphrodites.*
6. Sexual reproduction in the placental mammals involves the fertilization of a large, nonmotile ovum by a small, highly motile sperm.
7. The primary sex organs of the human male are the *testes,* which consist of the *seminiferous tubules* in which sperm are produced.
8. From the seminiferous tubules, sperm from each testis pass into the *epididymis, vas deferens,* and *seminal vesicles.* Sperm then enter the *urethra,* which extends downward through the penis. The *prostate gland* and *Cowper's glands* release fluids into the urethra.
9. *Semen* consists of sperm plus the secretions of the seminal vesicles, prostate gland, and Cowper's glands.
10. The testes are suspended within the *scrotum.* Failure of the testes to descend from the abdominal cavity of the fetus into the scrotum is known as *cryptorchidism.*
11. The production of active sperm requires a temperature several degrees cooler than normal body temperature (37° C). This cooler temperature is maintained within the scrotum.

12. The *primary and secondary sexual characteristics* of the human male are developed and maintained under the influence of the male sex hormone *testosterone*. Testosterone is produced by the *interstitial cells* of the testes.
13. The production and secretion of testosterone are controlled by *interstitial cell-stimulating hormone* (ICSH), released from the anterior pituitary gland. The secretion of ICSH is controlled by *luteinizing hormone releasing factor* (LRF), produced in the hypothalamus.
14. Spermatogenesis is initiated by *follicle-stimulating hormone* (FSH), secreted by the anterior pituitary gland. The hypothalamus produces *follicle-stimulating hormone releasing factor* (FRF), which stimulates the secretion of FSH.
15. Sexual disorders of the human male include sterility, impotence, and cancer of the prostate gland.
16. The primary sex organs of the human female are the *ovaries*, which contain the *primary follicles* in which the ova are produced. Other female reproductive organs include the *vagina, fallopian tubes*, and *uterus*.
17. Cells within the follicles—the *oogonia*—enlarge before birth to become *primary oocytes*. Prior to *ovulation*, a primary oocyte undergoes the first meiotic division to produce a *secondary oocyte* and a small *polar body*.
18. The *primary and secondary sexual characteristics* of the human female are developed and maintained by two types of hormones, the *estrogens* and *progesterone*.
19. The *menstrual cycle* involves a series of hormonal and physical activities that prepare the uterus for implantation of the embryo.
20. Secretion of the female sex hormones is controlled by releasing factors from the hypothalamus and tropic hormones from the anterior pituitary gland. The hypothalamus produces *follicle-stimulating hormone releasing factor* (FRF) and *luteinizing hormone releasing factor* (LRF). FRF stimulates the secretion of *follicle-stimulating hormone* (FSH) from the anterior pituitary, and LRF stimulates the secretion of *luteinizing hormone* (LH).
21. FSH stimulates maturation of the follicle and primary oocyte; a growing follicle releases increasing amounts of estrogens. The estrogens inhibit the hypothalamus, which reduces the secretion of FSH from the anterior pituitary. With maturation of the follicle, LH is released from the anterior pituitary.
22. FSH and LH act together to cause rupture of the follicle and release of the secondary oocyte into the fallopian tube, a process called *ovulation*. This normally occurs on the fourteenth day of a 28-day menstrual cycle.

23. If fertilization does not occur, the secondary oocyte degenerates within the fallopian tube; if the oocyte is fertilized, it undergoes the second meiotic division to produce the haploid *ovum*.
24. Under the influence of LH, an empty follicle is transformed into the *corpus luteum*, which secretes estrogens and progesterone. If fertilization has occurred, progesterone prepares the uterus for implantation of the embryo.
25. If fertilization does not occur, the corpus luteum degenerates, accompanied by a rapid decline in the levels of estrogens and progesterone. This is followed by disintegration of uterine tissues, which pass out of the vagina as the *menstrual flow*.
26. Low levels of estrogens and progesterone free the hypothalamus from feedback inhibition, and the anterior pituitary begins to secrete FSH and LH. In this manner, a new menstrual cycle is initiated.
27. The human sexual act involves a complex of physical, psychical, and physiological events, normally culminating in orgasm.
28. Fertilization of the human ovum usually occurs high in the fallopian tube. Within the fallopian tube, sperm normally remain viable for 48 hours; the secondary oocyte remains viable for 24 hours. Thus, fertilization must occur within a 3-day period.
29. Fertilization by a single sperm produces a "fertilization membrane" around the oocyte, which is then impenetrable by other sperm.
30. After entering the secondary oocyte, the head of the sperm forms the *male pronucleus*. Upon entry of the sperm, the oocyte undergoes the second meiotic division to produce the haploid *ovum* and a small polar body.
31. The ovum contains the *female pronucleus*, which fuses with the male pronucleus to form the diploid *zygote*.
32. After implantation, the embryo secretes *human chorionic gonadotropin* (HCG), which maintains the corpus luteum for up to 3 or 4 months. During this time, the corpus luteum secretes estrogens and progesterone, which maintain the uterine lining.
33. After 3 or 4 months, the corpus luteum gradually disintegrates, and the uterine lining is then maintained by estrogens and progesterone secreted by the *placenta*.
34. An embryo that becomes implanted in tissues other than the uterus results in an *ectopic pregnancy*.
35. *Contraception* refers to any method that will prevent pregnancy. Female contraceptives include *birth control pills*, *intrauterine devices*, and the *diaphragm*. The only male contraceptive is the *condom*. Other contraceptive measures include the *rhythm method*, *coitus interruptus*, and *sterilization*.

36. Sterilization methods include *vasectomy* in the male and *tubal ligation* in the female.
37. *Abortion* is the premature expulsion of a fetus from the uterus. A spontaneous abortion is called a miscarriage. *Induced abortions*, performed within the first 12 weeks of pregnancy, include the methods of *suction* and *dilation and curettage* (D & C). After the fifteenth week of pregnancy, abortions may be induced by injection of salt solutions into the amnion.

APPLYING THE CONCEPTS

1. Describe the types of asexual reproduction. What are the advantages and disadvantages of asexual reproduction?
2. Trace the path of sperm from the seminiferous tubules to the urethra.
3. Describe the production and secretion of testosterone in the human male. What is the function of testosterone?
4. How is spermatogenesis controlled in the human male?
5. Outline the steps of the human menstrual cycle.
6. What happens to a secondary oocyte if fertilization does *not* occur? What happens if fertilization *does* occur?
7. What is the fate of the corpus luteum if fertilization does *not* occur and if fertilization *does* occur?
8. What factors are responsible for initiating the menstrual cycle?
9. Describe the events that occur at fertilization in human beings.
10. What is the function of human chorionic gonadotropin? What is the significance of this function?
11. Describe the available methods of human contraception. What are the advantages and disadvantages of each?
12. What is the difference between sterilization and abortion?

development

THE ESSENTIAL OBJECTIVES

You have understood this chapter when you are able to:

1. Explain the concepts of growth, cellular differentiation, and morphogenesis as they apply to human embryonic development.
2. Trace the early embryonic development of the human embryo, including cleavage, formation of blastocyst, inner cell mass, amniotic cavity, and embryonic disc.
3. Describe the formation and general functions of the fetal membranes.
4. Describe the structure and function of the placenta.
5. Outline the general development of the human embryo and fetus from the first to the ninth month.
6. Describe the process of human birth.
7. Outline the circulation of blood in the fetus and describe the circulatory changes that occur in the fetus after birth.
8. Use experimental examples to discuss the role of the nucleus, cytoplasm, and tissues in cellular differentiation and control of development.
9. Describe the factors controlling development at the molecular level.
10. Summarize the factors thought to be involved in the aging and death of cells.

Embryonic development is the process by which the fertilized egg develops into a complete multicellular organism. Involved in the process are growth, cellular differentiation, and morphogenesis, each of which is influenced by several cellular activities. Development is actually a lifelong process, continuing into old age and culminating in death.

PREVIEW OF ESSENTIAL TERMS

cleavage A series of mitotic divisions of the fertilized ovum, or zygote.

morula In human embryonic development, a solid ball of about 16 embryonic cells.

blastocyst A hollow ball of cells resulting from formation of a cavity in the morula.

inner cell mass A specialized group of cells formed within the blastocyst; the inner cell mass gives rise to the body of the human embryo.

fetal membranes The four membranes that serve to protect and sustain the embryo and fetus; they are the amnion, yolk sac, allantois, and chorion.

placenta The organ in man and other mammals across which nutrients and wastes are exchanged between mother and embryo; it is formed from the lining of the uterus and the chorion of the fetus.

embryonic tissue One of the three basic types of tissue laid down in the embryo; they are ectoderm, endoderm, and mesoderm. All body structures arise from one or a combination of these tissues.

foramen ovale In the human fetus, the opening between the right atrium and left atrium of the heart. The foramen ovale normally closes after birth.

ductus arteriosus In the human fetus, a blood vessel connecting the pulmonary artery directly to the aorta. The ductus arteriosus normally becomes occluded after birth.

induction The developmental process of producing a specific effect—such as differentiation—through the influence of one tissue or structure on another.

gastrula A stage in embryonic development in which the blastula becomes a three-layered embryo composed of ectoderm, endoderm, and mesoderm.

organizer A part or region of an embryo that induces some other part or region to differentiate or develop in a specific manner.

Development

> *Embryonic development involves three basic processes: growth, cellular differentiation,* and *morphogenesis.*

INTRODUCTION

It is characteristic of life that, in whatever form, it almost always begins with a single cell. In most multicellular organisms, this original unit of life, called a *zygote*, undergoes a complex series of events referred to as **embryonic development**. As we shall examine it in human beings, embryonic development involves the interaction of three basic processes: **growth, cellular differentiation,** and **morphogenesis** (Fig. 21.1). The information required to direct these processes is largely contained within the zygote itself. The human zygote

Figure 21.1 The process of development involves growth, cellular differentiation, and morphogenesis.

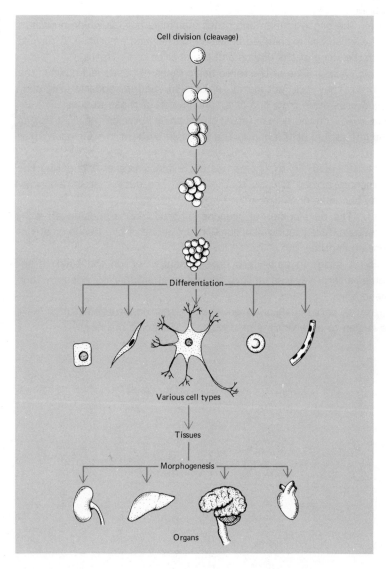

> *Fraternal twins* occur when two separate eggs are fertilized by different sperm. *Identical twins* always arise from a single egg fertilized by a single sperm.

receives most of this information in the form of 46 chromosomes, 23 from the father and 23 from the mother.

The process of growth involves an increase in the number, size, and complexity of cells. The single-celled zygote grows by mitotic cell division to produce a multicellular embryo. Subsequently, the cells of the embryo increase in size and become structurally and functionally more complex.

Cellular differentiation refers to the formation of different types of cells. As embryonic development proceeds, cells arise that are destined to become muscle cells, bone cells, skin cells, and so on. In human beings, more than 200 different cell types develop, each type having a specific task to perform in the life of the organism.

Morphogenesis means "development of form" and pertains to the movement and rearrangement of cells that give shape to the bodily structures and organs. As we shall discuss, the location to which an embryonic cell moves during development often determines its ultimate fate. Indeed, the arrangement of cells in relation to other types of cells is a critical factor in directing normal embryonic development.

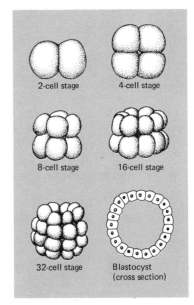

Figure 21.2 Cleavage of the zygote consists of a series of mitotic divisions, which give rise to the hollow blastocyst.

CLEAVAGE

About 24 hours after fertilization, the human zygote, normally located high in the fallopian tube, completes its first mitotic division. This is the first in a series of mitotic divisions known as *cleavage* (Figs. 21.2 and 21.3). Each of the cells from the first division undergoes mitosis to produce four cells, which in turn divide to produce eight cells, and so on. At the 16-cell stage, the embryo is in the form of a compact, spherical mass of cells called the *morula* (L. "mulberry").

During the cleavage process, the developing embryo slowly moves down the fallopian tube toward the uterus. About 3 days after fertilization, the morula arrives at the uterus but usually remains unattached for another 3 days. During this period, the morula undergoes reorganization to become a hollow ball of cells called the *blastocyst* (Fig. 21.4). The cellular wall of the blastocyst is called the *trophoblast* and is destined to become a part of the placenta. In an area along the inner wall of the blastocyst, a small group of cells arises to form the *inner cell mass.* The inner cell mass gives rise to the body of the human embryo.

Sometimes the inner cell mass separates into two groups of cells, each of which develops into a human embryo. Since both groups of cells arise from the same zygote, the embryos formed are identical twins. Fraternal (nonidentical) twins are produced when two sepa-

Figure 21.3 Four-cell cleavage stage of a "sea squirt" embryo.

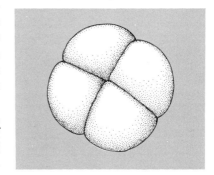

Development

> The embryo, suspended in the protective *amniotic fluid*, is surrounded by the *amnion* (first fetal membrane), which is housed within the *amniotic cavity*.

rate egg cells are fertilized by different sperm. In these instances, the woman ovulates two egg cells instead of only one. Identical twins, of course, are always of the same sex, whereas fraternal twins may be of opposite sex and are no more alike genetically than any other brothers or sisters. The frequency of twins is approximately 1 of every 88 births (Fig. 21.5).

IMPLANTATION AND THE FETAL MEMBRANES

By the seventh day of development, the embryo becomes attached to the lining of the uterus, and by the tenth day it has become implanted in the uterine wall. During the second week, the interior of the inner cell mass develops a fluid-filled hollow space called the *amniotic cavity*. Eventually, this cavity becomes lined with one of the

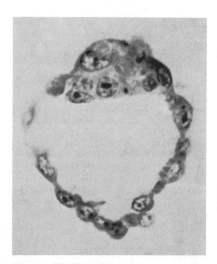

Figure 21.4 A 5-day human embryo (blastocyst), showing the inner cell mass (top) and the cellular wall (trophoblast). (From Villee, C. A.: *Biology*. 7th ed. Philadelphia: Saunders College Publishing, 1977, p. 600.)

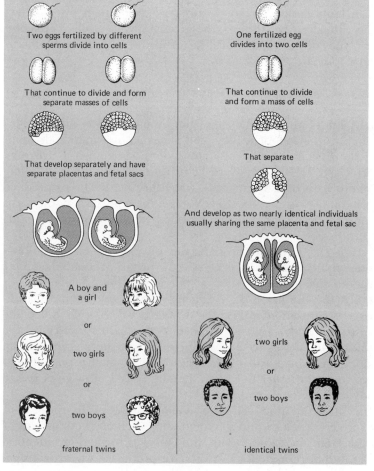

Figure 21.5 Fraternal twins *(left)*, developing from different eggs and sperms, are no more alike than other brothers and sisters. Identical twins *(right)* develop from a single egg and sperm and have the same genetic make up.

> The embryonic disk is composed of three embryonic tissues: *ectoderm, endoderm,* and *mesoderm.*

fetal membranes, the **amnion,** which completely surrounds the embryo. The fluid within the membrane, called **amniotic fluid,** suspends the embryo in a moist environment and serves as a shock absorber to protect the embryo against jarring blows.

Development of the amniotic cavity is followed by differentiation of cells in the inner cell mass to form the **embryonic disc.** This is the specific structure from which the body of the human embryo will develop. The embryonic disc is initially composed of two *embryonic tissues,* called **ectoderm** (Gk. "outer skin") and **endoderm** (Gk. "inner skin") (Fig. 21.6). Ectoderm eventually gives rise to the amnion, the membrane surrounding the embryo, and endoderm produces a second fetal membrane, the **yolk sac.** Since the ova of females contain but little yolk, the yolk sac in humans does not function as a site of yolk storage. Instead, it serves to absorb nu-

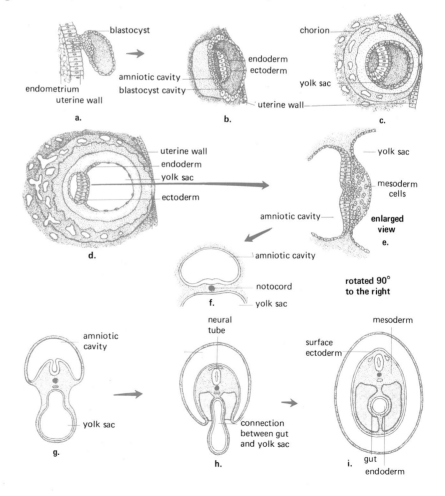

Figure 21.6 Early developmental stages of the human embryo. The embryonic disc is composed initially of ectoderm and endoderm; later, mesoderm forms between the ectoderm and endoderm. (From Applewhite, P., and Wilson, S.: *Understanding Biology.* New York: Holt, Rinehart and Winston, 1978, p. 291.)

> The *chorion* burrows into the uterus and eventually forms part of the *placenta,* an organ across which nutrients and oxygen pass into the embryo while allowing waste materials to pass from the embryo into the mother.

trients and forms the first blood cells. Later in development, the yolk sac eventually degenerates.

In the meantime, the outer cellular wall of the blastocyst, the trophoblast, has developed into the third fetal membrane, the **chorion.** The chorion burrows deep into the lining of the uterus until the blastocyst is completely enclosed by maternal tissue. The chorion and the surrounding tissues of the uterus eventually form the **placenta,** an organ across which nutrients and oxygen pass into the blood of the embryo, while waste materials from the embryo pass into the mother's blood to be excreted. Although the blood of the mother and the embryo are in close proximity, separated only by thin capillary walls, the two circulations never mix.

In addition to its function as an organ of exchange, the placenta is also an endocrine organ. You will recall from Chapter 19 that cells of the developing embryo secrete HCG (human chorionic gonadotropin), which stimulates the corpus luteum to continue its release of estrogens and progesterone during the early stages of pregnancy. After the first 3 months of pregnancy, however, the corpus luteum greatly reduces its secretion of these hormones, and the placenta, which by this time is well developed, secretes large amounts of its own estrogens and progesterone.

The fourth fetal membrane, the **allantois,** develops as an outpocketing of the primitive hindgut of the developing embryo. Part of the allantois supplies blood vessels to the placenta, and the stalk connecting the allantois to the embryo becomes the **umbilical cord** (Fig. 21.7). The blood vessels within the allantoic stalk become the umbilical arteries and vein. Eventually, the membranous portion of the allantois is reduced to a ligament attached to the urinary bladder.

The formation of the four fetal membranes—amnion, yolk sac, chorion, and allantois—early in development is essential because, as mentioned, the human ovum contains no yolk to nourish the embryo. Instead, nourishment is supplied through the blood of the mother, with the umbilical vessels and fetal membranes as avenues of transportation and exchange.

It was pointed out that the embryonic disc arises from the inner cell mass, which differentiates to form the disc from ectoderm and endoderm. Soon thereafter, the third embryonic tissue, the **mesoderm** (Gk. "middle skin"), is laid down between the ectoderm and endoderm. Each of these embryonic tissues eventually differentiates into the various cell types that constitute all of the organs and structures of the human body. For example, ectoderm will give rise to epidermal structures, such as hair and skin, and will form the tissues of the nervous system; mesoderm produces bone, muscle, blood,

Between the second and the eighth week of development the human organism is known as an *embryo*.

Embryonic Development: First Month

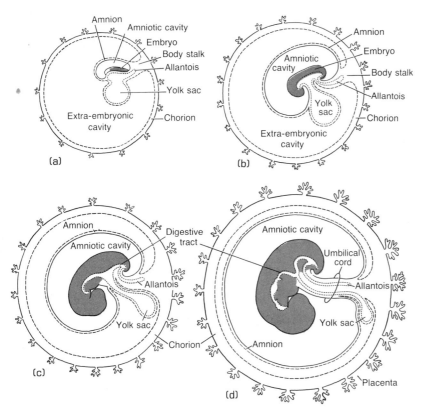

Figure 21.7 (a)–(d) Successive stages in the development of the umbilical cord and body form in the human embryo. The *solid lines* represent layers of ectoderm; the *dashed lines*, mesoderm; and the *dotted lines*, endoderm. (From Villee, C. A.: *Biology*. 7th ed. Philadelphia: Saunders College Publishing, 1977, p. 627).

and the lining of the body cavities; endoderm gives rise to the lining of the digestive and respiratory tracts and forms the tissues of the thyroid and parathyroid glands. This, of course, is only a partial listing and represents an oversimplified view of tissue origin. Actually, most of the organs and structures of the body consist of tissues derived from all three embryonic layers.

EMBRYONIC DEVELOPMENT: FIRST MONTH

Technically, the term "embryo" is applied to the human organism beginning at the second week of development and extending to the eighth week. During the second week, a rodlike structure called the **notochord** grows along the length of the embryo. The notochord is present in all chordate embryos, but in adult humans and other vertebrates, it is replaced by the vertebral column. Ectoderm overlying the notochord curves dorsally to form the **neural tube,** which will give rise to the central nervous system (brain and spinal cord) (Fig. 21.8). In addition, the rudiments of the skeletal and muscular systems are developing from the embryonic mesoderm.

Limb buds on the embryo develop into arms and legs.

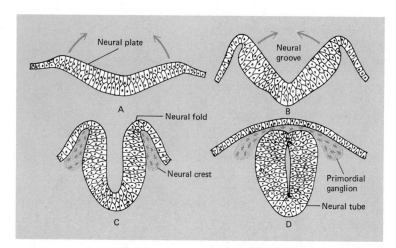

Figure 21.8 In human embryos, ectoderm overlying the notochord curves dorsally to form the neural tube. The neural crest gives rise to spinal and sympathetic ganglia.

By the end of the first month, the head of the embryo is quite large, and some of the organs of special sense—eyes, ears, and nose—have begun to form. The digestive system, including the liver and pancreas, is also developing, and the tiny heart is actively pumping blood. Present, too, are small *limb buds*, which will develop into arms and legs (Fig. 21.9*A,B*). Small slits and "gill pouches" resembling those of a fish embryo are evident along either side of the head and neck. In fish and some amphibians, the slits develop into gills, but in humans and other land vertebrates, they normally close before birth. However, tissues around the pouches develop into such human structures as the eardrum and eustachian tube.

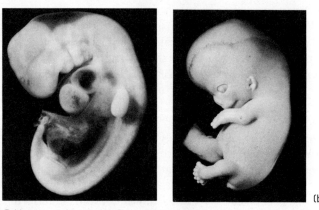

Figure 21.9 Human embryos: (a) One month. Note the limb buds and indentations of the gill pouches in the head region. (b) One and one-half months. Note the human characteristics and development of the limb buds. A remnant of a tail is present at this stage. (From Ebert, J. D., Loewy, A. G., Miller, R. S., and Schneiderman, H. A.: *Biology.* New York: Holt, Rinehart and Winston, 1973, p. 203.)

685
Embryonic
Development: Second
Month

Human embryos develop a small tail that normally degenerates before birth.

During the second month, morphogenesis continues as the embryo acquires somewhat more distinctive human characteristics. Organ development continues, and there is increasing prominence of facial features, limb buds, and the musculoskeletal system (Fig. 21.10 A,B). Internal sex organs—either ovaries or testes—may be distinguished, but the sex of the embryo cannot yet be determined from external structures. As in other vertebrate embryos, a small tail develops, but it regresses and finally disappears by the end of the second month. There have been rare cases, however, in which the tail was still present at birth. Fortunately this condition can be corrected surgically.

EMBRYONIC DEVELOPMENT: SECOND MONTH

Figure 21.10 Human embryos: (a) At 39 days of development, shown with its protective membranes. (b) About 2 months. (From Ebert, J. D., Loewy, A. G., Miller, R. S., and Schneiderman, H. A.: *Biology.* New York: Holt, Rinehart and Winston, 1973, pp. 180, 203.)

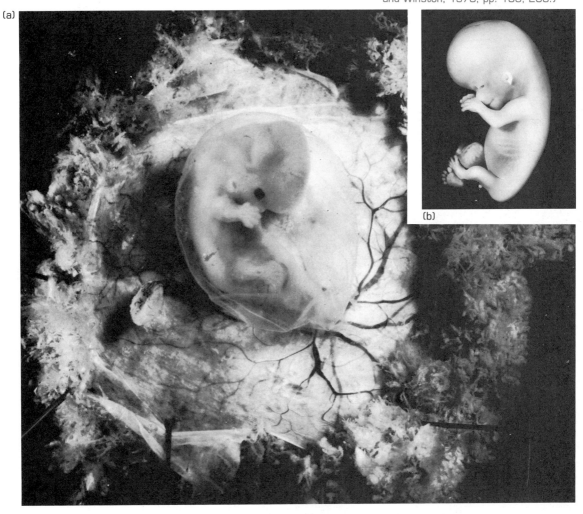

> From the ninth week of development until birth the human organism is known as a *fetus*.

FETAL DEVELOPMENT: THIRD MONTH

The term "fetus" (L. "offspring") is applied to the human organism from the ninth week of development until birth. By the end of the third month, the fetus possesses the rudiments of all the major organs. Facial expressions, movements of the arms and legs, and reflexes, including the sucking reflex, become quite evident during the third month. In addition, sex of the fetus can be determined externally, breathing movements are evident, and there is degeneration of the notochord. At the end of the third month, the fetus is slightly less than 3 inches long from head to rump and weighs about ½ oz.

FETAL DEVELOPMENT: FOURTH, FIFTH, AND SIXTH MONTHS

From the fourth to the end of the fifth month, the fetus more than doubles in length and increases in weight to about ½ lb. During the sixth month, it will attain a length of about 12 inches and will weigh about 1½ lb (Fig. 21.11 *A,B*). Its heart, which can be heard with a stethoscope, beats at a rate of about 140 times per minute. During

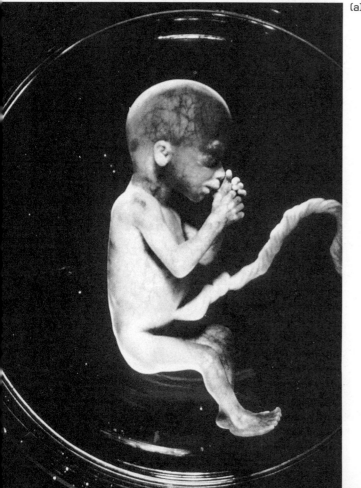
(a)

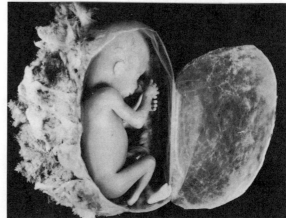
(b)

Figure 21.11 Human embryos: (a) At 4 months, the embryo has become a fetus, and all of its major organs are formed. Although it weighs only about 11 oz, its movements may already be felt by the mother. Further development involves growth and maturation of these organs. (b) Beginning of the fifth month. ((a) from Clark, M. E.: *Contemporary Biology*. 2nd ed. Philadelphia: Saunders College Publishing, 1979; (b) from Ebert, J. D., Loewy, A. G., Miller, R. S., and Schneiderman, H. A.: *Biology*. New York: Holt, Rinehart and Winston, 1973, p. 203.)

During the final 3 months of prenatal life there is rapid body growth and final development of the organ systems.

the fifth month, the mother can feel the movements of the fetus as it shifts positions within the amniotic cavity. By the sixth month of development, the fetus has acquired hair on its head, and its skin is red and wrinkled.

FETAL DEVELOPMENT: SEVENTH, EIGHTH, AND NINTH MONTHS

The final 3 months of prenatal life are marked by rapid growth of the body and final development of the organ systems. The brain and entire nervous system rapidly become more complex as an array of new nerve cells and tracts develop. The bones of the skull are soft and flexible, allowing for continued growth of the brain. By the end of the eighth month, the fetus is about 18 inches long and weighs approximately $2\frac{1}{2}$ lb. The baby also acquires a deposit of fat tissue, causing the skin to appear less wrinkled.

At the end of the final month of pregnancy, human fetuses average 20 inches in length and weigh about 7 lb (Fig. 21.12). Generally, the time required for complete development of a human baby—measured from conception until birth—is 266 days. However, many normal babies are born 2 or 3 weeks before or after this time period. Human fetuses born during the seventh or eighth month of development have a fair chance of survival; but a fetus born before the 26th week usually does not survive, primarily because of inadequate development of the respiratory system.

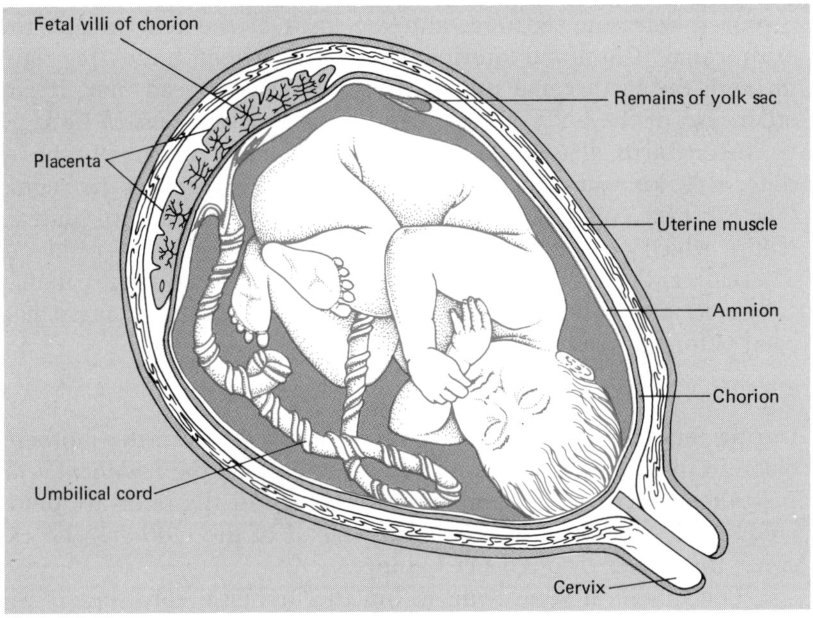

Figure 21.12 A diagrammatical section through the uterus, showing the placenta and the fetus shortly before birth.

> *Parturition* is the act of childbirth, during which the uterus undergoes strong contractions known as *labor*.

BIRTH

The act of childbirth, known as **parturition** (L. "to be in labor"), normally begins with strong contractions of the uterus. These contractions are thought to result from stretching of the uterus and cervix, caused by movements of the fetus, and from hormonal factors. It is probable that stretch of the cervix by the head of the fetus initiates muscular reflexes in the wall of the uterus. The resulting contractions of the uterine muscle help push the baby into the birth canal.

We have mentioned that the uterus secretes large amounts of estrogens and progesterone throughout most of pregnancy. The effect of progesterone is to inhibit contractions of the uterus, whereas estrogens increase the intensity of contractions. It has been suggested that increased uterine contractions late in pregnancy result in part from proportionately greater secretion of estrogens than progesterone. In addition, stretch of the uterus and cervix causes reflex stimulation of the posterior pituitary gland to release the hormone oxytocin. This hormone also acts to increase the rate and force of uterine contractions.

The exceptionally strong contractions of the uterus, known as **labor,** occur about once every 30 minutes as the birth process begins (Fig. 21.13). These contractions become progressively stronger as the process continues and soon occur every minute or so. The cervix progressively dilates during the first stage of labor and becomes fully expanded, usually within 24 hours. Once this occurs, the amnion ruptures, releasing its fluid, and the head of the fetus enters the birth canal. Continual uterine contractions help force the fetus through the birth canal until it emerges, usually head first. If any other part of the body is presented first, it is called a *breech birth*.

After birth of the infant, the placenta and fetal membranes—collectively known as the *afterbirth*—are expelled through the vagina (Fig. 21.14). For a short time, the uterus continues to contract mildly, which constricts local blood vessels, limiting the loss of blood. Thereafter, the uterus begins to revert to its normal prepregnancy condition and within 4 or 5 weeks usually returns to the size it had been before pregnancy.

FETAL CIRCULATION

During fetal development, oxygen and nutrients from the mother's blood in the placenta are supplied to the fetus via the **umbilical vein.** Waste materials that accumulate in the blood of the fetus are delivered via the **umbilical arteries** to the blood of the mother, who excretes the wastes through her kidneys.

The umbilical vein, lying within the umbilical cord, enters the body of the fetus at the navel area and carries oxygenated blood to

> In the fetus, the *foramen ovale* permits blood to be shunted from the right side of the heart to the left side, bypassing the lungs.

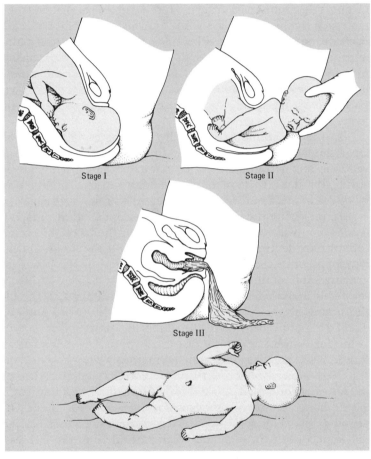

Figure 21.13 The stages of labor can be described as follows. The first stage is characterized by regular contractions, rupture of the membranes, and complete dilatation; the second stage extends from the time of complete cervical dilatation to delivery; during the third stage the "afterbirth" is delivered.

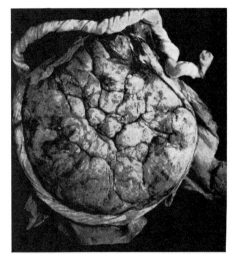
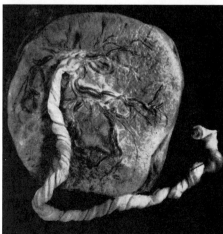

Figure 21.14 Photographs of the maternal (top) and fetal (bottom) surfaces of a human placenta at the end of a normal pregnancy. (From Greenhill, J. P.: *Obstetrics*. 13th ed. Philadelphia: W. B. Saunders Co., 1965.)

the inferior vena cava. Some of the deoxygenated blood already in the vena cava mixes with the oxygenated blood, and the mixture then travels to the right atrium of the heart. However, instead of passing into the right ventricle, out through the pulmonary artery, and into the lungs, most of the blood is shunted from the right atrium directly into the left atrium through an opening called the *foramen ovale* (L. "oval opening"). As mentioned, oxygen from the mother is supplied to the fetus through blood in the umbilical vein. This is necessary because the lungs of the fetus are nonfunctional until birth. Consequently, the foramen ovale permits blood to be shunted from the right side of the heart to the left side, while bypassing the lungs (Fig. 21.15).

However, a small portion of blood in the inferior vena cava, along with deoxygenated blood from the superior vena cava, *does* enter the right ventricle from the right atrium and then passes out

> In the unborn fetus, the lungs are nonfunctional, and respiratory gas exchange occurs across the placenta.

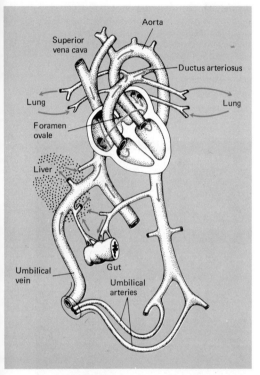

Figure 21.15 The fetal circulation. The foramen ovale and ductus arteriosus shunt blood away from the lungs.

through the pulmonary artery. Here again, most of this blood does not reach the lungs but is detoured through a small vessel called the **ductus arteriosus.** This vessel transports blood from the pulmonary artery directly across to the arch of the aorta. Some of this blood in the aorta is delivered by branching arteries to lower parts of the body; the rest of the blood enters the umbilical arteries to be transported to the placenta, where it is reoxygenated and cleared of waste materials.

In the meantime, the partially oxygenated blood that had entered the left atrium through the foramen ovale passes into the left ventricle and is pumped out through the aorta. Branches off the aortic arch deliver the blood to the heart muscle, forelimbs, and regions of the neck and head, including the brain.

In reviewing the circulation of blood in the adult human (Chapter 17), you will recall that oxygenation of the blood and removal of carbon dioxide are the primary functions of the lungs. During fetal development, these functions are carried out by the placenta. Oxygen from the mother's blood in the placental capillaries diffuses into the fetal capillaries and, as we have seen, is carried in the blood to the umbilical vein. Carbon dioxide and other wastes in the fetal blood are removed through the umbilical arteries. The blood travels to the fetal capillaries of the placenta, where the waste materials diffuse into the capillary blood of the mother.

To summarize the discussion of fetal circulation, remember that the lungs of the unborn fetus are nonfunctional and that respiratory gas exchange occurs in the placenta. Consequently, most of the oxygenated fetal blood is shunted away from the lungs, either from the right atrium through the foramen ovale into the left atrium or from the pulmonary artery to the aortic arch by way of the ductus arteriosus. The blood in the left atrium is delivered to the heart, forelimbs, and head, whereas some of the blood in the aorta is circulated to the lower parts of the body. The remaining portion in the aorta enters the umbilical arteries and is delivered to the placenta for reoxygenation and waste removal. Thus, the essential differences between fetal and adult circulations involve the rerouting of blood away from the lungs and the role of the umbilical vessels in respiratory gas exchange and in transporting fetal blood to and from the placenta.

Circulatory Changes at Birth. When a baby is born, cutting the umbilical cord terminates blood circulation through the placenta. The first breath of the infant inflates the lungs, which then must assume the task of respiratory gas exchange. Accordingly, the blood must now circulate through the entire pulmonary system, i.e., from the

> At birth, the foramen ovale and ductus arteriosus close, forcing the blood to circulate through the entire pulmonary system.

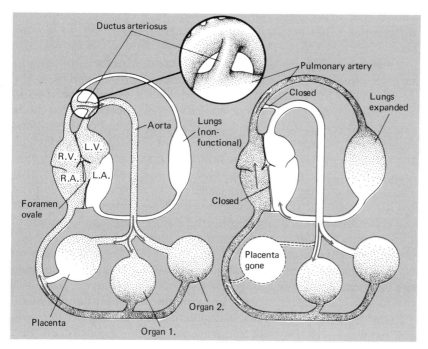

Figure 21.16 Changes in the human circulatory system at birth. Left, The circulatory system of the fetus. Right, The circulatory system of the newborn child. Aerated blood is shown in white, nonaerated blood in dark stippling, and a mixture of the two in lighter stippling. In the embryo most of the blood entering the right atrium reaches the aorta either via the oval window (foramen ovale) or via the arterial duct (ductus arteriosus) *(inset)*. The changes at birth are (1) the loss of the placenta, (2) the expansion of the lungs, (3) the closing of the foramen ovale, and (4) the closing and degeneration of the arterial duct.

heart to the lungs and back to the heart. At birth, the foramen ovale closes, preventing blood from passing from the right atrium directly into the left atrium. Instead, blood in the right atrium passes into the right ventricle and exits the heart through the pulmonary artery. In addition, the ductus arteriosus constricts and soon becomes occluded with tough, fibrous tissue. As a result, blood is prevented from passing out of the pulmonary artery directly into the aorta; instead, blood remains in the pulmonary artery and circulates to the lungs for oxygenation and carbon dioxide removal (Fig. 21.16).

Failure of the foramen ovale or ductus arteriosus to close at birth results in low blood oxygen levels (since some of the blood will not reach the lungs), and the baby's skin and nails acquire a bluish tint (cyanosis). An infant so affected is called a "blue baby." This condition can usually be corrected by surgical closure of the foramen ovale or ductus arteriosus.

CONTROL OF DEVELOPMENT

The Role of the Nucleus. We have seen that development proceeds from the division of a single cell, which ultimately gives rise to many different kinds of cells and a variety of organs. How, you might ask, is this possible? What factors might influence embryonic cells to "change," or differentiate? Although at least some of the factors are

> Experimental finds demonstrate that all the body cells of a multicellular organism are genetically identical; this indicates that cellular differentiation is a consequence of the expression of different genes in different cell types.

known, the exact mechanisms by which they operate to influence differentiation are not fully understood. We do, however, have some clues.

Cleavage of the zygote occurs by mitosis, a process that results in each daughter cell having the same set of chromosomes as the parent cell. Throughout childhood, adolescence, and adulthood in man, mitosis continues to provide the cells necessary for body growth and repair. Each of these cells is a descendant of the original single-celled zygote; it follows, then, that all of the body cells (i.e., all the cells except the gametes) of human beings and other multicellular organisms are genetically identical. It is obvious, however, that human skin cells do not look or function like muscle cells or that red blood cells do not resemble nerve cells. Nonetheless, each of these cell types arises from a single cell—the zygote—through a process of differentiation. Can it be demonstrated that these and other body cell types really are genetically identical? The answer to this question is important because it eliminates one of two possibilities underlying the basic mechanism of differentiation. First, if the answer is no, this means that each different cell type contains different genetic material and that some of the original genetic material in the zygote was lost from the cells somewhere during the development process. Second, if the answer is yes, we are led to conclude that in each different cell type only part of the genetic material is expressed, i.e., differentiation of cells depends upon different genes being expressed in each particular cell type.

Some of the classic experiments designed to answer the question involve the procedure of nuclear transplantation (Fig. 21.17). By destroying the nucleus of an unfertilized frog egg and transplanting a nucleus taken from a frog blastula or from an intestinal cell of a young tadpole, it was found that the egg will often develop into a fully grown frog. Thus, the nuclei of the blastula cell and intestinal cell contain the same genetic information, i.e., all the genetic instructions required to form a new adult organism. Similar experiments with other nucleated body cells have yielded essentially the same results. Still further evidence has been provided when it was found that a single cell taken from a carrot root and placed in a nutrient solution developed into a complete carrot plant. Thus, the root cell nucleus must contain the genetic information required to direct the development of the entire plant.

From these experimental findings, it is apparent that all the body cells of a multicellular organism are indeed genetically identical. Therefore, as stated earlier, we are led to conclude that cellular differentiation is a consequence of the expression of different genes in different cell types. But what influences gene expression in a

> Evidence suggests that the cytoplasm contains substances that affect the activity of certain genes.

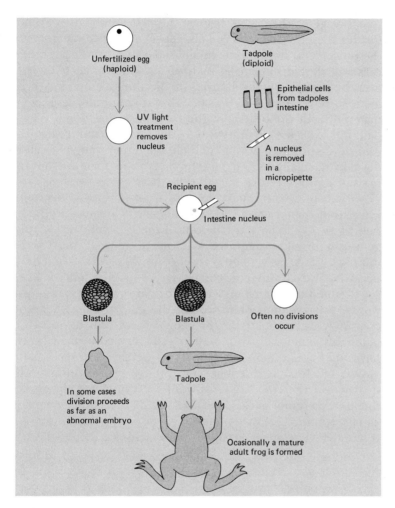

Figure 21.17 Nuclear transplant experiments have provided evidence that the nuclei of many types of body cells contain all the genetic information required to form a new adult organism. All of these cells, then, are genetically identical.

given embryonic cell? Actually, this is stating in a different form the original question: What factors might influence embryonic cells to "change," or differentiate? On the basis of the foregoing information, we now can discuss those factors in terms of their influence on gene expression.

The Role of the Cytoplasm. The immediate environment of a cell nucleus is the surrounding cytoplasm. There is evidence to suggest that the cytoplasm contains stimulatory and inhibitory substances that affect the activity of certain genes. One of the experiments that demonstrate cytoplasmic regulation of development involves the study of frog embryos. After fertilization, a small mass of material called the *gray crescent* appears in the cytoplasm of the frog egg cell.

As cleavage of the zygote begins, the first mitotic division normally divides the gray crescent equally between the two daughter cells. If these two cells are experimentally separated from each other, both cells will develop normally to form two individual tadpoles. However, by experimentally altering the first mitotic division, it is possible to produce one daughter cell that contains all the gray crescent material, with the other daughter cell receiving none. When these daughter cells are separated, only the one containing the gray crescent develops into a normal tadpole; the other daughter cell will not develop normally but divides for a time to form an unorganized mass of nondifferentiated cells (Fig. 21.18).

From this experiment, it is apparent that equal distribution of cytoplasmic materials at the first cleavage is involved in the production of two cells having the same genetic potential, i.e., the same genetic information required to form two normal embryos. Later in cleavage, however, a stage is reached at which embryonic cells separated from each other will not form complete embryos but rather are destined to become only a specific part of the organism. In other words, as development proceeds, the embryonic cells become increasingly differentiated. Why should these cells have such limited potentials while the two daughter cells formed at the first cleavage have full developmental potential? As we have just seen, part of the answer involves the distribution of the cytoplasm. Although the first cleavage of the frog zygote divides the cytoplasm equally between the two daughter cells, later cleavages divide the cytoplasm unequally among the various cells. Accordingly, the nuclei (with the contained genetic material) of these cells are exposed to varying

Figure 21.18 Influence of cytoplasmic constituents on development. If the first mitotic division divides the gray crescent material equally between the two daughter cells, a complete frog embryo will develop from each cell. However, if the first mitotic division is altered, only the cell receiving the gray crescent material will produce a complete embryo.

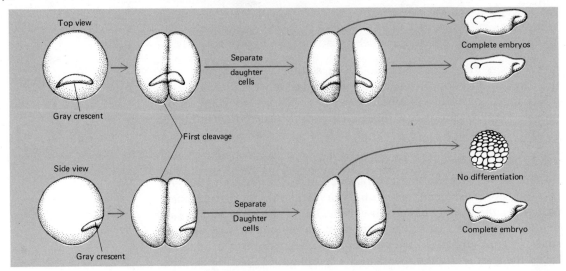

> In addition to the nucleus and cytoplasm, certain tissues also influence the course of embryonic development.

amounts of cytoplasm and, hence, to different cytoplasmic materials. In some manner, the particular cytoplasmic materials affect gene expression, which controls cellular differentiation and thereby directs the future course of embryonic development.

There are some animal groups in which the *first* cleavage does not divide the cytoplasmic materials equally. When the two daughter cells are experimentally divided, each one has a *different* developmental fate. The daughter cell receiving a particular cytoplasmic substance has the potential to develop into a normal embryo; the daughter cell lacking that substance has no such potential and fails to develop normally. This should not suggest that the daughter cell lacking the cytoplasmic material will not develop normally when it is included in the entire developing embryo; rather, the experiment indicates that each of the daughter cells has a different developmental potential as a result of unequal division of the cytoplasm. Therefore, since cleavage does not always divide the cytoplasmic materials equally, the developmental fate of various embryonic cells, i.e., the type of cells they will become, will be different.

Induction. Not only do the nucleus and cytoplasm influence the course of embryonic development, but also entire tissues have a crucial role to play. An illustration of this role involves another of the classic experiments in developmental biology. In experiments using amphibian embryos, it has been found that by dividing a blastula in two, *both* portions will develop into normal embryos. However, once development proceeds beyond the blastula stage, dividing the embryo in two produces a different result. The developmental stage immediately after the blastula is the **gastrula** stage, characterized in vertebrates by formation of ectoderm, endoderm, and mesoderm. In amphibians, development of the gastrula begins with the appearance of a small slit, the **blastopore,** on the surface of the blastula. The mass of cells just above the slit is called the **dorsal lip of the blastopore.** The dorsal lip is destined to give rise to mesoderm, which will form the various body tissues derived from mesoderm. The outer cell layer of the gastrula is ectoderm, whereas endoderm develops along the inner wall of the gastrula. Eventually, the mesoderm becomes situated between the ectoderm and endoderm.

When the amphibian *gastrula* is divided in two, only one embryo develops. This embryo forms only from the half containing the dorsal lip of the blastopore. In an attempt to determine the significance of this finding, grafting experiments have been carried out using the early gastrula stage of salamander embryos. Using two embryos, one light-colored and one dark-colored, the dorsal lip from the light-colored gastrula was removed and transplanted to the belly region

> *Induction* occurs when one type of embryonic tissue causes the differentiation of another type. Chemical substances from the inducing tissue stimulate or inhibit tissue development by activating or repressing certain genes.

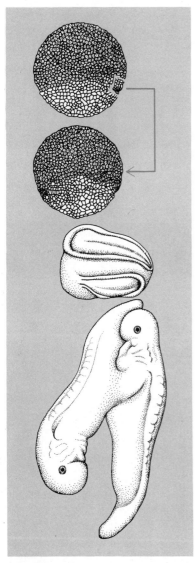

Figure 21.19 Role of the dorsal lip of the blastopore as an organizer of embryonic tissues. When transplanted to the belly region of another gastrula, the dorsal lip organizes the tissues of the host gastrula to form a second complete embryo.

of the dark-colored gastrula (Fig. 21.19). Normally, the tissues of the dorsal lip form the notochord and, as mentioned, give rise to other mesodermal structures. The mesoderm associated with the notochord also appears to influence formation of the neural tube, from which the brain and spinal cord arise. After the transplant, the host (dark-colored) gastrula continued development to form a notochord, central nervous system, and sometimes a complete embryo. In addition, a second gastrula formed at the site of the transplanted dorsal lip, eventually producing a second notochord and nervous system and sometimes a complete embryo. When two complete embryos were formed, they were joined together at the belly region. Except for the notochord, the other tissues in the second embryo were dark-colored like those of the host. What had happened? In some manner, the transplanted dorsal lip tissue interacted with and altered the host tissues, organizing them into a neural tube and eventually a complete second embryo.

The dorsal lip of the blastopore has since been referred to as the *organizer* and apparently is crucial during the early stages of embryonic development. As discussed in the transplantation experiment, the dorsal lip causes, or induces, the differentiation of certain tissues and organs. In similar experiments, transplanting tissues from other parts of the gastrula did not result in the formation of a second gastrula or embryo. From these initial experiments, it is now known that induction is actually a common process characteristic of practically all tissues of developing embryos. As evidenced by the dorsal lip experiment, one type of embryonic tissue can alter the fate of another type. This may be a result of physical contact between the tissues, but it is thought that induction also is controlled by chemical substances (in some cases, these are hormones).

By slightly separating two layers of tissue and inserting a porous filter between them so that they do not touch, induction will still take place. However, if a nonporous material is placed between the tissues, induction is inhibited. Apparently, then, some chemical substance diffuses from one tissue to the other to influence the induction process. More specifically, chemical substances influence induction by activating or repressing certain genes. It should be noted that chemical inducers inhibit as well as stimulate tissue development. Inhibition of development is important because once a certain tissue or organ develops in the body, it would be pointless to have another of the same tissue or organ form in the same area. We might wind up with a chest full of hearts or a couple of livers.

Gene Expression. We have seen that embryonic development is controlled in part by the interactions of the nucleus, cytoplasm, and

> Differentiation operates at the molecular level by activating or repressing certain genes that control the synthesis of specific proteins in a given cell or group of cells.

even entire tissues. Through the particular distribution of cytoplasmic materials and the process of induction, the nuclei of embryonic cells are exposed to varied environments and substances, which affect the expression of certain genes. The result is that cells "change," or differentiate. Each of these cells, however, still has the same genetic make-up as the original zygote from which it arose. Since differentiation depends ultimately on cellular changes at the molecular level, i.e., at the level of the deoxyribonucleic acid (DNA) molecule, we need to consider what those changes are and how they determine the expression of certain genes.

The process of differentiation operates at the molecular level by activating or repressing certain genes that control the synthesis of specific proteins in a given cell or group of cells. Obviously, if every body cell synthesized the same proteins, all the cells would be identical. This would mean that the DNA in each cell would be transcribing the same messenger ribonucleic acid (mRNA), which in turn would direct the synthesis of the same proteins (see Chapter 8). Thus, if cells are to differentiate, only specific segments of DNA in a given cell can be active in protein synthesis. It would be expected, then, that each kind of differentiating cell would have different kinds of mRNA. Observations of vertebrate oocyte chromosomes have revealed that at different stages of development, part of a chromosome may be tightly coiled, whereas another part may be unfolded, or unraveled, forming a lateral loop. Chromosomes arranged this way are called *lampbrush chromosomes* (Fig. 21.20).

The looped parts of such chromosomes are sites where DNA is actively synthesizing mRNA, i.e., sites where a certain sequence of DNA nucleotides (a gene) is initiating the production of a particular protein. Depending upon the stage of development of the tissue, location of the loops on the chromosome may be different, indicating that different proteins are being synthesized on each loop. The tightly coiled portions of a chromosome are sites where mRNA is not being transcribed, and, hence, no proteins are being synthesized. Giant chromosomes found in various insects form lateral loops called *chromosome puffs,* which can be stained to show the presence of mRNA (Fig. 21.21). Those parts of the chromosome not transcribing mRNA do not show the same staining pattern. As you would expect, the locations of puffs on a chromosome will differ in different kinds of tissues. The result is the synthesis of proteins specific for each type of tissue. Thus, even though differentiating cells have identical DNA, they can at given times produce different kinds of mRNA.

We now can visualize somewhat more clearly the influence of the cytoplasm and tissues on the differentiation process. The pres-

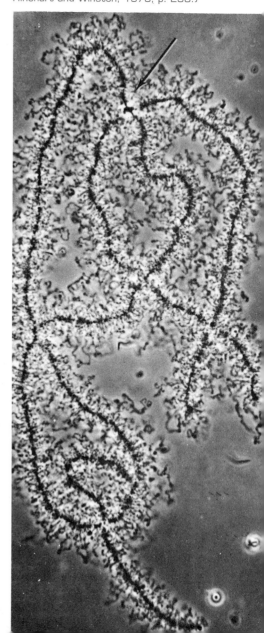

Figure 21.20 Lampbrush chromosomes from an amphibian oocyte. The feathery array of lateral loops are segments of DNA that are actively transcribing messenger RNA. (From Goodenough, U.: *Genetics.* 2nd ed. New York: Holt, Rinehart and Winston, 1978, p. 293.)

Physical environmental factors such as temperature and gravity also affect embryonic cells.

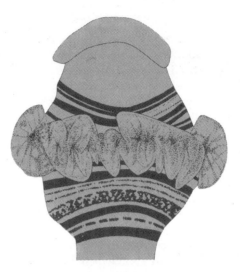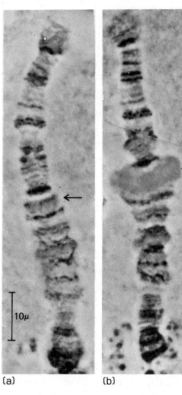

Figure 21.21 Left, A segment of an insect giant chromosome showing a puff. Right. A giant chromosome at two stages of development. (a) Small puff at arrow; (b) The same puff much larger 2 days later. (From Ebert, J. D., Loewy, A. G., Miller, R. S., and Schneiderman, H. A.: *Biology.* New York: Holt, Rinehart and Winston, 1973, p. 216.)

ence or absence of a certain cytoplasmic substance, such as the gray crescent in frog egg cells, affects the activity of the genes. Some genes may be activated to synthesize proteins, whereas other genes may be inhibited or repressed. The proteins synthesized are specific for each kind of cell undergoing differentiation.

In a somewhat similar manner, developing tissues release chemical inducers or inhibitors that influence gene activity. For example, some chemicals may act on the genetic machinery to induce the formation of ectoderm, which in turn releases chemicals that direct the development of skin tissue or nervous tissue. You will recall that this is the mechanism by which the dorsal lip of the blastopore induces formation of the neural tube. Thus, both the contents of the cytoplasm and the chemicals released by developing tissues exert their influence at the molecular level by activating or repressing certain genes. Although every body cell of the embryo may be equally endowed genetically, the factors discussed here ultimately determine which genes will be expressed in the course of differentiation.

This is not by any means the entire story of control of embryonic development. There are, for instance, physical environmental

> *Aging* is the biological process of cellular degeneration.

factors that affect embryonic cells. These factors include light, temperature, gravity, and a few others. In the plant kingdom, for example, it is known that most flowering plants produce blooms in response to the length and timing of the periods of light and darkness. In what are referred to as long-day (short-night) plants, development of the flower occurs only when the length of darkness is limited within a certain maximum number of hours. On the other hand, short-day (long-night) plants require a length of darkness beyond a certain minimum number of hurs. It is presumed by some scientists that plants exposed to light for a given period of time manufacture a hormone called *florigen* that induces flowering. However, this remains a hypothesis, since there has never been actual identification of florigen or any other specific flower-inducing hormone. Consequently, a comprehensive explanation of flowering is not yet possible.

AGING AND DEATH

We have discussed development primarily from an embryonic standpoint, but development is actually a lifetime process. Throughout its existence an organism continues to change until it dies. An integral part of that change is the gradual degeneration of body cells and tissues that cannot be replaced. The biological process of degeneration, known as aging, eventually renders the entire organism physiologically unfit for survival. Little is known about the process of aging other than the symptoms it produces; nonetheless, various theories have been proposed to explain at least certain aspects of the process.

Through experiments using cell cultures, it has been shown that human embryonic cells can undergo about 50 mitotic divisions, i.e., a cell can produce about 50 generations. This finding suggests that cells are "programmed" to divide only a certain number of times during the life of the organism (Fig. 21.22). After their allotted number of divisions, the cells eventually become nonfunctional. It is known, for example, that the number of functional nerve cells in the human body declines steadily after the age of 30 (Fig. 21.23).

Another hypothesis assumes that aging of cells results from gene mutations that alter the genetic information coded in DNA. Such changes might result in the synthesis of defective enzymes, which could upset the normal metabolism of various cells; or they may alter cells in such a manner that lysosomes release increased amounts of tissue-destroying enzymes.

As we saw in Chapter 14, after skeletal muscle cells die, they are not replaced by new cells. Instead, the dead cells are replaced by a tough connective tissue. With fewer functional cells, a working mus-

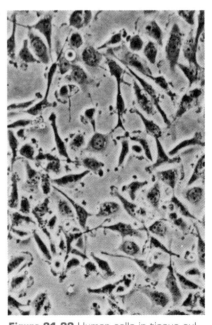

Figure 21.22 Human cells in tissue culture. Such cells can divide only a limited number of times before either dying or becoming transformed into cancer cells. (From Clark, M. E.: *Contemporary Biology*. 2nd ed. Philadelphia: Saunders College Publishing, 1979, p. 471.)

> The loss of natural immunity and hormonal imbalance have been suggested as factors contributing to aging.

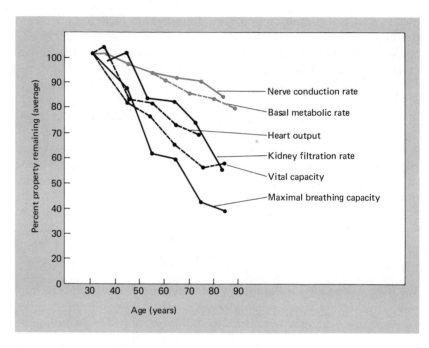

Figure 21.23 Decline in physiological functions after age 30.

cle may become overtaxed, a factor that may hasten the aging of the remaining muscle cells. Similarly, the amount of connective tissue in other organs and structures increases with time. Blood vessels infiltrated with excessive connective tissue become hardened and less elastic, thereby contributing to arteriosclerosis and hypertension, two common diseases of the aging.

Other factors, such as loss of natural immunity and hormonal imbalance, also have been suggested as contributing factors in the aging process. With a loss of immunity, the body becomes more susceptible to infection and disease, and the body may be attacked even by its own antibodies. A decline in the level of hormones alters the chemistry of the body, leading to malfunction of the endocrine organs and an upset of homeostasis.

In considering the factors just presented, perhaps the fundamental question is the following: Are these factors the *cause* of aging or simply the *result*? Or both? The truth is that we do not know. The science of the physiology of the aging—called **gerontology**—is a relatively new discipline in the biological sciences, and the questions it seeks to answer are formidable indeed. What is aging? Why do organisms age? Is aging inevitable? For now, the answers remain hidden from our view, locked somewhere within a living cell.

REVIEW OF ESSENTIAL CONCEPTS

1. Human *embryonic development* involves the processes of *growth*, *cellular differentiation*, and *morphogenesis*.
2. *Cleavage* is a series of mitotic divisions of the zygote. When cleavage reaches the 16-cell stage, the embryo is called a *morula*.
3. As cleavage proceeds, the human zygote moves down the fallopian tube toward the uterus. Prior to implantation, the morula becomes a hollow ball of cells called the *blastocyst*. The blastocyst contains the *inner cell mass* from which the body of the human embryo arises.
4. *Identical twins* arise from the same zygote; *fraternal twins* form from two separate egg cells fertilized by different sperm.
5. *Implantation* occurs by the tenth day of embryonic development. During the second week, the *amniotic cavity* forms within the inner cell mass. One of the fetal membranes, the *amnion*, lines the amniotic cavity. The *amniotic fluid* within the amnion suspends and protects the embryo.
6. The *embryonic disc*, which forms within the inner cell mass, is composed of the embryonic tissues *ectoderm* and *endoderm*. The amnion is formed from ectoderm, whereas the second fetal membrane, the *yolk sac*, arises from endoderm.
7. The third fetal membrane, the *chorion*, develops from the outer wall, or *trophoblast*, of the blastocyst.
8. The chorion and uterine tissues form the *placenta*, an organ across which respiratory gases, nutrients, and wastes are exchanged between the mother and fetus.
9. The fourth fetal membrane, the *allantois*, develops from the hindgut of the embryo. The blood vessels and stalk of the allantois eventually become incorporated into the *umbilical cord*.
10. The third embryonic tissue, *mesoderm*, forms between the ectoderm and endoderm. Most of the organs and structures of the human body consist of tissues derived from all three embryonic tissues.
11. The first month of embryonic development is characterized by the appearance of the *notochord* and *neural tube*. In addition, the skeletal and muscular systems begin developing, along with the organs of special sense, the heart, and the digestive system. *Limb buds* appear, and the head and neck area develop small slits and pouches.
12. During the second month of embryonic development, the embryo becomes more human-like; internal sex organs are distinguishable, and a small tail develops and regresses.
13. During the third month of development, the fetus possesses the rudiments of all the major organs. Sex can be determined externally, reflexes are evident, and the notochord degenerates.
14. The fourth, fifth, and sixth months of development result in a fourfold

increase in the length of the fetus and a weight gain of nearly a pound and a half. Hair is present on the head, and the skin is red and wrinkled.

15. The final 3 months of development are characterized by rapid body growth and final development of the organ systems. At the end of the ninth month, human fetuses average 20 inches in length and weigh about 7 lb.
16. Normal human development—measured from conception until birth—requires 266 days.
17. Childbirth, or *parturition,* results from a combination of factors that induce contractions of the uterus and cervix. Movement of the fetus and increased secretion of estrogens and oxytocin set up strong uterine contractions that help push the baby into the birth canal.
18. The birth process begins with strong uterine contractions known as *labor.* Next, the amnion ruptures, and the head of the fetus enters the birth canal for delivery. After birth, the *afterbirth* is expelled through the vagina.
19. The developing fetus receives oxygen and nutrients from the placenta via the *umbilical vein;* waste materials from the fetus are delivered to the placenta via the *umbilical arteries.*
20. Blood coming in to the fetus through the umbilical vein enters the right atrium of the heart. Most of the blood is shunted into the left atrium through the *foramen ovale.* From the left atrium, the blood passes into the left ventricle and out the aorta.
21. Some of the blood entering the right atrium of the fetal heart moves into the right ventricle and out through the pulmonary artery. Blood is then detoured through the *ductus arteriosus* into the aorta. Some of this blood circulates to the lower parts of the body, and some enters the umbilical arteries to be delivered to the placenta.
22. At birth, the foramen ovale and ductus arteriosus normally close, so that blood then circulates to the lungs. Failure of these structures to close results in *cyanosis,* or a "blue baby."
23. All human body cells arise by mitotic division of the zygote; therefore, these cells are genetically identical. That this is so has been demonstrated by nuclear transplant experiments and by the development of a complete carrot plant from a single carrot root cell.
24. There is evidence that the cytoplasm of embryonic cells contains stimulatory and inhibitory substances that affect gene activity. In frog eggs, the *gray crescent* normally is divided equally between the two daughter cells at the first mitotic division. These cells develop normally when experimentally separated from each other. If mitosis is altered, only the cell containing the gray crescent develops into a tadpole; the other daughter cell divides to form an unorganized cell mass.

25. As embryonic development proceeds, cells become increasingly differentiated. This is due in part to unequal division of the cytoplasm; the different amounts of cytoplasm and different kinds of cytoplasmic materials in some manner affect gene expression, which controls cellular differentiation.
26. Entire tissues influence embryonic development through the process of *induction*. Experiments using amphibian embryos have shown that the *dorsal lip of the blastopore* acts as an inducer, or *organizer*, to alter the fate of other embryonic tissues.
27. Induction is thought to be controlled by chemical substances that diffuse from one tissue to another. These chemicals stimulate or inhibit development by activating or repressing certain genes.
28. At the molecular level, cellular differentiation is controlled through the synthesis of a particular kind of messenger RNA. Only the unraveled, or looped, portions of a chromosome are the sites where DNA is synthesizing mRNA. The result is the synthesis of proteins specific for each type of tissue.
29. Development is a lifelong process of which *aging* is a part. Although little is fully understood concerning the process of aging, a variety of factors apparently contribute to the process.

APPLYING THE CONCEPTS

1. Explain what is meant by each of the following terms: cleavage, growth, cellular differentiation, morphogenesis, and blastocyst.
2. What embryonic structures arise from the blastocyst?
3. What are the functions of the four fetal membranes?
4. Describe the placenta.
5. What factors influence parturition?
6. How does fetal circulation differ from circulation after birth? Why is fetal circulation different?
7. Describe an experiment that indicates that all body cells are genetically identical.
8. What is the experimental evidence that the cytoplasm of a cell influences differentiation?
9. What is the significance of the fact that some embryonic cells undergo unequal cytoplasmic division?
10. What is induction? Give an example of this process.
11. Explain the basic mechanism of the control of cellular differentiation at the molecular level.
12. What are chromosome "puffs"?
13. Why is aging considered to be a part of the process of development?

22 animal behavior

THE ESSENTIAL OBJECTIVES

You have understood this chapter when you are able to:

1. Explain the concept of anthropomorphism as applied to behavior.
2. Describe the basic types of innate and learned behavior in animals.
3. Using examples, discuss the significance of social behavior in insect societies and vertebrate societies.

Fundamentally, any type of behavior is a response to a stimulus. The patterns of behavioral responses are influenced by genetic and environmental factors that are usually of survival value to the organism.

PREVIEW OF ESSENTIAL TERMS

anthropomorphism The practice of ascribing human characteristics to something that is not human.

taxis In animals, an automatic orientation of the body in response to an external stimulus.

releaser An environmental stimulus or "signal" that elicits specific behavioral responses in animals.

pheromone A chemical substance that, when released by an animal, is capable of causing behavioral changes in other animals of the same species.

circadian rhythm An activity by an organism that occurs on a 24-hour cycle.

society An organized group of animals living together to mutual advantage.

territoriality The defense of a section of the home range against intruders of the same species.

migration A seasonal expedition by an animal group from its home ground.

Animal Behavior

> *Anthropomorphism* is the practice of ascribing human characteristics to something that is not human.

INTRODUCTION

One of the more complex areas of biological study is that dealing with the causes and interpretation of the behavior of living organisms, particularly animals. Part of the difficulty lies in the fact that, as human beings, we often tend to take the **anthropomorphic** (Gk. "human form") view by placing ethical or volitional judgments on the behavior of organisms other than ourselves. Anthropomorphism is simply the practice of ascribing human characteristics to something that is not human. There is, then, the temptation to regard the behavior of a given organism as either "good" or "bad" or to assume that the organism is driven by some goal-directed purpose or motivation in a manner similar to that which we experience. These are not uncommon assumptions, considering our usual desire for easy "explanations" or a familiar frame of reference. Scientifically, however, we are obligated to remain within the boundaries of objectivity and experimental evidence. Accordingly, to state that a mother bird "loves" her nestlings (Fig. 22.1) or that a female spider devours her mate because she does not "care" for him anymore is not only speculation but also a venture into the realm of human sentiment. In other words, such speculation and sentiment constitute an anthropomorphic view of animal behavior.

How, then, are we to interpret the behavioral activities of animals? Essentially, we have to think in terms of the physical and

Figure 22.1 To assume that a mother bird cares for her nestlings because she "loves" them is to hold an anthropomorphic view of animal behavior. (From Orr, R. T.: *Vertebrate Biology*. 4th ed. Philadelphia: Saunders College Publishing, 1982.

> Analyzing behavior is a complex problem since it is often necessary to separate and understand the relative significance of *inheritance* versus *learning*.

chemical attributes of the organism itself, plus the influence of the environment in which it acts. This entails a realization of the structural and functional relationships that exist between the cells, organs, and systems of an organism and the manner in which these are affected by various external stimuli. Quite importantly, this involves a consideration of inheritance, since the behavior of any organism is restricted within certain limits as determined by its genetic make-up.

We have mentioned that the analysis of behavior is a complex problem. Much of the complexity lies in separating and understanding the relative significance of *inheritance* on the one hand and *learning* on the other. Although it is unnecessary for our purposes to analyze the behavioral theories concerning these two factors, it is important that you are aware that both inheritance and learning, particularly in the higher animals, are believed to play crucial roles in most behavior patterns. The two systems intricately bound to behavior in higher animals are the endocrine and nervous systems, both of which are inherited. Being highly developed in these animals, the two systems interact to provide a wide range of potential behavioral patterns. On the other hand, the behavioral patterns of lower animals are more limited as a result of the inheritance of less complex systems. This is particularly true in the case of the nervous system, since its extent of development largely determines the influence of learning on behavior.

Since the purpose of this book is to acquaint you with the essential biological principles, another point concerning the study of animal behavior is in order. You may be aware of attempts to extrapolate the behavior of some animals to that of human beings. This can be rather dangerous ground scientifically in that man, with his complex higher brain centers, diverse personal experiences, and varied environmental influences, is not readily disposed to comparative behavioral analyses. This does not necessarily mean, however, that *nothing* of value concerning human behavior can be gleaned from observing other animals. Nonetheless, we would be wise to scrutinize those behavioral theories mutually applied to mice and men.

TYPES OF ANIMAL BEHAVIOR
Innate Behavior

Innate behavior is genetically programmed, "inborn" behavior in which learning plays little, if any, part. We generally speak of this type of behavior as being *instinctive*, manifested as an automatic response to a stimulus. Such responses in lower animals are called **taxes** (Gk. "arrangement"), a term similar to "tropisms" in the plant world. An animal such as a planarian, for example, exhibits a **negative phototaxis** when exposed to light. This means, of course, that

Animal Behavior

> Another type of innate response is the *reflex*, an unlearned, automatic response to a stimulus involving the following nervous system components: receptor cell, conductor cell, and effector cell.

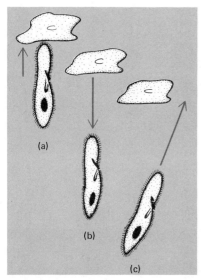

Figure 22.2 Avoidance behavior: negative chemotaxis in *Paramecium*. (a) When confronted with a noxious chemical, the paramecium backs off (b) and swims forward at a different angle (c).

Figure 22.3 A spider spinning its web. This type of behavior is innate, requiring no prior input of information.

the animal moves away from the light source. Similarly, a paramecium, when confronted with some noxious chemical, backs off, turns to one side, and swims forward again at a different angle (Fig. 22.2). In this avoidance reaction the paramecium exhibits a **negative chemotaxis,** or a movement away from the chemical stimulus. Many other protozoans are equally capable of this type of avoidance behavior.

Other examples of innate behavior include the spider's spinning of its web (Fig. 22.3) and nest building in birds. In either case, no prior information or observation is necessary for the organisms to perform their respective tasks. Such behavior is instinctive and usually serves to promote the welfare and survival of the species.

Among many lower and all higher animals is a type of innate response called a *reflex*. In lower animals, reflexes may account for the totality of the observed behavior, but in higher animals reflexes may be only a small part of the overall behavior pattern. A reflex is an unlearned, automatic response to a stimulus and involves some type of nervous system. Except for the sponges, all other animal groups have some type of nervous system and are capable, therefore, of reflex behavior. Involved in the nervous component of reflex behavior are (1) a *receptor* cell, which picks up the stimulus; (2) a *conductor* cell, which relays the nerve impulse; and (3) an *effector* cell, which makes the response (Fig. 22.4). When prodded or otherwise physically stimulated, a lower animal such as an earthworm will simply withdraw in an effort to escape the stimulus. Each time the earthworm encounters such a stimulus, its reflex behavior is pretty much the same. In those animals with higher, more complex nerve centers, a simple reflex may elicit a variety of behavioral patterns. In these cases, the reflex is influenced by nerve impulses emitted from the higher nerve centers. This latter instance is characteristic of you and all other human beings. To illustrate this, consider the following contrasting examples.

Assume that you experience a sudden painful stimulus, such as burning your finger with a lighted match. Your pain receptor cells are stimulated, and the resulting signal is transmitted by a conductor cell up the arm to the spinal cord. From the cord, the signal is transmitted back out to the effector (muscle) cells in your arm, and the contraction of the cells causes you to yank your hand away from the painful stimulus. This also is the fundamental type of reflex behavior observable in many lower organisms—even those such as the earthworm that lack a spinal cord.

Now, in another example, assume that you are in a grave situation, aware of impending danger. First of all, your *awareness* results from sensory stimuli picked up by special sense organs—the eyes and ears, for example—and from your higher brain centers, from

Releasers are specific environmental signals (sounds, odors, chemicals, colors, and actions) that evoke particular responses in animals.

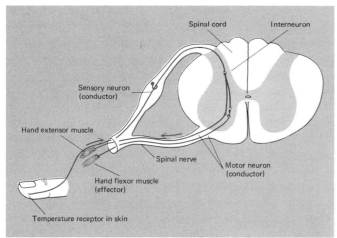

Figure 22.4 Reflex behavior involves some type of nervous system. In human beings and other vertebrates, nerve impulses from a receptor organ are relayed to the spinal cord and then to an effector by way of conducting nerves.

which comes your recognition of the *concept* of danger. In a matter of moments, all sensory stimuli are integrated and signals are relayed to the muscles that cause you to act. On one given occasion your reaction may be to run away in panic; on another you might freeze in your tracks. Although both of these examples in human beings illustrate the same basic reflex response, you can appreciate in the second example the influence of the higher nervous centers. Moreover, the reflex behavior pattern elicited by the burning match is almost always the same; with impending danger, one's behavior pattern is generally less stereotyped. In these and most other situations, reflexes basically serve as protective mechanisms for an animal.

There are many instances in which certain discrete signals from the environment evoke particular behavioral responses in animals. Such external "sign" stimuli are called **releasers**. Specific sounds, odors, chemicals, colors, actions, and so on may act as releasers. One of several classic experiments that elegantly demonstrate releasing behavior involves a small, three-spined fish appropriately called a stickleback. In the spring, the underside of the male stickleback becomes bright red. Fighting among the males apparently is associated with this red color. Male sticklebacks will attack wooden models that appear quite unlike real male sticklebacks except that the underside of the models has been painted a bright red color. Other models that are very fishlike in appearance but lack the red underside evoke little response from the combative males (Fig. 22.5). Ostensibly, the red underside acts as the releaser for the fighting behavior observed in these fish.

A quite interesting group of releasers comprises certain chemicals known as releaser **pheromones**. These substances, secreted by a

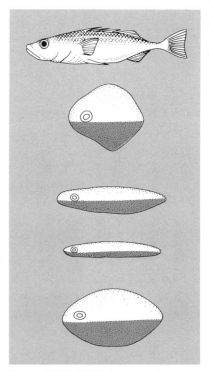

Figure 22.5 Stickleback models. The upper model, although fishlike in appearance, did not stimulate aggressive behavior in male sticklebacks. However, all of the cruder models with red undersides stimulated aggressive behavior.

> *Pheromones*, chemical substances secreted by many animals, cause immediate behavioral changes.

Figure 22.6 By secreting a "tracking" pheromone, ants lay down a chemical trail that can be followed by other members of the colony.

variety of different animals, are capable of causing immediate behavioral changes in other members of the same species. Various female insects, for example, utilize pheromones as sex attractants that the males are able to detect, often over long distances. Ants secrete a "tracking" pheromone as they wend their way to a food source and back to the nest (Fig. 22.6). Other members of the colony follow this chemical trail to the food source, and they also release pheromone on the return trip home. In this manner, the trail stays fresh as long as the food source is available. Ants that do not find food do not secrete the pheromone and eventually, when the food is gone, the trail disappears as the pheromone vaporizes.

A particularly interesting example of ant behavior controlled by pheromones involves disposal of the dead members of the colony. The dead, decomposing body of an ant releases certain chemical substances that act as pheromones. On detecting these pheromones, some of the workers of the colony drag their dead associate out of the nest and carry the carcass to the local dumping site. In experimental procedures, researchers extracted the decomposition pheromones from the bodies of dead ants and applied these chemicals to the bodies of living ants. Invariably, they, too, were hauled off to the dump, but shortly made their way back to the colony, only to be carted off again. From this example, it is evident that the releaser pheromone is the overriding stimulus directing the behavior of the ants in disposing of their dead. The fact that the living ants coated with pheromones are struggling or otherwise active while being carried is totally disregarded by the workers. This type of rigid, innate behavior is characteristic of many insect societies.

Throughout the plant and animal kingdoms there are numerous instances of behavioral and associated physiological activities that occur on a fairly regular schedule. From your own experiences, you have heard of or seen animals that are more active at night than during the day. These nocturnal animals leave their lairs and forage for food only once in a 24-hour period. Various flowers open their petals in the morning and close them at sundown (Fig. 22.7). Man, too, often functions according to a regular schedule. For example, your body temperature usually rises in the later afternoon and declines in the early morning. Many people can awake at the same hour every morning, regardless of their desire to do so. These and many other behavioral activities of plants and animals may occur on a fairly regular 24-hour cycle and are called *circadian* (L. "about a day") *rhythms.* There is considerable evidence to indicate that such cyclical behavior is influenced by an internal "biological clock." Believed to originate within the organism, such a clock is some type of internal timing device that controls the rhythm of certain bodily ac-

> *Biological clocks* are believed to control daily and seasonal behavioral cycles.

Figure 22.7 "Sleep" movements, shown here in a bean plant, occur in a rhythmic cycle. The leaves extend upward during the day and fold inward at night.

tivities. You are familiar with the yearly hibernation periods of squirrels, bears, and other mammals (Fig. 22.8) and the seasonal migrations of birds. These annual cycles, too, are believed to be controlled by innate biological clocks. It should be noted, however, that at present virtually nothing is known about the chemical or physical nature of these clocks.

It has been suggested that rhythmic behavior is, in actuality, a response to some environmental stimulus, such as gravity, cosmic radiation, and so on. This is not generally believed to be the case, however, although biological clocks most certainly are responsive to external stimuli. In fact, the clocks may be set or reset by environmental conditions. Consider, for example, the flowers previously mentioned that open in the morning and close at sundown. Such flowers are on a normal light-dark cycle of approximately 24 hours. If the flowers are placed in *constant* light, however, the same 24-hour rhythm of opening and closing continues for a time. Furthermore, by experimentally altering the periods of light and dark so that they do not conform to the normal periods, the flowers will continue to open and close on a 24-hour cycle. This cycle, though, may be out of phase with the normal one, i.e., the periods of opening and closing of the flowers may not occur at the same time of day or night as in the normal cycle. However, if the "artificial" cycle is too far out of phase, the plant may revert to its natural cycle. Moreover, if the plant is returned to its regular light-dark cycle, it will resume its normal 24-hour rhythm. Thus, the internal clock that controls this circadian rhythm can be reset by altering the environmental conditions of light and darkness.

The concept of biological clocks is of considerable significance in understanding rhythms in human beings. In addition to fluctua-

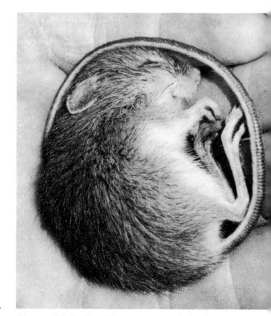

Figure 22.8 The hibernation period of this jumping mouse and other animals is thought to be controlled by an innate biological clock. (From Ebert, J. D., Loewy, A. G., Miller, R. S., and Schneiderman, H. A.: *Biology.* New York: Holt, Rinehart and Wilson, 1973, p. 396.)

> *Learning* is a relatively long-lasting modification in behavior resulting from previous experience or practice.

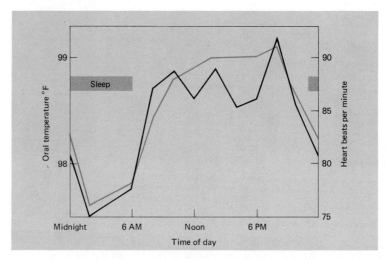

Figure 22.9 Daily changes in heart rate. Correlation of heart rate (black line) with daily temperature fluctuations (color line).

tions in body temperature, human beings have daily rhythms that result in regular sleep-wake cycles, variations in concentrations of hormones in the blood, changes in enzymatic activity, fluctuations in mental abilities, and alterations in many other physiological processes (Fig. 22.9). If these rhythms are altered significantly, the individual may become irritable, disoriented, or even ill. Today, we often hear of the disconcerting effects of "jet lag" resulting from long-distance flying. Anyone who has experienced this knows that it usually takes a day or two or longer for the body to adjust itself to the new time zone. In effect, one's biological clock has been reset.

Learned Behavior

Much of the behavior we observe in animals cannot be explained by any of the types of innate behavior we have just discussed. Innate behavior, however, can be modified by learning, which usually develops through experience. With some animals, it is difficult to determine if a change in their behavior is truly learned or if it is the result of maturation or physiological changes. A young animal, for example, may not be able to walk on its first try, but with maturation and development of its bones, muscles, nerves, and so forth, it soon accomplishes the task. Consequently, it is doubtful that walking in this case is a type of learned behavior. Learning, of course, can be quite complex, as is seen in the higher primates, particularly man. Different groups of animals have different learning potentials, attributable in part to the degree of development of the higher brain centers. As a working definition, we can consider learning to be a relatively long-lasting modification in behavior resulting from previous experience or practice.

> *Habituation* is a type of learning characterized by a loss of response to previously experienced stimuli.

Habituation. This type of learning is quite simple and is characterized by a loss of response to previously experienced stimuli. More exactly, an animal becomes habituated to certain unimportant stimuli to the extent that eventually it ignores them. A new puppy in the home may, at first, withdraw from or display fear at every household sound. In time, however, the pup becomes habituated to the sounds and, finding that the sound stimuli are unimportant to its welfare, finally ceases to respond at all. As anyone who has owned a dog can attest, this form of learning is long-lasting in that over a period of many years the animal may never respond to the stimuli again.

Conditioning. Basically, there are two types of conditioning behavior—*classic conditioning* and *operant conditioning*. The first of these, which also is the simplest type of conditioning, involves a form of reflex behavior. Classic conditioning was first demonstrated by the Russian physiologist I. P. Pavlov in his experiments with dogs (Fig. 22.10). As you may be aware, placing food in a dog's mouth causes the dog to salivate. Each time Pavlov fed his dogs he also rang a bell. Eventually, the dogs became conditioned by associating the sound of the bell with food. Thereafter, when Pavlov rang the bell, the dogs would salivate even if the food was not offered. The salivating response by the dogs to the sound of the bell is called a *conditioned reflex* and is prevalent in many animals. Some insects, for example, that are brightly colored or patterned smell or taste bad to potential predators. After a few unpleasant encounters, the predator becomes

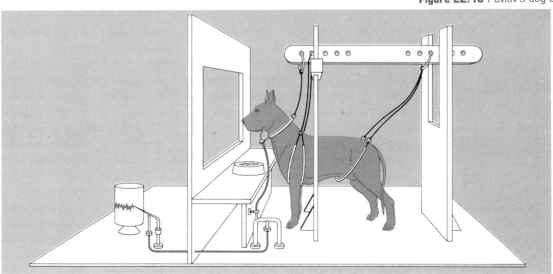

Figure 22.10 Pavlov's dog experiment.

> *Operant conditioning*, also known as trial-and-error learning, is based upon reward or punishment from a particular action.

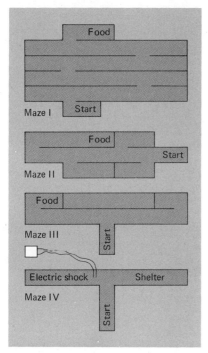

Figure 22.11 Different kinds of mazes used in trial-and-error learning experiments.

conditioned to avoid many other brightly colored insects, even though these may not be at all disagreeable. The predators, then, become conditioned to the particular color stimulus and associate it with all similarly colored insects.

Operant conditioning is based upon reward or punishment resulting from a particular action. Experimentally, an animal can be induced to perform some activity for which it is immediately rewarded, and it may be punished if it fails to perform correctly. This also is referred to as ***trial-and-error*** learning because the animal may have several options available in finding the one that is rewarded. Through trial-and-error experiments in the laboratory, animals such as ants, earthworms, planarians, and mice have navigated various mazes to be rewarded with food. If, however, they make a wrong turn in the maze, they are punished with a mild electric shock (Fig. 22.11). Over a period of time, each animal learns to avoid the disagreeable shock and seeks its reward instead.

Many of us are all too familiar with trial-and-error learning, since it is involved in much of what got us into trouble as children. Hurling unpleasantries at your friend next door often resulted in painful atonement at the hands of your friend, your parents, or both. Thereafter, you knew that hurling unpleasantries was an act to avoid—at least audibly. By trial and error we all learned, and continue to learn, which acts are to be avoided and which ones result in reward. Fortunately, most human beings often require only one trial to learn which acts are pleasant and which are not. In a lighter vein, your relative proficiency at sports and games comes about largely as a result of trial-and-error learning.

Imprinting. Early in the lives of many animals, there is a strong response and attachment to the first moving object they see. Ordinarily, of course, this object is their mother, but it has been found that these young animals will imprint upon any moving object, including a toy, some other animal, or even a human being. In baby chicks and ducklings, the critical period for imprinting is within about 36 hours after hatching (Fig. 22.12). Beyond this time, imprinting behavior is lost, even if the young animal encounters its own mother. Even in its adult life, a goose will direct its sexual drive toward the imprinted object, rather than toward another of its own kind. Apparently, the early attachment phase is necessary for the young animal to learn to recognize the imprinted object. Something no doubt similar to imprinting occurs in human infants, who form a strong maternal bond early in life. Imprinting obviously has survival value for a young animal, ensuring that it remains for a time under the protective custody of its mother.

> *Reasoning,* or insight learning, is the highest form of learning and can be defined as the ability to draw conclusions from known or presumed facts.

Figure 22.12 Through imprinting, young animals follow the first moving object they encounter. Usually this object is their mother.

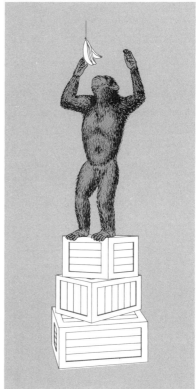

Reasoning. The highest form of learning is the result of reasoning, also referred to as *insight learning.* Reasoning is the ability to draw conclusions from known or presumed facts. It reaches its highest form in man, but some of the higher primates, particularly monkeys and chimpanzees, also demonstrate behavior based on the ability to reason. A classic example of reasoning has been observed in the chimpanzee. When confronted with the problem of reaching a bunch of bananas hanging from the ceiling, a chimp will begin to stack boxes lying about the room until it can climb up to reach the food (Fig. 22.13). Notice that trial-and-error learning is not involved in this instance but rather the ability to figure out a solution without prior stimuli and without being taught.

Associated with the process of reasoning in man are factors such as concept formation and abstraction. A certain animal, for example, may be able to distinguish the color red, but it is unable to comprehend the concept of redness. Similarly, animals cannot separate the attributes of an object from the object itself. Man may refer to a given object as "beautiful," but "beauty," as far as we know, is an abstraction impossible for the animal to make. As you are well aware from your own experiences, reasoning or insight learning can result in extremely complex behavioral patterns.

Figure 22.13 Insight learning. Confronted with the problem of reaching food hanging from the ceiling, the chimpanzee solves the problem by stacking boxes until it can climb up and reach the food.

GROUP BEHAVIOR

It is common knowledge that many species of animals live and interact together within their own groups. These include animals such as deer and zebra that run in herds, wolves that form packs, birds of a flock, and so forth. If the individuals in these groups interact to mu-

In *animal societies*, group members interact to mutual adaptive advantage. Characteristics of a society are interbreeding within the group, division of labor, and rigid organization.

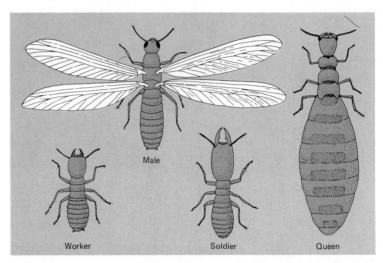

Figure 22.14 Castes of the North American termite.

tual adaptive advantage, as they usually do, the group is called a *society*. A society (or social behavior) is characterized by interbreeding within the group, division of labor among the members, and a rather rigid organization of the group. Nonsocial groupings of animals are referred to as *aggregations*. For example, animals of the same species that simply gather at the watering hole or, in the case of houseflies, converge on a hamburger patty obviously are aggregating, but they do not form a society.

The most highly developed societies are found in the insects and mammals. The caste society of the common termite is a good example (Fig. 22.14). Having been around for some 200 million years, termites constitute the oldest animal society known. Division of labor among the members is quite exact within the termite colony. The *reproductives* include only the queen and king. The queen exists solely to lay eggs, and the king usually remains in the chamber with her. The *workers*, which are sterile males or females, tend the colony, feed the young, and wait upon the king and queen. The *soldier* termites, which possess strong jaws and a thick body covering, serve to defend the colony against enemies. The reproductives and soldiers are both fed by the workers.

The eggs laid by the queen develop into either workers or soldiers, depending upon the needs of the colony at the time. Although the workers and soldiers are sterile, at certain times of the year, winged reproductive males and females develop and soon fly away to found a new colony. Most of the behavioral patterns of termites, as with other insect societies, are primarily instinctive.

An even more diverse division of labor is seen in honeybee societies. A honeybee colony has one queen; several hundred males,

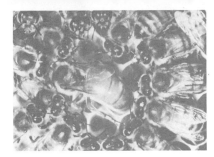

Figure 22.15 Worker honeybees tending the queen (center), their mother. The workers feed and groom the queen, raise the eggs she lays, and defend the hives from intruders. (From Arms, K., and Camp, P. S.: *Biology.* New York: Holt, Rinehart and Winston, 1979, p. 827.)

or drones; and thousands of sterile female workers (Fig. 22.15). Like her counterpart in the termite colony, the queen honeybee exists primarily to lay eggs. To prevent the production of rivals, the queen releases a pheromone known as "queen substance," which prevents the development of ovaries in the workers so that no new eggs are produced. However, in the early spring, the secretion of "queen substance" declines, a stimulus that causes the workers to begin to construct large specialized queen cells in the hive in anticipation of a new queen. The queen then lays a fertilized egg in each of these cells; later, she also lays an unfertilized egg in the smaller cells of the hive. The developing larvae are fed a whitish "brood food," secreted from head glands of worker bees. Some of the queen larvae continue to be fed on "brood food," also known as royal jelly, throughout their development. In honeybees, all fertilized eggs develop into diploid females, but only those fed exclusively on royal jelly can become queens. If several queens develop at the same time they fight it out until only one remains. After two or three days of being fed "brood food," all other larvae are fed honey and pollen. On this diet, the fertilized eggs develop into sterile female workers, whereas the unfertilized eggs develop parthenogenetically into haploid male drones.

When the new queen emerges, she leaves the hive on her nuptial flight, followed by the drones of the colony. She mates with several drones, accumulating enough sperm in her sperm sac to provide for the next 5 to 10 years of egglaying. The drones die almost instantly after mating. The queen returns to the hive to begin a new colony; in the meantime, the old queen, along with many of the workers, leave the old hive and swarm to a new location, where they, too, establish a new colony.

The role of the queen honeybee is limited almost exclusively to reproduction, but the workers perform a variety of tasks. Young workers feed the larvae and prepare the cells of the hive for the eggs laid by the queen. Other workers act as defenders of the hive and store food brought in from the outside by the older workers. The sole function of the drones is to fertilize the queen. Those not involved in mating during the nuptial flight are eventually driven out of the hive or killed by the workers before the new drones emerge.

Group behavior among the mammals and many other vertebrates assumes a variety of forms. One rather common form is that of ***territoriality,*** which is the defense of a section of the home range against intruders of the same species (Fig. 22.16). An animal's home range is the area in which it normally carries out its day-to-day activities; however, the home ranges of some species may overlap with

> *Dominance* in social hierarchies serves to reduce fighting and establish order.

Figure 22.16 Territories. Many sea birds, like these Adelie penguins in Antarctica, nest in crowded breeding groups on land. Each bird defends a small territory, which may include only the nest and standing room immediately around it. Three nests (depressions in the ground lined with stones) are visible only a few feet apart in this photograph: one on the right and one on the left in the foreground, and one under the bird sitting in the middle. (From Arms, K., and Camp, P. S.: *Biology*. New York: Holt, Rinehart and Winston, 1979, p. 611.)

those of another species. A territory is usually within, and therefore smaller than, the home range, and is the specific area most strongly defended by an animal group. The size of the home range varies with the size of the animal; in general, larger animals have more spacious home ranges than do smaller animals. Other than serving the obvious function of protecting the feeding and mating areas, territoriality also aids in social stabilization by preventing overcrowding. As a major result, competition for resources is minimized.

Within many vertebrate societies, the group is organized into a *social hierarchy*. Involved in social hierarchies is the concept of *dominance* (Fig. 22.17). Depending upon the society, there may be several dominant individuals, or there may be one or more dominant groups. The former instance is characteristic of chickens, in which the social hierarchy consists of a "pecking order." The dominant bird in the order pecks all the others, but they will not peck back. The bird second in rank pecks all the chickens of inferior rank in the order and is itself pecked only by the one dominant chicken. This is repeated at successive levels in the hierarchy, with the unfortunate chicken at the bottom being dominant over none of the others. There is also a pecking order of sorts found among various primate groups, such as monkeys and baboons. In these instances, groups of individual males or females may enjoy higher social status than do others. Such dominant positions may be in part inherited from dominant forebears, but age, strength, seniority, and other factors often are involved as well.

Within the social group, dominance serves to reduce fighting and establish order. This is of significant survival value to all members, in terms of both protection and mating.

Migrations, involving both innate and learned behavior, are expeditions of animals that periodically leave their home ground for extended periods of time, later returning to the original site.

Figure 22.17 Dominance among wolves. The dominant male in the center receives the "affectionate" attention of the pack. (From Ebert, J. D. Loewy, A. G., Miller, R. S., and Schneiderman, H. A.: *Biology*. New York: Holt, Rinehart and Winston, 1973, p. 452.)

Although group life assures the possibility of finding a mate, the dominant males and females usually select among each other, whereas those on the lower strata must choose among others of the same social status. Often, however, mating among members low in the order is much less frequent than among the more dominant animals. This, too, has survival value for the group by ensuring a strong and healthy posterity.

Migration. Some of the most complex behavior patterns are seen in animals that periodically leave their home ground for extended periods of time and later return to the original site. Such expeditions are known as ***migrations*** and involve both innate and learned behavior (Fig. 22.18). Animals such as birds, fish, salamanders, and butterflies annually migrate hundreds or thousands of miles en route to breeding and feeding grounds; many other herbivores and predators do not follow an annual migration pattern but usually migrate in response to a critical shortage of food.

The migratory impulse apparently is regulated by an internal biological clock that operates on a seasonal rhythm rather than on a 24-hour cycle. For example, in the winter, gray whales migrate from the Arctic Ocean south to their breeding grounds in the waters off the coast of Mexico (Fig. 22.19). Birds such as warblers migrate in autumn, leaving their feeding grounds in Canada and flying nonstop to South America. As pointed out earlier, biological clocks are influenced by external environmental stimuli. Changes in temperature, tidal rhythms, lunar cycles, and so on often act as cues to stimulate migratory behavior.

720
Animal Behavior

> *Navigation* is the act of determining the correct course or route of travel and is an important aspect of animal migrations.

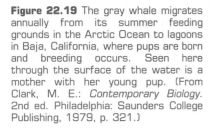

Figure 22.18 Aerial view of elk in migration. (From Orr, R. T.: *Vertebrate Biology.* 4th ed. Philadelphia: Saunders College Publishing, 1982.)

Migration is essentially another adaptive mechanism beneficial to survival of a species. Animals that migrate in groups are afforded protection by other members of the group and are in close contact with potential breeding partners. Furthermore, by moving to varied environments, migrating animals are not limited to only one source of food.

Perhaps the most remarkable aspect of animal migration is **navigation,** the act of determining the correct course or route of travel. It is known that certain animals use such navigational cues as visual

Figure 22.19 The gray whale migrates annually from its summer feeding grounds in the Arctic Ocean to lagoons in Baja, California, where pups are born and breeding occurs. Seen here through the surface of the water is a mother with her young pup. (From Clark, M. E.: *Contemporary Biology.* 2nd ed. Philadelphia: Saunders College Publishing, 1979, p. 321.)

> Many animals are capable of communicating with other members of their species through the use of pheromones, sounds, and visual signals.

landmarks, chemical stimuli, or sound. Many species of migrating birds use the sun and stars as visual landmarks (Fig. 22.20), whereas various land animals apparently recognize the surrounding landscape. Young salmon, after leaving their freshwater birthplace to swim to the ocean, return several years later to spawn in the very stream in which they were hatched. There is strong evidence that the salmon are guided to their birthplace by the odor of the water. Some of the mammals, most notably bats and dolphins, navigate by means of *echolocation*. This mechanism involves the detection of reflected sound waves originally emitted by the animal. Dolphins, for example, emit high-pitched squeaks and whistles, which are reflected off a submerged object; the returning echoes inform the dolphins of the location of the object, which can then be avoided. The echolocation system of bats functions in essentially the same way (Fig. 22.21).

Animal Communication. Many animal groups are capable of communicating with other members of the same species and usually with animals of another species (Fig. 22.22). In broadest terminology, communication involves any transmission or exchange of information. This may include information related to mating, food, antagonism, or danger. As we have seen in this chapter, ants and honeybees often communicate to members of their own species by pheromones. Insects also communicate by sound. Male crickets rub their wings together to produce calls that attract females, and male mosquitoes respond with apparent ardor to the sound vibrations made by the wings of the female mosquito during flight.

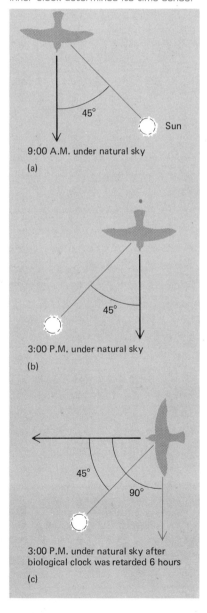

Figure 22.20 Solar navigation in birds. A fall migrant at 9:00 AM flies 45° left of the sun (a) and at 3:00 PM, 45° to the right (b). If its biological clock is retarded 6 hours, it "sees" the 3:00 PM sun as 9:00 AM and flies due west. Its inner clock determines its time sense.

Figure 22.21 A bat in flight avoids obstacles by a remarkably efficient echolocation sense. (From Camp, P. S., and Arms, K.: *Exploring Biology*. Philadelphia: Saunders College Publishing, 1981, p. 452.)

Figure 22.22 This white-sided dolphin, a member of the porpoise family, is capable of communicating several dozen distinct sounds. (From Clark, M. E.: *Contemporary Biology*. 2nd ed. Philadelphia: Saunders College Publishing, 1979, p. 359.)

Courtship among some animals, such as birds, for instance, often involves an elaborate—and sometimes amusing (to us)—system of communication. Male prairie chickens, for example, have enormous red throat pouches, which they inflate and display at mating time (Fig. 22.23). With a group of females looking on, the males first

Figure 22.23 Male sage grouse display their ruffs on a communal mating ground in Montana, attracting the less brightly colored females. (From Clark, M. E.: *Contemporary Biology*. 2nd ed. Philadelphia: Saunders College Publishing, 1979, p. 381.)

Figure 22.24 Threat display of a snapping turtle. By raising herself on her front legs, the turtle makes herself look larger than normal. She withdraws her head into a position where she is ready to lash out at an intruder. (From Arms, K., and Camp, P. S.: *Biology*. New York: Holt, Rinehart and Winston, 1979, p. 612.)

begin to strut, wings hanging to their sides, and then prance back and forth with their bills close to the ground. Adequately impressed, each female eventually chooses one of the males for her mate.

Antagonism among many animals is visually communicated through threat displays. Almost everyone has seen a dog bare its teeth and raise its hackles when confronting an intruder on its territory. Snapping turtles and toads, in an attempt to appear larger than normal to their antagonists, raise themselves on their legs as high as possible (Fig. 22.24).

Among animals of the same species, threatening behavior more often concludes with a "facing-down" encounter, rather than actual fighting. The "loser" in the encounter communicates its defeat by appeasement displays, such as turning its face away from the opponent or huddling in a crouch to make the body appear as small as possible. When fighting does occur between animals of the same species, each is usually careful not to maim or kill the other. Rattlesnakes never bite each other but fight by pushing their heads together until one submits. Bighorn sheep almost always attack each other head on instead of going for a vulnerable spot (Fig. 22.25). There are occasions, however, when animals of the same species inflict serious injury on each other. Ordinarily, this occurs when males fight to defend a territory or to acquire a female. With the exception of these two instances, fighting is of little advantage to the participants. Even in victory, an animal runs the risk of injury to itself, a condition that makes the animal more vulnerable to attacks by its enemies.

Figure 22.25 Animals of the same species, such as these bighorn rams, rarely attack each other with the intent to maim or kill.

REVIEW OF ESSENTIAL CONCEPTS

1. The interpretation of behavior involves consideration of the physical and chemical attributes of an organism, plus the influence of its environment.
2. Behavior in many animals involves a complex interplay between learning and inheritance. The complexity of animal behavior patterns is limited by the extent of development of the nervous system.
3. *Innate* or *instinctive* behavior involves an automatic response to a stimulus. In lower animals, such responses are called *taxes;* examples include *phototaxis* and *chemotaxis.*
4. Innate responses called *reflexes* are characteristic of many lower animals and all higher animals. Reflex behavior among higher animals is influenced by their more complex nerve centers; accordingly, higher animals generally elicit more complex behavioral patterns.
5. External sign stimuli known as *releasers* are signals from the environment that evoke particular behavioral responses in animals. Releasers include sounds, odors, chemicals, colors, actions, and *pheromones.*
6. Pheromones, which are chemicals secreted by many animals, cause immediate behavioral changes in other members of the same species. Examples include a "tracking" pheromone released by ants and a pheromone released from the bodies of dead, decaying ants.
7. Behavioral activities that occur on a regular 24-hour schedule are called *circadian rhythms.* Such rhythms are believed to be influenced by internal timing devices known as "biological clocks."
8. *Learning* is a relatively long-lasting modification in behavior resulting from previous experience or practice.
9. *Habituation* is a type of learning characterized by a loss of response to previously experienced stimuli.
10. *Classic conditioning* is a type of learning that involves an association response known as a *conditioned reflex.* Conditioned animals respond to a stimulus associated with the original stimulus.
11. *Operant conditioning,* or *trial-and-error learning,* is based upon reward or punishment resulting from a particular action by an animal. Through operant conditioning, animals learn which acts to avoid and which acts result in reward.
12. The attachment by young animals to the first moving object they see is called *imprinting.* Beyond a critical time period after birth, imprinting behavior is lost in many animals.
13. The highest form of learned behavior results from reasoning, or *insight learning.* Reasoning is the ability to draw conclusions from known or presumed facts; no teaching or prior stimuli are necessary.
14. A group in which the individual animals interact to mutual adaptive

advantage is called a *society*. Societies are characterized by interbreeding within the group, division of labor, and a rather rigid organization. Vertebrate societies, however, are generally far less rigid than insect societies.

15. Insect societies, such as those of termites and honeybees, are caste societies with a rigidly defined division of labor. Included in these societies are reproductives, soldiers, and workers.
16. One form of group behavior among many vertebrates is *territoriality*, the defense of a section of the home range against intruders of the same species. Territoriality protects the feeding and mating areas and prevents overcrowding.
17. Many vertebrate societies are organized into a *social hierarchy*, in which some individuals or groups exercise *dominance*. This is evident in the "pecking order" among chickens. Dominance is also prevalent in primate societies, such as those of monkeys and baboons. Dominance serves to establish order within a group and this is of significant survival value.
18. *Migration*, which involves both innate and learned behavior, is a seasonal expedition by an animal group from its home ground, to which it eventually returns. Migration apparently is regulated by an internal biological clock that operates on a seasonal rhythm.
19. *Navigation* by an animal is the act of determining the correct course of movement or route of travel. Navigational cues include visual landmarks, chemical stimuli, and sound.
20. Communication among animals involves any transmission or exchange of information. Communication methods include the use of pheromones, sound, courtship rituals, threat displays, and appeasement displays.

APPLYING THE CONCEPTS

1. Why is the anthropomorphic view of animal behavior scientifically indefensible?
2. Describe three examples of innate animal behavior.
3. Explain by example the concept of circadian rhythms.
4. List the types of learned behavior. How does learned behavior differ from innate behavior?
5. What are the basic differences between classic conditioning and operant conditioning?
6. What is an animal society? How would a society differ from a nonsocial group?
7. Explain what is meant by a "social hierarchy" in animal groups.
8. In what manner are innate and learned behavior involved in migration?

index

Abalone, 326
Abdomen, 332, 338
Abortion, 671
Abscisic acid (ABA), 286
Abscission, 286
Absorption, in small intestine, 599
Abstraction, 715
Accelerator nerves, 568
Acetic acid, 425
Acetylcholine, 473
Acetylcoenzyme A, 134
Achilles tendon, 520
Acid(s), definition, 38
   strong and weak, 42
Acquired characteristics, inheritance of, 435
Acromegaly, 629
Actin, 514
Active site, of enzyme, 60
Active transport, 83, 109
Adaptation, 9, 421
   role in evolution, 443
   short-term vs. long term, 18
Addison's disease, 634
Adenine, 62, 172
Adenosine diphosphate (ADP), 76
Adenosine monophosphate (AMP), 76
   cyclic (cAMP), 621, 624
Adenosine triphosphate (ATP), 47
   active transport and, 110
   energy storage in, 76–77
   formation in light reactions, 123
   in protein synthesis, 177
   in protocells, 428
   ion pumps and, 471
   muscular contraction and, 516
   ribose in, 51, 65
   yield from glycolysis, 136–137
Adenyl cyclase, 625
Adhesion, 68
Adipose tissue, 464
Adrenal glands, 612, 631
Adrenalin, 634
Adrenocorticotropic hormone (ACTH), 626, 633
Aerobic process, 130, 431
African sleeping sickness, 229
Afterbirth, 688
Agar, 252
Agglutination, 554
Aggregation, social, 716
Aging, 699
Agnatha, 348
Albino, 504
Alcohol, diuretic effect, 631
   nerve action of, 473
Aldosterone, 585, 612, 631
Alexia, 495
Algae, 251–259
   blue-green, 211, 221–223, 430
   colonial, 257
   corals and, 252
   unicellular, 223–227

Algin, 254
Alimentary canal, 586
Allantois, 358, 682
Allele, 181
   multiple, 192
Alligator, 358–359
All-or-none law, 471
Alpha cells, 640
Alternation of generations, 249, 250
Altitude, biomes and, 387
Alveoli, 525, 531
Amino acids, 57, 589
   destruction in liver, 600
   essential, 605
   in protein synthesis, 177
Ammocoete larva, 349
Ammonia, 425, 607
   bacterial use of, 215
   in nitrogen cycle, 404
Amniocentesis, 167
Amnion, 358, 681
Amniotic cavity, 680
Amoeba, 231
Amoeboid movement, 229
Amphetamines, action on nerves, 473
Amphibia, 354–357
Amphineura, 327
Amphioxus, 347
Amylase, pancreatic, 596
   salivary, 589
Anaerobes, obligate, 217
Anaerobic process, 130
   bacterial photosynthesis, 214
Anal pore, 234
Analogous structures, 295, 301
Anaphase, 147–150
   I and II in meiosis, 158, 160, 161
Anatomy, 462
Anaximander, 4, 433
Anemia, 535
   iron deficiency, 548
Anesthetics, 473
Angiosperms, 157, 272, 277
Angiotensin, 613
Angler fish, 353
Animalia, 202, 296
Annelida, 320
Anopheles, malaria and, 236
Ant, pheromones and behavior, 710
Antagonism, muscular, 501
Anteater, spiny, 366
Antenna, 334
   function of, 490
Anther, 278
Antheridia, 249, 261, 272
Anthropoid apes, 421, 449
Anthropomorphism, 705, 706
Antibiotics, penicillin, 242
   production by bacteria, 215
Antibody, 192, 545, 553
Anticodon, 174, 177
Antidiuretic hormone (ADH), 612, 630

Antigen, 192, 545, 553
Anti-inflammatory agent, 633
Antiserum, 579
Antithrombin, 553
Anus, 601
Anvil, in ear, 490
Aorta, 563
Apes, anthropoid, 421, 449
Aphasia, 495
Apical meristem, 249, 264
Appeasement display, 723
Appendicitis, 602
Appendicular skeleton, 507
Appendix, 602
Apraxia, 495
Arachnida, 332
Archegonia, 249, 261, 272
Archeopteryx, 362
Aristotle, 4, 84, 202
Arteriosclerosis, 570, 642, 700
Artery, 559
Arthritis, 521
   rheumatoid, 513
Arthropoda, 329–342
   circulatory system, 557
   vision, 486
Ascaris, 317
Aschelminthes, 316
Ascomycota, 240
Ascus, 240
Asexual reproduction, algae, 256, 650
Asthma, 540
Athlete's foot, 243
Atmosphere, modification by photosynthesis, 127, 430
   primitive, 423, 426, 430
   unit of pressure, 269
Atom, 11, 25, 28
Atomic number, 29
ATP, *see* Adenosine triphosphate.
Atrioventricular node, 567
Atrium, 324
Atrophy, 499, 517
Auditory nerve, 490
Aurelia, 307
Australopithecus, 453
Autoimmunity, 545, 579
Autonomic nervous system, 474, 480, 482–483
Autosomes, 143, 144, 188
Autotroph, 377, 393, 429
Autotrophic bacteria, 201, 214
Auxins, 284
Aves, 362–364
   migration, 719
Awareness, 708
Axial skeleton, 507
Axillary bud, 264
Axon, 461, 466

B lymphocytes, 577
Babinski reflex, 469
Bacilli, 212
Bacteria, 211–221
   chemosynthetic, 215, 393, 430
   facultative anaerobes, 217
   green sulfur, 214
   obligate anaerobes, 217
   pathogenic, 216
   purple, 214
   symbiotic, 215, 602
   vitamins synthesized by, 215, 602
Bacteriochlorophyll, 214
Bacteriophages, 207
Barnacle, 335
Base(s), definition, 38
   nitrogen, in nucleotides, 62
   strong and weak, 42
Base pairing, in DNA, 63, 172
Basidiomycota, 242
Basidium, 242
Bat, 368
   echolocation, 721
Bauhin, Kaspar, 202
Behavior, group, 715–723
   innate, 707
   learned, 712
Benthic zone, 387
Beta cells, 640
Bicarbonate buffer system, 41
Bicarbonate ion, 525, 536
Biceps muscle, 518
Bicuspid valve, 563
Bighorn sheep, fighting, 723
Bile, 56, 597, 598
Bile duct, common, 597
Bile salts, 598, 600
Bilirubin, 554, 598
Binary fission, 201, 211, 222, 229
Binomial taxonomy, 202
Biodegradable materials, 389
Biogenetic law, 3, 18
Biogeochemical cycles, 377, 400
Biological magnification, 408
Biological oxygen demand (BOD), 377, 391
Biology, definition, 9
Biomass, pyramid of, 396
Biomes, aquatic, 387
   terrestrial, 380–387
Biosphere, 377, 378
Biotic potential, 377, 409
Bird, *see* Aves.
Birth, 688
Birth control pill, 668
Biston betularia, 443
Bivalve, 324
Bladder, urinary, 608
Blastocele, 302
Blastocyst, 677, 679
Blastopore, 303, 695
Blastula, 302
Blind spot, 489
Blood, 547–557
   as tissue, 299
   clotting, 552
   glucose level, 599
   pH of, 612, 642
Blood group system, ABO, 192, 553
Blood pressure, 564
   regulation, 613, 630
Bolus, 590
Bond, chemical, 25, 35–36
   hydrogen, 36
Bone, 300, 506
Bony fishes, 351
Botulin, 473
Botulism, 216
Bowman's capsule, 608
Bracket fungi, 242
Brain, human, 475
   planarian, 485

Brain hormone, 342
Breathing, 531–533
Breech birth, 688
Brittle star, 343
Bronchi, bronchioles and, 530
Brood food, 717
Brown algae, 251, 253
Bryophyta, 260
Budding, in yeasts, 240, 650
Buffers, 40–42
Bursitis, 521

Caffeine, 473
Calcaneus, 512
Calcitonin, 635, 638
Calcium, in body, 48
   regulation of, 638
Calcium carbonate, 400
Calcium ions, in muscle, 515
Calorie, 73, 76, 604
Cambium, vascular, 249, 264
Camouflage, 353
Canaliculi, 507
Cancer, lung, 540
   prostate, 658
   viruses and, 211
Capillary, 559, 566
Carbaminohemoglobin, 525, 536
Carbohydrates, 50–54
   glycolysis of, 131
Carbon, 48–50
Carbon cycle, 400
Carbon dioxide, 102, 400
   in photosynthesis, 125
Carbonic acid, 41
Carboxypeptidase, 597
Carbuncle, 521
Cardiac muscle, 464
Cardiovascular system, 547, 557–571
Carnivores, 394
Carotenoid pigments, 221
Carpals, 512
Carpel, 249, 278
Carragheenin, 252
Carrier molecule, 108
Carrying capacity, 377, 409
Cartilage, 299, 464, 506
Cartilaginous fishes, 350
Catalyst, 59
Cataracts, 489
Catastrophism, 421, 435
Cecilians, 354
Cecum, 602
Cell(s), 11, 463
   animal vs. plant, 96
   haploid, 144
   structure of, 89–100
Cell body, neuron, 466
Cell membrane, 98
Cell plate, 152
Cell theory, 3, 12, 84
Cell wall, 96, 99
Cellular respiration, 527
Cellular slime mold, 227
Cellulose, 53, 99, 250
Centipede, 337
Central nervous system (CNS), 461, 474–479
Centriole, 94, 143, 147
Centromere, 143, 150, 159
Cephalization, 295, 312, 475, 485
Cephalochordata, 347
Cephalopoda, 327
Cephalothorax, 332
Cerebellum, 477
Cerebral cortex, 475
Cerebrospinal fluid, 479
Cerebrum, 475
Cerumen, 56, 506
Cervical vertebrae, 511

Cervix, 659
Chambered nautilus, 327
Chelicerae, 332
Chelicerata, 332
Chemical evolution, 425
Chemical symbol, 27
Chemoreception, 493, 525, 538
Chemosynthetic bacteria, 215, 393, 430
Chemotaxis, 493, 551
   negative, 708
Chestnut blight, 242
Chilopoda, 335, 337
Chimpanzee, 449
Chitin, 54, 329, 338
Chiton, 328
Chlamydomonas, 256
Chlorobium chlorophyll, 214
Chlorophyll, 96, 118, 122–123
   in algae, 253, 254
   types I and II, 123
Chlorophyta, 251, 256
Chloroplast, 96, 97, 117, 118, 256
Cholecystokinin, 596
Cholera, 603
Cholesterol, 56, 598
   heart disease and, 570
   steroids from, 623
   vitamin D from, 505
Chondrichthyes, 349
Chordata, 345–368
Chorion, 358, 682
Choroid, 488
Chromatid, 143, 150, 161
Chromatin, 147
Chromoplast, 98
Chromosome(s), 90
   DNA in, 65
   homologous, 143, 157
   lampbrush, 697
   mutations, 180, 441
   number of, 144–146
   types of, 144
Chromosome puffs, 697
Chrysophyta, 223, 224
Chyme, 585, 591
Chymotrypsin, 597
Cilia, 96
Ciliata, 229, 232
Circadian rhythm, 705, 710
Circulatory system, closed vs. open, 295, 320
   fetal, 688
   human, 547
Cirrhosis, 603
Clam, 324
Clam worm, 322
Class, taxonomic, 203
Classic conditioning, 713
Clavicle, 512
Cleavage, 302, 677, 679
Cleavage furrow, 150
Cleidoic egg, 296, 358
Climax community, 377, 414
Clitoris, 664
Cloaca, 351
Clock, biological, 710
Clostridium, 216
Clot, 552
Club fungi, 242
Cnidaria, 307
Coagulation, of proteins, 59
Cocci, 212
Coccidiosis, 236
Coccyx, 511
Cochlea, 490
Cocoon, 322
Codon, 174, 176
   termination, 178
Coelenterata, 307
   nervous system, 484
Coenzymes, 47, 65
Cohesion, 67

Cohesion-tension theory, 269
Coitus interruptus, 670
Cold sores, herpes virus and, 208
Colitis, 603
Collecting duct, urinary, 608
Colon, 601
Colonies, algal, 257
Color blindness, sex-linked, 188
Comb jellies, 309
Commensalism, 413
Communication, 721
Community, 12, 392
   climax, 377, 414
Competition, ecological, 410
Complementarity, of codon and anticodon, 177
Complementary bases, 63
Compound, chemical, 11, 25, 30
   organic and inorganic, 48
Compound eye, 486
Concentration, of solutions, 103
Concept formation, 715
Conch, 326
Conditional reflex, 713
Conditioning, behavioral, 713
Condom, 670
Conduction, nerve impulse, 470
Cone cells, 488
Conidia, 240
Coniferae, 272, 274
Conjugation, 201, 233–234, 258, 650
Connective tissue, 297, 463
Constipation, 603
Consumers, levels of, 394
Contraception, 667
Contractile vacuole, 92, 107, 224, 256
Contraction, muscular, 464, 514
Coral, 307
   algae and, 252
Cork cambium, 264
Cornea, 488
Coronary artery, 564
Coronary occlusion, 570
Corpus luteum, 649, 663
Cortex, 268
Cortisol, 632
Cortisone, 56
Cotyledon, 249, 277, 282
Cotylosaurs, 296, 359
Covalent bond, 36
Cowper, William, 654
Cowper's glands, 653, 654
Crab, 335
Cranial nerves, 348, 474, 479
Cranium, 507
Crayfish, 335
Creatine phosphate, 499, 517
Creation, 434
Crick, Francis, 172, 176
Crocodile, 358–359
Cro-Magnon man, 455
Cross, genetic, 184–185
Cross bridges, muscle fiber, 514
Cross-fertilization, 313
Crossing over, in meiosis, 143, 158
   linked genes and, 195
Crossopterygians, 295, 353
Cross-pollination, 280
Crust, of earth, 378
Crustacea, 335
Cryptorchidism, 655
Ctenophora, 309
Cushing's disease, 634
Cuticle, leaf, 119, 260
Cuttlefish, 327
Cuvier, Georges, 435
Cyanophyta, 221
Cyanosis, 540, 691
Cycle, carbon, 400
   nitrogen, 404
   oxygen, 403
   water, 407

Cyclic adenosine monophosphate (cAMP), 621, 624
Cytokinesis, 147
Cytokinins, 287
Cytoplasm, 89
   developmental control, 693
Cytoplasmic streaming, 93
Cytosine, 62, 172

da Vinci, Leonardo, 462
Dark reactions, 119, 125–128
Dart, Raymond A., 453
Darwin, Charles Robert, 436
Darwin-Wallace Theory, 437
Daughter cells, 145
DDT, 389
de Vries, Hugo, 439
Deactivating enzyme, neurotransmitter, 461, 473
Deciduous forest, 383
Decomposers, 394
Decomposition, bacteria and, 215
Decussation, 479
Defecation reflex, 602
Degeneration, 699
Dehydration synthesis, of disaccharides, 47, 52
Democritus, 4
Denatured enzyme, 61
Dendrite, 461, 466
Denitrification, 405
Density, ecological factors and, 409
Deoxyribonucleic acid (DNA), 61, 63–65, 89
   in chromosomes, 144
   in protocells, 428
   in viruses, 204
   replication, 172
   structure of, 172
Deoxyribose, 51, 63
Dermatitis, 521
Dermis, 499, 504
Desert, 383
Development, embryonic, 678
   environment and, 698
   regulation, 691–699
Diabetes insipidus, 614
Diabetes mellitus, 614, 642
Dialysis, 614
Diarrhea, 603
Diastole, 504, 545, 561
Diatoms, 224
Dicotyledon, 282
Diet, 603–606
Differentiation, 16, 697
   zone of, 266
Diffusion, 83, 101
   facilitated, 83, 108, 640
   in gas exchange, 533
Digestion, 586–601
Dihybrid cross, 185
Dilation and curettage, 671
Dinoflagellates, 226
Dinosaurs, 359, 361
   extinction of, 18
Dioecious organism, 295, 305
Dipeptide, 57
Diploblastic embryo, 303
Diplococci, 213
Diploid number, 143, 144
Diplopoda, 335, 337
Disaccharides, 51
Diseases, bacteria and, 216
   viruses and, 205
Dislocation, bone, 512
Display, communication by, 721–723
Dissociation, of acid and base in water, 38
Distal tubule, 608
Diversity, 20
Division, taxonomic, 203
DNA, *see* Deoxyribonucleic acid.
DNA ligase, 219

Dolphin, echolocation, 721
Dominance, incomplete, 190
   social, 718
Dominant allele, 171, 181
Double fertilization, 280
Down's syndrome, 164
Drosophila, 439
Dryopithecus, 449
Duckbill platypus, 366
Ducts, lymph, 573
Ductus arteriosus, 677, 690
Duodenum, 593
Dutch elm disease, 242
Dwarfism, 629
Dysbasia, 495

Ear, parts of, 490
Earth, origin of, 422
   structure of, 378
Earthworm, 320
Ecdysone, 342
Echinodermata, 343
Echolocation, 721
Ecological niche, 377, 392
Ecological succession, 377, 414
Ecology, 378
Ecosystem, 12, 377, 379
   aquatic, 387
Ectoderm, 295, 302, 681
Edema, 573
Effector organ, 467
Efficiency, 73, 78
   of metabolism, 137
Egg, 153
   cleidoic, 296, 358
Electrocardiogram, 568
Electron, 11, 28
   in atom, 31, 33
Electron carriers, 124
Electron configuration, 31
Electron shell, 33
Electron transport system, 129, 135–138
Element, 27
   in living matter, 27, 48, 49
Elephantiasis, 318, 573
Elongation, zone of, 266
Embolus, 552
Embryo, 302, 683
   development, 683–685
Embryo sac, 280
Embryology, 302
Embryonic disc, 581
Empedocles, 4, 433
Emphysema, 540
Emulsification, 585, 598
Endocrine gland, 621, 622
Endocrine system, 348, 620–647
Endocytosis, 111
Endoderm, 295, 302, 681
Endodermis, 268
Endonuclease, restriction, 201, 218
Endoplasmic reticulum, 91
Endorphins, 481
Endoskeleton, 345, 500
Endosperm, 250, 280
Energy, ATP synthesis and, 66
   controlled release of, 124
   in food chain, 393
Energy, kinetic and potential, 32
   law of conservation, 75
   light, 121
   muscular, 516
   nutrition and, 604
   origin of life and, 424
   storage in ATP, 76–77
Energy flow, 14, 78
Energy levels, electron, 31
Energy pyramid, 396

Energy use, 9, 12, 74
Enkephalins, 481
Enterogastric reflex, 593
Entomology, 338
Environmental resistance, 377, 409
Enzymes, 47, 59–61
　digestive, 587, 589, 591, 595
　in DNA replication, 172
　in genetic engineering, 218
　in protein synthesis, 177
　neurotransmitter deactivating, 461, 473
　pancreatic, 595
　pH dependence, 40
Epicotyl, 282
Epidermal cell, 119
Epidermis, 307, 499, 504
Epididymis, 653
Epiglottis, 528, 590
Epilepsy, 495
Epinephrine, 634
Epiphytes, 385
Epithelial tissue, 297, 463
Equatorial plate, 150
Equilibrium, ear and, 492
　genetic, 445
　in diffusion, 101
　osmotic, 105
Ergot, 240
Erythroblastosis fetalis, 556
Erythrocytes, 299, 547–549
Erythropoietin, 545, 548
Escherichia coli, 216
　in humans, 413
Esophagus, 586, 590
Estrogens, 649, 660
Ethyl alcohol, 132, 240
Ethylene, as plant hormone, 286
Euglena, 223
Euglenophyta, 223
Eukaryotic cells, 83, 85, 87, 89
　origin of, 431
Euphotic zone, 387
Eustachian tube, 345, 491
Eutrophication, 377, 405
Evolution, 18, 420–459
　chemical, 425
　crossing over and, 196
　synthetic theory of, 421, 440
Excitatory neuron, 473
Excretion, 606
Exercise, blood chemistry and, 539
　muscles in, 517
Exocytosis, 112
Exoskeleton, 500
Experimentation, 8
Expression, of gene, 696
Extensor, 520
Extracellular fluid, 100
Eye, compound, 330, 335, 486
　human, 487–490
Eyespots, flatworms, 312

Facilitated diffusion, 83, 108, 640
Fallopian tubes, 659
Families, taxonomic, 202
Farsightedness, 489
Fast-twitch fiber, 499, 515
Fats, 55
Fatty acids, 54, 588
　in bacterial photosynthesis, 214
　saturated, 55
Feathers, 362
Feces, 602
Feedback, negative, 545, 548
Felis domestica, 204
Femur, 512
Fermentation, 117, 132, 240, 428
Ferns, 272

Fertilization, 649, 652, 664
　double, 280
Fertilization membrane, 665
Fetus, circulation, 688
　definition, 686
　development, 676
Fibrin, 552
Fibrinogen, 552
Fibrocartilage, 506
Fibrous root, 266
Fibula, 512
"Fight or flight" response, 635
Fighting, 723
Filament, 278
Filaria worm, 318, 573
Filicineae, 272
Filter feeder, 326
Filtration, in kidney, 609
Fishes, bony, 351
　cartilaginous, 350
　circulatory system, 558
Fission, 650
　multiple, 201, 234
"Fixity of species" concept, 434
Flagellum, 96, 223, 229, 256
Flame cells, 311
Flavin adenine dinucleotide (FAD), 117, 134
　FAD and $FADH_2$, 65–66
Flemming, Walther, 147
Flexor, 520
Florigen, 699
Flounder, camouflage of, 353
Flower, 249, 278
Flukes, 310, 313
Follicle, primary, 658
Follicle-stimulating hormone (FSH), 626, 656, 662
Food, 603–606
Food chain, 393
Food vacuole, 92, 234
Food web, 394
Foot, in mollusks, 324
Foramen ovale, 677, 689
Foraminifera, 231
Forebrain, 475
Forest, deciduous, 383
　tropical rain, 384
Formaldehyde, 425
Fossil record, 433–444
Fox, Sidney W., 427
Fracture, bone, 512
Fragmentation, 650
Freshwater ecosystems, 389
Frog, 354
Fructose, 50, 588
Fructose phosphate, 131
Fruit, 250, 278, 282
Fruiting body, 227
Fucoxanthin, 254
Fungi, 202, 236–243
Fungi Imperfecti, 243
Fur, 365

Galactose, 50, 588
Galactosemia, 448
Gallbladder, 597, 598
Gallstones, 598
Gametes, 153, 650
Gametogenesis, 143, 155, 161
Gametophyte, 157, 254, 261
Ganglion, 312, 468, 485
Gas exchange, 533
Gastric juice, 591
Gastrin, 591
Gastrocnemius, 520
Gastrodermis, 307
Gastrointestinal tract, 586
Gastropoda, 324, 326
Gastrovascular cavity, 307

Gastrula, 303, 677, 695
Gecko, 360
Gemmule, 304
Gene, 176
　expression, 696
Gene flow, 421, 442
Gene linkage, 194
Gene mutation, 441
Gene pool, 421, 443
Genera, 202
Generation, first filial (F_1), 184
Generative cell, 275
Generative nucleus, 278
Genetic code, 175
Genetic engineering, 218
Genetic equilibrium, 445
Genetic recombination, 158, 442
Genetic variability, 433
Genetics, 180–197
Genotype, 171, 182
　inheritance of, 435
Genotypic ratio, 184
Geotropism, 285
Germanium, 50
Gerontology, 700
Gibberellins, 287
Gibbon, 449
Gigantism, 629
Gila monster, 360
Gill(s), in mollusks, 324
　in vertebrates, 348
Gill pouch, embryonic, 684
Gill slits, pharyngeal, 345
Giraffe, evolution of, 435, 438
Glaucoma, 489
Glomerulus, 608
Glottis, 528
Glucagon, 640
Glucose, 50, 127, 588
　blood level, 599
Gluteus maximus, 520
Glycerol, 54, 588
Glycogen, 53, 599
Glycolysis, 129–133
Goiter, 636
Golgi, Camillo, 91
Golgi body, 91
Gonorrhea, 657
Gorilla, 449
Gout, 614
Grand mal seizure, 495
Grasshopper, 338
　circulatory system, 558
Grasslands, 383
Gray crescent, 693
Green algae, 251, 256
Greenhouse effect, 402
Group behavior, 715–723
Growth, 9, 14
Growth hormone, 626, 629
Guanine, 62, 172
Guard cell, 119, 263
Gullet, in *Paramecium*, 234
Guttation, 249, 269
Gymnomycota, 227
Gymnosperms, 272, 274

Habitat, 378
Habituation, 713
Hagfish, 348
Hair, 365, 505
Hair follicle, 505
Hammer, in ear, 490
Hangover, 479
Haploid number, 143, 144
　meiosis and, 155
Hardy, G. H., 439
Hardy-Weinberg Law, 421, 439, 445–448

Harvey, Williams, 462
Haversian system, 499, 507
Hearing, 490
Heart, amphibians, 357
   birds, 364
   comparative structure, 559
   fibrillation, 521
   fishes, 351
   human, 560–569
   mammals, 365
   mollusks, 324
   muscle tissue of, 464
   reptiles, 358
Heart sounds, 561, 569
Heartbeat, regulation of, 567–569
Heartburn, 591
Heliozoa, 232
Helix, DNA structure, 65, 172
Hemoglobin, 98, 525, 535
   mutations and, 179
   oxygen transport, 548
Hemophilia, 190, 552
Henle, loop of, 608
Heparin, 553, 614
Hepatitis, 603
Herbaceous stem, 264
Herbivores, 394
Hermaphroditism, 651
Herpes simplex, 208
Heterotroph, 393, 428
Heterotrophic bacteria, 201, 214
Heterozygous, 181
Hibernation, 711
Hierarchy, social, 718
Hindbrain, 475, 477
Hirudin, 322
Histamine, 551
Home range, 717
Homeostasis, 9, 18, 461, 463
   of climax community, 416
   regulation, 622
Homeothermy, 296, 362
Homo erectus, 455
Homo habilis, 454
Homo sapiens, 204, 455
   evolution of, 449–457
Homologous chromosomes, 143, 157
Homologous structures, 295, 301
Homozygous, 181
Honeybee, society, 716
Hooke, Robert, 84
Hookworms, 317
Hormone(s), 17, 621, 622
   blood pressure and, 613, 630
   female, 660
   general, 624
   juvenile, 342
   local, 624
   male, 655
   mechanisms, 624
   neuropeptide, 481
   plant, 250, 283–288
   steroid, 56, 625
Hornworts, 260
Host, parasite and, 411
Human beings, bacteria and, 413
   classification of, 204, 368
   evolution of, 449–457
   inherited traits, 196
   meiosis in, 161–164
   symbiotic bacteria and, 215
Human chorionic gonadotropin (HCG), 666
Humerus, 511
Hyaline cartilage, 506
Hyaline membrane disease, 533
Hybrid, 153
Hydra, 307
   algae and, 254
   nerve net in, 484
Hydrochloric acid, 591, 595
Hydrogen bond, 36

Hydrogen cyanide, 425
Hydrogen ion (H^+), 38
Hydrogen sulfide, 214, 393
Hydrolysis, 585, 587, 591
   of disaccharides, 47, 52
   of fats, 55
   of proteins, 58
Hydroxyl ion (OH^-), 38
Hymen, 659
Hyperglycemia, 641
Hyperopia, 489
Hypertension, 570, 700
Hyperthyroidism, 636
Hypertonic solution, 83, 104
Hyperventilation, 540
Hyphae, 237
Hypocotyl, 249, 277, 282
Hypothalamus, 475
   functions of, 623
Hypothesis, 3, 7
Hypothyroidism, 636
Hypotonic solution, 83, 104

Ice, structure of, 69
Ichthyosaurs, 359
Ileocecal valve, 602
Ileum, 593
Immigration, in evolution, 442
Immune system, 576
Immunity, loss of, 700
   types of, 576–577
Implantation, fetal, 680
Impotence, 658
Imprinting, behavioral, 714
Impulse, nerve, 470
Incomplete dominance, 190
Independent assortment, principle of, 186
Induction, developmental, 677, 695
Industrial melanism, 443–445
Inert elements, 33
Inflammation, 551, 633
Inheritance, learning vs., 707
   of acquired characteristics, 435
Inhibitory factor, 623
Inhibitory neuron, 473
Inner cell mass, 677, 679
Insecta, 335, 338
Insight learning, 715
Instinct, 707
Insulin, 640
   bacterial synthesis, 219
Intercostal muscle, 520
Intercourse, sexual, 663
Interferon, 210
Interneurons, 467
Interphase, 147–149
   in meiosis, 158
Interstitial cells, 655
Interstitial cell-stimulating hormone, 627, 655
Intertidal zone, 389
Intervertebral disc, 510
Intestine, large, 586, 601
   small, 586, 593
Intracellular fluid, 100
Intrauterine device (IUD), 668
Intrinsic factor, 591
Invertebrates, 296, 303–345
Iodine, thyroid gland and, 636
Ion, definition, 35
Ion pump, 470
Ionic bond, 35
Iris, 488
   muscle of, 502
Iron, in hemoglobin, 548
Iron deficiency anemia, 548
Islets of Langerhans, 640
Isometric exercise, 517
Isotonic exercise, 517

Isotonic solution, 83, 104
Isotope, 29

Jaundice, 554
Jaws, 508
Jejunum, 593
Jellyfish, 307
Jenner, Edward, 579
Jet lag, 712
Joint, skeletal, 502
Junction, neuromuscular, 515
Juvenile hormone, 342

Kangaroo, 366
Kidney(s), 347, 607–614
   artificial, 614
   blood pressure and, 613
   urea excretion, 600, 610
Kidney stones, 614, 639
Kilocalorie, 76
Kinetic energy, 32
Kingdoms, 202
Klinefelter's syndrome, 166
Koala bear, 366
Krebs, Hans, 134
Krebs cycle, 129, 133–135
Kwashiorkor, 399, 605
Kyphosis, 521

Labor, birth, 688
   division of, 249, 258, 716
Lactase, 598
Lacteal, 594
Lactic acid, 113, 517
Lactose, 52, 588
Lacunae, 507
Lamarck, Jean Baptiste, 435
Lampbrush chromosomes, 697
Lamprey eels, 348
Lancelet, 347
Land plants, 259
Langerhans, islets of, 640
Larva, 295
   ammocoete, 349
   insect, 330
   sponge, 305
Laryngitis, 540
Larynx, 590
Latimeria, 353
Law(s), natural, 3, 8
   of conservation of matter, 377, 399
   of thermodynamics, 75
Leaf, structure of, 119
Learning, inheritance vs., 707
   insight, 715
   trial-and error, 714
Leeches, 322
Leeuwenhoek, Anton van, 84
Legumes, 404
Lens, 489
Lesion, 495
Leucoplast, 98
Leukemia, 551
Leukocytes, 299, 549
Leukocytosis, 551
Leukopenia, 551
Lewis, G. Edward, 449
Lichen, 222, 242, 412
   algae in, 254
Life, origin of, 426
Life cycle, 201, 227
Ligaments, 464, 502

Light, 117, 121
  reaction to, 284, 707
Light reactions, 119, 122–125
Lightning, early life and, 424
Lignin, 250
Limb buds, 684
Linkage group, genetic, 194
Linnaeus, Carolus, 202
Lipase, pancreatic, 597
Lipids, 47, 54, 588
  digestion, 599
Lipoproteins, 545, 570
Liver, 348, 587, 597
  amino acids in, 600
  cirrhosis of, 603
  urea formation, 607
Liver fluke, Chinese, 313
Liverworts, 260
Lizard, 358
Lobster, 335
Loop of Henle, 608
Lordosis, 521
Lumbar vertebrae, 511
Lumbricus terrestris, 320
Lungfish, 353
Lungs, 348
  book, 334
Luteinizing hormone, 626, 655, 662
Lyell, Charles, 437
Lymph, 299, 545, 572
  triglyceride absorption, 601
Lymph nodes and nodules, 574
Lymphatic system, 547, 571–575, 601
Lymphocytes, 545, 574
Lysergic acid, 242
Lysogenic cycle, 208
Lysosomes, 92
Lysozymes, 92
Lytic cycle, 208

Macronucleus, 232
Macronutrients, 603
Magnification, biological, 408
Malaria, 236
  sickle-cell anemia and, 192
Malnutrition, 605
Malpighian tubules, 329
Maltase, 598
Maltose, 52, 588
Mammalia, 365–368
Mammary gland, 365
Mandible, insect, 334
  human, 508
Mandibulata, 332, 334
Mania, 495
Mantle, in mollusks, 324
Marsupials, 366
Mass flow hypothesis, 271
Mass number, 28
Mastigophora, 229
Matter, 25, 26
  law of conservation of, 377, 399
Maxilla, 508
Mechanism, 3, 5
Medulla oblongata, 477
Medusa, 307
Megaspore, 249, 276
Meiosis, 153–164
  abnormalities of, 164
  origin of, 433
  stages I and II, 157
Melanin, 193, 499, 504
Melanism, industrial, 443–445
Melatonin, 643
Membrane, cell, 98
  fetal, 680
  nuclear, 90
Memory cells, 577

Mendel, Gregor, 180, 439
Meninges, 479
Meningitis, 479
Menstrual cycle, 661
Mercury, biological hazard, 408
Merozoite, 236
Mesoderm, 295, 302, 682
Mesoglea, 307
Mesophyll, 119
Messenger RNA, 173
Metabolism, 3, 12, 37
  cellular, 117, 129
  efficiency of, 137
Metacarpals, 512
Metamorphosis, 295, 330, 341
Metaphase, 147–150
  I and II in meiosis, 157, 159, 161
Metatarsals, 512
Metazoa, 296, 306
Methane, 425
Microfilament, 93
Micronucleus, 232
Micronutrients, 603
Microspores, 249, 275
Microtubule, 93, 94
Microvillus, 585, 594
Midbrain, 475, 477
Migration, 705, 719
Milk of magnesia, 42
Miller, Stanley, 426
Millipede, 337
Mimosa pudica, 288
Minerals, human requirement, 605
Mite, 332
Mitochondrion, 92, 93
  Krebs cycle in, 134
Mitosis, 144–152
  origin of, 433
  phases of, 147
Mitral valve, 563
Molecule, 11, 25, 30
Mollusca, 322
Molting, 295, 329, 341
Monera, 202, 211–223
Monocotyledon, 282
Monoecious organism, 295, 305
Monoglycerides, 588
Monohybrid cross, 184
Monosaccharides, 50, 588
Monotremes, 366
Morgan, Thomas Hunt, 439–440
Morphine, 481
Morphogenesis, 16, 678
Morula, 677, 679
Mosses, 206–263
  spores of, 157
Motor neurons, 467
Motor unit, muscle, 499, 514
Mouth, 586, 589
Movement, amoeboid, 229
Mucus, 589
Multinucleate cell, 147
Multiple alleles, 192
Multiple gene inheritance, 193
Multiple sclerosis, 495
Mumps, sterility and, 657
Muscle(s), antagonistic pairs, 501
  skeletal, 518–520
    anaerobic metabolism in, 132–133
Muscle fibers, 513–515
Muscle tissue, 300, 463, 464
  types of, 500–504
Muscle tone, 517
Muscular dystrophy, 521
Mushroom, 242
Mussel, 324
Mutation, 179, 439, 441
Mutualism, 201, 222, 229, 412
Myasthenia gravis, 521
Mycelium, 237
Mycobacterium tuberculosis, 540

Myelin sheath, 471
Myofibrils, 499, 513
Myoglobin, 499, 515
Myopia, 489
Myosin, 514

Nacre, 326
Nails, 505
Nastic movements, 288
Natural law, 3, 8
Natural selection, theory of, 421, 437
Navigation, 720
Neanderthal man, 455
Nearsightedness, 489
Nectar, 280
Nematocysts, 307
Nematoda, 316
Nephridia, 321
Nephron, 608
Nereis, 322
Nerve cord, 312, 345
Nerve gases, 473
Nerve impulse, 470
Nerve net, 484
Nervous system, disorders of, 495
  divisions of, 474
  human, 460–497
  ladder type, 485
Nervous tissue, 300, 464
Neural tube, 683
Neuromuscular junction, 499, 515
Neuron, 300, 461, 464–467
Neuropeptides, 481
Neurotransmitters, 461, 472
Neutron, 11, 28
Nicotinamide adenine dinucleotide (NAD), 117, 129
  NAD and $NADH_2$, 65–66
Nicotinamide adenine dinucleotide phosphate,
  reduced ($NADPH_2$), 123
Nicotine, 473
Nirenberg, Marshall, 176
Nitrate and nitrite, 404
Nitrification, 404
Nitrogen, fixation, 222, 404
  in metabolic waste, 607
Nitrogen cycle, 404
Nondisjunction, chromosomal, 143, 164
Norepinephrine, 634
Nose, 528
Notochord, 295, 345, 683
Nuclear membrane, 90
Nucleic acids, 47, 61–66, 428
Nucleoid, 211
Nucleolus, 90
Nucleoplasm, 89
Nucleotides, 61
Nucleus, atomic, 28
  cell, 89
  division in mitosis, 145
  generative, 278
  polar, 280
  transplantation, 692
Nutrition, 603–606

Obelia, 309
Obesity, 606
Ochoa, Severo, 176
Octopus, 327
  vision, 486
Oils, 55
Olfactory cells, 494
Ommatidium, 335
Omnivore, 398
Oocytes, primary and secondary, 163–164, 658
Oogenesis, 155, 163, 658
  hormones and, 662

Oogonia, 155, 238, 658
Oomycota, 238
Operant conditioning, 713
Opisthorchis sinensis, 313, 314
Optic nerve, 489
Orangutan, 449
Orders, taxonomic, 202
Organ, 11, 301, 465
Organelle, 83, 87
Organism, 11
Organization, 9, 11
Osculum, 303
Osmosis, 83, 104
  in plants, 269
Osmotic equilibrium, 105
Osteichthyes, 351
Osteoarthritis, 521
Osteoblast, 507
Osteocyte, 507
Ostracoderms, 295, 349
Otolith, 493
Ova, 153
Ovary, animal, 153
  plant, 278
Ovipositor, 341
Ovulation, 164, 663
Ovule, 277, 278
Oxidation, 35
Oxidation-reduction reaction, 35
Oxygen, 102
  formation in photosynthesis, 123
  production by phytoplankton, 222
  skin absorption, 526
Oxygen cycle, 403
Oxygen debt, 117, 133
Oxyhemoglobin, 525, 535
Oxytocin, 630, 631
Oyster, 324
Ozone, 31, 404, 431

P wave, 568
Pacemaker, 567
Panaphobia, 495
Pancreas, 348, 587, 595, 639–642
Pancreatic amylase, 596
Pancreatic lipase, 597
Pancreatitis, 597
Paralogia, 495
Paramecium, 234
  algae and, 254
  cytoplasm, 546
Paramylum, 223
Paraphrasia, 495
Parapodia, 322
Parasitism, 214, 410, 411
Parasympathetic nerves, 481, 483, 568
Parathyroid glands, 637
Parazoa, 303
Parkinson, James, 495
Parkinson's disease, 495
Parotid glands, 589
Parthenogenesis, 295, 319, 650
Parturition, 688
Patella, 512
Pathogenic bacteria, 216
Pavlov, I. P., 713
Pecking order, 718
Pectoral girdle, 512
Peking man, 455
Pelagic zone, 387
Pelecypoda, 324
Pelvis, 510
Penicillium, 242
Penis, 653
Pepsin, 591
Pepsinogen, 591
Peptide bond, 57
Peripheral nervous system (PNS), 461, 474, 479–481

Peristalsis, 585, 587, 591
Peritonitis, 602
Permeability, of nephron, 610
  of nerve membrane, 470
  selective, 98
Pernicious anemia, 592
Perspiration, 68
Pesticides, action on nerves, 473
Petal, 280
Petit mal seizure, 495
pH, 25, 38–40
  blood, 612, 642
  enzymes and, 61
Phaeophyta, 251, 253
Phage virus, 207
Phagocytosis, 83, 111, 549
Phalanges, 512
Pharynx, 528, 586, 590
Phenotype, 171, 182
  non-inheritance of, 436
Phenotypic ratio, 184
Pheromones, 332, 709–710
Phloem, 119, 263
Phobia, 495
Phosphate group, in nucleotides, 62
Phosphoglyceraldehyde (PGAL), 126
Phosphoglyceric acid (PGA), 126
Phospholipids, 56
  in cell membrane, 98
Photolysis, 117, 123
Photons, 121
Photosynthesis, 98, 117–128
  energy flow and, 78
  in bacteria, 214
  in carbon cycle, 401
  summary of, 126, 127
Phototaxis, negative, 707
Phototropism, 284
Phycocyanin, 221, 253
Phycoerythrin, 221, 253
Phylum, 203
Physiology, 462
Phytoplankton, 201, 222, 387
Pineal gland, 643
Pinocytosis, 83, 111
Pinworms, 317
Pioneer stage of succession, 416
Pith, 268
Pituitary gland, anterior, 625–630
  posterior, 630–631
Placenta, 296, 366, 667, 682, 689
Placoderms, 295, 349
Planarians, 310
  nervous system, 485
Planetesimal, 423
Plankton, *see* Phytoplankton, Zooplankton.
Plantae, 202, 250
Plantar reflex, 469
Plants, meiosis in, 156–157
  mineral requirements, 268
  vascular and nonvascular, 260
Planula, 307
Plaques, arteriosclerotic, 570
Plasma, 299, 547
  clearance by nephron, 608
Plasma cells, 578
Plasma membrane, 89
Plasmid, 201, 218
Plasmodial slime mold, 227
Plasmodium, 227
Plasmodium, 236
Plasmolysis, 83, 108
Plastid, 96
Platelets, 299, 551
Plato, 26
Platyhelminthes, 310
Platypus, duckbill, 366
Pleura, 531
Pneumonia, 540
Poikilothermy, 296, 363
Polar body, 164, 662

Polar molecule, 67
Polar nucleus, 280
Pollen, 157, 275
Pollen tube, 277
Pollination, 280
Pollution, 389–392
Polymers, 54
Polynomial taxonomy, 202
Polyp, 307
Polypeptide, 57, 588
Polyribosome, 177
Polysaccharides, 52, 588
Population, 12, 392
  regulation of size, 409
Population genetics, 445
Porifera, 303
Potassium ions, in nerves, 470
Potato blight, 238
Potential energy, 32
Prairie, 383
Prairie chicken, mating dance, 722
Praying mantis, 332
Prebiotic structures, 427
Predation, 410
Pregnancy, ectopic, 667
Primary consumer, 394
Primary electron acceptor, 123
Primary structure, of protein, 58
Primates, 20, 368
Principle of competitive exclusion, 392
Producer, 395
Productivity, net primary, 377, 397
Progesterone, 649, 660
Proglottids, 315
Prokaryote(s), 202, 211
  primitive, 429
Prokaryotic cells, 83, 85, 88
Prolactin, 626, 629
Pronucleus, 649, 665
Prophase, 147–150
  I and II in meiosis, 157, 158, 160
Prostaglandins, 643
Prostate gland, 653
Proteins, digestion, 588, 599
  structure, 57–59
  synthesis, 177–179
Proteinoid microsphere, 421, 427
Prothallus, 272
Prothrombin, 552
Protista, 202, 223–236
Protobiont, 429
Protocell, 421, 428
Proton, 11, 28
Protonema, 263
Protoplasm, 89
  water in, 100
Protozoa, 223, 228–236
Proximal tubule, 608
Pseudoplasmodium, 228
Pseudopodia, 229
Psoriasis, 521
Pterosaurs, 359
Ptyalin, 61, 589
Pubic bones, 512
Puffballs, 242
Pulmonary artery, 561
Pulmonary circulation, 357
Pulse, 564
Punnett square, 184
Pupa, 330
Pupil, 488
Purine, 62
  in gout, 614
Purkinje fibers, 567
Pyramid of biomass, 396
Pyrenoid, 223
Pyrimidine, 62
Pyrrophyta, 223, 226
Pyruvate, 131
Pyruvic acid, 129–132

Index 733

Quaternary consumer, 394
Queen substance, honeybee, 717
QRS complex, 568

Radicle, 249, 277, 282
Radioactivity, 29
Radiolaria, 232
Radius, 511
Ramapithecus, 449
Ray, 350
Reabsorption, in kidney, 609, 610
Reaction, chemical, 33
Reasoning, 715
Receptor, hormonal, 625
Recessive allele, 171, 181
Recombinant DNA, 218
Recombination, genetic, 158, 442
Rectum, 601
Red algae, 251
Red blood cells, 89, 98, 299, 547–549
"Red tide," 227
Reduction, chemical, 35
Reflex, 461, 468–470, 708
 conditional, 713
 defecation, 602
 inflation and deflation, 538
Reflex arc, 468
Regeneration, coelenterates, 309
 echinoderms, 343
 planarians, 313
 salamanders, 356
 sponges, 305
Releaser, environmental, 705, 709
Releasing factor, 623
 hypothalamus and, 477
Remora fish, 413
Renal artery, 608
Renin, 613
Reproduction, 9, 17, 648–675
Reproductive system, female, 659
 male, 653
Resistance, environmental, 377, 409
Respiration, cellular, 527
Respiratory center, 525, 538
Respiratory gases, 102
Respiratory system, 528
Response, 9, 16, 17
 stimulus and, 466
Resting state, of nerve, 470
Restriction endonuclease, 201, 218
Retina, 488
Retroviruses, 206
Reverse transcriptase, 201, 206
Rh factor, 555
Rhizoid, 238, 260
Rhizopus, 238, 650
Rhodophyta, 251
Rhythm, circadian, 705, 710
Rhythm method, 668
Rib, 511
Ribonucleic acid (RNA), 61, 65
 in protocells, 428
 in viruses, 204
 types of, 173
Ribose, 51, 63
Ribosomal RNA, 173, 175
Ribosomes, 91, 175
Ribulose diphosphate (RuDP), 125
Rickets, 521
Rigor mortis, 518
RNA, *see* Ribonucleic acid.
RNA transcriptases, 201, 206
Rod cells, 488
Root, 265
Root cap, 266, 282
Root hair, 266
Root pressure, 269
Roquefort cheese, 242

Rosa multiflora, 203
Rose, classification of, 203
Rotifera, 316, 318
Round window, in ear, 491
Royal jelly, 717
Ruminants, bacteria and, 215
Rust fungi, 242
Rye, ergot in, 240

Sac fungi, 240
Sacral vertebrae, 511
Sacrum, 511
Salamander, 354
Saliva, taste and, 493
Salivary amylase, 61, 589
Salivary glands, 587, 589
Salmon, navigation, 721
Salt, *see* Sodium chloride.
Sand dollar, 343
Saprophytes, 214
Sarcodina, 229
Satiety control center, 606
Scallop, 324
Scaphopoda, 327, 328
Scapula, 512
Schistosoma, 314
Schistosomiasis, 314
Schizomycetes, 211
Science, 3, 5
Scientific method, 3, 6
Sclera, 488
Scolex, 315
Scoliosis, 521
Scorpion, 332
Scrotum, 655
Sea anemone, 307
Sea cucumber, 343
Sea lily, 343
Sea squirt, 346
Sea urchin, 343
Seaweeds, 251, 253
Sebaceous glands, 499, 505
Seborrhea, 521
Sebum, 55
Secondary consumer, 394
Secondary structure, of protein, 58
Secretin, 596
Secretion, endocrine, 622
 kidney, 609, 612
Seed, 249, 274
Segregation, principle of, 183
Seizure, 495
Self-pollination, 280
Semen, 654
Semicircular canals, 492
Semilunar valves, 561, 563
Seminal vesicle, 653
Seminiferous tubule, 653
Sensation, 9, 16
Senses, special, 486–494
Sensory neurons, 467
Sepal, 280
Sequoia sempervirens, 204
Setae, 320
Sex chromosomes, 143, 144, 188
Sex linkage, 188
Sexual intercourse, 663
Sexual reproduction, 650
 origin of, 433
Shark, 350
Sheath, myelin, 471
Sheep, bighorn, 723
Shingles, 521
Shrew, pygmy, 368
Shrimp, 335
Sickle-cell anemia, 179, 535
 incomplete dominance in, 190
Silicon, 50

Silkworm moth, 342
Sinoatrial node, 567
Skate, 350
Skeletal muscle, 132, 464
Skin, 347, 500, 504–506
Skull, 507
Sliding filament theory, 514
Slime molds, 223, 227
Slow-twitch fiber, 499, 515
Slug, 326
Smell, 493
Smith, William, 434
Smooth muscle, 464
Smut fungi, 242
Snail, 326
Snake, 358, 360
Social hierarchy, 718
Society, 705, 716
Socrates, 4
Sodium bicarbonate, 41
 secretion by pancreas, 595
Sodium chloride, formation, 34
 kidney and, 610
Sodium ions, in nerves, 470
Solute and solvent, 38, 103
Solution, 25, 38, 103
Somatic nervous system, 474, 479
Sorus, 272
Sound, 490
Species, 202, 392
Specific heat, of water, 68
Specificity, of enzyme, 59–60
Sperm, 153
Spermatids, 161
Spermatocytes, primary and secondary, 161
Spermatogenesis, 155, 161
 regulation, 656
Spermatogonia, 155, 161
Spermicide, 668
Sphenodon, 360
Sphincter, 585, 591
 anal, 602
Spicules, 304
Spider, 332
Spinal cord, 467–469, 478, 511
Spinal nerves, 474, 479
Spindle, microtubules in, 94
Spindle fiber, 150, 159
Spine, 510
Spinneret, 332
Spiracle, 340
Spirilla, 212
Spirochetes, 213
Spirogyra, 258
Spleen, 574
Sponges, 303
 circulation, 546
Spongin, 304
Spongocoel, 303
Sporangium, 201, 227, 238
Spores, haploid, 144, 156, 650
Sporophyte, 157, 254, 261
Sporozoa, 229, 234
Sporozoite, 234
Sprain, 512
Squid, 327
 vision, 486
Stamen, 249, 278
Staphylococci, 212
Starch, 52, 127, 588
Starfish, 343
Starvation, 606
State, of matter, 26
Stegocephalians, 296, 354
Stems, herbaceous and woody, 264
Stereoscopic vision, 486
Sterility, male, 655
Sterilization, 670
Sternocleidomastoid muscle, 518
Sternum, 511
Steroids, 56, 623

Stethoscope, 569
Stickleback, behavior, 709
Stigma, in *Chlamydomonas*, 256
  in *Euglena*, 223
  in plants, 279
Stimulants, 473
Stimulus, 16, 466
  threshold, 471
Stirrup, in ear, 490
Stomach, 586, 591
Stomach acid, 42, 591, 595
Stomate, 119, 261
Stratification, vertical, 385
Streptococci, 213
Stretch receptor, 525, 538
Striated muscle, 464
Stroke, 570
Structure, correlation with function, 54
Style, 279
Subcutaneous layer, 499, 506
Sublingual glands, 589
Submaxillary glands, 589
Substrate, concentration of, 61
  of enzyme, 60
Sucrase, 598
Sucrose, 51, 588
  hydrolysis of, 587
Sugar(s), in nucleotides, 62
  simple, 50
Sun, energy from, 14, 74, 393
Surface area, volume and, 100
Surface tension, 67
Surfactant, 533
Sutherland, Earl W., 625
Sweat glands, 505
Swim bladder, 353
Symbiosis, 201, 215, 377, 411
Sympathetic nerves, 481, 482, 568, 634
Synapse, 461, 472, 515
Syndactylia, 521
Synovial fluid, 499, 513
Synthetic theory of evolution, 421, 440
Syphilis, 657
System, of organs, 11, 301, 466
Systemic circulation, 357
Systole, 504, 545, 561
Szent-Gyorgyi, Albert, 118

T lymphocytes, 575
T wave, 568
Tadpole, tail absorption, 92
Taiga, 381
Tail, 345, 685
Tapeworms, 310, 315
Taproot, 266
Target tissue, hormonal, 622
Tarsals, 512
Taste, 493
Taste bud, 493
Taxis, 705, 707
Taxonomy, 201, 202
  definition, 20
Teeth, 589
Telophase, 147–150
  I and II in meiosis, 158, 160, 161
Temperature, enzymes and, 61
  protein structure and, 58
  regulation in body, 68, 505, 566
Template, DNA as, 177
Ten Percent Rule, 397
Tendons, 464, 502
Tentacles, in coelenterates, 307
Termination codon, 178
Termite, society, 716
  zooflagellates and, 229, 412
Territoriality, 705, 717
Tertiary consumer, 394
Tertiary structure, of protein, 58

Test cross, 184–185
Testes, 153
Testosterone, 649, 655
Tetanus, 217
Tetany, 639
Thalamus, 475
Thallus, 251
Thecodonts, 296, 359
Theory, 3, 8
Therapsids, 296, 365
Thermal pollution, 391
Thermodynamics, first and second laws of, 75, 137
Thiamine, 602
Thigmotropism, 289
Thoracic vertebrae, 511
Thorax, 338
Threat display, 723
Threshold stimulus, 471
Thrombin, 552
Thrombocytes, 299, 551
Thrombocytopenia, 552
Thrombus, 552
Thymine, 62, 172
Thymosin, 575
Thymus, 575
Thyroid gland, 635–637
Thyroid-stimulating hormone (TSH), 626, 635
Thyroxine, 635
Tibia, 512
Tick, 332
Tissue, 11, 295, 297, 463
Tissue fluid, 572
Toad, 354
Toadstools, 242
Tongue, taste and, 493
Tortoise, 358
Toxins, 473
Trace elements, 48
Trachea, human, 530, 590
  insect, 340, 526
Tracheophyta, 260
Tranquilizers, 473
Transcription, mRNA, 171, 174
Transfer RNA, 173, 174
  charged, 177
Transfusion, blood, 554
Transition reaction, acetyl-CoA in, 134
Translation, in protein synthesis, 178
Translocation, 249, 268
Transpiration, 249, 263, 408
Transpiration theory, 271
Transplant, kidney, 614
Transport, active, 83, 109
Treponema pallidum, 214
Trial-and-error learning, 714
Triceps muscle, 518
Trichinella spiralis, 317, 521
Trichinosis, 317, 521
Trichocyst, 234
Tricuspid valve, 561
Triglycerides, 55, 588
  heart disease and, 570
Triploblastic embryo, 303
Trisomy-X, 166
Trochophore, 295, 322
Trophic levels, 395
Trophoblast, 679
Tropic hormone, 621, 626
Tropical rain forest, 384
Tropism, 250, 283
Truffles, 242
Trypanosoma, 229
Trypsin, 597
Tuatara, 360
Tubal ligation, 671
Tube cell, 275
Tube feet, 343
Tube nucleus, 278
Tuberculosis, 540
Tulipa montana, 204
Tumor, 211

Tundra, 380
Tunicates, 346
Turgor pressure, 107, 271
Turner's syndrome, 166
Turtle, 358
Twins, 679
Tympanum, 340
Tyrannosaurus, 361

Ulcer, duodenal, 596
  esophageal, 591
  gastric, 592
Ulna, 511
Ultraviolet radiation, 404, 424
Umbilical cord, 682, 688
Unity, of living things, 20
Uracil, 62 174
Urea, 505, 600, 607
Uremic syndrome, 614
Ureter, 608
Urethra, 608, 653
Urey, Harold, 426
Uric acid, 614
Urinalysis, 614
Urinary bladder, 608
Urinary system, human, 607–614
Urochordata, 346
Uterus, 659
Utricle, 493

Vaccination, 578
Vacuole, 92
Vagina, 659
Vagus nerve, 480
Valves, heart, 561
  venous, 565
Variability, genetic, 433
Varicose vein, 570
Vas deferens, 653
Vascular cambrium, 249, 264
Vascular system, 559
  in echinoderms, 343
Vascular tissue, 260
Vasectomy, 670
Vasoconstrictor, 613
Vasopressin, 612
Vegetative functions, 476
Vein, 559
  leaf, 119
Vena cava, 561
Venom, snake, 360, 473
Vent, 351
Ventricle, 324
Venus flytrap, 289
Vertebrae, 510
Vertebral column, 347, 510
Vertebrates, 296, 346–368
Vertical stratification, 385
Vesalius, Andreas, 462
Vesicle, 111
Vibration, sense of, 490
Villus, 585, 594, 599
Virchow, Rudolph, 144
Viruses, 204–211
  bacteriophage, 207
  cancer and, 211
  temperate, 208
  virulent, 208
Visceral mass, 324
Vision, 486–490
Vitalism, 3, 5
Vitamin(s), absorption, 601
  bacterial synthesis of, 215, 602
  list, 603, 604
Vitamin B_{12}, 592, 602

Vitamin D, 56
  in skin, 505
Vitamin deficiency, 604
Vitamin K, 602
Vitreous humor, 488
Vocal cords, 529
Volume, surface area and, 100
Volvox, 257
Vorticella, 234

Wallace, Alfred Russell, 436
Waste, metabolic, 607
Water, conservation by kidneys, 611, 630
  expansion on freezing, 69
  in body, 38
  in cells, 100
  motion in plants, 268–271
  purification by bacteria, 215
  role in life, 66–70
Water cycle, 407
Water flea, 335
Water pollution, 389
Water potential, 105
Water vascular system, 343
Watson, James, 172
Watson-Crick model of DNA, 172
Wavelength, 121, 122
Waxes, 56
Web, food, 394
Weight, nutrition and, 604
Weinberg, William, 439
Whale, 368
  migration, 719
Whelk, 326
White blood cells, 299, 549–551
Wigglesworth, V. B., 341
Withdrawal reflex, 468
Wombat, 366
Woody stem, 264

X-linked trait, 188
Xylem, 119, 263
XYY-syndrome, 167

Yeasts, 237, 240
  fermentation by, 132
Yolk sac, 358, 681

Zooflagellates, 229
Zooplankton, 335, 387
Zoospore, 249, 256
Zygomycota, 238
Zygospore, 201, 239
Zygote, 153, 649, 652

736 Index